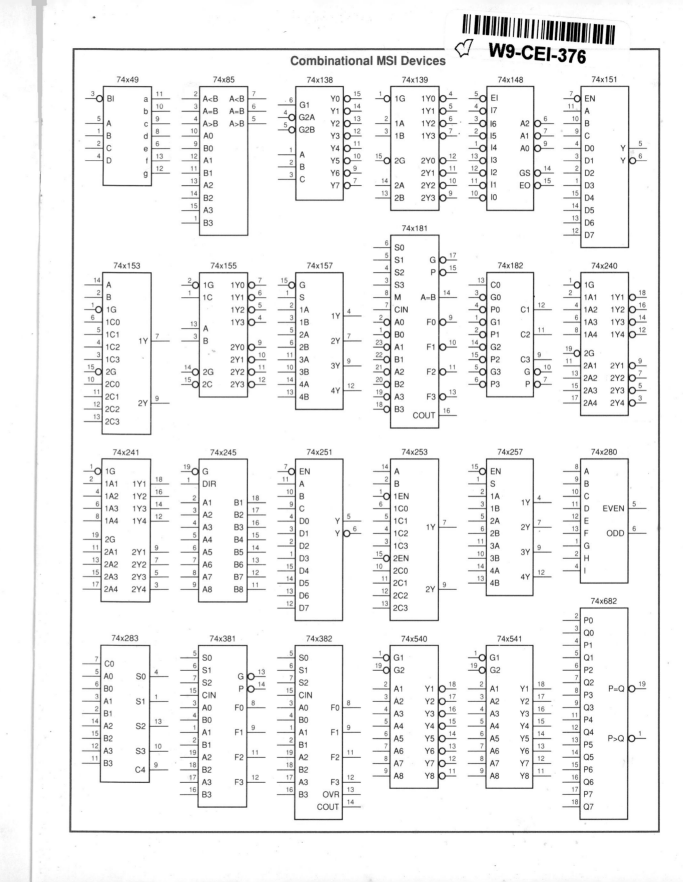

Combinational MSI Devices

W9-CEI-376

Digital Design
Principles and Practices

Digital Design
Principles and Practices

Second Edition

JOHN F. WAKERLY

Alantec Corporation

Stanford University

PRENTICE HALL, Englewood Cliffs, New Jersey 07632

Library of Congress Cataloging-in-Publication Data

Wakerly, John F.
  Digital design: principles and practices. / John F. Wakerly. --2nd ed.
  p. cm. -- (Prentice Hall series in computer engineering)
  Includes bibliographical references and index
  ISBN 0-13-211459-3
  1. Digital integrated circuits--Design and construction.
  I. Title. II. Series
TK7874.65.W34   1994
621.39`5--dc20   93-43333
                       CIP

Acquisitions editor: **Don Fowley**
Production editor: **Irwin Zucker**
Copy editor: **Brenda Melissaratos**
Production coordinator: **David Dickey**
Supplements editor: **Alice Dworkin**
Editorial assistant: **Jennifer Klein**
Cover design: **Glen Biren/Topdesk Pre-Press, Inc.**
Cover illustration: **Photo of circuit board, Bruce Forster/Allstock**
            Background texture, Allstock

© 1994, 1990 by Prentice-Hall, Inc.
A Paramount Communications Company
Englewood Cliffs, New Jersey 07632

TRADEMARK INFORMATION: PostScript is a trademark of Adobe Systems Incorporated. Macintosh is a trademark of Apple Computer, Inc. TRI-STATE is a trademark of National Semiconductor Corporation. PAL is a trademark of Advanced Micro Devices, Inc. ABEL is a trademark of Data I/O Corporation.

Printed in the United States of America

10 9 8 7 6 5 4 3 2 1

ISBN 0-13-211459-3

Prentice-Hall International (UK) Limited, *London*
Prentice-Hall of Australia Pty. Limited, *Sydney*
Prentice-Hall Canada Inc., *Toronto*
Prentice-Hall Hispanoamericana. S.A., *Mexico*
Prentice-Hall of India Private Limited, *New Delhi*
Prentice-Hall of Japan, Inc., *Tokyo*
Simon & Schuster Asia Pte. Ltd., *Singapore*
Editora Prentice-Hall do Brasil, Ltda., *Rio de Janeiro*

To Kate

It's about time!

CONTENTS

FOREWORD

John Wakerly has written a beautiful text, with grace and wit—and with love and respect for the field of digital design. The Principles and Practices subtitle of this text is not casually used. Rather it is the essence of this introduction to the field. The beginning student will find that the simple and concisely stated definitions and procedures provide an accurate exposure to the fundamentals of logical systems and the classic techniques for economically representing logic functions and their implementations in a digital circuit.

The design practice sections contain a tour de force of prevailing logic components and practical design techniques, with many examples of their use in a problem context. The text is replete with actual commercial component descriptions whose function and use in a system context naturally extends the fundamentals in the previous chapters. The sections on practical state-machine design methodology and design with programmable logic devices are outstanding and unique. This material provides survival skills that enable the novice designer to make the transition from simple classical textbook examples of components and design algorithms to practical design of complex systems with modern components. The range of straightforward, yet powerful design methodologies are representative of the state of the art and efficiently cut through the vast combinatorial sea of possible designs to achieve a rich variety of economical, logically valid design solutions.

Prevailing design practice has evolved greatly since John Wakerly and I first met during his graduate student days almost a quarter century ago, before the microprocessor was invented. In the intervening years, John has served continuously at the forefront of the field as a practicing designer and inventor, and as an author and editor. His first text helped define the national curriculum in undergraduate logic design laboratories which have become standard fare today. This text displayed his talent for pinpointing the simplest experimental environment required to illustrate the essence of a pedagogical point without sacrificing realistic, practical design considerations. In the 1980s, John took over the editorship of ACM's *Computer Architecture News*. His diligence and vision transformed the CAN into an open forum for timely discussion and debate of the salient philosophical, conceptual, and design issues of the day.

John Wakerly's *Digital Design Principles and Practices* bears the hallmark of the author's extensive experience as a practical designer, a gifted educator, and a master of expository writing with an incisive intellect and an abiding knowledge of the field. This book has set a new paradigm for an enlightened and enlightening first course in digital design. The instructor and the student alike can have full confidence in the relevance of the techniques presented, and in the meticulous attention to detail. It was a great pleasure for me to have this artfully executed textbook to guide me and my students through my return to the classroom after seven years of administrative assignments. I trust that others will also find this text to be both an excellent introductory tutorial and a valued reference handbook for years to come.

Edward S. Davidson
Ann Arbor, Michigan

PREFACE

This book is for everyone who wants to design and build real digital circuits. It is based on the idea that, in order to do this, you have to grasp the fundamentals, but at the same time you need to understand how things work in the real world. Hence, the "principles and practices" theme.

The material in this book is appropriate for introductory courses on digital logic design in electrical or computer engineering or computer science curricula. Computer science students who are unfamiliar with basic electronics concepts or who just aren't interested in the electrical behavior of digital devices may wish to skip Chapter 2; the rest of the book is written to be independent of this material as much as possible. On the other hand, *anyone* with a basic electronics background who wants to get up to speed on digital electronics can do so by reading Chapter 2. In addition, students with *no* electronics background can ask their instructor for a copy of "Electrical Circuits Review" by Bruce M. Fleischer, a freely reproducible 20-page electronics tutorial that appears in the Instructor's Manual for this book.

introductory courses

electronics concepts

Although this book's level is introductory, it contains much more material than can be taught in a typical introductory course. Once I started writing, I found that I had many important things to say that wouldn't fit into a one-quarter course at Stanford or a 400-page book. Therefore, I have followed my usual practice of including *everything* that I think is at least moderately important, and leaving it up to the instructor or reader to decide what is most important in a particular environment. To help these decisions along, though, I've marked the headings of *optional sections* with an asterisk. In general, these sections can be skipped without any loss of continuity in the non-optional sections that follow.

optional sections

Undoubtedly, some people will use this book in advanced courses and in laboratory courses. Advanced students will want to skip the basics, of course, and get right into the fun stuff. Once you know the basics, the most important

advanced courses
laboratory courses
fun stuff

*working digital
designers*

and fun stuff in this book is Chapters 6 and 10, where you'll discover that your programming courses actually helped prepare you to design hardware.

Another use of this book is as a self-study reference for a working digital designer, one of two kinds:

Novice If you're just getting started as a working digital designer, and you took a very "theoretical" logic design course in school, you should concentrate on Chapters 3, 5, 6, and 9–12 to get prepared for the real world.

Old pro If you're experienced, you may not need all of the "practices" material in this book, but the principles in Chapters 2, 4, and 7 can help you organize your thinking, and the discussions there of what's important and what's not might relieve the guilt you feel for not having used a Karnaugh map in 10 years. The design examples in Chapter 8 should give you additional insights into and appreciation for synchronous and state machine design methods, and the description of the IEEE standard in Appendix A will enable you to decipher those strange symbols that have appeared in the TI data books. Finally, Chapters 6 and 10 may serve as your first organized introduction to programmable logic devices and PLD-based design.

*marginal notes
marginal pun*

All readers should make good use of the comprehensive index and of the *marginal notes* throughout the text that call attention to definitions and important topics. Maybe the highlighted topics in *this* section were more marginal than important, but I just wanted to show off my text formatting system.

CHAPTER DESCRIPTIONS

What follows is a list of short descriptions of this book's twelve chapters and appendix. This may remind you of the "For People Who Hate Reading Manuals" section in software guides. If you read this list, then maybe you don't have to read the rest of the book.

- Chapter 1 gives a few basic definitions and lays down the ground rules for what we think is important, and what is not, in this book.

- Chapter 2 is an introduction to binary number systems and codes. Readers who are already familiar with binary number systems from a software course should still read Sections 2.10–2.13 to get an idea of how binary codes are used by hardware. Advanced students can get a nice introduction to error-detecting codes by reading Sections 1.14 and 1.15. The material in Section 1.16.1 should be read by everyone; it is used in some important design examples in Chapter 6.

- Chapter 3 describes "everything you ever wanted to know about" digital circuit operation, with primary emphasis placed on the external electrical characteristics of logic devices. The starting point is a basic electronics background including voltage, current, and Ohm's law; readers unfamiliar with these concepts may wish to consult the "Electrical Circuits Review" that appears in the Instructor's Manual. This chapter may be omitted by readers who aren't interested in how to make real circuits work, or who have the luxury of employing someone else to do the dirty work.

- Chapter 4 teaches combinational logic design principles, including switching algebra and combinational circuit analysis, synthesis, and minimization.

- Chapter 5 begins with a discussion of digital-system documentation standards, probably the most important practice for aspiring designers to start practicing. The rest of the chapter describes commonly used combinational logic functions and corresponding MSI devices and applications.

- Chapter 6 is an introduction to programmable logic devices. It describes their basic capabilities for combinational design, showing how they can realize both the basic MSI functions of the preceding chapter and more advanced, custom applications.

- Chapter 7 teaches sequential logic design principles, starting with latches and flip-flops. The main emphasis in this chapter is on the analysis and design of clocked synchronous state machines. However, for the brave and daring, the chapter ends with an introduction to fundamental-mode circuits and the analysis and design of feedback sequential circuits.

- Chapter 8 continues our coverage of clocked synchronous state-machine design. It begins with a look at the "big picture"—synchronous system structure—and then offers many examples of and approaches to practical state-machine design.

- Chapter 9 is all about the practical design of sequential circuits. Like Chapter 5 before it, this chapter focuses on commonly used functions, and gives examples using MSI devices and applications. Sections 9.7 and 9.8 discuss the inevitable impediments to the ideal of fully synchronous design, and address the important problem of how to live synchronously in an asynchronous world.

- Chapter 10 continues our PLD coverage with an introduction to sequential PLDs. Their application to both "standard" sequential functions such as counters and their use as state machines are discussed thoroughly. Experienced digital designers and novice students alike may feel that "if it wasn't hard, they wouldn't call it hardware." This chapter shows that there *is* an easier way—you can design a state machine by writing a program!

- Chapter 11 is an introduction to memory devices, including read-only memory and read-write memory. Both static and dynamic read-write memories are explained, from the points of view of both internal circuitry and functional behavior.

- Chapter 12 discusses several miscellaneous real-world topics that are of interest to digital designers. When I started writing what I thought would be a 300-page book, I included this chapter in the outline to pad out the "core" material to a more impressive length. Well, the book is obviously long enough without it, but this material is useful just the same.

- Appendix A is a concise and sometimes critical introduction to IEEE standard symbols for digital logic elements. In some environments, use of these symbols is important and necessary, but they have yet to replace "traditional" logic symbols in most of industry.

Most of the chapters contain references, drill problems, and exercises. Drill problems are typically short-answer or turn-the-crank questions that can be answered directly based on the text material, while exercises may require a little more thinking. The drill problems in Chapter 3 are particularly extensive, and are designed to allow non-EE types to ease into this material.

HOW THIS BOOK WAS PREPARED

The text of the second edition of this book was prepared on a string of Apple Macintosh computers, using the Textures implementation of Don Knuth's TEX manuscript preparation program, published by Blue Sky Research (Portland, OR). My TEX environment uses Leslie Lamport's LATEX macro package and my own book-format customizations. All of the figures in this book were created on Apple Macintosh computers using a very buggy drawing program from a now defunct software company. The `psfigtex` macro package, written by Trevor Darrell, allowed PostScript figure files created by the drawing program to be merged into the TEX output. Color separations were created from the page files by custom PostScript drivers written by the author, and the camera-ready final pages were run on a 1270-dpi PostScript imagesetter.

ERRORS

Warning: This book may contain errors. The author assumes no liability for any damage—incidental, brain, or otherwise—caused by errors.

There, that should make the lawyers happy. Now, to make *you* happy, let me assure you that a great deal of care has gone into the preparation of this manuscript to make it as error free as possible. I am anxious to learn of the remaining errors so that they may be fixed in future printings, editions, and spin-offs. Therefore I will pay $5 to the first finder of each undiscovered error, be it technical, typographical, or otherwise. Please send your comments to me at one of my electronic mail addresses, either wakerly@alantec.com or wakerly@shasta.stanford.edu. (I'm likely to remain associated with one or both of those places for a few years to come.) Failing that, you can write to me care of the Computer Engineering Editor, Prentice Hall, Inc., Englewood Cliffs, NJ 07632.

Any reader can request an up-to-date list of discovered errors sending me mail to one of the internet addresses above, or by writing to me at the postal address above and enclosing a stamped self-addressed envelope. Only one ounce of first-class postage will be needed, I hope.

FOR INSTRUCTORS

A set of transparency masters including full-size black-and-white versions of all the figures and tables in this book is available from the publisher. Also available is an Instructor's Manual containing answers to exercises, sample exams, and a set of three-hour introductory laboratory assignments. Finally, you can get a Macintosh diskette containing the text files (in Microsoft Word) for the exams and lab assignments, and a PC-compatible diskette containing the sources of all the Pascal and ABEL programs in Chapters 6, 10, and 11 of the book.

All of Motorola's data books are available at a much reduced price for classroom use through their outstanding University Support Program. For more information, contact the program manager, Fritz Wilson, at Motorola Semiconductor Products Sector, Mail Drop 56-242, P.O. Box 52073, Phoenix, AZ 85072.

ACKNOWLEDGEMENTS

Many people helped make this book possible. Most of them helped with the first edition, and are acknowledged there. Preparation of the second edition has been a lonelier and more difficult task, especially since I didn't have the luxury of taking a year off to work on it!

My sponsoring editor at Prentice Hall, Don Fowley, deserves great thanks for his patience. After being lured to Prentice Hall in part by the (falsely attractive) prospect of working on this project, only to find out on his arrival that it was over a year late, he nevertheless hunkered down and helped me enough to complete the project on time (at least, according to the Rev. 3 schedule!).

Production editor Irwin Zucker also deserves credit for putting up with sometimes uncooperative bureaucracies and suppliers. Ed Fletcher and Jerry Hill of Pizazz Printing, a San Jose, CA firm with a "can-do" attitude, worked with me to produce final negatives from my PostScript files in record time.

For the ideas on the "principles" side of this book, I still owe great thanks to my teacher, research advisor, and friend Ed McCluskey. On the "practices" side, I must continue to acknowledge the influence of several outstanding digital designers over the years (in chronological order): Ed Davidson, Jim McClure, Courtenay Heater, Sam Wood, Curt Widdoes, Prem Jain, and Ted Tracy. Special mention goes this time to long-time colleague Dave Raaum; how he can design a system with over 90 PLD-based state machines and still have it debugged in less than a month is still beyond me!

Several other professionals have assisted by suggesting changes for this edition and reviewing my efforts. Tops on the list is Larry Crum of Wright State University, one of the best teachers of my undergraduate days at Marquette University. Larry reviewed most of the first-draft manuscript and deserves credit for many improvements in the pedagogy of Chapter 3.

Another great helper was co-worker Rebecca Farley at Alantec, who reviewed much of the new technical material and occasionally encouraged me by saying that the new jokes weren't really so bad, after all. All of my colleagues at Alantec were very supportive of my need to "get this book out of me" while simultaneously trying to build a successful start-up company.

Since the first edition was published, I have received many helpful comments from readers. In addition to suggesting or otherwise motivating many improvements, readers have spotted dozens of typographical and technical errors whose fixes are incorporated in this second edition. The comments of several anonymous reviewers were also quite helpful, though I won't have time to incorporate all of their suggestions until the next edition.

Data I/O Corporation supplied the ABEL software that was used to develop and check the PLD examples in Chapters 6 and 10. I also found ABEL to be handy for minimizing and checking logic equations in other parts the text.

It seems like some disaster always strikes just as I am completing one of these projects. For the first edition, it was the World-Series earthquake of 1989. This time around, it was surgery four days ago for a ruptured and very yucky-looking appendix. Special thanks go to my surgeon, Dr. Tony Marzoni; this wasn't the first time he's had me under the knife. He did such a good job on this little surprise that I'll finish the book *sooner* than I would have without the emergency surgery; I feel great and I now have an excellent excuse to stay home from my other jobs for a week to clean up all the other loose ends. In any case, I'm very grateful that I'm able to put Dr. Marzoni's name in a digital design book, rather than having him put *my* name in some medical book!

Finally, I must once again thank my wife Kate for putting up with the late hours, frustration, crabbiness, preoccupation, and phone calls from weird people that occur when I'm engaged in a writing project like this. We hope you enjoy starting this book as much as we enjoy finishing it!

John F. Wakerly
Mountain View, California

Digital Design
Principles and Practices

Hi, I'm John....

1

INTRODUCTION

Welcome to the world of digital design. Perhaps you're an electrical engineering student who already knows something about analog electronics and circuit design, but you wouldn't know a bit if it bit you. Or perhaps you're a computer science student who knows all about computer software and programming, but you're still trying to figure out how all that fancy hardware could possibly work. No matter. Starting from a fairly basic level, this book will show you how to design digital circuits and subsystems.

We'll give you the basic principles that you need to figure things out, and we'll give you lots of examples. Along with principles, we'll try to convey the flavor of real-world digital design by discussing current, practical considerations whenever possible. And I, the author, will keep on referring to myself as "we" in the hope that you'll be drawn in and feel that we're walking through the learning process together.

1.1   ABOUT DIGITAL DESIGN

Some people call it "logic design." That's OK, but ultimately the goal of design is to build systems. To that end, we'll cover a lot more in this text than just logic equations and theorems.

This book claims to be about principles and practices. Most of the principles that we present will continue to be important years from now; some may be applied in ways that have not even been discovered yet. As for practices, the practices may be a little different from what's presented here by the time you start working in the field, and they will certainly continue to change throughout your career. So you should treat the "practices" material in this book as a way to reinforce principles, and as a way to learn design methods by example.

Listed in the box on this page, there are several key points that you should learn through your studies with this text. Most of these items probably make no sense to you right now, but you should come back and review them later.

Digital design is engineering, and engineering means "problem solving."

IMPORTANT
THEMES IN
DIGITAL DESIGN

- Good tools do not guarantee good design, but they help a lot by taking the pain out of doing things right.

- Digital circuits have analog characteristics.

- Know when to worry and when not to worry about the analog aspects of digital design.

- Always document your designs to make them understandable by yourself and others.

- Associate active levels with signal names and practice bubble-to-bubble logic design.

- Understand and use standard functional building blocks.

- Design for minimum cost at the system level, including your own engineering effort as part of the cost.

- State-machine design is like programming; approach it that way.

- Use programmable logic to simplify designs, reduce cost, and accommodate last-minute modifications.

- Avoid asynchronous design. Practice synchronous design until a better methodology comes along.

- Pinpoint the unavoidable asynchronous interfaces between different subsystems and the outside world, and provide reliable synchronizers.

- Catching a glitch in time saves nine.

My experience is that only 5%–10% of digital design is "the fun stuff"—the creative part of design, the flash of insight, the invention of a new approach. Much of the rest is just "turning the crank." To be sure, turning the crank is much easier now than it was 20 or even 10 years ago, but you still can't spend 100% or even 50% of your time on the fun stuff.

Besides the fun stuff and turning the crank, there are many other areas in which a successful digital designer must be competent, including the following:

- *Debugging.* It's next to impossible to be a good designer without being a good troubleshooter. Successful debugging takes advance planning, a systematic approach, patience, and logic: if you can't discover where a problem *is*, find out where it *is not*!

- *Business requirements and practices.* A digital designer's work is dependent on a variety of non-engineering factors, including component availability, documentation standards, feature definitions, target specifications, task scheduling, office politics, and going to lunch with vendors.

- *Risk-taking.* When you begin a design project you must carefully balance risks against potential rewards and consequences, in areas ranging from new-component selection (will it be available when I'm ready to build the first prototype?) to schedule commitments (will I still have a job if it slips?).

- *Communication.* Eventually, you'll hand off your successful designs to other engineers, other departments, and customers. Without good communication skills, you'll never complete this step successfully. (In other words, try not to be *too* nerdy.)

In the rest of this chapter, and throughout the text, I'll continue to state some opinions about what's important and what is not. I think I'm entitled to do so as a moderately successful practioner of digital design (tens of millions of dollars worth of products incorporating my designs have shipped in just the last year). Of course, you are always welcome to share your own opinions and experience (an e-mail address is given in the Preface).

1.2 ANALOG VERSUS DIGITAL

Analog devices and systems process time-varying signals that can take on any value across a continuous range of voltage, current, or other metric. So do *digital* circuits and systems; the difference is that we can pretend that they don't! A digital signal is modeled as taking on, at any time, only one of two discrete values, which we call *0* and *1* (or LOW and HIGH, FALSE and TRUE, negated and asserted, Sam and Fred, or whatever).

analog
digital

0
1

Digital computers have been around since the 1940s, and have been in widespread commercial use since the 1960s. Yet only in the past 10 to 20 years has the "digital revolution" spread to many other aspects of life. Examples of once-analog systems that have now "gone digital" include the following:

- *Audio recordings.* Once made exclusively by impressing analog waveforms onto vinyl or magnetic tape, audio recordings now commonly use digital compact discs (CDs) and digital audio tape (DAT). A CD or DAT stores music as a sequence of 16-bit numbers corresponding to samples of the original analog waveform, one sample every 22 microseconds or so. A typical CD contains over five billion bits of information.

- *Automobile carburetors.* Once controlled strictly by mechanical linkages (including clever "analog" mechanical devices that sensed temperature, pressure, etc.), automobile engines are now controlled by embedded microprocessors. Various electronic and electromechanical sensors convert engine conditions into numbers that the microprocessor can examine to determine how to control the flow of fuel and oxygen to the engine. The microprocessor's output is a time-varying sequence of numbers that operate electromechanical actuators which, in turn, control the engine.

- *The telephone system.* It started out a hundred years ago with analog microphones and receivers connected to the ends of a pair of copper wires (or was it string?). Even today, most homes and businesses still use analog telephones, which transmit analog signals to the phone company's central office (CO). However, in the majority of COs, these analog signals are converted into a digital format before they are routed to their destinations, be they in the same CO or across the continent or the world.

- *Traffic lights.* Stop lights used to be controlled by electromechanical timers that would give the green light to each direction for a predetermined amount of time. Later, relays were used in controllers that could activate the lights according to the pattern of traffic detected by sensors embedded in the pavement. Today's controllers use microprocessors, and can control the lights in ways that maximize vehicle throughput or, in some California cities, frustrate drivers in all kinds of creative ways.

- *Movie effects.* Special effects used to be made exclusively with miniature clay models, stop action, trick photography, and numerous overlays of film on a frame-by-frame basis. Today, spaceships, creatures, other-worldly scenes, and even babies from hell (in Pixar's animated feature *Tin Toy*) are synthesized entirely using digital computers. Might the stunt man or woman someday no longer be needed, either?

The electronics revolution has been going on for quite some time now, and the "solid-state" revolution began with analog devices and applications like transistors and transistor radios. So why has there now been a *digital* revolution? There are in fact many reasons to favor digital circuits over analog ones:

- *Reproducibility of results.* Given the same set of inputs (in both value and time sequence), a properly designed digital circuit always produces precisely the same results. The outputs of an analog circuit may vary with temperature, power-supply voltage, component aging, and other factors.

- *Ease of design.* Digital design, often called "logic design," is logical. No special math skills are needed, and the behavior of small logic circuits can be visualized mentally without any special insights about the operation of capacitors, transistors, or other devices that require calculus to model.

- *Flexibility and functionality.* Once a problem has been reduced to digital form, it can be solved using a set of logical steps in space and time. For example, you can design a digital circuit that scrambles your recorded voice so that it is absolutely indecipherable by anyone who does not have your "key" (password), but can be heard virtually undistorted by anyone who does. Try doing that with an analog circuit.

- *Programmability.* You're probably already quite familiar with digital computers and the ease with which you can design, write, and debug programs for them. Well, guess what? Many of the computer and digital-system components themselves, called *programmable logic devices (PLDs)*, can be programmed to perform different low-level logic functions! Furthermore, the "programs" in these devices can be easily modified during circuit debugging, which greatly simplifies and speeds product development.

  programmable logic device (PLD)

- *Speed.* Today's digital devices are fast. A typical, primitive digital logic element can examine its inputs and produce an output in less than 10 nanoseconds. This means that such a element can produce 100 million or more results per second.

- *Economy.* Digital circuits can provide a lot of functionality in a small space. Circuits that are used repetitively can be "integrated" into a single "chip" and mass-produced at extremely attractive costs, making possible throwaway items like calculators, digital watches, and singing birthday cards. (You may ask, "Is this such a good thing?" Never mind.)

A *microsecond (μs)* is 10^{-6} second. A *nanosecond (ns)* is just 10^{-9} second. In a vacuum, light travels about a 12 inches in a nanosecond.

SHORT TIMES

- *Steadily advancing technology.* When you design a digital system, you almost always know that there will be a faster, cheaper, or otherwise better technology for it in a few years. Clever designers can accommodate these expected advances right in the initial design of a system, to forestall system obsolesence and to add value for customers. For example, desktop computers often have "expansion sockets" to accommodate faster processors or larger memories than are available at the time of the computer's introduction.

So, that's enough of a sales pitch. The rest of this chapter will give you a bit more technical background to prepare you for the rest of the book.

1.3 DIGITAL DEVICES

gate

The most basic digital devices are called *gates*. These devices have one or more inputs and produce an output that is a function of the current input value(s). While the inputs and outputs may be analog conditions such as voltage, current, even hydraulic pressure, they are modeled as taking on just two discrete values, 0 and 1.

AND gate

Figure 1–1 shows symbols for the three most important kinds of gates. A 2-input *AND gate*, shown in (a), produces a 1 output if both of its inputs are 1; otherwise it produces a 0 output. The figure shows the same gate four times, with the four possible combinations of inputs that may be applied to it and the resulting outputs. A gate is called a *combinational* circuit because its output depends only on the current input combination.

combinational

OR gate

A 2-input *OR gate*, shown in (b), produces a 1 output if one or both of its inputs are 1; it produces a 0 output only if both inputs are 0. Once again, there are four possible input combinations, which result in the outputs shown in the figure.

NOT gate
inverter

A *NOT gate*, more commonly called an *inverter*, produces an output value that is the opposite of the input value, as shown in (c).

We called these three gates the most important for good reason. Any digital function can be realized using just these three kinds of gates. In Chapter 3 we'll show how gates are realized using transistor circuits. You should know, however, that gates have been built (and are hoped to be buildable) using other technologies, such as relays, vacuum tubes, hydraulics, and molecular structures.

flip-flop
state

A *flip-flop* is a device that stores either a 0 or 1. The *state* of a flip-flop is the value that it currently stores. The stored value can be changed only at certain times determined by a "clock" input, and the new value may further depend on the the flip-flop's current state and its "control" inputs.

sequential circuit

A digital circuit that contains flip-flops is called a *sequential circuit* because

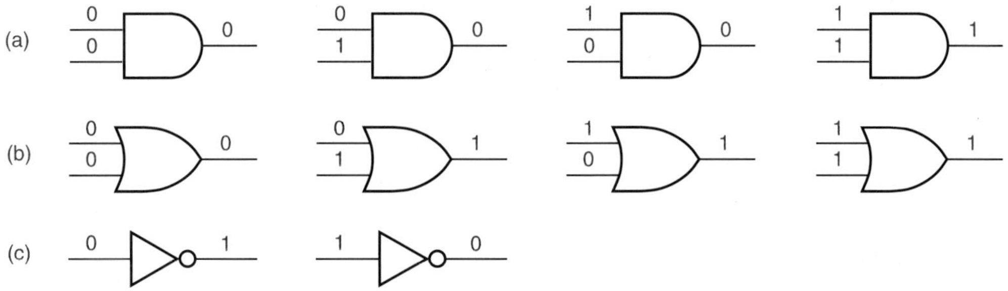

Figure 1–1   Digital devices: (a) AND gate; (b) OR gate; (c) NOT gate or inverter.

its output at any time depends not only on its current input, but also on the past sequence of inputs that have been applied to it. In other words, a sequential circuit has *memory* of past events.   *memory*

Besides gates, most designers think of "flip-flops" as fundamental building blocks of digital design. However, we show in Chapter 7 that flip-flops in most digital technologies are actually constructed using gates. In fact, some chip manufacturers design their large chips as a "sea of gates" containing tens of thousands of individual gates. Such a chip is customized for a particular customer's application during the last few manufacturing steps, which establish interconnections between the gates themselves and the external input/output pins of the device.

1.4   ELECTRONIC ASPECTS OF DIGITAL DESIGN

Digital circuits are not exactly a binary version of alphabet soup—with all due respect to Figure 1–1, they don't have little 0s and 1s floating around in them. As we'll see in Chapter 3, digital circuits deal with analog voltages and currents, and are built with analog components. The "digital abstraction" allows analog behavior to be ignored in most cases, so circuits can be modeled as if they really did process 0s and 1s.

One important aspect of the digital abstraction is to associate a *range* of voltages with each logic value (0 or 1). As shown in Figure 1–2, a typical gate is not guaranteed to have a precise voltage level for a logic 0 output. Rather, it may produce a voltage somewhere in a range that is a *subset* of the range guaranteed to be recognized as a 0 by other gate inputs. The difference between the range boundaries is called *noise margin*—in a real circuit, a gate's output   *noise margin* can be corrupted by this much noise and still be correctly recognized by other inputs.

Behavior for logic 1 outputs is similar. Note in the figure that there is an "invalid" region between the input ranges for logic 0 and logic 1. Although any given digital device operating at a particular voltage and temperature will have

Figure 1–2
Logic values and noise margins.

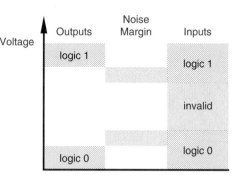

a fairly well defined boundary (or threshold) between the two ranges, different devices may have different boundaries. However, all properly operating devices have their boundary *somewhere* in the undefined range. Therefore, any signal that is within the defined ranges for 0 and 1 will be interpreted identically by different devices.

It is the job of an *electronic* circuit designer to ensure that logic gates produce and recognize logic signals that are within the appropriate ranges. This is an an analog circuit-design problem; we touch upon some aspects of this in Chapter 3. Of course, it is not possible to design a circuit that has the desired behavior under all possible conditions of power-supply voltage, temperature, loading, and other factors; the analog designer (or device manufacturer) provides *specifications* that define the conditions under which correct behavior is guaranteed.

specifications

As a *digital* designer, then, you need not delve into the detailed analog behavior of a digital device to ensure its correct operation. Rather, you need only examine enough about the device's operating environment to determine that it is operating within its published specifications. Granted, some analog knowledge is needed to perform this examination, but not nearly what you'd need to design a digital device starting from scratch. In Chapter 3, we'll give you just what you need.

1.5 SOFTWARE ASPECTS OF DIGITAL DESIGN

Digital design need not involve any software tools. For example, Figure 1–3 shows the primary tool of the "old school" of digital design—a plastic template for drawing logic symbols in schematic diagrams by hand (the designer's name was engraved into the plastic with a soldering iron).

Today, however, software tools are an important part of digital design. In *computer-aided engineering (CAE)* various software tools improve the designer's productivity and help to improve the correctness and quality of designs. In a competitive world, the use of software tools is almost mandatory to obtain high-

computer-aided engineering (CAE)

Figure 1–3
A logic-design template.

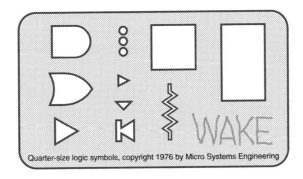

quality results on agressive schedules. Some examples of software tools for digital design are listed below:

- *Schematic entry.* This is the digital designer's equivalent of a word processor. It allows schematic diagrams to be drawn "on-line," instead of with paper and pencil. The more advanced schematic-entry programs also check for common, easy-to-spot errors, such as shorted outputs, signals that don't go anywhere, and so on. Such programs are discussed in greater detail in Section 12.1.

- *Compilers for programmable-logic devices.* Programmable logic devices are customized by the digital designer to perform specific logic functions as required in a system design. The designer specifies the functions using "programs" in a specialized language, and the compiler produces a set of "fuse" patterns that can be programmed into the device to obtain the desired functions. One such language is ABEL, described in Chapters 6 and 10.

- *Timing analyzers and verifiers.* The time dimension is very important in digital design. All digital circuits take time to produce a new output value in response to an input change, and much of a designer's effort is spent ensuring that such output changes occur quickly enough (or, in some cases, not too quickly). Programs such as TimingDesigner (Chronology Corp., Redmond, WA) can automate the tedious task of drawing timing diagrams and verifying the timing relationships between different signals in a complex system.

- *Simulators.* During debugging it's very difficult, often impossible, to change the gates and interconnections in a digital circuit that has been integrated into a single chip. Usually, changes must be made in the original design database and a new chip must be manufactured to incorporate the required changes. Since this process can take days to months to complete, chip designers are highly motivated to "get it right" (or almost right) on the very first try.

Simulators help designers predict the electrical and functional behavior of a chip without actually building it, allowing most bugs to be found before the chip is fabricated. Simulation is discussed in greater detail in Section 12.1.

Simulators are also sometimes used in the design of systems that incorporate many chips, but they are less critical in this case because it's easier for the designer to make changes in components and interconnections on a printed-circuit board. (Also, see box on this page.)

In addition to using the tools above, designers may sometimes write specialized programs in a high-level language like C or Pascal to solve a particular

PROGRAMMABLE
LOGIC DEVICES
VERSUS
SIMULATION

Later in this book you'll learn how programmable logic devices allow you to design a circuit or subsystem by writing a sort of program. PLDs are now available with tens of thousands of gates, and the capabilities of PLD technology are ever increasing. If a PLD-based design doesn't work the first time, you can often fix it by changing the PLD program and physically reprogramming the PLD, without changing any components or interconnections at the system level. The ease of prototyping and modifying PLD-based systems usually eliminates the need for simulation in board-level design; simulation is required only for chip-level designs.

One view of industry trends says that as chip technology advances, more and more design will be done at the chip level, rather than the board level. Therefore, the ability to perform complete and accurate simulation will become increasingly important to the average digital designer.

However, another view is possible. If we extrapolate trends in PLD capabilities, in the next decade we will witness the emergence of PLDs that include not only gates and flip-flops as building blocks, but also higher-level functions such as processors, memories, and input/output controllers. At this point, most digital designers will use on-chip components and interconnections whose basic functions have already been tested by the PLD manufacturer.

In this future view, it is still possible to misapply high-level programmable functions, but it is also possible to fix mistakes simply by changing a program; detailed simulation of a design before simply "trying it out" could be a waste of time. Another, compatible view is that the PLD is merely a full-speed simulator for the program, and this full-speed simulator is what gets shipped in the product!

Does this extreme view have any validity? To guess the answer, ask yourself the following question. How many software programmers do you know who debug a new program by "simulating" its operation rather than just trying it out?

In any case, modern digital systems are much too complex for a designer to have any chance of testing every possible input condition, with or without simulation. As in software, correct operation of digital systems is best accomplished through practices that ensure that the systems are "correct by design." It is a goal of this text to encourage such practices.

design problem. For example, Section 11.1 gives a few examples of Pascal programs that generate the "truth tables" for complex combinational logic functions.

While CAE tools are important, they don't make or break a digital designer. To take an analogy from another field, you couldn't consider yourself to be a great writer just because you're a fast typist or very handy with a word processor.

During your study of digital design, you are encouraged to use any specific tools that might be available to you, such as schematic-entry programs, simulators, and PLD compilers. However, remember that learning to use tools is no guarantee that you'll be able produce good results with them. Please pay attention to what you're producing with them.

1.6 INTEGRATED CIRCUITS

A collection of one or more gates fabricated on a single silicon chip is called an *integrated circuit (IC)*. Large ICs with millions of transistors may be half an inch or more on a side, while small ICs may be less than one-tenth of an inch on a side. Figure 1–4 shows a large microprocessor IC. The manufacturer has labeled the photo with rectangles and legends showing the uses of the silicon area in this particular IC. Around the periphery of the IC are small squares called em bonding pads, which serve as the electrical contact points for tiny wires that connect the chip to the pins of the package that eventually houses it.

integrated circuit (IC)

bonding pad

Regardless of its size, an IC is initially part of a much larger *wafer*, up to eight inches in diameter, containing dozens to hundreds of replicas of the same IC. All of the IC chips on the wafer are fabricated at the same time, like pizzas that are eventually sold by the slice, except in this case, each piece (IC chip) is called a *die*. After the wafer is fabricated, all of the dice are tested, and the defective ones are marked. Then the wafer is sliced up to produce the individual dice, and the marked ones are discarded. (Compare with the the pizza-maker who

wafer

die

Figure 1–4
A microprocessor IC containing about 1.3 million transistors, the IDT R3081. (Courtesy of Integrated Device Technology, Inc.)

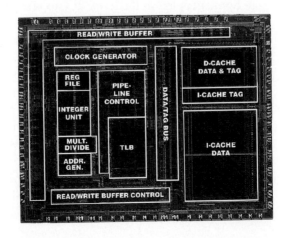

sells all the pieces, even the ones without enough pepperoni!) Each unmarked die is mounted in a package, its pads are connected to the package pins, and the packaged IC is subjected to a final test and shipped to a customer.

To give you an idea of the scale of things, Figure 1–5 shows four packaged microprocessor ICs sitting on top of a wafer full of the same ICs; each package is about an inch and a half on a side. The rectangular outlines of individual dice can be seen on the wafer.

Figure 1–5 Part of a wafer and four packaged ICs. (Courtesy of Integrated Device Technology, Inc.)

Some people use the term "IC" to refer to a silicon die. Some use "chip" to refer to the same thing. Still others use "IC" or "chip" to refer to the combination of a silicon die and its package. Digital designers tend to use the two terms interchangeably, and they really don't care what they're talking about. They don't require a precise definition, since they're only looking at the functional and electrical behavior of these things. In the balance of this text, we'll use the term *IC* to refer to a packaged die (not to be confused with icee, a drink that you buy at a convenience store).

IC

ICs can be classified by how many gates they contain. The simplest type of commercially available ICs are called *small-scale integration (SSI)*, and contain the equivalent of 1 to 20 gates. SSI ICs typically contain a handful of gates or flip-flops, the basic building blocks of digital design. One might also define SSI ICs in terms of cost: Inflation willing, an SSI IC costs about 25 cents.

small-scale integration (SSI)

Figure 1–6
Dual in-line pin (DIP) packages :
(a) 14-pin; (b) 20-pin; (c) 28-pin.

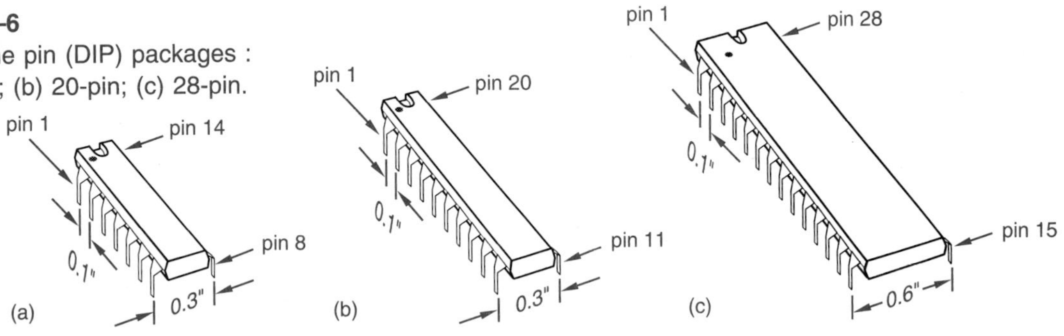

Most SSI ICs come in a 14-pin *dual in-line pin (DIP) package*, where the spacing between pins in a column is 0.1 inch and the spacing between columns is 0.3 inch, as shown in Figure 1–6(a). A *pin diagram* shows the assignment of device signals to package pins, or *pinout*. Figure 1–7 shows the pin diagrams for commonly used SSI ICs. All 7400-series ICs with the same part number have the pinout, even if the alphabetic portion of the part name is different. For example, a 74LS00 and a 74ACT00 have the same pinout.

dual in-line pin (DIP) package

pin diagram
pinout

Note that pin diagrams given in Figure 1–7 are used only for mechanical reference, when a designer needs to determine the pin numbers for a particular IC. In the schematic diagram for a digital circuit, pin diagrams are not used. Instead, the various gates are grouped functionally, as we'll show in Section 5.1. In particular, see Figure 5–19 on page 290.

Although SSI ICs are still sometimes used as "glue" to tie together larger-scale elements in complex systems, they have been largely supplanted by programmable logic devices, which we'll study in Chapters 6 and 10.

The next larger commercially available ICs are called *medium-scale integration (MSI)*, and contain the equivalent of about 20 to 200 gates. An MSI IC typically contains a functional building block, such as a decoder, register, or counter, as we'll show in Chapters 5 and 9. One might also define MSI ICs in terms of cost: The Federal Reserve willing, an MSI IC costs under a dollar.

medium-scale integration (MSI)

Many standard "high-level" functions appear over and over as building blocks in digital design. Historically, these functions were first integrated in MSI circuits. Subsequently, they have appeared as components in the "macro" libraries for ASIC design, as "standard cells" in VLSI design, as "canned" functions in PLD programming languages, and as library functions in digital-design languages such as VHDL.

Standard logic functions are introduced in Chapters 5 and 9 as 74-series MSI parts. However, the discussion and examples in these chapters provide a basis for understanding and using these functions in any form.

STANDARD
LOGIC
FUNCTIONS

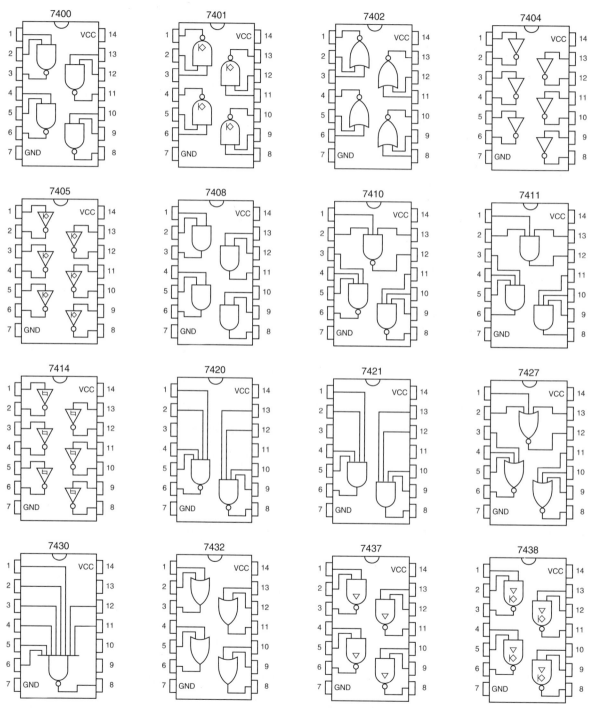

Figure 1–7 Commonly used 7400-series SSI ICs.

Large-scale integration (LSI) ICs are bigger still, containing the equivalent of 200 to 200,000 gates or more. LSI parts include memories, microprocessors, and programmable logic devices; their costs can range from a dollar to over $20.

large-scale integration (LSI)

The dividing line between LSI and *very large-scale integration (VLSI)* is fuzzy, and tends to be stated in terms of transistor count rather than gate count. Any IC with over 500,000 transistors is definitely VLSI. Below that, the name depends on how well the marketing folks have promoted the product. Examples of VLSI devices include 1-Mbit and larger memories and 32-bit microprocessors. Prices of VLSI devices can range from well under $10 (for 1-Mbit memories) to hundreds of dollars (for the 3-million-transistor Intel Pentium microprocessor.) Perhaps the next level of integration will be called RLSI—ridiculously large-scale integration.

very large-scale integration (VLSI)

1.7 APPLICATION-SPECIFIC ICS

Probably the most interesting developments in IC technology for the average logic designer are not the ever-increasing chip sizes, but the ever-increasing opportunities to "design your own chip." Chips designed for a particular, limited product or application are called *semicustom ICs* or *application-specific ICs (ASICs).* ASICs generally reduce the total component and manufacturing cost of a product by reducing chip count, physical size, and power consumption, and they often provide higher performance.

semicustom IC

application-specific IC (ASIC)

The *nonrecurring engineering (NRE) cost* for designing an ASIC can exceed the cost of a discrete design by $5,000 to $250,000 or more. NRE charges are paid to the IC manufacturer and others who are responsible for designing the internal structure of the chip, creating tooling such as the metal masks for manufacturing the chips, developing tests for the manufactured chips, and actually making the first few sample chips.

nonrecurring engineering (NRE) cost

The NRE cost for a typical, medium-complexity ASIC with about 20,000 gates is $30–$50,000. An ASIC design normally makes sense only when the NRE cost can be amortized by the per-unit savings over the expected sales volume of the product.

The NRE cost to design a *custom LSI* chip—a chip whose functions, internal architecture, and detailed transistor-level design is tailored for a specific customer—is very high, $250,000 or more. Thus, full custom LSI design is done only for chips that have general commercial application or that will enjoy very high sales volume in a specific application (e.g., a digital watch chip, or a bus-interface circuit for a popular laptop computer).

custom LSI

To reduce NRE cost, IC manufacturers have developed libraries of *standard cells* including commonly used MSI functions such as decoders, registers, and counters, and commonly used LSI functions such as memories, programmable logic arrays, and microprocessors. In a *standard-cell design*, the logic designer

standard cells

standard-cell design

interconnects functions in much the same way as in a multichip MSI/LSI design. Custom cells are created (at added cost, of course) only if absolutely necessary. All of the cells are then laid out on the chip, optimizing the layout to reduce propagation delays and minimize the size of the chip. Minimizing the chip size reduces the per-unit cost of the chip, since it increases the number of chips that can be fabricated on a single wafer. The NRE cost for a standard-cell design is typically on the order of $100,000.

gate array

Well, $100,000 is still a lot of money for most folks, so IC manufacturers have gone one step further to bring ASIC design capability to the masses. A *gate array* is an IC whose internal structure is an array of gates whose interconnections are initially unspecified. The logic designer specifies the gate types and interconnections. Even though the chip design is ultimately specified at this very low level, the designer typically works with "macrocells," the same high-level functions used in multichip MSI/LSI and standard-cell designs; software expands the high-level design into a low-level one.

The main difference between standard-cell and gate-array design is that the macrocells and the chip layout of a gate array are not as highly optimized as those in a standard-cell design, so the chip may be 25% or more larger, and therefore may cost more. Also, there is no opportunity to create custom cells in the gate-array approach. On the other hand, a gate-array design can be completed faster and at lower NRE cost, ranging from about $5000 (what the marketing folks tell you initially) to $50,000 (what you find you've spent when you're all done).

field-programmable gate array (FPGA)

One of the most recent developments in ASIC technology allows an ASIC to be created overnight (or even during the day, if that's when you like to work). In all of the ASIC technologies that we've described so far, gate-level interconnections are established when the device is manufactured. A *field-programmable gate array (FPGA)* contains "fuses" that allow the interconnection pattern to be loaded and changed *after* the device is manufactured.

Since an FPGA is a standard, in-stock part, its major appeal is that it can be programmed and tested in an application as soon as the design is completed, compared with 2–6 weeks to manufacture a fixed gate array. For designers who make mistakes, an even bigger appeal is that modifications can be made with very little time and cost, compared with many weeks and big bucks in the fixed gate-array and standard-cell approaches. A downside is that programmable gate arrays are typically slower and more costly than their fixed-interconnect cousins.

GATE-ARRAY
TECHNOLOGY

> Most gate-array technologies don't really contain an array of gates. Rather, they use an array of uncommitted CMOS transistors. Different gate types (NAND, NOR, etc.) and flip-flops (D, J-K, etc.) are created by interconnecting the transistors in different ways.

The basic digital design methods that you'll study throughout this book apply very well to the functional design of ASICs. However, there are additional opportunities, constraints, and steps in ASIC design, which usually depend on the particular ASIC vendor and design environment.

1.8 PRINTED-CIRCUIT BOARDS

An IC is normally mounted on a *printed-circuit board (PCB)* [sometimes called a *printed-wiring board (PWB)*] that connects it to other ICs in a system. The multilayer PCBs used in typical digital systems have copper wiring etched on multiple, thin layers of fiberglass that are laminated into a single board typically about 1/16 inch thick. Figure 1–8 shows a PCB with many IC packages and other components.

Individual wire connections, or *PCB traces* are usually quite narrow, 10 to 25 mils in typical PCBs. (A *mil* is one-thousandth of an inch.) In *fine-line* PCB technology, the traces are extremely narrow, as little as 5 mils wide with 5-mil spacing between adjacent traces. Thus, up to 100 connections may be routed in

printed-circuit board (PCB)

printed-wiring board (PWB)

PCB traces
mil
fine-line

Figure 1–8 A PCB with ICs, connectors, and other components. (Courtesy of Alantec, Inc.)

a one-inch-wide band on a single layer of the PCB. If higher connection density is needed, then more layers are used.

surface-mount technology (SMT)

Most of the components in Figure 1–8 use *surface-mount technology (SMT)*. Instead of having the long pins of Figure 1–6 that poke through the board and are soldered to the underside, the leads of SMT IC packages are bent to make flat contact with the top surface of the PCB. Before such components are mounted on the PCB, a special "solder paste" is applied to contact pads on the PCB using a stencil whose hole pattern matches the contact pads to be soldered. Then the SMT components are placed (by hand or by machine) on the pads, where they are held in place by the solder paste (or in some cases, by glue). Finally, the entire assembly is passed through an oven to melt the solder paste, which then solidifies when cooled.

Surface-mount component technology, coupled with fine-line PCB technology, allows extremely dense packing of integrated circuits and other components on a PCB. This dense packing does more than save space. For very high-speed circuits, dense packing goes a long way toward minimizing adverse analog phenomena, including transmission-line effects and speed-of-light limitations.

multichip module (MCM)

To satisfy the most stringent requirements for speed and density, *multichip modules (MCMs)* have been developed in the past few years. In this technology, IC dice are not mounted in individual plastic or ceramic packages. Instead, the IC dice for a high-speed subsystem (say, a processor and its cache memory) are bonded directly to a substrate that contains the required interconnections on multiple layers. The MCM is hermetically sealed and has its own external pins for power, ground, and just those signals that are required by the system that contains it.

WHERE ARE THE CAPS?

Astute readers may notice the PCB in Figure 1–8 appears to have remarkably few decoupling capacitors and other analog components, and no SSI chips. The dirty little secret is that such components are actually mounted on the underside of the board.

With SMT, it is relatively easy to place components on both sides of the PCB, thereby increasing the amount a circuitry that can be crammed into a given board area. The PCB in Figure 1–8 has all of its decoupling capacitors, pull-up and pull-down resistors, and SSI gates mounted on the underside of the board, what would be considered to be the "solder side" of the board with through-hole technology. SMT components can be soldered onto both sides on the board, but any through-hole components (such as the connectors and the fiber-optic interfaces in the figure) are normally mounted on one side only.

1.9 THE NAME OF THE GAME

Given the functional and performance requirements for a digital system, the name of the game in practical digital design is to minimize cost. For *board-level designs*—systems that are packaged on a single PCB—this usually means minimizing the number of IC packages. If too many ICs are required, they won't all fit on the PCB. "Well, just use a bigger PCB," you say. Unfortunately, PCB sizes are usually constrained by factors such as pre-existing standards (e.g., add-in boards for PCs), packaging constraints (e.g., it has to fit in a toaster), or edicts from above (e.g., in order to get the project approved three months ago, you foolishly told your manager that it would all fit on a 3×5 inch PCB, and now you've got to deliver!). In each of these cases, the cost of using a larger PCB or multiple PCBs may be unacceptable.

board-level design

Minimizing the number of ICs is usually the rule even though individual IC costs vary. For example, a typical SSI IC may cost 15 cents, while an MSI IC may cost 50 cents. It may be possible to perform a particular function with two SSI ICs (30 cents) or one MSI IC (50 cents). In most situations, the more expensive MSI solution is used, not because the designer owns stock in the IC company, but because the MSI solution uses less PCB area.

We'll see this idea repeated when we design with PLDs in Chapters 6 and 10. One PLD, costing perhaps $2, may replace half a dozen 15–50-cent SSI and MSI ICs. The PLD may cost a little more or a little less than the SSI/MSI total, but in any case it saves substantial design time and PCB space.

In *ASIC design*, the name of the game is a little different, but the importance of MSI (i.e., structured, functional) design techniques is the same. Although it's easy to burn hours and weeks creating custom macrocells and minimizing the total gate count of an ASIC, only rarely is this advisable. The per-unit cost reduction achieved by having a 10% smaller chip is negligible except in high-volume applications. In applications with low to medium volume (the majority), two other factors are more important: design time and NRE cost.

ASIC design

A shorter design time allows a product to reach the market sooner, increasing revenues over the lifetime of the product. A lower NRE cost also flows right to the "bottom line," and in small companies may be the only way the project can be completed before the company runs out of money (believe me, I've been there!). If the product is successful, it's always possible and profitable to "tweak" the design later to reduce per-unit costs. The need to minimize design time and NRE cost argues in favor of a structured, as opposed to highly optimized, approach to ASIC design, using standard MSI-type functions provided in the ASIC manufacturer's library.

1.10 GOING FORWARD

This concludes the introductory chapter. As you continue reading this book, keep in mind two things. First, the ultimate goal of digital design is to build systems that solve problems for people; while this book will give you the basic tools for design, it's still your job to keep "the big picture" in the back of your mind. Second, cost is an important factor in every design decision; and you must consider not only the cost of digital components, but also the cost of the design activity itself.

Finally, as you get deeper into the text, if you encounter something that you think you've seen before but don't remember where, please consult the index. I've tried to make it as helpful and complete as possible.

D R I L L   P R O B L E M S

1.1   Suggest some better-looking chapter-opening artwork to put on page 1 of the next edition of this book.

1.2   Give three different definitions for the word "bit" as used in this chapter.

1.3   Define the following acronyms: ASIC, CAE, CD, CO, DAT, DIP, FPGA, IC, LSI, MCM, MSI, NRE, OK, PCB, PLD, PWB, RLSI, SMT, SSI, VLSI. (Is OK really an acronym?)

1.4   Excluding the topics in Section 1.2, list three once-analog systems that have "gone digital" since you were born.

1.5   Draw a digital circuit consisting of a 2-input AND gate and three inverters, where an inverter is connected to each of the AND gate's inputs and its output. For each of the four possible combinations of inputs applied to the two primary inputs of this circuit, determine the value produced at the primary output. Is there a simpler circuit that gives the same input/output behavior?

1.6   When should you use the pin diagrams of Figure 1–7 in the schematic diagram of a circuit?

1.7   What is the relationship between "die" and "dice"?

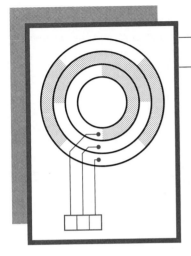

NUMBER SYSTEMS AND CODES

Digital systems are built from circuits that process binary digits—0s and 1s—yet very few real-life problems are based on binary numbers or any numbers at all. Therefore, a digital system designer must establish some correspondence between the binary digits processed by digital circuits and real-life numbers, events, and conditions. The purpose of this chapter is to show you how familiar numeric quantities can be represented and manipulated in a digital system, and how nonnumeric data, events, and conditions also can be represented.

The first nine sections describe binary number systems and show how addition, subtraction, multiplication, and division are performed in these systems. Sections 2.10–2.13 show how other things, such as decimal numbers, text characters, mechanical positions, and arbitrary conditions, can be encoded using strings of binary digits.

Section 2.14 introduces "n-cubes," which provide a way to visualize the relationship between different bit strings. The n-cubes are especially useful in the study of error-detecting codes in Section 2.15. We conclude the chapter with an introduction to codes for transmitting and storing data one bit at a time.

2.1 POSITIONAL NUMBER SYSTEMS

positional number system

weight

The traditional number system that we learned in school and use every day in business is called a *positional number system.* In such a system, a number is represented by a string of digits where each digit position has an associated *weight.* The value of a number is a weighted sum of the digits, for example:

$$1734 = 1 \cdot 1000 + 7 \cdot 100 + 3 \cdot 10 + 4 \cdot 1$$

Each weight is a power of 10 corresponding to the digit's position. A decimal point allows negative as well as positive powers of 10 to be used:

$$5185.68 = 5 \cdot 1000 + 1 \cdot 100 + 8 \cdot 10 + 5 \cdot 1 + 6 \cdot 0.1 + 8 \cdot 0.01$$

In general, a number D of the form $d_1 d_0.d_{-1} d_{-2}$ has the value

$$D = d_1 \cdot 10^1 + d_0 \cdot 10^0 + d_{-1} \cdot 10^{-1} + d_{-2} \cdot 10^{-2}$$

base

radix

Here, 10 is called the *base* or *radix* of the number system. In a general positional number system, the radix may be any integer $r \geq 2$, and a digit in position i has weight r^i. The general form of a number in such a system is

$$d_{p-1} d_{p-2} \cdots d_1 d_0.d_{-1} d_{-2} \cdots d_{-n}$$

radix point

where there are p digits to the left of the point and n digits to the right of the point, called the *radix point.* If the radix point is missing, it is assumed to be to the right of the rightmost digit. The value of the number is the sum of each digit multiplied by the corresponding power of the radix:

$$D = \sum_{i=-n}^{p-1} d_i \cdot r^i$$

high-order digit

most significant digit

low-order digit

least significant digit

Except for possible leading and trailing zeroes, the representation of a number in a positional number system is unique. (Obviously, 0185.6300 equals 185.63, and so on.) The leftmost digit in such a number is called the *high-order* or *most significant digit*; the rightmost is the *low-order* or *least significant digit.*

binary digit

bit

binary radix

As we'll learn in Chapter 3, digital circuits have signals that are normally in one of only two conditions—low or high, charged or discharged, off or on. The signals in these circuits are interpreted to represent *binary digits* (or *bits*) that have one of two values, 0 and 1. Thus, the *binary radix* is normally used to represent numbers in a digital system. The general form of a binary number is

$$b_{p-1} b_{p-2} \cdots b_1 b_0.b_{-1} b_{-2} \cdots b_{-n}$$

and its value is

$$B = \sum_{i=-n}^{p-1} b_i \cdot 2^i$$

In a binary number, the radix point is called the *binary point*. When dealing *binary point* with binary and other nondecimal numbers, we use a subscript to indicate the radix of each number, unless the radix is clear from the context. Examples of binary numbers and their decimal equivalents are given below.

$$100011_2 = 1 \cdot 16 + 0 \cdot 8 + 0 \cdot 4 + 1 \cdot 2 + 1 \cdot 1 = 19_{10}$$

$$1000010_2 = 1 \cdot 32 + 0 \cdot 16 + 0 \cdot 8 + 0 \cdot 4 + 1 \cdot 2 + 0 \cdot 1 = 34_{10}$$

$$101.001_2 = 1 \cdot 4 + 0 \cdot 2 + 1 \cdot 1 + 0 \cdot 0.5 + 0 \cdot 0.25 + 1 \cdot 0.125 = 5.125_{10}$$

The leftmost bit of a binary number is called the *high-order* or *most significant bit (MSB)*; the rightmost is the *low-order* or *least significant bit (LSB)*.

MSB
LSB

2.2 OCTAL AND HEXADECIMAL NUMBERS

Radix 10 is important because we use it in everyday business, and radix 2 is important because binary numbers can be processed directly by digital circuits. Numbers in other radices are not often processed directly, but may be important for documentation or other purposes. In particular, the radices 8 and 16 provide convenient shorthand representations for multibit numbers in a digital system.

The *octal number system* uses radix 8, while the *hexadecimal number system* uses radix 16. Table 2–1 shows the binary integers from 0 to 1111 and their octal, decimal, and hexadecimal equivalents. The octal system needs 8 digits, so it uses digits 0–7 of the decimal system. The hexadecimal system needs 16 digits, so it supplements decimal digits 0–9 with the letters *A–F*.

octal number system
hexadecimal number system

hexadecimal digits A–F

The octal and hexadecimal number systems are useful for representing multibit numbers because their radices are powers of 2. Since a string of three bits can take on eight different combinations, it follows that each 3-bit string can be uniquely represented by one octal digit, according to the third and fourth columns of Table 2–1. Likewise, a 4-bit string can be represented by one hexadecimal digit according to the fifth and sixth columns of the table.

Thus, it is very easy to convert a binary number to octal. Starting at the binary point and working left, we simply separate the bits into groups of three and replace each group with the corresponding octal digit:

binary to octal conversion

$$100011001110_2 = 100\ 011\ 001\ 110_2 = 4316_8$$

$$1110110111010101_2 = 011\ 101\ 101\ 110\ 101\ 001_2 = 355651_8$$

The procedure for binary to hexadecimal conversion is similar, except we use groups of four bits:

binary to hexadecimal conversion

$$100011001110_2 = 1000\ 1100\ 1110_2 = 8CE_{16}$$

$$1110110111010101_2 = 0001\ 1101\ 1011\ 1010\ 1001_2 = 1DBA9_{16}$$

Table 2–1
Binary, decimal, octal, and
hexadecimal numbers.

Binary	Decimal	Octal	3-Bit String	Hexadecimal	4-Bit String
0	0	0	000	0	0000
1	1	1	001	1	0001
10	2	2	010	2	0010
11	3	3	011	3	0011
100	4	4	100	4	0100
101	5	5	101	5	0101
110	6	6	110	6	0110
111	7	7	111	7	0111
1000	8	10	—	8	1000
1001	9	11	—	9	1001
1010	10	12	—	A	1010
1011	11	13	—	B	1011
1100	12	14	—	C	1100
1101	13	15	—	D	1101
1110	14	16	—	E	1110
1111	15	17	—	F	1111

In these examples we have freely added zeroes on the left to make the total number of bits a multiple of 3 or 4 as required.

If a binary number contains digits to the right of the binary point, we can convert them to octal or hexadecimal by starting at the binary point and working right. Both the left-hand and right-hand sides can be padded with zeroes to get multiples of three or four bits, as shown in the example below:

$$10.1011001011_2 \ = \ 010.101\ 100\ 101\ 100_2 \ = \ 2.5454_8$$
$$= \ 0010.1011\ 0010\ 1100_2 \ = \ 2.B2C_{16}$$

octal or hexadecimal to binary conversion   Converting in the reverse direction, from octal or hexadecimal to binary, is very easy. We simply replace each octal or hexadecimal digit with the corresponding 3- or 4-bit string, as shown below:

$$1357_8 \ = \ 001\ 011\ 101\ 111_2$$

$$2046.17_8 \ = \ 010\ 000\ 100\ 110.001\ 111_2$$

$$BEAD_{16} \ = \ 1011\ 1110\ 1010\ 1101_2$$

$$9F.46C_{16} \ = \ 1001\ 1111.0100\ 0110\ 1100_2$$

The octal number system was quite popular twenty years ago because of certain minicomputers that had their front-panel lights and switches arranged in groups of three. However, the octal number system is not used much today,

because of the preponderence of machines that process 8-bit *bytes*. It is difficult   *byte*
to extract individual byte values in multibyte quantities in the octal representation;
for example, what are the octal values of the four 8-bit bytes in the 32-bit number
with octal representation 12345670123_8?

In the hexadecimal system, two digits neatly represent an 8-bit byte, and
$2n$ digits represent an n-byte word; each pair of digits constitutes exactly one
byte. For example, the 32-bit hexadecimal number $5678ABCD_{16}$ consists of
four bytes with values 56_{16}, 78_{16}, AB_{16}, and CD_{16}. In this context, a 4-bit
hexadecimal digit is sometimes called a *nibble*; a 32-bit (4-byte) number has   *nibble*
eight nibbles. Hexadecimal numbers are often used to describe a computer's
memory address space. For example, a computer with 16-bit addresses might
be described as having read/write memory installed at addresses 0–$EFFF_{16}$, and
read-only memory at addresses $F000$–$FFFF_{16}$.

2.3 GENERAL POSITIONAL NUMBER SYSTEM CONVERSIONS

In general, conversion between two radices cannot be done by simple substitutions; arithmetic operations are required. In this section, we show how to convert
a number in any radix to radix 10 and vice versa, using radix-10 arithmetic.

In Section 2.1, we indicated that the value of a number in any radix is given   *radix-r to decimal*
by the formula   *conversion*

$$D = \sum_{i=-n}^{p-1} d_i \cdot r^i$$

where r is the radix of the number and there are p digits to the left of the
radix point and n to the right. Thus, the value of the number can be found by
converting each digit of the number to its radix-10 equivalent and expanding the
formula using radix-10 arithmetic. Some examples are given below:

$$1CE8_{16} = 1 \cdot 16^3 + 12 \cdot 16^2 + 14 \cdot 16^1 + 8 \cdot 16^0 = 7400_{10}$$

$$F1A3_{16} = 15 \cdot 16^3 + 1 \cdot 16^2 + 10 \cdot 16^1 + 3 \cdot 16^0 = 61859_{10}$$

$$436.5_8 = 4 \cdot 8^2 + 3 \cdot 8^1 + 6 \cdot 8^0 + 5 \cdot 8^{-1} = 286.625_{10}$$

$$132.3_4 = 1 \cdot 4^2 + 3 \cdot 4^1 + 2 \cdot 4^0 + 3 \cdot 4^{-1} = 30.75_{10}$$

A shortcut for converting whole numbers to radix 10 is obtained by rewriting the expansion formula as follows:

$$D = ((\cdots((d_{p-1}) \cdot r + d_{p-2}) \cdot r + \cdots) \cdot r + d_1) \cdot r + d_0$$

That is, we start with a sum of 0; beginning with the leftmost digit, we multiply the sum by r and add the next digit to the sum, repeating until all digits have been processed. For example, we can write

$$F1AC_{16} = (((15) \cdot 16 + 1) \cdot 16 + 10) \cdot 16 + 12$$

decimal to radix-r conversion

Although this formula is not too exciting in itself, it forms the basis for a very convenient method of converting a decimal number D to a radix r. Consider what happens if we divide the formula by r. Since the parenthesized part of the formula is evenly divisible by r, the quotient will be

$$Q = (\cdots((d_{p-1}) \cdot r + d_{p-2}) \cdot r + \cdots) \cdot r + d_1$$

and the remainder will be d_0. Thus, d_0 can be computed as the remainder of the long division of D by r. Furthermore, the quotient Q has the same form as the original formula. Therefore, successive divisions by r will yield successive digits of D from right to left, until all the digits of D have been derived. Examples are given below:

$$179 \div 2 = 89 \text{ remainder } 1 \quad \text{(LSB)}$$
$$\div 2 = 44 \text{ remainder } 1$$
$$\div 2 = 22 \text{ remainder } 0$$
$$\div 2 = 11 \text{ remainder } 0$$
$$\div 2 = 5 \text{ remainder } 1$$
$$\div 2 = 2 \text{ remainder } 1$$
$$\div 2 = 1 \text{ remainder } 0$$
$$\div 2 = 0 \text{ remainder } 1 \quad \text{(MSB)}$$
$$179_{10} = 10110011_2$$

$$467 \div 8 = 58 \text{ remainder } 3 \quad \text{(least significant digit)}$$
$$\div 8 = 7 \text{ remainder } 2$$
$$\div 8 = 0 \text{ remainder } 7 \quad \text{(most significant digit)}$$
$$467_{10} = 723_8$$

$$3417 \div 16 = 213 \text{ remainder } 9 \quad \text{(least significant digit)}$$
$$\div 16 = 13 \text{ remainder } 5$$
$$\div 16 = 0 \text{ remainder } 13 \quad \text{(most significant digit)}$$
$$3417_{10} = D59_{16}$$

Table 2–2 summarizes methods for converting among the most common radices.

Table 2–2 Conversion methods for common radices.

Conversion	Method	Example
Binary to		
Octal	Substitution	$10111011001_2 = 10\ 111\ 011\ 001_2 = 2731_8$
Hexadecimal	Substitution	$10111011001_2 = 101\ 1101\ 1001_2 = 5D9_{16}$
Decimal	Summation	$10111011001_2 = 1 \cdot 1024 + 0 \cdot 512 + 1 \cdot 256 + 1 \cdot 128 + 1 \cdot 64$ $+ 0 \cdot 32 + 1 \cdot 16 + 1 \cdot 8 + 0 \cdot 4 + 0 \cdot 2 + 1 \cdot 1 = 1497_{10}$
Octal to		
Binary	Substitution	$1234_8 = 001\ 010\ 011\ 100_2$
Hexadecimal	Substitution	$1234_8 = 001\ 010\ 011\ 100_2 = 0010\ 1001\ 1100_2 = 29C_{16}$
Decimal	Summation	$1234_8 = 1 \cdot 512 + 2 \cdot 64 + 3 \cdot 8 + 4 \cdot 1 = 668_{10}$
Hexadecimal to		
Binary	Substitution	$C0DE_{16} = 1100\ 0000\ 1101\ 1110_2$
Octal	Substitution	$C0DE_{16} = 1100\ 0000\ 1101\ 1110_2 = 1\ 100\ 000\ 011\ 011\ 110_2 = 140336_8$
Decimal	Summation	$C0DE_{16} = 12 \cdot 4096 + 0 \cdot 256 + 13 \cdot 16 + 14 \cdot 1 = 49374_{10}$
Decimal to		
Binary	Division	$108_{10} = 1101100_2$ $108_{10} \div 2 = 54 \text{ remainder } 0 \quad \text{(LSB)}$ $\div 2 = 27 \text{ remainder } 0$ $\div 2 = 13 \text{ remainder } 1$ $\div 2 = 6 \text{ remainder } 1$ $\div 2 = 3 \text{ remainder } 0$ $\div 2 = 1 \text{ remainder } 1$ $\div 2 = 0 \text{ remainder } 1 \quad \text{(MSB)}$
Octal	Division	$108_{10} = 154_8$ $108_{10} \div 8 = 13 \text{ remainder } 4 \quad \text{(least significant digit)}$ $\div 8 = 1 \text{ remainder } 5$ $\div 8 = 0 \text{ remainder } 1 \quad \text{(most significant digit)}$
Hexadecimal	Division	$108_{10} = 6C_{16}$ $108_{10} \div 16 = 6 \text{ remainder } 12 \quad \text{(least significant digit)}$ $\div 16 = 0 \text{ remainder } 6 \quad \text{(most significant digit)}$

2.4 ADDITION AND SUBTRACTION OF NONDECIMAL NUMBERS

Addition and subtraction of nondecimal numbers by hand uses the same technique that we learned in grammar school for decimal numbers; the only catch is that the addition and subtraction tables are different.

Table 2–3
Binary addition and subtraction table.

c_{in} or b_{in}	x	y	c_{out}	s	b_{out}	d
0	0	0	0	0	0	0
0	0	1	0	1	1	1
0	1	0	0	1	0	1
0	1	1	1	0	0	0
1	0	0	0	1	1	1
1	0	1	1	0	1	0
1	1	0	1	0	0	0
1	1	1	1	1	1	1

binary addition   Table 2–3 is the addition and subtraction table for binary digits. To add two binary numbers X and Y, we add together the least significant bits with an initial carry (c_{in}) of 0, producing carry (c_{out}) and sum (s) bits according to the table. We continue processing bits from right to left, including the carry out of each column in the next column's sum.

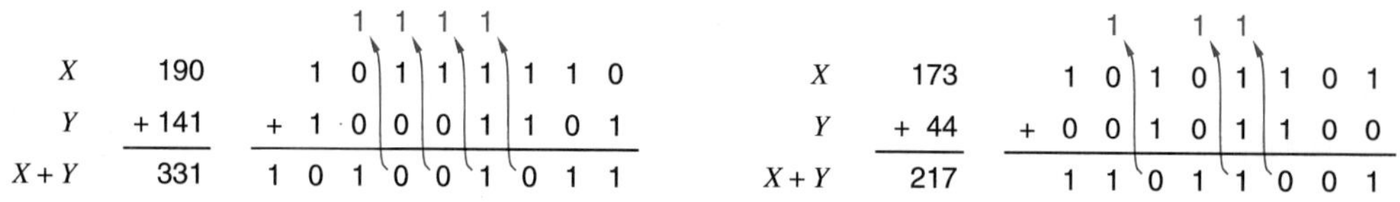

$$
\begin{array}{rrl}
X & 190 & 1\ 0\ 1\ 1\ 1\ 1\ 1\ 0 \\
Y & +\ 141 & +\ 1\ 0\ 0\ 0\ 1\ 1\ 0\ 1 \\ \hline
X+Y & 331 & 1\ 0\ 1\ 0\ 0\ 1\ 0\ 1\ 1
\end{array}
\qquad
\begin{array}{rrl}
X & 173 & 1\ 0\ 1\ 0\ 1\ 1\ 0\ 1 \\
Y & +\ 44 & +\ 0\ 0\ 1\ 0\ 1\ 1\ 0\ 0 \\ \hline
X+Y & 217 & 1\ 1\ 0\ 1\ 1\ 0\ 0\ 1
\end{array}
$$

Figure 2–1   Examples of decimal and corresponding binary additions.

Two examples of decimal additions and the corresponding binary additions are shown in Figure 2–1, using a colored arrow to indicate a carry of 1. The same examples are repeated below along with two more, with the carries shown as a bit string C:

$$
\begin{array}{rrr}
C & & 101111000 \\
X & 190 & 10111110 \\
Y & +\ 141 & +\ 10001101 \\ \hline
X+Y & 331 & 101001011
\end{array}
\qquad
\begin{array}{rrr}
C & & 001011000 \\
X & 173 & 10101101 \\
Y & +\ 44 & +\ 00101100 \\ \hline
X+Y & 217 & 11011001
\end{array}
$$

$$
\begin{array}{rrr}
C & & 011111110 \\
X & 127 & 01111111 \\
Y & +\ 63 & +\ 00111111 \\ \hline
X+Y & 190 & 10111110
\end{array}
\qquad
\begin{array}{rrr}
C & & 000000000 \\
X & 170 & 10101010 \\
Y & +\ 85 & +\ 01010101 \\ \hline
X+Y & 255 & 11111111
\end{array}
$$

binary subtraction   Binary subtraction is performed similarly, using borrows (b_{in} and b_{out}) instead of carries between steps, and producing a difference bit d. Two examples of decimal subtractions and the corresponding binary subtractions are shown in

Figure 2–2. As in decimal subtraction, the binary minuend values in the columns *minuend* are modified when borrows occur, as shown by the colored arrows and bits. The *subtrahend* examples of the figure are repeated below along with two more, this time showing the borrows as a bit string B:

B		001111100	B		011011010
X	229	11100101	X	210	11010010
Y	− 46	− 00101110	Y	− 109	− 01101101
$X-Y$	183	10110111	$X-Y$	101	01100101

B		010101010	B		000000000
X	170	10101010	X	221	11011101
Y	− 85	− 01010101	Y	− 76	− 01001100
$X-Y$	85	01010101	$X-Y$	145	10010001

A very common use of subtraction in computers is to compare two numbers. *comparing numbers* For example, if the operation $X-Y$ produces a borrow out of the most significant bit position, then X is less than Y; otherwise, X is greater than or equal to Y. The relationship between carries and borrows in adders and subtracters will be explored in Section 5.9.

Addition and subtraction tables can be developed for octal and hexadecimal digits, or any other desired radix. However, few computer engineers bother to memorize these tables. If you must manipulate nondecimal numbers only infrequently, then it's easy enough to convert them to decimal, compute results, and convert back. On the other hand, if you must perform calculations in binary, octal, or hexadecimal frequently, then you should ask Santa for a programmer's "hex calculator" from Texas Instruments or Casio.

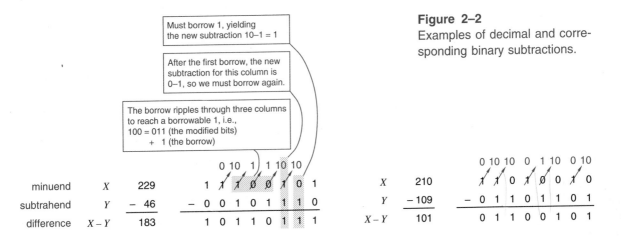

Must borrow 1, yielding the new subtraction 10–1 = 1

After the first borrow, the new subtraction for this column is 0–1, so we must borrow again.

The borrow ripples through three columns to reach a borrowable 1, i.e., 100 = 011 (the modified bits) + 1 (the borrow)

Figure 2–2
Examples of decimal and corresponding binary subtractions.

minuend	X	229	0 10 1 1 10 10
			1 1 1 0 0 1 0 1
subtrahend	Y	− 46	− 0 0 1 0 1 1 1 0
difference	$X-Y$	183	1 0 1 1 0 1 1 1

X	210	0 10 10 0 1 10 0 10
		1 1 0 1 0 0 1 0
Y	− 109	− 0 1 1 0 1 1 0 1
$X-Y$	101	0 1 1 0 0 1 0 1

If the calculator's battery wears out, some mental shortcuts can be used to facilitate nondecimal arithmetic. In general, each column addition (or subtraction) can be done by converting the column digits to decimal, adding in decimal, and converting the result to corresponding sum and carry digits in the nondecimal radix. (A carry is produced whenever the column sum equals or exceeds the radix.) Since the addition is done in decimal, we rely on our knowledge of the decimal addition table; the only new thing that we need to learn is the conversion from decimal to nondecimal digits and vice versa. The sequence of steps for mentally adding two hexadecimal numbers is shown below:

hexadecimal addition

C	$1\ 1\ 0\ 0$	1	1	0	0
X	$1\ 9\ B\ 9_{16}$	1	9	11	9
Y	$+\ C\ 7\ E\ 6_{16}$	$+$ 12	7	14	6
$X + Y$	$E\ 1\ 9\ F_{16}$	14	17	25	15
		14	16+1	16+9	15
		E	1	9	F

2.5   REPRESENTATION OF NEGATIVE NUMBERS

So far, we have dealt only with positive numbers, but there are many ways to represent negative numbers. In everyday business, we use the signed-magnitude system, discussed next. However, most computers use one of the complement number systems that we introduce later.

2.5.1   Signed-Magnitude Representation

signed-magnitude system

In the *signed-magnitude system*, a number consists of a magnitude and a symbol indicating whether the magnitude is positive or negative. Thus, we interpret decimal numbers +98, −57, +123.5, and −13 in the usual way, and we also assume that the sign is "+" if no sign symbol is written. There are two possible representations of zero, "+0" and "−0", but both have the same value.

sign bit

The signed-magnitude system is applied to binary numbers by using an extra bit position to represent the sign (the *sign bit*). Traditionally, the most significant bit (MSB) of a bit string is used as the sign bit (0 = plus, 1 = minus), and the lower-order bits contain the magnitude. Thus, we can write several 8-bit signed-magnitude integers and their decimal equivalents:

$$01010101_2\ =\ +85_{10} \qquad\qquad 11010101_2\ =\ -85_{10}$$
$$01111111_2\ =\ +127_{10} \qquad\qquad 11111111_2\ =\ -127_{10}$$
$$00000000_2\ =\ +0_{10} \qquad\qquad 10000000_2\ =\ -0_{10}$$

The signed-magnitude system contains an equal number of positive and negative integers. An n-bit signed-magnitude integer lies within the range $-(2^{n-1}-1)$ through $+(2^{n-1}-1)$, with two possible representations of zero.

Now suppose that we wanted to build a digital logic circuit that adds signed-magnitude numbers. The circuit must examine the signs of the addends to determine what to do with the magnitudes. If the signs are the same, it must add the magnitudes and give the result the same sign. If the signs are different, it must compare the magnitudes, subtract the smaller from the larger, and give the result the sign of the larger. All of these "ifs," "adds," "subtracts," and "compares" translate into a lot of logic-circuit complexity. Adders for complement number systems are much simpler, as we'll show next. Perhaps the one redeeming feature of a signed-magnitude system is that, once we know how to build a signed-magnitude adder, a signed-magnitude subtractor is almost trivial to build—it need only change the sign of the subtrahend and pass it along with the minuend to an adder. *signed-magnitude adder*

signed-magnitude subtractor

2.5.2 Complement Number Systems

While the signed-magnitude system negates a number by changing its sign, a *complement number system* negates a number by taking its complement as defined by the system. Taking the complement is more difficult than changing the sign, but two numbers in a complement number system can be added or subtracted directly without the sign and magnitude checks required by the signed-magnitude system. We shall describe two complement number systems, called the "radix complement" and the "diminished radix-complement." *complement number system*

In any complement number system, we normally deal with a fixed number of digits, say n. (However, we can increase the number of digits by "sign extension" as shown in Exercise 2.22, and decrease the number by truncating high-order digits as shown in Exercise 2.23.) We further assume that the radix is r, and that numbers have the form

$$D = d_{n-1}d_{n-2}\cdots d_1 d_0.$$

The radix point is on the right and so the number is an integer. If an operation produces a result that requires more than n digits, we throw away the extra high-order digit(s). If a number D is complemented twice, the result is D.

2.5.3 Radix-Complement Representation

In a *radix-complement system*, the complement of an n-digit number is obtained by subtracting it from r^n. In the decimal number system, the radix complement is called the *10's complement*. Some examples using 4-digit decimal numbers (and subtraction from 10,000) are shown in Table 2–4. *radix-complement system*

10's complement

Table 2–4
Examples of 10's and 9s'
complements.

Number	10's complement	9s' complement
1849	8151	8150
2067	7933	7932
100	9900	9899
7	9993	9992
8151	1849	1848
0	10000 (= 0)	9999

By definition, the radix complement of an n-digit number D is obtained by subtracting it from r^n. If D is between 1 and $r^n - 1$, this subtraction produces another number between 1 and $r^n - 1$. If D is 0, the result of the subtraction is r^n, which has the form $100\cdots00$, where there are a total of $n + 1$ digits. We throw away the extra high-order digit and get the result 0. Thus, there is only one representation of zero in a radix-complement system.

computing the radix complement

It seems from the definition that a subtraction operation is needed to compute the radix complement of D. However, this subtraction can be avoided by rewriting r^n as $(r^n - 1) + 1$ and $r^n - D$ as $((r^n - 1) - D) + 1$. The number $r^n - 1$ has the form $mm\cdots mm$, where $m = r - 1$ and there are n m's. For example, 10,000 equals 9,999 + 1. If we define the complement of a digit d as $r - 1 - d$, then $(r^n - 1) - D$ is obtained by complementing the digits of D. Therefore, the radix

Table 2–5
Digit complements.

Digit	Complement			
	Binary	Octal	Decimal	Hexadecimal
0	1	7	9	F
1	0	6	8	E
2	–	5	7	D
3	–	4	6	C
4	–	3	5	B
5	–	2	4	A
6	–	1	3	9
7	–	0	2	8
8	–	–	1	7
9	–	–	0	6
A	–	–	–	5
B	–	–	–	4
C	–	–	–	3
D	–	–	–	2
E	–	–	–	1
F	–	–	–	0

numbers is somewhat trickier than a two's-complement adder (see Exercise 7.53). Also, zero-detecting circuits in a ones'-complement system must check for both representations of zero, or must always convert $11 \cdots 11$ to $00 \cdots 00$.

* 2.5.7 Excess Representations

Yes, the number of different systems for representing negative numbers *is* excessive, but there's just one more for us to cover. In *excess-B representation*, an *m*-bit string whose unsigned integer value is M ($0 \le M < 2^m$) represents the signed integer $M - B$, where B is called the *bias* of the number system.

excess-B representation
bias

      For example, an *excess-2^{m-1} system* represents a number X in the range -2^{m-1} through $+2^{m-1} - 1$ by the *m*-bit binary representation of $X + 2^{m-1}$ (which is always nonnegative and less than 2^m). The range of this representation is exactly the same as that of *m*-bit two's-complement numbers. In fact, the representations of any number in the two systems are identical except for the sign bits, which are always opposite. (Note that this is true only when the bias is 2^{m-1}.)

excess-2^{m-1} system

      The most common use of excess representations is in floating-point number systems (see References).

2.6 TWO'S-COMPLEMENT ADDITION AND SUBTRACTION

2.6.1 Addition Rules

A table of decimal numbers and their equivalents in different number systems, Table 2–6, reveals why the two's complement is preferred for arithmetic operations. If we start with 1000_2 (-8_{10}) and count up, we see that each successive two's-complement number all the way to 0111_2 ($+7_{10}$) can be obtained by adding 1 to the previous one, ignoring any carries beyond the fourth bit position. The same cannot be said of signed-magnitude and ones'-complement numbers. Because ordinary addition is just an extension of counting, two's-complement numbers can thus be added by ordinary binary addition, ignoring any carries beyond the MSB. The result will always be the correct sum as long as the range of the number system is not exceeded. Some examples of decimal addition and the corresponding 4-bit two's-complement additions confirm this:

two's-complement addition

+3	0011	−2	1110
+ +4	+ 0100	+ −6	+ 1010
+7	0111	−8	1 1000

+6	0110	+4	0100
+ −3	+ 1101	+ −7	+ 1001
+3	1 0011	−3	1101

Table 2–6

Decimal and 4-bit numbers.

Decimal	Two's Complement	Ones' Complement	Signed Magnitude	Excess 2^{m-1}
−8	1000	—	—	0000
−7	1001	1000	1111	0001
−6	1010	1001	1110	0010
−5	1011	1010	1101	0011
−4	1100	1011	1100	0100
−3	1101	1100	1011	0101
−2	1110	1101	1010	0110
−1	1111	1110	1001	0111
0	0000	1111 or 0000	1000 or 0000	1000
1	0001	0001	0001	1001
2	0010	0010	0010	1010
3	0011	0011	0011	1011
4	0100	0100	0100	1100
5	0101	0101	0101	1101
6	0110	0110	0110	1110
7	0111	0111	0111	1111

2.6.2 A Graphical View

Another way to view the two's-complement system uses the 4-bit "counter" shown in Figure 2–3. Here we have shown the numbers in a circular or "modular" representation. The operation of this counter very closely mimics that of a real up/down counter circuit, which we'll study in Section 9.3. Starting with the arrow pointing to any number, we can add +n to that number by counting up n times, that is, by moving the arrow n positions clockwise. It is also evident that we can subtract n from a number by counting down n times, that is, by moving the arrow n positions counterclockwise. Of course, these operations give correct results only if n is small enough that we don't cross the discontinuity between −8 and +7.

What is most interesting is that we can also subtract n (or add −n) by moving the arrow $16 - n$ positions clockwise. Notice that the quantity $16 - n$ is what we defined to be the 4-bit two's complement of n, that is, the two's-complement representation of −n. This graphically supports our earlier claim that a negative number in two's-complement representation may be added to another number simply by adding the 4-bit representations using ordinary binary addition. Adding a number in Figure 2–3 is equivalent to moving the arrow a corresponding number of positions clockwise.

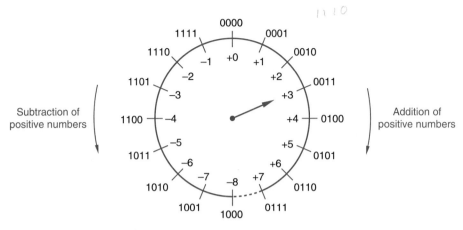

Figure 2–3 A modular counting representation of 4-bit two's-complement numbers.

2.6.3 Overflow

If an addition operation produces a result that exceeds the range of the number system, *overflow* is said to occur. In the modular counting representation of Figure 2–3, overflow occurs during addition of positive numbers when we count past +7. Addition of two numbers with different signs can never produce overflow, but addition of two numbers of like sign can, as shown by the following examples: *overflow*

$$
\begin{array}{rl}
-3 & 1101 \\
+\ -6 & +\ 1010 \\
\hline
-9 & 1\,0111 = +7
\end{array}
\qquad\qquad
\begin{array}{rl}
+5 & 0101 \\
+\ +6 & +\ 0110 \\
\hline
+11 & 1011 = -5
\end{array}
$$

$$
\begin{array}{rl}
-8 & 1000 \\
+\ -8 & +\ 1000 \\
\hline
-16 & 1\,0000 = +0
\end{array}
\qquad\qquad
\begin{array}{rl}
+7 & 0111 \\
+\ +7 & +\ 0111 \\
\hline
+14 & 1110 = -2
\end{array}
$$

Fortunately, there is a simple rule for detecting overflow in addition: An addition overflows if the signs of the addends are the same and the sign of the sum is different from the addends' sign. The overflow rule is sometimes stated in terms of carries generated during the addition operation: An addition overflows if the carry bits c_{in} into and c_{out} out of the sign position are different. Close examination of Table 2–3 on page 28 shows that the two rules are equivalent—there are only two cases where $c_{in} \neq c_{out}$, and these are the only two cases where $x = y$ and the sum bit is different. *overflow rules*

2.6.4   Subtraction Rules

two's-complement
subtraction

Two's-complement numbers may be subtracted as if they were ordinary unsigned binary numbers, and appropriate rules for detecting overflow may be formulated. However, most subtraction circuits for two's-complement numbers do not perform subtraction directly. Rather, they negate the subtrahend by taking its two's complement, and then add it to the minuend using the normal rules for addition.

Negating the subtrahend and adding the minuend can be accomplished with only one addition operation as follows: Perform a bit-by-bit complement of the subtrahend and add the complemented subtrahend to the minuend with an initial carry (c_{in}) of 1 instead of 0. Examples are given below:

$$
\begin{array}{rrr}
 & & 1 \longrightarrow c_{in} \\
+4 & 0100 & 0100 \\
-\ +3 & -\ 0011 & +\ 1100 \\
\hline
+1 & & 1\ 0001
\end{array}
\qquad
\begin{array}{rrr}
 & & 1 \longrightarrow c_{in} \\
+3 & 0011 & 0011 \\
-\ +4 & -\ 0100 & +\ 1011 \\
\hline
-1 & & 1111
\end{array}
$$

$$
\begin{array}{rrr}
 & & 1 \longrightarrow c_{in} \\
+3 & 0011 & 0011 \\
-\ -4 & -\ 1100 & +\ 0011 \\
\hline
+7 & & 0111
\end{array}
\qquad
\begin{array}{rrr}
 & & 1 \longrightarrow c_{in} \\
-3 & 1101 & 1101 \\
-\ -4 & -\ 1100 & +\ 0011 \\
\hline
+1 & & 1\ 0001
\end{array}
$$

Overflow in subtraction can be detected by examining the signs of the minuend and the *complemented* subtrahend, using the same rule as in addition. Or, using the technique in the preceding examples, the carries into and out of the sign position can be observed and overflow detected irrespective of the signs of inputs and output, again using the same rule as in addition.

An attempt to negate the "extra" negative number results in overflow according to the rules above, when we add 1 in the complementation process:

$$
\begin{array}{rl}
-(-8) = -1000 = & 0111 \\
+ & 0001 \\
\hline
& 1000 = -8
\end{array}
$$

However, this number can still be used in additions and subtractions as long as the final result does not exceed the number range:

$$
\begin{array}{rr}
+4 & 0100 \\
+\ -8 & +\ 1000 \\
\hline
-4 & 1100
\end{array}
\qquad
\begin{array}{rrr}
 & & 1 \longrightarrow c_{in} \\
-3 & 1101 & 1101 \\
-\ -8 & -\ 1000 & +\ 0111 \\
\hline
+5 & & 1\ 0101
\end{array}
$$

2.6.5 Two's-Complement and Unsigned Binary Numbers

Since two's-complement numbers are added and subtracted by the same basic binary addition and subtraction algorithms as unsigned numbers of the same length, a computer or other digital system can use the same adder circuit to handle numbers of both types. However, the results must be interpreted differently depending on whether the system is dealing with signed numbers (e.g., –8 through +7) or unsigned numbers (e.g., 0 through 15). *signed vs. unsigned numbers*

   We introduced a graphical representation of the 4-bit two's-complement system in Figure 2–3. We can relabel this figure as shown in Figure 2–4 to obtain a representation of the 4-bit unsigned numbers. The binary combinations occupy the same positions on the wheel, and a number is still added by moving the arrow a corresponding number of positions clockwise, and subtracted by moving the arrow counterclockwise.

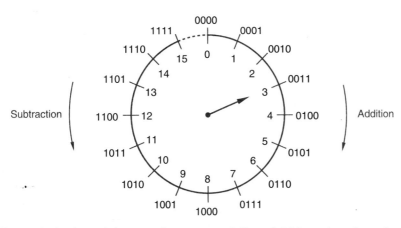

Figure 2–4   A modular counting representation of 4-bit unsigned numbers.

   An addition operation exceeds the range of the 4-bit unsigned number system in Figure 2–4 if the arrow moves clockwise through the discontinuity between 0 and 15. In this case a *carry* out of the most significant bit position is said to occur. *carry*

   Likewise a subtraction operation exceeds the range of the number system if the arrow moves counterclockwise through the discontinuity. In this case a *borrow* out of the most significant bit position is said to occur. *borrow*

   From Figure 2–4 it is also evident that we may subtract an unsigned number n by counting *clockwise* $16 - n$ positions. This is equivalent to *adding* the 4-bit two's-complement of n. The subtraction produces a borrow if the corresponding addition of the two's complement *does not* produce a carry.

   In summary, in unsigned addition the carry or borrow in the most significant bit position indicates an out-of-range result. In signed, two's-complement addition the overflow condition defined earlier indicates an out-of-range result. The carry from the most significant bit position is irrelevant in signed addition in the sense that overflow may or may not occur independently of whether or not a carry occurs.

* 2.7 ONES'-COMPLEMENT ADDITION AND SUBTRACTION

Another look at Table 2–6 helps to explain the rule for adding ones'-complement numbers. If we start at 1000_2 (-7_{10}) and count up, we obtain each successive ones'-complement number by adding 1 to the previous one, *except* at the transition from 1111_2 (negative 0) to 0001_2 ($+1_{10}$). To maintain the proper count, we must add 2 instead of 1 whenever we count past 1111_2. This suggests a technique for adding ones'-complement numbers: Perform a standard binary addition, but add an extra 1 whenever we count past 1111_2.

ones'-complement addition

Counting past 1111_2 during an addition can be detected by observing the carry out of the sign position. Thus, the rule for adding ones'-complement numbers can be stated quite simply:

- Perform a standard binary addition; if there is a carry out of the sign position, add 1 to the result.

end-around carry

This rule is often called *end-around carry*. Examples of ones'-complement addition are given below; the last three include an end-around carry:

$$
\begin{array}{rl}
+3 & 0011 \\
+\ +4 & +\ 0100 \\
\hline
+7 & 0111
\end{array}
\qquad
\begin{array}{rl}
+4 & 0100 \\
+\ -7 & +\ 1000 \\
\hline
-3 & 1100
\end{array}
\qquad
\begin{array}{rl}
+5 & 0101 \\
+\ -5 & +\ 1010 \\
\hline
-0 & 1111
\end{array}
$$

$$
\begin{array}{rl}
-2 & 1101 \\
+\ -5 & +\ 1010 \\
\hline
-7 & 1\,0111 \\
 & +\quad 1 \\
\hline
 & 1000
\end{array}
\qquad
\begin{array}{rl}
+6 & 0110 \\
+\ -3 & +\ 1100 \\
\hline
+3 & 1\,0010 \\
 & +\quad 1 \\
\hline
 & 0011
\end{array}
\qquad
\begin{array}{rl}
-0 & 1111 \\
+\ -0 & +\ 1111 \\
\hline
-0 & 1\,1110 \\
 & +\quad 1 \\
\hline
 & 1111
\end{array}
$$

Following the two-step addition rule above, the addition of a number and its ones' complement produces negative 0. In fact, an addition operation using this rule can never produce positive 0 unless both addends are positive 0.

ones'-complement subtraction

As with two's complement, the easiest way to do ones'-complement subtraction is to complement the subtrahend and add. Overflow rules for ones'-complement addition and subtraction are the same as for two's complement.

Table 2–7 summarizes the rules that we presented in this and previous sections for negation, addition, and subtraction in binary number systems.

Table 2–7 Summary of addition and subtraction rules for binary numbers.

Number System	Addition Rules	Negation Rules	Subtraction Rules
Unsigned	Add the numbers. Result is out of range if a carry out of the MSB occurs.	Not applicable	Subtract the subtrahend from the minuend. Result is out of range if a borrow out of the MSB occurs.
Signed magnitude	(same sign) Add the magnitudes; overflow occurs if a carry out of MSB occurs; result has the same sign. (opposite sign) Subtract the smaller magnitude from the larger; overflow is impossible; result has the sign of the larger.	Change the sign bit.	Change the sign bit of the subtrahend and proceed as in addition.
Two's complement	Add, ignoring any carry out of the MSB. Overflow occurs if the carries into and out of MSB are different.	Complement all bits of the subtrahend; add 1 to the result.	Complement all bits of the subtrahend and add to the minuend with an initial carry of 1.
Ones' complement	Add; if there is a carry out of the MSB, add 1 to the result. Overflow if carries into and out of MSB are different.	Complement all bits of the subtrahend.	Complement all bits of the subtrahend and proceed as in addition.

* 2.8 BINARY MULTIPLICATION

In grammar school we learned to multiply by adding a list of shifted multiplicands computed according to the digits of the multiplier. The same method can be used to obtain the product of two unsigned binary numbers. Forming the shifted multiplicands is trivial in binary multiplication, since the only possible values of the multiplier digits are 0 and 1. An example is shown below:

shift-and-add multiplication

unsigned binary multiplication

```
       11              1011     multiplicand
    ×  13           ×  1101     multiplier
   ───────          ───────
       33              1011  ⎫
       11              0000  ⎪
   ───────             1011  ⎬ shifted multiplicands
      143              1011  ⎭
                   ─────────
                   10001111     product
```

partial product

Instead of listing all the shifted multiplicands and then adding, in a digital system it is more convenient to add each shifted multiplicand as it is created to a *partial product*. Applying this technique to the previous example, four additions and partial products are used to multiply 4-bit numbers:

$$
\begin{array}{rl}
11 & \\
\times\ 13 & \\
\hline
\end{array}
\qquad
\begin{array}{rl}
1011 & \text{multiplicand} \\
\times\quad 1101 & \text{multiplier} \\
\hline
0000 & \text{partial product} \\
1011 & \text{shifted multiplicand} \\
\hline
01011 & \text{partial product} \\
0000\!\downarrow & \text{shifted multiplicand} \\
\hline
001011 & \text{partial product} \\
1011\!\downarrow\downarrow & \text{shifted multiplicand} \\
\hline
0110111 & \text{partial product} \\
1011\!\downarrow\downarrow\downarrow & \text{shifted multiplicand} \\
\hline
10001111 & \text{product}
\end{array}
$$

In general, when we multiply an n-bit number by an m-bit number, the resulting product requires at most $n + m$ bits to express. The shift-and-add algorithm requires m partial products and additions to obtain the result, but the first addition is trivial, since the first partial product is zero. Although the first partial product has only n significant bits, after each addition step the partial product gains one more significant bit, since each addition may produce a carry. At the same time, each step yields one more partial product bit, starting with the rightmost and working toward the left, that does not change for the rest of the computation. The shift-and-add algorithm can be performed by a digital circuit that includes a shift register, an adder, and control logic, as shown in Section 8.1.2.

signed multiplication

Multiplication of signed numbers can be accomplished using unsigned multiplication and the usual grammar school rules: Perform an unsigned multiplication of the magnitudes and make the product positive if the operands had the same sign, negative if they had different signs. This is very convenient in signed-magnitude systems, since the sign and magnitude are separate.

In the two's-complement system, obtaining the magnitude of a negative number and negating the unsigned product are nontrivial operations. This leads us to seek a more efficient way of performing two's-complement multiplication, described next.

two's-complement multiplication

Conceptually, unsigned multiplication is accomplished by a sequence of unsigned additions of the shifted multiplicands; at each step, the shift of the multiplicand corresponds to the weight of the multiplier bit. The bits in a two's-complement number have the same weights as in an unsigned number, except for the MSB, which has a negative weight (see Section 2.5.4). It follows that we can perform two's-complement multiplication by a sequence of two's-complement additions of shifted multiplicands, except for the last step, in which the shifted multiplicand corresponding to the MSB of the multiplier must be negated before it is added to the partial product. Our previous example is repeated below, this time interpreting the multiplier and multiplicand as two's-complement numbers:

$$
\begin{array}{rl}
-5 & \\
\times\ -3 & \\
\end{array}
\qquad
\begin{array}{rl}
1011 & \text{multiplicand} \\
\times\quad 1101 & \text{multiplier} \\
\hline
00000 & \text{partial product} \\
11011 & \text{shifted multiplicand} \\
\hline
111011 & \text{partial product} \\
00000\!\downarrow & \text{shifted multiplicand} \\
\hline
1111011 & \text{partial product} \\
11011\!\downarrow\!\downarrow & \text{shifted multiplicand} \\
\hline
11100111 & \text{partial product} \\
00101\!\downarrow\!\downarrow\!\downarrow & \text{shifted and negated multiplicand} \\
\hline
00001111 & \text{product} \\
\end{array}
$$

Handling the MSBs is a little tricky because we gain one significant bit at each step and we are working with signed numbers. Therefore, before adding each shifted multiplicand and k-bit partial product, we change them to $k+1$ significant bits by sign extension, as shown in color above. Each resulting sum has $k+1$ bits; any carry out of the MSB of the $k+1$-bit sum is ignored.

* 2.9   BINARY DIVISION

The simplest binary division algorithm is based on the shift-and-subtract method that we learned in grammar school. Table 2–8 gives examples of this method for unsigned decimal and binary numbers. In both cases, we mentally compare the reduced dividend with multiples of the divisor to determine which multiple of the shifted divisor to subtract. In the decimal case, we first pick 11 as the greatest multiple of 11 less than 21, and then pick 99 as the greatest multiple less than 107. In the binary case, the choice is somewhat simpler, since the only two choices are zero and the divisor itself. *shift-and-subtract division* *unsigned division*

Division methods for binary numbers are somewhat complementary to binary multiplication methods. A typical division algorithm accepts an $n+m$-bit dividend and an n-bit divisor, and produces an m-bit quotient and an n-bit remainder. A division *overflows* if the divisor is zero or the quotient would take more than m bits to express. In most computer division circuits, $n = m$. *division overflow*

Division of signed numbers can be accomplished using unsigned division *signed division*

Table 2–8
Example of long division.

	19		10011	quotient
11)̄	217	1011)̄	11011001	dividend
	11		1011	shifted divisor
	107		0101	reduced dividend
	99		0000	shifted divisor
	8		1010	reduced dividend
			0000	shifted divisor
			10100	reduced dividend
			1011	shifted divisor
			10011	reduced dividend
			1011	shifted divisor
			1000	remainder

and the usual grammar school rules: Perform an unsigned division of the magnitudes and make the quotient positive if the operands had the same sign, negative if they had different signs. The remainder should be given the same sign as the dividend. As in multiplication, there are special techniques for performing division directly on two's-complement numbers; these techniques are often implemented in computer division circuits (see References).

2.10 BINARY CODES FOR DECIMAL NUMBERS

Even though binary numbers are the most appropriate for the internal computations of a digital system, most people still prefer to deal with decimal numbers. As a result, the external interfaces of a digital system may read or display decimal numbers, and some digital devices actually process decimal numbers directly.

The human need to represent decimal numbers doesn't change the basic nature of digital electronic circuits—they still process signals that take on one of only two states that we call 0 and 1. Therefore, a decimal number is represented in a digital system by a string of bits, where different combinations of bit values in the string represent different decimal numbers. For example, if we use a 4-bit string to represent a decimal number, we might assign bit combination 0000 to decimal digit 0, 0001 to 1, 0010 to 2, and so on.

code
code word
A set of n-bit strings in which different bit strings represent different numbers or other things is called a *code*. A particular combination of n bit-values is called a *code word*. As we'll see in the examples of decimal codes in this section, there may or may not be an arithmetic relationship between the bit values in a code word and the thing that it represents. Furthermore, a code that uses n-bit strings need not contain 2^n valid code words.

Table 2–9
Decimal codes.

Decimal digit	BCD (8421)	2421	Excess-3	Biquinary	1-out-of-10
0	0000	0000	0011	0100001	1000000000
1	0001	0001	0100	0100010	0100000000
2	0010	0010	0101	0100100	0010000000
3	0011	0011	0110	0101000	0001000000
4	0100	0100	0111	0110000	0000100000
5	0101	1011	1000	1000001	0000010000
6	0110	1100	1001	1000010	0000001000
7	0111	1101	1010	1000100	0000000100
8	1000	1110	1011	1001000	0000000010
9	1001	1111	1100	1010000	0000000001
Unused code words					
	1010	0101	0000	0000000	0000000000
	1011	0110	0001	0000001	0000000011
	1100	0111	0010	0000010	0000000101
	1101	1000	1101	0000011	0000000110
	1110	1001	1110	0000101	0000000111
	1111	1010	1111	. . .	. . .

At least four bits are needed to represent the ten decimal digits. There are billions and billions of different ways to choose ten 4-bit code words, but some of the more common decimal codes are listed in Table 2–9.

Perhaps the most "natural" decimal code is *binary-coded decimal (BCD)*, which encodes the digits 0 through 9 by their 4-bit unsigned binary representations, 0000 through 1001. The code words 1010 through 1111 are not used. Conversions between BCD and decimal representations are trivial, a direct substitution of four bits for each decimal digit. Some computer programs place two BCD digits in one 8-bit byte in *packed-BCD representation*; thus, one byte may represent the values from 0 to 99 as opposed to 0 to 255 for a normal unsigned 8-bit binary number. BCD numbers with any desired number of digits may be obtained by using one byte for each two digits.

As with binary numbers, there are many possible representations of negative

binary-coded decimal (BCD)

packed-BCD representation

The number of different ways to choose m items out of a set of n items is given by a *binomial coefficient*, denoted $\binom{n}{m}$, whose value is $\frac{n!}{m!\cdot(n-m)!}$. For a 4-bit decimal code, there are $\binom{16}{10}$ different ways to choose 10 out of 16 4-bit code words, and 10! ways to assign each different choice to the 10 digits. So there are $\frac{16!}{10!\cdot6!} \cdot 10!$ or 29,059,430,400 different 4-bit decimal codes.

BINOMIAL
COEFFICIENTS

BCD numbers. Signed BCD numbers have one extra digit position for the sign. Both the signed-magnitude and 10's-complement representations are popular. In signed-magnitude BCD, the encoding of the sign bit string is arbitrary; in 10's-complement, 0000 indicates plus and 1001 indicates minus.

BCD addition

Addition of BCD digits is similar to adding 4-bit unsigned binary numbers, except that a correction must be made if a result exceeds 1001. The result is corrected by adding 6; examples are shown below:

$$
\begin{array}{rl}
5 & 0101 \\
+\ \ 9 & +\ 1001 \\
\hline
14 & \quad 1110 \\
& +\ 0110 \ \text{— correction} \\
\hline
10{+}4 & 1\,0100
\end{array}
\qquad
\begin{array}{rl}
4 & 0100 \\
+\ \ 5 & +\ 0101 \\
\hline
9 & \quad 1001
\end{array}
$$

$$
\begin{array}{rl}
8 & 1000 \\
+\ \ 8 & +\ 1000 \\
\hline
16 & 1\,0000 \\
& +\ 0110 \ \text{— correction} \\
\hline
10{+}6 & 1\,0110
\end{array}
\qquad
\begin{array}{rl}
9 & 1001 \\
+\ \ 9 & +\ 1001 \\
\hline
18 & 1\,0010 \\
& +\ 0110 \ \text{— correction} \\
\hline
10{+}8 & 1\,1000
\end{array}
$$

Notice that the addition of two BCD digits produces a carry into the next digit position if either the initial binary addition or the correction factor addition produces a carry. Many computers perform packed-BCD arithmetic using special instructions that handle the carry correction automatically.

weighted code

Binary-coded decimal is a *weighted code* because each decimal digit can be obtained from its code word by assigning a fixed weight to each code-word bit. The weights for the BCD bits are 8, 4, 2, and 1, and for this reason the code is sometimes called the *8421 code*. Another set of weights results in the *2421 code* shown in Table 2–9. This code has the advantage that it is *self-complementing*, that is, the code word for the 9s' complement of any digit may be obtained by complementing the individual bits of the digit's code word.

8421 code
2421 code
self-complementing code

excess-3 code

Another self-complementing code shown in Table 2–9 is the *excess-3 code*. Although this code is not weighted, it has an arithmetic relationship with the BCD code—the code word for each decimal digit is the corresponding BCD code word plus 0011_2. Because the code words follow a standard binary counting sequence, standard binary counters can easily be made to count in excess-3 code, as we'll show in Section 9.3.

biquinary code

Decimal codes can have more than four bits; for example, the *biquinary code* in Table 2–9 uses seven. The first two bits in a code word indicate whether the number is in the range 0–4 or 5–9, and the last five bits indicate which of the five numbers in the selected range is represented.

One potential advantage of using more than the minimum number of bits in a code is an error-detecting property. In the biquinary code, if any one bit in a code word is accidentally changed to the opposite value, the resulting code word does not represent a decimal digit and can therefore be flagged as an error. Out of 128 possible 7-bit code words, only 10 are valid and recognized as decimal digits; the rest can be flagged as errors if they appear.

A *1-out-of-10 code* such as the one shown in the last column of Table 2–9   *1-out-of-10 code* is the sparsest encoding for decimal digits, using 10 out of 1024 possible 10-bit code words.

2.11 GRAY CODE

In electromechanical applications of digital systems—such as machine tools, automotive braking systems, and copiers—it is sometimes necessary for an input sensor to produce a digital value that indicates a mechanical position. For example, Figure 2–5 is a conceptual sketch of an encoding disk and a set of contacts that produce one of eight 3-bit binary-coded values depending on the rotational position of the disk. The dark areas of the disk are connected to a signal source corresponding to logic 1, and the light areas are unconnected, which the contacts interpret as logic 0.

The encoder in Figure 2–5 has a problem when the disk is positioned at certain boundaries between the regions. For example, consider the boundary between the 001 and 010 regions of the disk; two of the encoded bits change here. What value will the encoder produce if the disk is positioned right on the theoretical boundary? Since we're on the border, both 001 and 010 are acceptable. However, because the mechanical assembly is not perfect, the two right-hand contacts may both touch a "1" region, giving an incorrect reading of 011. Likewise, a reading of 000 is possible. In general, this sort of problem can

Figure 2–5
A mechanical encoding disk using a 3-bit binary code.

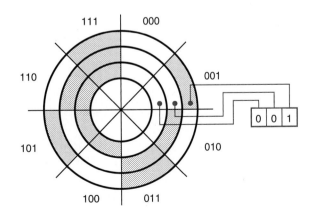

Table 2–10
A comparison of 3-bit binary code and Gray code.

Decimal number	Binary code	Gray code
0	000	000
1	001	001
2	010	011
3	011	010
4	100	110
5	101	111
6	110	101
7	111	100

occur at any boundary where more than one bit changes. The worst problems occur when all three bits are changing, as at the 000–111 and 011–100 boundaries.

Gray code

The encoding-disk problem can be solved by devising a digital code in which only one bit changes between each pair of successive code words. Such a code is called a *Gray code*; a 3-bit Gray code is listed in Table 2–10. We've redesigned the encoding disk using this code as shown in Figure 2–6. Only one bit of the new disk changes at each border, so borderline readings give us a value on one side or the other of the border.

reflected code

There are two convenient ways to construct a Gray code with any desired number of bits. The first method is based on the fact that Gray code is a *reflected code*; it can be defined (and constructed) recursively using the following rules:

(1) A 1-bit Gray code has two code words, 0 and 1.

(2) The first 2^n code words of an $n + 1$-bit Gray code equal the code words of an n-bit Gray code, written in order with a leading 0 appended.

(3) The last 2^n code words of an $n + 1$-bit Gray code equal the code words of an n-bit Gray code, written in reverse order with a leading 1 appended.

Figure 2–6
A mechanical encoding disk using a 3-bit Gray code.

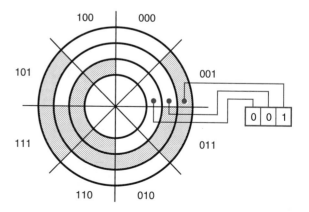

If we draw a line between rows 3 and 4 of Table 2–10, we can see that rules 2 and 3 are true for the 3-bit Gray code. Of course, to construct an n-bit Gray code for an arbitrary value of n with this method, we must also construct a Gray code of each length smaller than n.

The second method allows us to derive an n-bit Gray-code code word directly from the corresponding n-bit binary code word:

(1) The bits of an n-bit binary or Gray-code code word are numbered from right to left, from 0 to $n - 1$.

(2) Bit i of a Gray-code code word is 0 if bits i and $i + 1$ of the corresponding binary code word are the same, else bit i is 1. (When $i + 1 = n$, bit n of the binary code word is considered to be 0.)

Again, inspection of Table 2–10 shows that this is true for the 3-bit Gray code.

* 2.12 CHARACTER CODES

As we showed in the preceding section, a string of bits need not represent a number, and in fact most of the information processed by computers is nonnumeric. The most common type of nonnumeric data is *text*, strings of characters *text* from some character set. Each character is represented in the computer by a bit string according to an established convention.

The most commonly used character code is *ASCII* (pronounced *ASS key*), *ASCII* the American Standard Code for Information Interchange. ASCII represents each character by a 7-bit string, a total of 128 different characters shown in Table 2–11. The code contains the uppercase and lowercase alphabet, numerals, punctuation, and various nonprinting control characters. Thus, the text string "Yeccch!" is represented by a rather innocuous-looking list of seven 7-bit numbers:

1011001   1100101   1100011   1100011   1100011   1101000   0100001

2.13 CODES FOR ACTIONS, CONDITIONS, AND STATES

The codes that we've described so far are generally used to represent things that we would probably consider to be "data"—things like numbers, positions, and characters. Programmers know that dozens of different data types can be used in a single computer program.

In digital system design, we often encounter nondata applications where a string of bits must be used to control an action, to flag a condition, or to represent the current state of the hardware. Probably the most commonly used type of code

Table 2–11 American Standard Code for Information Interchange (ASCII), Standard No. X3.4–1968 of the American National Standards Institute.

$b_3b_2b_1b_0$	Row (hex)	000 / 0	001 / 1	010 / 2	011 / 3	100 / 4	101 / 5	110 / 6	111 / 7
0000	0	NUL	DLE	SP	0	@	P	`	p
0001	1	SOH	DC1	!	1	A	Q	a	q
0010	2	STX	DC2	"	2	B	R	b	r
0011	3	ETX	DC3	#	3	C	S	c	s
0100	4	EOT	DC4	$	4	D	T	d	t
0101	5	ENQ	NAK	%	5	E	U	e	u
0110	6	ACK	SYN	&	6	F	V	f	v
0111	7	BEL	ETB	'	7	G	W	g	w
1000	8	BS	CAN	(	8	H	X	h	x
1001	9	HT	EM	)	9	I	Y	i	y
1010	A	LF	SUB	*	:	J	Z	j	z
1011	B	VT	ESC	+	;	K	[	k	{
1100	C	FF	FS	,	<	L	\	l	|
1101	D	CR	GS	–	=	M	]	m	}
1110	E	SO	RS	.	>	N	^	n	~
1111	F	SI	US	/	?	O	_	o	DEL

$b_6b_5b_4$ (column)

Control codes

NUL	Null	DLE	Data link escape
SOH	Start of heading	DC1	Device control 1
STX	Start of text	DC2	Device control 2
ETX	End of text	DC3	Device control 3
EOT	End of transmission	DC4	Device control 4
ENQ	Enquiry	NAK	Negative acknowledge
ACK	Acknowledge	SYN	Synchronize
BEL	Bell	ETB	End transmitted block
BS	Backspace	CAN	Cancel
HT	Horizontal tab	EM	End of medium
LF	Line feed	SUB	Substitute
VT	Vertical tab	ESC	Escape
FF	Form feed	FS	File separator
CR	Carriage return	GS	Group separator
SO	Shift out	RS	Record separator
SI	Shift in	US	Unit separator
SP	Space	DEL	Delete or rubout

for such an application is a simple binary code. If there are n different actions, conditions, or states, we can represent them with a binary code with $b = \lceil \log_2 n \rceil$ bits. (The brackets $\lceil \ \rceil$ denote the *ceiling function*—the smallest integer greater than or equal to the bracketed quantity. Thus, b is the smallest integer such that $2^b \geq n$.)

$\lceil \ \rceil$
ceiling function

For example, consider a simple traffic-light controller. The signals at the intersection of a north-south (N-S) and an east-west (E-W) street might be in any of the six states listed in Table 2–12. These states can be encoded in three bits, as shown in the last column of the table. Only six of the eight possible 3-bit code words are used, and the assignment of the six chosen code words to states is arbitrary, so many other encodings are possible. An experienced digital designer chooses a particular encoding to minimize circuit cost or to optimize some other parameter (like design time—there's no need to try billions and billions of possible encodings).

Table 2–12   States in a traffic-light controller.

| | Lights | | | | | | |
State	N-S green	N-S yellow	N-S red	E-W green	E-W yellow	E-W red	Code word
N-S go	ON	off	off	off	off	ON	000
N-S wait	off	ON	off	off	off	ON	001
N-S delay	off	off	ON	off	off	ON	010
E-W go	off	off	ON	ON	off	off	100
E-W wait	off	off	ON	off	ON	off	101
E-W delay	off	off	ON	off	off	ON	110

Another application of a binary code is illustrated in Figure 2–7(a). Here, we have a system with n devices, each of which can perform a certain action. The characteristics of the devices are such that they may be enabled to operate only one at a time. The control unit produces a binary-coded "device select" word with $\lceil \log_2 n \rceil$ bits to indicate which device is enabled at any time. The "device select" code word is applied to each device, which compares it with its own "device ID" to determine whether it is enabled.

Although its code words have the minimum number of bits, a binary code isn't always the best choice for encoding actions, conditions, or states. Figure 2–7(b) shows how to control n devices with a *1-out-of-n code*, an n-bit code in which valid code words have one bit equal to 1 and the rest of the bits equal to 0. Each bit of the 1-out-of-n code word is connected directly to the enable input of a corresponding device. This simplifies the design of the devices, since they no longer have device IDs; they need only a single "enable" input bit.

1-out-of-n code

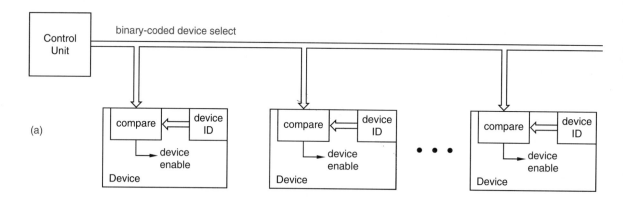

(a)

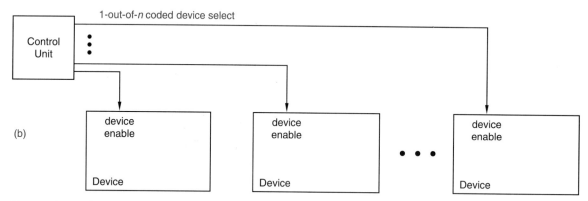

(b)

Figure 2–7   Control structure for a digital system with n devices: (a) using a binary code; (b) using a 1-out-of-n code.

inverted 1-out-of-n code

m-out-of-n code

The code words of a 1-out-of-10 code were listed in Table 2–9. Sometimes an all-0s word may also be included in a 1-out-of-n code, to indicate that no device is selected. Another common code is an *inverted 1-out-of-n code*, in which valid code words have one 0 bit and the rest of the bits equal to 1.

An *m-out-of-n code* is a generalization of the 1-out-of-n code in which valid code words have m bits equal to 1 and the rest of the bits equal to 0. A valid m-out-of-n code word can be detected with an m-input AND gate, which produces a 1 output if all of its inputs are 1. This is fairly simple and inexpensive to do, yet for most values of m, an m-out-of-n code typically has far more valid code words than a 1-out-of-n code. The total number of code words is given by the binomial coefficient $\binom{n}{m}$, which has the value $\frac{n!}{m! \cdot (n-m)!}$. Thus, a 2-out-of-4 code has 6 valid code words, and a 3-out-of-10 code has 120.

In complex systems, a combination of coding techniques may be used in a single code. For example, consider a system similar to Figure 2–7(b), in which each of the n devices contains up to s subdevices. The control unit could produce a device select code word with a 1-out-of-n coded field to select a device, and a $\lceil \log_2 s \rceil$-bit binary-coded field to select one of the s subdevices of the selected device. Or, if it were possible to enable m subdevices at once, an m-out-of-s code might be used to select subdevices.

* 2.14 *N*-CUBES AND DISTANCE

An n-bit string can be visualized geometrically, as a vertex of an object called an n-cube. Figure 2–8 shows n-cubes for $n = 1, 2, 3, 4$. An n-cube has 2^n vertices, each of which is labeled with an n-bit string. Edges are drawn so that each vertex is adjacent to n other vertices whose labels differ from the given vertex in only one bit. Beyond $n = 4$, n-cubes are really tough to draw.   *n-cube*

Figure 2–8
n-cubes for n = 1, 2, 3, and 4.

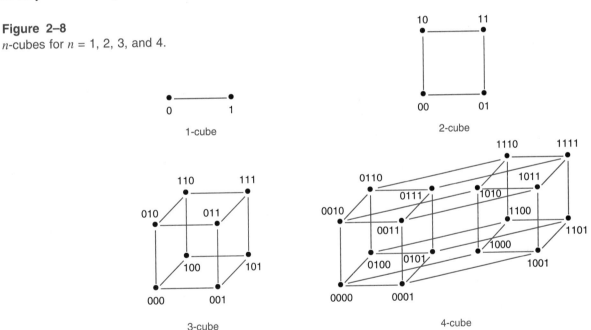

For reasonable values of n, n-cubes make it easy to visualize certain coding and logic minimization problems. For example, the problem of designing an n-bit Gray code is equivalent to finding a path along the edges of an n-cube, a path that visits each vertex exactly once. The paths for 3- and 4-bit Gray codes are shown in Figure 2–9.

Cubes also provide a geometrical interpretation for the concept of *distance*, also called *Hamming distance*. The distance between two n-bit strings is the number of bit positions in which they differ. In terms of an n-cube, the distance   *distance*

Hamming distance

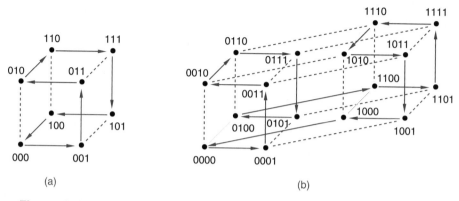

Figure 2–9   Traversing n-cubes in Gray-code order: (a) 3-cube; (b) 4-cube.

is the minimum length of a path between the two corresponding vertices. Two adjacent vertices have distance 1; vertices 001 and 100 in the 3-cube have distance 2. The concept of distance is crucial in the design and understanding of error-detecting codes, discussed in the next section.

m-subcube   An *m-subcube* of an n-cube is a set of 2^m vertices in which $n - m$ of the bits have the same value at each vertex, and the remaining m bits take on all 2^m combinations. For example, the vertices (000, 010, 100, 110) form a 2-subcube of the 3-cube. This subcube can also be denoted by a single string, xx0, where

don't-care   "x" denotes that a particular bit is a *don't-care*; any vertex whose bits match in the non-x positions belongs to this subcube. The concept of subcubes is particularly useful in visualizing algorithms that minimize the cost of combinational logic functions, as we'll show in Section 4.4.

* 2.15   CODES FOR DETECTING AND CORRECTING ERRORS

error   An *error* in a digital system is the corruption of data from its correct value to
failure   some other value. An error is caused by a physical *failure*. Failures can be either
temporary failure   temporary or permanent. For example, a cosmic ray or alpha particle can cause
permanent failure   a temporary failure of a memory circuit, changing the value of a bit stored in it. Letting a circuit get too hot or zapping it with static electricity can cause a permanent failure, so that it never works correctly again.

error model   The effects of failures on data are predicted by *error models*. The simplest
independent error   error model, which we consider here, is called the *independent error model*. In
  model   this model, a single physical failure is assumed to affect only a single bit of data;
single error   the corrupted data is said to contain a *single error*. Multiple failures may cause
multiple error   *multiple errors*—two or more bits in error—but multiple errors are normally assumed to be less likely than single errors.

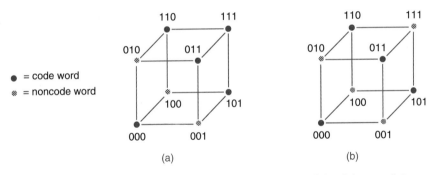

Figure 2–10   Code words in two different 3-bit codes: (a) minimum distance
= 1, does not detect all single errors; (b) minimum distance =
2, detects all single errors.

2.15.1   Error-Detecting Codes

Recall from our definitions in Section 2.10 that a code that uses n-bit strings
need not contain 2^n valid code words; this is certainly the case for the codes that
we now consider. An *error-detecting code* has the property that corrupting or   *error-detecting code*
garbling a code word will likely produce a bit string that is not a code word (a
noncode word).   *noncode word*

    A system that uses an error-detecting code generates, transmits, and stores
only code words. Thus, errors in a bit string can be detected by a simple rule—if
the bit string is a code word, it is assumed to be correct; if it is a noncode word,
it contains an error.

    An n-bit code and its error-detecting properties under the independent error
model are easily explained in terms of an n-cube. A code is simply a subset of
the vertices of the n-cube. In order for the code to detect all single errors, no
code-word vertex can be immediately adjacent to another code-word vertex.

    For example, Figure 2–10(a) shows a 3-bit code with five code words.
Code word 111 is immediately adjacent to code words 110 and 011. Since a
single failure could change 111 to 110 or 011, this code does not detect all single
errors. If we make 111 a noncode word, we obtain a code that does have the
single-error-detecting property, as shown in (b). No single error can change one
code word into another.

    The ability of a code to detect single errors can be stated in terms of the
concept of distance introduced in the preceding section:

- A code detects all single errors if the *minimum distance* between all possible   *minimum distance*
  pairs of code words is 2.

    In general, we need $n+1$ bits to construct a single-error-detecting code with
2^n code words. The first n bits of a code word, called *information bits*, may be   *information bit*

parity bit

even-parity code

odd-parity code
1-bit parity code

any of the 2^n n-bit strings. To obtain a minimum-distance-2 code, we add one more bit, called a *parity bit*, that is set to 0 if there are an even number of 1s among the information bits, and to 1 otherwise. This is illustrated in the first two columns of Table 2–13 for a code with three information bits. A valid $n+1$-bit code word has an even number of 1s, and this code is called an *even-parity code*. We can also construct a code in which the total number of 1s in a valid $n+1$-bit code word is odd; this is called an *odd-parity code* and is shown in the third column of the table. These codes are also sometimes called *1-bit parity codes*, since they each use a single parity bit.

Table 2–13
Distance-2 codes with three information bits.

Information Bits	Even-parity Code	Odd-parity Code
000	000 0	000 1
001	001 1	001 0
010	010 1	010 0
011	011 0	011 1
100	100 1	100 0
101	101 0	101 1
110	110 0	110 1
111	111 1	111 0

The 1-bit parity codes do not detect 2-bit errors, since changing two bits does not affect the parity. However, the codes can detect errors in any *odd* number of bits. For example, if three bits in a code word are changed, then the resulting word has the wrong parity and is a noncode word. This doesn't help us much, though. Under the independent error model, 3-bit errors are much less likely than 2-bit errors, which are not detectable. Thus, practically speaking, the 1-bit parity codes' error detection capability stops after 1-bit errors. Other codes, with minimum distance greater than 2, can be used to detect multiple errors.

2.15.2 Error-Correcting and Multiple-Error-Detecting Codes

By using more than one parity bit according to some well-chosen rules, we can create a code whose minimum distance is greater than 2. Before showing how this can be done, let's look at how such a code can be used to correct single errors or detect multiple errors.

Suppose that a code has a minimum distance of 3. Figure 2–11 shows a fragment of the n-cube for such a code. As shown, there are at least two noncode words between each pair of code words. Now suppose we transmit code words and assume that failures affect at most one bit of each received code word. Then a received noncode word with a 1-bit error will be closer to the originally transmitted code word than to any other code word. Therefore, when we receive

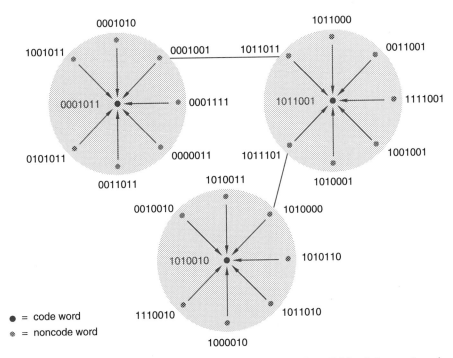

Figure 2–11   Some code words and noncode words in a 7-bit, distance-3 code.

a noncode word, we can *correct* the error by changing the received noncode *error correction* word to the nearest code word, as indicated by the arrows in the figure. Deciding which code word was originally transmitted to produce a received word is called *decoding decoding*, and the hardware that does this is an error-correcting *decoder*. *decoder*

A code that is used to correct errors is called an *error-correcting code*. In *error-correcting* general, if a code has minimum distance $2c + 1$, it can be used to correct errors *code* that affect up to c bits ($c = 1$ in the preceding example). If a code's minimum distance is $2c + d + 1$, it can be used to correct errors in up to c bits and to detect errors in up to d additional bits.

For example, Figure 2–12(a) shows a fragment of the *n*-cube for a code with minimum distance 4 ($c = 1$, $d = 1$). Single-bit errors that produce non-code words 00101010 and 11010011 can be corrected. However, an error that produces 10100011 cannot be corrected, because no single-bit error can produce this noncode word, and either of two 2-bit errors could have produced it. So the code can detect a 2-bit error, but it cannot correct it.

| The names *decoding* and *decoder* make sense, since they are are just distance-1 perturbations of *deciding* and *decider*. | DECISIONS, DECISIONS |

Figure 2–12
Some code words and
noncode words in an
8-bit, distance-4 code:
(a) correcting 1-bit and
detecting 2-bit errors;
(b) incorrectly "correcting" a
3-bit error; (c) correcting no
errors but detecting up to
3-bit errors.

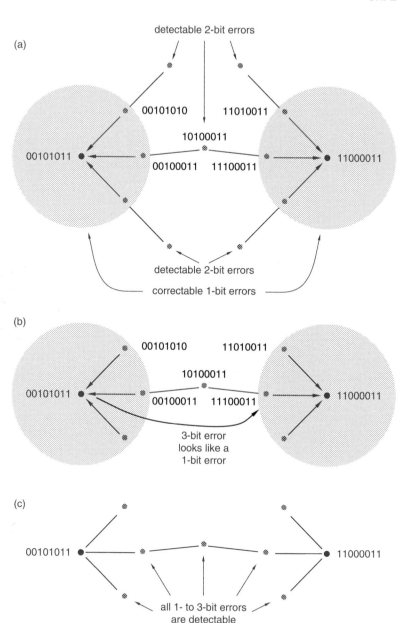

When a noncode word is received, we don't know which code word was originally transmitted; we only know which code word is closest to what we've received. Thus, as shown in Figure 2–12(b), a 3-bit error may be "corrected" to the wrong value. The possibility of making this kind of mistake may be acceptable if 3-bit errors are very unlikely to occur. On the other hand, if we are concerned about 3-bit errors, we can change the decoding policy for the code. Instead of trying to correct errors, we just flag all noncode words as uncorrectable errors. Thus, as shown in (c), we can use the same distance-4 code to detect up to 3-bit errors but correct no errors ($c = 0$, $d = 3$).

Stopped here

2.15.3 Hamming Codes

In 1950, R. W. Hamming described a general method for constructing codes with a minimum distance of 3, now called *Hamming codes*. For any value of i, *Hamming code* his method yields a $2^i - 1$-bit code with i parity bits and $2^i - 1 - i$ information bits. Distance-3 codes with a smaller number of information bits are obtained by deleting information bits from a Hamming code with a larger number of bits.

The bit positions in a Hamming code word are numbered from 1 to $2^i - 1$. Any position whose number is a power of 2 is a parity bit, and the remaining positions are information bits. Each parity bit is grouped with a subset of the information bits, as specified by a *parity-check matrix*. As shown in *parity-check matrix* Figure 2–13(a), each parity bit is grouped with the information positions whose numbers have a 1 in the same bit when expressed in binary. For example, parity

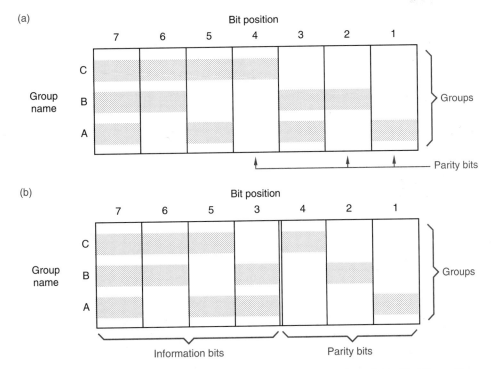

Figure 2–13 Parity-check matrices for 7-bit Hamming codes: (a) with bit positions in numerical order; (b) with parity bits and information bits separated.

Table 2-14
Code words in distance-3 and distance-4 Hamming codes with four information bits.

Minimum-distance-3 code		Minimum-distance-4 code	
Information Bits	Parity Bits	Information Bits	Parity Bits
0000	000	0000	0000
0001	011	0001	0111
0010	101	0010	1011
0011	110	0011	1100
0100	110	0100	1101
0101	101	0101	1010
0110	011	0110	0110
0111	000	0111	0001
1000	111	1000	1110
1001	100	1001	1001
1010	010	1010	0101
1011	001	1011	0010
1100	001	1100	0011
1101	010	1101	0100
1110	100	1110	1000
1111	111	1111	1111

bit 2 (010) is grouped with information bits 3 (011), 6 (110), and 7 (111). For a given combination of information-bit values, each parity bit is chosen so that the total number of 1s in its group is even.

Traditionally, the bit positions of a parity-check matrix and the resulting code words are rearranged so that all of the parity bits are on the right, as in Figure 2–13(b). The first two columns of Table 2–14 list the resulting code words.

We can prove that the minimum distance of a Hamming code is 3 by proving that at least a 3-bit change must be made to a code word to obtain another code word. That is, we'll prove that a 1-bit or 2-bit change in a code word yields a noncode word.

If we change one bit of a code word, in position j, then we change the parity of every group that contains position j. Since every information bit is contained in at least one group, at least one group has incorrect parity, and the result is a noncode word.

What happens if we change two bits, in positions j and k? Parity groups that contain both positions j and k will still have correct parity, since parity is unaffected when an even number of bits are changed. However, since j and k are different, their binary representations differ in at least one bit, corresponding to one of the parity groups. This group has only one bit changed, resulting in incorrect parity and a noncode word.

If you understand this proof, you should also see how the position numbering rules for constructing a Hamming code are a simple consequence of the proof. For the first part of the proof (1-bit errors), we required that the position numbers be nonzero. And for the second part (2-bit errors), we required that no two positions have the same number. Thus, with an i-bit position number, you can construct a Hamming code with up to $2^i - 1$ bit positions.

The proof also suggests how we can design an *error-correcting decoder* for a received Hamming code word. First, we check all of the parity groups; if all have even parity, then the received word is assumed to be correct. If one or more groups have odd parity, then a single error is assumed to have occurred. The pattern of groups that have odd parity (called the *syndrome*) must match one of the columns in the parity-check matrix; the corresponding bit position is assumed to contain the wrong value and is complemented. For example, using the code defined by Figure 2–13(b), suppose we receive the word 0101011. Groups B and C have odd parity, corresponding to position 6 of the parity-check matrix (the syndrome is 110, or 6). By complementing the bit in position 6 of the received word, we determine that the correct word is 0001011.

error-correcting decoder

syndrome

A distance-3 Hamming code can easily be modified to increase its minimum distance to 4. We simply add one more parity bit, chosen so that the parity of all the bits, including the new one, is even. As in the 1-bit even-parity code, this bit ensures that all errors affecting an odd number of bits are detectable. In particular, any 3-bit error is detectable. We already showed that 1- and 2-bit errors are detected by the other parity bits, so the minimum distance of the modified code must be 4.

Distance-3 and distance-4 Hamming codes are commonly used to detect and correct errors in computer memory systems, especially in large mainframe computers where memory circuits account for the bulk of the system's failures. These codes are especially attractive for very wide memory words, since the required number of parity bits grows slowly with the width of the memory word, as shown in Table 2–15.

Table 2–15   Word sizes of distance-3 and distance-4 Hamming codes.

Information Bits	Minimum-distance-3 Codes		Minimum-distance-4 Codes	
	Parity Bits	Total Bits	Parity Bits	Total Bits
1	2	3	3	4
≤ 4	3	≤ 7	4	≤ 8
≤ 11	4	≤ 15	5	≤ 16
≤ 26	5	≤ 31	6	≤ 32
≤ 57	6	≤ 63	7	≤ 64
≤ 120	7	≤ 127	8	≤ 128

2.15.4 Two-Dimensional Codes

two-dimensional code

Another way to obtain a code with large minimum distance is to construct a *two-dimensional code*, as illustrated in Figure 2–14(a). The information bits are conceptually arranged in a two-dimensional array, and parity bits are provided to check both the rows and the columns. A code C_{row} with minimum distance d_{row} is used for the rows, and a possibly different code C_{col} with minimum distance d_{col} is used for the columns. That is, the row-parity bits are selected so that each row is a code word in C_{row} and the column-parity bits are selected so that each column is a code word in C_{col}. (The "corner" parity bits can be chosen according to either code.) The minimum distance of the two-dimensional code is the product of d_{row} and d_{col}; in fact, two-dimensional codes are sometimes called *product codes*.

product code

As shown in Figure 2–13(b), the simplest two-dimensional code uses 1-bit even-parity codes for the rows and columns, and has a minimum distance of $2 \cdot 2$, or 4. You can easily prove that the minimum distance is 4 by convincing yourself that any pattern of one, two, or three bits in error causes incorrect parity of a row or a column or both. In order to obtain an undetectable error, at least four bits must be changed in a rectangular pattern as in (c).

Figure 2–14
Two-dimensional codes:
(a) general structure; (b) using even parity for both the row and column codes to obtain minimum distance 4; (c) typical pattern of an undetectable error.

(a)

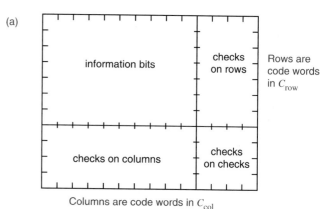

Columns are code words in C_{col}

(b)

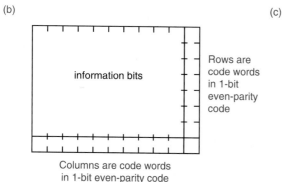

Columns are code words in 1-bit even-parity code

(c)

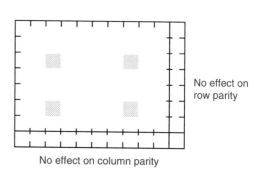

No effect on column parity

To obtain an even larger minimum distance, a distance-3 or -4 Hamming code can be used for the row or column code or both. It is also possible to construct a code in three or more dimensions, with minimum distance equal to the product of the minimum distances in each dimension.

Two-dimensional codes are sometimes used to check computer memory systems, which are inherently two-dimensional structures—n words $\times$ b bits. Suppose we provide an even-parity check bit for each word in memory; this is a row code. We can use one extra memory word to create a column code. Each bit of the column check word stores the parity of the corresponding bit of all the words in memory. The bit-by-bit Exclusive OR instruction available in typical computer processors can be used $n-1$ times to form the check word.

The error detecting and correcting procedures for the two-dimensional code described above are straightforward. As we read each byte from the memory, we check its row parity, and thereby detect any single-bit error. If an error is detected, we can't tell which bit is wrong from the row check alone. However, we can reconstruct the erroneous word by forming the bit-by-bit Exclusive OR of all of the words in memory, omitting the erroneous word, but including the column check word.

2.15.5   Checksum Codes

The parity-checking operation that we've used in the previous subsections is essentially modulo-2 addition of bits—the sum modulo 2 of a group of bits is 0 if the number of 1s in the group is even, and 1 if it is odd. The technique of modular addition can be extended to other bases besides 2 to form check digits.

For example, a computer stores information as a set of 8-bit bytes. Each byte may be considered to have a decimal value from 0 to 255. Therefore, we can use modulo-256 addition to check the bytes. We form a single check byte, called a *checksum*, that is the sum modulo 256 of all the information bytes. The resulting *checksum code* can detect any single *byte* error, since such an error will cause a recomputed sum of bytes to disagree with the checksum.

checksum

checksum code

A checksum code can be combined with a 1-bit even-parity code to form a two-dimensional, distance-4, error-correcting code for a computer memory system. This code is similar to the code described at the end of the preceding subsection, except that the column check word is formed by adding all of the memory words, rather than forming their bit-by-bit Exclusive OR.

2.15.6   *m*-out-of-*n* Codes

The 1-out-of-*n* and *m*-out-of-*n* codes that we introduced in Section 2.13 have a minimum distance of 2, since changing only one bit changes the total number of 1s in a code word and therefore produces a noncode word.

unidirectional error

These codes have another useful error-detecting property—they detect unidirectional multiple errors. In a *unidirectional error*, all of the erroneous bits change in the same direction (0s change to 1s, or vice versa). This property is very useful in systems where the predominant error mechanism tends to change all bits in the same direction.

2.16 CODES FOR SERIAL DATA TRANSMISSION AND STORAGE

2.16.1 Parallel and Serial Data

parallel data

Most computers and other digital systems transmit and store data in a *parallel* format. In parallel data transmission, a separate signal line is provided for each bit of a data word. In parallel data storage, all of the bits of a data word can be written or read simultaneously.

Parallel formats are not cost-effective for some applications. For example, parallel transmission of data bytes over the telephone network would require eight phone lines, and parallel storage of data bytes on a magnetic disk would require a disk drive with eight separate read/write heads. *Serial* formats allow data to be transmitted or stored one bit at a time, and reduce system cost in many applications.

serial data

Figure 2–15 illustrates some of the basic ideas in serial data transmission. A repetitive clock signal, named CLOCK in the figure, defines the rate at which bits are transmitted, one bit per clock cycle. Thus, the *bit rate* in bits per second (bps) numerically equals the clock frequency in cycles per second (hertz, or Hz).

bit rate

bit time

The reciprocal of the bit rate is called the *bit time* and numerically equals the clock period in seconds (s). This amount of time is reserved on the serial data line (named SERDATA in the figure) for each bit that is transmitted. The

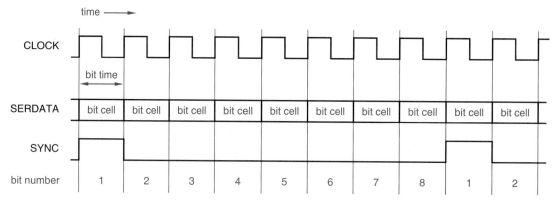

Figure 2–15 Basic concepts for serial data transmission.

time occupied by each bit is sometimes called a *bit cell*. The format of the actual signal that appears on the line during each bit cell depends on the *line code*. In the simplest line code, called *Non-Return-to-Zero (NRZ)*, a 1 is transmitted by placing a 1 on the line for the entire bit cell, and a 0 is transmitted as a 0. However, more complex line codes have other rules, as discussed in the next subsection.

bit cell
line code
Non-Return-to-Zero (NRZ)

Regardless of the line code, a serial data transmission or storage system needs some way of identifying the significance of each bit in the serial stream. For example, suppose that 8-bit bytes are transmitted serially. How can we tell which is the first bit of each byte? A *synchronization signal*, named SYNC in Figure 2–15, provides the necessary information; it is 1 for the first bit of each byte.

synchronization signal

Evidently, we need a minimum of three signals to recover a serial data stream: a clock to define the bit cells, a synchronization signal to define the word boundaries, and the serial data itself. In some applications, like the interconnection of modules in a computer or telecommunications system, a separate wire is used for each of these signals, since reducing the number of wires per connection from *n* to three is savings enough. We'll give an example of a 3-wire serial data system in Section 9.4.4.

In many applications, the cost of having three separate signals is still too high (e.g., three phone lines, three read/write heads). Such systems typically combine all three signals into a single serial data stream and use sophisticated analog and digital circuits to recover the clock and synchronization information from the data stream.

* 2.16.2   Serial Line Codes

The most commonly used line codes for serial data are illustrated in Figure 2–16. In the NRZ code, each bit value is sent on the line for the entire bit cell. This is the simplest and most reliable coding scheme for short distance transmission. However, it generally requires a clock signal to be sent along with the data to define the bit cells. Otherwise, it is not possible for the receiver to determine how many 0s or 1s are represented by a continuous 0 or 1 level. Without a clock to define the bit cells, the NRZ waveform in Figure 2–15 might be erroneously interpreted as 01010.

A *digital phase-locked loop (DPLL)* is an analog/digital circuit that can be used to recover a clock signal from a serial data stream. The DPLL works only if the serial data stream contains enough 0-to-1 and 1-to-0 transitions to give the DPLL "hints" about when the original clock transitions took place. With NRZ-coded data, the DPLL works only if the data does not contain any long, continuous streams of 1s or 0s.

digital phase-locked loop (DPLL)

Some serial transmission and storage media are *transition sensitive*; they cannot transmit or store absolute 0 or 1 levels, only transitions between two

transition-sensitive media

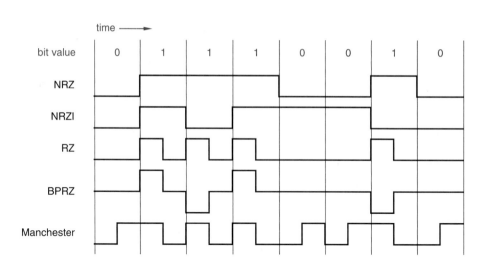

Figure 2–16   Commonly used line codes for serial data.

discrete levels. For example, a magnetic disk or tape stores information by
changing the polarity of the medium's magnetization in regions corresponding to
the stored bits. When the information is recovered, it is not feasible to determine
the absolute magnetization polarity of a region, only that the polarity changes
between one region and the next.

Non-Return-to-Zero
Invert-on-1s
(NRZI)

   Data stored in NRZ format on transition-sensitive media cannot be recovered unambiguously; the data in Figure 2–16 might be interpreted as 01110010
or 10001101. The *Non-Return-to-Zero Invert-on-1s (NRZI)* code overcomes this
limitation by sending a 1 as the opposite of the level that was sent during the
previous bit cell, and a 0 as the same level. A DPLL can recover the clock
from NRZI-coded data as long as the data does not contain any long, continuous
streams of 0s.

Return-to-Zero (RZ)

   The *Return-to-Zero (RZ)* code is similar to NRZ except that, for a 1 bit, the
1 level is transmitted only for a fraction of the bit time, usually 1/2. With this
code, data patterns that contain a lot of 1s create lots of transitions for a DPLL
to use to recover the clock. However, as in the other line codes, a string of 0s
has no transitions, and a long string of 0s makes clock recovery impossible.

Bipolar
Return-to-Zero
(BPRZ)

Alternate Mark
Inversion (AMI)

   All of the preceding codes transmit or store only two signal levels. The
Bipolar Return-to-Zero (BPRZ) code transmits three signal levels: +1, 0, and −1.
The code is like RZ except that 1s are alternately transmitted as +1 and −1; for
this reason, the code is also known as *Alternate Mark Inversion (AMI)*.

   The biggest advantage of BPRZ over the other codes is that for any data
pattern, the average signal level of the BPRZ serial data stream is zero. This
improves the noise immunity of the analog circuits that receive highly attenuated
BPRZ signals sent over a long distance. In fact, the BPRZ code has been used
in the telephone network for decades, where analog speech signals are converted

into streams of 8000 8-bit digital samples per second that are transmitted in BPRZ format on 64 Kbps serial channels.

As with RZ, it is possible to recover the clock signal from a BPRZ stream as long as there aren't too many 0s in a row. Although TPC (The Phone Company) has no control over what you say (at least, not yet), they have a neat way of getting around this problem. If one of the 8-bit bytes that results from sampling your analog speech pattern is all 0s, they simply change one of its bits to 1! I'll bet you never noticed.

The last code in Figure 2–16 is called *Manchester* or *diphase* code. The major strength of this code is that, regardless of the transmitted data pattern, it provides at least one transition per bit cell, making it very easy to recover the clock. As shown in the figure, a 0 is encoded as a 0-to-1 transition in the middle of the bit cell, and a 1 is encoded as a 1-to-0 transition. The Manchester code's major strength is also its major weakness. Since it has more transitions per bit cell than other codes, it also requires more media bandwidth to transmit a given bit rate. Bandwidth is not a problem in coaxial cable, however, which is used in Ethernet local area networks to carry Manchester-coded serial data at the rate of 10 Mbps (megabits per second).

Manchester
diphase

ABOUT TPC

> Watch the 20-year old James Coburn movie, *The President's Analyst*, for an amusing view of TPC. With the growing pervasiveness of digital technology and cheap wireless communications, the concept of universal, *personal* connectivity to the phone network preesented in the movie's conclusion is becoming less and less far-fetched.

R E F E R E N C E S

The presentation in the first nine sections of this chapter is based on Chapter 4 of *Microcomputer Architecture and Programming*, by John F. Wakerly (Wiley, 1981). Precise, thorough, and entertaining discussions of these topics can also be found in Donald E. Knuth's *Seminumerical Algorithms* (Addison-Wesley, 1969). Mathematically inclined readers will find Knuth's analysis of the properties of number systems and arithmetic to be excellent, and all readers should enjoy the insights and history sprinkled throughout the text.

Descriptions of digital logic circuits for arithmetic operations, as well as an introduction to properties of various number systems, appear in *Computer Arithmetic* by Kai Hwang (Wiley, 1979). *Decimal Computation* by Hermann Schmid (Wiley, 1974) contains a thorough description of techniques for BCD arithmetic.

An introduction to algorithms for binary multiplication and division and to floating-point arithmetic appears in *Microcomputer Architecture and Program-*

ming: The 68000 Family by John F. Wakerly (Wiley, 1989). A more thorough discussion of arithmetic techniques and floating-point number systems can be found in *Introduction to Arithmetic for Digital Systems Designers* by Shlomo Waser and Michael J. Flynn (Holt, Rinehart and Winston, 1982).

The classic book on error-detecting and error-correcting codes is *Error-Correcting Codes* by W. W. Peterson and E. J. Weldon, Jr. (MIT Press, 1972, 2nd ed.); however, this book is recommended only for mathematically sophisticated readers. A more accessible introduction can be found in *Error Control Coding: Fundamentals and Applications* by S. Lin and D. J. Costello, Jr. (Prentice Hall, 1983). Another recent, communication-oriented introduction to coding theory can be found in *Error-Control Techniques for Digital Communication* by A. M. Michelson and A. H. Levesque (Wiley-Interscience, 1985). Hardware applications of codes in computer systems are discussed in *Error-Detecting Codes, Self-Checking Circuits, and Applications* by John F. Wakerly (Elsevier/North-Holland, 1978).

An introduction to coding techniques for serial data transmission, including mathematical analysis of the performance and bandwidth requirements of several codes, appears in *Introduction to Communications Engineering* by R. M. Gagliardi (Wiley-Interscience, 1988, 2nd ed.). A nice introduction to the serial codes used in magnetic disks and tapes is given in *Computer Storage Systems and Technology* by Richard Matick (Wiley-Interscience, 1977).

DRILL PROBLEMS

2.1 Perform the following number system conversions:

(a) $01101011_2 = ?_{16}$ (b) $174003_8 = ?_2$

(c) $10110111_2 = ?_{16}$ (d) $67.24_8 = ?_2$

(e) $10100.1101_2 = ?_{16}$ (f) $F3A5_{16} = ?_2$

(g) $11011001_2 = ?_8$ (h) $AB3D_{16} = ?_2$

(i) $101111.0111_2 = ?_8$ (j) $15C.38_{16} = ?_2$

2.2 Convert the following octal numbers into binary and hexadecimal:

(a) $1023_8 = ?_2 = ?_{16}$ (b) $761302_8 = ?_2 = ?_{16}$

(c) $163417_8 = ?_2 = ?_{16}$ (d) $552273_8 = ?_2 = ?_{16}$

(e) $5436.15_8 = ?_2 = ?_{16}$ (f) $13705.207_8 = ?_2 = ?_{16}$

2.3 Convert the following hexadecimal numbers into binary and octal:

(a) $1023_{16} = ?_2 = ?_8$ (b) $7E6A_{16} = ?_2 = ?_8$

(c) $ABCD_{16} = ?_2 = ?_8$ (d) $C350_{16} = ?_2 = ?_8$

(e) $9E36.7A_{16} = ?_2 = ?_8$ (f) $DEAD.BEEF_{16} = ?_2 = ?_8$

128 + 64 + 16
1 + 2 + 16 + 64 + 128 = 213

2.4 Convert the following numbers into decimal:

(a) $1101011_2 = ?_{10}$ (b) $174003_8 = ?_{10}$

(c) $10110111_2 = ?_{10}$ (d) $67.24_8 = ?_{10}$

(e) $10100.1101_2 = ?_{10}$ (f) $F3A5_{16} = ?_{10}$

(g) $12010_3 = ?_{10}$ (h) $AB3D_{16} = ?_{10}$

(i) $7156_8 = ?_{10}$ (j) $15C.38_{16} = ?_{10}$

2.5 Perform the following number system conversions:

(a) $125_{10} = ?_2$ (b) $3489_{10} = ?_8$

(c) $209_{10} = ?_2$ (d) $9714_{10} = ?_8$

(e) $132_{10} = ?_2$ (f) $23851_{10} = ?_{16}$

(g) $727_{10} = ?_5$ (h) $57190_{10} = ?_{16}$

(i) $14351_{10} = ?_8$ (j) $651131_{10} = ?_{16}$

2.6 Add the following pairs of binary numbers, showing all carries:

(a) 110101   (b) 101110   (c) 11011101   (d) 1110010
     + 11001   + 100101   + 1100011   + 1101101

1001110

2.7 Repeat Drill 2.6 using subtraction instead of addition, and showing borrows instead of carries.

(a)

100101
11001
11000

2.8 Add the following pairs of octal numbers:

(a) 1372   (b) 47135   (c) 175214   (d) 110321
     + 4631   + 5125   + 152405   + 56573

6223

2.9 Add the following pairs of hexadecimal numbers:

(a) 1372   (b) 4F1A5   (c) F35B   (d) 1B90F
     + 4631   + B8D5   + 27E6   + C44E

59A3

2.10 Write the 8-bit signed-magnitude, two's-complement, and ones'-complement representations for each of these decimal numbers: +18, +115, +79, −49, −3, −100.

2.11 Indicate whether or not overflow occurs when adding the following 8-bit two's-complement numbers:

(a) 11010100   (b) 10111001   (c) 01011101   (d) 00100110
     + 10101011   + 11010110   + 00100001   + 01011010

(1)01111111

2.12 How many errors can be detected by a code with minimum distance d?

2.13 What is the minimum number of parity bits required to obtain a distance-4, two-dimensional code with n information bits?

$2\overline{)18}/2 = 9$ r 0
$/2 = 4$ r 1
$/2 = 2$ r 0
$/2 = 0$ r 0
$1/2 = r 1$

000 1 0010
1 1 1 0 1 1 0 1   1's
1 1 1 0 1 1 1 0   2's

EXERCISES

2.14 Here's a problem to whet your appetite. What is the hexadecimal equivalent of 61453_{10}?

2.15 Each of the following arithmetic operations is correct in at least one number system. Determine possible radices of the numbers in each operation.

(a) $1234 + 5432 = 6666$ (b) $41/3 = 13$

(c) $33/3 = 11$ (d) $23 + 44 + 14 + 32 = 223$

(e) $302/20 = 12.1$ (f) $\sqrt{41} = 5$

2.16 The first expedition to Mars found only the ruins of a civilization. From the artifacts and pictures, the explorers deduced that the creatures who produced this civilization were four-legged beings with a tentacle that branched out at the end with a number of grasping "fingers." After much study, the explorers were able to translate Martian mathematics. They found the following equation:

$$5x^2 - 50x + 125 \;=\; 0$$

with the indicated solutions $x = 5$ and $x = 8$. The value $x = 5$ seemed legitimate enough, but $x = 8$ required some explanation. Then the explorers reflected on the way in which Earth's number system developed, and found evidence that the Martian system had a similar history. How many fingers would you say the Martians had? (From *The Bent of Tau Beta Pi*, February, 1956.)

2.17 Suppose a $4n$-bit number B is represented by an n-digit hexadecimal number H. Prove that the two's complement of B is represented by the 16's complement of H. Make and prove true a similar statement for octal representation.

2.18 Repeat Exercise 2.17 using the ones' complement of B and the 15s' complement of H.

2.19 Given an integer x in the range $-2^{n-1} \le x \le 2^{n-1} - 1$, we define $[x]$ to be the two's-complement representation of x, expressed as a positive number: $[x] = x$ if $x \ge 0$ and $[x] = 2^n - |x|$ if $x < 0$, where $|x|$ is the absolute value of x. Let y be another integer in the same range as x. Prove that the two's-complement addition rules given in Section 2.6 are correct by proving that the following equation is always true:

$$[x + y] \;=\; [x] + [y] \text{ modulo } 2^n$$

(*Hints:* Consider four cases based on the signs of x and y. Without loss of generality, you may assume that $|x| \ge |y|$.)

2.20 Repeat Exercise 2.19 using appropriate expressions and rules for ones'-complement addition.

2.21 State an overflow rule for addition of two's-complement numbers in terms of counting operations in the modular representation of Figure 2–3.

2.22 Show that a two's-complement number can be converted to a representation with more bits by *sign extension*. That is, given an n-bit two's-complement number X, show that the m-bit two's-complement representation of X, where $m > n$, can be obtained by appending $m - n$ copies of X's sign bit to the left of the n-bit representation of X.

sign extension

2.23 Show that a two's-complement number can be converted to a representation with fewer bits by removing higher-order bits. That is, given an n-bit two's-complement number X, show that the m-bit two's-complement number Y obtained by discarding the d leftmost bits of X represents the same number as X if and only if the discarded bits all equal the sign bit of Y.

2.24 Why is the punctuation of "two's complement" and "ones' complement" inconsistent? (See the first two citations in the References.)

2.25 A n-bit binary adder can be used to perform an n-bit unsigned subtraction operation $X - Y$, by performing the operation $X + \overline{Y} + 1$, where X and Y are n-bit unsigned numbers and $\overline{Y}$ represents the bit-by-bit complement of Y. Demonstrate this fact as follows. First, prove that $(X - Y) = (X + \overline{Y} + 1) - 2^n$. Second, prove that the carry out of the n-bit adder is the opposite of the borrow from the n-bit subtraction. That is, show that the operation $X - Y$ produces a borrow out of the MSB position if and only if the operation $X + \overline{Y} + 1$ *does not* produce a carry out of the MSB position.

2.26 In most cases, the product of two n-bit two's-complement numbers requires $2n - 1$ or fewer bits to represent. In fact, there is only one case in which $2n$ bits are needed—find it.

2.27 Prove that a two's-complement number can be multiplied by 2 by shifting it one bit position to the left, with a carry of 0 into the least significant bit position and disregarding any carry out of the most significant bit position, assuming no overflow. State the rule for detecting overflow.

2.28 State and prove correct a technique similar to the one described in Exercise 2.27, for multiplying a ones'-complement number by 2.

2.29 Show how to subtract BCD numbers, by stating the rules for generating borrows and applying a correction factor. Show how your rules apply to each of the following subtractions: $9 - 3$, $5 - 7$, $4 - 9$, $1 - 8$.

2.30 How many different 3-bit binary state encodings are possible for the traffic-light controller of Table 2–12?

2.31 List all of the "bad" boundaries in the mechanical encoding disc of Figure 2–5, where an incorrect position may be sensed.

2.32 As a function of n, how many "bad" boundaries are there in a mechanical encoding disc that uses an n-bit binary code?

2.33 On-board altitude transponders on commercial and private aircraft use Gray code to encode the altitude readings that are transmitted to air traffic controllers. Why?

2.34   An incandescent light bulb is stressed every time it is turned on, so in some applications the lifetime of the bulb is limited by the number of on/off cycles rather than the total time it is illuminated. Use your knowledge of codes to suggest a way to double the lifetime of 3-way bulbs in such applications.

2.35   As a function of n, how many different distinct subcubes of an n-cube are there?

2.36   Find a way to draw a 3-cube on a sheet of paper (or other two-dimensional object) so that none of the lines cross, or prove that it's impossible.

2.37   Repeat Exercise 2.36 for a 4-cube.

2.38   Write a formula that gives the number of m-subcubes of an n-cube for a specific value of m. (Your answer should be a function of n and m.)

2.39   Define parity groups for a distance-3 Hamming code with 11 information bits.

2.40   Write the code words of a Hamming code with one information bit.

2.41   Exhibit the pattern for a 3-bit error that is not detected if the "corner" parity bits are not included in the two-dimensional codes of Figure 2–14.

code rate

2.42   The *rate of a code* is the ratio of the number of information bits to the total number of bits in a code word. High rates, approaching 1, are desirable for efficient transmission of information. Construct a graph comparing the rates of distance-2 parity codes and distance-3 and -4 Hamming codes for up to 100 information bits.

2.43   Which type of distance-4 code has a higher rate—a two-dimensional code or a Hamming code? Support your answer with a table in the style of Table 2–15, including the rate as well as the number of parity and information bits of each code for up to 100 information bits.

2.44   Show how to construct a distance-6 code with four information bits. Write a list of its code words.

2.45   In the style of Figure 2–16, draw the waveforms for the bit pattern 10101110 when sent serially using the NRZ, NRZI, RZ, BPRZ, and Manchester codes, assuming that the bits are transmitted in order from left to right.

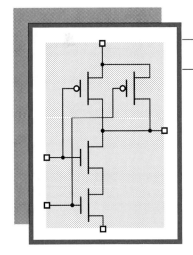

3

DIGITAL CIRCUITS

Marketing hype notwithstanding, we live in an analog world, not a digital one. Voltages, currents, and other physical quantities in real circuits take on values that are infinitely variable, depending on properties of the real devices that comprise the circuits. Because real values are continuously variable, we could use a physical quantity such as a signal voltage in a circuit to represent a real number (e.g., 3.14159265358979 volts represents the mathematical constant *pi* to 14 decimal digits of precision).

Unfortunately, stability and accuracy in physical quantities are difficult to obtain in real circuits. They are affected by manufacturing tolerances, temperature, power-supply voltage, cosmic rays, and noise created by other circuits, among other things. If we used an analog voltage to represent *pi*, we might find that instead of being an absolute mathematical constant, *pi* varied over a range of 10% or more.

Also, many mathematical and logical operations are difficult or impossible to perform with analog quantities. While it is possible with some cleverness to build an analog circuit whose output voltage is the square root of its input voltage, no one has ever built a 100-input, 100-output analog circuit whose outputs are a set of voltages identical to the set of input voltages, but sorted arithmetically.

3.1   LOGIC SIGNALS AND GATES

digital logic

logic values

Digital logic hides the pitfalls of the analog world by mapping the infinite set of real values for a physical quantity into two subsets corresponding to just two possible numbers or *logic values*—0 and 1. As a result, digital logic circuits can be analyzed and designed functionally, using tables and other abstract means to describe the operation of well-behaved 0s and 1s in a circuit.

binary digit

bit

A logic value, 0 or 1, is often called a *binary digit*, or *bit*. If an application requires more than two discrete values, additional bits may be used, with a set of n bits representing 2^n different values.

Examples of the physical phenomena used to represent bits in some modern (and not-so-modern) digital technologies are given in Table 3–1. With most phenomena, there is an undefined region between the 0 and 1 states (e.g., voltage = 1.8 V, dim light, capacitor slightly charged, etc.). This undefined region is needed so that the 0 and 1 states can be unambiguously defined and reliably detected. Noise can more easily corrupt results if the boundaries separating the 0 and 1 states are too close.

LOW

HIGH

When discussing electronic logic circuits such as CMOS and TTL, digital designers often use the words "LOW" and "HIGH" in place of "0" and "1" to remind them that they are dealing with real circuits, not abstract quantities:

LOW   A signal in the range of algebraically lower voltages, which is interpreted as a logic 0.

HIGH   A signal in the range of algebraically higher voltages, which is interpreted as a logic 1.

Table 3–1   Physical states representing bits in different computer logic and memory technologies.

	State Representing Bit	
Technology	0	1
Relay logic	Circuit open	Circuit closed
Complementary metal-oxide semiconductor (CMOS) logic	0–1.5 V	3.5–5.0 V
Transistor-transistor logic (TTL)	0–0.8 V	2.0–5.0 V
Fiber optics	Light off	Light on
Dynamic memory	Capacitor discharged	Capacitor charged
Nonvolatile, erasable memory	Electrons trapped	Electrons released
Bipolar read-only memory	Fuse blown	Fuse intact
Bubble memory	No magnetic bubble	Bubble present
Magnetic tape or disk	No flux reversal	Flux reversal
Polymer memory	Molecule in state A	Molecule in state B
Compact disc	No pit	Pit

Note that the assignments of 0 and 1 to LOW and HIGH are somewhat arbitrary. Assigning 0 to LOW and 1 to HIGH seems most natural, and is called *positive logic*. The opposite asignment, 1 to LOW and 0 to HIGH, is not often used, and is called *negative logic*.

positive logic

negative logic

    Because a wide range of physical values represent the same binary value, digital logic is highly immune to component and power supply variations and noise. Furthermore, *buffer amplifier* circuits can be used to regenerate "weak" values into "strong" ones, so that digital signals can be transmitted over arbitrary distances without loss of information. For example, a buffer amplifier for CMOS logic converts any HIGH input voltage into a 5.0-V output, and any LOW input voltage into a 0.0-V output.

buffer amplifier

    A logic circuit can be represented with a minimum amount of detail simply as a "black box" with a certain number of inputs and outputs. For example, Figure 3–1 shows a logic circuit with three inputs and one output. However, this representation does not describe how the circuit responds to input signals.

Figure 3–1
"Black box" representation of a three-input, one-output logic circuit.

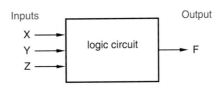

    From the point of view of electronic circuit design, it takes a lot of information to describe the precise electrical behavior of a circuit. However, since the inputs of a digital logic circuit can be viewed as taking on only discrete 0 and 1 values, the circuit's "logical" operation can be described with a table that ignores electrical behavior and lists only discrete 0 and 1 values.

    A logic circuit whose outputs depend only on its current inputs is called a *combinational circuit*. Its operation is fully described by a *truth table* that lists all combinations of input values and the output value(s) produced by each one. Table 3–2 is the truth table for a logic circuit with three inputs X, Y, and Z and a single output F.

combinational circuit

truth table

    A circuit with memory, whose outputs depend on the current input *and* the sequence of past inputs, is called a *sequential circuit*. The behavior of such a circuit may be described by a *state table* that specifies its output and next state as functions of its current state and input. Sequential circuits will be introduced in Chapter 7.

sequential circuit

state table

    As we'll show in Section 4.1, just three basic logic functions, AND, OR, and NOT, can be used to build any combinational digital logic circuit. Figure 3–2 shows the truth tables and symbols for logic "gates" that perform these functions. The symbols and truth tables for AND and OR may be extended to gates with any number of inputs. The gates' functions are easily defined in words:

Table 3–2
Truth table for a combinational logic circuit.

X	Y	Z	F
0	0	0	0
0	0	1	1
0	1	0	0
0	1	1	0
1	0	0	0
1	0	1	0
1	1	0	1
1	1	1	1

AND gate

- An *AND gate* produces a 1 output only if all of its inputs are 1.

OR gate

- An *OR gate* produces a 1 only if any one or more of its inputs are 1.

NOT gate
complement gate
inverter

- A *NOT gate*, often called a *complement gate* or *inverter*, produces an output value that is the opposite of its input value.

inversion bubble

The circle on the inverter symbol's output is called an *inversion bubble*, and is used in this and other gate symbols to denote this "inverting" behavior.

In the definitions of AND and OR functions above, we only had to state the input conditions for which the output is 1, because there is only possibility when the output is not 1—it must be 0.

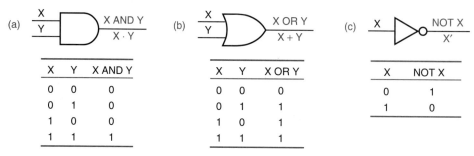

X	Y	X AND Y
0	0	0
0	1	0
1	0	0
1	1	1

X	Y	X OR Y
0	0	0
0	1	1
1	0	1
1	1	1

X	NOT X
0	1
1	0

Figure 3–2   Basic logic elements: (a) AND; (b) OR; (c) NOT.

Two more logic functions are obtained by combining NOT with an AND or OR function in a single gate. Figure 3–3 shows the truth tables and symbols for these gates; Their functions are also easily described in words:

NAND gate

- A *NAND gate* produces the opposite of an AND gate's output, a 0 output only if all of its inputs are 1.

NOR gate

- A *NOR gate* produces the opposite of an OR gate's output, 0 if any one or more of its inputs are 1.

Figure 3–3
Inverting gates: (a) NAND;
(b) NOR.

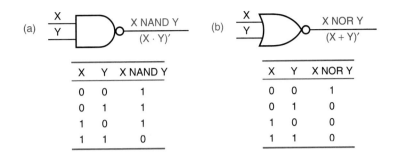

X	Y	X NAND Y
0	0	1
0	1	1
1	0	1
1	1	0

X	Y	X NOR Y
0	0	1
0	1	0
1	0	0
1	1	0

As with AND and OR gates, the symbols and truth tables for NAND and NOR may be extended to gates with any number of inputs.

Figure 3–4 is a logic circuit using AND, OR, and NOT gates that functions according to the truth table of Table 3–2. In Chapter 4 you'll learn how to go from a truth table to a logic circuit, and vice versa, and you'll also learn about the switching-algebra notation that was used in the preceding figures.

Figure 3–4
Logic circuit with the truth table
of Table 3–2.

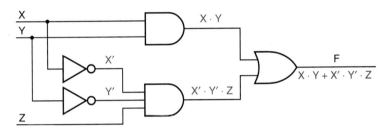

Real logic circuits also function in another analog dimension—time. For example, Figure 3–5 is a *timing diagram* that shows how the circuit of Figure 3–4 might respond to a time-varying pattern of input signals. The timing diagram shows that the logic signals do not change between 0 and 1 instantaneously, and also that there is a lag between an input change and the corresponding output change. Later in this chapter, you'll learn some of the reasons for this timing behavior, and how it is specified and handled in real circuits. And once again,

timing diagram

Figure 3–5
Timing diagram for a logic circuit.

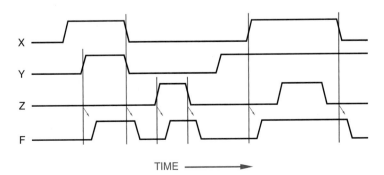

TIME

you'll learn in a later chapter how this analog timing behavior can be generally ignored in most sequential circuits, and instead the circuit can be viewed as moving between discrete states at precise intervals defined by a clock signal.

Thus, even if you know nothing about analog electronics, you should be able to understand the logical behavior of digital circuits. However, there comes a time in design and debugging when every digital logic designer must temporarily throw out "the digital abstraction" and consider the analog phenomena that limit or disrupt digital performance. The purpose of the rest of this chapter is to prepare you for that day by discussing the electrical characteristics of digital logic circuits.

THERE'S HOPE
FOR NON-EE'S

> If all of this electrical "stuff" bothers you, don't worry, at least for now. The rest of this book is written to be as independent of this stuff as possible. But you'll need it later, if you ever have to design and build digital systems in the real world.

3.2   LOGIC FAMILIES

There are many, many ways to design an electronic logic circuit. The first electrically controlled logic circuits, developed at Bell Laboratories in 1930s, were based on relays. In the mid-1940s, the first electronic digital computer, the Eniac, used logic circuits based on vacuum tubes. The Eniac had about 18,000 tubes and a similar number of logic gates, not a lot by today's standards of microprocessor chips with millions of transistors. However, the Eniac could hurt you a lot more than a chip could if it fell on you—it was 100 feet long, 10 feet high, 3 feet deep, and consumed 140,000 watts of power!

semiconductor diode
bipolar junction transistor
integrated circuit (IC)

The inventions of the *semiconductor diode* and the *bipolar junction transistor* allowed the development of smaller, faster, and more capable computers in the late 1950s. In the 1960s, the invention of the *integrated circuit (IC)* allowed multiple diodes, transistors, and other components to be fabricated on a single chip, and computers got still better.

logic family

The 1960s also saw the introduction of the first integrated-circuit logic families. A *logic family* is a collection of different integrated-circuit chips that have similar input, output, and internal circuit characteristics, but that perform different logic functions. Chips from the same family can be interconnected to perform any desired logic function. On the other hand, chips from differing families may not be compatible, in that they may use different power-supply voltages or may use different input and output conditions to represent logic values.

bipolar logic family
transistor-transistor logic (TTL)

The most successful *bipolar logic family* (one based on bipolar junction transistors) is *transistor-transistor logic (TTL)*. First introduced in the 1960s,

TTL now is actually a family of logic families that are compatible with each other but differ in speed, power consumption, and cost. Digital systems can mix components from several different TTL families, according to design goals and constraints in different parts of the system. We describe TTL families beginning in Section 3.10.

Ten years *before* the bipolar junction transistor was invented, the principles of operation were patented for another type of transistor, called the *metal-oxide semiconductor field effect transistor (MOSFET)*, or simply *MOS transistor*. However, MOS transistors were difficult to fabricate in the early days, and it wasn't until the 1960s that a wave of developments made MOS-based logic and memory circuits practical. Even then, MOS circuits lagged bipolar circuits considerably in speed, and were attractive only in selected applications because of their lower power consumption and higher levels of integration.

metal-oxide semiconductor field effect transistor (MOSFET)
MOS transistor

In recent years, advances in the design of MOS circuits, in particular *complementary MOS (CMOS)* circuits, have vastly increased their performance and popularity. By far the majority of new large-scale integrated circuits, such as microprocessors and memories, use CMOS. Likewise, small- to medium-scale applications, for which TTL was once the logic family of choice, are now likely to use CMOS devices with equivalent functionality but higher speed and lower power consumption. CMOS circuits now account for well over half of the worldwide IC market.

complementary MOS (CMOS)

CMOS logic is both the most capable and the easiest to understand commercial digital logic technology. Beginning in the next section, we describe the basic structure of CMOS logic circuits and introduce the most commonly used commercial CMOS logic families.

Because of TTL's continued popularity, most CMOS families are designed to be at least somewhat, and often fully, compatible with TTL. In Section 3.13, we show how TTL and CMOS families can be mixed within a single system.

Nowadays, the acronym "MOS" is usually spoken as "moss," rather than spelled out. And "CMOS" has always been spoken as "sea moss."

GREEN STUFF

3.3 CMOS LOGIC

The functional behavior of a CMOS logic circuit is fairly easy to understand, even if your knowledge of analog electronics is not particularly deep. The basic (and typically only) building blocks in CMOS logic circuits are MOS transistors, described shortly. Before introducing MOS transistors and CMOS logic circuits, we must say a few words about logic levels.

3.3.1   CMOS Logic Levels

Abstract logic elements process binary digits, 0 and 1. However, real logic circuits process electrical signals such as voltage levels.

In any logic circuit, there is a range of voltages (or other circuit conditions) that is interpreted as a logic 0, and another, nonoverlapping range that is interpreted as a logic 1. A typical CMOS logic circuit may interpret any voltage in the range 0–1.5 V as a logic 0, and in the range 3.5–5.0 V as a logic 1. Thus, the definitions of LOW and HIGH for CMOS logic are as shown in Figure 3–6. Voltages in the intermediate range (1.5–3.5 V) are not expected to occur except during signal transitions, and yield undefined logic values (i.e., a circuit may interpret them as either 0 or 1).

Figure 3–6
Logic levels for typical CMOS logic circuits.

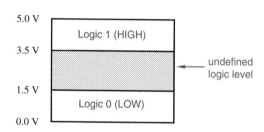

3.3.2   MOS Transistors

A MOS transistor can be modeled as a 3-terminal device that acts like a voltage-controlled resistance. As suggested by Figure 3–7, an input voltage applied to one terminal controls the resistance between the remaining two terminals. In digital logic applications, a MOS transistor is operated so its resistance is always either very high (and the transistor is "off") or very low (and the transistor is "on").

n-channel MOS (NMOS) transistor

gate

source

drain

There are two types of MOS transistors, *n*-channel and *p*-channel; the names refer to the type of semiconductor material used for the resistance-controlled terminals. The circuit symbol for an *n-channel MOS (NMOS) transistor* is shown in Figure 3–8. The terminals are called *gate*, *source*, and *drain*. (Note that the "gate" of a MOS transistor has nothing to do with a "logic gate.") As you might guess from the orientation of the circuit symbol, the drain is normally at a higher voltage than the source.

Figure 3–7
The MOS transistor as a voltage-controlled resistance.

Figure 3–8
Circuit symbol for an *n*-channel
MOS (NMOS) transistor.

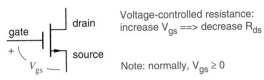

The voltage from gate to source (V_{gs}) in an NMOS transistor is normally zero or positive. If $V_{gs} = 0$, then the resistance from drain to source (R_{ds}) is very high, on the order of a megohm (10^6 ohms) or more. As we increase V_{gs} (i.e., increase the voltage on the gate), R_{ds} decreases to a very low value, 10 ohms or less in some devices.

The circuit symbol for a *p-channel MOS (PMOS) transistor* is shown in Figure 3–9. Operation is analogous to that of an NMOS transistor, except that the source is normally at a higher voltage than the drain, and V_{gs} is normally zero or negative. If V_{gs} is zero, then the resistance from source to drain (R_{ds}) is very high. As we algebraically decrease V_{gs} (i.e., *decrease* the voltage on the gate), R_{ds} decreases to a very low value.

*p-channel MOS
(PMOS) transistor*

Figure 3–9
Circuit symbol for a *p*-channel
MOS (PMOS) transistor.

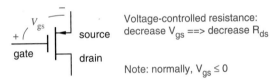

The gate of a MOS transistor has a very high impedance. That is, the gate is separated from the source and the drain by an insulating material with a very high resistance. However, the gate voltage creates an electric field that enhances or retards the flow of current between source and drain. This is the "field effect" in the "MOSFET" name.

Regardless of gate voltage, almost no current flows from the gate to source, or from the gate to drain for that matter. The resistance between the gate and the other terminals of the device is extremely high, well over a megohm. The small amount of current that flows across this resistance is very small, typically less than one microampere (10^{-6} A), and is called a *leakage current*.

leakage current

The MOS transistor symbol itself reminds us that there is no connection between the gate and the other two terminals of the device. However, the gate of a MOS transistor is capacitively coupled to the source and drain, as the symbol might suggest. In high-speed circuits, the power needed to charge and discharge

Technically, there's a difference between "impedance" and "resistance," but electrical engineers often use the terms interchangeably. So do we in this text.

IMPEDANCE VS.
RESISTANCE

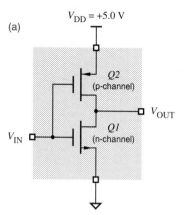

(a)

$V_{DD} = +5.0$ V

Q2
(p-channel)

V_{OUT}

Q1
(n-channel)

V_{IN}

Figure 3–10
CMOS inverter: (a) circuit
diagram; (b) functional behavior;
(c) logic symbol.

(b)

V_{IN}	Q1	Q2	V_{OUT}
0.0 (L)	off	on	5.0 (H)
5.0 (H)	on	off	0.0 (L)

(c)

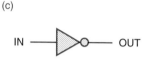

IN — OUT

this capacitance on each input-signal transition accounts for a nontrivial portion of a circuit's power consumption.

3.3.3   Basic CMOS Inverter Circuit

CMOS logic

NMOS and PMOS transistors are used together in a complementary way to form *CMOS logic*. The simplest CMOS circuit, a logic inverter, requires only one of each type of transistor, connected as shown in Figure 3–10. The power supply voltage, V_{DD}, typically may be in the range 2–6 V, and is most often set at 5.0 V for compatibility with TTL circuits.

Ideally, the functional behavior of the CMOS inverter circuit can be characterized by just two cases tabulated in Figure 3–10(b):

(1) V_{IN} is 0.0 V. In this case, the bottom, *n*-channel transistor *Q1* is off, since its V_{gs} is 0, but the top, *p*-channel transistor *Q2* is on, since its V_{gs} is a large negative value (−5.0 V). Therefore, *Q2* presents only a small resistance between the power supply terminal (V_{DD}, +5.0 V) and the output terminal (V_{OUT}), and the output voltage is 5.0 V.

(2) V_{IN} is 5.0 V. Here, *Q1* is on, since its V_{gs} is a large positive value (+5.0 V), but *Q2* is off, since its V_{gs} is 0. Thus, *Q1* presents a small resistance between the output terminal and ground, and the output voltage is 0 V.

With the foregoing functional behavior, the circuit clearly behaves as a logical inverter, since a 0-volt input produces a 5-volt output, and vice versa.

Another way to visualize CMOS operation uses switches. As shown in Figure 3–11(a), the *n*-channel (bottom) transistor is modeled by a normally-open switch, and the *p*-channel (top) transistor by a normally-closed switch. Applying a HIGH voltage changes each switch to the opposite of its normal state, as shown in (b).

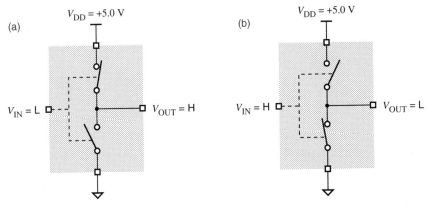

$V_{DD} = +5.0$ V

(a)

$V_{IN} = $ L

$V_{OUT} = $ H

$V_{DD} = +5.0$ V

(b)

$V_{IN} = $ H

$V_{OUT} = $ L

Figure 3–11   Switch model for CMOS inverter: (a) LOW input; (b) HIGH input.

The switch model gives rise to a way of drawing CMOS circuits that makes their logical behavior more readily apparent. As shown in Figure 3–12, different symbols are used for the *p*- and *n*-channel transistors to reflect their logical behavior. The *n*-channel transistor (*Q1*) is switched "on," and current flows between source and drain, when a HIGH voltage is applied to its gate; this seems

Figure 3–12
CMOS inverter logical operation.

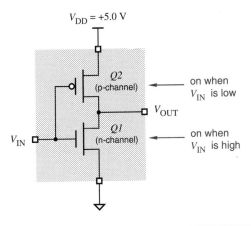

$V_{DD} = +5.0$ V

Q2
(p-channel)

on when
V_{IN} is low

V_{OUT}

Q1
(n-channel)

on when
V_{IN} is high

V_{IN}

The "DD" in the name "V_{DD}" refers to the *drain* terminal of an MOS transistor. This may seem strange, since in the CMOS inverter V_{DD} is actually connected to the *source* terminal of a PMOS transistor. However, CMOS logic circuits evolved from NMOS logic circuits, where the supply *was* connected to the drain of an NMOS transistor through a load resistor, and the name "V_{DD}" stuck. Also note that ground is sometimes referred to as "V_{SS}" in CMOS and NMOS circuits. Some authors and circuit manufacturers use "V_{CC}" as the symbol for the CMOS supply voltage, since this name is used in TTL circuits, which historically preceded CMOS.

WHAT'S IN
A NAME?

natural enough. The *p*-channel transistor (*Q2*) has the opposite behavior. It is "on" when a LOW voltage is applied; the inversion bubble on its gate indicates this inverting behavior.

3.3.4 CMOS NAND and NOR Gates

Both NAND and NOR gates can be constructed using CMOS. A *k*-input gate uses *k* *p*-channel and *k* *n*-channel transistors. Figure 3–13 shows a 2-input CMOS NAND gate. If either input is LOW, the output Z has a low-impedance connection to V_{DD} through the corresponding "on" *p*-channel transistor, and the path to ground is blocked by the corresponding "off" *n*-channel transistor. If both inputs are HIGH, the path to V_{DD} is blocked, and Z has a low-impedance connection to ground. Figure 3–14 shows the switch model for the NAND gate's operation.

Figure 3–15 shows a CMOS NOR gate. If both inputs are LOW, the output Z has a low-impedance connection to V_{DD} through the "on" *p*-channel transistors, and the path to ground is blocked by the "off" *n*-channel transistors. If either input is HIGH, the path to V_{DD} is blocked, and Z has a low-impedance connection to ground.

(a)

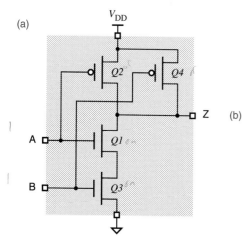

Figure 3–13
CMOS 2-input NAND gate:
(a) circuit diagram; (b) function table; (c) logic symbol.

(b)

A	B	*Q1*	*Q2*	*Q3*	*Q4*	Z
L	L	off	on	off	on	H
L	H	off	on	on	off	H
H	L	on	off	off	on	H
H	H	on	off	on	off	L

(c)

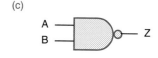

NAND vs. NOR

CMOS NAND and NOR gates do not have identical performance. For a given silicon area, an *n*-channel transistor has lower "on" resistance than a *p*-channel transistor. Therefore, when transistors are put in series, *k* *n*-channel transistors have lower "on" resistance than do *k* *p*-channel ones. As a result, a *k*-input NAND gate is generally faster than and preferred over a *k*-input NOR gate.

3.3.5   Fan-In

The number of inputs that a gate can have in a particular logic family is called
the logic family's *fan-in*. CMOS gates with more than two inputs can be obtained   *fan-in*
by extending series-parallel designs on Figures 3–13 and 3–15 in the obvious
manner. For example, Figure 3–16 shows a 3-input CMOS NAND gate.

In principle, you could design a CMOS NAND or NOR gate with a very
large number of inputs. In practice, however, the additive "on" resistance of

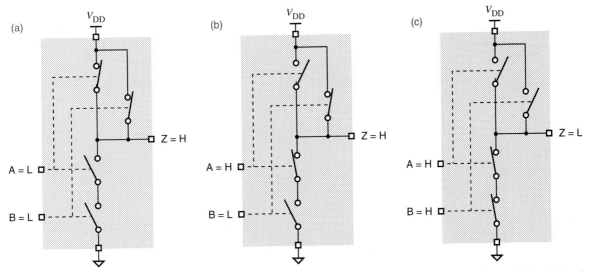

Figure 3–14   Switch model for CMOS 2-input NAND gate: (a) both inputs LOW; (b) one input HIGH; (c) both inputs HIGH.

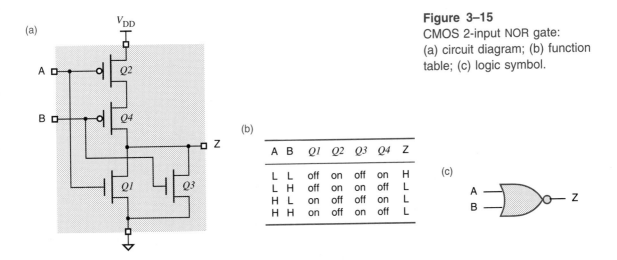

Figure 3–15
CMOS 2-input NOR gate:
(a) circuit diagram; (b) function
table; (c) logic symbol.

A	B	Q1	Q2	Q3	Q4	Z
L	L	off	on	off	on	H
L	H	off	on	on	off	L
H	L	on	off	off	on	L
H	H	on	off	on	off	L

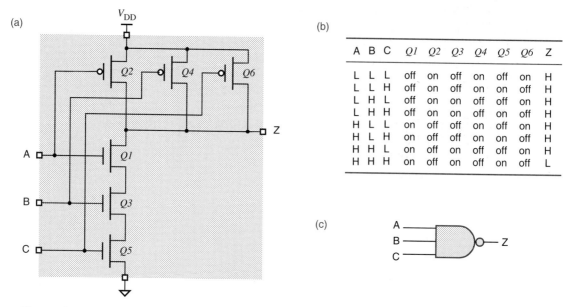

(a)

(b)

A	B	C	Q1	Q2	Q3	Q4	Q5	Q6	Z
L	L	L	off	on	off	on	off	on	H
L	L	H	off	on	off	on	on	off	H
L	H	L	off	on	on	off	off	on	H
L	H	H	off	on	on	off	on	off	H
H	L	L	on	off	off	on	off	on	H
H	L	H	on	off	off	on	on	off	H
H	H	L	on	off	on	off	off	on	H
H	H	H	on	off	on	off	on	off	L

(c)

Figure 3–16 CMOS 3-input NAND gate: (a) circuit diagram; (b) function table; (c) logic symbol.

series transistors limits the fan-in of CMOS gates, typically to 4 for NOR gates and 6 for NAND gates.

As the number of inputs is increased, CMOS gate designers may compensate by increasing the size of the series transistors to reduce their resistance and the corresponding switching delay. However, at some point this becomes inefficient or impractical. Gates with a large number of inputs can be made faster and smaller by cascading gates with fewer inputs. For example, Figure 3–17 shows the logical structure of an 8-input CMOS NAND gate. The total delay through a 4-input NAND, a 2-input NOR, and an inverter is typically less than the delay of a one-level 8-input NAND circuit.

3.3.6 Noninverting Gates

In CMOS, and in most other logic families, the simplest gates are inverters, and the next simplest are NAND gates and NOR gates. A logical inversion comes "for free," and it typically is not possible to design a noninverting gate with a smaller number of transistors than an inverting one.

CMOS noninverting buffers and AND and OR gates are obtained by connecting an inverter to the output of the corresponding inverting gate. Thus, Figure 3–18 shows a noninverting buffer and Figure 3–19 shows an AND gate.

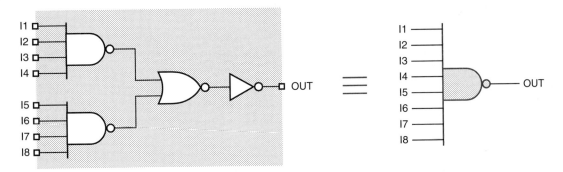

Figure 3–17   Logic diagram equivalent to the internal structure of an 8-input CMOS NAND gate.

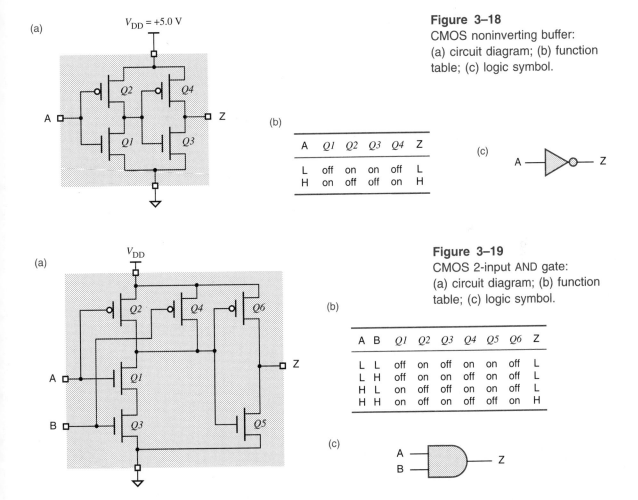

Figure 3–18
CMOS noninverting buffer:
(a) circuit diagram; (b) function
table; (c) logic symbol.

(b)

A	$Q1$	$Q2$	$Q3$	$Q4$	Z
L	off	on	on	off	L
H	on	off	off	on	H

Figure 3–19
CMOS 2-input AND gate:
(a) circuit diagram; (b) function
table; (c) logic symbol.

(b)

A	B	$Q1$	$Q2$	$Q3$	$Q4$	$Q5$	$Q6$	Z
L	L	off	on	off	on	on	off	L
L	H	off	on	on	off	on	off	L
H	L	on	off	off	on	on	off	L
H	H	on	off	on	off	off	on	H

3.3.7 CMOS AND-OR-INVERT and OR-AND-INVERT Gates

AND-OR-INVERT
(AOI) gate

CMOS circuits can perform two levels of logic with just a single "level" of transistors. For example, the circuit in Figure 3–20 is a two-wide, two-input CMOS *AND-OR-INVERT (AOI) gate*. The function table for this circuit is shown in Figure 3–21(a), and a logic diagram for this function using AND and NOR gates is shown in (b). Transistors can be added to or removed from this circuit to obtain an AOI function with a different number of ANDs or a different number of inputs per AND.

The contents of each of the $Q1$–$Q8$ columns in Figure 3–21(a) depends only on the input signal connected to the corresponding transistor's gate. The

Figure 3–20
Circuit diagram for CMOS AND-OR-INVERT gate.

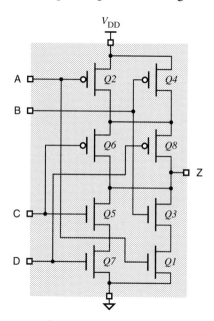

(a)

A	B	C	D	$Q1$	$Q2$	$Q3$	$Q4$	$Q5$	$Q6$	$Q7$	$Q8$	Z
L	L	L	L	off	on	off	on	off	on	off	on	H
L	L	L	H	off	on	off	on	off	on	on	off	H
L	L	H	L	off	on	off	on	on	off	off	on	H
L	L	H	H	off	on	off	on	on	off	on	off	L
L	H	L	L	off	on	on	off	off	on	off	on	H
L	H	L	H	off	on	on	off	off	on	on	off	H
L	H	H	L	off	on	on	off	on	off	off	on	H
L	H	H	H	off	on	on	off	on	off	on	off	L
H	L	L	L	on	off	off	on	off	on	off	on	H
H	L	L	H	on	off	off	on	off	on	on	off	H
H	L	H	L	on	off	off	on	on	off	off	on	H
H	L	H	H	on	off	off	on	on	off	on	off	L
H	H	L	L	on	off	on	off	off	on	off	on	L
H	H	L	H	on	off	on	off	off	on	on	off	L
H	H	H	L	on	off	on	off	on	off	off	on	L
H	H	H	H	on	off	on	off	on	off	on	off	L

Figure 3–21
CMOS AND-OR-INVERT gate:
(a) function table; (b) logic diagram.

(b)

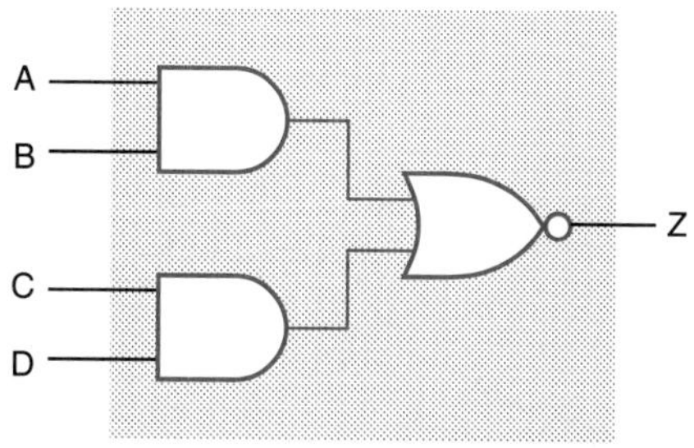

last column is constructed by examining each input combination and determining whether Z is connected to V_{DD} or ground by "on" transistors for that input combination. Note that Z is never connected to *both* V_{DD} and ground for any input combination; in such a case the output would be a non-logic value somewhere between LOW and HIGH, and the output structure would consume excessive power due to the low-impedance connection between V_{DD} and ground.

A circuit can also be designed to perform an OR-AND-INVERT function. For example, Figure 3–22 is a two-wide, two-input CMOS *OR-AND-INVERT (OAI) gate*. The function table for this circuit is shown in Figure 3–23(a); the values

OR-AND-INVERT (OAI) gate

Figure 3–22
Circuit diagram for CMOS OR-AND-INVERT gate.

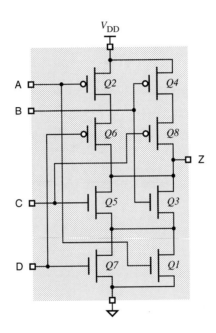

(a)

A	B	C	D	Q1	Q2	Q3	Q4	Q5	Q6	Q7	Q8	Z
L	L	L	L	off	on	off	on	off	on	off	on	H
L	L	L	H	off	on	off	on	off	off	on	on	H
L	L	H	L	off	on	off	on	on	on	off	off	H
L	L	H	H	off	on	off	on	on	off	on	off	L
L	H	L	L	off	on	on	off	off	on	off	on	H
L	H	L	H	off	on	on	off	off	off	on	on	L
L	H	H	L	off	on	on	off	on	on	off	off	H
L	H	H	H	off	on	on	off	on	off	on	off	L
H	L	L	L	on	off	off	on	off	on	off	on	H
H	L	L	H	on	off	off	on	off	off	on	on	H
H	L	H	L	on	off	off	on	on	on	off	off	L
H	L	H	H	on	off	off	on	on	off	on	off	L
H	H	L	L	on	off	on	off	off	on	off	on	L
H	H	L	H	on	off	on	off	off	off	on	on	L
H	H	H	L	on	off	on	off	on	on	off	off	L
H	H	H	H	on	off	on	off	on	off	on	off	L

Figure 3–23
CMOS OR-AND-INVERT gate:
(a) function table; (b) logic diagram.

(b)

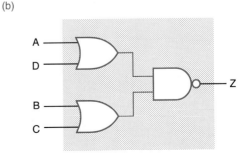

in each column are determined just as we did for the CMOS AOI gate. A logic diagram for the OAI function using OR and NAND gates is shown in (b).

The speed and other electrical characteristics of a CMOS AOI or OAI gate are quite comparable to those of a single CMOS NAND or NOR gate. As a result, these gates are very appealing because they can perform two levels of logic (AND-OR or OR-AND) with just one level of delay. Although most logic designers don't bother with AOI gates in their discrete designs, many larger-scale CMOS devices use these gates internally. For example, if you examine the internal logic diagrams of larger-scale devices in Chapter 5, you'll find that the AOI structure appears quite often. In addition, AOI and OAI gates are usually available as building blocks in CMOS gate-array and VLSI design.

3.4   ELECTRICAL BEHAVIOR OF CMOS CIRCUITS

The next three sections discuss electrical, not logical, aspects of CMOS circuit operation. It's important to understand this material when you design real circuits using CMOS or other logic families. Most of this material is aimed at providing a framework for ensuring that the "digital abstraction" is really valid for a given circuit. In particular, a circuit or system designer must provide in a number of areas adequate engineering design margins—insurance that the circuit will work properly even under the worst of conditions.

engineering design
margins

3.4.1   Overview

The topics that we discuss in Sections 3.5–3.7 include the following:

- *DC noise margins.* Depending on loading and other factors, a CMOS output signal has a certain voltage in the LOW state, and another voltage in the HIGH state. Circuits that observe such an output signal, depending on their environment, are capable of interpreting a certain range of input voltages as LOW, and another range as HIGH. Nonnegative DC noise margins ensure that the highest LOW voltage produced by an output is always lower than the highest voltage that an input can reliably interpret as LOW, and that the lowest HIGH voltage produced by an output is always higher than the lowest voltage that an input can reliably interpret as HIGH.

- *Fanout.* This refers to the number and type of inputs that are connected to a given output. If too many inputs are connected to an output, the DC noise margins of the circuit may be inadequate. Fanout may also affect the speed at which the output changes from one state to another.

- *Speed.* The time that it takes a CMOS output to change from the LOW state to the HIGH state, or vice versa, depends on both the internal structure of the device and the characteristics of the other devices that it drives, even to the extent of being affected by the wire or printed-circuit-board traces connected to the output. We'll look at two separate components of "speed"—transition time and propagation delay.

- *Power consumption.* The power consumed by a CMOS device depends on a number of factors, including not only its internal structure, but also the input signals that it receives, the other devices that it drives, and how often its output changes between LOW and HIGH.

- *Noise.* The main reason for providing engineering design margins is to ensure proper circuit operation in the presence of noise. Noise can be generated by a number of sources; several of them are listed below, from the least likely to the (perhaps surprisingly) most likely:

  - Cosmic rays.
  - Magnetic fields from nearby machinery.
  - Power-supply disturbances.
  - The switching action of the logic circuits themselves.

- *Electrostatic discharge.* Would you believe that you can destroy a CMOS device just by touching it?

- *Open-drain outputs.* Some CMOS outputs omit the usual *p*-channel pull-up transistors. In the HIGH state, such outputs are effectively a "no-connection," which is useful in some applications.

- *Three-state outputs.* Some CMOS devices have an extra "output enable" control input that can be used to disable both the *p*-channel pull-up transistors and the *n*-channel pull-down transistors. Many such device outputs can be tied together to create a multisource bus, as long as the control logic is arranged so that at most one output is enabled at a time.

3.4.2 A Real-World CMOS Data Sheet

The manufacturers of real-world devices provide *data sheets* that specify the devices' logical and electrical characteristics. The electrical specifications portion of a typical CMOS data sheet is shown in Table 3–3, and the test circuits used by the manufacturer to define various parameters are shown in Figure 3–24.

data sheet

Most of the terms in the data sheet and the waveforms in the figure are probably meaningless to you at this point. In fact, different manufacturers may

Table 3–3 Specifications for a typical CMOS device, based on the manufacturer's data sheet for the IDT74FCT257T multiplexer (Integrated Device Technology, Inc., Santa Clara, CA).

DC ELECTRICAL CHARACTERISTICS OVER OPERATING RANGE
The following conditions apply unless otherwise specified:
Commercial: $T_A = 0°C$ to $+70°C$, $V_{CC} = 5.0V \pm 5\%$;   Military: $T_A = -55°C$ to $+125°C$, $V_{CC} = 5.0V \pm 10\%$

Sym.	Parameter	Test Conditions[1]		Min.	Typ.[2]	Max.	Unit
V_{IH}	Input HIGH level	Guaranteed logic HIGH level		2.0	—	—	V
V_{IL}	Input LOW level	Guaranteed logic LOW level		—	—	0.8	V
I_{IH}	Input HIGH current	V_{CC} = Max.	V_I = 2.7 V	—	—	5	μA
I_{IL}	Input LOW current	V_{CC} = Max.	V_I = 0.5 V	—	—	−5	μA
I_{OZH}	High-impedance output current	V_{CC} = Max.	V_O = 2.7 V	—	—	10	μA
I_{OZL}			V_O = 0.5 V	—	—	−10	
I_I	Input HIGH current	V_{CC} = Max., $V_I = V_{CC}$ (Max.)		—	—	20	μA
V_{IK}	Clamp diode voltage	V_{CC} = Min., I_N = −18 mA		—	−0.7	−1.2	V
I_{IOS}	Short-circuit current	V_{CC} = Max.,[3] V_O = GND		−60	−120	−225	mA
V_{OH}	Output HIGH voltage	V_{CC} = Min. $V_{IN} = V_{IH}$ or V_{IL}	I_{OH} = −6 mA Mil. I_{OH} = −8 mA Com'l.	2.4	3.3	—	V
			I_{OH} = −12 mA Mil. I_{OH} = −15 mA Com'l.	2.0	3.0	—	V
V_{OL}	Output LOW voltage	V_{CC} = Min. $V_{IN} = V_{IH}$ or V_{IL}	I_{OL} = 32 mA Mil. I_{OL} = 48 mA Com'l.	—	0.3	0.5	V
V_H	Input hysteresis	—		—	200	—	mV
I_{CC}	Quiescent power supply current	V_{CC} = Max. V_{IN} = GND or V_{CC}		—	0.2	1.5	mA

CAPACITANCE ($T_A = 25°C$, f = 1.0 MHz)

Sym.	Parameter[4]	Test Conditions	Min.	Typ.	Max.	Unit
C_{IN}	Input capacitance	V_{IN} = 0 V	—	6	10	pF
C_{IN}	Output capacitance	V_{OUT} = 0 V	—	8	12	pF

NOTES:
1. For conditions shown as Max. or Min., use appropriate value specified under Electrical Characteristics.
2. Typical values are at V_{CC} = 5.0 V, +25°C ambient, and maximum loading.
3. Not more than one output should be shorted at a time. Duration of short-circuit test should not exceed one second.
4. This parameter is guaranteed but not tested.

Table 3–3 (continued)

POWER SUPPLY CHARACTERISTICS							
Sym.	Parameter	Test Conditions[1]		Min.	Typ.[2]	Max.	Unit
ΔI_{CC}	Quiescent power supply current, TTL inputs HIGH	V_{CC} = Max. V_{IN} = 3.4 V[3]		—	0.5	2.0	mA
I_{CCD}	Dynamic power supply current[4]	V_{CC} = Max. Outputs open $\overline{E}$ or $\overline{OE}$ = GND One bit toggling 50% duty cycle	$V_{IN} = V_{CC}$ V_{IN} = GND	—	0.15	0.25	mA/MHz
I_C	Total power supply current[5]	V_{CC} = Max. Outputs open f_i = 10 MHz	$V_{IN} = V_{CC}$ V_{IN} = GND	—	1.7	4.0	mA
		$\overline{E}$ or $\overline{OE}$ = GND One bit toggling 50% duty cycle	V_{IN} = 3.4 V V_{IN} = GND	—	2.0	4.0	
		V_{CC} = Max. Outputs open f_i = 2.5 MHz	$V_{IN} = V_{CC}$ V_{IN} = GND	—	1.7	4.0[6]	
		$\overline{E}$ or $\overline{OE}$ = GND Four bits toggling 50% duty cycle	V_{IN} = 3.4 V V_{IN} = GND	—	2.7	8.0	

NOTES:

1. For conditions shown as Max. or Min., use appropriate value specified under Electrical Characteristics.
2. Typical values are at V_{CC} = 5.0 V, +25°C ambient.
3. Per TTL driven input (V_{IN} = 3.4 V); all other inputs at V_{CC} or GND.
4. This parameter is not directly testable, but is derived for use in Total Power Supply Current calculations.
5. $I_C = I_{QUIESCENT} + I_{INPUTS} + I_{DYNAMIC}$
   $I_C = I_{CC} + \Delta I_{CC} D_H N_T + I_{CCD}(f_{CP}/2 + f_i N_i)$
   I_{CC} = quiescent power suppy current
   ΔI_{CC} = power supply current for a TTL HIGH input (V_{IN} = 3.4 V)
   D_H = duty cycle for TTL inputs HIGH
   N_T = number of TTL inputs at D_H
   I_{CCD} = dynamic current caused by an output transition pair (LHL or HLH)
   f_{CP} = clock frequency for register devices (zero for non-register devices)
   f_i = input clock frequency
   N_i = number of inputs at f_i
   All currents are in milliamperes and all frequencies are in megahertz.
6. Values for these conditions are examples of the I_C formula. These limits are guaranteed but not tested.

Table 3–3 (continued)

SWITCHING CHARACTERISTICS OVER OPERATING RANGE										
			54/74FCT257T				54/74FCT257AT			
			Com'l.		Mil.		Com'l.		Mil.	
Sym.	Parameter	Condition[1]	Min.[2]	Max.	Min.[2]	Max.	Min.[2]	Max.	Min.[2]	Max.	Unit
t_{PLH} t_{PHL}	Propagation delay In to Zn	$C_L = 50$ pF $R_L = 500\ \Omega$	1.5	6.0	1.5	7.0	1.5	5.0	1.5	5.8	ns
t_{PLH} t_{PHL}	Propagation delay S to Zn		1.5	10.5	1.5	12.0	1.5	7.0	1.5	8.1	ns
t_{PLH} t_{PHL}	Output enable time		1.5	8.5	1.5	10.0	1.5	7.0	1.5	8.0	ns
t_{PLH} t_{PHL}	Output disable time		1.5	6.0	1.5	8.0	1.5	5.5	1.5	5.8	ns

NOTES:
1. See test circuit and waveforms.
2. Minimum limits on propagation delays are guaranteed but not tested.

use slightly different terms and symbols to convey the same information in their data sheets. However, after reading the next three sections you should know enough about the electrical characteristics of CMOS circuits that you'll be able to understand the salient points of this or any other data sheet. As a logic designer, you'll need this knowledge to create reliable and robust real-world circuits and systems.

WHAT'S IN
A NUMBER?

> Two different prefixes, "74" and "54," are used in the part numbers of CMOS and TTL devices. These prefixes simply distinguish between commercial and military versions as explained in Section 3.8.

BE NOT AFRAID

> Computer science students and other non-EE readers should not have undue fear of the material in the next three sections. Only a basic understanding of electronics, at about the level of Ohm's law, is required.

TEST CIRCUIT FOR ALL OUTPUTS

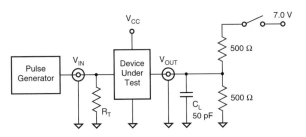

SWITCH POSITION

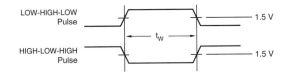

Test	Switch
Open Drain Three-State Disable from LOW Three-State Enable from LOW	Closed
All Other Tests	Open

DEFINITIONS:
C_L = Load capacitance, includes jig and probe capacitance.
R_T = Termination resistance, should equal Z_{OUT} of the Pulse Generator.

SETUP, HOLD, AND RELEASE TIMES

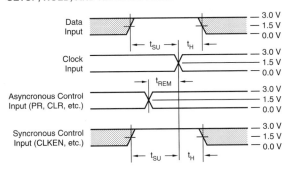

PULSE WIDTH

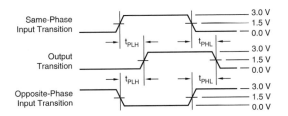

PROPAGATION DELAY

THREE-STATE ENABLE AND DISABLE TIMES

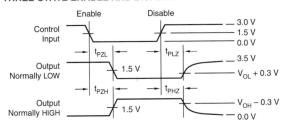

Figure 3–24　Test circuits and waveforms for 74FCT logic, based on the manufacturer's data sheet for the IDT74FCT257T multiplexer (Integrated Device Technology, Inc., Santa Clara, CA).

3.5　CMOS STEADY-STATE ELECTRICAL BEHAVIOR

This section discusses the steady-state behavior of CMOS circuits, that is, the circuits' behavior when inputs and outputs are not changing. The next section discusses dynamic behavior, including speed and power dissipation.

3.5.1　Logic Levels and Noise Margins

The table in Figure 3–10(b) on page 82 defined the CMOS inverter's behavior only at two discrete input voltages; other input voltages may yield different output

Figure 3–25
Typical input-output transfer
characteristic of a CMOS
inverter.

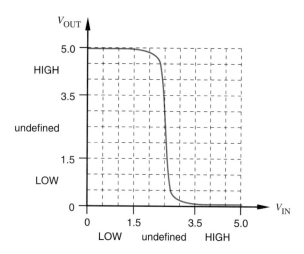

voltages. The complete input-output transfer characteristic can be described by a graph such as Figure 3–25. In this graph, the input voltage is varied from 0 to 5 V, as shown on the X axis; the Y axis plots the output voltage.

If we believed the curve in Figure 3–25, we could define a CMOS LOW input level as any voltage under 2.4 V, and a HIGH input level as anything over 2.6 V. Only when the input is between 2.4 and 2.6 V does the inverter produce a nonlogic output voltage under this definition.

Unfortunately, the typical transfer characteristic shown in Figure 3–25 is just that—typical, but not guaranteed. It varies greatly under different conditions of power supply voltage, temperature, and output loading. The transfer characteristic may even vary depending on when the device was fabricated. For example, after months of trying to figure out why gates made on some days were good and on other days were bad, one manufacturer discovered that the bad gates were victims of airborne contamination by a particularly noxious perfume worn by one of its production-line workers!

Sound engineering practice dictates that we use more conservative specifications for LOW and HIGH. The conservative specs for a typical CMOS logic family (HC-series) are depicted in Figure 3–26. These parameters are specified by CMOS device manufacturers in a "data sheet," and are defined as follows:

V_{OHmin} The minimum output voltage in the HIGH state, 4.9 V for HC-series CMOS with a 5-volt power supply.

V_{IHmin} The minimum input voltage guaranteed to be recognized as a HIGH, 3.5 V for HC-series CMOS with a 5-volt power supply.

V_{ILmax} The maximum input voltage guaranteed to be recognized as a LOW, 1.5 V for HC-series CMOS with a 5-volt power supply.

Figure 3–26
Logic levels and noise margins
for the HC-series CMOS logic
family.

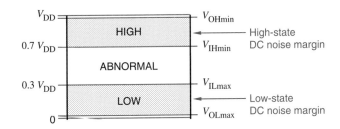

V_{OLmax} The maximum output voltage in the LOW state, 0.1 V for HC-series
CMOS.

The input parameters are determined mainly by switching thresholds of the two
transistors, while the output parameters are determined mainly by the "on" resistance of the transistors. These parameters apply when the device inputs and
outputs are connected only to other CMOS devices. Different parameters may
apply when CMOS devices drive other logic families or resistive loads, as explained later.

    At 1.5 V, V_{ILmax} exceeds V_{OLmax} by 1.4 V, so HC-series CMOS has a *DC*   *DC noise margin*
noise margin of 1.4 V in the LOW state. That is, it takes at least 1.4 V of noise
to corrupt a worst-case LOW output into a voltage that is not guaranteed to be
recognizable by another gate as a LOW input. There is a DC noise margin of
1.4 V in the HIGH state also. Compared with other logic families that we'll study,
HC-series CMOS has excellent DC noise margins.

    Regardless of the voltage applied to the input of a CMOS inverter, the input
consumes very little current, only the leakage current of the two transistors' gates.
The maximum amount of leakage current that can flow is another parameter
specified by the device manufacturer:

I_{Imax} The maximum current that flows into the input in any state, ±1 μA for
HC-series CMOS with a 5-volt power supply.

Thus, it takes very little power to maintain a CMOS inverter input in one state
or the other. This is in sharp constrast to bipolar logic circuits, whose inputs
may consume significant current (and power) in one or both states.

    All of the parameters in Figure 3–26 are guaranteed by CMOS manufac-
turers over a range of temperature and output loading. Parameters are also guaranteed over a range of power-supply voltage V_{DD}, typically 5.0 V±10%. Note
that some parameters are a function of the actual power-supply voltage:

V_{ILmax} 30% of V_{DD}

V_{IHmin} 70% of V_{DD}

V_{OHmin} $V_{DD} - 0.1$ V

power-supply rails   The power-supply voltage V_{DD} and ground are often called the *power-supply rails*. CMOS output voltages are usually close to the power-supply rails, typically 0 V and 5 V for LOW and HIGH, respectively.

3.5.2 Circuit Behavior with Resistive Loads

When the output of a CMOS circuit is connected to TTL gate inputs or other resistive loads, the output behavior is not as ideal as we described previously, for a number of reasons. In either logic state, the CMOS output transistor that is "on" has a nonzero resistance, and a load connected to the output terminal will cause a voltage drop across this resistance. Thus, in the LOW state, the output voltage may be somewhat higher than 0.1 V, and in the HIGH state it may be lower than 4.9 V. The easiest way to see how this happens is look at a resistive model of the CMOS circuit and load.

Figure 3–27(a) shows the resistive model. The *p*-channel and *n*-channel transistors have resistances R_p and R_n, respectively. In normal operation, one resistance is high (> 1 MΩ) and the other is low (perhaps 100 Ω), depending on whether the input voltage is HIGH or LOW. The load in this circuit consists of two resistors attached to the supply rails; a real circuit may have any resistor values, or an even more complex resistive network. Such a load may be attached to the CMOS output for any of a variety of reasons:

- Discrete resistors may be included to provide transmission-line termination, discussed in Section 12.4.

- Discrete resistors may not really be present in the circuit, but the load presented by one or more TTL or other non-CMOS inputs may be modeled by a simple resistor network.

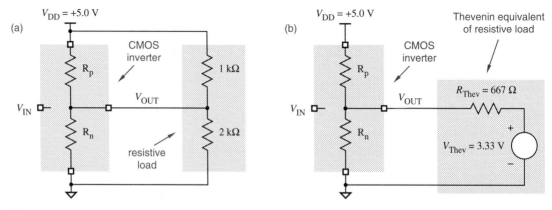

Figure 3–27   Resistive model of a CMOS inverter with a resistive load: (a) showing actual load circuit; (b) using Thévenin equivalent of load.

Any two-terminal circuit consisting of only voltage sources and resistors can be modeled by a *Thévenin equivalent* consisting of a single voltage source in series with a single resistor. The *Thévenin voltage* is the open-circuit voltage of the original circuit, and the *Thévenin resistance* is the Thévenin voltage divided by the short-circuit current of the original circuit.

In the example of Figure 3–27, the Thévenin voltage of the resistive load, including its connection to V_{DD}, is established by the 1-kΩ and 2-kΩ resistors, which form a voltage divider:

$$V_{Thev} = \frac{2\ k\Omega}{(2\ k\Omega + 1\ k\Omega)} \cdot 5.0\ V = 3.33\ V$$

The short-circuit current is $(5.0\ V)/(1\ k\Omega) = 5$ mA, so the Thévenin resistance is $(3.33\ V)/(5\ mA) = 667\ \Omega$. Experienced readers may recognize this as the parallel resistance of the 1-kΩ and 2-kΩ resistors.

- The resistors may be part of or may model a current-consuming device such as a light-emitting diode (LED) or a relay coil.

In any case, a resistive load, consisting only of resistors and voltage sources, can always be modeled by a Thévenin equivalent network, such as the one shown in Figure 3–27(b). Such a load is often called a *DC load*.   *DC load*

When the CMOS inverter has a HIGH input, the output should be LOW; the actual output voltage can be predicted using the resistive model shown in Figure 3–28. The *p*-channel transistor is "off" and has a very high resistance, high enough to be negligible in the calculations that follow. The *n*-channel transistor is "on" and has a low resistance, which we assume to be 100 Ω. (The actual "on" resistance depends on the CMOS family and other characteristics such as operating temperature and whether or not the device was manufactured on a good day.) The "on" transistor and the Thévenin-equivalent resistor R_{Thev}

Figure 3–28
Resistive model for CMOS LOW output with resistive load.

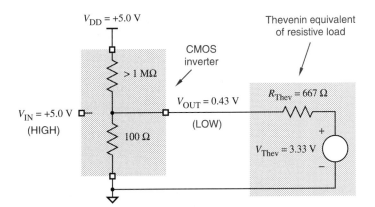

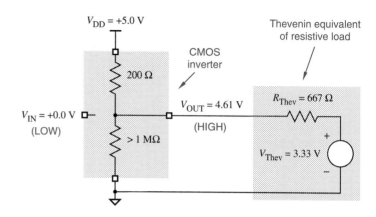

Figure 3–29
Resistive model for CMOS HIGH
output with resistive load.

in Figure 3–28 form a simple voltage divider. The resulting output voltage can
be calculated as follows:

$$V_{OUT} = 3.33 \text{ V} \cdot [100/(100+667)]$$
$$= 0.43 \text{ V}$$

Similarly, when the inverter has a LOW input, the output should be HIGH,
and the actual output voltage can be predicted with the model in Figure 3–29.
We'll assume that the *p*-channel transistor's "on" resistance is 200 Ω. Once
again, the "on" transistor and the Thévenin-equivalent resistor R_{Thev} in the figure
form a simple voltage divider, and the resulting output voltage can be calculated
as follows:

$$V_{OUT} = 3.33 \text{ V} + (5 \text{ V} - 3.33 \text{ V}) \cdot [667/(200+667)]$$
$$= 4.61 \text{ V}$$

In practice, it's seldom necessary to calculate output voltages as in the
preceding examples. In fact, IC manufacturers normally don't specify the equivalent resistances of the "on" transistors, so you wouldn't have the necessary
information to make the calculation anyway. Instead, IC manufacturers specify
a maximum load for the output in each state (HIGH or LOW), and guarantee a
worst-case output voltage for that load. The load is specified in terms of current:

I_{OLmax} The maximum current that the output can sink in the LOW state while
still maintaining an output voltage no greater than V_{OLmax}.

I_{OHmax} The maximum current that the output can source in the HIGH state while
still maintaining an output voltage no less than V_{OHmin}.

sinking current

These definitions are illustrated in Figure 3–30. A device output is said to *sink
current* when current flows from the power supply, through the load, and through

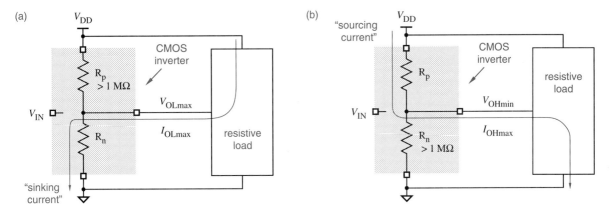

Figure 3–30 Circuit definitions of (a) I_{OLmax}; (b) I_{OHmax}.

the device output to ground as in (a). The output is said to *source current* when   *sourcing current*
current flows from the power supply, out of the device output, and through the
load to ground as in (b).

Most CMOS devices have two sets of loading specifications. One set is
for "CMOS" loads, meaning that the device output is connected to other CMOS
inputs, which consume very little current. The other set is for "TTL" loads,
meaning that the output is connected to TTL inputs or other devices that consume
significant current. For example, the specifications for HC-series CMOS outputs
are shown in Table 3–4.

Notice in the table that the output current in the HIGH state is shown as a
negative number. By convention, the current measured at a device terminal is
positive if current flows *into* the device; in the HIGH state, current flows *out* of
the output terminal.

As the table shows, with CMOS loads, the CMOS gate's output voltage is
maintained within 0.1 V of the power-supply rail. With TTL loads, the output
voltage may be degraded quite a bit. Also notice in this case that the maximum

Table 3–4 Output loading specifications for HC-series CMOS with a 5-volt supply.

	CMOS load		TTL load	
Parameter	Name	Value	Name	Value
Maximum LOW-state output current (mA)	I_{OLmaxC}	0.02	I_{OLmaxT}	4.0
Maximum LOW-state output voltage (V)	V_{OLmaxC}	0.1	V_{OLmaxT}	0.33
Maximum HIGH-state output current (mA)	I_{OHmaxC}	−0.02	I_{OHmaxT}	−4.0
Minimum HIGH-state output voltage (V)	V_{OHminC}	4.9	V_{OHminT}	4.3

voltage drop with respect to the power-supply rail is twice as much in the HIGH state (0.7 V) as in the LOW state (0.33 V). This suggests that the *p*-channel transistors in HC-series CMOS have a higher "on" resistance than the *n*-channel transistors do. This is natural, since in any CMOS circuit, a *p*-channel transistor has over twice the "on" resistance of an *n*-channel transistor with the same area. Equal voltage drops in both states could be obtained only by making the *p*-channel transistors much larger than the *n*-channel transistors.

It is usually quite straightforward to determine how much current an output sources or sinks in a given situation. In Figure 3–28 on page 99, the "on" *n*-channel transistor modeled by a 100-Ω resistor has a 0.43-V drop across it; therefore it sinks $(0.43\text{ V})/(100\ \Omega) = 4.3$ mA of current. Similarly, the "on" *p*-channel transistor in Figure 3–29 sources $(0.39\text{ V})/(200\ \Omega) = 1.95$ mA.

The actual "on" resistances of CMOS output transistors usually aren't published, so it's not always possible to use the exact models of the previous paragraphs. However, you can estimate "on" resistances using the following equations, which rely on specifications that are always published:

$$R_{p(on)} = \frac{V_{DD} - V_{OHminT}}{|I_{OHmaxT}|}$$

$$R_{n(on)} = \frac{V_{OLmaxT}}{I_{OLmaxT}}$$

These equations use Ohm's law to compute the "on" resistance as the voltage drop across the "on" transistor divided by the current through it with a worst-case resistive load. Using the numbers given for HC-series CMOS in Table 3–4, we can calculate $R_{p(on)} = 175\ \Omega$ and $R_{n(on)} = 82.5\ \Omega$.

Very good *worst-case* estimates of output current can be made by assuming that there is *no* voltage drop across the "on" transistor. This assumption

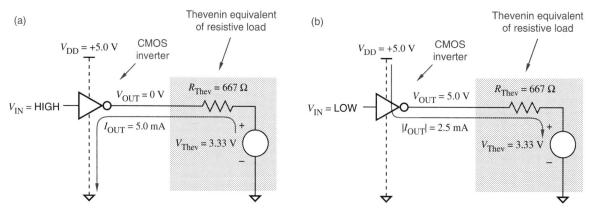

Figure 3–31   Estimating sink and source current: (a) output LOW; (b) output HIGH.

simplifies the analysis, and yields a conservative result that is almost always good enough for practical purposes. For example, Figure 3–31 shows a CMOS inverter driving the same Thévenin-equivalent load that we've used in previous examples. The resistive model of the output structure is not shown, because it is no longer needed; we assume that there is no voltage drop across the "on" CMOS transistor. In (a), with the output LOW, the entire 3.33-V Thévenin-equivalent voltage source appears across R_{Thev}, and the estimated sink current is (3.33 V)/(667 Ω) = 5.0 mA. In (b), with thc output HIGH and assuming a 5.0-V supply, the voltage drop across R_{Thev} is 1.67 V, and the estimated source current is (1.67 V)/(667 Ω) = 2.5 mA.

An important feature of the CMOS inverter (or any CMOS circuit) is that the output structure by itself consumes very little current in either state, HIGH or LOW. In either state, one of the transistors is in the high-impedance "off" state. All of the current flow that we've been talking about occurs when a resistive load is connected to the CMOS output. If there's no load, then there's no current flow, and the power consumption is zero. With a load, however, current flows through both the load and the "on" transistor, and power is consumed in both.

As we've stated elsewhere, an "off" transistor's resistance is over one megohm, but it's not infinite. Therefore, a very tiny leakage current actually does flow in "off" transistors and the CMOS output structure does have a correspondingly tiny but nonzero power consumption. In most applications, this power consumption is tiny enough to ignore. It is usually significant only in "standby mode" in battery-powered devices, such as the laptop computer on which this chapter was prepared.

THE TRUTH
ABOUT POWER
CONSUMPTION

3.5.3 Circuit Behavior with Nonideal Inputs

So far, we have assumed that the HIGH and LOW inputs to a CMOS circuit are ideal voltages, very close to the power supply rails. However, the behavior of a real CMOS inverter circuit depends on the input voltage as well as on the characteristics of the load.

If the input voltage is not close to the power-supply rail, then the "on" transistor may not be fully "on" and its resistance may increase. Likewise, the "off" transistor may not be fully "off' and its resistance may be quite a bit less than one megohm. These two effects combine to move the output voltage away from the power-supply rail.

For example, Figure 3–32(a) shows a CMOS inverter's possible behavior with a 1.5-V input. The p-channel transistor's resistance has doubled at this point, and that the n-channel transistor is beginning to turn on. (These values are simply assumed for the purposes of illustration; the actual values depend on the detailed characteristics of the transistors.)

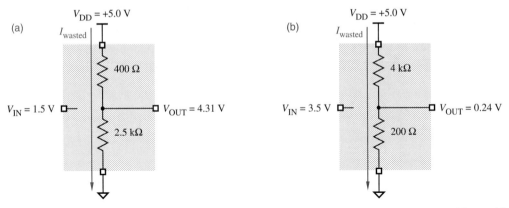

Figure 3–32   CMOS inverter with nonideal input voltages: (a) equivalent circuit with 1.5-V
input; (b) equivalent circuit with 3.5-V input.

In the figure, the output at 4.31 V is still well within the valid range for a
HIGH signal, but not quite the ideal of 5.0 V. Similarly, with a 3.5-V input in (b),
the LOW output is 0.24 V, not 0 V. The slight degradation of output voltage is
generally tolerable; what's worse is that the output structure is now consuming
a nontrivial amount of power. The current flow with the 1.5-V input is

$$I_{wasted} \;=\; 5.0 \text{ V} / (400 \text{ }\Omega + 2.5 \text{ k}\Omega) \;=\; 1.72 \text{ mA}$$

and the power consumption is

$$P_{wasted} \;=\; 5.0 \text{ V} \cdot I_{wasted} \;=\; 8.62 \text{ mW}$$

The output voltage of a CMOS inverter deteriorates further if a resistive
load is attached to it. Such a load may be attached for any of a variety of
reasons given in the preceding subsection. Figure 3–33 shows a CMOS inverter's
possible behavior with a resistive load. With a 1.5-V input, the output at 3.98 V

Figure 3–33
CMOS inverter with load and
nonideal 1.5-V input.

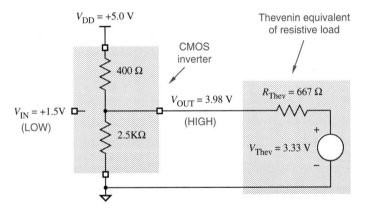

Figure 3–34
CMOS inverter with load and
nonideal 3.5-V input.

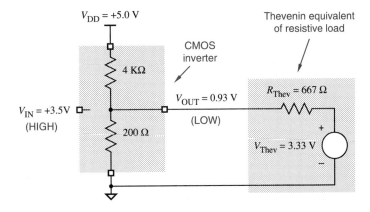

is still within the valid range for a HIGH signal, but it is far from the ideal of
5.0 V. Similarly, with a 3.5-V input as shown in Figure 3–34, the LOW output is
0.93 V, not 0 V.

In "pure" CMOS systems, all of the logic devices in a circuit are CMOS.
Since CMOS inputs have a very high impedance, they present very little resistive
load to the CMOS outputs that drive them. Therefore, the CMOS output levels
all remain very close to the power-supply rails (0 V and 5 V), and none of the
devices waste power in their output structures. On the other hand, if TTL outputs
or other nonideal logic signals are connected to CMOS inputs, then the CMOS
outputs use power in the way depicted in this subsection. In addition, if TTL
inputs or other resistive loads are connected to CMOS outputs, then the CMOS
outputs use power in the way depicted in the preceding subsection.

3.5.4 Fanout

The *fanout* of a logic gate is the number of inputs that the gate can drive without   *fanout*
exceeding its worst-case loading specifications. The fanout depends not only on
the characteristics of the output, but also on the inputs that it is driving. Fanout
must be examined for both possible output states, HIGH and LOW.

For example, we showed in Table 3–4 on page 101 that the maximum
LOW-state output current I_{OLmaxC} for an HC-series CMOS gate driving CMOS
inputs is 0.02 mA (20 μA). We also stated previously that the maximum input
current I_{Imax} for an HC-series CMOS input in any state is ±1 μA. Therefore,
the *LOW-state fanout* for an HC-series output driving HC-series inputs is 20.   *LOW-state fanout*
Table 3–4 also showed that the maximum HIGH-state output current I_{OHmaxC} is
−0.02 mA (−20 μA). Therefore, the *HIGH-state fanout* for an HC-series output   *HIGH-state fanout*
driving HC-series inputs is also 20.

overall fanout

Note that the HIGH-state and LOW-state fanouts of a gate are not necessarily equal. In general, the *overall fanout* of a gate is the minimum of its HIGH-state and LOW-state fanouts, 20 in the foregoing example.

In the fanout example that we just completed, we assumed that we needed to maintain the gate's output at CMOS levels, that is, within 0.1 V of the power-supply rails. If we were willing to live with somewhat degraded, TTL output levels, then we could use I_{OLmaxT} and I_{OHmaxT} in the fanout calculation. According to Table 3–4, these specifications are 4.0 mA and −4.0 mA, respectively. Therefore, the fanout of an HC-series output driving HC-series inputs at TTL levels is 4000, virtually unlimited, apparently.

DC fanout

Well, not quite. The calculations that we've just carried out give the *DC fanout*, defined as the number of inputs that an output can drive *with the output in a constant state* (HIGH or LOW). Even if the DC fanout specification is met, a CMOS output driving a large number of inputs may not behave satisfactorily on transitions, LOW-to-HIGH or vice versa.

During transitions, the CMOS output must charge or discharge the stray capacitance associated with the inputs that it drives. If this capacitance is too large, the transition from LOW to HIGH (or vice versa) may be too slow, causing improper system operation.

AC fanout

The ability of an output to charge and discharge stray capacitance is sometimes called *AC fanout*, though it is seldom calculated as precisely as DC fanout. As you'll see in Section 3.6.1, it's more a matter of deciding how much speed degradation you're willing to live with.

3.5.5 Effects of Loading

Loading an output with more than its rated fanout has several effects:

- In the LOW state, the output voltage (V_{OL}) may increase beyond V_{OLmax}.

- In the HIGH state, the output voltage (V_{OH}) may fall below V_{OHmin}.

- Propagation delay to the output may increase beyond specifications.

- Output rise and fall times may increase beyond their specifications.

- The operating temperature of the device may increase, thereby reducing the reliability of the device and eventually causing device failure.

The first four effects reduce the DC noise margins and timing margins of the circuit. Thus, a slightly overloaded circuit may work properly in ideal conditions, but experience says that it will fail once it's out of the friendly environment of the engineering lab.

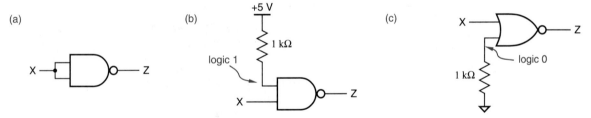

Figure 3–35   Unused inputs: (a) tied to another input; (b) NAND pulled up; (c) NOR pulled down.

3.5.6   Unused Inputs

Sometimes not all of the inputs of a logic gate are used. In a real design problem, you may need an n-input gate but have only an $n+1$-input gate available. Tying together two inputs of the $n+1$-input gate gives it the functionality of an n-input gate. You can convince yourself of this fact intuitively now, or use switching algebra to prove it after you've studied Section 4.1. Figure 3–35(a) shows a NAND gate with its inputs tied together.

You can also tie unused inputs to a constant logic value. An unused AND or NAND input should be tied to logic 1, as in (b), and an unused OR or NOR input should be tied to logic 0, as in (c). In high-speed circuit design, it's usually better to use method (b) or (c) rather than (a), which increases the capacitive load on the driving signal and may slow things down. In (b) and (c), a resistor value in the range 1–10 kΩ is typically used, and a single pull-up or pull-down resistor can serve multiple unused inputs. It is also possible to tie unused inputs directly to the appropriate power-supply rail.

Unused CMOS inputs should never be left unconnected (or *floating*). On *floating input* one hand, such an input will behave as if it had a LOW signal applied to it and will normally show a value of 0 V when probed with an oscilloscope or voltmeter. So you might think that an unused OR or NOR input can be left floating, because it will act as if a logic 0 is applied and not affect the gate's output. However, since CMOS inputs have such high impedance, it takes only a small amount of circuit noise to temporarily make a floating input look HIGH, creating some very nasty intermittent circuit failures.

Floating CMOS inputs are often the cause of mysterious circuit behavior, as an   SUBTLE BUGS
unused input erratically changes its effective state based on noise and conditions
elsewhere in the circuit. When you're trying to debug such a problem, the extra
capacitance of an oscilloscope probe touched to the floating input is often enough
to damp out the noise and make the problem go away. This can be especially
baffling if you don't realize that the input is floating!

3.5.7 Current Spikes and Decoupling Capacitors

When a CMOS output switches between LOW and HIGH, current flows from V_{DD} to ground through the partially-on p- and n-channel transistors. These currents, often called *current spikes* because of their brief duration, may show up as noise on the power-supply and ground connections in a CMOS circuit, especially when multiple outputs are switched simultaneously.

current spike

For this reason, systems that use CMOS circuits require *decoupling capacitors* between V_{DD} and ground. These capacitors must be distributed throughout the circuit, at least one within an inch or so of each chip, to supply current during transitions. The large *filtering capacitors* typically found in the power supply itself don't satisfy this requirement, because stray wiring inductance prevents them from supplying the current fast enough, hence the need for a *physically distributed* system of decoupling capacitors.

decoupling capacitors

filtering capacitors

3.5.8 How to Destroy a CMOS Device

Hit it with a sledge hammer. Or simply walk across a carpet and then touch an input pin with your finger. Because CMOS device inputs have such high impedance, they are subject to damage from *electrostatic discharge (ESD)*.

electrostatic discharge (ESD)

ESD occurs when a buildup of charge on one surface arcs through a dielectric to another surface with the opposite charge. In the case of a CMOS input, the dielectric is the insulation between an input transistor's gate and its source and drain. ESD may damage this insulation, causing a short-circuit between the device's input and output.

The input structures of modern CMOS devices use various measures to reduce their susceptibility to ESD damage, but no device is completely immune. Therefore, to protect CMOS devices from ESD damage during shipment and handling, manufacturers normally package their devices in conductive bags, tubes, or foam. To prevent ESD damage when handling loose CMOS devices, circuit assemblers and technicians usually wear conductive wrist straps that are connected by a coil cord to earth ground; this prevents a static charge from building up on their bodies as they move around the factory or lab.

latch-up

Once a CMOS device is installed in a system, another possible source of damage is *latch-up*. The physical input structure of just about any CMOS device contains parasitic bipolar transistors between V_{DD} and ground configured as a "silicon-controlled rectifier (SCR)." In normal operation, this "parasitic SCR" has no effect on device operation. However, an input voltage that is less than ground or more than V_{DD} can "trigger" the SCR, creating a virtual short-circuit between V_{DD} and ground. Once the SCR is triggered, the only way to turn it off is to turn off the power supply. Before you have a chance to do this, enough power may be dissipated to destroy the device (i.e., you may see smoke).

Some design engineers consider themselves above such inconveniences, but to be safe you should follow several ESD precautions in the lab:

- Before handling a CMOS device, touch the grounded metal case of a plugged-in instrument or another source of earth ground.

- Before transporting a CMOS device, insert it in conductive foam.

- When carrying a circuit board containing CMOS devices, handle the board by the edges, and touch a ground terminal on the board to earth ground before poking around with it.

- When handing over a CMOS device to a partner, especially on a dry winter day, touch the partner first. He or she will thank you for it.

ELIMINATE RUDE, SHOCKING BEHAVIOR

One possible trigger for latch-up is "undershoot" on high-speed HIGH-to-LOW signal transitions, discussed in Section 12.4. In this situation, the input signal may go several volts below ground for several nanoseconds before settling into the normal LOW range. However, modern CMOS logic circuits are fabricated with special structures that prevent latch-up in this transient case.

Latch-up can also occur when CMOS inputs are driven by the outputs of another system or subsystem with a separate power supply. If a HIGH input is applied to a CMOS gate before power is present, the gate may come up in the "latched-up" state when power is applied. Again, modern CMOS logic circuits are fabricated with special structures that prevent this in most cases. However, if the driving output is capable of sourcing lots of current (e.g., tens of mA), latch-up is still possible. One solution to this problem is to apply power before hooking up input cables.

3.6 CMOS DYNAMIC ELECTRICAL BEHAVIOR

Both the speed and the power consumption of a CMOS device depend to a large extent on AC or dynamic characteristics of the device and its load, that is, what happens when the output changes between states. In the design of gate arrays and other semicustom CMOS circuits, logic designers must carefully examine the effects of output loading. Even in board-level design, the effects of loading must be considered for clocks, buses, and other signals that have high fanout or long interconnections.

Speed depends on two characteristics, transition time and propagation delay, discussed in the next two subsections. Power dissipation is discussed in the third subsection.

3.6.1 Transition Time

transition time

The amount of time that the output of a logic circuit takes to change from one state to another is called the *transition time*. Figure 3–36(a) shows how we might like outputs to change state—in zero time. However, real outputs cannot change instantaneously, because they need time to charge the stray capacitance of the wires and other components that they drive. A more realistic view of a circuit's output is shown in (b). An output takes a certain time, called the *rise time (t_r)*, to change from LOW to HIGH, and a possibly different time, called the *fall time (t_f)*, to change from HIGH to LOW.

rise time (t_r)
fall time (t_f)

Figure 3–36
Transition times: (a) ideal case of zero-time switching; (b) a more realistic approximation; (c) actual timing, showing rise and fall times.

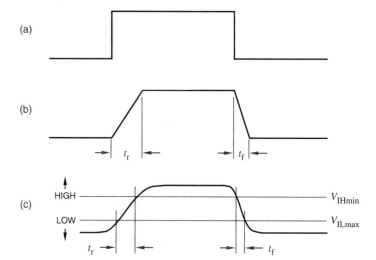

Even Figure 3–36(b) is not quite accurate, because the rate of change of the output voltage does not change instantaneously, either. Instead, the beginning and the end of a transition are smooth, as shown in (c). To avoid difficulties in defining the endpoints, rise and fall times are normally measured at the boundaries of the valid logic levels as indicated in the figure.

With the convention in (c), the rise and fall times indicate how long an output voltage takes to pass through the "undefined" region between LOW and HIGH. The initial part of a transition is not included in the rise- or fall-time number. Instead, the initial part of a transition contributes to the "propagation delay" number discussed in the next subsection.

The rise and fall times of a CMOS output depend mainly on two factors, the "on" transistor resistance and the load capacitance. A large capacitance increases transition times; since this is undesirable, it is extremely rare for logic designer to purposely connect a capacitor to a logic circuit's output. However, *stray capacitance* is present in every circuit; it comes from at least three sources:

stray capacitance

(1) Output circuits, including a gate's output transistors, internal wiring, and packaging, have some capacitance associated with them, on the order of 2–10 picofarads (pF) in typical logic families, including CMOS.

(2) The wiring that connects an output to other inputs has capacitance, about 1 pF per inch or more, depending on the wiring technology.

(3) Input circuits, including transistors, internal wiring, and packaging, have capacitance, from 2 to 15 pF per input in typical logic families.

Stray capacitance is sometimes called a *capacitive load* or an *AC load*.

capacitive load
AC load

A CMOS output's rise and fall times can be analyzed using the equivalent circuit shown in Figure 3–37. As in the preceding section, the p-channel and n-channel transistors are modeled by resistances R_p and R_n, respectively. In normal operation, one resistance is high and the other is low, depending on the output's state. The output's load is modeled by an *equivalent load circuit* with three components:

equivalent load circuit

R_L, V_L These two components represent the DC load and determine the voltages and currents that are present when the output has settled into a stable HIGH or LOW state. The DC load doesn't have too much effect on transition times when the output changes states.

C_L This capacitance represents the AC load and determines the voltages and currents that are present while the output is changing, and how long it takes to change from one state to the other.

When a CMOS output drives only CMOS inputs, the DC load is negligible. To simplify matters, we'll analyze only this case, with $R_L = \infty$ and $V_L = 0$, in the remainder of this subsection. The presence of a nonnegligible DC load would affect the results, but not dramatically (see Exercise 3.68).

Figure 3–37
Equivalent circuit for analyzing transition times of a CMOS output.

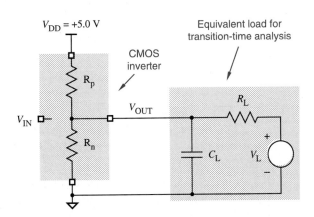

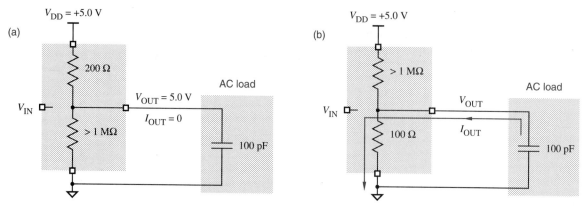

Figure 3–38   Model of a CMOS HIGH-to-LOW transition: (a) in the HIGH state; (b) after *p*-channel transistor turns off and *n*-channel transistor turns on.

We can now analyze the transition times of a CMOS output. For the purposes of this analysis, we'll assume $C_L = 100$ pF, a moderate capacitive load. Also, we'll assume that the "on" resistances of the *p*-channel and *n*-channel transistors are 200 Ω and 100 Ω, respectively, as in the preceding subsection. The rise and fall times depend on how long it takes to charge or discharge the capacitive load C_L.

First, we'll look at fall time. Figure 3–38(a) shows the electrical conditions in the circuit when the output is in a steady HIGH state. (R_L and V_L are not drawn; they have no effect, since we assume $R_L = \infty$.) For the purposes of our analysis, we'll assume that when CMOS transistors change between "on" and "off," they do so instantaneously. We'll assume that at time $t = 0$ the CMOS output changes to the LOW state, resulting in the situation depicted in (b).

At time $t = 0$, V_{OUT} is still 5.0 V. (A useful electrical engineering maxim is that the voltage across a capacitor cannot change instantaneously.) Yet at time $t = \infty$, the capacitor must be fully discharged and V_{OUT} will be 0 V. In between, the value of V_{OUT} is governed by an exponential law:

$$
\begin{aligned}
V_{OUT} &= V_{DD} \cdot e^{-t/R_n C_L} \\
&= 5.0 \cdot e^{-t/(100 \cdot 100 \cdot 10^{-12})} \text{ V} \\
&= 5.0 \cdot e^{-t/(10 \cdot 10^{-9})} \text{ V}
\end{aligned}
$$

RC time constant

The factor $R_n C_L$ has units of seconds, and is called an *RC time constant*. The preceding calculation shows that the *RC* time constant for HIGH-to-LOW transitions is 10 nanoseconds (ns).

Figure 3–39 plots V_{OUT} as a function of time. To calculate fall time, recall that 1.5 V and 3.5 V are the defined boundaries for LOW and HIGH levels for

CMOS inputs being driven by the CMOS output. To obtain the fall time, we must solve the preceding equation for $V_{OUT} = 3.5$ and $V_{OUT} = 1.5$, yielding

$$t = -R_n C_L \cdot \ln \frac{V_{OUT}}{V_{DD}}$$

$$= -10 \cdot 10^{-9} \cdot \ln \frac{V_{OUT}}{5.0}$$

$$t_{3.5} = 3.57 \text{ ns}$$

$$t_{1.5} = 12.04 \text{ ns}$$

The fall time t_f is the difference between these two numbers, or about 8.5 ns.

Rise time can be calculated in a similar manner. Figure 3–40(a) shows the conditions in the circuit when the output is in a steady LOW state. If at time $t = 0$

Figure 3–39
Fall time for a HIGH-to-LOW
transition of a CMOS output.

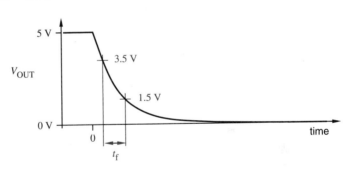

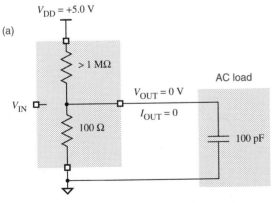

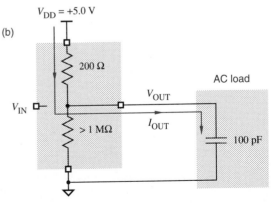

Figure 3–40   Model of a CMOS LOW-to-HIGH transition: (a) in the LOW state; (b) after *n*-channel transistor turns off and *p*-channel transistor turns on.

Figure 3–41
Rise time for a LOW-to-HIGH
transition of a CMOS output.

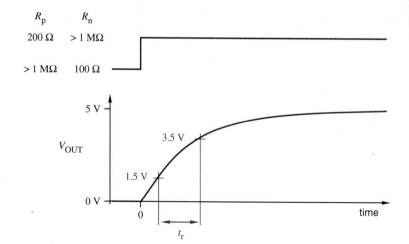

the CMOS output changes to the HIGH state, the situation depicted in (b) results. Once again, V_{OUT} cannot change instantly, but at time $t = \infty$, the capacitor will be fully charged and V_{OUT} will be 5.0 V. Once again, the value of V_{OUT} in between is governed by an exponential law:

$$
\begin{aligned}
V_{OUT} &= V_{DD} \cdot (1 - e^{-t/R_p C_L}) \\
&= 5.0 \cdot (1 - e^{-t/(200 \cdot 100 \cdot 10^{-12})}) \text{ V} \\
&= 5.0 \cdot (1 - e^{-t/(20 \cdot 10^{-9})}) \text{ V}
\end{aligned}
$$

The RC time constant in this case is 20 ns. Figure 3–41 plots V_{OUT} as a function of time. To obtain the rise time, we must solve the preceding equation for $V_{OUT} = 1.5$ and $V_{OUT} = 3.5$, yielding

$$
\begin{aligned}
t &= -RC \cdot \ln \frac{V_{DD} - V_{OUT}}{V_{DD}} \\
&= -20 \cdot 10^{-9} \cdot \ln \frac{5.0 - V_{OUT}}{5.0} \\
t_{1.5} &= 7.13 \text{ ns} \\
t_{3.5} &= 24.08 \text{ ns}
\end{aligned}
$$

The rise time t_r is the difference between these two numbers, or about 17 ns.

The foregoing example assumes that the p-channel transistor has twice the resistance of the n-channel one, and as a result the rise time is twice as long as the fall time. It takes longer for the "weak" p-channel transistor to pull the output up than it does for the "strong" n-channel transistor to pull it down; the output's drive capability is "asymmetric." High-speed CMOS devices are sometimes

> Calculated transition times are actually quite sensitive to the choice of logic levels. In the examples in this subsection, if we used 2.0 V and 3.0 V instead of 1.5 V and 3.5 V as the thresholds for LOW and HIGH, we would calculate shorter transition times. On the other hand, if we used 0.0 and 5.0 V, the calculated transition times would be infinity! You should also be aware that in some logic families (most notably TTL), the thresholds are not symmetric around the voltage midpoint. Still, it is the author's experience that the "time-constant-equals-transition-time" rule of thumb usually works for practical circuits.

THERE'S A CATCH!

fabricated with larger *p*-channel transistors to make the transition times more nearly equal and output drive more symmetric.

Regardless of the transistors' characteristics, an increase in the load capacitance cause an increase in the *RC* time constant, and a corresponding increase in the transition times of the output. Thus, it is a goal of high-speed circuit designers to minimize load capacitance, especially on the most timing-critical signals. This can be done by minimizing the number of inputs driven by the signal, by creating multiple copies of the signal, and by careful physical layout of the circuit.

When dealing with real digital circuits, it's often useful to estimate transition times, without going through a detailed analysis. A useful rule of thumb is that the transition time approximately equals the *RC* time constant of the charging or discharging circuit. For example, estimates of 10 and 20 ns for fall and rise time in the preceding example would have been pretty much on target, especially considering that most assumptions about load capacitance and transistor "on" resistances are approximate to begin with.

Manufacturers of commercial CMOS circuits typically do not specify transistor "on" resistances on their data sheets. If you search carefully, you might find this information published in the manufacturers' application notes. In any case, you can estimate an "on" resistance as the voltage drop across the "on" transistor divided by the current through it with a worst-case resistive load, as we showed in Section 3.5.2:

$$R_{p(on)} = \frac{V_{DD} - V_{OHminT}}{|I_{OHmaxT}|}$$

$$R_{n(on)} = \frac{V_{OLmaxT}}{I_{OLmaxT}}$$

3.6.2 Propagation Delay

Rise and fall times only partially describe the dynamic behavior of a logic element; we need additional parameters to relate output timing to input timing. A

signal path

propagation delay t_p

signal path is the electrical path from a particular input signal to a particular output signal of a logic element. The *propagation delay* t_p of a signal path is the amount of time that it takes for a change in the input signal to produce a change in the output signal.

A complex logic element with multiple inputs and outputs may specify a different value of t_p for each different signal path. Also, different values may be specified for a particular signal path, depending on the direction of the output change. Ignoring rise and fall times, Figure 3–42(a) shows two different propagation delays for the input-to-output signal path of a CMOS inverter, depending on the direction of the output change:

t_{pHL}

t_{pHL} The time between an input change and the corresponding output change when the output is changing from HIGH to LOW.

t_{pLH}

t_{pLH} The time between an input change and the corresponding output change when the output is changing from LOW to HIGH.

Several factors lead to nonzero propagation delays. In a CMOS device, the rate at which transistors change state is influenced both by the semiconductor physics of the device and by the circuit environment, including input-signal transition rate, input capacitance, and output loading. Multistage devices such as noninverting gates or more complex logic functions may require several internal transistors to change state before the output can change state. And even when the output begins to change state, with nonzero rise and fall times it takes quite some time to cross the region between states, as we showed in the preceding subsection. All of these factors are included in propagation delay.

To factor out the effect of rise and fall times, manufacturers usually specify propagation delays at the midpoints of input and output transitions, as shown in

Figure 3–42
Propagation delays for a CMOS inverter: (a) ignoring rise and fall times; (b) measured at midpoints of transitions.

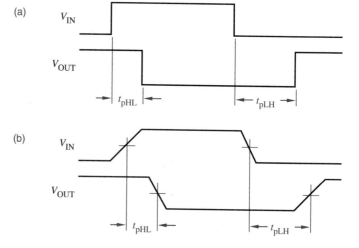

Figure 3–43
Worst-case timing specified using
logic-level boundary points.

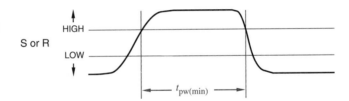

Figure 3–42(b). However, sometimes the delays are specified at the logic-level boundary points, especially if the device's operation may be adversely affected by slow rise and fall times. For example, Figure 3–43 shows how the minimum input pulse width for an S-R latch (discussed in Section 7.2.1) might be specified.

   In addition, a manufacturer may specify absolute maximum input rise and fall times that must be satisfied to guarantee proper operation. High-speed CMOS circuits may consume excessive current or oscillate if their input transitions are too slow.

3.6.3 Power Consumption

The power consumption of a CMOS circuit whose output is not changing is called *static power dissipation* or *quiescent power dissipation*. (The words *consumption* and *dissipation* are used pretty much interchangeably when discussing how much power a device uses.) Most CMOS circuits have very low static power dissipation. This is what makes them so attractive for laptop computers and other low-power applications—when computation pauses, very little power is consumed. A CMOS circuit consumes significant power only during transitions; this is called *dynamic power dissipation*.

*static power
dissipation*

*quiescent power
dissipation*

*dynamic power
dissipation*

   One source of dynamic power dissipation is the partial short-circuiting of the CMOS output structure. When the input voltage is not close to one of the power supply rails (0 V or V_{DD}), both the *p*-channel and *n*-channel output transistors may be partially "on," creating a series resistance of 600 Ω or less. In this case, current flows through the transistors from V_{DD} to ground. The amount of power consumed in this way depends on both the value of V_{DD} and the rate at which output transitions occur, according to the formula

$$P_T \ = \ C_{PD} \cdot V_{DD}^2 \cdot f$$

The following variables are used in the formula:

P_T The circuit's internal power dissipation due to output transitions.

V_{DD} The power supply voltage. As all electrical engineers know, power dissipation across a resistive load (the partially-on transistors) is proportional to the *square* of the voltage.

transition frequency
f The *transition frequency* of the output signal. This specifies the number of power-consuming output transitions per second. (But note that frequency is defined as the number of transitions divided by 2.)

power dissipation capacitance
C_{PD} The *power dissipation capacitance*. This constant, normally specified by the device manufacturer, completes the formula. C_{PD} turns out to have units of capacitance, but does not represent an actual output capacitance. Rather, it embodies the dynamics of current flow through the changing output-transistor resistances during a single pair of output transitions, HIGH-to-LOW and LOW-to-HIGH. For example, C_{PD} for HC-series CMOS is typically 24 pF, even though the actual output capacitance is much less.

The P_T formula is valid only if input transitions are fast enough, leading to fast output transitions. If the input transitions are too slow, then the output transistors stay partially on for a longer time, and power consumption increases. Device manufacturers usually recommend a maximum input rise and fall time, below which the value specified for C_{PD} is valid.

A second, and often more significant, source of CMOS power consumption is the capacitive load (C_L) on the output. During a LOW-to-HIGH transition, current flows through a *p*-channel transistor to charge C_L. Likewise, during a HIGH-to-LOW transition, current flows through an *n*-channel transistor to discharge C_L. In each case, power is dissipated in the "on" resistance of the transistor. We'll use P_L to denote the total amount power dissipated by charging and discharging C_L.

P_L
The units of P_L are power, or energy usage per unit time. The energy for one transition could be determined by calculating the current through the charging transistor as a function of time (using the RC time constant as in Section 3.6.1), squaring this function, multiplying by the "on" resistance of the charging transistor, and integrating over time. An easier way is described below.

During a transition, the voltage across the load capacitance C_L changes by $\pm V_{DD}$. According to the definition of capacitance, the total amount of charge that must flow to make a voltage change of V_{DD} across C_L is $C_L \cdot V_{DD}$. The total amount of energy used in one transition is charge times the average voltage change, or $C_L \cdot V_{DD}^2 / 2$. (The first little bit of charge makes a voltage change of V_{DD}, while the last bit of charge makes a vanishingly small voltage change, hence the average change is $V_{DD}/2$). If there are $2f$ transitions per second, the total power dissipated due to the capacitive load is

$$
\begin{aligned}
P_L &= C_L \cdot (V_{DD}^2 / 2) \cdot 2f \\
    &= C_L \cdot V_{DD}^2 \cdot f
\end{aligned}
$$

The total dynamic power dissipation of a CMOS circuit is the sum of P_T and P_L:

$$
\begin{aligned}
P_D &= P_T + P_L \\
&= C_{PD} \cdot V_{DD}^2 \cdot f + \cdot C_L \cdot V_{DD}^2 \cdot f \\
&= (C_{PD} + C_L) \cdot V_{DD}^2 \cdot f
\end{aligned}
$$

Based on this formula, dynamic power dissipation is often called *CV^2f power*.   *CV^2f power*
In most applications of CMOS circuits, *CV^2f* power is the main type of power
dissipation. Note that *CV^2f* power is also consumed by bipolar logic circuits
like TTL, but at low to moderate frequencies it is insignificant compared to the
static (DC or quiescent) power dissipation of bipolar circuits.

3.7   OTHER CMOS INPUT AND OUTPUT STRUCTURES

Circuit designers have modified the basic CMOS circuit in many ways to produce
gates that are tailored for specific applications. This section describes some of
the more common variations in CMOS input and output structures.

3.7.1   Schmitt-Trigger Inputs

The input-output transfer characteristic for a typical CMOS gate was shown in
Figure 3–25 on page 96. The corresponding transfer characteristic for a gate with
Schmitt-trigger inputs is shown in Figure 3–44(a). A Schmitt trigger is a special   *Schmitt-trigger*
circuit that uses feedback internally to shift the switching threshold depending   *input*
on whether the input is changing from LOW to HIGH or from HIGH to LOW.

For example, suppose the input of a Schmitt-trigger inverter is initially at
0 V, a solid LOW. Then the output is HIGH, close to 5.0 V. If the input voltage is
increased, the output will not go LOW until the input voltage reaches about 2.9 V.

Figure 3–44
A Schmitt-trigger inverter:
(a) input-output transfer characteristic; (b) logic symbol.

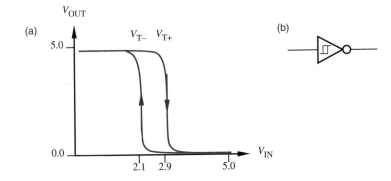

However, once the output is LOW, it will not go HIGH again until the input is decreased to about 2.1 V. Thus, the switching threshold for positive-going input changes, denoted V_{T+}, is about 2.9 V, and for negative-going input changes, denoted V_{T-}, is about 2.1 V. The difference between the two thresholds is called *hysteresis*. The Schmitt-trigger inverter provides about 0.8 V of hysteresis.

hysteresis

To demonstrate the usefulness of hysteresis, Figure 3–45(a) shows an input signal with long rise and fall times and about 0.5 V of noise on it. An ordinary inverter, without hysteresis, has the same switching threshold for both positive-going and negative-going transitions, $V_T \approx 2.5$ V. Thus, the ordinary inverter responds to the noise as shown in (b), producing multiple output changes each time the noisy input voltage crosses the switching threshold. However, as shown in (c), a Schmitt-trigger inverter does not respond to the noise, because its hysteresis is greater than the noise amplitude.

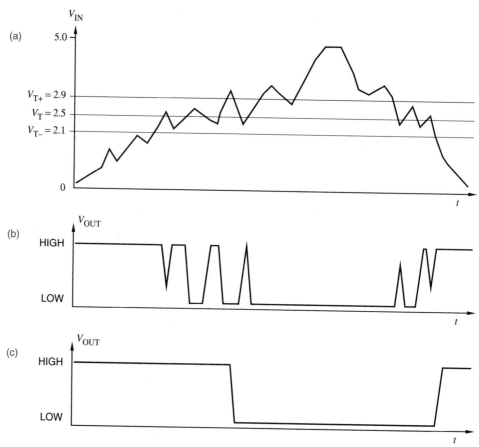

Figure 3–45 Device operation with slowly changing inputs: (a) a noisy, slowly changing input; (b) output produced by an ordinary inverter; (c) output produced by an inverter with 0.8 V of hysteresis.

Schmitt-trigger inputs have better noise immunity than ordinary gate inputs for signals that contain transmission-line reflections, discussed in Section 12.4, or that have long rise and fall times. Such signals typically occur in physically long connections, such as input-output buses and computer interface cables. Noise immunity is important in these applications because long signal lines are more likely to have reflections or to pick up noise from adjacent signal lines, circuits, and appliances.

FIXING YOUR TRANSMISSION

3.7.2 Three-State Outputs

Logic outputs have two normal states, LOW and HIGH, corresponding to logic values 0 and 1. However, some outputs have a third electrical state that is not a logic state at all, called the *high impedance, Hi-Z, or floating state*. In this state, the output behaves as if it isn't even connected to the circuit, except for a small leakage current that may flow into or out of the output pin. Thus, an output can have one of three states—logic 0, logic 1, and Hi-Z.

high impedance state
Hi-Z state
floating state

An output with three possible states is called (surprise!) a *three-state output* or, sometimes, a *tristate output*. Three-state devices have an extra input, usually called "output enable" or "output disable," for placing the device's output(s) in the high-impedance state.

three-state output
tristate output

A *three-state bus* is created by wiring several three-state outputs together. Control circuitry for the "output enables" must ensure that at most one output is enabled (not in its Hi-Z state) at any time. The single enabled device can transmit logic levels (HIGH and LOW) on the bus. Examples of three-state bus design are given in Section 5.4.

three-state bus

A circuit diagram for a CMOS *three-state buffer* is shown in Figure 3–46(a). To simplify the diagram, the internal NAND, NOR, and inverter functions are shown in functional rather than transistor form; they actually use a total of 10 transistors (see Exercise 3.78). As shown in the function table (b), when the

three-state buffer

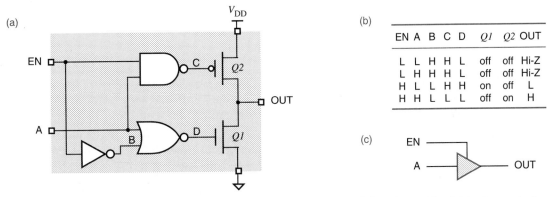

EN	A	B	C	D	Q1	Q2	OUT
L	L	H	H	L	off	off	Hi-Z
L	H	H	H	L	off	off	Hi-Z
H	L	L	H	H	on	off	L
H	H	L	L	L	off	on	H

Figure 3–46 CMOS three-state buffer: (a) circuit diagram; (b) function table; (c) logic symbol.

enable (EN) input is LOW, both output transistors are off, and the output is in the Hi-Z state. Otherwise, the output is HIGH or LOW as controlled by the "data" input A. Logic symbols for three-state buffers and gates are normally drawn with the enable input coming into the top, as shown in (c).

In practice, the three-state control circuit may be different from what we have shown, in order to provide proper dynamic behavior of the output transistors during transitions to and from the Hi-Z state. In particular, devices with three-state outputs are normally designed so that the output-enable delay (Hi-Z to LOW or HIGH) is somewhat longer than the output-disable delay (LOW or HIGH to Hi-Z). Thus, if a control circuit activates one device's output-enable input at the same time that it deactivates a second's, the second device is guaranteed to enter the Hi-Z state before the first places a HIGH or LOW level on the bus.

If two three-state outputs on the same bus are enabled at the same time and try to maintain opposite states, the situation is similar to tying standard active-pull-up outputs together as in Figure 3–53 on page 127—a nonlogic voltage is produced on the bus. The devices probably will not be damaged, but the large current drain through the tied outputs can produce noise pulses that affect circuit behavior elsewhere in the system.

There is a leakage current of up to 10 μA associated with a CMOS three-state output in its Hi-Z state. This current, as well as the input currents of receiving gates, must be taken into account when calculating the maximum number of devices that can be placed on a three-state bus. That is, in the LOW or HIGH state, an enabled three-state output must be capable of sinking or sourcing 10 μA of leakage current for every other three-state output on the bus, as well as sinking the current required by every input on the bus. As with standard CMOS logic, separate LOW-state and HIGH-state calculations must be made to ensure that the fanout requirements of a particular circuit configuration are met.

LEGAL NOTICE

"TRI-STATE" is a trademark of National Semiconductor Corporation. Their lawyers thought you'd like to know. And my copy editor would like you to know that their use of a hyphen is incorrect.

* 3.7.3 Open-Drain Outputs

active pull-up

open-drain output

The *p*-channel transistors in CMOS output structures are said to provide *active pull-up*, since they actively pull up the output voltage on a LOW-to-HIGH transition. These transistors are omitted in gates with *open-drain outputs*, such as the NAND gate in Figure 3–47(a). The drain of the topmost *n*-channel transistor is left unconnected internally, so if the output is not LOW it is "open," as indicated in (b). The underscored diamond in the symbol in (c) is sometimes used

---

*Throughout this book, optional sections are marked with an asterisk.

(a)

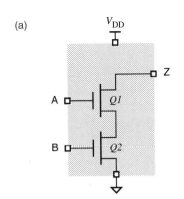

(b)

A	B	$Q1$	$Q2$	Z
L	L	off	off	open
L	H	off	on	open
H	L	on	off	open
H	H	on	on	L

Figure 3–47
Open-drain CMOS NAND gate:
(a) circuit diagram; (b) function
table; (c) logic symbol.

(c)

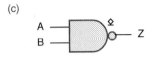

to indicate an open-drain output. (A similar structure, called an "open-collector output," is provided in TTL logic families; see Section 3.11.4.)

An open-drain output requires an external *pull-up resistor* to provide *passive pull-up* to the HIGH level. For example, Figure 3–48 shows an open-drain CMOS NAND gate, with its pull-up resistor, driving a load.

pull-up resistor
passive pull-up

For the highest possible speed, an open-drain output's pull-up resistor should be as small as possible; this minimizes the RC time constant for LOW-to-HIGH transitions (rise time). However, the pull-up resistance cannot be arbitrarily small; the minimum resistance is determined by the open-drain output's maximum sink current, I_{OLmax}. For example, in HC- and HCT-series CMOS, I_{OLmax} is 4 mA, and the pull-up resistor can be no less than 5.0 V/4 mA, or 1.25 kΩ. Since this is an order of magnitude greater than the "on" resistance of the *p*-channel transistors in a standard gate, the LOW-to-HIGH output transitions are much slower for an open-drain gate than for standard gate with active pull-up.

As an example, let us assume that the open-drain gate in Figure 3–48 is HC-series CMOS, the pull-up resistance is 1.5 kΩ, and the load capacitance is 100 pF. We showed in Section 3.5.2 that the "on" resistance of an HC-series CMOS output in the LOW state is about 80 Ω, Thus, the RC time constant for a HIGH-to-LOW transition is about 80 $\Omega \cdot$ 100 pF = 8 ns, and the output's fall time

Figure 3–48
Open-drain CMOS NAND gate
driving a load.

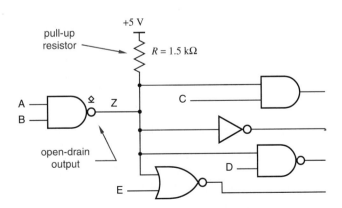

is about 8 ns. However, the RC time constant for a LOW-to-HIGH transition is about 1.5 kΩ · 100 pF = 150 ns, and the rise time is about 150 ns. This relatively slow rise time is contrasted with the much faster fall time in Figure 3–49. A friend of the author calls such slow rising transitions *ooze*.

ooze

So why use open-drain outputs? Despite slow rise times, they can be useful in at least three applications: driving light-emitting diodes (LEDs) and other devices; performing wired logic; and driving multisource buses.

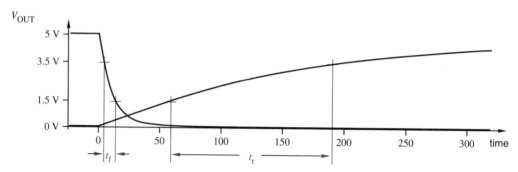

Figure 3–49   Rising and falling transitions of an open-drain CMOS output.

* 3.7.4   Driving LEDs

An open-drain output can drive an LED as shown in Figure 3–50. If either input X or Y is LOW, the corresponding *n*-channel transistor is off and the LED is off. When X and Y are both HIGH, both transistors are on, the output Z is LOW, and the LED is on. The value of the pull-up resistor R is chosen so that the proper amount of current flows through the LED in the "on" state.

Typical LEDs require 10 mA for normal brightness. HC- and HCT-series CMOS outputs are only specified to sink or source 4 mA and are not normally used to drive LEDs. However, the outputs in advanced CMOS families such as 74AC and 74ACT can sink 24 mA or more, and can be used quite effectively to drive LEDs.

Three pieces of information are needed to calculate the proper value of the pull-up resistor R:

(1) The LED current I_{LED} needed for the desired brightness, 10 mA for typical LEDs.

(2) The voltage drop V_{LED} across the LED in the "on" condition, about 1.6 V for typical LEDs.

(3) The output voltage V_{OL} of the open-drain output that sinks the LED current. In the 74AC and 74ACT CMOS families, V_{OLmax} is 0.37 V. If an output

Figure 3–50
Driving an LED with an open-
drain output.

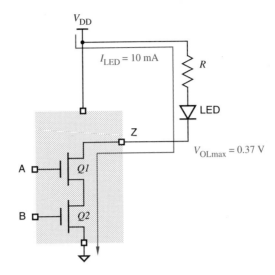

can sink I_{LED} and maintain a lower voltage, say 0.2 V, then the calculation below yields a resistor value that is a little too low, but normally with no harm done. A little more current than I_{LED} will flow and the LED will be a little brighter than expected.

Using the above information, we can write the following equation:

$$V_{OL} + V_{LED} + (I_{LED} \cdot R) = V_{DD}$$

Assuming $V_{DD} = 5.0$ V and the other typical values above, we can solve for the required value of R:

$$R = \frac{V_{DD} - V_{OL} - V_{LED}}{I_{LED}}$$

$$= (5.0 - 0.37 - 1.6) \text{ V} / 10 \text{ mA} = 303 \text{ } \Omega$$

Note that you don't have to use an open-drain output to drive an LED. Figure 3–51(a) shows an LED driven by ordinary CMOS NAND-gate output with active pull-up. If both inputs are HIGH, the bottom (*n*-channel) transistors pull the output LOW as in the open-drain version. If either input is LOW, the output is HIGH; although one or both of the top (*p*-channel) transistors is on, no current flows through the LED.

In most applications, the precise value of LED series resistors is unimportant, as long as groups of nearby LEDs have similar drivers and resistors to give equal apparent brightness. In the example in this subsection, one might use an off-the-shelf resistor value of 270, 300, or 330 ohms, whatever is readily available.

RESISTOR
VALUES

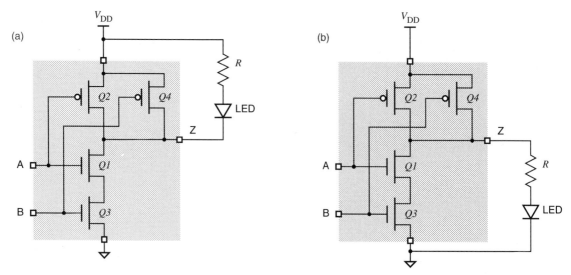

Figure 3–51   Driving an LED with an ordinary CMOS output: (a) sinking current, "on" in the LOW state; (b) sourcing current, "on" in the HIGH state.

With some CMOS families, you can turn an LED "on" when the output is in the HIGH state, as shown in Figure 3–51(b). This is possible if the output can source enough current to satisfy the LED's requirements. However, method (b) isn't used as often as method (a), because most CMOS and TTL outputs cannot source as much current in the HIGH state as they can sink in the LOW state.

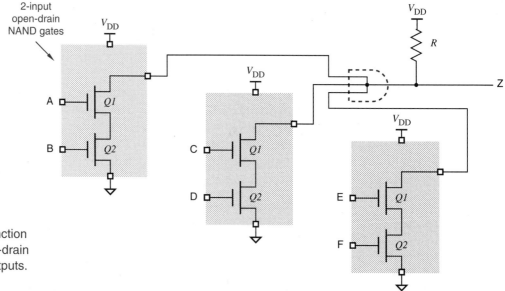

Figure 3–52
Wired-AND function on three open-drain NAND-gate outputs.

* 3.7.5 Wired Logic

If the outputs of several open-drain gates are tied together with a single pull-up resistor, then *wired logic* is performed. (That's *wired*, not *wierd*!) An AND   *wired logic* function is obtained, since the wired output is HIGH if and only if all of the individual gate outputs are HIGH (actually, open); any output going LOW is sufficient to pull the wired output LOW. For example, a three-input *wired AND*   *wired AND* function is shown in Figure 3–52. If any of the individual 2-input NAND gates has both inputs HIGH, it pulls the wired output LOW; otherwise, the pull-up resistor R pulls the wired output HIGH.

Note that wired logic cannot be performed using gates with active pull-up. Two such outputs wired together and trying to maintain opposite logic values result in a very high current flow and an abnormal output voltage. Figure 3–53

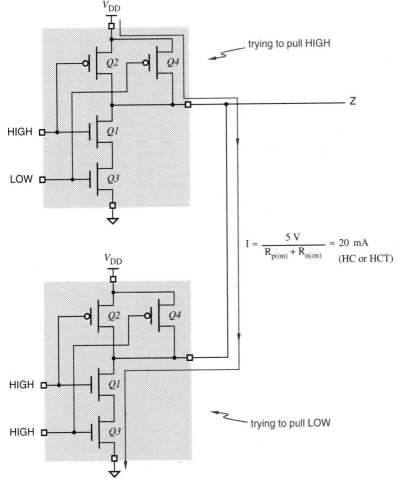

$$I \approx \frac{5\text{ V}}{R_{p(on)} + R_{n(on)}} \approx 20\text{ mA}$$
(HC or HCT)

Figure 3–53
Two CMOS outputs trying to maintain opposite logic values on the same line.

fighting

shows this situation, which is sometimes called *fighting*. The exact output voltage depends on the relative "strengths" of the fighting transistors, but is typically about 1–2 V, almost always a nonlogic voltage.

* 3.7.6   Multisource Buses

open-drain bus

Open-drain outputs can be tied together to allow several devices, one at a time, to put information a common bus. At any time all but one of the outputs on the bus are in their HIGH (open) state. The remaining output either stays in the HIGH state or pulls the bus LOW, depending on whether it wants to transmit a logical 1 or a logical 0 on the bus. Control circuitry selects the particular device that is allowed to drive the bus at any time.

For example, in Figure 3–54, eight 2-input open-drain NAND-gate outputs drive a common bus. The top input of each NAND gate is a data bit, and the bottom input of each is a control bit. At most one control bit is HIGH at any time, enabling the corresponding data bit to be passed through to the bus. (Actually, the complement of the data bit is placed on the bus.) The other gate outputs are HIGH, that is, "open," so the data input of the enabled gate determines the value on the bus.

Figure 3–54
Eight open-drain outputs driving a bus.

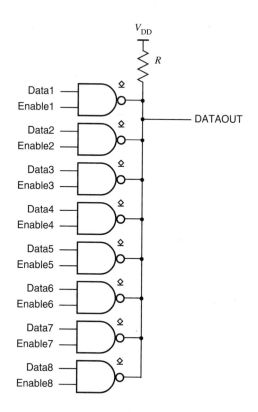

* 3.7.7 Pull-Up Resistors

A proper choice of value for the pull-up resistor R must be made in open-drain   *pull-up resistor*
applications. Two calculations are made to bracket the allowable values of R: *calculation*

Minimum The sum of the current through R in the LOW state and the LOW-
state input currents of the gates driven by the wired outputs must not
exceed the LOW-state driving capability of the active output, 4 mA
for HC and HCT, 24 mA for AC and ACT.

Maximum The voltage drop across R in the HIGH state must not reduce the output
voltage below V_{OHmin}, the guaranteed minimum HIGH output voltage
(2.4 V for HCT, ACT, and FCT; 4.4 V for HC and AC). This drop
is produced by the HIGH-state output leakage current of the wired
outputs and the HIGH-state input currents of the driven gates.

For example, suppose that four HCT open-drain outputs are wired together
and drive two LS-TTL inputs as shown in Figure 3–55. A LOW output must sink
0.4 mA from each LS-TTL input as well as sink the current through the pull-up
resistor R. For the total current to stay within the HCT I_{OLmax} spec of 4 mA, the
current through R may be no more than

$$I_{R(max)} = 4 - (2 \cdot 0.4) = 3.2 \text{ mA}$$

Assuming that V_{OL} of the open-drain output is 0.0 V, the minimum value of R is

$$R_{min} = (5.0 - 0.0)/I_{R(max)} = 1562.5 \ \Omega$$

Figure 3–55
Four open-drain outputs driving
two inputs in the LOW state.

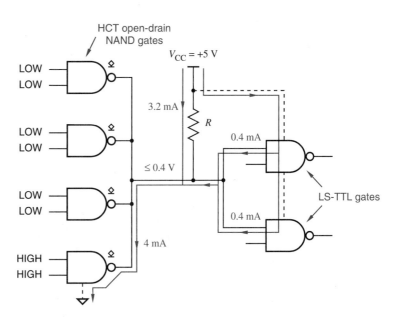

Figure 3–56

Four open-drain outputs driving
two inputs in the HIGH state.

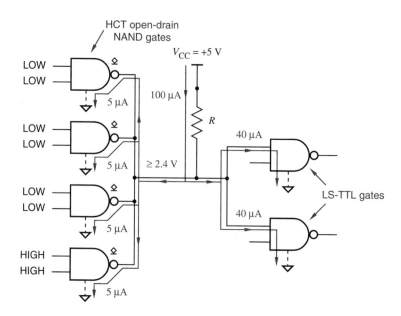

In the HIGH state, typical open-drain outputs have a maximum leakage current of 5 μA, and LS-TTL inputs require 40 μA of source current. Hence, the HIGH-state current requirement as shown in Figure 3–56 is

$$I_{R(\text{leak})} = (4 \cdot 5) + (2 \cdot 40) = 100 \ \mu\text{A}$$

This current produces a voltage drop across R, and must not lower the output voltage below $V_{\text{OHmin}} = 2.4$ V; thus the maximum value of R is

$$R_{\text{max}} = (5.0 - 2.4) / I_{R(\text{leak})} = 26 \ \text{k}\Omega$$

Hence, any value of R between 1562.5 Ω and 26 kΩ may be used. Higher values reduce power consumption and improve the LOW-state noise margin, while lower values increase power consumption but improve both the HIGH-state noise margin and the speed of LOW-to-HIGH output transitions.

OPEN-DRAIN
ASSUMPTION

In our open-drain resistor calculations, we assume that the output voltage can be as low as 0.0 V rather than 0.4 V (V_{OLmax}) in order to obtain a worst-case result. That is, even if the open-drain output is so strong that it can pull the output voltage all the way down to 0.0 V (it's only required to pull down to 0.4 V), we'll never allow it to sink more than 4 mA, so it doesn't get overstressed. Some designers prefer to use 0.4 V in this calculation, figuring that if the output is so good that it can pull lower than 0.4 V, a little bit of excess sink current beyond 4 mA won't hurt it.

* 3.7.8 Transmission Gates

A *p*-channel and *n*-channel transistor pair can be used as a logic-controlled switch. Shown in Figure 3–57(a), this circuit is called a CMOS *transmission gate*.   *transmission gate*

A transmission gate is operated so that its input signals EN and /EN are always at opposite levels. When EN is HIGH and /EN is LOW, there is a low-impedance connection (as low as 5 Ω) between points A and B. When EN is LOW and /EN is HIGH, points A and B are disconnected.

Once a transmission gate is enabled, the propagation delay from A to B (or vice versa) is very short. Because of their short delays and conceptual simplicity, trransmission gates are often used internally in larger-scale CMOS devices such as multiplexers and flip-flops. For example, Figure 3–58 shows how transmission gates can be used to create a "2-input multiplexer." When S is LOW, the X "input" is connected to the Z "output"; when S is HIGH, Y is connected to Z.

At least one commercial manufacturer (Quality Semiconductor) makes a variety of logic functions based on transmission gates. In their multiplexer devices, it takes several nanoseconds for a change in the "select" inputs (such as S in Figure 3–58) to affect the input-output path (X or Y to Z). Once a path is set up, however, the propagation delay from input to output is specified to be at most 0.25 ns; this is the fastest CMOS multiplexer you can buy.

Figure 3–57
CMOS transmission gate.

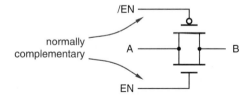

Figure 3–58
Two-input multiplexer using
CMOS transmission gates.

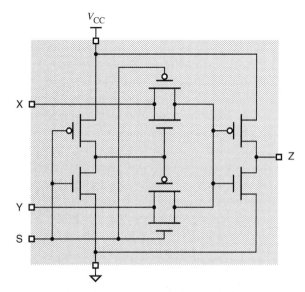

3.8 CMOS LOGIC FAMILIES

4000-series CMOS

The first commercially successful CMOS family was *4000-series CMOS*. Although 4000-series circuits offered the benefit of low power dissipation, they were fairly slow and were not easy to interface with the most popular logic family of the time, bipolar TTL. Thus, the 4000 series was supplanted in most applications by the more capable CMOS families discussed in this section.

All of the CMOS devices that we discuss have part numbers of the form "74FAM*nn*," where "FAM" is an alphabetic family mnemonic and *nn* is a numeric function designator. Devices in different families with the same value of *nn* perform the same function. For example, the 74HC30, 74HCT30, 74AC30, and 74ACT30 are all 8-input NAND gates.

The prefix "74" is a simply a number that was used by an early, popular supplier of TTL devices, Texas Instruments. The prefix "54" is used for identical parts that are specified for operation over a wider range of temperature and power-supply voltage, for use in military applications. Such parts are usually fabricated in the same way as their 74-series counterparts, except that they are tested, screened, and marked differently, a lot of extra paperwork is generated, and a higher price is charged, of course.

3.8.1 HC and HCT

HC (High-speed CMOS)
HCT (High-speed CMOS, TTL compatible)

The first two 74-series CMOS families are *HC (High-speed CMOS)* and *HCT (High-speed CMOS, TTL compatible)*. Compared with the original 4000 family, HC and HCT both have higher speed and better current sinking and sourcing capability. The HCT family uses a power supply voltage V_{DD} of 5 V and can be intermixed with TTL devices, which also use a 5-V supply.

The HC family is optimized for use in systems that use CMOS logic exclusively, and can use any power supply voltage between 2 and 6 V. A higher voltage is used for higher speed, and a lower voltage for lower power dissipation. Lowering the supply voltage is especially effective, since most CMOS power dissipation is proportional to the square of the voltage (CV^2f power).

Even when used with a 5-V supply, HC devices are not quite compatible with TTL. In particular, HC circuits are designed to recognize CMOS input levels. Assuming a supply voltage of 5.0 V, Figure 3–59(a) shows the input and output levels of HC devices. The output levels produced by TTL devices do not quite match this range, so HCT devices use the different input levels shown in (b). These levels are established in the fabrication process by making transistors with different switching thresholds, producing the different transfer characteristics shown in Figure 3–60.

We'll have more to say about CMOS/TTL interfacing in Section 3.13. For

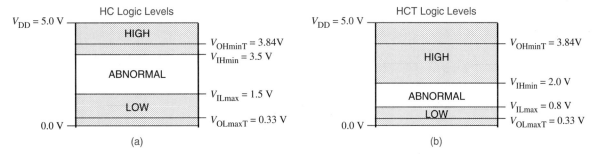

Figure 3–59 Input and output levels for CMOS devices using a 5-V supply: (a) HC; (b) HCT.

now, it is useful simply to note that HC and HCT are essentially identical in their output specifications; only their input levels differ.

3.8.2 AC and ACT

Introduced in the mid-1980s, a pair of more advanced CMOS families are aptly named—*AC (Advanced CMOS)* and *ACT (Advanced CMOS, TTL compatible)*. These families are very fast, comparable to ALS TTL (Section 3.12), and they can source or sink gobs of current, more than most TTL circuits can. Like HC and HCT, the AC and ACT families differ only in the input levels that they recognize; their output characteristics are the same.

AC (Advanced CMOS)

ACT (Advanced CMOS, TTL compatible)

Also like HC/HCT, AC/ACT outputs have *symmetric output drive*. That is, an output can sink or source equal amounts of current; the output is just as "strong" in both states. Other logic families, including the FCT and TTL families introduced later, have *asymmetric output drive*; they can sink much more current in the LOW state than they can source current in the HIGH state.

symmetric output drive

asymmetric output drive

Device outputs in the AC and ACT families have very fast rise and fall times. In fact, they are so fast that they are often a major source of "analog" problems, including switching noise and "ground bounce."

Figure 3–60
Transfer characteristics of HC and HCT circuits under typical conditions.

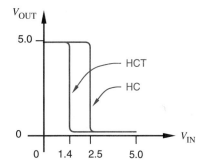

3.8.3   HC, HCT, AC, and ACT Electrical Characteristics

Electrical characteristics of the HC, HCT, AC, and ACT families are summarized in this subsection. The specifications assume that the devices are used with a nominal 5-V power supply, although (derated) operation is possible with any supply voltage in the range 2–6 V.

Commercial (74-series) parts are intended to be operated at temperatures between 0°C and 70°C, while military (54-series) parts are characterized for operation between −55°C and 125°C. The specs in Table 3–5 assume an operating temperature of 25°C. A full manufacturer's data sheet provides additional specifications for device operation over the entire temperature range.

Most devices within a given logic family have the same electrical specifications for inputs and outputs, typically differing only in power consumption and propagation delay. Table 3–5 includes specifications for a 74x00 two-input NAND gate and a 74x138 3-to-8 decoder in the HC, HCT, AC, and ACT families. The '00 NAND gate is included as the smallest logic-design building block in each family, while the '138 is a "medium-scale" part containing the equivalent of about 15 NAND gates. (The '138 spec is included to allow comparison with the faster FCT family in a later subsection; '00 gates are not manufactured in the FCT family.)

The first row of Table 3–5 specifies propagation delay. As discussed in Section 3.6.2, two numbers, t_{pHL} and t_{pLH} may be used to specify delay; the number in the table is the average of these two. Skipping ahead to Table 3–11 on page 171, you can see that HC and HCT are somewhat slower than LS TTL, but that AC and ACT have about the same speed as ALS. The propagation delay for the '138 is somewhat longer than for the '00, since signals must travel through three or four levels of gates internally.

A big difference between CMOS and TTL is power dissipation. A TTL gate dissipates power in resistors and "on" transistors whether or not the gate's output is changing. In CMOS, the quiescent power dissipation is practically nil, only a few microwatts (μW) if the inputs have CMOS levels—0 V for LOW and V_{DD} for HIGH. (Note that in the table, the quiescent power dissipation numbers given for the '00 are per gate, while for the '138 they apply to the entire MSI device.)

HCT and ACT circuits can also be driven by TTL devices, which may produce HIGH output levels as low as 2.4 V. The HCT/ACT output specs are

NOTE ON
NOTATION

> The "x" in the notation "74x00" takes the place of a family designator such as HC, HCT, AC, ACT, FCT, LS, ALS, AS, or F. We may also refer to such a generic part simply as a " '00 " and leave off the "74x."

Table 3–5   Speed and power characteristics of CMOS families operating at 5 V.

Description	Part	Symbol	Condition	Family HC	HCT	AC	ACT
Typical propagation delay (ns)	'00			18	18	5.25	4.75
	'138			36	36	9.25	10.5
Quiescent power dissipation	'00		$V_{in} = 0$ or V_{DD}	0.0025	0.0025	0.005	0.005
(mW)	'00		$V_{in} = 2.4$ V	n/a	12	n/a	7.5
	'138		$V_{in} = 0$ or V_{DD}	0.04	0.04	0.04	0.04
	'138		$V_{in} = 2.4$ V	n/a	12	n/a	7.5
Power dissipation capacitance	'00	C_{PD}		24	24	30	30
(pF)	'138	C_{PD}		85	85	60	60
Dynamic power dissipation	'00			0.60	0.60	0.75	0.75
(mW/MHz)	'138			2.1	2.1	1.5	1.5
Total power dissipation (mW)	'00		$f = 100$ kHz	0.0625	0.0625	0.080	0.080
	'00		$f = 1$ MHz	0.6025	0.6025	0.755	0.755
	'00		$f = 10$ MHz	6.0025	6.0025	7.505	7.505
	'138		$f = 100$ kHz	0.25	0.25	0.19	0.19
	'138		$f = 1$ MHz	2.14	2.14	1.54	1.54
	'138		$f = 10$ MHz	21.04	21.04	15.04	15.04
Speed-power product (pJ)	'00		$f = 100$ kHz	1.1	1.1	0.4	0.4
	'00		$f = 1$ MHz	10.8	10.8	3.9	3.6
	'00		$f = 10$ MHz	108	108	39	36
	'138		$f = 100$ kHz	9.0	9.0	1.8	2.0
	'138		$f = 1$ MHz	77	77	14.2	16.2
	'138		$f = 10$ MHz	757	757	139	158

still met in this condition, since the switching threshold is well below 2.4 volts, and the bottom output transistors are fully "on." However, the top transistors are also partially "on," resulting in additional current flow across the output transistors and power dissipation in the amount listed for the "$V_{in} = 2.4$ V" line in the table. Fortunately, TTL outputs are normally able to drive CMOS inputs to a HIGH level well above the 2.4-volt V_{OHmin} spec, usually to 3.5 V or more, so this extra power dissipation is usually not nearly as bad as the worst-case spec given in the table. In any case, the dynamic (CV^2f) power dissipation is much more significant in most applications.

As we discussed in Section 3.6.3, the dynamic power dissipation of a CMOS gate depends on the voltage swing of the output (usually V_{DD}), the output transition frequency (f), and the capacitance that is being charged and discharged on transitions, according to the formula

$$P_D = (C_L + C_{PD}) \cdot V_{DD}^2 \cdot f$$

Here, C_{PD} is the power dissipation capacitance of the device and C_L is the capacitance of the load attached to the CMOS output in a given application. The table lists both C_{PD} and an equivalent dynamic power dissipation factor in units of milliwatts per megahertz, assuming that $C_L = 0$. Using this factor, the total power dissipation is computed at various frequencies as the sum of the dynamic power dissipation at that freqency and the quiescent power dissipation.

speed-power product

Shown next in the table, the *speed-power product* is simply the product of the propagation delay and power consumption of a typical gate; the result is measured in picojoules (pJ). Recall from physics that the joule is a unit of energy, so the speed-power product measures a sort of efficiency—how much energy a logic gate uses to switch its output. In this day and age, it's obvious that the lower the energy usage, the better.

At high transition frequencies (large f), CMOS families actually use more power than TTL. For example, compare HCT CMOS at $f = 10$ MHz with LS TTL (Table 3–11); a CMOS gate uses three times as much power as a TTL gate at this frequency. Both HCT and LS may be used in systems with maximum "clock" frequencies of up to about 20 MHz, so you might think that CMOS is not so good for high-speed systems. However, the transition frequencies of most outputs in typical systems are much less than the maximum frequency present in the system (e.g., see Exercise 3.75). Thus, typical CMOS systems have a lower total power dissipation than they would have if they were built with TTL.

Table 3–6 gives the input specs of typical CMOS devices in each of the families. Some of the specs assume that the 5-V supply has a ±10% margin; that is, V_{DD} can be anywhere between 4.5 and 5.5 V. These parameters were discussed in previous sections, but for reference purposes their meanings are summarized here:

I_{Imax} The maximum input current for any value of input voltage. This spec states that the current flowing into or out of a CMOS input is 1 μA or less for any value of input voltage. In other words, CMOS inputs create almost no DC load on the circuits that drive them.

SAVING ENERGY

> There are practical as well as geopolitical reasons for saving energy in digital systems. Lower energy consumption means lower cost of power supplies and cooling systems. Also, a digital system's reliability is improved more by running it cooler than by any other single reliability improvement strategy.

Table 3–6 Input specifications for CMOS families with V_{DD} between 4.5 and 5.5 V.

			Family			
Description	Symbol	Condition	HC	HCT	AC	ACT
Input leakage current (μA)	I_{Imax}	V_{in} = any	±1	±1	±1	±1
Typical input capacitance (pF)	C_{INtyp}		6.5	6.5	4.5	4.5
LOW-level input voltage (V)	V_{ILmax}		1.35	0.8	1.35	0.8
HIGH-level input voltage (V)	V_{IHmin}		3.85	2.0	3.85	2.0

C_{INtyp} The typical capacitance of an input. This number can be used when figuring the AC load on an output that drives this and other inputs. Some manufacturers may also specify a maximum input capacitance, perhaps 10 pF, but designers normally use the "typical" spec to estimate AC load.

V_{ILmax} The maximum voltage that an input is guaranteed to recognize as LOW. Notice that the values are different for HC/AC versus HCT/ACT. The "CMOS" value, 1.35 V, is 30% of the minimum power-supply voltage, while the "TTL" value is 0.8 V for compatibility with all TTL families.

V_{IHmin} The minimum voltage that an input is guaranteed to recognize as HIGH. The "CMOS" value, 3.85 V, is 70% of the maximum power-supply voltage, while the "TTL" value is 2.0 V for compatibility with all TTL families. (Unlike CMOS levels, TTL input levels are not symmetric with respect to the power-supply rails.)

The specifications for TTL-compatible CMOS outputs usually have two sets of output parameters; one set or the other is used depending on how an output is loaded. A *CMOS load* is one that requires the output to sink and source very little DC current, 20 μA for HC/HCT and 50 μA for AC/ACT. This is, of course, the case when the CMOS outputs drive only CMOS inputs. With CMOS loads, CMOS outputs maintain an output voltage within 0.1 V of the supply rails, 0 and V_{DD}. (Recall that a worst-case V_{DD} = 4.5 V is used for some table entries; hence, V_{OHminC} = 4.4 V.)

CMOS load

A *TTL load* requires much more sink and source current, up to 4 mA for HC/HCT and 24 mA for AC/ACT. In this case, a higher voltage drop occurs across the "on" transistors in the CMOS output circuit, but the output voltage is still guaranteed to be within the normal range of TTL output levels.

TTL load

Table 3–7   Output specifications for CMOS families operating with V_{DD} between 4.5 and 5.5 V.

Description	Symbol	Condition	Family			
			HC	HCT	AC	ACT
LOW-level output current (mA)	I_{OLmaxC}	CMOS load	0.02	0.02	0.05	0.05
	I_{OLmaxT}	TTL load	4.0	4.0	24.0	24.0
LOW-level output voltage (V)	V_{OLmaxC}	$I_{out} \leq I_{OLmaxC}$	0.1	0.1	0.1	0.1
	V_{OLmaxT}	$I_{out} \leq I_{OLmaxT}$	0.33	0.33	0.37	0.37
HIGH-level output current (mA)	I_{OHmaxC}	CMOS load	−0.02	−0.02	−0.05	−0.05
	I_{OHmaxT}	TTL load	−4.0	−4.0	−24.0	−24.0
HIGH-level output voltage (V)	V_{OHminC}	$|I_{out}| \leq |I_{OHmaxC}|$	4.4	4.4	4.4	4.4
	V_{OHminT}	$|I_{out}| \leq |I_{OHmaxT}|$	3.84	3.84	3.76	3.76

Table 3–7 lists CMOS output specifications for both CMOS and TTL loads. These parameters have the following meanings:

I_{OLmaxC} The maximum current that an output can supply in the LOW state while driving a CMOS load. Since this is a positive value, current flows *into* the output pin.

I_{OLmaxT} The maximum current that an output can supply in the LOW state while driving a TTL load.

V_{OLmaxC} The maximum voltage that a LOW output is guaranteed to produce while driving a CMOS load, that is, as long as I_{OLmaxC} is not exceeded.

V_{OLmaxT} The maximum voltage that a LOW output is guaranteed to produce while driving a TTL load, that is, as long as I_{OLmaxT} is not exceeded.

I_{OHmaxC} The maximum current that an output can supply in the HIGH state while driving a CMOS load. Since this is a negative value, positive current flows *out of* the output pin.

I_{OHmaxT} The maximum current that an output can supply in the HIGH state while driving a TTL load.

V_{OHminC} The minimum voltage that a HIGH output is guaranteed to produce while driving a CMOS load, that is, as long as I_{OHmaxC} is not exceeded.

V_{OHminT} The minimum voltage that a HIGH output is guaranteed to produce while driving a TTL load, that is, as long as I_{OHmaxT} is not exceeded.

The voltage parameters above determine DC noise margins. The LOW-state DC noise margin is the difference between V_{OLmax} and V_{ILmax}. This depends on the characteristics of both the driving output and the driven inputs. For example, the LOW-state DC noise margin of a HCT driving a few HCT inputs (a CMOS load) is $0.8 - 0.1 = 0.7$ V. If a TTL load is added (perhaps a few LS-TTL inputs), then the noise margin for the HCT inputs drops to $0.8 - 0.33 = 0.47$ V. Similarly, the HIGH-state DC noise margin is the difference between V_{OHmin} and V_{IHmin}. In general, when different families are interconnected, you have to compare the appropriate V_{OLmax} and V_{OHmin} of the driving gate with V_{ILmax} and V_{IHmin} of all the driven gates to determine the worst-case noise margins. (Exercise 3.73 asks you to create a complete matrix of noise margins for different interconnections.)

The I_{OLmax} and I_{OHmax} parameters in the table determine fanout capability, and are especially important when an output drives inputs in one or more different families. Two calculations must be performed to determine whether an output is operating within its rated fanout capability:

HIGH-state fanout The I_{IHmax} values for all of the driven inputs are added. The sum must be less than I_{OHmax} of the driving output.

LOW-state fanout The I_{ILmax} values for all of the driven inputs are added. The sum must be less than I_{OLmax} of the driving output.

Note that the input and output characteristics of specific components may vary from the representative values given in Table 3–7, so you must always consult the manufacturers' data books when analyzing a real design.

3.8.4 FCT and FCT-T

In the early 1990s, yet another CMOS family was launched. The *FCT (Fast CMOS, TTL compatible)* family combines circuit innovations with smaller transistor geometries to produce devices that are even faster than AC and ACT while reducing power consumption and maintaining full compatibility with TTL. A variation of this family, *FCT-T (Fast CMOS, TTL compatible with TTL V_{OH})*, has additional circuit innovations that reduce the HIGH-level output voltage, thereby reducing both power consumption and switching noise while maintaining the same high operating speed as standard FCT. A suffix of "T" is used on part numbers to denote the FCT-T output structure, for example, 74FCT138T versus 74FCT138.

FCT (Fast CMOS, TTL compatible)

FCT-T (Fast CMOS, TTL compatible with TTL V_{OH})

3.8.5 FCT and FCT-T Electrical Characteristics

Electrical characteristics of the FCT families are summarized in Table 3–8. The FCT family is specifically designed to be intermixed with TTL devices, so its operation is only specified with a nominal 5-V supply and TTL logic levels.

Table 3–8   Specifications for a 74x138 decoder in FCT and FCT-T logic families.

Description	Symbol	Condition	Family FCT	Family FCT-T
Typical propagation delay (ns)	t_p		9	9
Quiescent power dissipation (mW)		$V_{in} = 0$ or V_{DD}	7.5	7.5
		$V_{in} = 3.4$ V	10.0	10.0
Dynamic power suppy current (mA/MHz)	I_{CCD}		0.3	0.3
Total power dissipation (mW)		$f = 100$ kHz	0.080	0.080
		$f = 1$ MHz	0.755	0.755
		$f = 10$ MHz	7.505	7.505
Speed-power product (pJ)		$f = 100$ kHz	0.4	0.4
		$f = 1$ MHz	3.9	3.6
		$f = 10$ MHz	39	36
Input leakage current (μA)	I_{Imax}	$V_{in} =$ any	±5	±5
Typical input capacitance (pF)	C_{INtyp}		6	6
LOW-level input voltage (V)	V_{ILmax}		0.8	0.8
HIGH-level input voltage (V)	V_{IHmin}		2.0	2.0
LOW-level output current (mA)	I_{OLmax}		48	48
LOW-level output voltage (V)	V_{OLmax}	$I_{out} \le I_{OLmax}$	0.5	0.5
HIGH-level output current (mA)	I_{OHmax}		−15	−8
HIGH-level output voltage (V)	V_{OHmin}	$|I_{out}| \le |I_{OHmax}|$	2.4	2.4
	V_{OHtyp}	$|I_{out}| \le |I_{OHmax}|$	4.3	3.3

Note that the specifications for FCT and FCT-T almost identical, the only differences being FCT-T's reduced maximum I_{OH} and typical V_{OH}. The reduced V_{OH} is an important advantage because it reduces switching noise and power dissipation due to transitions.

Individual logic gates are not manufactured in the FCT families. Just about the simplest FCT logic element is a 74FCT138/74FCT138T decoder, which has six inputs, eight outputs, and contains the equivalent of about a dozen 4-input gates internally. (This function is described later, in Section 5.3.4.) Comparing its propagation delay and power consumption in Table 3–8 with the corresponding HCT and ACT numbers in Table 3–5 on page 135, you can see that the FCT families are superior in both speed and power dissipation.

Unlike other CMOS families, FCT does not have a C_{PD} specification. However, the I_{CCD} specification in the table gives the same information in a different way. The circuit's internal power dissipation due to transitions at a given frequency f can be calculated by the formula

$$P_T \ = \ V_{CC} \cdot I_{CCD} \cdot f$$

Thus, I_{CCD} / V_{CC} is algebraically equivalent to the C_{PD} specification of other CMOS families (see Exercise 3.82).

Table 3–3 on page 92 showed portions of the manufacturer's data sheet for an FCT logic element, the 74FCT257T 4-bit, 2-input multiplexer with three-state outputs. The "Switching Characteristics" portion of the table also showed timing for a 74FCT257AT. The "A" designation in the suffix indicates a faster speed grade, as reflected in the timing table. Even faster (and more expensive) "C" and "D" speed grades are available.

3.9 BIPOLAR LOGIC

Bipolar logic families use semiconductor diodes and bipolar junction transistors as the basic building blocks of logic circuits. TTL and ECL are the most commonly used bipolar logic families.

The simplest bipolar logic elements use diodes and resistors to perform logic operations; this is called *diode logic* and is described first. Many TTL logic gates, described next, use diode logic internally and boost their output drive capability using transistor circuits. Other TTL gates that we'll study use parallel configurations of transistors to perform logic functions. ECL gates, described last, use transistors as current switches to achieve very high speed.

diode logic

3.9.1 Diodes

A *semiconductor diode* is fabricated from two types of semiconductor material, called *p*-type and *n*-type, that are brought into contact with each other as shown in Figure 3–61(a). This is basically the same type of material that is used in *p*-channel and *n*-channel MOS transistors. The point of contact between the *p* and *n* materials is called a *pn junction*. (Actually, a diode is normally fabricated from

semiconductor diode
p-type material
n-type material

pn junction

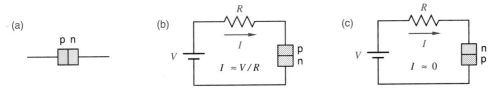

Figure 3–61   Semiconductor diodes: (a) the *pn* junction; (b) forward-biased junction allowing current flow; (c) reverse-biased junction blocking current flow.

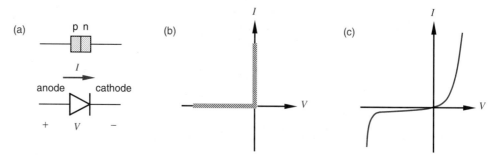

Figure 3–62 Diodes: (a) symbol; (b) transfer characteristic of an ideal diode; (c) transfer characteristic of a real diode.

a single monolithic crystal of semiconductor material in which the two halves are "doped" with different impurities to give them p-type and n-type properties.)

The physical properties of a pn junction are such that positive current can easily flow from the p-type material to the n-type. Thus, if we build the circuit shown in Figure 3–61(b), the pn junction acts almost like a short circuit. However, the physical properties also make it very difficult for positive current to flow in the opposite direction, from n to p. Thus, in the circuit of Figure 3–61(c), the pn junction behaves almost like an open circuit. This is called *diode action*.

diode action

Although it's possible to build vacuum tubes and other devices that exhibit diode action, modern systems use pn junctions—semiconductor diodes—which we'll henceforth call simply *diodes*. Figure 3–62(a) shows the schematic symbol for a diode. As we've shown, in normal operation significant amounts of current can flow only in the direction indicated by the two arrows, from *anode* to *cathode*. In effect, the diode acts like a short circuit as long as the voltage across the anode-to-cathode junction is nonnegative. If the anode-to-cathode voltage is negative, the diode acts like an open circuit and no current flows.

diode

anode

cathode

The transfer characteristic of an ideal diode shown in Figure 3–62(b) further illustrates this principle. If the anode-to-cathode voltage, V, is negative, the diode is said to be *reverse biased* and the current I through the diode is zero.

reverse-biased diode

YES, THERE ARE
TWO ARROWS

...in Figure 3–62(a). The second arrow is built into the diode symbol to help you remember the direction of current flow. Once you know this, there are many ways to remember which end is called the anode and which is the cathode. Afficianados of vacuum-tube hi-fi amplifiers may remember that electrons travel from the hot cathode to the anode, and therefore positive current flow is from anode to cathode. Those of us who were still watching "Sesame Street" when most vacuum tubes went out of style might like to think in terms of the alphabet—current flows alphabetically from A to C.

However, if V is nonnegative, the diode is said to be *forward biased* and I can   *forward-biased*
be an arbitrarily large positive value. In fact, V can never get larger than zero,   *diode*
because an ideal diode acts like a zero-resistance short circuit when forward
biased.

A nonideal, real diode has a resistance that is less than infinity when reverse
biased, and greater than zero when forward biased, so the transfer characteristic
looks like Figure 3–62(c). When the diode is forward biased, the diode acts
like a small nonlinear resistance, in that the voltage drop increases as current
increases, although not strictly proportionally. When the diode is reverse biased,
a small amount of negative *leakage current* flows. If the voltage is made too   *leakage current*
negative, the diode *breaks down*, and large amounts of negative current can flow;   *diode breakdown*
in most applications, this type of operation is avoided.

A real diode can be modeled very simply as shown in Figure 3–63(a) and
(b). When the diode is reverse biased, it acts like an open circuit; we ignore
leakage current. When the diode is forward biased, it acts like a small resistance,
R_f, in series with V_d, a small voltage source. R_f is called the *forward resistance*   *forward resistance*
of the diode, and V_d is called a *diode-drop*.   *diode-drop*

Careful choice of values for R_f and V_d yields a reasonable piecewise-linear
approximation to the real diode transfer characteristic, as in Figure 3–63(c). In
a typical small-signal diode such as a 1N914, the forward resistance R_f is about
25 Ω and the diode-drop V_d is about 0.6 V.

In order to get a feel for diodes, you should remember that a real diode
does not actually contain the 0.6-V source that appears in the model. It's just

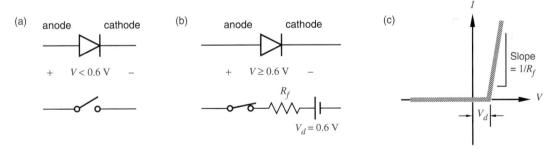

Figure 3–63   Model of a real diode: (a) reverse biased; (b) forward biased; (c) transfer characteristic of forward-biased diode.

Zener diodes take advantage of diode breakdown, in particular the steepness of   ZENER DIODES
the *V–I* slope in the breakdown region. A Zener diode can function as a voltage
regulator when used with a resistor to limit the breakdown current. A wide variety
of Zeners with different specified breakdown voltages are produced for voltage-
regulator applications.

that, due to the nonlinearity of the real diode's transfer characteristic, significant amounts of current do not begin to flow until the diode's forward voltage V has reached about 0.6 V. Also note that in typical applications, the 25-Ω forward resistance of the diode is small compared to other resistances in the circuit, so that very little additional voltage drop occurs across the forward-biased diode once V has reached 0.6 V. Thus, for practical purposes, a forward-biased diode may be considered to have a fixed drop of 0.6 V or so.

3.9.2   Diode Logic

LOW

HIGH

Diode action can be exploited to perform logical operations. Consider a logic system with a 5-V power supply and the characteristics shown in Table 3–9. Within the 5-volt range, signal voltages are partitioned into two ranges, LOW and HIGH, with a 1-volt noise margin between. A voltage in the LOW range is considered to be a logic 0, and a voltage in the HIGH range is a logic 1.

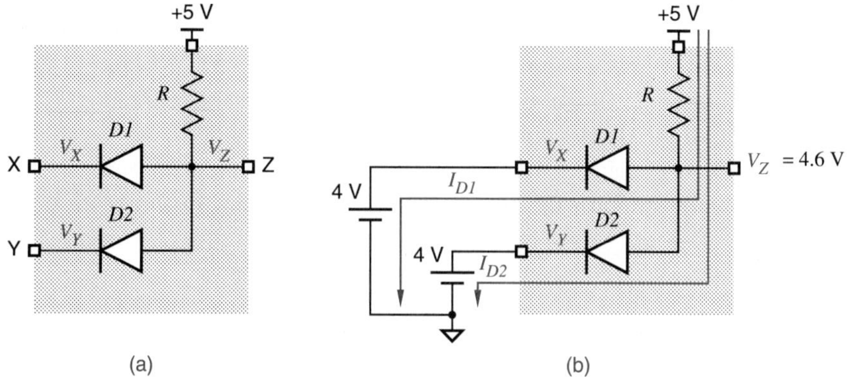

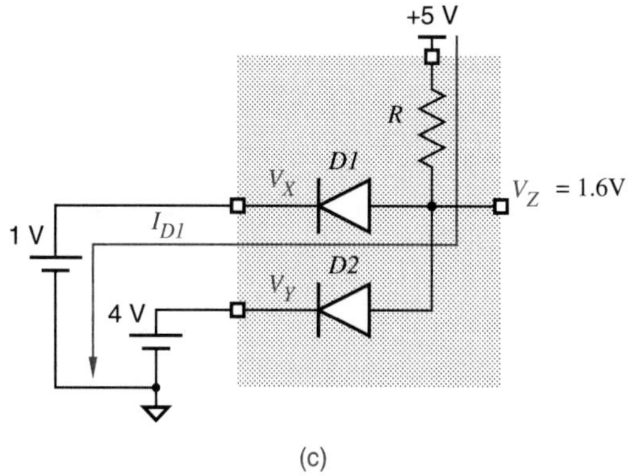

V_X	V_Y	V_Z		X	Y	Z
low	low	low		0	0	0
low	high	low		0	1	0
high	low	low		1	0	0
high	high	high		1	1	1

(c)   (d)   (e)

Figure 3–64   Diode AND gate: (a) electrical circuit; (b) both inputs HIGH; (c) one input HIGH, one LOW; (d) function table; (e) truth table.

Table 3–9
Logic levels in a simple diode
logic system.

Signal Level	Designation	Binary Logic Value
0–2 volts	LOW	0
2–3 volts	noise margin	undefined
3–5 volts	HIGH	1

With these definitions, a *diode AND gate* can be constructed as shown in   *diode AND gate*
Figure 3–64(a). In this circuit, suppose that both inputs X and Y are connected
to HIGH voltage sources, say 4 V, so that V_X and V_Y are both 4 V as in (b). Then
both diodes are forward biased, and the output voltage V_Z is one diode-drop
above 4 V, or about 4.6 V. A small amount of current, determined by the value
of R, flows from the 5-V supply through the two diodes and into the 4-V sources.
The colored arrows in the figure show the path of this current flow.

Now suppose that V_X drops to 1 V as in Figure 3–64(c). In the diode AND
gate, the output voltage equals the *lower* of the two input voltages plus a diode-
drop. Thus, V_Z drops to 1.6 V, and diode D2 is reverse biased (the anode is at
1.6 V and the cathode is still at 4 V). The single LOW input "pulls down" the
output of the diode AND gate to a LOW value. Obviously, two LOW inputs create
a LOW output as well. This functional operation is summarized in (d) and is
repeated in terms of binary logic values in (e); clearly, this is an AND gate.

Figure 3–65(a) shows a logic circuit with two AND gates connected together;
Figure 3–65(b) shows the equivalent electrical circuit with a particular set of
input values. This example shows the necessity of diodes in the AND circuit: *D3*

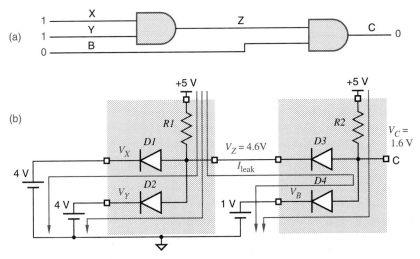

Figure 3–65 Two AND gates: (a) logic diagram; (b) electrical circuit.

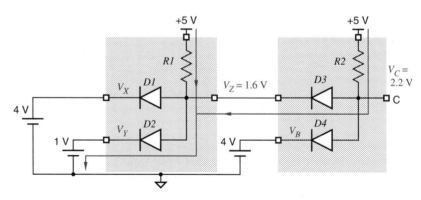

Figure 3–66   Two AND gates with a different set of input values.

allows the output Z of the first AND gate to remain HIGH while the output C of the second AND gate is being pulled LOW by input signal B through $D4$.

Also notice in Figure 3–65(b) that a leakage current, I_{leak}, flows through the reverse-biased diode $D3$. Suppose that $R1$ is 1000 Ω and I_{leak} is 100 μA (microamperes), both plausible values for real circuits. Then the leakage current creates a 100 mV drop across resistor $R1$, even if the X and Y inputs to the gate are left unconnected. This is no problem, given the noise margin in the circuit. But now suppose that twenty other gate inputs, not just one, are connected to output Z. Then the combined leakage currents create a voltage drop of 20 · 100 mV = 2 V, and the voltage at point Z drops to 3 V, the minimum logic 1 value in our system. So we see that real logic gates have a maximum *fanout*—the maximum number of gate inputs that a gate output can drive before logic levels are degraded beyond worst-case specifications.

fanout

Figure 3–66 shows the same circuit with a different set of input values. In this case, both gate outputs Z and C are LOW. Notice that since the logic LOW voltage of a gate is one diode-drop higher than its lowest input voltage, the logic LOW voltage creeps up as gates are cascaded. That is, an input voltage source of 1.0 V on input Y creates an output of 1.6 V at point Z, which in turn creates an output of 2.2 V at point C. This is outside the range we defined for valid logic LOW signals! To get a valid signal, we must either start out with a lower voltage at input Y, or insert a buffer amplifier on signal line Z to restore the logic level to a really low LOW voltage. In fact, all modern logic families contain such a buffer amplifier on the output of each gate.

Diode logic forms a good theoretical basis for understanding bipolar logic gates, but it is seldom used in practice today because of the wide availability, high performance, and low cost of logic families with active buffer amplifiers in every gate. However, logic designers are occasionally tempted to use diode logic under special circumstances; for example, see Exercise 3.93.

3.9.3 Bipolar Junction Transistors

A *bipolar junction transistor* is a three-terminal device that, in most logic circuits, acts like a current-controlled switch. If we put a small current into one of the terminals, called the *base*, then the switch is "on"—current may flow between the other two terminals, called the *emitter* and the *collector*. If no current is put into the base, then the switch is "off"—no current flows between the emitter and the collector. *bipolar junction transistor*

base

emitter

collector

To study the operation of a transistor, we first consider the operation of a pair of diodes connected as shown in Figure 3–67(a). In this circuit, current can flow from node B to node C or node E, when the appropriate diode is forward biased. However, no current can flow from C to E, or vice versa, since for any choice of voltages on nodes B, C, and E, one or both diodes will be reverse biased. The *pn* junctions of the two diodes in this circuit are shown in (b).

Now suppose that we fabricate the back-to-back diodes so that they share a common *p*-type region, as shown in Figure 3–67(c). The resulting structure is called an *npn transistor* and has an amazing property. (At least, the physicists working on transistors back in the 1950s thought it was amazing!) If we put current across the base-to-emitter *pn* junction, then current is also enabled to flow across the collector-to-base *np* junction (which is normally impossible) and from there to the emitter. *npn transistor*

The circuit symbol for the *npn* transistor is shown in Figure 3–67(d). Notice that the symbol contains a subtle arrow in the direction of positive current flow. This also reminds us that the base-to-emitter junction is a *pn* junction, the same as a diode whose symbol has an arrow pointing in the same direction.

It is also possible to fabricate a *pnp transistor*, as shown in Figure 3–68. However, *pnp* transistors are seldom used in digital circuits, so we won't discuss them any further. *pnp transistor*

The current I_e flowing out of the emitter of an *npn* transistor is the sum

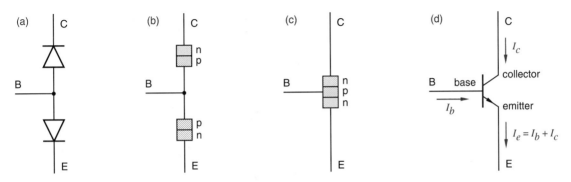

Figure 3–67 Development of an *npn* transistor: (a) back-to-back diodes; (b) equivalent *pn* junctions; (c) structure of an *npn* transistor; (d) *npn* transistor symbol.

Figure 3–68
A *pnp* transistor: (a) structure;
(b) symbol.

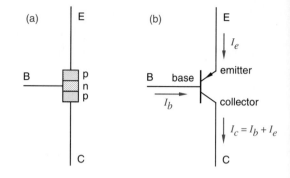

amplifier

active region

common-emitter
configuration

cut off (OFF)

of the currents I_b and I_c flowing into the base and the collector. A transistor is often used as a signal *amplifier*, because over a certain operating range (the *active region*) the collector current is equal to a fixed constant times the base current ($I_c = \beta \cdot I_b$). However, in digital circuits, we normally use a transistor as a simple switch that's always fully "on" or fully "off," as explained next.

Figure 3–69 shows the *common-emitter configuration* of an *npn* transistor, which is most often used in digital switching applications. This configuration uses two discrete resistors, *R1* and *R2*, in addition to a single *npn* transistor. In this circuit, if V_{IN} is 0 or negative, then the base-to-emitter diode junction is reverse biased, and no base current (I_b) can flow. If no base current flows, then no collector current (I_c) can flow, and the transistor is said to be *cut off (OFF)*.

Since the base-to-emitter junction is a *real* diode, as opposed to an ideal one, V_{IN} must reach at least +0.6 V (one diode-drop) before any base current can flow. Once this happens, Ohm's law tells us that

$$I_b = (V_{IN} - 0.6) / R1$$

Figure 3–69
Common-emitter configuration of
an *npn* transistor.

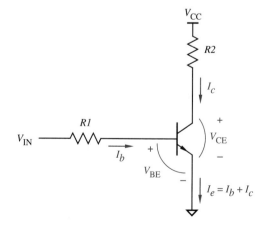

(We ignore the forward resistance R_f of the forward-biased base-to-emitter junction, which is usually small compared to the base resistor $R1$.) When base current flows, then collector current can flow in an amount proportional to I_b, that is,

$$I_c = \beta \cdot I_b$$

The constant of proportionality, β, is called the *gain* of the transistor, and is in the range of 10 to 100 for typical transistors.

β
gain

    Although the base current I_b controls the collector current flow I_c, it also indirectly controls the voltage V_{CE} across the collector-to-emitter junction, since V_{CE} is just the supply voltage V_{CC} minus the voltage drop across resistor $R2$:

$$\begin{aligned} V_{CE} &= V_{CC} - I_c \cdot R2 \\ &= V_{CC} - \beta \cdot I_b \cdot R2 \\ &= V_{CC} - \beta \cdot (V_{IN} - 0.6) \cdot R2/R1 \end{aligned}$$

Table 3–10   A Pascal program that simulates the function of an *npn* transistor in the common-emitter configuration.

```
PROGRAM CommonEmitterNPNtransistor;
CONST {Transistor parameters}
  diodedrop = 0.6; {volts}
  beta = 10;
  VCEsat = 0.2; {volts}
  RCEsat = 50; {ohms}
VAR
  Vcc, Vin, R1, R2: real; {circuit parameters}
  Ib, Ic, Vce: real; {circuit conditions}
BEGIN
  IF (Vin < diodedrop) THEN BEGIN {cut off}
      Ib := 0.0;
      Ic := 0.0;
      Vce := Vcc;
    END
  ELSE BEGIN {active or saturated}
      Ib := (Vin - diodedrop) / R1;
      IF (Vcc - ((beta * Ib) * R2)) >= VCEsat THEN {active}
        BEGIN
          Ic := beta * Ib;
          Vce := Vcc - (Ic * R2);
        END
      ELSE BEGIN {saturated}
          Vce := VCEsat;
          Ic := (Vcc - Vce) / (R2 + RCEsat);
        END
    END;
END.
```

However, in an ideal transistor V_{CE} can never be less than zero (the transistor cannot create a negative potential out of thin air), and in a real transistor V_{CE} can never be less than $V_{CE(sat)}$, a transistor parameter that is typically about 0.2 V.

If the values of V_{IN}, β, $R1$, and $R2$ in a given circuit configuration are such that the preceding equation predicts a value of V_{CE} that is less than $V_{CE(sat)}$, then the transistor cannot be operating in the active region and the equation does not apply. Instead, the transistor is operating in the *saturation region*, and is said to be *saturated (ON)*. No matter how much current I_b we put into the base, V_{CE} cannot drop below $V_{CE(sat)}$, and the collector current I_c is determined mainly by the load resistor $R2$:

saturation region
saturated (ON)

$$ I_c \;=\; (V_{CC} - V_{CE(sat)}) / (R2 + R_{CE(sat)}) $$

saturation resistance

Here, $R_{CE(sat)}$ is the *saturation resistance* of the transistor. Typically, $R_{CE(sat)}$ is 50 Ω or less and is insignificant compared with $R2$.

transistor simulation

Computer scientists might like to imagine an *npn* transistor as a device that continuously looks at its environment and executes the program in Table 3–10 on the preceding page.

3.9.4 Transistor Logic Inverter

Figure 3–70 shows that we can make a logic inverter from an *npn* transistor in the common-emitter configuration. When the input voltage is LOW, the output voltage is HIGH, and vice versa.

In digital switching applications, bipolar transistors are often operated so they are always either cut off or saturated. That is, digital circuits such as the inverter in Figure 3–70 are designed so that their transistors are always (well, almost always) in one of the states depicted in Figure 3–71. When the input

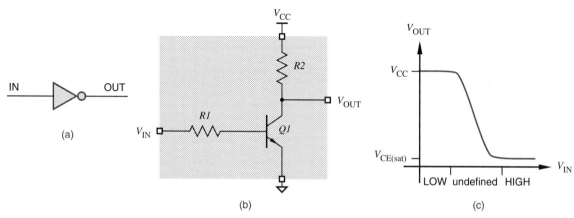

Figure 3–70   Transistor inverter: (a) logic symbol; (b) circuit diagram; (c) transfer characteristic.

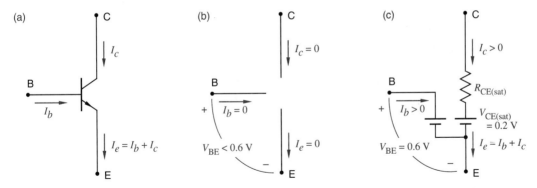

Figure 3–71 Normal states of an *npn* transistor in a digital switching circuit: (a) transistor symbol and currents; (b) equivalent circuit for a cut-off (OFF) transistor; (c) equivalent circuit for a saturated (ON) transistor.

voltage V_{IN} is LOW, it is low enough that I_b is zero and the transistor is cut off; the collector-emitter junction looks like an open circuit. When V_{IN} is HIGH, it is high enough (and $R1$ is low enough and β is high enough) that the transistor will be saturated for any reasonable value of $R2$; the collector-emitter junction looks almost like a short circuit. Input voltages in the undefined region between LOW and HIGH are not allowed, except during transitions. This undefined region corresponds to the noise margin that we discussed in conjunction with Table 3–1.

Another way to visualize the operation of a transistor inverter is shown in Figure 3–72. When V_{IN} is HIGH, the transistor switch is closed, and the output terminal is connected to ground, definitely a LOW voltage. When V_{IN} is LOW, the transistor switch is open and the output terminal is pulled to +5 V through

Figure 3–72
Switch model for a transistor inverter.

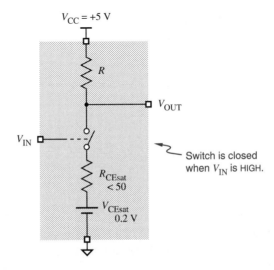

Figure 3–73
Schottky-clamped transistor:
(a) circuit; (b) symbol.

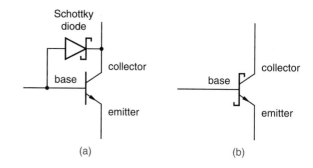

(a)   (b)

a resistor; the output voltage is HIGH unless the output terminal is too heavily loaded (i.e., improperly connected through a low impedance to ground).

3.9.5 Schottky Transistors

storage time

When the input of a saturated transistor is changed, the output does not change immediately; it takes extra time, called *storage time*, to come out of saturation. In fact, storage time accounts for a significant portion of the propagation delay in the original TTL logic family.

Schottky diode
Schottky-clamped transistor
Schottky transistor

　　　　Storage time can be eliminated and propagation delay can be reduced by ensuring that transistors do not saturate in normal operation. Modern TTL logic familes do this by placing a *Schottky diode* between the base and collector of each transistor that might saturate, as shown in Figure 3–73. The resulting transistors, which do not saturate, are called *Schottky-clamped transistors* or *Schottky transistors* for short.

　　　　When forward biased, a Schottky diode's voltage drop is much less than a standard diode's, 0.25 V vs. 0.6 V. In a standard saturated transistor, the base-to-collector voltage is 0.4 V, as shown in Figure 3–74(a). In a Schottky transistor, the Schottky diode shunts current from the base into the collector before the

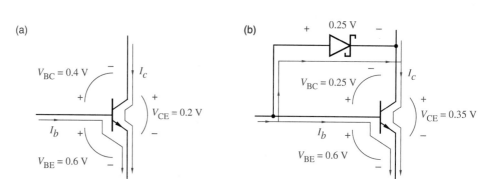

Figure 3–74 Operation of a transistor with large base current: (a) standard saturated transistor; (b) transistor with Schottky diode to prevent saturation.

transistor goes into saturation, as shown in (b). Figure 3–75 is the circuit diagram of a simple inverter using a Schottky transistor.

An undesirable side effect of using Schottky-clamped transistors is that the collector-to-emitter voltage of the "almost saturated" transistor is 0.35 V, which is 0.15 V higher than 0.2-V drop of a standard saturated transistor. Thus, the LOW-state noise margin provided by a Schottky output is less than that provided by a saturated output.

Figure 3–75
Inverter using Schottky transistor.

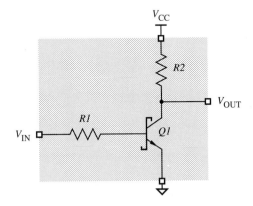

3.10 TRANSISTOR-TRANSISTOR LOGIC

By far the most commonly used bipolar logic family is transistor-transistor logic. Actually, there are many different TTL families, with a range of speed, power consumption, and other characteristics. The circuit examples in this section are based on a representative TTL family, Low-power Schottky (LS or LS-TTL).

TTL families use basically the same logic levels as the TTL-compatible CMOS families in previous sections. We'll use the following definitions of LOW and HIGH in our discussions of TTL circuit behavior:

LOW 0–0.8 volts.

HIGH 2.0–5.0 volts.

3.10.1 Basic TTL NAND Gate

The circuit diagram for a two-input LS-TTL NAND gate made by Texas Instruments, part number 74LS00, is shown in Figure 3–76. The NAND function is obtained by combining a diode AND gate with an inverting buffer amplifier. (Note that there is no transistor $Q1$ in this circuit; $Q1$ appears in a variant of the circuit discussed later.) The circuit's operation is best understood by dividing it

Figure 3–76
Circuit diagram of two-input
LS-TTL NAND gate.

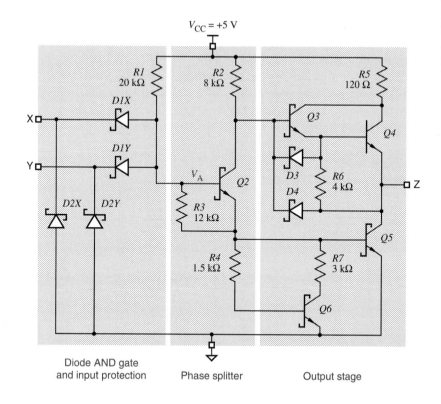

into the three parts that are shown in the figure and discussed in the paragraphs
that follow:

- Diode AND gate and input protection.

- Phase splitter.

- Output stage.

diode AND gate

Diodes *D1X* and *D1Y* and resistor *R1* in Figure 3–76 form a *diode AND
gate*, as in Section 3.9.2. Diodes *D2X* and *D2Y* do nothing in normal operation,
but limit undesirable negative excursions on the inputs to a single diode drop.
Such negative excursions may occur on HIGH-to-LOW input transitions as a result
of transmission-line effects, discussed in Section 12.4.

output stage

totem-pole output

push-pull output

The *output stage* has two transistors, *Q4* and *Q5*, only one of which is on
at any time. The TTL output stage is sometimes called a *totem-pole* or *push-pull
output*. Similar to the *p*-channel and *n*-channel transistors in CMOS, *Q4* and *Q5*
provide active pull-up and pull-down to the HIGH and LOW states, respectively.

phase splitter

Transistor *Q2* and the surrounding resistors form a *phase splitter* that con-

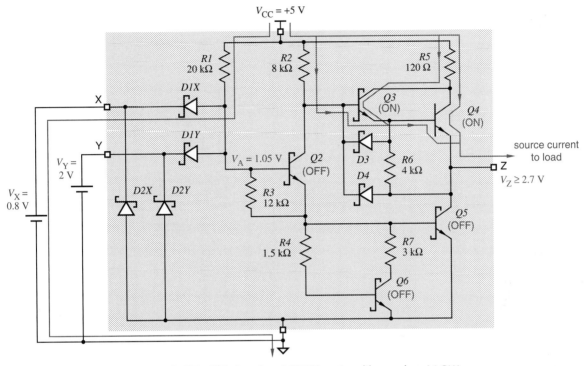

Figure 3–77   TTL two-input NAND gate with one input LOW.

trols the output stage. Depending on whether the diode AND gate produces a
"low" or a "high" voltage at V_A, $Q2$ is either cut off or turned on.

The first case is shown in Figure 3–77. When either input is LOW (0.8 V
or less), V_A is 1.05 V or less ($D1X$ and $D1Y$ are Schottky diodes and have a
voltage drop of only 0.25 V). In this case, $Q2$, $Q6$, and $Q5$ are all cut off (to turn
them on, a minimum of 1.2 V would be required at V_A, due to the two 0.6-V
base-emitter drops through $Q2$ and $Q5$). At the same time, current flows from
V_{CC} through $R2$ and into the base of $Q3$, turning on $Q4$. The output Z is HIGH
in this case.

Diodes $D3$ and $D4$ don't appear to perform any useful function when the TTL circuit is in a steady state, and in fact, they don't. However, when Z changes from HIGH to LOW, they temporarily provide a path for discharging both $Q4$'s base capacitance and the capacitance of the load, thereby speeding up the HIGH-to-LOW transition.	MORE USELESS DIODES?

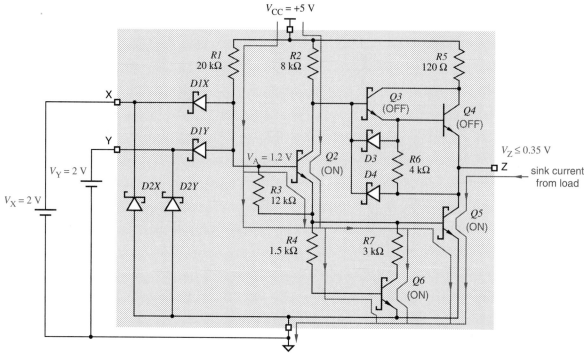

Figure 3–78   TTL two-input NAND gate with both inputs HIGH.

Figure 3–78 shows the situation when both inputs are HIGH. A turned-on transistor has a 0.6-V diode-drop in its base-to-emitter junction, so $Q2$ and $Q5$ are turned on when V_A is 1.2 V. This occurs when the voltages at the X and Y inputs are 0.95 V or more. In this case, current from V_{CC} flows through $R2$ and $Q2$ into the base of $Q5$, turning on $Q5$. However, the voltage at the base of $Q3$ is insufficient to turn on $Q3$, and $Q4$ is turned off. The output Z is LOW in this case.

The functional operation of the TTL NAND gate as we just described it is summarized in Figure 3–79(a). The gate does indeed perform the NAND function, with the truth table and logic symbol shown in (b) and (c). TTL NAND gates can be designed with any desired number of inputs simply by changing the number of

OH DARLING!

Older TTL families use just a single transistor in place of transistors $Q3$ and $Q4$ in Figure 3–76. The combination of $Q3$ and $Q4$ is called a *Darlington transistor pair*. It is used in newer TTL families because it can source more current to the load than a single transistor can. Also note that $Q4$ is not a Schottky transistor. The circuit parameters are such that $Q4$ cannot saturate, so an ordinary transistor is used.

Figure 3–79
Functional operation of a TTL
two-input NAND gate: (a) function
table; (b) truth table; (c) logic
symbol.

(a)

X	Y	V_A	$Q2$	$Q3$	$Q4$	$Q5$	$Q6$	V_Z	Z
L	L	≤1.05	off	on	on	off	off	≥2.7	H
L	H	≤1.05	off	on	on	off	off	≥2.7	H
H	L	≤1.05	off	on	on	off	off	≥2.7	H
H	H	1.2	on	off	off	on	on	≤0.35	L

(b)

X	Y	Z
0	0	1
0	1	1
1	0	1
1	1	0

(c)

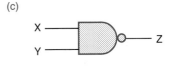

diodes in the diode AND gate in the figure. Commercially available TTL NAND
gates have as many as 13 inputs. A TTL inverter is designed as a 1-input NAND
gate, omitting diodes *D1Y* and *D2Y* in Figure 3–76 on page 154.

Since the output transistors *Q3* and *Q4* are normally complementary—one
ON and the other OFF—you might question the purpose of the 120 Ω resistor *R5*
in the output stage. A value of 0 Ω would give even better driving capability in
the HIGH state. This is certainly true from a DC point of view. However, when
the TTL output is changing from HIGH to LOW or vice versa, there is a short
time when both transistors may be on. The purpose of *R5* is to limit the amount
of current that flows from V_{CC} to ground during this time.

The combination of *Q6*, *R4*, and *R7* is called a *squaring circuit* because it "squares
up" the transfer characteristic of the LS-TTL circuit compared with older TTL
families that use a single pull-down resistor in its place. In the older families,
both *Q2* and *Q5* begin to turn on when V_A is as low as 0.6 V. The squaring circuit
diverts the current from *Q2*'s emitter away from the base of *Q5* until *Q2* is able to
supply enough current to turn on *Q5* fully.

SQUARING
CIRCUIT

Even with a 120 Ω resistor in the TTL output stage, higher-than-normal currents
called *current spikes* flow when TTL outputs are switched. These are similar to
the current spikes that occur when high-speed CMOS outputs switch.

Current spikes can show up as noise on the power-supply and ground connections in TTL and CMOS circuits, especially when multiple outputs are switched
simultaneously. For this reason, reliable circuits require decoupling capacitors between V_{CC} and ground, distributed throughout the circuit so that there is a capacitor
within an inch or so of each chip. Decoupling capacitors supply the instantaneous
current needed during transitions.

CURRENT SPIKES
AGAIN

Figure 3–80
A TTL output driving a TTL input LOW.

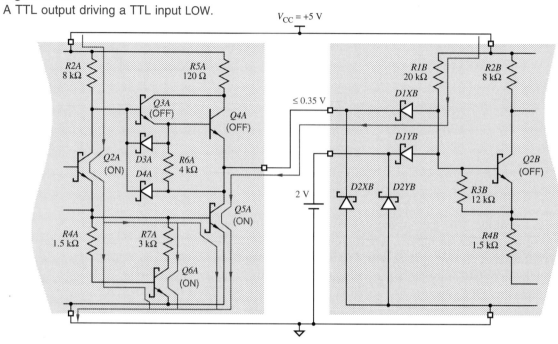

Figure 3–81
A TTL output driving a TTL input HIGH.

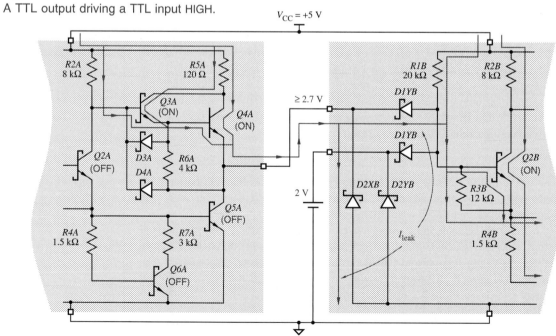

Figure 3–82
Two-input TTL NAND gate using
multi-emitter transistor.

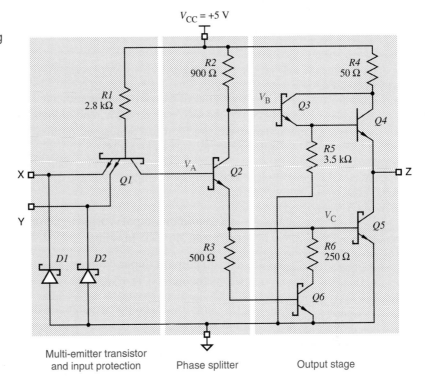

So far we have shown the input signals to a TTL gate as ideal voltage sources. Figure 3–80 shows the situation when a TTL input is driven LOW by the output of another TTL gate. Transistor $Q3A$ in the driving gate is ON, and thereby provides a path to ground for the current flowing out of the emitter of $Q1B$ in the driven gate. When current flows *into* a TTL output in the LOW state, as in this case, the output is said to be *sinking current*.   *sinking current*

Figure 3–81 shows the same circuit with a HIGH output. In this case, $Q4A$ in the driving gate is turned on enough to supply the small amount of leakage current required by $Q1B$ and reverse-biased protection diode in the driven gate. When current flows *out of* a TTL output in the HIGH state, the output is said to be *sourcing current*.   *sourcing current*

The details of TTL circuit design vary both by TTL family and by manufacturer. For example, Figure 3–82 shows the internal circuit design of a two-input Schottky TTL (S-TTL) NAND gate from Texas Instruments (part number 74S00).   *multiple-emitter*
The most striking variation in this circuit is the use of a *multiple-emitter transistor* ($Q1$) instead of diodes to perform the AND operation. $Q1$ might be thought   *transistor*
of as the "AND" of multiple single-emitter transistors. If any of its emitters is low with respect to the base, then current flows from the base to that emitter and from the collector to that emitter; thus, the collector voltage may be pulled

low. If all emitters are high, then the multiple-emitter transistor is cut off and the collector voltage may be pulled high.

3.10.2 Logic Levels and Noise Margins

At the beginning of this section, we indicated that we would consider TTL signals between 0 and 0.8 V to be LOW, and signals between 2.0 and 5.0 V to be HIGH. Actually, we can be more precise by defining TTL input and output levels in the same way as we did for CMOS:

V_{OHmin} The minimum output voltage in the HIGH state, 2.7 V for most TTL families.

V_{IHmin} The minimum input voltage guaranteed to be recognized as a HIGH, 2.0 V for all TTL families.

V_{ILmax} The maximum input voltage guaranteed to be recognized as a LOW, 0.8 V for most TTL families.

V_{OLmax} The maximum output voltage in the LOW state, 0.5 V for most TTL families.

These noise margins are illustrated in Figure 3–83.

DC noise margin In the HIGH state, the V_{OHmin} specification of most TTL families exceeds V_{IHmin} by 0.7 V, so TTL has a *DC noise margin* of 0.7 V in the HIGH state. That is, it takes at least 0.7 V of noise to corrupt a worst-case HIGH output into a voltage that is not guaranteed to be recognizable as a HIGH input. In the LOW state, however, V_{ILmax} exceeds V_{OLmax} by only 0.3 V, so the DC noise margin in the LOW state is only 0.3 V. In general, TTL and TTL-compatible circuits tend to be more sensitive to noise in the LOW state than in the HIGH state.

3.10.3 Fanout

fanout As we defined it previously in Section 3.5.4, fanout is a measure of the number of gate inputs that are connected to (and driven by) a single gate output. As we showed in that section, the DC fanout of CMOS outputs driving CMOS inputs

Figure 3–83
Noise margins for popular TTL logic families (74LS, 74S, 74ALS, 74AS, 74F).

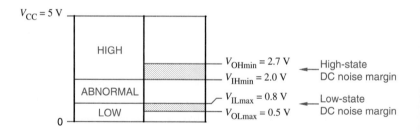

is virtually unlimited, because CMOS inputs require almost no current in either state, HIGH or LOW. This is not the case with TTL inputs. As a result, there are very definite limits on the fanout of TTL or CMOS outputs driving TTL inputs, as you'll see in the paragraphs that follow.

By convention, the *current flow* in a TTL input or output lead is defined   *current flow*
to be positive if the current actually flows *into* the lead, and negative if current flows *out of* the lead. As a result, when an output is connected to one or more inputs, the algebraic sum of all the input and output currents is 0.

The amount of current required by a TTL input depends on whether the input is HIGH or LOW, and is specified by two parameters:

I_{ILmax} The maximum current that an input requires to pull it LOW. Recall from the discussion of Figure 3–80 on page 158 that positive current is actually flowing from V_{CC}, through $R1B$, through diode $D1XB$, *out of* the input lead, through the driving output transistor $Q5A$, and into ground.

Since current flows *out of* a TTL input in the LOW state, I_{ILmax} has a negative value. Most LS-TTL inputs have $I_{\text{ILmax}} = -0.4$ mA, which is sometimes called a *LOW-state unit load* for LS-TTL.   *LOW-state unit load*

I_{IHmax} The maximum current that an input requires to pull it HIGH. As shown in Figure 3–81 on page 158, positive current flows from V_{CC}, through $R5A$ and Q4A of the driving gate, and *into* the driven input, where it leaks to ground through reversed-biased diodes $D1XB$ and $D2XB$.

Since current flows *into* a TTL input in the HIGH state, I_{IHmax} has a positive value. Most LS-TTL inputs have $I_{\text{IHmax}} = 20$ μA, which is sometimes called a *HIGH-state unit load* for LS-TTL.   *HIGH-state unit load*

Like CMOS outputs, TTL outputs can source or sink a certain amount of current depending on the state, HIGH or LOW:

I_{OLmax} The maximum current an output can sink in the LOW state while maintaining an output voltage no more than V_{OLmax}. Since current flows into the output, I_{OLmax} has a positive value, 8 mA for most LS-TTL outputs.

I_{OHmax} The maximum current an output can source in the HIGH state while maintaining an output voltage no less than V_{OHmin}. Since current flows out of the output, I_{OHmax} has a negative value, -400 μA for most LS-TTL outputs.

Notice that the value of I_{OLmax} for typical LS-TTL outputs is exactly 20 times the absolute value of I_{ILmax}. As a result, LS-TTL is said to have a *LOW-state fanout* of 20, because an output can drive up to 20 inputs in the LOW   *LOW-state fanout*

**TTL OUTPUT
ASYMMETRY**

> Although LS-TTL's numerical fanouts for HIGH and LOW states are equal, LS-TTL and other TTL families have a pronounced asymmetry in current driving capability—an LS-TTL output can sink 8 mA in the LOW state, but can source only 400 μA in the HIGH state. This asymmetry is no problem when TTL outputs drive other TTL inputs, because it is matched by a corresponding asymmetry in TTL input current requirements (I_{ILmax} is large, while I_{IHmax} is small). However, it *is* a problem when TTL is used to drive LEDs, relays, solenoids, or other devices requiring large amounts of current, often tens of milliamperes. Circuits using these devices must be designed so that current flows (and the driven device is "on") when the TTL output is in the LOW state, and so little or no current flows in the HIGH state. Special TTL buffer/driver gates are made that can sink up to 60 mA in the LOW state, but that still have a rather puny current sourcing capability in the HIGH state (2.4 mA).

HIGH-state fanout
overall fanout

state. Similarly, the absolute value of I_{OHmax} is exactly 20 times I_{IHmax}, so LS-TTL is said to have a *HIGH-state fanout* of 20 also. The *overall fanout* is the lesser of the LOW- and HIGH-state fanouts.

If an output drives more than its rated fanout, we might say that it is "overloaded." The logic levels of an overloaded output may be outside the normal guaranteed range, reducing the system's immunity to noise. The output's transition time and propagation delay are also likely to increase beyond their rated maximums, causing improper system operation. Severely overloaded outputs may produce nonlogic values and overloaded chips may also overheat, degrading system reliability.

The overall fanout of a TTL family can often be used as a rule of thumb to determine whether outputs are being overloaded. However, some inputs of some TTL devices, including flip-flops and certain MSI parts, require more current than one unit load. Also, different TTL families have different input current requirements and output current capabilities. Thus, to determine whether outputs are being overloaded when different logic families and device types are mixed, you may have to look up the exact specifications for the output and all of the

**BURNED
FINGERS**

> If a TTL output is forced to sink a lot more than I_{OLmax} (say, 50 mA or more), the device may be damaged, especially if high current is allowed to flow for more than a second or so. For example, suppose that a TTL output in the LOW state is short-circuited directly to the 5 V supply. The ON resistance, $R_{\text{CE(sat)}}$, of the saturated $Q5$ transistor in a typical TTL output stage is less than 10 Ω. Thus, $Q5$ must dissipate about $5^2/10$ or 2.5 watts. Don't try this yourself unless you're prepared to deal with the consequences! That's enough heat to destroy the device (and burn your finger) in a very short time.

> Loading a TTL output with more than its rated fanout has the same deleterious effects that were described for CMOS devices in Section 3.5.5 on page 106. That is, DC noise margins may be reduced or eliminated, transition times and delays may increase, and the device may overheat.

DON'T
OVERLOAD!

inputs that it drives. Specifications for different TTL families are given later, in Section 3.12.

In general, two calculations must be carried out to confirm that an output is not being overloaded:

HIGH state   The I_{IHmax} values for all of the driven inputs are added. This sum must be less than or equal to the absolute value of I_{OHmax} for the driving output.

LOW state   The I_{ILmax} values for all of the driven inputs are added. The absolute value of this sum must be less than or equal to I_{OLmax} for the driving output.

For example, suppose you designed a system in which a certain LS-TTL output drives ten LS-TTL and three S-TTL gate inputs. In the HIGH state, a total of $10 \cdot 20 + 3 \cdot 50\ \mu A = 350\ \mu A$ is required. This is within an LS-TTL output's HIGH-state current-sourcing capability of $400\ \mu A$. In the LOW state, a total of $10 \cdot 0.4 + 3 \cdot 2.0\ mA = 10.0\ mA$ is required. This more that an LS-TTL output's LOW-state current-sinking capability of 8 mA, so the output is overloaded.

3.10.4   Unused Inputs

Unused inputs of TTL gates can be handled in the same way as we described for CMOS gates in Section 3.5.6 on page 107. That is, unused inputs may be tied to used ones, or unused inputs may be pulled HIGH or LOW as is appropriate for the logic function.

The resistance value of a pull-up or pull-down resistor is more critical with TTL gates than CMOS gates, because TTL inputs draw significantly more current, especially in the LOW state. If the resistance is too large, the voltage

> Analysis of the TTL input structure shows that unused inputs left unconnected (or *floating*) behave as if they have a HIGH voltage applied—they are pulled HIGH by base resistor *R1* in Figure 3–76 on page 154. However, *R1*'s pull-up is much weaker than that of a TTL output driving the input. As a result, a small amount of circuit noise, such as that produced by other gates when they switch, can make a floating input spuriously behave like it's LOW. Therefore, for the sake of reliability, unused TTL inputs should be tied to a stable HIGH or LOW voltage source.

FLOATING
TTL INPUTS

Figure 3–84
Pull-down resistor for TTL inputs.

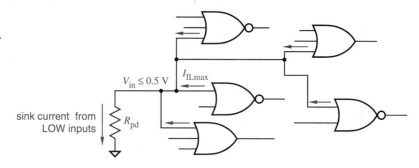

drop across the resistor may result in a gate input voltage beyond the normal LOW or HIGH range.

For example, consider the pull-down resistor shown in Figure 3–84. The pull-up resistor must sink 0.4 mA of current from each of the unused LS-TTL inputs that it drives. Yet the voltage drop across the resistor must be no more 0.5 V in order to have a LOW input voltage no worse than that produced by a normal gate output. If the resistor drives n LS-TTL inputs, then we must have

$$n \cdot 0.4 \text{ mA} \cdot R_{pd} < 0.5 \text{ V}$$

Thus, if the resistor must pull 10 LS-TTL inputs LOW, then we must have $R_{pd} < 0.5/(10 \cdot 4 \cdot 10^{-3})$, or $R_{pd} < 125 \ \Omega$.

Similarly, consider the pull-up resistor shown in Figure 3–85. It must source 20 μA of current to each unused inputs while producing a HIGH voltage no worse than that produced by a normal gate output, 2.7 V. Therefore, the voltage drop across the resistor must be no more 2.3 V; if n LS-TTL inputs are driven, we must have

$$n \cdot 20 \ \mu\text{A} \cdot R_{pu} < 2.3 \text{ V}$$

Thus, if 10 LS-TTL inputs are pulled up, then $R_{pu} < 2.3/(10 \cdot 20 \cdot 10^{-6})$, or $R_{pu} < 11.5 \text{ K}\Omega$.

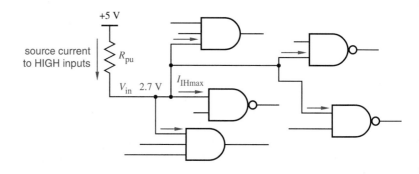

Figure 3–85
Pull-up resistor for TTL inputs.

3.11 ADDITIONAL TTL GATE TYPES

Although the NAND gate is the "workhorse" of the TTL family, other types of gates can be built with the same general circuit structure.

3.11.1 NOR Gates

The circuit diagram for an LS-TTL NOR gate is shown in Figure 3–86. If either input X or Y is HIGH, the corresponding phase-splitter transistor $Q2X$ or $Q2Y$ is turned on, which turns off $Q3$ and $Q4$ while turning on $Q5$ and $Q6$, and the output is LOW. If both inputs are LOW, then both phase-splitter transistors are off, and the output is forced HIGH. This functional operation is summarized in Figure 3–87.

The LS-TTL NOR gate's input circuits, phase splitter, and output stage are almost identical to those of an LS-TTL NAND gate. The difference is that an LS-TTL NAND gate uses diodes to perform the AND function, while an LS-TTL NOR gate uses parallel transistors in the phase splitter to perform the OR function.

The speed, input, and output characteristics of a TTL NOR gate are comparable to those of a TTL NAND. However, an n-input NOR gate uses more transistors and resistors and is thus more expensive in silicon area than an n-input NAND. Also, internal leakage current limits the number of $Q2$ transistors that can be placed in parallel, so NOR gates have poor fan-in. (The largest discrete TTL NOR gate has only 5 inputs, compared with a 13-input NAND.) As a result, NOR gates are less commonly used than NAND gates in TTL designs.

3.11.2 AND-OR-INVERT Gates

The diode AND function and the OR function of parallel phase-splitter transistors   *AND-OR-INVERT*
are combined in a TTL *AND-OR-INVERT (AOI) gate*. As shown in Figure 3–88,   *(AOI) gate*
the circuit for a two-wide, two-input LS-TTL AOI gate is identical that of the

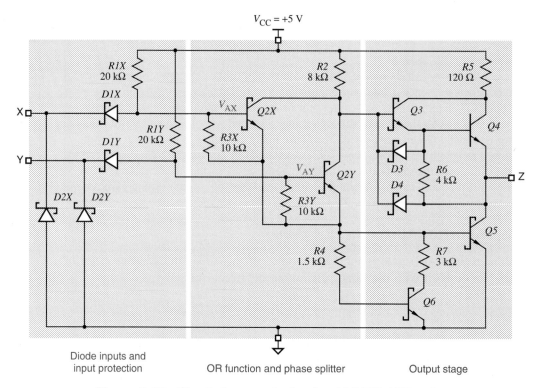

Diode inputs and
input protection

OR function and phase splitter

Output stage

Figure 3–86   Circuit diagram of a two-input LS-TTL NOR gate.

(a)

X	Y	V_{AX}	Q2X	V_{AY}	Q2Y	Q3	Q4	Q5	Q6	V_Z	Z
L	L	≤1.05	off	≤1.05	off	on	on	off	off	≥2.7	H
L	H	≤1.05	off	1.2	on	off	off	on	on	≤0.35	L
H	L	1.2	on	≤1.05	off	off	off	on	on	≤0.35	L
H	H	1.2	on	1.2	on	off	off	on	on	≤0.35	L

(b)

X	Y	Z
0	0	1
0	1	0
1	0	0
1	1	0

(c)

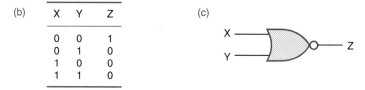

Figure 3–87   Two-input LS-TTL NOR gate: (a) function table; (b) truth
table; (c) logic symbol.

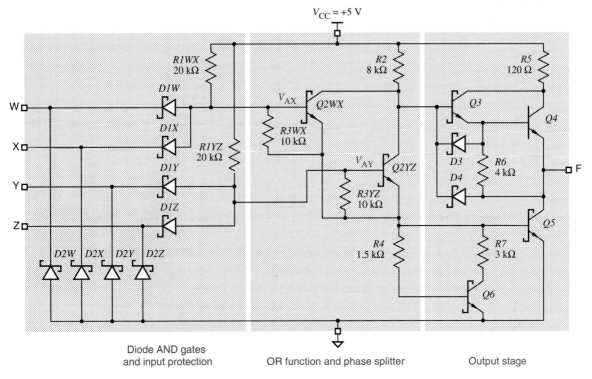

Figure 3–88   Circuit diagram for a two-wide, two-input LS-TTL AOI gate.

two-input NOR gate in Figure 3–86, except that each input is replaced by a pair of inputs driving a diode AND gate.

The truth table for the AND-OR-INVERT gate is shown in Figure 3–89(a). The x's in the truth table indicate that the value of the specified input may take on either value, 0 or 1, without affecting the output for that row of the truth table. The overall logic function of the AOI gate is equivalent to a pair of AND gates followed by a NOR, as suggested by the symbol in (b).

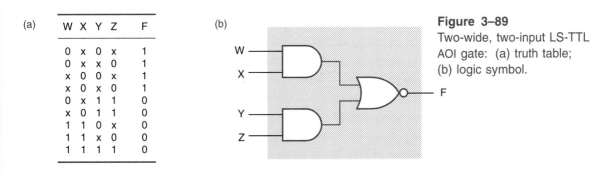

(a)

W	X	Y	Z	F
0	x	0	x	1
0	x	x	0	1
x	0	0	x	1
x	0	x	0	1
0	x	1	1	0
x	0	1	1	0
1	1	0	x	0
1	1	x	0	0
1	1	1	1	0

(b)

Figure 3–89
Two-wide, two-input LS-TTL AOI gate: (a) truth table; (b) logic symbol.

Like CMOS AOI gates, TTL AOI gates have speed, input, and output characterstics that are quite comparable to those of individual NAND or NOR gates. Therefore, many larger-scale TTL devices use these gates internally to perform two levels of logic (AND-OR) with just one level of delay.

3.11.3   Noninverting Gates

The most "natural" TTL gates are inverting gates like NAND and NOR. Noninverting TTL gates include an extra inverting stage, typically between the input stage and the phase splitter. As a result, noninverting TTL gates are typically larger and slower than the inverting gates on which they are based.

While inverting gates are faster and smaller than noninverting ones, they are often more difficult to deal with conceptually, especially when lots of them are interconnected. However, in Section 4.3.2, we'll show how certain designs using noninverting gates can easily be modified to use inverting ones, and in Section 5.1 we'll describe documentations standards that make inverted signals a lot easier to understand.

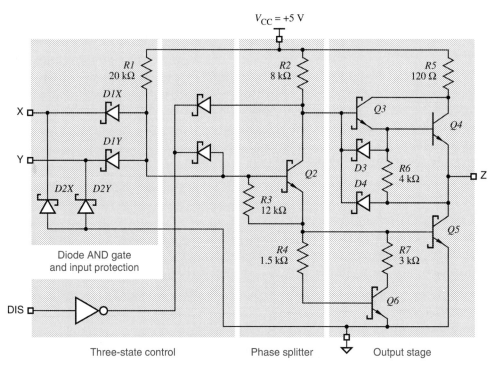

Figure 3–90   Conceptual circuit diagram for a two-input LS-TTL NAND gate with three-state output.

3.11.4 Three-State and Open-Collector Outputs

Like CMOS, TTL gates can be designed with three-state outputs. Such gates have an "output-enable" or "output-disable" input that allows the output to be placed in a high-impedance state where neither output transistor is turned on.

Figure 3–90 shows how the LS-TTL two-input NAND gate of Figure 3–76 on page 154 can be modified to achieve three-state operation. When the DIS input is HIGH, the output is supposed to be in the Hi-Z state. This is accomplished by inverting DIS, and using the resulting signal to pull down the bases of both $Q2$ and $Q3$ LOW through diodes when DIS is HIGH. As a result, both $Q4$ and $Q5$ are turned off.

Some TTL gates are also available with *open-collector outputs*. Such gates omit the entire upper half of the output stage in Figure 3–76, so that only passive pull-up to the HIGH state is provided by an external resistor. The applications and required calculations for TTL open-collector gates are similar to those for CMOS gates with open-drain outputs.

open-collector output

3.12 TTL FAMILIES

TTL families have evolved over the years in response to the demands of digital designers for better performance. As a result, three TTL families have come and gone, and today's designers have at least four newer families from which to choose. All of the TTL families are compatible in that they use the same power supply voltage and logic levels, but each family has its own advantages in terms of speed, power consumption, and cost.

3.12.1 Early TTL Families

The original TTL family of logic gates was introduced by Sylvania in 1963. It was popularized by Texas Instruments, whose "7400-series" part numbers for gates and other TTL components have become an industry standard.

As in 7400-series CMOS, devices in a given TTL family have part numbers of the form 74FAM*nn*, where "FAM" is an alphabetic family mnemonic and *nn* is a numeric function designator. Devices in different families with the same value of *nn* perform the same function. In the original TTL family, "FAM" is null and the family is called *74-series TTL.*

74-series TTL

Resistor values in the original TTL circuit were changed to obtain two more TTL families with different performance characteristics. The *74H (High-speed TTL)* family used lower resistor values to reduce propagation delay at the expense of increased power consumption. The *74L (Low-power TTL)* family used higher resistor values to reduce power consumption at the expense of propagation delay.

74H (High-speed TTL)

74L (Low-power TTL)

The availability of three TTL families allowed digital designers in the 1970s to make a choice between high speed and low power consumption for their circuits. However, like many people in the 1970s, they wanted to "have it all, now." The development of Schottky transistors provided this opportunity, and made 74, 74H, and 74L TTL obsolete. The characteristics of better-performing, contemporary TTL families are discussed in the rest of this section.

3.12.2   Schottky TTL Families

74S (Schottky TTL)

Historically, the first family to make use of Schottky transistors was *74S (Schottky TTL)*. With Schottky transistors and low resistor values, this family has much higher speed, but higher power consumption, than the original 74-series TTL.

74LS (Low-power Schottky TTL)

Perhaps the most widely used TTL family is *74LS (Low-power Schottky TTL)*, introduced shortly after 74S. By combining Schottky transistors with higher resistor values, 74LS TTL matches the speed of 74-series TTL but has about one-fifth of its power consumption. Thus, 74LS is a preferred logic family for new TTL logic designs.

74AS (Advanced Schottky TTL)

74ALS (Advanced Low-power Schottky TTL)

74F (Fast TTL)

Subsequent IC processing and circuit innovations gave rise to two more Schottky logic families. The *74AS (Advanced Schottky TTL)* family offers speeds approximately twice as fast as 74S with approximately the same power consumption. The *74ALS (Advanced Low-power Schottky TTL)* family offers both lower power and higher speeds than 74LS, and rivals 74LS in popularity for general-purpose requirements in new TTL designs. The *74F (Fast TTL)* family is positioned between 74AS and 74ALS in the speed/power tradeoff, and is probably the most popular choice for high-speed requirements in new TTL designs.

3.12.3   Characteristics of TTL Families

The important characteristics of contemporary TTL families are summarized in Table 3–11. The first two rows of the table list the propagation delay (in nanoseconds) and the power consumption (in milliwatts) of a typical 2-input NAND gate in each family.

speed-power product

One figure of merit of a logic family is its *speed-power product* listed in the third row of the table. This is simply the product of the propagation delay and power consumption of a typical gate; the result is measured in picojoules (pJ). Recall from physics that the joule is a unit of energy, so the speed-power product measures a sort of efficiency—how much energy a logic gate uses to switch its output. In our age of PC (political correctness), it's obvious that the lower the energy usage, the better.

The remaining rows in Table 3–11 describe the input and output parameters of typical TTL gates in each of the families. Using this information, you can

Table 3–11   Characteristics of contemporary TTL families.

Description	Symbol	74S	74LS	74AS	74ALS	74F
				Family		
Typical propagation delay (ns)		3	9	1.7	4	3
Power consumption per gate (mW)		19	2	8	1.2	4
Speed-power product (pJ)		57	18	13.6	4.8	12
LOW-level input voltage (V)	V_{ILmax}	0.8	0.8	0.8	0.8	0.8
LOW-level output voltage (V)	V_{OLmax}	0.5	0.5	0.5	0.5	0.5
HIGH-level input voltage (V)	V_{IHmin}	2.0	2.0	2.0	2.0	2.0
HIGH-level output voltage (V)	V_{OHmin}	2.7	2.7	2.7	2.7	2.7
LOW-level input current (mA)	I_{ILmax}	−2.0	−0.4	−0.5	−0.2	−0.6
LOW-level output current (mA)	I_{OLmax}	20	8	20	8	20
HIGH-level input current (μA)	I_{IHmax}	50	20	20	20	20
HIGH-level output current (μA)	I_{OHmax}	−1000	−400	−2000	−400	−1000

analyze the external behavior of TTL gates without knowing the details of the
internal TTL circuit design. These parameters were defined and discussed in
Sections 3.10.2 and 3.10.3. Note that the input and output characteristics of
specific components may vary from the representative values given in Table 3–11,
so you must always consult the manufacturer's data book when analyzing a real
design.

3.12.4   A TTL Data Sheet

Table 3–12 shows the part of a typical manufacturer's data sheet for the 74LS00.
The 54LS00 listed in the data sheet is identical to the 74LS00, except that it is
specified to operate over the full "military" temperature and voltage range, and
it costs more. Most TTL parts have corresponding 54-series (military) versions.
Three sections of the data sheet are shown in the table:

- *Recommended operating conditions* specify power-supply voltage, input-
  voltage ranges, DC output loading, and temperature values under which the
  device is normally operated.

*recommended
operating
conditions*

Table 3–12 Typical manufacturer's data sheet for the 74LS00.

RECOMMENDED OPERATING CONDITIONS

Parameter	Description	SN54LS00 Min.	SN54LS00 Nom.	SN54LS00 Max.	SN74LS00 Min.	SN74LS00 Nom.	SN74LS00 Max.	Unit
V_{CC}	Supply voltage	4.5	5.0	5.5	4.75	5.0	5.25	V
V_{IH}	High-level input voltage	2.0			2.0			V
V_{IL}	Low-level input voltage			0.7			0.8	V
I_{OH}	High-level output current			−0.4			−0.4	mA
I_{OL}	Low-level output current			4			4	mA
T_A	Operating free-air temperature	−55		125	0		70	°C

ELECTRICAL CHARACTERISTICS OVER RECOMMENDED FREE-AIR TEMPERATURE RANGE

Parameter	Test Conditions[1]	SN54LS00 Min.	SN54LS00 Typ.[2]	SN54LS00 Max.	SN74LS00 Min.	SN74LS00 Typ.[2]	SN74LS00 Max.	Unit
V_{IK}	V_{CC} = Min., I_N = −18 mA			−1.5			−1.5	V
V_{OH}	V_{CC} = Min., V_{IL} = Max., I_{OH} = −0.4 mA	2.5	3.4		2.7	3.4		V
V_{OL}	V_{CC} = Min., V_{IH} = 2.0 V, I_{OL} = 4 mA		0.25	0.4		0.25	0.4	V
	V_{CC} = Min., V_{IH} = 2.0 V, I_{OL} = 8 mA					0.35	0.5	
I_I	V_{CC} = Max., V_I = 7.0 V			0.1			0.1	mA
I_{IH}	V_{CC} = Max. V_I = 2.7 V			20			20	μA
I_{IL}	V_{CC} = Max. V_I = 0.4 V			−0.4			−0.4	mA
I_{IOS}[3]	V_{CC} = Max.	−20		−100	−20		−100	mA
I_{CCH}	V_{CC} = Max., V_I = 0 V		0.8	1.6		0.8	1.6	mA
I_{CCL}	V_{CC} = Max., V_I = 4.5 V		2.4	4.4		2.4	4.4	mA

SWITCHING CHARACTERISTICS, V_{CC} = 5.0 V, T_A = 25°C

Parameter	From (Input)	To (Output)	Test Conditions	Min.	Typ.	Max.	Unit
t_{PLH}	A or B	Y	R_L = 2 kΩ, C_L = 15 pF		9	15	ns
t_{PHL}					10	15	

NOTES:
1. For conditions shown as Max. or Min., use appropriate value specified under Recommended Operating Conditions.
2. All typical values are at V_{CC} = 5.0 V, T_A = 25°C.
3. Not more than one output should be shorted at a time; duration of short-circuit should not exceed one second.

- *Electrical characteristics* specify additional DC voltages and currents that are observed at the device inputs and output when it is operated under the recommended conditions: *electrical characteristics*

  I_I Maximum input current for a very high HIGH input voltage.

  I_{OS} Output current with HIGH output shorted to ground.

  I_{CCH} Power-supply current when all outputs (on four NAND gates) are HIGH. (The number given is for the entire package, which contains four NAND gates, so the current per gate is one-fourth of the specified amount.)

  I_{CCL} Power-supply current when all outputs (on four NAND gates) are LOW.

- *Switching characteristics* give maximum and typical propagation delays under "typical" operating conditions of $V_{CC} = 5$ V and $T_A = 25°C$. A conservative designer must increase these delays by 5%–10% to account for different power-supply voltages and temperatures, and even more under heavy loading conditions. *switching characteristics*

A fourth section is also included in the manufacturer's data book:

- *Absolute maximum ratings* indicate the worst-case conditions for operating or storing the device without damage. *absolute maximum ratings*

A complete data book also shows test circuits that are used to measure the parameters when the device is manufactured, and graphs that show how the typical parameters vary with operating conditions such as power-supply voltage (V_{CC}), ambient temperature (T_A), and load (R_L, C_L).

3.13 CMOS/TTL INTERFACING

A digital designer selects a "default" logic family to use in a system, based on general requirements of speed, power, cost, and so on. However, the designer may select devices from other families in some cases because of availability or other special requirements. (For example, not all 74LS part numbers are available in 74HCT, and vice versa.) Thus, it's important for a designer to understand the implications of connecting TTL outputs to CMOS inputs, and vice versa.

There are several factors to consider in TTL/CMOS interfacing, and the first is noise margin. The LOW-state DC noise margin depends on V_{OLmax} of the driving output and V_{ILmax} of the driven input, and equals $V_{ILmax} - V_{OLmax}$. Similarly, the HIGH-state DC noise margin equals $V_{OHmin} - V_{IHmin}$. Figure 3–91 shows the relevant numbers for TTL and CMOS families.

Figure 3–91

Output and input levels for interfacing TTL and CMOS families. (Note that HC and AC inputs are not TTL-compatible.)

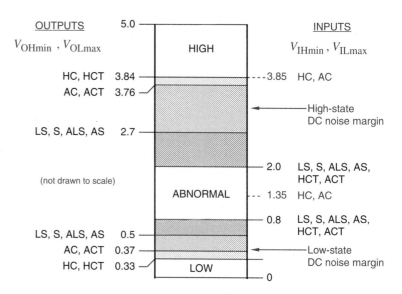

For example, the LOW-state DC noise margin of HC or HCT driving TTL is $0.8 - 0.33 = 0.47$ V. On the other hand, the HIGH-state DC noise margin of TTL driving HC or AC is $2.7 - 3.85 = -1.15$ V. In other words, TTL driving HC or AC doesn't work unless the TTL HIGH output happens to be higher and the CMOS HIGH threshold happens to be lower by a total of 1.15 V compared to their worst-case specs. (To drive a CMOS subsystem from TTL outputs, you should use HCT or ACT devices rather than HC or AC devices.)

The next factor to consider is fanout. As discussed in Section 3.10.3 for TTL, a designer must sum the input current requirements of devices driven by an output and compare with the output's capabilities in both states. Fanout is not a problem when TTL drives CMOS, since CMOS inputs require almost no current in either state. On the other hand, TTL inputs, especially in the LOW state, require substantial current, especially compared to 74HC and 74HCT output capabilities. For example, a 74HC or 74HCT output can drive 10 74LS or only two 74S TTL inputs.

The last factor is capacitive loading. We've seen that load capacitance increases both the delay and the power dissipation of logic circuits. Increases in delay are especially noticeable with 74HC and 74HCT outputs, whose transition times increase about 1 ns for each 5 pF of load capacitance. The transistors in 74AC and 74ACT outputs have very low "on" resistances, so their transition times increase only about 0.1 ns for each 5 pF of load capacitance.

For a given load capacitance, power-supply voltage, and application, all of the CMOS families have similar dynamic power dissipation, since each variable in the CV^2f equation is the same. On the other hand, TTL outputs have substantially lower dynamic power dissipation, since the voltage swing between TTL HIGH and LOW levels is smaller.

Although you can sometimes get away with allowing unused TTL inputs to float, you shouldn't do this with CMOS. The TTL input structure contains a resistor that provides some pull-up to the HIGH state. The CMOS input structure contains no such resistor and has extremely high impedance. Therefore, a very small amount of noise coupled into a CMOS input can make the device behave as if the input were HIGH or LOW. Worse, if the noise voltage is halfway between 0 and 1, both the top and bottom transistors may be "on," and excessive current may flow. In some cases, this current is sufficient to affect the power supply voltage seen by the device, and may actually create a sustained high-frequency oscillation at the device output. Thus, unused CMOS inputs must always be tied to V_{CC}, ground, or another HIGH or LOW signal source appropriate to the device's function.

> Two potential pitfalls of using TTL levels to drive TTL-compatible CMOS inputs are a reduction in noise margin and a possible increase in power consumption in driven gates (if their "off" transistors begin to turn on).

TTL-TO-CMOS CAVEATS

* 3.14 CURRENT-MODE LOGIC

The key to reducing propagation delay in a bipolar logic family is to prevent a gate's transistors from saturating. In Section 3.9.5, we learned how Schottky diodes prevent saturation in TTL gates. However, it is also possible to prevent saturation by using a radically different circuit structure, called *current-mode logic (CML)*.

current-mode logic (CML)

Unlike the other logic families in this chapter, CML does not produce a large voltage swing between the LOW and HIGH levels. Instead, it has a small voltage swing, less than a volt, and it internally switches current between two possible paths, depending on the output state.

The first CML logic family was introduced by General Electric in 1961. The concept was refined by Motorola and others to produce today's 10K and 100K *emitter-coupled logic (ECL)* families. These ECL families are extremely fast, offering propagation delays as short as 1 ns. In fact, throughout the evolution of digital circuit technology, some type of CML has always been the fastest commercial logic family.

emitter-coupled logic (ECL)

Still, commercial ECL families aren't nearly as popular as TTL and CMOS, mainly because they consume much more power. In fact, high power consumption has made the design of ECL supercomputers, such as the Cray-1 and Cray-2, as much of a challenge in cooling technology as in digital design. Also, ECL has a poor speed-power product, does not provide a high level of integration, has fast edge rates requiring design for transmission-line effects in most applications, and is not directly compatible with TTL and CMOS. Nevertheless, proponents of ECL believe that, as the various digital techologies evolve, some kind of ECL will continue to be the family of choice for those who require maximum performance regardless of cost.

* 3.14.1 Basic CML Circuit

differential amplifier

The basic idea of current-mode logic is illustrated by the inverter/buffer circuit in Figure 3–92. This circuit has both an inverting output (OUT1) and a noninverting output (OUT2). Two transistors are connected as a *differential amplifier* with a common emitter resistor. The supply voltages for this example are V_{CC} = 5.0, V_{BB} = 4.0, and V_{EE} = 0 V, and the input LOW and HIGH levels are defined to be 3.6 and 4.4 V. This circuit produces output LOW and HIGH levels that are 0.6 V higher (4.2 and 5.0 V), but we'll deal with that problem later.

When V_{IN} is HIGH, as shown in the figure, transistor $Q1$ is on, but not saturated, and transistor $Q2$ is OFF. This is true because of a careful choice of resistor values and voltage levels, as shown in the following analysis. Since $Q1$ is on, V_E is one diode-drop lower than V_{IN}, or 3.8 V. Therefore, the current through $R3$ is 3.8 / 1.3 kΩ = 2.92 mA. If $Q1$ has a β of about 10, then about 2.65 mA of this current comes through the collector and $R1$, so the voltage drop across $R1$ is about 0.8 V, and V_{OUT1} is about 4.2 V (LOW). Since the collector-

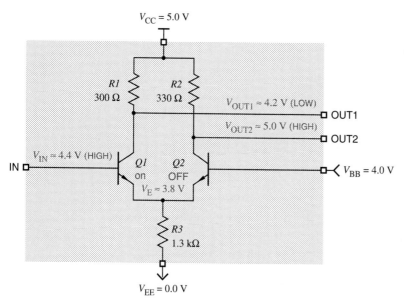

V_{CC} = 5.0 V

$R1$
300 Ω

$R2$
330 Ω

V_{OUT1} ≈ 4.2 V (LOW)

☐ OUT1

V_{OUT2} ≈ 5.0 V (HIGH)

☐ OUT2

V_{IN} ≈ 4.4 V (HIGH)

IN ☐

$Q1$
on

$Q2$
OFF

V_E ≈ 3.8 V

☐ V_{BB} = 4.0 V

$R3$
1.3 kΩ

V_{EE} = 0.0 V

Figure 3–92

Basic CML inverter/buffer circuit with input HIGH.

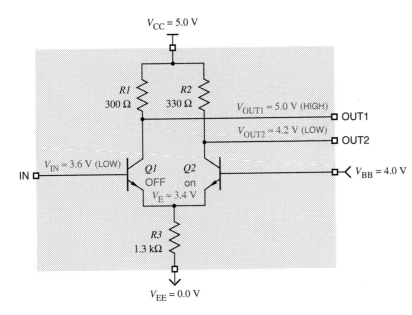

Figure 3–93
Basic CML inverter/buffer circuit with input LOW.

to-emitter voltage of $Q1$ ($4.2 - 3.8 = 0.4$ V) is greater than $V_{CE(sat)}$, $Q1$ is not saturated. $Q2$ is OFF because its base-to-emitter voltage ($4.0 - 3.8 = 0.2$ V) is less than 0.6 V. Thus, V_{OUT2} is pulled to 5.0 V (HIGH) through $R2$.

When V_{IN} is LOW, as shown in Figure 3–93, transistor $Q2$ is on, but not saturated, and transistor $Q1$ is OFF, as we'll show by an analysis similar to the foregoing. Since $Q2$ is on, V_E is one diode-drop lower than V_{BB}, or 3.4 V. Therefore, the current through $R3$ is $3.4 / 1.3$ kΩ = 2.62 mA. If $Q2$ has a β of about 10, then about 2.38 mA of this current comes through the collector and $R2$, so the voltage drop across $R2$ is about 0.8 V, and V_{OUT2} is about 4.2 V. Since the collector-to-emitter voltage of $Q2$ ($4.2 - 3.4 = 0.8$ V) is greater than $V_{CE(sat)}$, $Q2$ is not saturated. $Q1$ is OFF because its base-to-emitter voltage ($3.6 - 3.4 = 0.2$ V) is less than 0.6 V. Thus, V_{OUT1} is pulled to 5.0 V through $R1$.

To perform logic with the basic circuit of Figure 3–93, we simply place additional transistors in parallel with $Q1$. For example, Figure 3–94 shows a 2-input OR/NOR gate. If any input is HIGH, the corresponding input transistor is active, and V_{OUT1} is LOW (NOR output). At the same time, $Q3$ is OFF, and V_{OUT2} is HIGH (OR output).

Recall that the input levels for the inverter/buffer are defined to be 3.6 and 4.4 V, while the output levels that it produces are 4.2 and 5.0 V. This is obviously a problem. We could put a diode in series with each output to lower it by 0.6 V to match the input levels, but that still leaves another problem—the outputs have poor fanout. A HIGH output must supply base current to the inputs that it drives, and this current creates an additional voltage drop across $R1$ or $R2$, reducing the

(a)

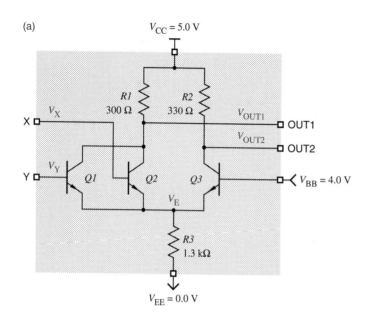

Figure 3-94
CML 2-input OR/NOR gate:
(a) circuit diagram; (b) function
table; (c) truth table; (d) logic
symbol.

(d)

X ———
Y ——— ⟩o— OUT1
         —— OUT2

(b)

X	Y	V_X	V_Y	Q1	Q2	Q3	V_E	V_{OUT1}	V_{OUT2}	OUT1	OUT2
L	L	3.6	3.6	OFF	OFF	on	3.4	5.0	4.2	H	L
L	H	3.6	4.4	OFF	on	OFF	3.8	4.2	5.0	L	H
H	L	4.4	3.6	on	OFF	OFF	3.8	4.2	5.0	L	H
H	H	4.4	4.4	on	on	OFF	3.8	4.2	5.0	L	H

(c)

X	Y	OUT1	OUT2
0	0	1	0
0	1	0	1
1	0	0	1
1	1	0	1

output voltage (and we don't have much margin to work with). These problems
are solved in commercial ECL families, such as the 10K family that we describe
next.

* 3.14.2 ECL 10K Family

ECL 10K family

The gates in today's most popular ECL family have 5-digit part numbers of
the form "10xxx" (e.g., 10102, 10181, 10209), so the family is called *ECL 10K*.
This family has the following improvements over the basic ECL circuit described
previously:

- An emitter-follower output stage shifts the output levels to match the input
  levels and provides very high current-driving capability, up to 50 mA per
  output. It is also responsible for the family's name, "emitter-coupled" logic.

- An internal, temperature- and voltage-compensated bias network provides
  V_{BB} without the need for a separate, external power supply.

- The family is designed to operate with $V_{CC} = 0$ (ground) and $V_{EE} = -5.2$ V. This improves the family's immunity to power-supply noise because noise on V_{EE} is a "common-mode" signal that is rejected by the input structure's differential amplifier.

Logic LOW and HIGH levels are defined in the ECL 10K family as shown in Figure 3–95. Note that even though the power supply is negative, ECL assigns the names LOW and HIGH to the *algebraically* lower and higher voltages, respectively.

DC noise margins in ECL 10K are much less than in TTL, only 0.155 V in the LOW state and 0.125 V in the HIGH state. However, ECL gates do not need as much noise margin as TTL. Unlike TTL, an ECL gate generates very little power-supply and ground noise when it changes state; its current requirement remains constant as it merely steers current from one path to another. Also, ECL's emitter-follower outputs have very low impedance in either state, and it is difficult to couple noise from an external source into a signal line driven by such a low-impedance output.

Figure 3–96(a) is the circuit for an ECL OR/NOR gate, one section of a quad OR/NOR gate with part number 10102. A pull-down resistor on each input ensures that if the input is left unconnected, it is treated as LOW. The bias network has component values selected to generate $V_{BB} = -1.29$ V for proper operation of the differential amplifier. Each output transistor, using the emitter-follower configuration, maintains its emitter voltage at one diode-drop below its base voltage, thereby achieving the required output level shift. Figure 3–96(b) summarizes the electrical operation of the gate.

Figure 3–95
ECL 10K logic levels.

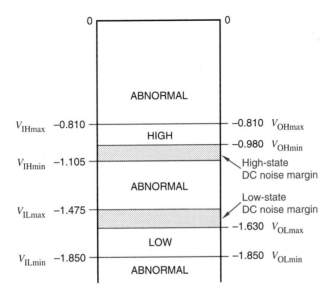

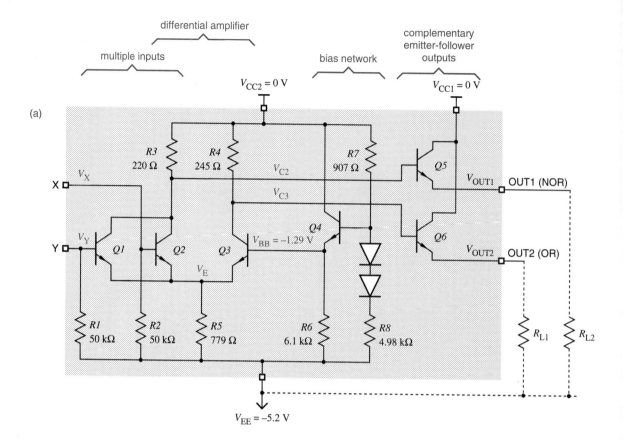

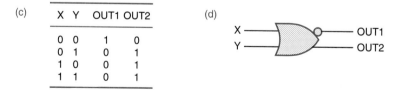

Figure 3–96 Two-input 10K ECL OR/NOR gate: (a) circuit diagram; (b) function table; (c) truth table; (d) logic symbol.

The emitter-follower outputs used in ECL 10K require external pull-down resistors, as shown in the figure. The 10K family is designed to use external rather than internal pull-down resistors for good reason. The rise and fall times of ECL output transitions are so fast (typically 2 ns) that any connection longer than a few inches must be treated as a transmission line, and must be terminated as discussed in Section 12.4. Rather than waste power with an internal pull-down resistor, ECL 10K allows the designer to select an external resistor that satisfies both pull-down and transmission-line termination requirements. The simplest termination, sufficient for short connections, is to connect a resistor in the range of 270 Ω to 2 kΩ from each output to V_{EE}.

A typical ECL 10K gate has a propagation delay of 2 ns, comparable to 74AS TTL. With its outputs left unconnected, a 10K gate consumes about 26 mW of power, also comparable to a 74AS TTL gate, which consumes about 20 mW. However, the termination required by ECL 10K also consumes power, from 10 to 150 mW per output depending on the termination circuit configuration. A 74AS TTL output may or may not require a power-consuming termination circuit, depending on the physical characteristics of the application.

* 3.14.3 ECL 100K Family

Members of the *ECL 100K family* have 6-digit part numbers of the form "100xxx"   *ECL 100K family*
(e.g., 100101, 100117, 100170), but in general have functions different than 10K parts with similar numbers. The 100K family has the following major differences from the 10K family:

- Reduced power-supply voltage, $V_{EE} = -4.5$ V.

- Shorter propagation delays, typically 0.75 ns.

- Shorter transition times, typically 0.70 ns.

- Higher power consumption, typically 40 mW per gate.

R E F E R E N C E S

Students who need to study the basics may wish to consult "Electrical Circuits Review" by Bruce M. Fleischer. This 20-page tutorial covers all of the basic circuit concepts that are used in this chapter. It appears both as an appendix in the first edition of this book, and as a reproducible section in the Instructor's Manual for the current edition.

If you're interested in history, a nice introduction to all of the early bipolar logic families can be *Logic Design with Integrated Circuits* by William E. Wickes

(Wiley-Interscience, 1968). The classic introduction to TTL electrical characteristics appeared in *The TTL Applications Handbook*, edited by Peter Alfke and Ib Larsen (Fairchild Semiconductor, 1973). Early logic designers also enjoyed *The TTL Cookbook* by Don Lancaster.

For another perspective on the electronics material in this chapter, you can consult almost any modern electronics text. Many contain a much more analytical discussion of digital circuit operation; for example, see *Microelectronics* by J. Millman and A. Grabel (McGraw-Hill, 1987, 2nd ed.). Another good introduction to ICs and important logic families can be found in *VLSI System Design* by Saburo Muroga (Wiley, 1982). For NMOS and CMOS circuits in particular, two good books are *Introduction to VLSI Systems* by Carver Mead and Lynn Conway (Addison-Wesley, 1980) and *Principles of CMOS VLSI Design* by Neil H. E. Weste and Kamran Eshraghian (Addison-Wesley, 1993).

Characteristics of today's logic families can be found in the data books published by the device manufacturers. Texas Instruments' data books for TTL and CMOS devices are listed in Table 5–20 on page 377. Motorola provides an excellent introduction to the HC and HCT families in *High-Speed CMOS Logic Data* (publ. DL129, rev. 3, 1988), and to the more recent AC and ACT families in *FACT Data* (publ. DL138, 1988).

Motorola also has a data book that covers two popular TTL families, LS and F, in a single volume (*FAST and LSTTL Data*, publ. DL121, rev. 4, 1989). Motorola covers ECL parts in the *MECL Device Data* book (publ. DL122, rev. 4, 1989) and in the *MECL System Design Handbook* (publ. HB205, rev. 1, 1988).

DRILL PROBLEMS

3.1 A particular logic family defines a LOW signal to be in the range 0.0–0.8 V, and a HIGH signal to be in the range 2.0–3.3 V. Under a positive-logic convention, indicate the logic value associated with each of the following signal levels:

   (a) 0.0 V   (b) 3.0 V   (c) 0.8 V   (d) 1.9 V
   (e) 2.0 V   (f) 5.0 V   (g) −0.7 V   (h) −3.0 V

3.2 Repeat Drill 3.1 using a negative-logic convention.

3.3 Discuss how a logic buffer amplifier is different from an audio amplifier.

3.4 Is a buffer amplifier equivalent to a 1-input AND gate or a 1-input OR gate?

3.5 True or false: For a given set of inputs, a NAND gate produces the opposite output as a NOR gate.

3.6 True or false: The Simpsons are a bipolar logic family.

3.7 Write two completely different (but valid) definitions of "gate."

3.8   What kind of transistors are used in CMOS gates?

3.9   (Electrical engineers only.) Draw an equivalent circuit for a CMOS inverter using a single-pole, double-throw relay.

3.10 For a given silicon area, which is likely to be faster, a CMOS NAND gate or a CMOS NOR?

3.11 Define "fan-in" and "fanout." Which one are you likely to have to calculate?

3.12 Draw the circuit diagram, function table, and logic symbol for a 3-input CMOS NOR gate in the style of Figure 3–16.

3.13 Draw switch models in the style of Figure 3–14 for a 2-input CMOS NOR gate for all four input combinations.

3.14 Draw a circuit diagram, function table, and logic symbol for a CMOS OR gate in the style of Figure 3–19.

3.15 Which has fewer transistors, a CMOS inverting gate or a noninverting gate?

3.16 Name and draw the logic symbols of four different 4-input CMOS gates that each use 8 transistors.

3.17 The circuit in Figure X3.17(a) is a type of CMOS AND-OR-INVERT gate. Write a function table for this circuit in the style of Figure 3–15(b), and a corresponding logic diagram using AND and OR gates and inverters.

Figure X3.17

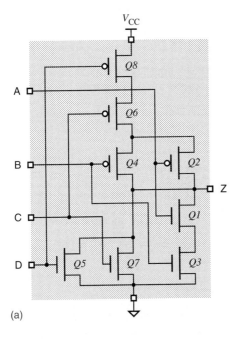

(a)

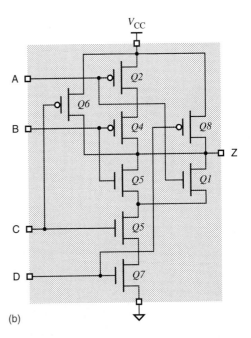

(b)

3.18 The circuit in Figure X3.17(b) is a type of CMOS OR-AND-INVERT gate. Write function table for this circuit in the style of Figure 3–15(b), and a corresponding logic diagram using AND and OR gates and inverters.

3.19 How is it that perfume can be bad for digital designers?

3.20 How much high-state DC noise margin is available in a CMOS inverter whose transfer characteristic under worst-case conditions looks like Figure 3–25? How much low-state DC noise margin is available? (Assume standard 1.5-V and 3.5-V thresholds for LOW and HIGH.)

3.21 Using the data sheet in Table 3–3, determine the worst-case LOW-state and HIGH-state DC noise margins of the 74FCT257T. State any assumptions required by your answer.

3.22 Section 3.5 defines seven different electrical parameters for CMOS circuits. Using the data sheet in Table 3–3, determine the worst-case value of each of these for the 74FCT257T. State any assumptions required by your answer.

3.23 Based on the conventions and definitions in Section 3.4, if the current at a device output is specified as a negative number, is the output sourcing current or sinking current?

3.24 For each of the following resistive loads, determine whether the output drive specifications of the 74FCT257T over the commercial operating range are exceeded. (Refer to Table 3–3, and use $V_{OHmin} = 2.4$ V and $V_{CC} = 5.0$ V.)

  (a) 120 Ω to V_{CC} (b) 270 Ω to V_{CC} and 330 Ω to GND
  (c) 1 KΩ to GND   (d) 150 Ω to V_{CC} and 150 Ω to GND
  (e) 100 Ω to V_{CC} (f) 75 Ω to V_{CC} and 150 Ω to GND
  (g) 75 Ω to GND   (h) 75 Ω to V_{CC} and 150 Ω to GND

3.25 Across the range of valid HIGH input levels, 2.0–5.0 V, at what input level would you expect the 74FCT257T (Table 3–3) to consume the most power?

3.26 Determine the LOW-state and HIGH-state DC fanout of the 74FCT257T when it drives 74LS00-like inputs. (Refer to Tables 3–3 and 3–12.)

3.27 Estimate the "on" resistances of the p-channel and n-channel output transistors of the 74FCT257T using information in Table 3–3.

3.28 Under what circumstances is it safe to allow an unused CMOS input to float?

3.29 Explain "latch up" and the circumstances under which it occurs.

3.30 Explain why putting all the decoupling capacitors in one corner of a printed-circuit board is not a good idea.

3.31 When is it important to hold hands with a friend?

3.32 Name the two components of CMOS logic gate's delay. Which one is most affected by load capacitance?

3.33 Determine the *RC* time constant for each of the following resistor-capacitor combinations:

(a) $R = 100 \; \Omega$, $C = 50$ pF   (b) $R = 330 \; \Omega$, $C = 150$ pF

(c) $R = 1 \; K\Omega$, $C = 30$ pF   (d) $R = 4.7 \; K\Omega$, $C = 100$ pF

3.34 Besides delay, what other characteristic(s) of a CMOS circuit are affected by load capacitance?

3.35 Explain the I_C formula in footnote 6 in Table 3–3 in terms of concepts presented in Sections 3.5 and 3.6.

3.36 It is possible to operate 74AC CMOS devices with a 3.3-volt power supply. How much power does this typically save, compared to 5-volt operation?

3.37 A particular Schmitt-trigger inverter has $V_{ILmax} = 0.8$ V, $V_{IHmin} = 2.0$ V, $V_{T+} = 1.6$ V, and $V_{T-} = 1.3$ V. How much hysteresis does it have?

3.38 Why are three-state outputs usually designed to "turn off" faster than they "turn on"? Discuss the pros and cons of larger versus smaller pull-up resistors for open-drain CMOS outputs or open-collector TTL outputs.

3.39 A particular LED has a voltage drop of about 2.0 V in the "on" state, and requires about 5 mA of current for normal brightness. Determine an appropriate value for the pull-up resistor when the LED is connected as shown in Figure 3–50.

3.40 How does the answer for Drill 3.39 change if the LED is connected as shown in Figure 3–51(a)?

3.41 A wired-AND function is obtained simply by typing two open-drain or open-collector outputs together, without going through another level of transistor circuitry. How is it, then, that a wired-AND function can actually be slower than a discrete AND gate? (*Hint:* Recall the title of a Teenage Mutant Ninja Turtles movie.)

3.42 Which CMOS or TTL logic family in this chapter has the strongest output driving capability?

3.43 Concisely summarize the difference between HC and HCT logic families. The same concise statement should apply to AC versus ACT.

3.44 Why don't the specifications for FCT devices include parameters like V_{OLmaxC} that apply to CMOS loads, as HCT and ACT specifications do?

3.45 How does FCT-T devices reduce power consumption compared to FCT devices?

3.46 How many diodes are required for an *n*-input diode AND gate?

3.47 True or false: A TTL NOR gate uses diode logic.

3.48 Are TTL outputs more capable of sinking current or sourcing current?

3.49 Compute the maximum fanout for each of the following cases of a TTL output driving multiple TTL inputs. Also indicate how much "excess" driving capability is available in the LOW or HIGH state for each case.

(a) 74LS driving 74LS   (b) 74LS driving 74S
(c) 74S driving 74AS   (d) 74F driving 74S
(e) 74AS driving 74AS   (f) 74AS driving 74F
(g) 74ALS driving 74F   (h) 74AS driving 74ALS

3.50 Which resistor dissipates more power, the pull-down for an unused LS-TTL NOR-gate input, or the pull-up for an unused LS-TTL NAND-gate input? Use the minimum allowable resistor value in each case.

3.51 Which would you expect to be faster, a TTL AND gate or a TTL AND-OR-INVERT gate? Why?

3.52 Describe the main benefit and the main drawback of TTL gates that use Schottky transistors.

3.53 Using the data sheet in Table 3–12, determine the worst-case LOW-state and HIGH-state DC noise margins of the 74LS00.

3.54 Section 3.10 defines eight different electrical parameters for TTL circuits. Using the data sheet in Table 3–12, determine the worst-case value of each of these for the 74LS00.

3.55 For each of the following resistive loads, determine whether the output drive specifications of the 74LS00 over the commercial operating range are exceeded. (Refer to Table 3–12, and use $V_{OLmax} = 0.5$ V and $V_{CC} = 5.0$ V.)

(a) 470 Ω to V_{CC} (b) 330 Ω to V_{CC} and 470 Ω to GND
(c) 10 KΩ to GND   (d) 390 Ω to V_{CC} and 390 Ω to GND
(e) 600 Ω to V_{CC} (f) 510 Ω to V_{CC} and 510 Ω to GND
(g) 4.7 KΩ to GND   (h) 220 Ω to V_{CC} and 330 Ω to GND

3.56 Compute the LOW-state and HIGH-state DC noise margins for each of the following cases of a TTL output driving a TTL-compatible CMOS input, or vice versa.

(a) 74HCT driving 74LS   (b) 74ACT driving 74AS
(c) 74LS driving 74HCT   (d) 74S driving 74ACT

3.57 Compute the maximum fanout for each of the following cases of a TTL-compatible CMOS output driving multiple inputs in a TTL logic family. Also indicate how much "excess" driving capability is available in the LOW or HIGH state for each case.

(a) 74HCT driving 74LS   (b) 74HCT driving 74S
(c) 74ACT driving 74AS   (d) 74ACT driving 74LS

3.58 For a given load capacitance and transition rate, which logic family in this chapter has the lowest dynamic power dissipation?

EXERCISES

3.59 Design a CMOS circuit that has the functional behavior shown in Figure X3.59. (*Hint:* Only six transistors are required.)

Figure X3.59

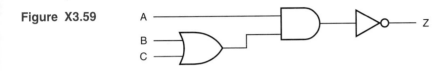

3.60 Design a CMOS circuit that has the functional behavior shown in Figure X3.60. (*Hint:* Only six transistors are required.)

Figure X3.60

3.61 Draw a circuit diagram, function table, and logic symbol in the style of Figure 3–19 for a CMOS gate with two inputs A and B and an output Z, where Z = 1 if A = 0 and B = 1, and Z = 0 otherwise. (*Hint:* Only six transistors are required.)

3.62 Draw a circuit diagram, function table, and logic symbol in the style of Figure 3–19 for a CMOS gate with two inputs A and B and an output Z, where Z = 0 if A = 1 and B = 0, and Z = 1 otherwise. (*Hint:* Only six transistors are required.)

3.63 Draw a figure showing the logical structure of an 8-input CMOS NOR gate, assuming that at most 4-input gate circuits are practical. Using your general knowledge of CMOS electrical characteristics, select a circuit structure that minimizes the NOR gate's propagation delay for a given area of silicon, and explain why this is so.

3.64 The circuit designers of TTL-compatible CMOS families presumably could have made the voltage drop across the "on" transistor under load in the HIGH state as little as it is in the LOW state, simply by making the *p*-channel transistors bigger. Why do you suppose they didn't bother to do this?

3.65 How much current and power are "wasted" in Figure 3–32(b)?

3.66 Perform a detailed calculation of V_{OUT} in Figures 3–34 and 3–33. (*Hint:* Create a Thévenin equivalent for the CMOS inverter in each figure.)

3.67 Consider the dynamic behavior of a CMOS output driving a given capacitive load. If the resistance of the charging path is double the resistance of the discharging path, is the rise time exactly twice the fall time? If not, what other factors affect the transition times?

3.68   Analyze the fall time of the CMOS inverter output of Figure 3–37, assuming that $R_L = 1$ kΩ and $V_L = 2.5$ V. Compare your answer with the results of Section 3.6.1 and explain.

3.69   Repeat Exercise 3.68 for rise time.

3.70   Assuming that the transistors in an FCT CMOS three-state buffer are perfect, zero-delay on-off devices that switch at an input threshold of 1.5 V, determine the value of t_{PLZ} for the test circuit and waveforms in Figure 3–24. (*Hint:* You have to determine the time using an RC time constant.) Explain the difference between your result and the specifications in Table 3–3.

3.71   Repeat Exercise 3.70 for t_{PHZ}.

3.72   Using the specifications in Table 3–6, estimate the "on" resistances of the p-channel and n-channel transistors in 74AC-series CMOS logic.

3.73   Create a 4×4×2×2 matrix of worst-case DC noise margins for the following CMOS interfacing situations: an (HC, HCT, AC, or ACT) output driving an (HC, HCT, AC, or ACT) input with a (CMOS, TTL) load in the (LOW, HIGH) state; Figure X3.73 illustrates. (*Hints:* There are 64 different combinations to examine, but many give identical results. Some combinations yield negative margins.)

		Input							
Output		HC		HCT		AC		ACT	
HC		CL	TL	CL	TL	CL	TL	CL	TL
		CH	TH	CH	TH	CH	TH	CH	TH
HCT		CL	TL	CL	TL	CL	TL	CL	TL
		CH	TH	CH	TH	CH	TH	CH	TH
AC		CL	TL	CL	TL	CL	TL	CL	TL
		CH	TH	CH	TH	CH	TH	CH	TH
ACT		CL	TL	CL	TL	CL	TL	CL	TL
		CH	TH	CH	TH	CH	TH	CH	TH

Key:
CL = CMOS load, LOW
CH = CMOS load, HIGH
TL = TTL load, LOW
TH = TTL load, HIGH

Figure X3.73

3.74   In the LED example in Section 3.7.4, a designer chose a resistor value of 300 Ω, and found that the open-drain gate was able to maintain its output at 0.1 V while driving the LED. How much current flows through the LED, and how much power is dissipated by the pull-up resistor in this case?

3.75   Consider a CMOS 8-bit binary counter (Section 9.3) clocked at 16 MHz. For the purposes of computing dynamic power dissipation, what is the transition frequency of least significant bit? Of the most significant bit? For the purposes of determining the dynamic power dissipation of the eight output bits, what frequency should be used?

3.76 Using only AND and NOR gates, draw a logic diagram for the logic function performed by the circuit in Figure 3–52.

3.77 Calculate the approximate output voltage at Z in Figure 3–53, assuming that the gates are HCT-series CMOS.

3.78 Redraw the circuit diagram of a CMOS 3-state buffer in Figure 3–46 using actual transistors instead of NAND, NOR, and inverter symbols. Can you find a circuit for the same function that requires a smaller total number of transistors? If so, draw it.

3.79 Modify the CMOS 3-state buffer circuit in Figure 3–46 so that the output is in the High-Z state when the enable input is HIGH. The modified circuit should require no more transistors than the original.

3.80 Using information in Table 3–3, estimate how much current can flow through each output pin when the outputs of two different 74FCT257Ts are fighting.

3.81 A computer system made by the Green PC Company had ten LED "status OK" indicators, each of which was turned on by an open-collector output in the style of Figure 3–50. However, in order to save a few cents, the logic designer connected the anodes of all ten LEDs together and replaced the ten, now parallel, 300-Ω pull-up resistors with a single 30-Ω resistor. This worked fine in the lab, but a big problem was found after volume shipments began. Explain.

3.82 Show that at a given power-supply voltage, an FCT-type I_{CCD} specification can be derived from an HCT/ACT-type C_{PD} specification, and vice versa.

3.83 If both V_Z and V_B in Figure 3–65(b) are 4.6 V, can we get $V_C = 5.2$ V? Explain.

3.84 Modify the program in Table 3–10 to account for leakage current in the OFF state.

3.85 Assuming "ideal" conditions, what is the minimum voltage that will be recognized as a HIGH in the TTL NAND gate in Figure 3–77?

3.86 Assuming "ideal" conditions, what is the maximum voltage that will be recognized as a LOW in the TTL NAND gate in Figure 3–78?

3.87 Find a commercial TTL part that can source 40 mA in the HIGH state. What is its application?

3.88 What happens if you try to drive an LED with its cathode grounded and its anode connected to a TTL totem-pole output, analogous to Figure 3–51 for CMOS?

3.89 What happens if you try to drive a 12-volt relay with a TTL totem-pole output?

3.90 Suppose that a single pull-up resistor to +5 V is used to provide a constant-1 logic source to 15 different 74LS00 inputs. What is the maximum value of this resistor? How much HIGH-state DC noise margin are you providing in this case?

3.91 The circuit in Figure X3.91 uses open-collector NAND gates to perform "wired logic." Write a truth table for output signal F and, if you've read Section 4.2, a logic expression for F as a function of the circuit inputs.

Figure X3.91

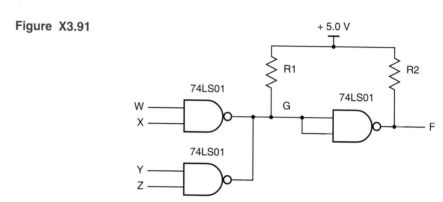

3.92 What is the maximum allowable value for *R1* in Figure X3.91? Assume that a 0.7 V HIGH-state noise margin is required. The 74LS01 has the specs shown in the 74LS column of Table 3–11, except that I_{OHmax} is 100 μA, a leakage current that flows *into* the output in the HIGH state.

3.93 A logic designer found a problem in a certain circuit's function after the circuit had been released to production and 1000 copies of it built. A portion of the circuit is shown in Figure X3.93 in black; all of the gates are 74LS00 NAND gates. The logic designer fixed the problem by adding the two diodes shown in color. What do the diodes do? Describe both the logical effects of this change on the circuit's function and the electrical effects on the circuit's noise margins.

Figure X3.93

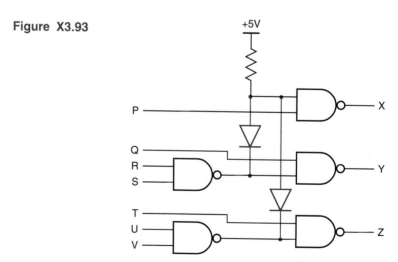

3.94 A *Thévenin termination* for an open-collector or three-state bus has the structure *Thévenin*
shown in Figure X3.94(a). The idea is that, by selecting appropriate values of *R1*   *termination*
and *R2*, a designer can obtain a circuit equivalent to the termination in (b) for any
desired values of V and R. The value of V determines the voltage on the bus when
no device is driving it, and the value of R is selected to match the characteristic
impedance of the bus for transmission-line purposes (Section 12.4). For each of
the following pairs of V and R, determine the required values of *R1* and *R2*.

(a)   V = 2.75, R = 148.5   (b)   V = 2.7, R = 180
(c)   V = 3.0, R = 130   (d)   V = 2.5, R = 75

Figure X3.94

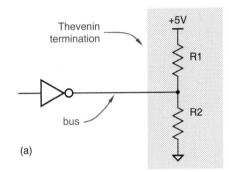
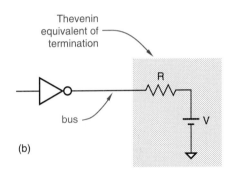

3.95 For each of the *R1* and *R2* pairs in Exercise 3.94, determine whether the termination can be properly driven by a three-state output in each of the following logic
families: 74LS, 74S, 74ACT. For proper operation, the family's I_{OL} and I_{OH} specs
must not be exceeded when $V_{OL} = V_{OLmax}$ and $V_{OH} = V_{OHmin}$, respectively.

3.96 Suppose that the output signal F in Figure X3.91 drives the inputs of two 74S04
inverters. Compute the minimum and maximum allowable values of *R2*, assuming
that a 0.7 V HIGH-state noise margin is required.

3.97 A 74LS125 is a buffer with a three-state output. When enabled, the output can sink
24 mA in the LOW state and source 2.6 mA in the HIGH state. When disabled, the
output has a leakage current of ±20 μA (the sign depends on the output voltage—
plus if the output is pulled HIGH by other devices, minus if it's LOW). Suppose a
system is designed with multiple modules connected to a bus, where each module
has a single 74LS125 to drive the bus, and one 74LS04 to receive information on
the bus. What is the maximum number of modules that can be connected to the
bus without exceeding the 74LS125's specs?

3.98 Repeat Exercise 3.97, this time assuming that a single pull-up resistor is connected
from the bus to +5 V to guarantee that the bus is HIGH when no device is driving
it. Calculate the maximum possible value of the pull-up resistor, and the number
of modules that can be connected to the bus.

3.99 Find the circuit design in a TTL data book for an actual three-state gate, and explain how it works.

3.100 Using the graphs in a TTL data book, develop some rules of thumb for derating the maximum propagation delay specification of LS-TTL under nonoptimal conditions of power-supply voltage, temperature, and loading.

3.101 Determine the total power dissipation of the circuit in Figure X3.101 as function of transition frequency f for two realizations: (a) using 74LS gates; (b) using 74HC gates. Assume that input capacitance is 3 pF for a TTL gate and 7 pF for a CMOS gate, that a 74LS gate has an internal power dissipation capacitance of 20 pF, and that there is an additional 20 pF of stray wiring capacitance in the circuit. Also assume that the X, Y, and Z inputs are always HIGH, and that input C is driven with a CMOS-level square wave with frequency f. Other information that you need for this problem can be found in Tables 3–5 and 3–11. State any other assumptions that you make. At what frequency does the TTL circuit dissipate less power than the CMOS circuit?

Figure X3.101

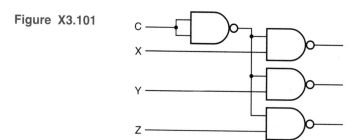

3.102 It is possible to drive one or more 74AC or 74HC inputs reliably with a 74LS TTL output by providing an external resistor to pull the TTL output all the way up to V_{CC} in the HIGH state. What are the design issues in choosing a value for this pull-up resistor?

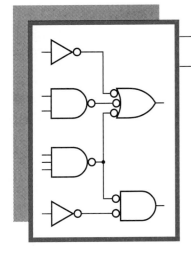

COMBINATIONAL LOGIC DESIGN PRINCIPLES

Logic circuits are classified into two types, "combinational" and "sequential."
A *combinational logic circuit* is one whose outputs depend only on its current
inputs. The rotary channel selector knob on an inexpensive TV is like a combinational circuit—its "output" selects a channel based only on the current position
of the knob ("input").

    The outputs of a *sequential logic circuit* depend not only on the current
inputs, but also on the past sequence of inputs, possibly arbitrarily far back in
time. The channel selector controlled by the up and down pushbuttons on a TV or
VCR is a sequential circuit—the channel selection depends on the past sequence
of up/down pushes, at least since when you started viewing 10 hours before, and
perhaps as far back as when you first powered-up the device. Sequential circuits
are discussed in Chapters 7 through 10.

    A combinational circuit may contain an arbitrary number of logic gates and
inverters but no feedback loops. A *feedback loop* is a signal path of a circuit
that allows the output of a gate to propagate back to the input of that same gate;
such a loop generally creates sequential circuit behavior.

    In combinational circuit *analysis*, we start with a logic diagram, and proceed
to a formal description of the function performed by that circuit, such as a truth

*combinational logic
circuit*

*sequential logic
circuit*

feedback loop

analysis

SYNTHESIS
VS. DESIGN

> Logic circuit design is a superset of synthesis, since in a real design problem we usually start out with an informal (word or thought) description of the circuit. Often the most challenging and creative part of design is to formalize the circuit description, defining the circuit's input and output signals and specifying its functional behavior by means of truth tables and equations. Once we've created the formal circuit description, we can usually follow a "turn-the-crank" synthesis procedure to obtain a logic diagram for a circuit with the required functional behavior. The material in this chapter is the basis for "turn-the-crank" procedures, whether the crank is turned by hand or by a computer. In Chapters 5 and 6 we'll encounter some examples of the real design process.

synthesis

table or a logic expression. In *synthesis*, we do the reverse, starting with a formal description and proceeding to a logic diagram.

Combinational circuits may have one or more outputs. Most analysis and synthesis techniques can be extended in an obvious way from single-output to multiple-output circuits (e.g., "Repeat these steps for each output"). However, some techniques can also be extended in a not-so-obvious way for improved effectiveness in the multiple-output case. For example, in multiple-output synthesis, common subexpressions can sometimes be shared among different outputs. We will revisit these ideas later, but we will not discuss formal multiple-output synthesis techniques in detail (see References).

The purpose of this chapter is to give you a solid theoretical foundation for the analysis and synthesis of combinational logic circuits, a foundation that will be doubly important later when we move on to sequential circuits. Although many of the analysis and synthesis procedures in this chapter are automated nowadays by computer-aided design tools, you need a basic understanding of the fundamentals to use the tools and to figure out what's wrong when you get unexpected or undesirable results.

Before launching into a discussion of combinational logic circuits, we must introduce switching algebra, the fundamental mathematical tool for analyzing and synthesizing logic circuits of all types.

4.1   SWITCHING ALGEBRA

Formal analysis techniques for digital circuits have their roots in the work of an English mathematician, George Boole. In 1854, he invented a two-valued

Boolean algebra

algebraic system, now called *Boolean algebra*, to "give expression ... to the fundamental laws of reasoning in the symbolic language of a Calculus." Using this system, a philosopher, logician, or inhabitant of the planet Vulcan can formulate propositions that are true or false, combine them to make new propositions, and determine the truth or falsehood of the new propositions. For example, if we agree that "People who haven't studied this material are either failures or not nerds," and "No computer designer is a failure," then we can answer questions like "If you're a nerdy computer designer, then have you already studied this?"

Long after Boole, in 1938, Bell Labs researcher Claude E. Shannon showed how to adapt Boolean algebra to analyze and describe the behavior of circuits built from relays, the most commonly used digital logic elements of that time. In Shannon's *switching algebra*, the condition of a relay contact, open or closed, is represented by a variable X that can have one of two possible values, 0 or 1. In today's logic technologies, these values correspond to a wide variety of physical conditions—voltage HIGH or LOW, light off or on, capacitor discharged or charged, fuse blown or intact, and so on—as we detailed in Table 3–1 on page 74.

switching algebra

In the remainder of this section, we develop the switching algebra directly, using "first principles" and what we already know about the behavior of logic elements (gates and inverters). For more historical and/or mathematical treatments of this material, consult the References.

4.1.1 Axioms

In switching algebra we use a symbolic variable, such as X, to represent the condition of a logic signal. A logic signal is in one of two possible conditions—low or high, off or on, and so on, depending on the technology. We say that X has the value "0" for one of these conditions and "1" for the other.

For example, with the CMOS and TTL logic circuits in Chapter 3, the *positive-logic convention* dictates that we associate the value "0" with a LOW voltage and "1" with a HIGH voltage. The *negative-logic convention* makes the opposite association: 0 = HIGH and 1 = LOW. However, the choice of positive or negative logic has no effect on our ability to develop a consistent algebraic description of circuit behavior; it only affects details of the physical-to-algebraic abstraction, as we'll explain later in our discussion of "duality." For the moment, we may ignore the physical realities of logic circuits and pretend that they operate directly on the logic symbols 0 and 1.

positive-logic convention
negative-logic convention

The *axioms* (or *postulates*) of a mathematical system are a minimal set of basic definitions that we assume to be true, from which all other information about the system can be derived. The first two axioms of switching algebra embody the "digital abstraction" by formally stating that a variable X can take on only one of two values:

axiom
postulate

$$(A1) \quad X = 0 \quad \text{if } X \neq 1 \qquad (A1') \quad X = 1 \quad \text{if } X \neq 0$$

Notice that we stated these axioms as a pair, with the only difference between A1 and A1′ being the interchange of the symbols 0 and 1. This is a characteristic of all the axioms of switching algebra, and is the basis of the "duality" principle that we'll study later.

In Section 3.3.3 we showed the design of an inverter, a logic circuit whose output signal level is the opposite (or *complement*) of its input signal level. We use a *prime (')* to denote an inverter's function. That is, if a variable X denotes an inverter's input signal, then X' denotes the value of a signal on the inverter's output. This notation is formally specified in the second pair of axioms:

complement
prime (')

$$\text{(A2)} \quad \text{If } X = 0, \text{ then } X' = 1 \qquad \text{(A2')} \quad \text{If } X = 1, \text{ then } X' = 0$$

As shown in Figure 4–1, the output of an inverter with input signal X may have an arbitrary signal name, say Y. However, algebraically, we write Y = X' to say "signal Y always has the opposite value as signal X." The prime (') is an *algebraic operator*, and X' is an *expression*, which you can read as "X prime" or "NOT X." This usage is analogous to what you've learned in programming languages, where if J is an integer variable, then -J is an expression whose value is $0 - J$. Although this may seem like a small point, you'll learn that the distinction between signal names (X, Y), expressions (X'), and equations (Y = X') is very important when we study documentation standards and logic design tools. In the logic diagrams in this book, we maintain this distinction by writing signal names in black and expressions in color.

algebraic operator
expression
NOT operation

Figure 4–1
Signal naming and algebraic notation for an inverter.

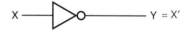
$X \longrightarrow \!\!\!\!\!\! \triangleright\!\!\circ \longrightarrow Y = X'$

In Section 3.3.6 we showed how to build a 2-input CMOS AND gate, a circuit whose output is 1 if both of its inputs are 1. The function of a 2-input AND gate is sometimes called *logical multiplication* and is symbolized algebraically by a *multiplication dot (·)*. That is, an AND gate with inputs X and Y has an output signal whose value is X·Y, as shown in Figure 4–2(a). Some

logical
multiplication
multiplication dot (·)

NOTE ON
NOTATION

> The notations $\overline{X}$, ~X, and ¬X are also used by some authors to denote the complement of X. The overbar notation is probably the most widely used and the best looking typographically. However, we use the prime notation to get you used to writing logic expressions on a single text line without the more graphical overbar, and to force you to parenthesize complex complemented subexpressions—because this is what you'll have to do when you start using computer-aided logic design languages and systems.

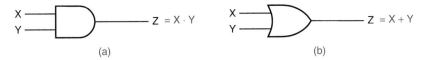

Figure 4–2   Signal naming and algebraic notation: (a) AND gate; (b) OR gate.

authors, especially mathematicians and logicians, denote logical multiplication with a wedge (X ∧ Y). We follow standard engineering practice by using the dot (X·Y).

    We also showed in Section 3.3.6 how to build a 2-input CMOS OR gate, a circuit whose output is 1 if either of its inputs is 1. The function of a 2-input OR gate is sometimes called *logical addition* and is symbolized algebraically *logical addition* by a plus sign (+). An OR gate with inputs X and Y has an output signal whose value is X + Y, as shown in Figure 4–2(b). Some authors denote logical addition with a vee (X ∨ Y), but we follow the standard engineering practice of using the plus sign (X + Y). By convention, in a logic expression involving both multiplication and addition, multiplication has *precedence*, just as in integer *precedence* expressions in conventional programming languages. That is, the expression W·X + Y·Z is equivalent to (W·X) + (Y·Z).

    The last three pairs of axioms state the formal definitions of the AND and *AND operation* OR operations by listing the output produced by each gate for each possible input *OR operation* combination:

(A3)	0·0 = 0	(A3′)	1 + 1 = 1
(A4)	1·1 = 1	(A4′)	0 + 0 = 0
(A5)	0·1 = 1·0 = 0	(A5′)	1 + 0 = 0 + 1 = 1

The five pairs of axioms, A1–A5 and A1′–A5′, completely define switching algebra. All other facts about the system can be proved using these axioms as a starting point.

4.1.2   Single-Variable Theorems

During the analysis or synthesis of logic circuits, we often write algebraic expressions that characterize a circuit's actual or desired behavior. Switching-algebra *theorems* are statements, known to be always true, that allow us to manipulate *theorem* algebraic expressions to allow simpler analysis or more efficient synthesis of the corresponding circuits. For example, the theorem X + 0 = X allows us to substitute every occurrence of X + 0 in an expression with X.

Table 4–1
Switching-algebra theorems with one variable.

(T1)	X + 0 = X	(T1′)	X·1 = X	(Identities)
(T2)	X + 1 = 1	(T2′)	X·0 = 0	(Null elements)
(T3)	X + X = X	(T3′)	X·X = X	(Idempotency)
(T4)	(X′)′ = X			(Involution)
(T5)	X + X′ = 1	(T5′)	X·X′ = 0	(Complements)

Table 4–1 lists switching-algebra theorems involving a single variable X. How do we know that these theorems are true? We can either prove them ourselves or take the word of someone who has. OK, we're in college now, let's learn how to prove them.

perfect induction

Most theorems in switching algebra are exceedingly simple to prove using a technique called *perfect induction*. Axiom A1 gives us the key to this technique—since a switching variable can take on only two different values, 0 and 1, we can prove a theorem involving a single variable X by proving that it is true for both X = 0 and X = 1. For example, to prove theorem T1, we make two substitutions:

$$[X = 0] \qquad 0 + 0 = 0 \qquad \text{true, according to axiom A4′}$$

$$[X = 1] \qquad 1 + 0 = 1 \qquad \text{true, according to axiom A5′}$$

All of the theorems in Table 4–1 can be proved using perfect induction, as requested in the Drills 4.2 and 4.3.

4.1.3   Two- and Three-Variable Theorems

Switching-algebra theorems with two or three variables are listed in Table 4–2. Each of these theorems is easily proved by perfect induction, by evaluating the theorem statement for the four possible combinations of X and Y, or the eight possible combinations of X, Y, and Z.

The first two theorem pairs are identical to the commutative and associative laws for addition and multiplication of integers and reals. Taken together, they

JUXT A
MINUTE...

> Older texts use simple *juxtaposition* (XY) to denote logical multiplication, but we don't. In general, juxtaposition is a clear notation only when signal names are limited to a single character. Otherwise, is XY a logical product or a 2-character signal name? One-character variable names are common in algebra, but in real digital design problems, we prefer to use multicharacter signal names that mean something. Thus, we need a separator between names, and the separator might just as well be a multiplication dot rather than a space. The multiplication dot or equivalent (typically * or &) is also required when logic formulas are written in computer-aided logic design languages and systems.

Table 4–2 Switching-algebra theorems with two or three variables.

(T6)	$X + Y = Y + X$	(T6′)	$X \cdot Y = Y \cdot X$	(Commutativity)
(T7)	$(X + Y) + Z = X + (Y + Z)$	(T7′)	$(X \cdot Y) \cdot Z = X \cdot (Y \cdot Z)$	(Associativity)
(T8)	$X \cdot Y + X \cdot Z = X \cdot (Y + Z)$	(T8′)	$(X + Y) \cdot (X + Z) = X + Y \cdot Z$	(Distributivity)
(T9)	$X + X \cdot Y = X$	(T9′)	$X \cdot (X + Y) = X$	(Covering)
(T10)	$X \cdot Y + X \cdot Y' = X$	(T10′)	$(X + Y) \cdot (X + Y') = X$	(Combining)
(T11)	$X \cdot Y + X' \cdot Z + Y \cdot Z = X \cdot Y + X' \cdot Z$			(Consensus)
(T11′)	$(X + Y) \cdot (X' + Z) \cdot (Y + Z) = (X + Y) \cdot (X' + Z)$			

indicate that the parenthesization or order of terms in a logical sum or logical product is irrelevant. For example, from a strictly algebraic point of view, an expression such as $W \cdot X \cdot Y \cdot Z$ is ambiguous; it should be written as $(W \cdot (X \cdot (Y \cdot Z)))$ or $(((W \cdot X) \cdot Y) \cdot Z)$ or $(W \cdot X) \cdot (Y \cdot Z)$ (see Exercise 4.28). But the theorems tell us that the ambiguous form of the expression is OK because we get the same results in any case. We even could have rearranged the order of the variables (e.g., $X \cdot Z \cdot Y \cdot W$) and gotten the same results.

As trivial as this discussion may seem, it is very important because it forms the theoretical basis for using logic gates with more than two inputs. We defined $\cdot$ and $+$ as *binary operators*—operators that combine *two* variables. Yet we use 3-input, 4-input, and larger AND and OR gates in practice. The theorems tell us we can connect gate inputs in any order (some printed-circuit-board layout programs actually take advantage of this). We can use either one n-input gate or $(n-1)$ 2-input gates interchangeably, though propagation delay and cost are likely to be higher with multiple 2-input gates.

binary operator

Theorem T8 is identical to the distributive law for integers and reals— that is, logical multiplication distributes over logical addition. Hence, we can "multiply out" an expression to obtain a sum-of-products form, as in the example below:

$$V \cdot (W + X) \cdot (Y + Z) \quad = \quad V \cdot W \cdot Y + V \cdot W \cdot Z + V \cdot X \cdot Y + V \cdot X \cdot Z$$

However, switching algebra also has the unfamiliar property that the reverse is true—logical addition distributes over logical multiplication—as demonstrated by theorem T8′. Thus, we can also "add out" an expression to obtain a product-of-sums form:

$$(V \cdot W \cdot X) + (Y \cdot Z) \quad = \quad (V + Y) \cdot (V + Z) \cdot (W + Y) \cdot (W + Z) \cdot (X + Y) \cdot (X + Z)$$

Theorems T9 and T10 are used extensively in the minimization of logic functions. For example, if the subexpression $X + X \cdot Y$ appears in a logic expression, the *covering theorem* T9 says that we need only include X in the expression;

covering theorem

cover   X is said to *cover* X·Y. The *combining theorem* T10 says that if the subexpression
combining theorem   X·Y + X·Y′ appears in an expression, we can replace it with X. Since Y must be
0 or 1, either way the original subexpression is 1 if and only if X is 1.

Although we could easily prove T9 by perfect induction, the truth of T9 is
more obvious if we prove it using the other theorems that we've proved so far:

$$
\begin{aligned}
X + X \cdot Y &= X \cdot 1 + X \cdot Y & \text{(according to T1$'$)} \\
&= X \cdot (1 + Y) & \text{(according to T8)} \\
&= X \cdot 1 & \text{(according to T2)} \\
&= X & \text{(according to T1$'$)}
\end{aligned}
$$

Likewise, the other theorems can be used to prove T10, where the key step is to
use T8 to rewrite the left-hand side as X·(Y + Y′).

consensus theorem   Theorem T11 is known as the *consensus theorem*. The Y·Z term is called
consensus   the *consensus* of X·Y and X′·Z. The idea is that if the Y·Z term is 1, then either
X·Y or X′·Z must also be 1, so the Y·Z term is redundant and may be dropped.
The consensus theorem has two important applications. It forms the basis of
the iterative-consensus method of finding prime implicants, as we'll show in
Section 4.4.3. And it can also be used to eliminate certain timing hazards in
combinational logic circuits, as we'll see in Section 4.5.

In all of the theorems, it is possible to replace each variable with an arbitrary
logic expression. A simple replacement is to complement one or more variables:

$$
(X + Y') + Z' = X + (Y' + Z') \qquad \text{(based on T7)}
$$

But more complex expressions may be substituted as well:

$$
(V' + X) \cdot (W \cdot (Y' + Z)) + (V' + X) \cdot (W \cdot (Y' + Z))' = V' + X \qquad \text{(based on T10)}
$$

4.1.4 *n*-Variable Theorems

Several important theorems, listed in Table 4–3, are true for an arbitrary number
of variables, n. Most of these theorems can be proved using a two-step method
finite induction   called *finite induction*—first proving that the theorem is true for $n = 2$ (the *basis*
basis step   *step*) and then proving that if the theorem is true for $n = i$, then it is also true for
induction step $n = i+1$ (the *induction step*). For example, consider the generalized idempotency
theorem T12. For $n = 2$, T12 is equivalent to T3 and is therefore true. If it is
true for a logical sum of i X's, then it is also true for a sum of $i+1$ X's, according
to the following reasoning:

$$
\begin{aligned}
X + X + X + \cdots + X &= X + (X + X + \cdots + X) & (i + 1 \text{ X's on either side}) \\
&= X + (X) & (\text{if T12 is true for } n = i) \\
&= X & (\text{according to T3})
\end{aligned}
$$

Thus, the theorem is true for all finite values of n.

Table 4–3 Switching-algebra theorems with n variables.

(T12)	$X + X + \cdots + X = X$	(Generalized idempotency)
(T12')	$X \cdot X \cdot \cdots \cdot X = X$	
(T13)	$(X_1 \cdot X_2 \cdot \cdots \cdot X_n)' = X_1' + X_2' + \cdots + X_n'$	(DeMorgan's theorems)
(T13')	$(X_1 + X_2 + \cdots + X_n)' = X_1' \cdot X_2' \cdot \cdots \cdot X_n'$	
(T14)	$[F(X_1,X_2,\ldots,X_n,+,\cdot)]' = F(X_1',X_2',\ldots,X_n',\cdot,+)$	(Generalized DeMorgan's theorem)
(T15)	$F(X_1,X_2,\ldots,X_n) = X_1 \cdot F(1,X_2,\ldots,X_n) + X_1' \cdot F(0,X_2,\ldots,X_n)$	(Shannon's expansion theorems)
(T15')	$F(X_1,X_2,\ldots,X_n) = [X_1 + F(0,X_2,\ldots,X_n)] \cdot [X_1' + F(1,X_2,\ldots,X_n)]$	

DeMorgan's theorems (T13 and T13′) are probably the most commonly used of all the theorems of switching algebra. Theorem T13 says that an n-input AND gate whose output is complemented is equivalent to an n-input OR gate whose inputs are complemented. That is, the circuits of Figure 4–3(a) and (b) are equivalent.

DeMorgan's theorems

In Section 3.3.4 we showed how to build a CMOS NAND gate. The output of a NAND gate for any set of inputs is the complement of an AND gate's output for the same inputs, so a NAND gate can have the logic symbol in Figure 4–3(c). However, the CMOS NAND circuit is not designed as an AND gate followed by a transistor inverter (NOT gate); it's just a collection of transistors that happens to perform the AND-NOT function. In fact, theorem T13 tells us that the logic symbol in (d) denotes the same logic function (bubbles on the OR-gate inputs indicate logical inversion). That is, a NAND gate may be viewed as performing a NOT-OR function.

By observing the inputs and output of a NAND gate, it is impossible to determine whether it has been built internally as an AND gate followed by an inverter, as inverters followed by an OR gate, or as a direct CMOS realization, *because all NAND circuits perform precisely the same logic function.* Although

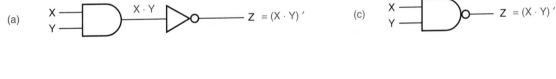

Figure 4–3 Equivalent circuits according to DeMorgan's theorem T13: (a) AND-NOT; (b) NOT-OR; (c) logic symbol for a NAND gate; (d) equivalent symbol for a NAND gate.

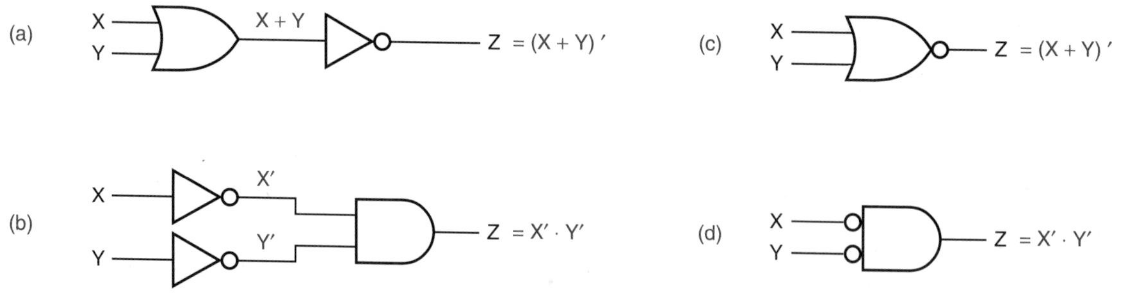

(a)   X, Y → [OR gate] → X + Y → [inverter] → Z = (X + Y)′

(c)   X, Y → [NOR gate] → Z = (X + Y)′

(b)   X → [inverter] → X′, Y → [inverter] → Y′ → [AND gate] → Z = X′ · Y′

(d)   X, Y → [equivalent NOR symbol] → Z = X′ · Y′

Figure 4–4 Equivalent circuits according to DeMorgan's theorem T13′: (a) OR-NOT; (b) NOT-AND; (c) logic symbol for a NOR gate; (d) equivalent symbol for a NOR gate.

the choice of symbol has no bearing on the functionality of a circuit, we'll show in Section 5.1 that the proper choice can make the circuit's function much easier to understand.

A similar symbolic equivalence can be inferred from theorem T13′. As shown in Figure 4–4, a NOR gate may be realized as an OR gate followed by an inverter, or as inverters followed by an AND gate.

generalized DeMorgan's theorem

complement of a logic expression

Theorems T13 and T13′ are just special cases of a *generalized DeMorgan's theorem*, T14, that applies to an arbitrary logic expression F. By definition, the *complement of a logic expression*, denoted (F)′, is an expression whose value is the opposite of F's for every possible input combination. Theorem T14 is very important because it gives us a way to manipulate and simplify the complement of an expression.

Theorem T14 states that, given any n-variable logic expression, its complement can be obtained by swapping $+$ and $\cdot$ and complementing all variables. For example, suppose that we have

$$F(W,X,Y,Z) \;=\; (W'{\cdot}X) + (X{\cdot}Y) + (W{\cdot}(X' + Z'))$$
$$=\; ((W)'{\cdot}X) + (X{\cdot}Y) + (W{\cdot}((X)' + (Z)'))$$

In the second line we have enclosed complemented variables in parentheses to remind you that the ′ is an operator, not part of the variable name. Applying theorem T14, we obtain

THE WORLD'S MOST EXPENSIVE NAND GATE

A synthesis procedure introduced Section 11.1 reinforces the notion that you can't tell anything about a gate's internal realization from its logic symbol. We'll see that any n-input logic function, including an n-input NAND gate, can be realized as a table of 2^n bits in a read-only memory (ROM) where, for each possible input combination, the ROM "looks up" the correct output value. This is a very expensive realization of an n-input NAND gate, but once again either Figure 4–3(c) or (d) is an acceptable logic symbol, independent of the NAND gate's realization.

$$[F(W,X,Y,Z)]' = ((W')' + X') \cdot (X' + Y') \cdot (W' + ((X')' \cdot (Z')'))$$

Using theorem T4, this can be simplified to

$$[F(W,X,Y,Z)]' = (W + X') \cdot (X' + Y') \cdot (W' + (X \cdot Z))$$

In general, we can use theorem T14 to complement a parenthesized expression by swapping + and ·, complementing all uncomplemented variables, and uncomplementing all complemented ones.

    The generalized DeMorgan's theorem T14 can be proved by showing that all logic functions can be written as either a sum or a product of subfunctions, and then applying T13 and T13' recursively. However, a much more enlightening and satisfying proof can be based on the principle of duality, explained next.

4.1.5 Duality

We stated all of the axioms of switching algebra in pairs. The primed version of each axiom (e.g., A5') is obtained from the unprimed version (e.g., A5) by simply swapping 0 and 1 and, if present, · and +. As a result, we can state the following *metatheorem*, a theorem about theorems:

metatheorem

Principle of Duality Any theorem or identity in switching algebra remains true if 0 and 1 are swapped and · and + are swapped throughout.

The metatheorem is true because the duals of all the axioms are true, so duals of all switching-algebra theorems can be proved using duals of the axioms.

    After all, what's in a name, or in a symbol for that matter? If the software that was used to typeset this book had a bug, one that swapped $0 \leftrightarrow 1$ and $\cdot \leftrightarrow +$ throughout this chapter, you still would have learned exactly the same switching algebra; only the nomenclature would have been a little weird, using words like "product" to describe an operation that uses the symbol "+".

    Duality is important because it doubles the usefulness of everything that you learn about switching algebra and manipulation of switching functions. Stated more practically, from a student's point of view, it halves the amount that you have to learn! For example, once you learn how to synthesize two-stage AND-OR logic circuits from sum-of-products expressions, you automatically know a dual technique to synthesize OR-AND circuits from product-of-sums expressions.

    There is just one convention in switching algebra where we did not treat · and + identically, so duality does not necessarily hold true—can you figure out what it is before reading the answer on the next page? Consider the following statement of theorem T9 and its clearly absurd "dual":

$X + X \cdot Y$	=	X	(theorem T9)
$X \cdot X + Y$	=	X	(after applying the principle of duality)
$X + Y$	=	X	(after applying theorem T3')

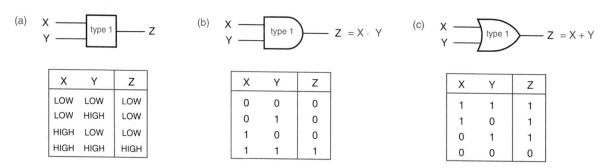

Figure 4–5   A "type-1" logic gate: (a) electrical function table; (b) logic function table and symbol with positive logic; (c) logic function table and symbol with negative logic.

Good point

Obviously the last line above is false—where did we go wrong? The problem is in operator precedence. We were able to write the left-hand side of the first line without parentheses because of our convention that · has precedence. However, once we applied the principle of duality, we should have given precedence to + instead, or written the second line as $X \cdot (X + Y) = X$. The best way to avoid problems like this is to parenthesize an expression fully before taking its dual.

dual of a logic expression

Let us formally define the *dual of a logic expression*. If $F(X_1, X_2, \ldots, X_n, +, \cdot, ')$ is a fully parenthesized logic expression involving the variables $X_1, X_2, \ldots, X_n$ and the operators +, ·, and ', then the dual of F, written F^D, is the same expression with + and · swapped:

$$F^D(X_1, X_2, \ldots, X_n, +, \cdot, ') \;=\; F(X_1, X_2, \ldots, X_n, \cdot, +, ')$$

You already knew this, of course, but we wrote the definition in this way just to highlight the similarity between duality and the generalized DeMorgan's theorem T14, which may now be restated as follows:

$$[F(X_1, X_2, \ldots, X_n)]' \;=\; F^D(X_1', X_2', \ldots, X_n')$$

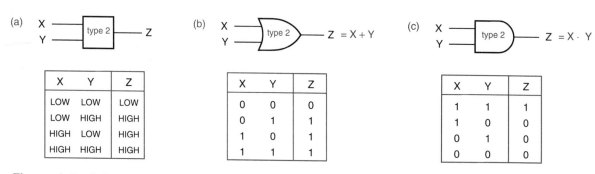

Figure 4–6   A "type-2" logic gate: (a) electrical function table; (b) logic function table and symbol with positive logic; (c) logic function table and symbol with negative logic.

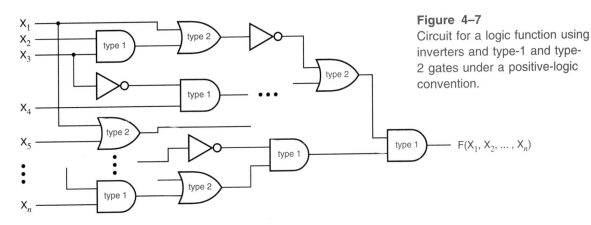

Figure 4–7
Circuit for a logic function using inverters and type-1 and type-2 gates under a positive-logic convention.

Let's examine this statement in terms of a physical network.

Figure 4–5(a) shows the electrical function table for a logic element that we'll simply call a "type-1" gate. Under the positive-logic convention (LOW = 0 and HIGH = 1), this is an AND gate, but under the negative-logic convention (LOW = 1 and HIGH = 0), it is an OR gate, as shown in (b) and (c). We can also imagine a "type-2" gate, shown in Figure 4–6, that is a positive-logic OR or a negative-logic AND. Similar tables can be developed for gates with more than two inputs.

Suppose that we are given an arbitrary logic expression, $F(X_1, X_2, \ldots, X_n)$. Following the positive-logic convention, we can build a circuit corresponding to this expression using inverters for NOT operations, type-1 gates for AND, and type-2 gates for OR, as shown in Figure 4–7. Now suppose that, without changing this circuit, we simply change the logic convention from positive to negative. Then we should redraw the circuit as shown in Figure 4–8. Clearly, for every possible combination of input voltages (HIGH and LOW), the circuit still produces the same output voltage. However, from the point of view of switching

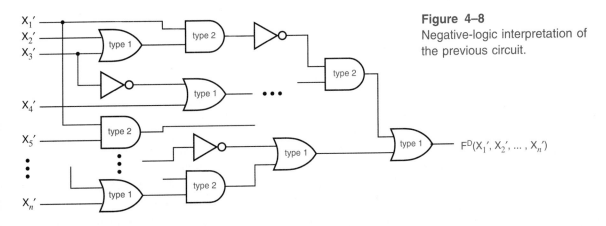

Figure 4–8
Negative-logic interpretation of the previous circuit.

algebra, the output value—0 or 1—is the opposite of what it was under the positive-logic convention. Likewise, each input value is the opposite of what it was. Therefore, for each possible input combination to the circuit in Figure 4–7, the output is the opposite of that produced by the opposite input combination applied to the circuit in Figure 4–8:

$$F(X_1, X_2, \ldots, X_n) \;=\; [F^D(X_1', X_2', \ldots, X_n')]'$$

By complementing both sides, we get the generalized DeMorgan's theorem:

$$[F(X_1, X_2, \ldots, X_n)]' \;=\; F^D(X_1', X_2', \ldots, X_n')$$

Amazing!

So, we have seen that duality is the basis for the generalized DeMorgan's theorem. Going forward, duality will halve the number of methods you must learn to manipulate and simplify logic functions.

MIXED-UP LOGIC

> Some authors exploit duality even further, allowing the logic convention (positive or negative) to be changed at arbitrary points in a circuit, so that AND gates can be used as OR gates, and vice versa. The goal of this system, called *mixed logic*, is to promote more effective use of available gates and more understandable logic diagrams.
>
> Getting correct results with mixed logic requires proper identification of the convention-change points, careful naming of signals, and using inverters as non-logic elements called *oops functions*. In this book, we take advantage of duality in a more customary and less confusing way. We use the positive-logic convention throughout this book, but as you'll see in Section 5.1, we choose signal names and gate symbols in a way that promotes better understanding of circuit functionality.

4.1.6 Standard Representations of Logic Functions

Before moving on to analysis and synthesis of combinational logic functions we'll introduce some necessary nomenclature and notation.

truth table

The most basic representation of a logic function is the *truth table*. Similar in philosophy to the perfect-induction proof method, this brute-force representation simply lists the output of the circuit for every possible input combination. Traditionally, the input combinations are arranged in rows in ascending binary counting order, and the corresponding output values are written in a column next to the rows. The general structure of a 3-variable truth table is shown in Table 4–4. The rows are numbered 0–7 corresponding to the binary input combinations, but this numbering is not an essential part of the truth table. The truth table for a particular 3-variable logic function is shown in Table 4–5. Each distinct pattern of 0s and 1s in the output column yields a different logic function;

Table 4–4
General truth table structure for a
3-variable logic function, F(X,Y,Z).

Row	X	Y	Z	F
0	0	0	0	F(0,0,0)
1	0	0	1	F(0,0,1)
2	0	1	0	F(0,1,0)
3	0	1	1	F(0,1,1)
4	1	0	0	F(1,0,0)
5	1	0	1	F(1,0,1)
6	1	1	0	F(1,1,0)
7	1	1	1	F(1,1,1)

there are 2^8 such patterns. Thus, the logic function in Table 4–5 is one of 2^8 different logic functions of three variables.

The truth table for an *n*-variable logic function has 2^n rows. Obviously, truth tables are practical to write only for logic functions with a small number of variables, say, 10 for students and about 5 for everyone else.

The information contained in a truth table can also be conveyed algebraically. To do so, we first need some definitions:

- A *literal* is a variable or the complement of a variable. Examples: X, Y, X′, Y′. *literal*

- A *product term* is a single literal or a logical product of two or more literals. *product term* Examples: Z′, W·X·Y, X·Y′·Z, W′·Y′·Z.

- A *sum-of-products expression* is a logical sum of product terms. Example: *sum-of-products* Z′ + W·X·Y + X·Y′·Z + W′·Y′·Z. *expression*

- A *sum term* is a single literal or a logical sum of two or more literals. *sum term* Examples: Z′, W + X + Y, X + Y′ + Z, W′ + Y′ + Z.

- A *product-of-sums expression* is a logical product of sum terms. Example: *product-of-sums* Z′·(W + X + Y)·(X + Y′ + Z)·(W′ + Y′ + Z). *expression*

Table 4–5
Truth table for a particular 3-
variable logic function, F(X,Y,Z).

Row	X	Y	Z	F
0	0	0	0	1
1	0	0	1	0
2	0	1	0	0
3	0	1	1	1
4	1	0	0	1
5	1	0	1	0
6	1	1	0	1
7	1	1	1	1

Table 4–6
Minterms and maxterms for a 3-variable logic function, F(X,Y,Z).

Row	X	Y	Z	F	Minterm	Maxterm
0	0	0	0	F(0,0,0)	X'·Y'·Z'	X + Y + Z
1	0	0	1	F(0,0,1)	X'·Y'·Z	X + Y + Z'
2	0	1	0	F(0,1,0)	X'·Y·Z'	X + Y' + Z
3	0	1	1	F(0,1,1)	X'·Y·Z	X + Y' + Z'
4	1	0	0	F(1,0,0)	X·Y'·Z'	X' + Y + Z
5	1	0	1	F(1,0,1)	X·Y'·Z	X' + Y + Z'
6	1	1	0	F(1,1,0)	X·Y·Z'	X' + Y' + Z
7	1	1	1	F(1,1,1)	X·Y·Z	X' + Y' + Z'

normal term

- A *normal term* is a product or sum term in which no variable appears more than once. A nonnormal term can always be simplified to a constant or a normal term using one of theorems T3, T3', T5, or T5'. Examples of nonnormal terms: W·X·X·Y', W + W + X' + Y, X·X'·Y. Examples of normal terms: W·X·Y', W + X' + Y.

minterm

- An n-variable *minterm* is a normal product term with n literals. There are 2^n such product terms. Examples of 4-variable minterms: W'·X·Y'·Z', W·X·Y'·Z, W'·X'·Y·Z'.

maxterm

- An n-variable *maxterm* is a normal sum term with n literals. There are 2^n such sum terms. Examples of 4-variable maxterms: W' + X' + Y' + Z', W + X' + Y' + Z, W' + X' + Y + Z'.

There is a close correspondence between the truth table and minterms and maxterms. A minterm can be defined as a product term that is 1 in exactly one row of the truth table. Similarly, a maxterm can be defined as a sum term that is 0 in exactly one row of the truth table. Table 4–6 shows this correspondence for a 3-variable truth table.

minterm number
minterm i

An n-variable minterm can be represented by an n-bit integer, the *minterm number*. We'll use the name *minterm i* to denote the minterm corresponding to row i of the truth table. In minterm i, a particular variable appears complemented if the corresponding bit in the binary representation of i is 0; otherwise, it is uncomplemented. For example, row 5 has binary representation 101 and the corresponding minterm is X·Y'·Z. As you might expect, the correspondence for

maxterm i

maxterms is just the opposite: in *maxterm i*, a variable appears complemented if the corresponding bit in the binary representation of i is 1. Thus, maxterm 5 (101) is X' + Y + Z'.

Based on the correspondence between the truth table and minterms, we can easily create an algebraic representation of a logic function from its truth table.

canonical sum

The *canonical sum* of a logic function is a sum of the minterms corresponding to truth-table rows (input combinations) for which the function produces a 1 output. For example, the canonical sum for the logic function in Table 4–5 on page 207 is

$$F = \Sigma_{X,Y,Z}(0,3,4,6,7)$$
$$= X'\cdot Y'\cdot Z' + X'\cdot Y\cdot Z + X\cdot Y'\cdot Z' + X\cdot Y\cdot Z' + X\cdot Y\cdot Z$$

Here, the notation $\Sigma_{X,Y,Z}(0,3,4,6,7)$ is a *minterm list* and means "the sum of   *minterm list* minterms 0, 3, 4, 6, and 7 with variables X, Y, and Z." You can visualize that each minterm "turns on" the output for exactly one input combination. Any logic function can be written as a canonical sum.

The *canonical product* of a logic function is a product of the maxterms   *canonical product* corresponding to input combinations for which the function produces a 0 output. For example, the canonical product for the logic function in Table 4–5 is

$$F = \Pi_{X,Y,Z}(1,2,5)$$
$$= (X + Y + Z')\cdot(X + Y' + Z)\cdot(X' + Y + Z')$$

Here, the notation $\Pi_{X,Y,Z}(1,2,5)$ is a *maxterm list* and means "the product of   *maxterm list* maxterms 1, 2, and 5 with variables X, Y, and Z." You can visualize that each maxterm "turns off" the output for exactly one input combination. Any logic function can be written as a canonical product.

It's easy to convert between a minterm list and a maxterm list. For a function of n variables, the possible minterm and maxterm numbers are in the set $\{0, 1, \ldots, 2^n - 1\}$; a minterm or maxterm list contains a subset of these numbers. To switch between list types, take the set complement, for example,

$$\Sigma_{A,B,C}(0,1,2,3) = \Pi_{A,B,C}(4,5,6,7)$$
$$\Sigma_{X,Y}(1) = \Pi_{X,Y}(0,2,3)$$
$$\Sigma_{W,X,Y,Z}(0,1,2,3,5,7,11,13) = \Pi_{W,X,Y,Z}(4,6,8,9,10,12,14,15)$$

[handwritten: Compliment every thing including the set list]

We have now learned five possible representations for a combinational logic function:

(1) A truth table.

(2) An algebraic sum of minterms, the canonical sum.

(3) A minterm list using the Σ notation.

(4) An algebraic product of maxterms, the canonical product.

(5) A maxterm list using the Π notation.

Each one of these representations specifies exactly the same information; given any one of them, we can derive the other four using a simple mechanical process.

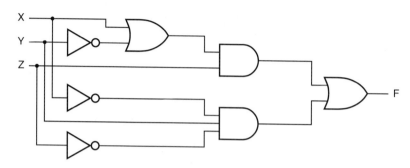

Figure 4–9
A three-input, one-output logic circuit.

4.2   COMBINATIONAL CIRCUIT ANALYSIS

We analyze a combinational logic circuit by obtaining a formal description of its logic function. Once we have a description of the logic function, a number of other operations are possible:

- We can determine the behavior of the circuit for various input combinations.

- We can manipulate an algebraic description to suggest different circuit structures for the logic function.

- We can transform an algebraic description into a standard form corresponding to an available circuit structure. For example, a sum-of-products expression corresponds directly to a circuit structure called a "programmable logic array."

- We can use an algebraic description of the circuit's functional behavior in the analysis of a larger system that includes the circuit.

Given a logic diagram for a combinational circuit, such as Figure 4–9, there are a number of ways to obtain a formal description of the circuit's function. The most primitive functional description is the truth table.

Using only the basic axioms of switching algebra, we can obtain the truth table of an n-input circuit by working our way through all 2^n input combinations. For each input combination, we determine all of the gate outputs produced by

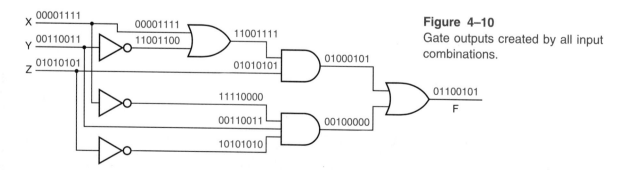

Figure 4–10
Gate outputs created by all input combinations.

Table 4–7
Truth table for the logic circuit of
Figure 4–9.

Row	X	Y	Z	F
0	0	0	0	0
1	0	0	1	1
2	0	1	0	1
3	0	1	1	0
4	1	0	0	0
5	1	0	1	1
6	1	1	0	0
7	1	1	1	1

that input, propagating information from the circuit inputs to the circuit outputs. Figure 4–10 applies this "exhaustive" technique to our example circuit. Written on each signal line in the circuit is a sequence of eight logic values, the values present on that line when the circuit inputs XYZ are 000, 001, ..., 111. The truth table can be written by transcribing the output sequence of the final OR gate, as shown in Table 4–7. Once we have the truth table for the circuit, we can also directly write a logic expression—the canonical sum or product—if we wish.

The number of input combinations of a logic circuit grows exponentially with the number of inputs, so the exhaustive approach can quickly become exhausting. Instead, we normally use an algebraic approach whose complexity is more linearly proportional to the size of the circuit. The method is simple—we build up a parenthesized logic expression corresponding to the logic operators and structure of the circuit. We start at the circuit inputs and propagate expressions through gates toward the output. Using the theorems of switching algebra, we may simplify the expressions as we go, or we may defer all algebraic manipulations until an output expression is obtained.

Figure 4–11 applies the algebraic technique to our example circuit. The output function is given on the output of the final OR gate:

$$F = ((X + Y') \cdot Z) + (X' \cdot Y \cdot Z')$$

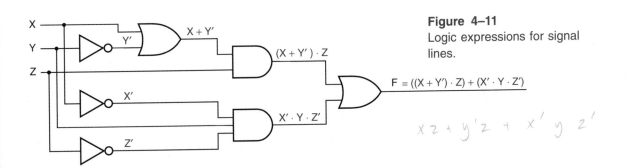

Figure 4–11
Logic expressions for signal lines.

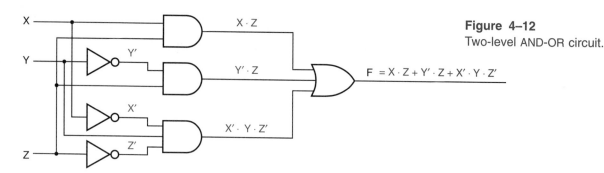

Figure 4–12
Two-level AND-OR circuit.

No switching-algebra theorems were used in obtaining this expression. However, we can use theorems to transform this expression into another form. For example, a sum of products can be obtained by "multiplying out":

$$F = X \cdot Z + Y' \cdot Z + X' \cdot Y \cdot Z'$$

The new expression corresponds to a different circuit for the same logic function, as shown in Figure 4–12.

Similarly, we can "add out" the original expression to obtain a product of sums:

$$
\begin{aligned}
F &= ((X + Y') \cdot Z) + (X' \cdot Y \cdot Z') \\
&= (X + Y' + X') \cdot (X + Y' + Y) \cdot (X + Y' + Z') \cdot (Z + X') \cdot (Z + Y) \cdot (Z + Z') \\
&= 1 \cdot 1 \cdot (X + Y' + Z') \cdot (X' + Z) \cdot (Y + Z) \cdot 1 \\
&= (X + Y' + Z') \cdot (X' + Z) \cdot (Y + Z)
\end{aligned}
$$

The corresponding logic circuit is shown in Figure 4–13.

Our next example of algebraic analysis uses a circuit with NAND and NOR gates, shown in Figure 4–14. This analysis is a little messier than the previous example, because each gate produces a complemented subexpression, not just

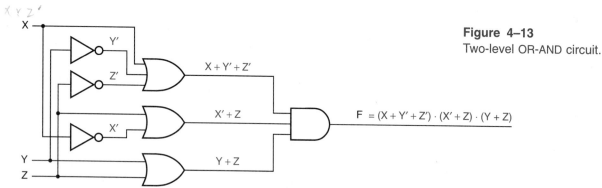

Figure 4–13
Two-level OR-AND circuit.

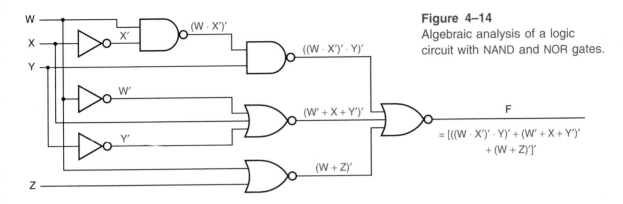

Figure 4–14
Algebraic analysis of a logic
circuit with NAND and NOR gates.

a simple sum or product. However, the output expression can be simplified by
repeated application of the generalized DeMorgan's theorem:

$$
\begin{aligned}
F &= [((W \cdot X')' \cdot Y)' + (W' + X + Y')' + (W + Z)']' \\
  &= ((W' + X)' + Y')' \cdot (W \cdot X' \cdot Y)' \cdot (W' \cdot Z')' \\
  &= ((W \cdot X')' \cdot Y) \cdot (W' + X + Y') \cdot (W + Z) \\
  &= ((W' + X) \cdot Y) \cdot (W' + X + Y') \cdot (W + Z)
\end{aligned}
$$

Quite often, DeMorgan's theorem can be applied *graphically* to simplify
algebraic analysis. Recall from Figures 4–3 and 4–4 that NAND and NOR gates
each have two equivalent symbols. By judiciously redrawing Figure 4–14, we
make it possible to cancel out some of the inversions during the analysis by
using theorem T4 [$(X')' = X$], as shown in Figure 4–15. This manipulation leads
us to a simplified output expression directly:

$$
F = ((W' + X) \cdot Y) \cdot (W' + X + Y') \cdot (W + Z)
$$

Figures 4–14 and 4–15 were just two different ways of drawing the same
physical logic circuit. However, when we simplify a logic expression using the

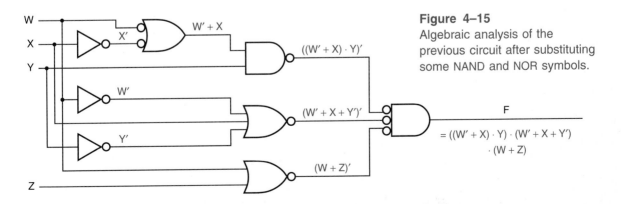

Figure 4–15
Algebraic analysis of the
previous circuit after substituting
some NAND and NOR symbols.

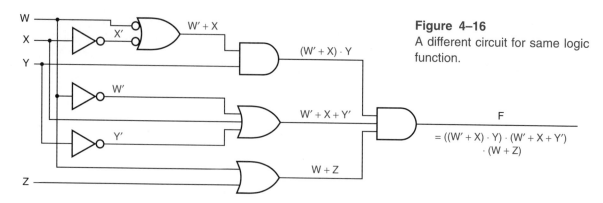

Figure 4–16
A different circuit for same logic function.

theorems of switching algebra, we get an expression corresponding to a different physical circuit. For example, the simplified expression above corresponds to the circuit of Figure 4–16, which is physically different from the one in the previous two figures. Furthermore, we could multiply out and add out the expression to obtain sum-of-products and product-of-sums expressions corresponding to two more physically different circuits for the same logic function.

Although we used logic expressions above to convey information about the physical structure of a circuit, we don't always do this. For example, we might use the expression $G(W,X,Y,Z) = W \cdot X \cdot Y + Y \cdot Z$ to describe any one of the circuits in Figure 4–17. Normally, the only sure way to determine a circuit's structure is to look at its schematic drawing. However, for certain restricted classes of circuits, structural information can be inferred from logic expressions. For example, the circuit in (a) could be described without reference to the drawing as "the two-level AND-OR circuit for $W \cdot X \cdot Y + Y \cdot Z$," while the circuit in (b) could be described as "the two-level NAND-NAND circuit for $W \cdot X \cdot Y + Y \cdot Z$."

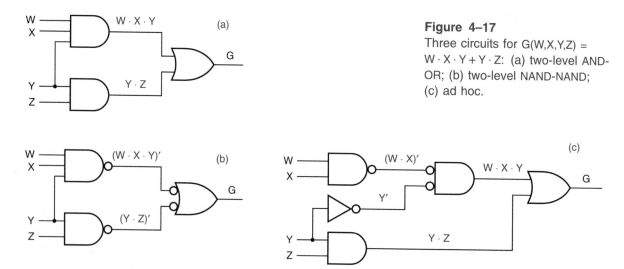

Figure 4–17
Three circuits for $G(W,X,Y,Z) = W \cdot X \cdot Y + Y \cdot Z$: (a) two-level AND-OR; (b) two-level NAND-NAND; (c) ad hoc.

4.3 COMBINATIONAL CIRCUIT SYNTHESIS

4.3.1 Circuit Descriptions and Designs

What is the starting point for designing combinational logic circuits? Usually, we are given a word description of a problem or we develop one ourselves. Occasionally, the description is a list of input combinations for which a signal should be on or off, the verbal equivalent of a truth table or the Σ or Π notation introduced in the preceding section. For example, the description of a 4-bit prime-number detector might be, "Given a 4-bit input combination $N = N_3N_2N_1N_0$, this function produces a 1 output for $N = 1,2,3,5,7,11$, and 13, and 0 otherwise." A logic function described in this way can be designed directly from the canonical sum or product expression. For the prime-number detector, we have

$$
\begin{aligned}
F &= \Sigma_{N_3,N_2,N_1,N_0}(1,2,3,5,7,11,13) \\
&= N_3'N_2'N_1'N_0 + N_3'N_2'N_1N_0' + N_3'N_2'N_1N_0 + N_3'N_2N_1'N_0 \\
&\quad + N_3'N_2N_1N_0 + N_3N_2'N_1N_0 + N_3N_2N_1'N_0
\end{aligned}
$$

The corresponding circuit is shown in Figure 4–18.

More often, we describe a logic function using the English-language connectives "and," "or," and "not." For example, we might describe an alarm circuit by saying, "The ALARM output is 1 if the PANIC input is 1, or if the ENABLE input is 1, the EXITING input is 0, and the house is not secure; the house is secure

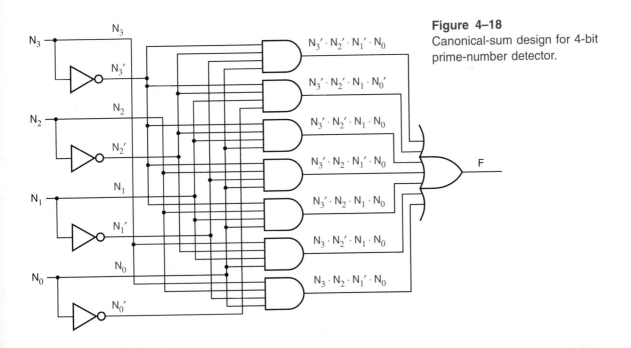

Figure 4–18
Canonical-sum design for 4-bit prime-number detector.

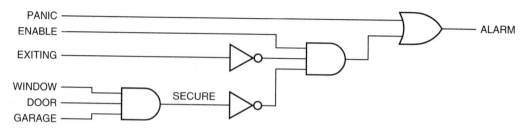

Figure 4–19   Alarm circuit derived from logic expression.

if the WINDOW, DOOR, and GARAGE inputs are all 1." Such a description can be translated directly into algebraic expressions:

$$ALARM = PANIC + ENABLE \cdot EXITING' \cdot SECURE'$$
$$SECURE = WINDOW \cdot DOOR \cdot GARAGE$$
$$ALARM = PANIC + ENABLE \cdot EXITING' \cdot (WINDOW \cdot DOOR \cdot GARAGE)'$$

Notice that we used the same method in switching algebra as in ordinary algebra to formulate a complicated expression—we defined an auxiliary variable SECURE to simplify the first equation, developed an expression for SECURE, and used substitution to arrive at the final expression. We can easily draw a circuit using AND, OR, and NOT gates that realizes the final expression, as shown in Figure 4–19. (We say that a circuit *realizes* ["makes real"] an expression if its output function equals that expression, and the circuit is called a *realization* of the function.)

realize
realization

Once we have an expression, any expression, for a logic function, we can do other things besides building a circuit directly from the expression. We can manipulate the expression to get different circuits. For example, the ALARM expression above can be multiplied out to get the sum-of-products circuit in Figure 4–20. Or, if the number of variables is not too large, we can construct the truth table for the expression and use any of the synthesis methods that apply to truth tables, including the canonical sum or product method described earlier and the minimization methods described later.

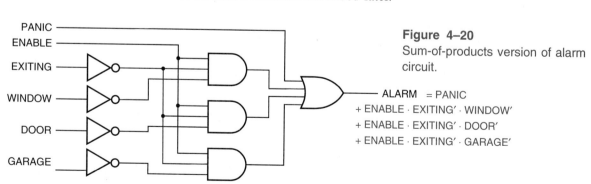

Figure 4–20
Sum-of-products version of alarm circuit.

$ALARM = PANIC$
$+ ENABLE \cdot EXITING' \cdot WINDOW'$
$+ ENABLE \cdot EXITING' \cdot DOOR'$
$+ ENABLE \cdot EXITING' \cdot GARAGE'$

In general, it's easier to describe a circuit in words using logical connectives and to write the corresponding logic expressions than it is to write a complete truth table, especially if the number of variables is large. However, sometimes we have to work with imprecise word descriptions of logic functions, for example, "The ERROR output should be 1 if the GEARUP, GEARDOWN, and GEARCHECK inputs are inconsistent." In this situation, the truth-table approach is best because it allows us to determine the output required for every input combination, based on our knowledge and understanding of the problem environment (e.g., the brakes cannot be applied unless the gear is down).

4.3.2 Circuit Manipulations

The design methods that we've described so far use AND, OR, and NOT gates. We might like to use NAND and NOR gates, too—they're faster than ANDs and ORs in most technologies. However, most people don't develop logical propositions in terms of NAND and NOR connectives. That is, you probably wouldn't say, "I won't go out with you if you're not clean or not wealthy and also you're not smart or not friendly." It would be more natural for you to say, "I'll go out with you if you're clean and wealthy, or if you're smart and friendly." So, given a "natural" logic expression, we need ways to translate it into other forms.

We can translate any logic expression into an equivalent sum-of-products expression, simply by multiplying it out. As shown in Figure 4–21, such an expression may be realized directly with AND and OR gates. The inverters required for complemented inputs are not shown.

Figure 4–21
Alternative sum-of-products
realizations: (a) AND-OR;
(b) AND-OR with extra inverter
pairs; (c) NAND-NAND.

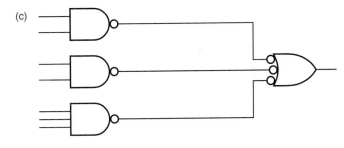

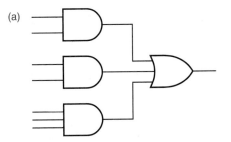

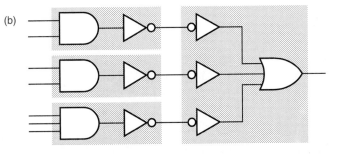

As shown in Figure 4–21(b), we may insert a pair of inverters between each AND-gate output and the corresponding OR-gate input in a two-level AND-OR circuit. According to theorem T4, these inverters have no effect on the output function of the circuit. In fact, we've drawn the second inverter of each pair with its inversion bubble on its input to provide a graphical reminder that the inverters cancel. However, if these inverters are absorbed into the AND and OR gates, we wind up with AND-NOT gates at the first level and a NOT-OR gate at the second level. These are just two different symbols for the same type of gate—a NAND gate. Thus, a two-level *AND-OR circuit* may be converted to a two-level *NAND-NAND circuit* simply by substituting gates.

AND-OR circuit
NAND-NAND circuit

If any of the product terms in the sum-of-products expression contain just a single literal, then we may gain or lose inverters in the transformation from AND-OR to NAND-NAND. For example, Figure 4–22 is an example where an inverter on the W input is no longer needed, but an inverter must be added to the Z input.

We have shown that any sum-of-products expression can be realized in either of two ways—as an AND-OR circuit or as a NAND-NAND circuit. The dual of this statement is also true: any product-of-sums expression can be realized as an *OR-AND circuit* or as a *NOR-NOR circuit*. Figure 4–23 shows an example. Any logic expression can be translated into an equivalent product-of-sums expression by adding it out, and hence has both OR-AND and NOR-NOR circuit realizations.

OR-AND circuit
NOR-NOR circuit

The same kind of manipulations can be applied to arbitrary logic circuits. For example, Figure 4–24(a) shows a circuit built from AND and OR gates. After adding pairs of inverters, we obtain the circuit in (b). However, one of the gates, a 2-input AND gate with a single inverted input, is not a standard type. We can use a discrete inverter as shown in (c) to obtain a circuit that uses only standard gate types—NAND, AND, and inverters. Actually, a better way to use

Figure 4–22
Another two-level sum-of-products circuit: (a) AND-OR; (b) AND-OR with extra inverter pairs; (c) NAND-NAND.

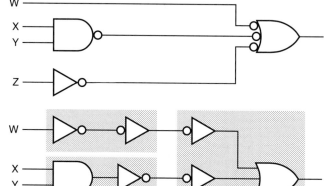
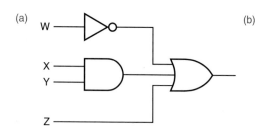

the inverter is shown in (d); one level of gate delay is eliminated, and the bottom
gate becomes a NOR instead of AND. In most logic technologies, inverting gates
like NAND and NOR are faster than noninverting gates like AND and OR.

Figure 4–23
Realizations of a product-of-sums
expression: (a) OR-AND; (b) OR-
AND with extra inverter pairs;
(c) NOR-NOR.

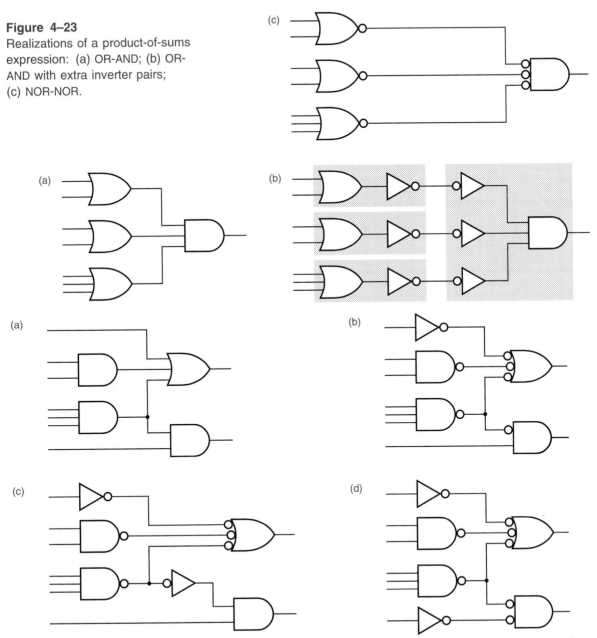

Figure 4–24   Logic-symbol manipulations: (a) original circuit; (b) transformation with a nonstandard
gate; (c) inverter used to eliminate nonstandard gate; (d) preferred inverter placement.

4.3.3 Combinational Circuit Minimization

It's often uneconomical to build a logic circuit directly from the first logic expression that pops into your head. Canonical sum and product expressions are especially expensive because the number of possible minterms or maxterms (and hence gates) grows exponentially with the number of variables. We *minimize* a combinational circuit by reducing the number and size of gates used to build it.

minimize

The traditional combinational circuit minimization methods that we'll study have as their starting point a truth table or, equivalently, a minterm list or maxterm list. If we are given a logic function that is not expressed in this form, then we must convert it to an appropriate form before using these methods. For example, if we are given an arbitrary logic expression, then we can evaluate it for every input combination to construct the truth table. The minimization methods reduce the cost of a two-level AND-OR, OR-AND, NAND-NAND, or NOR-NOR circuit in three ways:

(1) By minimizing the number of first-level gates.

(2) By minimizing the number of inputs on each first-level gate.

(3) By minimizing the number of inputs on the second-level gate. This is actually a side effect of the first reduction.

However, the minimization methods do not consider the cost of input inverters; they assume that both true and complemented versions of all input variables are available. While this is not the case in SSI-based design, it's quite appropriate for modern circuits built with programmable logic devices; such devices have both true and complemented versions of all input variables available "for free."

Most minimization methods are based on a generalization of the combining theorems, T10 and T10':

$$\text{given product term} \cdot Y + \text{given product term} \cdot Y' \;=\; \text{given product term}$$
$$(\text{given sum term} + Y) \cdot (\text{given sum term} + Y') \;=\; \text{given sum term}$$

That is, if two product or sum terms differ only in the complementing or not of one variable, we can combine them into a single term with one less variable. So we save one gate and the remaining gate has one fewer input.

We can apply this algebraic method repeatedly to combine minterms 1, 3, 5, and 7 of the prime-number detector shown in Figure 4–18 on page 215:

$$
\begin{aligned}
F \;&=\; \Sigma_{N_3,N_2,N_1,N_0}(1,3,5,7,2,11,13) \\
&=\; N_3'N_2'N_1'N_0 + N_3'N_2'N_1N_0 + N_3'N_2N_1'N_0 + N_3'N_2N_1N_0 + \ldots \\
&=\; (N_3'N_2'N_1'N_0 + N_3'N_2'N_1N_0) + (N_3'N_2N_1'N_0 + N_3'N_2N_1N_0) + \ldots \\
&=\; N_3'N_2'N_0 + N_3'N_2N_0 + \ldots \\
&=\; N_3'N_0 + \ldots
\end{aligned}
$$

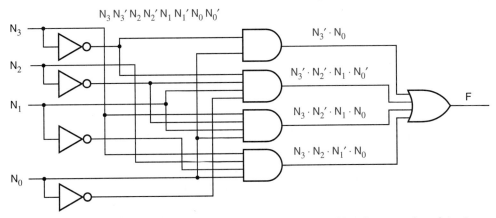

$N_3 N_3' N_2 N_2' N_1 N_1' N_0 N_0'$

$N_3' \cdot N_0$

$N_3' \cdot N_2' \cdot N_1 \cdot N_0'$

$N_3 \cdot N_2' \cdot N_1 \cdot N_0$

$N_3 \cdot N_2 \cdot N_1' \cdot N_0$

Figure 4–25 Simplified sum-of-products realization for 4-bit prime-number detector.

The resulting circuit is shown in Figure 4–25; it has three fewer gates and one of the remaining gates has only two inputs.

   If we had worked a little harder on the preceding expression, we could have saved a couple more first-level gate inputs, though not any gates. It's difficult to find terms that can be combined in a jumble of algebraic symbols. In the next subsection, we'll begin to explore a minimization method that is more fit for human consumption. Our starting point will be the graphical equivalent of a truth table.

4.3.4 Karnaugh Maps

A *Karnaugh map* is a graphical representation of a logic function's truth table.   *Karnaugh map*
Figure 4–26 shows Karnaugh maps for logic functions of 2, 3, and 4 variables.
The map for an *n*-input logic function is an array with 2^n cells, one for each possible input combination or minterm.

   The rows and columns of a Karnaugh map are labeled so that the input

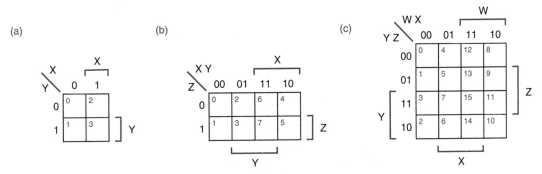

Figure 4–26 Karnaugh maps: (a) 2-variable; (b) 3-variable; (c) 4-variable.

combination for any cell is easily determined from the row and column headings for that cell. The small number inside each cell is the corresponding minterm number in the truth table, assuming that the truth table inputs are labeled alphabetically from left to right (e.g., X, Y, Z) and the rows are numbered in binary counting order, like all the examples in this text. For example, cell 13 in the 4-variable map corresponds to the truth-table row in which W X Y Z = 1101.

When we draw the Karnaugh map for a given function, each cell of the map contains the information from the like-numbered row of the function's truth table—a 0 if the function is 0 for that input combination, a 1 otherwise.

In this text, we use two redundant labelings for map rows and columns. For example, consider the 4-variable map in Figure 4–26(c). The columns are labeled with the four possible combinations of W and X, W X = 00, 01, 11, and 10. Similarly, the rows are labeled with the Y Z combinations. These labels give us all the information we need. However, we also use brackets to associate four regions of the map with the four variables. Each bracketed region is the part of the map in which the indicated variable is 1. Obviously, the brackets convey the same information that is given by the row and column labels.

When we draw a map by hand, it is much easier to draw the brackets than to write out all of the labels. So feel free to do so in your homework assignments. However, we retain the labels in the text's Karnaugh maps as an additional aid to understanding. In any case, you must be sure to label the rows and columns in the proper order to preserve the correspondence between map cells and truth table row numbers shown in Figure 4–26.

To represent a logic function on a Karnaugh map, we simply copy 1s and 0s from the truth table or equivalent to the corresponding cells of the map. Figures 4–27(a) and (b) show the truth table and Karnaugh map for a logic function that we analyzed (beat to death?) in Section 4.2. From now on, we'll reduce the clutter in maps by copying only the 1s or the 0s, not both.

4.3.5   Minimizing Sums of Products

By now you must be wondering about the "strange" ordering of the row and column numbers in a Karnaugh map. There is a very important reason for this ordering—each cell corresponds to an input combination that differs from each of its immediately adjacent neighbors in only one variable. For example, cells 5 and 13 in the 4-variable map differ only in the value of W. In the 3- and 4-variable maps, corresponding cells on the left/right or top/bottom borders are less obvious neighbors; for example, cells 12 and 14 in the 4-variable map are adjacent because they differ only in the value of Y.

Each input combination with a "1" in the truth table corresponds to a minterm in the logic function's canonical sum. Since pairs of adjacent "1" cells in the Karnaugh map have minterms that differ in only one variable, the minterm pairs can be combined into a single product term using the generalization of

Figure 4–27
$F = \Sigma_{X,Y,Z}(1,2,5,7)$: (a) truth
table; (b) Karnaugh map;
(c) combining adjacent 1-cells;
(d) AND-OR circuit.

X	Y	Z	F
0	0	0	0
0	0	1	1
0	1	0	1
0	1	1	0
1	0	0	0
1	0	1	1
1	1	0	0
1	1	1	1

(a)

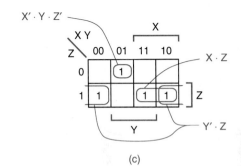

(b)

(c)

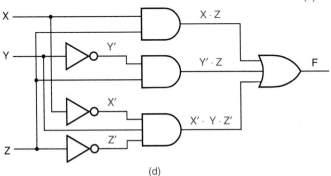

(d)

theorem T10, $\text{term} \cdot Y + \text{term} \cdot Y' = \text{term}$. Thus, we can use a Karnaugh map to simplify the canonical sum of a logic function.

For example, consider cells 5 and 7 in Figure 4–27(b), and their contribution to the canonical sum for this function:

$$
\begin{aligned}
F &= \dots + X \cdot Y' \cdot Z + X \cdot Y \cdot Z \\
&= \dots + (X \cdot Z) \cdot Y' + (X \cdot Z) \cdot Y \\
&= \dots + X \cdot Z
\end{aligned}
$$

Remembering wraparound, we see that cells 1 and 5 in Figure 4–27(b) are also adjacent and can be combined:

$$
\begin{aligned}
F &= X' \cdot Y' \cdot Z + X \cdot Y' \cdot Z + \dots \\
&= X' \cdot (Y' \cdot Z) + X \cdot (Y' \cdot Z) + \dots \\
&= Y' \cdot Z + \dots
\end{aligned}
$$

In general, we can simplify a logic function by combining pairs of adjacent 1-cells (minterms) whenever possible, and writing a sum of product terms that cover all of the 1-cells. Figure 4–27(c) shows the result for our example logic function. We circle a pair of 1s to indicate that the corresponding minterms are combined into a single product term. The corresponding AND/OR circuit is shown in (d).

In many logic functions, the cell-combining procedure can be extended to combine more than two 1-cells into a single product term. For example, consider the canonical sum for the logic function $F = \Sigma_{X,Y,Z}(0,1,4,5,6)$. We can use the algebraic manipulations of the previous examples iteratively to combine four of the five minterms:

$$
\begin{aligned}
F &= X'{\cdot}Y'{\cdot}Z' + X'{\cdot}Y'{\cdot}Z + X{\cdot}Y'{\cdot}Z' + X{\cdot}Y'{\cdot}Z + X{\cdot}Y{\cdot}Z' \\
&= [(X'{\cdot}Y'){\cdot}Z' + (X'{\cdot}Y'){\cdot}Z] + [(X{\cdot}Y'){\cdot}Z' + (X{\cdot}Y'){\cdot}Z] + X{\cdot}Y{\cdot}Z' \\
&= X'{\cdot}Y' + X{\cdot}Y' + X{\cdot}Y{\cdot}Z' \\
&= [X'{\cdot}(Y') + X{\cdot}(Y')] + X{\cdot}Y{\cdot}Z' \\
&= Y' + X{\cdot}Y{\cdot}Z'
\end{aligned}
$$

In general, 2^i 1-cells may be combined to form a product term containing $n - i$ literals, where n is the number of variables in the function.

A precise mathematical rule determines how 1-cells may be combined and the form of the corresponding product term:

A set of 2^i 1-cells may be combined if there are i variables of the logic function that take on all 2^i possible combinations within that set, while the remaining $n - i$ variables have the same value throughout that set. The corresponding product term has $n - i$ literals, where a variable is complemented if it appears as 0 in all of the 1-cells, and uncomplemented if it appears as 1.

rectangular sets of 1s

Graphically, this rule means that we can circle *rectangular* sets of 2^n 1s, literally as well as figuratively stretching the definition of rectangular to account for wraparound at the edges of the map. We can determine the literals of the corresponding product terms directly from the map; for each variable we make the following determination:

- If a circle covers only areas of the map where the variable is 0, then the variable is complemented in the product term.

- If a circle covers only areas of the map where the variable is 1, then the variable is uncomplemented in the product term.

- If a circle covers both areas of the map where the variable is 0 and areas where it is 1, then the variable does not appear in the product term.

A sum-of-products expression for a function must contain product terms (circled sets of 1-cells) that cover all of the 1s and none of the 0s on the map.

The Karnaugh map for our most recent example, $F = \Sigma_{X,Y,Z}(0,1,4,5,6)$, is shown in Figure 4–28(a) and (b). [Cell numbers are shown in (a) to show you

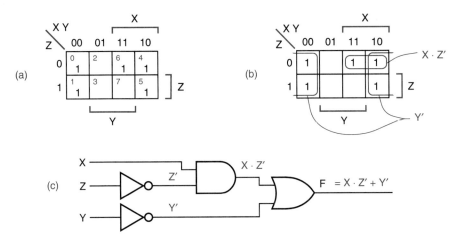

Figure 4–28 $F = \Sigma_{X,Y,Z}(0,1,4,5,6)$: (a) initial Karnaugh map; (b) Karnaugh map with circled product terms; (c) AND/OR circuit.

the correspondence between cell positions and minterm numbers, while they are omitted in (b) to reduce clutter.] We have circled one set of four 1s, corresponding to the product term Y', and a set of two 1s corresponding to the product term $X \cdot Z'$. Notice that the second product term has one less literal than the corresponding product term in our algebraic solution ($X \cdot Y \cdot Z'$). By circling the largest possible set of 1s containing cell 6, we have found a less expensive realization of the logic function, since a 2-input AND gate should cost less than a 3-input one. The fact that two different product terms now cover the same 1-cell (4) does not affect the logic function, since for logical addition $1 + 1 = 1$, not 2! The corresponding two-level AND/OR circuit is shown in (c).

A Karnaugh map and minimized circuit for our prime-number detector are shown in Figure 4–29 on the next page. As mentioned earlier, some of the gates in the minimal solution have fewer inputs than we (or at least I) could come up with in the algebraic approach.

At this point, we need some more definitions to clarify what we're doing:

- A *minimal sum* of a logic function $F(X_1,\ldots,X_n)$ is a sum-of-products ex-   *minimal sum*
  pression for F such that no sum-of-products expression for F has fewer
  product terms, and any sum-of-products expression with the same number
  of product terms has at least as many literals.

That is, the minimal sum has the fewest possible product terms (first-level gates and second-level gate inputs) and, within that constraint, the fewest possible literals (first-level gate inputs). Thus, among our three prime-number detector circuits, only the one in Figure 4–29 on the next page realizes a minimal sum.

(a)

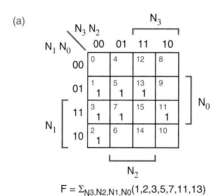

$$F = \Sigma_{N3,N2,N1,N0}(1,2,3,5,7,11,13)$$

(b)

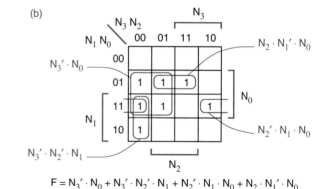

$$F = N_3' \cdot N_0 + N_3' \cdot N_2' \cdot N_1 + N_2' \cdot N_1 \cdot N_0 + N_2 \cdot N_1' \cdot N_0$$

(c)

Figure 4–29   Prime-number detector: (a) initial Karnaugh map;   (b) circled product terms; (c) minimized circuit.

The next definition says precisely what the word "imply" means when you're talking about logic functions:

imply
- A logic function $P(X_1,\ldots,X_n)$ *implies* a logic function $F(X_1,\ldots,X_n)$ if for every input combination such that $P = 1$, then $F = 1$ also.

That is, if P implies F, then F is 1 for every input combination that P is 1, and maybe some more. We may write the shorthand $P \Rightarrow F$. We may also say that "F *includes* P," or that "F *covers* P."

includes
covers
prime implicant
- A *prime implicant* of a logic function $F(X_1,\ldots,X_n)$ is a normal product term $P(X_1,\ldots,X_n)$ that implies F, such that if any variable is removed from P, then the resulting product term does not imply F.

In terms of a Karnaugh map, a prime implicant of F is a circled set of 1-cells satisfying our combining rule, such that if we try to make it larger (covering twice as many cells), it covers one or more 0s.

Now comes the most important part, a theorem that limits how much work we must do to find a minimal sum for a logic function:

Prime-Implicant Theorem A minimal sum is a sum of prime implicants.

That is, to find a minimal sum, we need not consider any product terms that are not prime implicants. This theorem is easily proved by contradiction. Suppose that a product term P in a "minimal" sum is *not* a prime implicant. Then according to the definition of prime implicant, if P is not one, it is possible to remove some literal from P to obtain a new product term P* that still implies F. If we replace P with P* in the "minimal" sum, the resulting sum still equals F but has one fewer literal. Therefore, the presumed "minimal" sum was not minimal after all.

Another minimization example, this time a 4-variable function, is shown in Figure 4–30. There are just two prime implicants, and it's quite obvious that both of them must be included in the minimal sum in order to cover all of the 1-cells on the map. We didn't draw the logic diagram for this example because you should know how to do that yourself by now.

The sum of all the prime implicants of a logic function is called the *complete sum*. Although the complete sum is always a legitimate way to realize a logic function, it's not always minimal. For example, consider the logic function shown in Figure 4–31. It has five prime implicants, but the minimal sum includes only three of them. So, how can we systematically determine which prime implicants to include and which to leave out? Two more definitions are needed:

complete sum ✓

- A *distinguished 1-cell* of a logic function is an input combination that is covered by only one prime implicant.

distinguished 1-cell ✓

- An *essential prime implicant* of a logic function is a prime implicant that covers one or more distinguished 1-cells.

essential prime implicant ✓

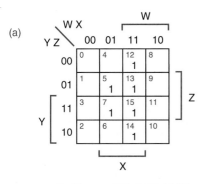

F = Σ_{W,X,Y,Z}(5,7,12,13,14,15)

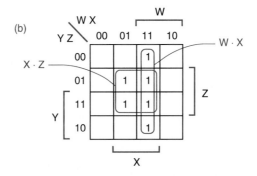

F = X · Z + W · X

Figure 4–30 $F = \Sigma_{W,X,Y,Z}(5,7,12,13,14,15)$: (a) Karnaugh map; (b) prime implicants.

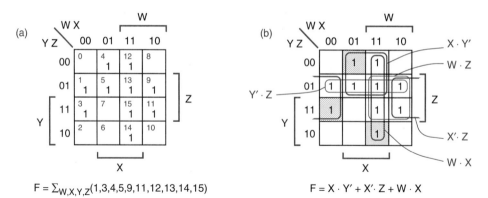

Figure 4–31 $F = \Sigma_{W,X,Y,Z}(1,3,4,5,9,11,12,13,14,15)$: (a) Karnaugh map; (b) prime implicants and distinguished 1-cells.

Since an essential prime implicant is the *only* prime implicant that covers some 1-cell, it *must* be included in every minimal sum for the logic function. So, the first step in the prime implicant selection process is simple—we identify distinguished 1-cells and the corresponding prime implicants, and include the essential prime implicants in the minimal sum. Then we need only determine how to cover the 1-cells, if any, that are not covered by the essential prime implicants. In the example of Figure 4–31, the three distinguished 1-cells are shaded, and the corresponding essential prime implicants are circled with heavier lines. All of the 1-cells in this example are covered by essential prime implicants, so we need go no further. Likewise, Figure 4–32 shows an example where all of the prime implicants are essential, and so all are included in the minimal sum.

A logic function in which not all the 1-cells are covered by essential prime implicants is shown in Figure 4–33. By removing the essential prime implicants and the 1-cells they cover, we obtain a reduced map with only a single 1-cell

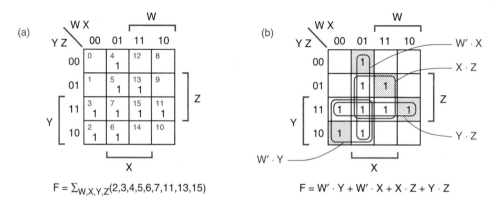

Figure 4–32 $F = \Sigma_{W,X,Y,Z}(2,3,4,5,6,7,11,13,15)$: (a) Karnaugh map; (b) prime implicants and distinguished 1-cells.

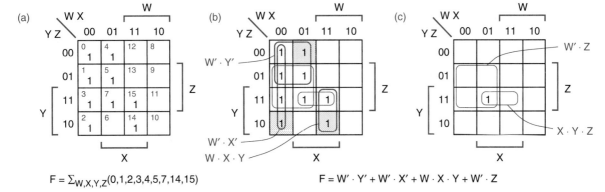

$$F = \Sigma_{W,X,Y,Z}(0,1,2,3,4,5,7,14,15)$$

$$F = W' \cdot Y' + W' \cdot X' + W \cdot X \cdot Y + W' \cdot Z$$

Figure 4–33 $F = \Sigma_{W,X,Y,Z}(0,1,2,3,4,5,7,14,15)$: (a) Karnaugh map; (b) prime implicants and distinguished 1-cells; (c) reduced map after removal of essential prime implicants and covered 1-cells.

and two prime implicants that cover it. The choice in this case is simple—we use the $W' \cdot Z$ product term because it has fewer inputs and therefore lower cost.

For more complex cases, we need yet another definition:

- Given two prime implicants P and Q in a reduced map, P is said to *eclipse*   *eclipse*
  Q (written P $\supseteq$ Q) if P covers at least all the 1-cells covered by Q.

If P costs no more than Q and eclipses Q, then removing Q from consideration cannot prevent us from finding a minimal sum; that is, P is at least as good as Q.

An example of eclipsing is shown in Figure 4–34. After removing essential prime implicants, we are left with two 1-cells, each of which is covered by two prime implicants. However, $X \cdot Y \cdot Z$ eclipses the other two prime implicants, which therefore may be removed from consideration. The two 1-cells are then   *secondary essential* covered only by $X \cdot Y \cdot Z$, which is a *secondary essential prime implicant* that must   *prime implicant* be included in the minimal sum.

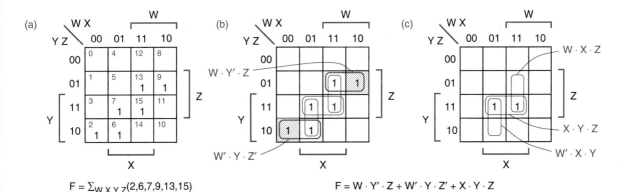

$$F = \Sigma_{W,X,Y,Z}(2,6,7,9,13,15)$$

$$F = W \cdot Y' \cdot Z + W' \cdot Y \cdot Z' + X \cdot Y \cdot Z$$

Figure 4–34 $F = \Sigma_{W,X,Y,Z}(2,6,7,9,13,15)$: (a) Karnaugh map; (b) prime implicants and distinguished 1-cells; (c) reduced map after removal of essential prime implicants and covered 1-cells.

Figure 4–35 shows a more difficult case—a logic function with no essential prime implicants. By trial and error we can find two different minimal sums for this function.

branching method   We can also approach the problem systematically using the *branching method*. Starting with any 1-cell, we arbitrarily select one of the prime implicants that covers it, and include it as if it were essential. This simplifies the remaining problem, which we can complete in the usual way to find a tentative minimal sum. We repeat this process starting with all other prime implicants that cover the starting 1-cell, generating a different tentative minimal sum from each starting point. We may get stuck along the way and have to apply the branching method recursively. Finally, we examine all of the tentative minimal sums generated in this way and select one that is truly minimal.

Obviously, prime-implicant selection can be a fairly involved process. Automatic programs for logic circuit minimization, discussed in Section 4.4, apply additional optimizations and heuristics not discussed here. In real logic-design applications, you are likely to encounter only two kinds of minimization problems: functions of a few variables that you can "eyeball" using the methods

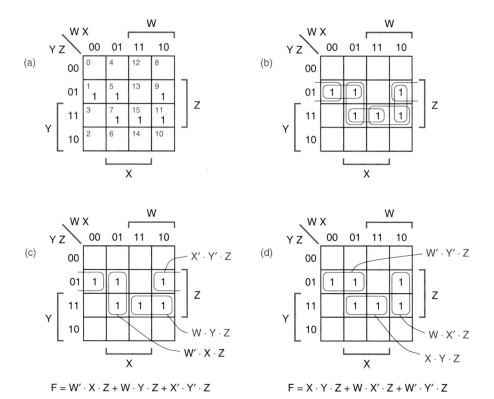

$$F = W' \cdot X \cdot Z + W \cdot Y \cdot Z + X' \cdot Y' \cdot Z$$

$$F = X \cdot Y \cdot Z + W \cdot X' \cdot Z + W' \cdot Y' \cdot Z$$

Figure 4–35 $F = \Sigma_{W,X,Y,Z}(1,5,7,9,11,15)$: (a) Karnaugh map; (b) prime implicants; (c) a minimal sum; (d) another minimal sum.

of this section, and more complex, multiple-output functions that are hopeless without the use of a minimization program.

If a minimal sum-of-products expression for a logic function has many terms, then a minimal product-of-sums expression for the same function may have few terms. As a trivial example of this tradeoff, consider a 4-input OR function:

$$F \;=\; (W) + (X) + (Y) + (Z) \quad \text{(a sum of four trivial product terms)}$$
$$=\; (W + X + Y + Z) \quad \text{(a product of one sum term)}$$

The opposite of this tradeoff is also sometimes true, as trivially illustrated by a 4-input AND:

$$F \;=\; (W){\cdot}(X){\cdot}(Y){\cdot}(Z) \quad \text{(a product of four trivial sum terms)}$$
$$=\; (W{\cdot}X{\cdot}Y{\cdot}Z) \quad \text{(a sum of one product term)}$$

However, for some logic functions, both expression forms are equally costly. For example, consider a 3-input "exclusive OR" function:

$$F \;=\; \Sigma_{X,Y,Z}(1,2,4,7)$$
$$=\; (X'{\cdot}Y'{\cdot}Z) + (X'{\cdot}Y{\cdot}Z') + (X{\cdot}Y'{\cdot}Z') + (X{\cdot}Y{\cdot}Z)$$
$$=\; (X + Y + Z){\cdot}(X + Y' + Z'){\cdot}(X' + Y + Z'){\cdot}(X' + Y' + Z)$$

Both expressions have four terms, and each term has three literals.

If you have the ability to use either expression form to realize a logic function (as in certain programmable logic devices discussed in Section 6.12), then in general you must work out both forms to discover which has the minimum number of terms. Some logic minimization programs do this for you automatically.

* 4.3.6 Simplifying Products of Sums

Using the principle of duality, we can minimize product-of-sums expressions by looking at the 0s on a Karnaugh map. Each 0 on the map corresponds to a maxterm in the canonical product of the logic function. The 0s can be combined to form sum terms with a smaller number of literals using the following rule:

A set of 2^i 0-cells may be combined if there are i variables of the logic function such that the i variables take on all 2^i possible combinations within that set, while the remaining $n - i$ variables take on the same combination within that set. The resulting sum term has $n - i$ literals, where a literal is complemented if its variable appears as 1 in all of the 0-cells, and uncomplemented if its variable appears as 0.

*Throughout this book, optional sections are marked with an asterisk.

The rules for including variables in the sum terms are as follows:

- If a circle covers only areas of the map where the variable is 1, then the variable is complemented in the sum term.

- If a circle covers only areas of the map where the variable is 0, then the variable is uncomplemented in the sum term.

- If a circle covers both areas of the map where the variable is 1 and areas where it is 0, then the variable does not appear in the sum term.

Notice that these rules are the precise duals of our rules for combining 1-cells and including variables in the corresponding product terms when forming sums of products. The circled sets of 0s are called "prime implicates" and are formally defined below:

prime implicate
- A *prime implicate* of a logic function $F(X_1,\ldots,X_n)$ is a normal sum term $S(X_1,\ldots,X_n)$ implied by F, such that if any variable is removed from S, then the resulting sum term is not implied by F.

In terms of a Karnaugh map, a prime implicate of F is a circled set of 0-cells satisfying our combining rule, such that if we try to make it larger (covering twice as many cells), it covers one or more 1s.

Finally, we have a dual for the prime-implicant theorem:

Prime-Implicate Theorem   A minimal product is a product of prime implicates.

Figure 4–36 shows the product-of-sums procedure for our favorite logic function, $F = \Sigma_{X,Y,Z}(1,2,5,7)$. Since this function is specified as a list of minterms, the first step is to write the corresponding maxterm list, $F = \Pi_{X,Y,Z}(0,3,4,6)$, so we can mark all of the function's 0s on the map. Next, we circle and choose prime implicates, using the same kind of rules as in the sum-of-products procedure. The only notable difference is in the way we form sum terms corresponding to circled sets of 0s—each sum term is precisely the complement of the product term we would have written if the 0s were 1s. Another example is shown in Figure 4–37.

In general, to find the lowest-cost two-level realization of a logic function, we have to find both a minimal sum of products and a minimal product of sums, and compare them. Both product-of-sums examples above were worked previously as sums of products, in Figures 4–27 and 4–28 on pages 223 and 225. In the first example, the OR-AND circuit has exactly the same cost (number of gates and gate inputs) as our earlier AND-OR circuit; in the second case, the

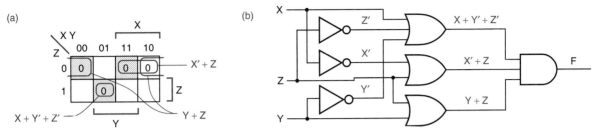

Figure 4–36 $F = \Sigma_{X,Y,Z}(1,2,5,7) = \Pi_{X,Y,Z}(0,3,4,6)$: (a) Karnaugh map; (b) OR/AND circuit.

AND-OR circuit is cheaper. An example where the OR-AND circuit is cheaper is shown in Figure 4–38; the product of sums has just two sum terms. The corresponding sum of products, which we obtained in Figure 4–32 on 228, has four product terms.

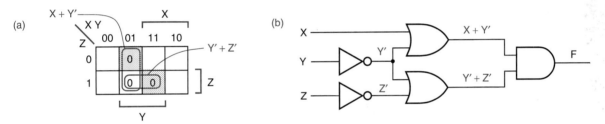

Figure 4–37 $F = \Sigma_{X,Y,Z}(0,1,4,5,6) = \Pi_{X,Y,Z}(2,3,7)$: (a) Karnaugh map; (b) OR/AND circuit.

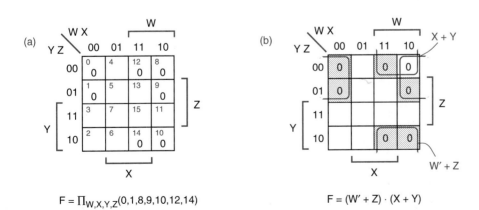

$$F = \Pi_{W,X,Y,Z}(0,1,8,9,10,12,14)$$

$$F = (W' + Z) \cdot (X + Y)$$

Figure 4–38 $F = \Sigma_{W,X,Y,Z}(2,3,4,5,6,7,11,13,15) = \Pi_{W,X,Y,Z}(0,1,8,9,10,12,14)$:
(a) Karnaugh map; (b) prime implicates and distinguished 0-cells.

PRODUCTS
OF SUMS
MADE EASY

If you find the product-of-sums minimization procedure to be a little bit disorient-ing, you can always use a simple trick. Complement the given function and find a minimal sum-of-products expression using the methods of the preceding sub-section. Then complement the result using the generalized DeMorgan's theorem, to obtain a minimal product-of-sums expression for the original function. (Note that if you simply "add out" a minimal sum-of-products expression for the original function, the resulting product-of-sums expression is not necessarily minimal; for example, see Exercise 4.54.)

* 4.3.7 "Don't-Care" Input Combinations

don't-care

Sometimes the specification of a combinational circuit is such that its output doesn't matter for certain input combinations, called *don't-cares*. This may be true because the outputs really don't matter when these input combinations oc-cur, or because these input combinations never occur in normal operation. For example, suppose we wanted to build a prime-number detector whose 4-bit input $N = N_3 N_2 N_1 N_0$ is always a BCD digit; then minterms 10–15 should never occur. A prime BCD-digit detector function may therefore be written as follows:

$$F = \Sigma_{N_3, N_2, N_1, N_0}(1,2,3,5,7) + d(10,11,12,13,14,15)$$

The d(...) term lists the don't-care input combinations for the function. Here F must be 1 for input combinations corresponding to the minterm set (1,2,3,5,7), F can have any values for the don't-care set (10,11,12,13,14,15), and F must be 0 for all other input combinations.

Figure 4–39 shows how to find a minimal sum-of-products realization for the prime BCD-digit detector, including don't-cares. The d's in the map denote

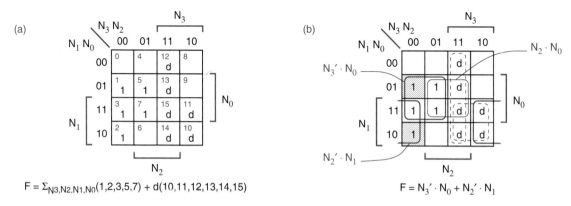

Figure 4–39 Prime BCD-digit detector: (a) initial Karnaugh map; (b) Karnaugh map with prime im-plicants and distinguished 1-cells.

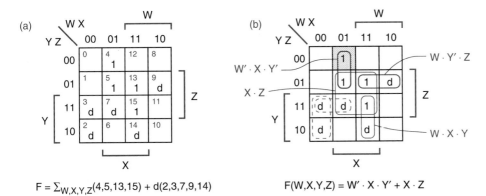

$$F = \Sigma_{W,X,Y,Z}(4,5,13,15) + d(2,3,7,9,14)$$

$$F(W,X,Y,Z) = W \cdot X \cdot Y' + X \cdot Z$$

Figure 4–40 $F = \Sigma_{W,X,Y,Z}(4,5,13,15) + d(2,3,7,9,14)$: (a) Karnaugh map; (b) prime implicants and distinguished 1-cells.

the don't-care input combinations. We modify the procedure for circling sets of 1s (prime implicants) as follows:

- Allow d's to be included when circling sets of 1s, to make the sets as large as possible. This reduces the number of variables in the corresponding prime implicants. Two such prime implicants ($N_2 N_0$ and $N_2' N_1$) appear in the example.

- Do not circle any sets that contain only d's. Including the corresponding product term in the function would unnecessarily increase its cost. Two such product terms ($N_3 N_2$ and $N_3 N_1$) are circled with dashed lines in the example.

- Just a reminder: As usual, do not circle any 0s.

The remainder of the procedure is the same. In particular, we look for distinguished 1-cells and *not* distinguished d-cells, and we include only the corresponding essential prime implicants and any others that are needed to cover all the 1s on the map. In the example, the two essential prime implicants are sufficient to cover all of the 1s on the map. Two of the d's also happen to be covered, so F will be 1 for don't-care input combinations 10 and 11, and 0 for the other don't-cares.

   Figure 4–40 shows another minimization problem with don't-cares. Once again, d's are included to make the circled sets as large as possible, and sets containing only d's are not circled. The single essential prime implicant $W' \cdot X \cdot Y'$ does not cover all the 1s. However, prime implicant $X \cdot Z$ eclipses the other two remaining prime implicants (considering only 1s, not d's), and is added to the first to obtain the minimal sum.

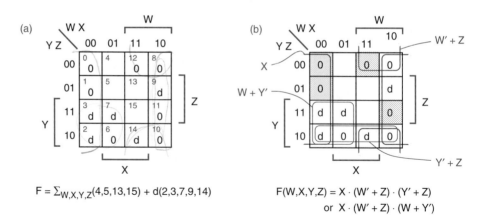

$$F = \Sigma_{W,X,Y,Z}(4,5,13,15) + d(2,3,7,9,14)$$

Figure 4–41 $F = \Sigma_{W,X,Y,Z}(4,5,13,15) + d(2,3,7,9,14)$: (a) Karnaugh map; (b) prime implicates and distinguished 0-cells.

Figure 4–41 shows how to find a minimal product of sums for the function in the previous example. Here, d's are included to make the circled sets of 0s as large as possible. In this case, including a d in the minimal product of sums forces the function to 0 for that input combination, while including a d in the minimal sum of products forced the function to 1. Interestingly, two different minimal sums exist. The first one algebraically equals the minimal sum of products derived earlier, while the second is different—it produces the opposite output for input combinations 7 and 14 (which are don't-cares, of course).

* 4.3.8 Multiple-Output Minimization

Most practical combinational logic circuits require more than one output. We can always handle a circuit with *n* outputs as *n* independent single-output design problems. However, in doing so, we may miss some opportunities for optimization. For example, consider the following two logic functions:

$$F = \Sigma_{X,Y,Z}(3,6,7)$$
$$G = \Sigma_{X,Y,Z}(0,1,3)$$

Figure 4–42 shows the design of F and G as two independent single-output functions. However, as shown in Figure 4–43, we can also find a pair of sum-of-products expressions that share a product term, such that the resulting circuit has one fewer gate than our original design.

When we design multiple-output combinational circuits using discrete gates, product-term sharing obviously reduces circuit size and cost. Of course, most modern designs don't use discrete gates; they use the larger-scale elements of Chapter 5 and the programmable logic devices (PLDs) of Chapter 6. However,

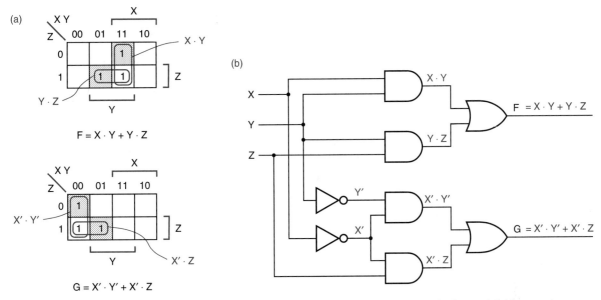

Figure 4–42 Treating a 2-output design as two independent single-output designs: (a) Karnaugh maps; (b) "minimal" circuit.

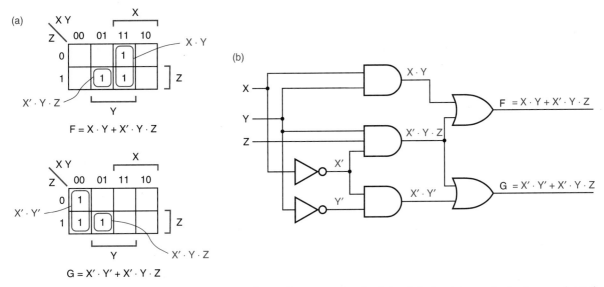

Figure 4–43 Multiple-output minimization for a 2-output circuit: (a) minimized maps including a shared term; (b) minimal multiple-output circuit.

as we'll see in Chapter 6, PLDs contain precisely the sum-of-products structure that we've been learning how to minimize, and they allow product terms to be shared among multiple outputs. Thus, the ideas introduced in this subsection are used extensively in modern logic minimization programs for multiple-output PLDs.

You probably could have "eyeballed" the Karnaugh maps for F and G in Figure 4–43, and discovered the minimal solution. However, larger circuits, such as might be built with a PLD, can be mimimized only with a formal multiple-output minimization algorithm. We'll outline the ideas in such an algorithm here; details can be found in the References.

The key to successful multiple-output minimization of a set of n functions is to consider not only the n original single-output functions, but also "product functions." An *m-product function* of a set of n functions is the product of m of the functions, where $2 \leq m \leq n$. There are $2^n - n - 1$ such functions. Fortunately, $n = 2$ in our example and there is only one product function, F·G, to consider. The Karnaugh maps for F, G, and F·G are shown in Figure 4–44; in general, the map for an m-product function is obtained by ANDing the maps of its m components.

m-product function

A *multiple-output prime implicant* of a set of n functions is a prime implicant of one of the n functions or of one of the product functions. The first step in multiple-output minimization is to find all of the multiple-output prime implicants. Each prime implicant of an m-product function is a possible term

multiple-output prime implicant

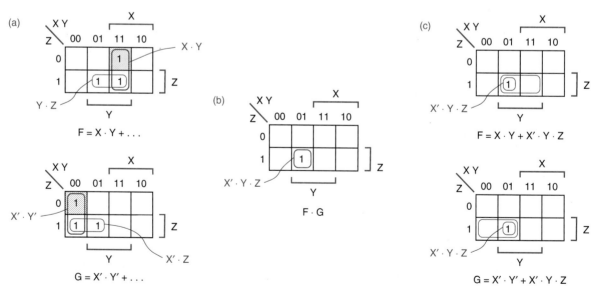

Figure 4–44 Karnaugh maps for a set of two functions: (a) maps for F and G; (b) 2-product map for F·G; (c) reduced maps for F and G after removal of essential prime implicants and covered 1-cells.

to include in the corresponding m outputs of the circuit. If we were trying to minimize a set of 8 functions, we would have to find the prime implicants for $2^8 - 8 - 1 = 247$ product functions as well as for the 8 given functions. Obviously, multiple-output minimization is not for the faint-hearted!

Once we have found the multiple-output prime implicants, we try to simplify the problem by identifying the essential ones. A *distinguished 1-cell* of a particular single-output function F is a 1-cell that is covered by exactly one prime implicant of F or of the product functions involving F. The distinguished 1-cells in Figure 4–44 are shaded. An *essential prime implicant* of a particular single-output function is one that contains a distinguished 1-cell. As in single-output minimization, the essential prime implicants must be included in a mimimum-cost solution. Only the 1-cells that are not covered by essential prime implicants are considered in the remainder of the algorithm.

distinguished 1-cell ✓

essential prime implicant ✓

The final step is to select a minimal set of prime implicants to cover the remaining 1-cells. In this step we must consider all n functions simultaneously, including the possibility of sharing; details of this procedure are discussed in the References. In the example of Figure 4–44(c), we see that there exists a single, shared product term that covers the remaining 1-cell in both F and G.

* 4.4   PROGRAMMED MINIMIZATION METHODS

We know that minimization operations can be performed visually for functions of a few variables using the Karnaugh-map method. We'll show in this section that the same operations can be performed for functions of an arbitrarily large number of variables (at least in principle) using a tabular method called the *Quine-McCluskey algorithm*. Like all algorithms, the Quine-McCluskey algorithm can be translated into a computer program. And like the map method, the algorithm has two steps: (a) finding all prime implicants of the function, and (b) selecting a minimal set of prime implicants that covers the function.

Quine-McCluskey algorithm

Most "real" combinational logic functions are far too big to be minimized by the Karnaugh-map approach, but it's still important to minimize them in many situations. That's because "random" combinational logic functions are most often built nowadays using programmable logic devices. As we'll show in Chapter 6, these devices contain a two-level AND-OR structure internally. Minimization isn't just a matter of saving 10% or 20% of the gates in a design. A given PLD has a limited, fixed number of AND gates available, so minimization can often determine whether or not a design "fits" at all in an available PLD. As a result, most software packages for PLD design include one or more combinational-logic minimization procedures.

WHY MINIMIZE?

The Quine-McCluskey algorithm is often described in terms of handwritten tables and manual check-off procedures. However, since no logic designer ever performs these procedures manually, it's more appropriate for us to discuss the algorithm in terms of data structures and procedures in a high-level programming language. The goal of this section is to give you an appreciation for computational complexity involved in a large minimization problem, and to mention a few alternatives to the traditional algorithm that are used in practice. We consider only fully specified, single-output functions; don't-cares and multiple-output functions can be handled by fairly straightforward modifications to the single-output algorithms, as discussed in the References.

* 4.4.1 Representation of Product Terms

The starting point for the Quine-McCluskey minimization algorithm is the truth table or, equivalently, the minterm list of a function. If the function is specified differently, it must first be converted into this form. For example, an arbitrary n-variable logic expression can be multiplied out (perhaps using De-Morgan's theorem along the way) to obtain a sum-of-products expression. Once we have a sum-of-products expression, each p-variable product term produces 2^{n-p} minterms in the minterm list.

We showed in Section 4.1.6 that a minterm of an n-variable logic function can be represented by an n-bit integer (the minterm number), where each bit indicates whether the corresponding variable is complemented or uncomplemented. However, a minimization algorithm must also deal with product terms that are not minterms, where some variables do not appear at all. Thus, we must represent three possibilities for each variable in a general product term:

1 Uncomplemented.

0 Complemented.

x Doesn't appear.

cube representation   These possibilities are represented by a string of n of the above digits in the *cube representation* of a product term. For example, if we are working with product terms of up to eight variables, X7, X6, ..., X1, X0, we can write the following product terms and their cube representations:

$$X7' \cdot X6 \cdot X5 \cdot X4' \cdot X3 \cdot X2 \cdot X1 \cdot X0' \equiv 01101110$$
$$X3 \cdot X2 \cdot X1 \cdot X0' \equiv xxxx1110$$
$$X7 \cdot X5' \cdot X4 \cdot X3 \cdot X2' \cdot X1 \equiv 1x01101x$$
$$X6 \equiv x1xxxxxx$$

Notice that for convenience, we named the variables just like the bit positions in n-bit binary integers.

In terms of the n-cube and m-subcube nomenclature of Section 2.14, the string 1x01101x represents a 2-subcube of an 8-cube, and the string 01101110 represents a 0-subcube of an 8-cube. However, in the miminimization literature, the maximum dimension n of a cube or subcube is usually implicit, and an m-subcube is simply called an "m-cube" or a "cube" for short; we'll follow this practice in this section.

To represent a product term in a computer program, we can use a data structure with n elements, each of which has three possible values. In Pascal, we might make the following declarations:

```
TYPE
  trit = (complemented, uncomplemented, doesntappear);
  cube = ARRAY [0..15] OF trit; {Represents a single product
                               term with up to 16 variables.}
VAR
  P1, P2, P3: cube;  {Allocate three cubes for use by program.}
```

However, these declarations do not lead to a particularly efficient internal representation of cubes. As we'll see, cubes are easier to manipulate using conventional computer instructions if an n-variable product term is represented by two n-bit computer words, as suggested by the following declarations:

```
CONST
  maxVars = 16;      {Maximum # of variables in a product term.}
  maxCint = 65535;   {2**(maxVars)-1}
TYPE
  word = 0..maxCint;     {This is a maxVars-bit integer.}
  cube = RECORD t: word; {Bits 1 for uncomplemented variables.}
              f: word; {Bits 1 for complemented variables.}
        END;
VAR
  P1, P2, P3: cube;  {Allocate three cubes for use by program.}
```

Here, a word is a 16-bit integer, and a 16-variable product term is represented by a record with two words, as shown in Figure 4–45(a). The first word in a cube has a 1 for each variable in the product term that appears uncomplemented (or "true," .t), and the second has a 1 for each variable that appears complemented (or "false," .f). If a particular bit position has 0s in both words, then the corresponding variable does not appear, while the case of a particular bit position having 1s in both words is not used. Thus, the program variable P1 in (b) represents the product term $P1 = X15 \cdot X12' \cdot X10' \cdot X9 \cdot X4' \cdot X1 \cdot X0$. If we wished to

Figure 4–45
Internal representation
of 16-variable product
terms in a Pascal program:
(a) general format; (b) P1 =
X15·X12′·X10′·X9·X4′·X1·X0.

(a)

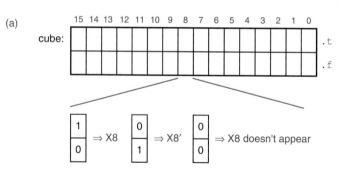

(b)

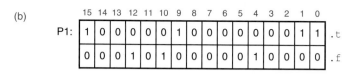

represent a logic function F of up to 16 variables, containing up to 100 product terms, we could declare an array of 100 cubes:

```
VAR
  F: ARRAY [1..100] OF cube; {Storage for a logic function
                              with up to 100 product terms.}
```

Using the foregoing cube representation, it is possible to write short, efficient procedures that manipulate product terms in useful ways. For example, the following Pascal function compares two cubes for equality:

```
FUNCTION EqualCubes(C1, C2: cube): boolean;
BEGIN
  EqualCubes := (C1.t = C2.t) AND (C1.f = C2.f);
END;
```

Figure 4–46 depicts how two cubes can be compared and combined if possible using theorem T10, term·X + term·X′ = term. This theorem says that two product terms can be combined if they differ in only one variable that appears complemented in one term and uncomplemented in the other. Combining two m-cubes yields an $(m + 1)$-cube. Using cube representation, we can apply the combining theorem to a few examples:

$$010 + 000 = 0x0$$
$$00111001 + 00111000 = 0011100x$$
$$101xx0x0 + 101xx1x0 = 101xxxx0$$
$$x111xx00110x000x + x111xx00010x000x = x111xx00x10x000x$$

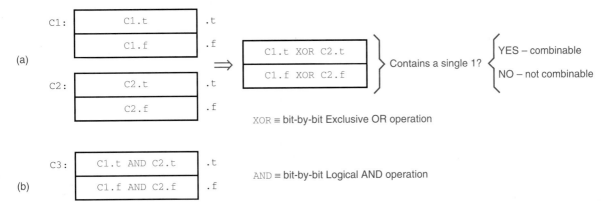

Figure 4–46 Cube manipulations: (a) determining whether two cubes are combinable using theorem T10, term·X + term·X′ = term; (b) combining cubes using theorem T10.

In the Pascal algorithm in the next subsection, we assume the existence of a Pascal function and procedure that support the theorem:

```
FUNCTION Combinable(C1, C2: cube): boolean;
{Returns true if C1 and C2 differ in only one variable,
 which appears true in one and false in the other.}
```

```
PROCEDURE Combine(C1, C2: cube; VAR C3: cube);
{Combines C1 and C2 using theorem T10, and stores the
 result in C3.  Assumes Combinable(C1,C2) is true.}
```

These procedures can be written in a few lines of assembly language using the bit-by-bit "logical operations" provided in almost all computers and in the C language. Unfortunately, bit-by-bit logical operations are not provided in standard Pascal, so we have given a pictorial description of the procedures.

* 4.4.2 Finding Prime Implicants by Combining Product Terms

The first step in the Quine-McCluskey algorithm is to determine all of the prime implicants of the logic function. With a Karnaugh map, we do this visually by identifying "largest possible rectangular sets of 1s." In the algorithm, this is done by systematic, repeated application of theorem T10 to combine minterms, then 1-cubes, 2-cubes, and so on, creating the largest possible cubes (smallest possible product terms) that cover only 1s of the function.

The Pascal program in Table 4–8 applies the algorithm to functions with up to 16 variables. It declares two 2-dimensional arrays, cubes[m,j] and covered[m,j], to keep track of m-cubes for $m = 0$ to maxVars. The 0-cubes (minterms) are supplied by the user. Starting with the 0-cubes, the program examines every pair of cubes at each level and combines them when possible

Table 4–8 A Pascal program that finds prime implicants using the Quine-McCluskey algorithm.

```
PROGRAM FindPIsQuineMcCluskey(input,output);
CONST
   maxVars = 16;          {Max # of variables in a product term.}
   maxCint = 65535;       {2**(maxVars)-1}
   maxCubes = 1000;       {Max # of cubes to allocate per level.}
TYPE
   word = 0..maxCint;     {This is a maxVars-bit integer.}
   cube = RECORD t: word; {Bits 1 for uncomplemented variables.}
                 f: word; {Bits 1 for complemented variables.}
          END;
VAR
   cubes: ARRAY [0..maxVars, 1..maxCubes] OF cube;
   covered: ARRAY [0..maxVars, 1..maxCubes] OF boolean;
   numCubes: ARRAY [0..maxVars] OF 0..maxCubes;
   m: 0..maxVars;              {Value of m in an m-cube, i.e., ''level m.''}
   j, k, p: 1..maxCubes;   {Index into the cubes or covered array.}
   tempCube: cube;
   found: boolean;

{Procedure and function definitions should be placed here.}
```

into cubes at the next level. Cubes that are combined into a next-level cube are marked as "covered"; cubes that are not covered are prime implicants.

Even though the program is short, an experienced programmer might become extremely pessimistic just looking at its structure. The inner FOR loop is nested four levels deep, and the number of times it might be executed is on the order of $maxVars \cdot maxCubes^3$. That's right, that's an exponent, not a footnote. We picked the value $maxCubes = 1000$ somewhat arbitrarily (in fact, too optimistically for many functions), but if you believe this number, then the inner loop can be executed *billions and billions* of times.

The maximum number of minterms of an n-variable function is 2^n, of course, and so by all rights the program in Table 4–8 should declare $maxCubes$ to be 2^{16}, at least to handle the maximum possible number of 0-cubes. Such a declaration would not be overly pessimistic. If an n-variable function has a product term equal to a single variable, then 2^{n-1} minterms are in fact needed to cover that product term.

For larger cubes, the situation is actually worse. The number of possible m-subcubes of an n-cube is $\binom{n}{m} \times 2^{n-m}$, where the binomial coefficient $\binom{n}{m}$ is the number of ways to choose m variables to be x's, and 2^{n-m} is the number of ways to assign 0s and 1s to the remaining variables. For 16-variable functions, the worst case occurs with $m = 5$; there are 8,945,664 possible 5-subcubes of a 16-cube. The total number of distinct m-subcubes of an n-cube, over all values

Table 4–8   (continued)

```
BEGIN
  {Initialize number of m-cubes at each level m.}
  FOR m := 0 to maxVars DO numCubes[m] := 0;

  {Read a list of minterms (0-cubes) supplied by the user, storing them
   in the cubes[0,j] subarray, setting covered[0,j] to false for each minterm,
   and setting numCubes[0] to the total number of minterms read.}
  ReadMinterms;

  FOR m := 0 to maxVars-1 DO {For all levels except the last...}
    FOR j := 1 TO numCubes[m] DO {For all cubes at this level...}
      FOR k := j+1 TO numCubes[m] DO {For other cubes at this level...}
        BEGIN
          IF Combinable(cubes[m,j], cubes[m,k]) THEN
            BEGIN
              {Mark the cubes as covered.}
              covered[m,j] := true; covered[m,k] := true;
              {Combine into an (m+1)-cube, store in tempCube.}
              Combine(cubes[m,j], cubes[m,k], tempCube);
              found := false; {See if we've generated this one before.}
              FOR p := 1 TO numCubes[m+1] DO
                IF EqualCubes(cubes[m+1,p],tempCube) THEN found := true;
              IF NOT found THEN {Add the new cube to the next level.}
                BEGIN
                  numCubes[m+1] := numCubes[m+1] + 1;
                  cubes[m+1,numCubes[m+1]] := tempCube;
                  covered[m+1,numCubes[m+1]] := false;
                END;
            END;
        END;
  FOR m := 0 to maxVars DO {For all levels...}
    FOR j := 1 TO numCubes[m] DO {For all cubes at this level...}
      {Print uncovered cubes -- these are the prime implicants.}
      IF NOT covered[m,j] THEN PrintCube(cubes[m,j]);
END.
```

of m, is 3^n. So a general minimization program might require a *lot* more memory than we've allocated in Table 4–8.

There *are* a few things that we can do to optimize the storage space and execution time required in Table 4–8 (see Exercises 4.70–4.73), but they are piddling compared to the overwhelming combinatorial complexity of the problem. Thus, even with today's fast computers and huge memories, direct application of the Quine-McCluskey algorithm for generating prime implicants is generally limited to functions with only a few variables (fewer than 15–20).

* 4.4.3   Finding Prime Implicants by Iterative Consensus

Programmable logic devices have a lot of inputs, typically 16 or more, and PLD-based functions often use most or all of the available inputs. Therefore, a minimization algorithm whose storage requirements and computational complexity grow with the number of inputs, such as the Quine-McCluskey algorithm of the preceding subsection, is not too useful in PLD-based design.

On the other hand, the number of product terms available in a typical PLD is comparatively limited, typically 4–16 per PLD output. Therefore, if a function is going to fit in a PLD at all, it must have only a small number of terms in its minimal sum-of-products expression. Also, designers tend to specify logic functions using a small number of product terms to begin with, far fewer than a canonical sum would require.

iterative consensus algorithm

Thus, logic functions intended for PLD realization might be minimized more effectively by an algorithm whose computational complexity grows with the number of product terms, rather than with the number of input variables. *Iterative consensus* is such an algorithm. The starting point for the iterative consensus algorithm is a sum-of-products expression, or equivalently, a list of cubes. The product terms *need not* be minterms or prime implicants, but *may* be either or anything in between. In other words, the cubes in the list may have any and all dimensions, from 0 to n in an n-variable function. Starting with the list of cubes, the algorithm generates a list of all the prime-implicant cubes of the function, without ever having to generate a full minterm list.

existence of consensus

Before describing the iterative consensus algorithm, we'll revisit two theorems of switching algebra that it uses heavily. The first is the covering theorem T9, termA + termA·termB = termA, which can be applied to a pair of cubes as shown in Figure 4–47. The iterative consensus algorithm uses this theorem to eliminate covered cubes; only the larger cube must be retained.

The second theorem, as you must have guessed, is the consensus theorem T11, X·term1 + X'·term2 = X·term1 + X'·term2 + term1·term2. In general, given two product terms, termA and termB, their consensus is said to *exist* if there is a variable X such that we can write

$$\text{termA} \ = \ \text{X} \cdot \text{term1}$$
$$\text{termB} \ = \ \text{X}' \cdot \text{term2}$$

consensus of two terms

and if no variable other than X appears complemented in one term and uncomplemented in the other. The product term1·term2 is called the consensus of termA and termB.

As shown in the References, it can be proved that starting with any sum-of-products expression, the process of iteratively adding consensus terms and removing covered terms eventually results in a complete sum. That is, all of the

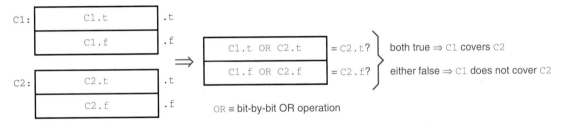

Figure 4–47   Determining whether one cube covers another cube using the covering theorem T9,
termA + termA·termB = termA.

prime implicants of the given function are generated, all other product terms are
removed, and the procedure eventually terminates. One version of the iterative
consensus algorithm, stated in words, is as follows:

(1) Make a list of the product terms in the given expression.

(2) Starting at the beginning of the list, compare each product term with every
product term above it.

(3) If any product term is found to be covered by another, delete the covered
term from the list.

(4) If the consensus of two product terms exists, compare the consensus term
with all product terms in the list, and add it to the bottom of the list if it is
not covered by any other term.

(5) Keep doing this until every product term in the list has been compared with
all the ones above it. The remaining terms are all the prime implicants of
the function.

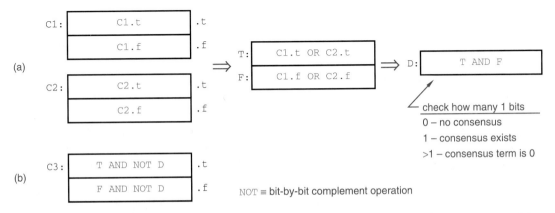

Figure 4–48   Consensus operations on cubes: (a) determining whether the consensus of two
terms exists; (b) deriving the consensus term.

Table 4-9   A Pascal program that finds prime implicants using iterative consensus.

```
PROGRAM FindPIsIterativeConsensus(input,output);
CONST
    maxVars = 16;              {Max # of variables in a product term.}
    maxCint = 65535;           {2**(maxVars)-1}
    maxCubes = 1000;           {Max # of cubes to allocate.}
TYPE
    word = 0..maxCint;         {This is a maxVars-bit integer.}
    cube = RECORD t: word;     {Bits 1 for uncomplemented variables.}
                 f: word;      {Bits 1 for complemented variables.}
           END;
VAR
    cubes: ARRAY [1..maxCubes] OF cube;
    covered: ARRAY [1..maxCubes] OF boolean;
    numCubes: 0..maxCubes;     {Number of cubes actually used in arrays.}
    j, k, p: 1..maxCubes;      {Index into the cubes or covered array.}
    tempCube: cube;
    found: boolean;

{The following procedures operate as shown in the previous two figures.}

FUNCTION Covers(C1, C2: cube): boolean;
{Returns true if C1 covers C2, that is, if C2 has 1s
 everywhere that C1 does.  Note that it is possible for
 Covers(X,Y) and Covers(Y,X) both to return true only if X=Y.}

FUNCTION ConsensusExists(C1, C2: cube): boolean;
{Returns true if the consensus of C1 and C2 exists.}

PROCEDURE StoreConsensus(C1, C2: cube; VAR C3: cube);
{Computes the consensus of C1 and C2, and stores the
 result in C3.  Assumes ConsensusExists(C1,C2) is true.}
```

Figure 4-48(a) on the preceding page shows how a program can determine whether the consensus exists for two product terms given in cube representation. The consensus term, if it exists, can be formed as shown in (b).

A Pascal program that embodies the iterative consensus algorithm is shown in Table 4-9. The program uses a 1-dimensional array, cubes[j], to store the initial list of cubes, and it adds cubes at the end of the array as it goes along. The program "deletes" a covered cube by simply marking it in a corresponding array, covered[j]; this array is checked at each step so that covered cubes are not considered.

Compared with the Quine-McCluskey program of Table 4-8, the iterative consensus program has two favorable characteristics. First, the number of iterations of the innermost FOR-loop is slightly less, on the order of maxCubes^3

Table 4–9   (continued)

```
BEGIN
  {Read a list of product terms (m-cubes) supplied by the user, storing them
   in the cubes[j] array, setting covered[j] to false for each product term,
   and setting numCubes to the total number of product terms read.}
  ReadProductTerms;

  j := 2;  {Start with the second cube in the array.}
  WHILE j <= numCubes DO    {For each cube...}
    BEGIN
        FOR k := 1 TO j-1 DO  {For all cubes before it in the array.}
          {Look only at pairs of uncovered cubes.}
          IF covered[j] OR covered[k] THEN BEGIN END {Don't bother.}
          {Else mark cubes that are covered, which are subsequently ignored.}
          ELSE IF Covers(cubes[j],cubes[k]) THEN covered[k] := true
          ELSE IF Covers(cubes[k],cubes[j]) THEN covered[j] := true
          {Else check for consensus.}
          ELSE IF ConsensusExists(cubes[j],cubes[k]) THEN
            BEGIN {Consensus exists, store it in tempCube}
              StoreConsensus(cubes[j],cubes[k],tempCube);
              found := false;           {Determine whether tempCube is  }
              FOR p := 1 TO numCubes DO  {  covered by an existing cube.  }
                IF Covers(cubes[p],tempCube) THEN found := true;
              IF NOT found THEN {Add tempCube to the end of array.}
                BEGIN
                  numCubes := numCubes + 1;
                  cubes[numCubes] := tempCube;
                  covered[numCubes] := false;
                END;
            END;
      j := j+1;
    END;
  FOR j := 1 TO numCubes DO {For all cubes...}
    {Print uncovered cubes -- these are the prime implicants.}
    IF NOT covered[j] THEN PrintCube(cubes[j]);
END.
```

instead of numVars · maxCubes^3. Second and more important, the number of
cubes generated is typically far fewer; in fact, maxCubes = 100 would work for
the majority of practical design problems with PLD realizations.

As in the Quine-McCluskey algorithm, several enhancements can be made
to the iterative consensus algorithm to improve its performance. One improvement has been used enough to have a name; it is called the *generalized consensus*
algorithm. The basic idea is to execute a shorter procedure maxVars times, where
execution *i* generates all possible consensus terms that result from input variable

*generalized
consensus
algorithm*

i appearing true in one product term and complemented in another. To be more specific, the generalized consensus algorithm consists of executing the following steps for each value of i:

(1) Partition the cubes array into three parts, C1, C0, and Cx, according to whether each cube contains a 1, 0, or x for variable i.

(2) Compare every possible pair of cubes that results from taking one cube from C1 and the other from C0. If a consensus exists, add it to a list CC of consensus cubes.

(3) Compare each consensus cube in CC with all cubes in CC and Cx, and remove any consensus cubes that are covered by other cubes. (Consensus cubes cannot be covered by cubes in C0 or C1 [Exercise 4.74]).

(4) Compare the remaining consensus cubes in CC with the cubes in C0, C1, and Cx, removing any of the latter that are covered by a consensus cube.

(5) Combine all four arrays and iterate on the next variable.

The innermost loop of this procedure has a worst-case execution time on the order of numVars $\cdot (.5s \cdot$ maxCubes$)^2 \cdot$ maxCubes, where s is the fraction of all the cubes that go into the C0 and C1 arrays on each step. The maximum value of s is 1, but a more typical value for large circuits might be 0.5 or less.

* 4.4.4 Finding a Minimal Cover Using Prime-Implicant Tables

prime-implicant table

The second step in minimizing a combinational logic function, once we have a list of all its prime implicants, is to select a minimal subset of them to cover all the 1s of the function. The Quine-McCluskey algorithm uses a two-dimensional array called a *prime-implicant table* to do this. Figure 4–49(a) shows a small but representative prime-implicant table, corresponding to the Karnaugh-map minimization problem of Figure 4–34. There is one column for each minterm of the function, and one row for each prime implicant. Each entry is a bit that is 1 if and only if the prime implicant for that row covers the minterm for that column (shown in the figure as a check).

The steps for selecting prime implicants with the table are analogous to the steps that we used in Section 4.3.5 with Karnaugh maps:

(1) Identify distinguished 1-cells. These are easily identified in the table as columns with a single 1, as shown in Figure 4–49(b).

(2) Include all essential prime implicants in the minimal sum. A row that contains a check in one or more distinguished-1-cell columns corresponds to an essential prime implicant.

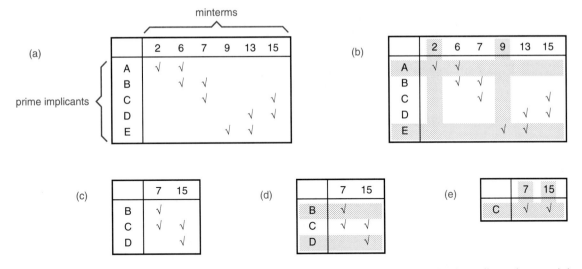

Figure 4–49   Prime-implicant tables: (a) original table; (b) showing distinguished 1-cells and essential prime implicants; (c) after removal of essential prime implicants; (d) showing eclipsed rows; (e) after removal of eclipsed rows, showing secondary essential prime implicant.

(3) Remove from consideration the essential prime implicants and the 1-cells (minterms) that they cover. In the table, this is done by deleting the corresponding rows and columns, marked in color in Figure 4–49(b). If any rows have no checks remaining, they are also deleted; the corresponding prime implicants are *redundant*, that is, completely covered by essential prime implicants. This step leaves the reduced table shown in (c).   *redundant prime implicant*

(4) Remove from consideration any prime implicants that are eclipsed by others with equal or lesser cost. In the table, this is done by deleting any rows whose checked columns are a proper subset of another row's, and deleting all but one of a set of rows with identical checked columns. This is shown in color in (d), and leads to the further reduced table in (e).

When a function is minimized for a PLD-based design, all of its prime implicants may be considered to have equal cost, because all of the AND gates in a PLD have all of the inputs available. Otherwise, the prime implicants must be sorted and selected according to the number of AND-gate inputs.

(5) Identify distinguished 1-cells and include all secondary essential prime implicants in the minimal sum. As before, any row that contains a check in one or more distinguished-1-cell columns corresponds to a secondary essential prime implicant.

(6) If all remaining columns are covered by the secondary essential prime implicants, as in (e), we're done. Otherwise, if any secondary essential prime implicants were found in the previous step, we go back to step 3 and iterate. Otherwise, the branching method must be used, as described in Section 4.3.5. This involves picking rows one at a time, treating them as if they were essential, and recursing (and cursing) on steps 3–6.

Although a prime-implicant table allows a fairly straightforward prime-implicant selection algorithm, the data structure required in a corresponding computer program is huge, since it requires on the order of $p \cdot 2^n$ bits, where p is the number of prime implicants and n is the number of input bits (assuming that the given function produces a 1 output for most input combinations). Worse, executing the steps that we so blithely described in a few sentences above requires a huge amount of computation.

* 4.4.5   Finding a Minimal Cover Using Iterative Consensus

As we've said before, practical PLD-based designs can have lots of inputs, so an applicable minimization algorithm must not generate or use a minterm list of the function; that rules out prime-implicant tables. However, the iterative consensus algorithm works at least for the first part of the minimization procedure, finding all prime implicants. With a little thought, we can design the second part of the procedure, finding a minimal cover, using iterative consensus as the basic operation, and hope thereby to obtain reasonable performance.

The first step in selecting prime implicants for the minimal cover is to identify essential prime implicants, as in steps 1 and 2 of the prime-implicant-table algorithm in the preceding subsection. And we must do this without generating a minterm list.

essential-test procedure

Let C denote the set of *all* the prime implicants of the given logic function, and let P denote a prime implicant that we would like to test for being essential. Suppose we temporarily delete P from C and run the iterative consensus algorithm on the remaining terms in C. If P is not essential, then all of the minterms it covers are also covered by other terms that remain in C; therefore, the iterative consensus algorithm will recreate P. On the other hand, if P is essential, it will not reappear. This procedure can be applied to all of the prime implicants in C to find all of the essential ones.

remove-redundant procedure

Let E be the set of essential prime implicants found above. Now suppose we run the iterative consensus algorithm on E; we'll generate some additional product terms that aren't necessarily prime implicants. However, any of the new terms that *are* prime implicants (i.e., that appear in C) are redundant, since they are completely covered by essential prime implicants. The essential and the redundant prime implicants are removed from C, corresponding to step 3 in the preceding subsection.

At this point, if there are no prime implicants left in C, we're done; the function is completely covered by just the essential prime implicants. Otherwise, we must go on to find secondary essential prime implicants. To do this, we must first identify and remove from C any prime implicants that are eclipsed by others, corresponding to step 4 in the preceding subsection. We do this as follows.

remove-eclipsed procedure

For each prime implicant P remaining in C, we want to know if P plus the terms in E completely cover any other equal- or lower-cost prime implicant Q in C. Any such Q may be deleted from C, since P covers at least as much of what's left to cover as Q does. We can perform this test for each P by temporarily adding P to E and running the iterative consensus algorithm on E. If any Q in C appears in the resulting set of product terms, it is eclipsed by P and is deleted from C, as in step 5 in the preceding subsection.

If the remove-eclipsed procedure removed one or more terms from C, then we have a chance of finding secondary essential prime implicants. So we go back to the essential-test procedure and iterate. Otherwise, we're stuck with the branching method, in which we curse and recurse.

* 4.4.6   Other Minimization Methods

Although the previous subsections form a good introduction to logic minimization algorithms, the methods they describe are by no means the latest and greatest. Spurred on by the ever increasing density of VLSI chips, many researchers have discovered more effective ways to minimize combinational logic functions. Their results fall roughly into three categories:

(1) *Computational improvements.* Improved algorithms typically use clever data structures or rearrange the order of the steps to reduce the memory requirements and execution time of the classical algorithms.

(2) *Heuristic methods.* Some minimization problems are just too big to be solved using an "exact" algorithm. These problems can be attacked using shortcuts and well-educated guesses to reduce memory size and execution time to a fraction of what an "exact" algorithm would require. However, instead of finding a provably minimal expression for a logic function, heuristic methods attempt to find an "almost minimal" expression. Even for problems that can be solved by an "exact" method, a heuristic method typically finds a good solution ten times faster. The most successful heuristic program, Espresso-II, does in fact produce minimal or near-minimal results for the majority of problems (within one or two product terms), including problems with dozens of inputs and hundreds of product terms.

(3) *Looking at things differently.* As we mentioned earlier, multiple-output minimization can be handled by straightforward, fairly mechanical modifications to single-output minimization methods. However, by looking at multiple-output minimization as a problem in multivalued (nonbinary) logic, the designers of the Espresso-MV algorithm were able to make substantial performance improvements over Espresso II.

More information on these methods can be found in the References.

* 4.5 TIMING HAZARDS

The analysis methods that we developed in Section 4.2 ignore circuit delay and predict only the *steady-state behavior* of combinational logic circuits. That is, they predict a circuit's output as a function of its inputs under the assumption that the inputs have been stable for a long time, relative to the delays in the circuit's electronics. However, we showed in Section 3.6 that the actual delay from an input change to the corresponding output change in a real logic circuit is nonzero and depends on many factors.

√ *steady-state behavior*

Because of circuit delays, the *transient behavior* of a logic circuit may differ from what is predicted by a steady-state analysis. In particular, a circuit's output may produce a short pulse, often called a *glitch*, at a time when steady-state analysis predicts that the output should not change. A *hazard* is said to exist when a circuit has the possibility of producing such a glitch. Whether or not the glitch actually occurs depends on the exact delays and other electrical characteristics of the circuit. Since such parameters are difficult to control in production circuits, a logic designer must be prepared to eliminate hazards (the *possibility* of a glitch) even though a glitch may occur only under a worst-case combination of logical and electrical conditions.

√ *transient behavior*

√ *glitch*

hazard

* 4.5.1 Static Hazards

A *static-1 hazard* is the possibility of a circuit's output producing a 0 glitch when we would expect the output to remain at a nice steady 1 based on a static analysis of the circuit function. A formal definition is given below.

√ *static-1 hazard*

> Any combinational circuit can be analyzed for the presence of hazards. However, a well-designed, *synchronous* digital system is structured so that hazard analysis is not needed for most of its circuits. In a synchronous system, all of the inputs to a combinational circuit are changed at a particular time, and the outputs are not "looked at" until they have had time to settle to a steady-state value. Hazard analysis and elimination are typically needed only in the design of asynchronous sequential circuits, such as the feedback sequential circuits discussed in Section 7.5. You'll rarely have reason to design such a circuit, but if you do, an understanding of hazards will be absolutely essential for a reliable result.

MOST HAZARDS
ARE NOT
HAZARDOUS!

Definition: A static-1 hazard is a pair of input combinations that: (a) differ in only one input variable and (b) both give a 1 output; such that it is possible for a momentary 0 output to occur during a transition in the differing input variable.

For example, consider the logic circuit in Figure 4–50(a). Suppose that X and Y are both 1 and Z is changing from 1 to 0. Then (b) shows the timing diagram assuming that the propagation delay through each gate or inverter is one unit time. Even though "static" analysis predicts that the output is 1 for both input combinations X,Y,Z = 111 and X,Y,Z = 110, the timing diagram shows that F goes to 0 for one unit time during a 1-0 transition on Z, because of the delay in the inverter that generates Z'.

A *static-0 hazard* is the possibility of a 1 glitch when we expect the circuit to have a steady 0 output:

static-0 hazard ✓

Definition: A static-0 hazard is a pair of input combinations that: (a) differ in only one input variable and (b) both give a 0 output; such that it is possible for a momentary 1 output to occur during a transition in the differing input variable.

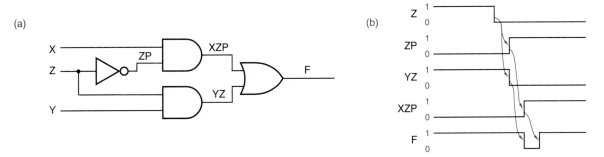

Figure 4–50   Circuit with a static-1 hazard: (a) logic diagram; (b) timing diagram.

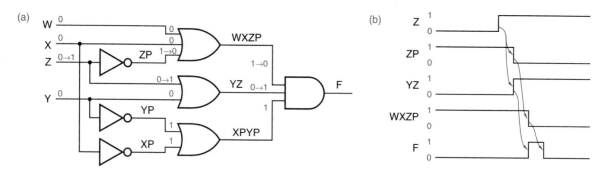

Figure 4–51   Circuit with static-0 hazards: (a) logic diagram; (b) timing diagram.

Since a static-0 hazard is just the dual of a static-1 hazard, an OR-AND circuit that is the dual of Figure 4–50(a) would have a static-0 hazard.

An OR-AND circuit with four static-0 hazards is shown in Figure 4–51(a). One of the hazards occurs when W,X,Y = 000 and Z is changed, as shown in (b). You should be able to find the other three hazards and eliminate all of them after studying the next subsection.

* 4.5.2   Finding Static Hazards Using Maps

A Karnaugh map can be used to detect static hazards in a two-level sum-of-products or product-of-sums circuit. The existence or nonexistence of static hazards depends on the circuit design for a logic function.

A properly designed two-level sum-of-products (AND-OR) circuit has no static-0 hazards. A static-0 hazard would exist in such a circuit only if both a variable and its complement were connected to the same AND gate, which would be silly. However, the circuit *may* have static-1 hazards. Their existence can be predicted from a Karnaugh map where the product terms corresponding to the AND gates in the circuit are circled.

Figure 4–52(a) shows the Karnaugh map for the circuit of Figure 4–50. It is clear from the map that there is no single product term that covers both input combinations X,Y,Z = 111 and X,Y,Z = 110. Thus, intuitively, it is possible for the output to "glitch" momentarily to 0 if the AND gate output that covers one of the combinations goes to 0 before the AND gate output covering the other input combination goes to 1. The way to eliminate the hazard is also quite apparent: Simply include an extra product term (AND gate) to cover the hazardous input pair, as shown in Figure 4–52(b). The extra product term, it turns out, is the *consensus* of the two original terms; in general, we must add consensus terms to eliminate hazards. The corresponding hazard-free circuit is shown in Figure 4–53.

consensus

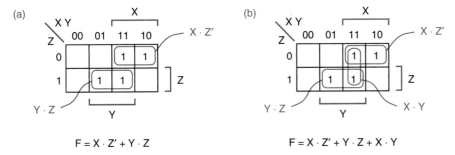

Figure 4–52   Karnaugh map for the circuit of Figure 4–50: (a) as originally
designed; (b) with static-1 hazard eliminated.

Another example is shown in Figure 4–54. In this example, three product
terms must be added to eliminate the static-1 hazards.

A properly designed two-level product-of-sums (OR-AND) circuit has no
static-1 hazards. It *may* have static-0 hazards, however. These hazards can be

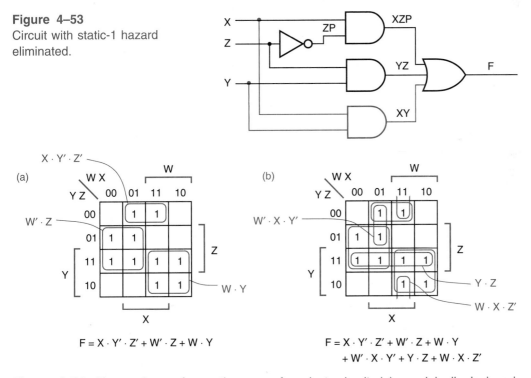

Figure 4–53
Circuit with static-1 hazard
eliminated.

Figure 4–54   Karnaugh map for another sum-of-products circuit: (a) as originally designed;
(b) with extra product terms to cover static-1 hazards.

detected and eliminated by studying the adjacent 0s in the Karnaugh map, in a manner dual to the foregoing.

* 4.5.3 Dynamic Hazards

dynamic hazard

A *dynamic hazard* is the possibility of an output changing more than once as the result of a single input transition. Multiple output transitions can occur if there are multiple paths with different delays from the changing input to the changing output.

For example, consider the circuit in Figure 4–55; it has three different paths from input X to the output F. One of the paths goes through a slow OR gate, and another goes through an OR gate that is even slower. If the input to the circuit is W,X,Y,Z = 0,0,0,1, then the output will be 1, as shown. Now suppose we change the X input to 1. Assuming that all of the gates except the two marked "slow" and "slower" are very fast, the transitions shown in black occur next, and the output goes to 0. Eventually, the output of the "slow" OR gate changes, creating the transitions shown in nonitalic color, and the output goes to 1. Finally, the output of the "slower" OR gate changes, creating the transitions shown in italic color, and the output goes to its final state of 0.

Dynamic hazards do not occur in a properly designed two-level AND-OR or OR-AND circuit, that is, one in which no variable and its complement are connected to the same first-level gate. In multilevel circuits, dynamic hazards can be discovered using a method described in the References.

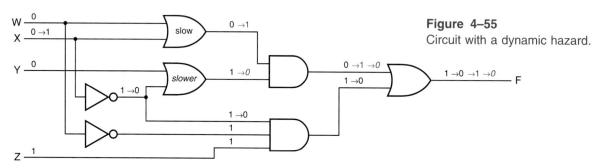

Figure 4–55
Circuit with a dynamic hazard.

* 4.5.4 Designing Hazard-Free Circuits

There are only a few situations, such as the design of feedback sequential circuits, that require hazard-free combinational circuits. Techniques for finding hazards in arbitrary circuits, described in the References, are rather difficult to use. So, when you require a hazard-free design, it's best to use a circuit structure that is easy to analyze.

cants" (*IRE Trans. Electron. Computers*, Vol. EC-9, No. 2, 1960, pp. 245–252). The generalized consensus algorithm was published by Pierre Tison in "Generalization of Consensus Theory and Application to the Minimization of Boolean Functions" (*IEEE Trans. Electron. Computers*, Vol. EC-16, No. 4, 1967, pp. 446–456). All of these algorithms are described by Thomas Downs in *Logic Design with Pascal* (Van Nostrand Reinhold, 1988).

As we explained in Section 4.4.6, the huge number of prime implicants in some logic functions makes it impractical or impossible deterministically to find them all or select a minimal cover. However, efficient heuristic methods can find solutions that are close to minimal, at least close enough for government work. The Espresso-II method is described in *Logic Minimization Algorithms for VLSI Synthesis* by R. K. Brayton, C. McMullen, G. D. Hachtel, and A. Sangiovanni-Vincentelli (Kluwer Academic Publishers, 1984). The more recent Espresso-MV and Espresso-EXACT algorithms are described in "Multiple-Valued Minimization for PLA Optimization" by R. L. Rudell and A. Sangiovanni-Vincentelli (*IEEE Trans. CAD*, Vol. CAD-6, No. 5, 1987, pp. 727–750).

In this chapter we described a map method for finding static hazards in two-level AND-OR and OR-AND circuits, but any combinational circuit can be analyzed for hazards. In both his 1965 and 1986 books, McCluskey defines the *0-sets* and *1-sets* of a circuit and shows how they can be used to find static hazards. He also defines *P-sets* and *S-sets* and shows how they can be used to find dynamic hazards.

0-set
1-set
P-set
S-set

Many deeper and varied aspects of switching theory have been omitted from this book, but have been beaten to death in other books and literature. A good starting point for an academic study of classical switching theory is Zvi Kohavi's book, *Switching and Finite Automata Theory*, 2nd ed. (McGraw-Hill, 1978), which includes material on set theory, symmetric networks, functional decomposition, threshold logic, fault detection, and path sensitization. Another area of great academic interest (but little commercial activity) is nonbinary *multiple-valued logic*, in which each signal line can take on more than two values. In his 1986 book, McCluskey gives a good introduction to multiple-valued logic, explaining its pros and cons and why it has seen little commercial development.

multiple-valued logic

DRILL PROBLEMS

4.1 Using variables NERD, DESIGNER, FAILURE, and STUDIED, write a boolean expression that is 1 for successful designers who never studied and for nerds who studied all the time.

4.2 Prove theorems T2–T5 using perfect induction.

4.3   Prove theorems T1′–T3′ and T5′ using perfect induction.

4.4   Prove theorems T6–T9 using perfect induction.

4.5   According to DeMorgan's theorem, the complement of $X + Y \cdot Z$ is $X' \cdot Y' + Z'$. Yet both functions are 1 for $XYZ = 110$. How can both a function and its complement be 1 for the same input combination? What's wrong here?

4.6   Use the theorems of switching algebra to simplify each of the following logic functions:

(a) $F = W \cdot X \cdot Y \cdot Z \cdot (W \cdot X \cdot Y \cdot Z' + W \cdot X' \cdot Y \cdot Z + W' \cdot X \cdot Y \cdot Z + W \cdot X \cdot Y' \cdot Z)$

(b) $F = A \cdot B + A \cdot B \cdot C' \cdot D + A \cdot B \cdot D \cdot E' + A \cdot B \cdot C' \cdot E + C' \cdot D \cdot E$

(c) $F = M \cdot N \cdot O + Q' \cdot P' \cdot N' + P \cdot R \cdot M + Q' \cdot O \cdot M \cdot P' + M \cdot R$

4.7   Write the truth table for each of the following logic functions:

(a) $F = X' \cdot Y + X' \cdot Y' \cdot Z$ (b) $F = W' \cdot X + Y' \cdot Z' + X' \cdot Z$

(c) $F = W + X' \cdot (Y' + Z)$ (d) $F = A \cdot B + B' \cdot C + C' \cdot D + D' \cdot A$

(e) $F = V \cdot W + X' \cdot Y' \cdot Z$ (f) $F = (A' + B' + C \cdot D) \cdot (B + C' + D' \cdot E')$

(g) $F = (W \cdot X)' \cdot (Y' + Z')'$ (h) $F = (((A + B)' + C')' + D)'$

(i) $F = (A' + B + C) \cdot (A + B' + D) \cdot (B + C' + D') \cdot (A + B + C + D)$

4.8   Write the truth table for each of the following logic functions:

(a) $F = X' \cdot Y' \cdot Z' + X \cdot Y \cdot Z + X \cdot Y' \cdot Z$ (b) $F = M' \cdot N' + M \cdot P + N' \cdot P$

(c) $F = A \cdot B + A \cdot B' \cdot C' + A' \cdot B \cdot C$ (d) $F = A' \cdot B \cdot (C \cdot B \cdot A' + B \cdot C')$

(e) $F = X \cdot Y \cdot (X' \cdot Y \cdot Z + X \cdot Y' \cdot Z + X \cdot Y \cdot Z' + X' \cdot Y' \cdot Z)$ (f) $F = M \cdot N + M' \cdot N' \cdot P'$

(g) $F = (A + A') \cdot B + B \cdot A \cdot C' + C \cdot (A + B') \cdot (A' + B)$ (h) $F = X \cdot Y' + Y \cdot Z + Z' \cdot X$

4.9   Write the canonical sum and product for each of the following logic functions:

(a) $F = \Sigma_{X,Y}(1,2)$ (b) $F = \Pi_{A,B}(0,1,2)$

(c) $F = \Sigma_{A,B,C}(2,4,6,7)$ (d) $F = \Pi_{W,X,Y}(0,1,3,4,5)$

(e) $F = X + Y' \cdot Z'$ (f) $F = V' + (W' \cdot X)'$

4.10 Write the canonical sum and product for each of the following logic functions:

(a) $F = \Sigma_{X,Y,Z}(0,3)$ (b) $F = \Pi_{A,B,C}(1,2,4)$

(c) $F = \Sigma_{A,B,C,D}(1,2,5,6)$ (d) $F = \Pi_{M,N,P}(0,1,3,6,7)$

(e) $F = X' + Y \cdot Z' + Y \cdot Z'$ (f) $F = A'B + B'C + A$

4.11 If the canonical sum for an n-input logic function is also a minimal sum, how many literals are in each product term of the sum? Might there be any other minimal sums in this case?

4.12 Give two reasons why the cost of inverters is not included in the definition of "minimal" for logic minimization.

4.13 Using Karnaugh maps, find a minimal sum-of-products expression for each of the following logic functions. Indicate the distinguished 1-cells in each map.

(a) $F = \Sigma_{X,Y,Z}(1,3,5,6,7)$ (b) $F = \Sigma_{W,X,Y,Z}(1,4,5,6,7,9,14,15)$

(c) $F = \Pi_{W,X,Y}(0,1,3,4,5)$ (d) $F = \Sigma_{W,X,Y,Z}(0,2,5,7,8,10,13,15)$

(e) $F = \Pi_{A,B,C,D}(1,7,9,13,15)$ (f) $F = \Sigma_{A,B,C,D}(1,4,5,7,12,14,15)$

4.14 Find a minimal product-of-sums expression for each function in Drill 4.13. Indicate the distinguished 0-cells in each map.

4.15 Using Karnaugh maps, find a minimal sum-of-products expression for each of the following logic functions. Indicate the distinguished 1-cells in each map.

(a) $F = \Sigma_{A,B,C}(0,1,2,4)$ (b) $F = \Sigma_{W,X,Y,Z}(1,4,5,6,11,12,13,14)$

(c) $F = \Pi_{A,B,C}(1,2,6,7)$ (d) $F = \Sigma_{W,X,Y,Z}(0,1,2,3,7,8,10,11,15)$

(e) $F = \Sigma_{W,X,Y,X}(1,2,4,7,8,11,13,14)$ (f) $F = \Pi_{A,B,C,D}(1,3,4,5,6,7,9,12,13,14)$

4.16 Find a minimal product-of-sums expression for each function in Drill 4.15. Indicate the distinguished 0-cells in each map.

4.17 Find the complete sum for the logic functions in Drill 4.15(d) and (e).

4.18 Using Karnaugh maps, find a minimal sum-of-products expression for each of the following logic functions. Indicate the distinguished 1-cells in each map.

(a) $F = \Sigma_{W,X,Y,Z}(0,1,3,5,14) + d(8,15)$ (b) $F = \Sigma_{W,X,Y,Z}(0,1,2,8,11) + d(3,9,15)$

(c) $F = \Sigma_{A,B,C,D}(1,5,9,14,15) + d(11)$ (d) $F = \Sigma_{A,B,C,D}(1,5,6,7,9,13) + d(4,15)$

(e) $F = \Sigma_{W,X,Y,Z}(3,5,6,7,13) + d(1,2,4,12,15)$

4.19 Repeat Drill 4.18, finding a minimal product-of-sums expression for each logic function.

4.20 For each logic function in the two preceding exercises, determine whether the minimal sum-of-products expression equals the minimal product-of-sums expression. Also compare the circuit cost for realizing each of the two expressions.

4.21 For each of the following logic expressions, find all of the static hazards in the corresponding two-level AND-OR or OR-AND circuit, and design a hazard-free circuit that realizes the same logic function.

(a) $F = W \cdot X + W' Y'$ (b) $F = W \cdot X' \cdot Y' + X \cdot Y' \cdot Z + X \cdot Y$

(c) $F = W' \cdot Y + X' \cdot Y' + W \cdot X \cdot Z$ (d) $F = W' \cdot X + Y' \cdot Z + W \cdot X \cdot Y \cdot Z + W \cdot X' \cdot Y \cdot Z'$

(e) $F = (W + X + Y) \cdot (X' + Z')$ (f) $F = (W + Y' + Z') \cdot (W' + X' + Z') \cdot (X' + Y + Z)$

(g) $F = (W + Y + Z') \cdot (W + X' + Y + Z) \cdot (X' + Y') \cdot (X + Z)$

EXERCISES

4.22 Design a non-trivial-looking logic circuit that contains a feedback loop but whose output depends only on its current input.

4.23 Prove the combining theorem T10 without using perfect induction, but assuming that theorems T1–T9 and T1′–T9′ are true.

4.24 Show that the combining theorem, T10, is just a special case of consensus (T11) used with covering (T9).

4.25 Prove that $(X + Y') \cdot Y = X \cdot Y$ *without* using perfect induction. You may assume that theorems T1–T11 and T1′–T11' are true.

4.26 Prove that $(X + Y) \cdot (X' + Z) = X \cdot Z + X' \cdot Y$ *without* using perfect induction. You may assume that theorems T1–T11 and T1′–T11' are true.

4.27 Show that an n-input AND gate can be replaced by $n - 1$ 2-input AND gates. Can the same statement be made for NAND gates? Justify your answer.

4.28 How many physically different ways are there to realize $V \cdot W \cdot X \cdot Y \cdot Z$ using four 2-input AND gates (4/4 of a 74LS08)? Justify your answer.

4.29 Use switching algebra to prove that tying together two inputs of an $n + 1$-input AND or OR gate gives it the functionality of an n-input gate.

4.30 Prove DeMorgan's theorems (T13 and T13′) using finite induction.

4.31 Which logic symbol more closely approximates the internal realization of a TTL NOR gate, Figure 4–4(c) or (d)? Why?

4.32 Use the theorems of switching algebra to rewrite the following expression using as few inversions as possible (complemented parentheses are allowed):

$$B' \cdot C + A \cdot C \cdot D' + A' \cdot C + E \cdot B' + E \cdot (A + C) \cdot (A' + D')$$

4.33 Prove or disprove the following propositions:

(a) Let A and B be switching-algebra *variables*. Then $A \cdot B = 0$ and $A + B = 1$ implies that $A = B'$.

(b) Let X and Y be switching-algebra *expressions*. Then $X \cdot Y = 0$ and $X + Y = 1$ implies that $X = Y'$.

4.34 Prove Shannon's expansion theorems. (*Hint:* Don't get carried away; it's easy.)

generalized Shannon expansion theorems

4.35 Shannon's expansion theorems can be generalized to "pull out" not just one but i variables so that a logic function can be expressed as a sum or product of 2^i terms. State the generalized Shannon expansion theorems.

4.36 Show how the generalized Shannon expansion theorems lead to the canonical sum and canonical product representations of logic functions.

4.37 An *Exclusive OR (XOR) gate* is a 2-input gate whose output is 1 if and only if exactly one of its inputs is 1. Write a truth table, sum-of-products expression, and corresponding AND-OR circuit for the Exclusive OR function.   *Exclusive OR (XOR) gate*

4.38 From the point of view of switching algebra, what is the function of a 2-input XOR gate whose inputs are tied together? How might the output behavior of a real XOR gate differ?

4.39 After completing the design and fabrication of a digital system, a designer finds that one more inverter is required. However, the only spare gates in the system are a 3-input OR, a 2-input AND, and a 2-input XOR. How should the designer realize the inverter function without adding another IC?

4.40 Any set of logic-gate types that can realize any logic function is called a *complete set* of logic gates. For example, 2-input AND gates, 2-input OR gates, and inverters are a complete set, because any logic function can be expressed as a sum of products of variables and their complements, and AND and OR gates with any number of inputs can be made from 2-input gates. Do 2-input NAND gates form a complete set of logic gates? Prove your answer.   *complete set*

4.41 Do 2-input NOR gates form a complete set of logic gates? Prove your answer.

4.42 Do 2-input XOR gates form a complete set of logic gates? Prove your answer.

4.43 Define a two-input gate, other than NAND, NOR, or XOR, that forms a complete set of logic gates if the constant inputs 0 and 1 are allowed. Prove your answer.

4.44 Some people think that there are *four* basic logic functions, AND, OR, NOT, and *BUT*. Figure X4.44 is a possible symbol for a 4-input, 2-output *BUT gate*. Invent a useful, nontrivial function for the BUT gate to perform. The function should have something to do with the name (BUT). Keep in mind that, due to the symmetry of the symbol, the function should be symmetric with respect to the A and B inputs of each section and with respect to sections 1 and 2. Describe your BUT's function and write its truth table. (Personal to RR: No offense intended.)   *BUT*   *BUT gate*

Figure X4.44
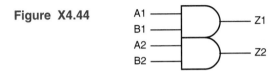

4.45 Write logic expressions for the Z1 and Z2 outputs of the BUT gate you designed in the preceding exercise, and draw a corresponding logic diagram using AND gates, OR gates, and inverters.

4.46 Most students have no problem using theorem T8 to "multiply out" logic expressions, but many develop a mental block if they try to use theorem T8′ to "add out" a logic expression. How can duality be used to overcome this problem?

4.47 How many different logic functions are there of n variables?

4.48 How many different 2-variable logic functions $F(X, Y)$ are there? Write a simplified algebraic expression for each of them.

self-dual logic function
$\oplus$

4.49 A *self-dual logic function* is a function F such that $F = F^D$. Which of the following functions are self-dual? (The symbol $\oplus$ denotes the XOR operation.)

(a) $F = X$

(b) $F = \Sigma_{X,Y,Z}(0, 3, 5, 6)$

(c) $F = X \cdot Y' + X' \cdot Y$

(d) $F = W \cdot (X \oplus Y \oplus Z) + W' \cdot (X \oplus Y \oplus Z)'$

(e) A function F of 7 variables such that $F = 1$ if and only if 4 or more of the variables are 1

(f) A function F of 10 variables such that $F = 1$ if and only if 5 or more of the variables are 1

4.50 How many self-dual logic functions of n input variables are there? (*Hint:* Consider the structure of the truth table of a self-dual function.)

4.51 Prove that any n-input logic function $F(X_1, \ldots, X_n)$ that can be written in the form $F = X_1 \cdot G(X_2, \ldots, X_n) + X_1' \cdot G^D(X_2, \ldots, X_n)$ is self-dual.

4.52 Assuming that an inverting gate has a propagation delay of 5 ns, and a noninverting gate has a propagation delay of 8 ns, compare the speeds of the circuits in Figure 4–24(a), (c), and (d).

4.53 Using switching algebra, show that the logic functions for the AND-OR circuits in Figures 4–27 and 4–28 equal their OR-AND counterparts in Figures 4–36 and 4–37.

4.54 Determine whether the product-of-sums expressions obtained by "adding out" the minimal sums in Figure 4–27 and 4–28 are minimal.

4.55 Prove that the rule for combining 2^i 1-cells in a Karnaugh map is true, using the axioms and theorems of switching algebra.

irredundant sum

4.56 An *irredundant sum* for a logic function F is a sum of prime implicants for F such that if any prime implicant is deleted, the sum no longer equals F. This sounds a lot like a minimal sum, but an irredundant sum is not necessarily minimal. For example, the minimal sum of the function in Figure 4–34 has only three product terms, but there is an irredundant sum with four product terms. Find the irredundant sum and draw a map of the function, circling only the prime implicants in the irredundant sum.

4.57 Find another logic function in Section 4.3 that has one or more nonminimal irredundant sums, and draw its map, circling only the prime implicants in the irredundant sum.

4.58 Derive the minimal product-of-sums expression for the prime BCD-digit detector function of Figure 4–39. Determine whether or not the expression algebraically equals the minimal sum-of-products expression and explain your result.

4.59 Draw a Karnaugh map and assign variables to the inputs of the AND-XOR circuit in Figure X4.59 so that its output is $F = \Sigma_{W,X,Y,Z}(6,7,12,13)$. Note that the output gate is a 2-input XOR rather than an OR.

Figure X4.59

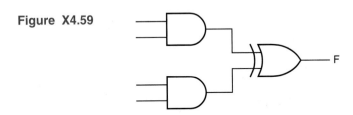

4.60 The text indicates that a truth table or equivalent is the starting point for traditional combinational minimization methods. A Karnaugh map itself contains the same information as a truth table. Given a sum-of-products expression, it is possible to write the 1s corresponding to each product term directly on the map without developing an explicit truth table or minterm list, and then proceed with the map minimization procedure. Find a minimal sum-of-products expression for each of the following logic functions in this way:

(a) $F = X' \cdot Z + X \cdot Y + X \cdot Y' \cdot Z$ (b) $F = A' \cdot C' \cdot D + B' \cdot C \cdot D + A \cdot C' \cdot D + B \cdot C \cdot D$
(c) $F = W \cdot X \cdot Z' + W \cdot X' \cdot Y \cdot Z + X \cdot Z$ (d) $F = (X' + Y') \cdot (W' + X' + Y) \cdot (W' + X + Z)$
(e) $F = A \cdot B \cdot C' \cdot D' + A' \cdot B \cdot C' + A \cdot B \cdot D + A' \cdot C \cdot D + B \cdot C \cdot D'$

4.61 Repeat Exercise 4.60, finding a minimal product-of-sums expression for each logic function.

4.62 A Karnaugh map for a 5-variable function can be drawn as shown in Figure X4.62. In such a map, cells that occupy the same relative position in the $V = 0$ and $V = 1$ submaps are considered to be adjacent. (Many worked examples of 5-variable Karnaugh maps appear in Sections 7.4.4 and 7.4.5.) Find a minimal sum-of-products expression for each of the following functions using a 5-variable map: *5-variable Karnaugh map*

(a) $F = \Sigma_{V,W,X,Y,Z}(5,7,13,15,16,20,25,27,29,31)$
(b) $F = \Sigma_{V,W,X,Y,Z}(0,7,8,9,12,13,15,16,22,23,30,31)$
(c) $F = \Sigma_{V,W,X,Y,Z}(0,1,2,3,4,5,10,11,14,20,21,24,25,26,27,28,29,30)$
(d) $F = \Sigma_{V,W,X,Y,Z}(0,2,4,6,7,8,10,11,12,13,14,16,18,19,29,30)$
(e) $F = \Pi_{V,W,X,Y,Z}(4,5,10,12,13,16,17,21,25,26,27,29)$
(f) $F = \Sigma_{V,W,X,Y,Z}(4,6,7,9,11,12,13,14,15,20,22,25,27,28,30) + d(1,5,29,31)$

Figure X4.62

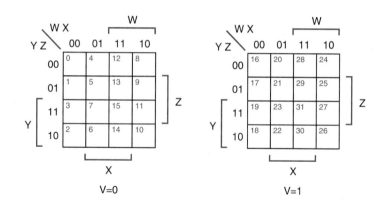

4.63 Repeat Exercise 4.62, finding a minimal product-of-sums expression for each logic function.

6-variable Karnaugh map

4.64 A Karnaugh map for a 6-variable function can be drawn as shown in Figure X4.64. In such a map, cells that occupy the same relative position in adjacent submaps are considered to be adjacent. Minimize the following functions using 6-variable maps:

(a) $F = \Sigma_{U,V,W,X,Y,Z}(1,5,9,13,21,23,29,31,37,45,53,61)$

(b) $F = \Sigma_{U,V,W,X,Y,Z}(0,4,8,16,24,32,34,36,37,39,40,48,50,56)$

(c) $F = \Sigma_{U,V,W,X,Y,Z}(2,4,5,6,12{-}21,28{-}31,34,38,50,51,60{-}63)$

4.65 A 3-bit "comparator" circuit receives two 3-bit numbers, $P = P_2P_1P_0$ and $Q = Q_2Q_1Q_0$. Design a minimal sum-of-products circuit that produces a 1 output if and only if $P > Q$.

4.66 Find minimal multiple-output sum-of-products expressions for $F = \Sigma_{X,Y,Z}(0,1,2)$, $G = \Sigma_{X,Y,Z}(1,4,6)$, and $H = \Sigma_{X,Y,Z}(0,1,2,4,6)$.

4.67 Prove whether or not the following expression is a minimal sum. Do it the easiest way possible (algebraically, not using maps).

$$F = T'{\cdot}U{\cdot}V{\cdot}W{\cdot}X + T'{\cdot}U{\cdot}V'{\cdot}X{\cdot}Z + T'{\cdot}U{\cdot}W{\cdot}X{\cdot}Y'{\cdot}Z$$

4.68 There are $2n$ m-subcubes of an n-cube for the value $m = n - 1$. Show their text representations and the corresponding product terms. (You may use ellipses as required, e.g., $1, 2, \ldots, n$.)

4.69 There is just one m-subcube of an n-cube for the value $m = n$; its text representation is xx...xx. Write the product term corresponding to this cube.

4.70 The Pascal program in Table 4–8 uses memory inefficiently because it allocates memory for a maximum number of cubes at each level, even if this maximum is

Figure X4.64

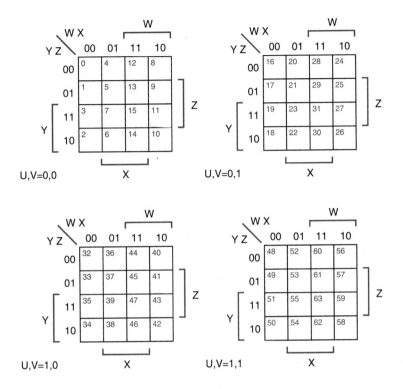

never used. Redesign the program so that the cubes and used arrays are one-dimensional arrays, and each level uses only as many array entries as needed. (*Hint:* You can still allocate cubes sequentially, but keep track of the starting point in the array for each level.)

4.71 As a function of m, how many times is each distinct m-cube rediscovered in Table 4–8, only to be found in the inner loop and thrown away? Suggest some ways to eliminate this inefficiency.

4.72 The third FOR-loop in Table 4–8 tries to combine all m-cubes at a given level with all other m-cubes at that level. In fact, only m-cubes with x's in the same positions can be combined, so it is possible to reduce the number of loop iterations by using a more sophisticated data structure. Design a data structure that segregates the cubes at a given level according to the position of their x's, and determine the maximum size required for various elements of the data structure. Rewrite Table 4–8 accordingly.

4.73 Estimate whether the savings in inner-loop iterations achieved in Exercise 4.72 outweighs the overhead of maintaining a more complex data structure. Try to make reasonable assumptions about how cubes are distributed at each level, and indicate how your results are affected by these assumptions.

4.74 Explain why, in the generalized consensus algorithm, cubes in CC cannot be covered by cubes in C0 and C1.

4.75 Write a Pascal program that selects a minimal cover for a logic function using the iterative consensus method of Section 4.4.4. Your program should behave reasonably if it given a function for which the branching method is required. Reasonable behaviors include halting with an error message, picking a pseudo-essential prime implicant at random, and actually executing the branching method. Getting stuck in an infinite loop is not reasonable.

4.76 *(Hamlet circuit.)* Complete the timing diagram and explain the function of the circuit in Figure X4.76. Where does the circuit get its name?

Figure X4.76

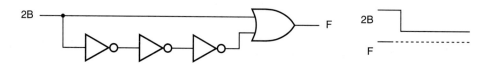

4.77 Prove that a two-level AND-OR circuit corresponding to the complete sum of a logic function is always hazard free.

4.78 Find a four-variable logic function whose minimal sum-of-products realization is not hazard free, but where there exists a hazard-free sum-of-products realization with fewer product terms than the complete sum.

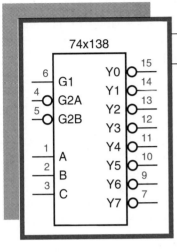

74x138

6	G1	Y0	15
4	G2A	Y1	14
5	G2B	Y2	13
		Y3	12
1	A	Y4	11
2	B	Y5	10
3	C	Y6	9
		Y7	7

5

COMBINATIONAL LOGIC DESIGN PRACTICES

The preceding chapter described the theoretical principles used in combinational logic design. In this chapter, we'll build on that foundation and describe many of the devices, structures, and methods used by engineers to solve practical digital design problems.

A practical combinational circuit may have dozens of inputs and outputs and could require hundreds, thousands, even millions of terms to describe as a sum of products, and *billions and billions* of rows to describe in a truth table. Thus, most real combinational logic design problems are too large to solve by "brute-force" application of theoretical techniques.

But wait, you say, how could any human being conceive of such a complex logic circuit in the first place? The key is structured thinking. A complex circuit or system is conceived as a collection of smaller subsystems, each of which has a much simpler description.

In combinational logic design, there are several straightforward structures—decoders, multiplexers, comparators, and the like—that turn up quite regularly as building blocks in larger systems. The most important of these structures appear in this chapter. We describe each structure generally and then give examples and applications using 74-series ICs. But first, we start this chapter by presenting documentation standards that are used by digital designers to ensure that their designs are correct, manufacturable, and maintainable.

THE IMPORTANCE
OF 74-SERIES
LOGIC

Later in this chapter, we'll look at commonly used 74-series ICs that perform well-structured logic functions. These parts are the most important building blocks in a digital designer's toolbox because their level of functionality matches a designer's level of thinking when partitioning a large problem into smaller chunks.

Even for PLD and ASIC designers, understanding 74-series MSI functions is important. In PLD-based design, standard MSI functions can be used as a starting point for developing the logic equations to perform more specialized functions. And in ASIC design, the basic building blocks (or "standard cells" or "macros") provided by the ASIC manufacturer may actually be defined as 74-series MSI functions, even to the extent of having similar descriptive numbers.

5.1 DOCUMENTATION STANDARDS

Good documentation is essential for correct design and efficient maintenance of digital systems. In addition to being accurate and complete, documentation must be somewhat instructive, so that a test engineer, maintenance technician, or even the original design engineer (six months after designing the circuit) can figure out how the system works just by reading the documentation.

Although the type of documentation depends on system complexity and the engineering and manufacturing environments, a documentation package should generally contain at least the following five items:

block diagram

(1) A *block diagram* is an informal pictorial description of the system's major functional modules and their basic interconnections.

schematic diagram

(2) A *schematic diagram* is a formal specification of the electrical components of the system, their interconnections, and all of the details needed to construct the system, including IC types, reference designators, and pin numbers. We've been using the term *logic diagram* for an informal drawing that does not have quite this level of detail.

logic diagram

timing diagram

(3) A *timing diagram* shows the values of various logic signals as a function of time, including the cause-and-effect delays between critical signals.

(4) A *structured logic description* describes the internal function of a structured logic device, such as a programmable logic device (PLD) or a read-only memory (ROM). Such a description may be in the form of logic equations, state tables or diagrams, or program listings.

structured logic description

(5) A *circuit description* is a narrative text document that, in conjunction with the other documentation, describes concisely how the circuit is intended to be used and explains how it works internally. The circuit description should list any assumptions and potential pitfalls in the circuit's design and operation, and point out the use of any nonobvious design "tricks."

circuit description

You've probably already seen block diagrams in many contexts. We present a few rules for drawing them in the next subsection, and then we concentrate on schematics for combinational logic circuits in the rest of this section. Section 5.2.1 introduces timing diagrams. Structured logic descriptions, such as PLD equations, are covered in later chapters.

The last area of documentation, the circuit description, is very important in practice. Just as an experienced programmer creates a program design document before beginning to write code, an experienced logic designer starts writing the circuit description before drawing the schematic. Unfortunately, the circuit description is sometimes the last document to be created, and sometimes it's never written at all. A circuit without a description is difficult to debug, manufacture, test, maintain, modify, and enhance.

In order to create great products, logic designers must develop their language and writing skills, especially in the area of *logical* outlining and organization. The most successful logic designers (and later, project leaders, system architects, and entrepreneurs) are the ones who communicate their ideas, proposals, and decisions effectively to others. Even though it's a lot of fun to tinker in the digital design lab, don't use that as an excuse to shortchange your writing and communications courses and projects.

DON'T FORGET
TO WRITE!

5.1.1 Block Diagrams

A *block diagram* shows the inputs, outputs, functional modules, internal data paths, and important control signals of a system. In general, it should not be so detailed that it occupies more than one page, yet it must not be too vague. A small block diagram may have three to six blocks, while a large one may have 10 to 15, depending on system complexity. In any case, the block diagram must show the most important system elements and how they "play together." Large systems may require additional block diagrams of individual subsystems, but there should always be a "top-level" diagram showing the entire system.

block diagram

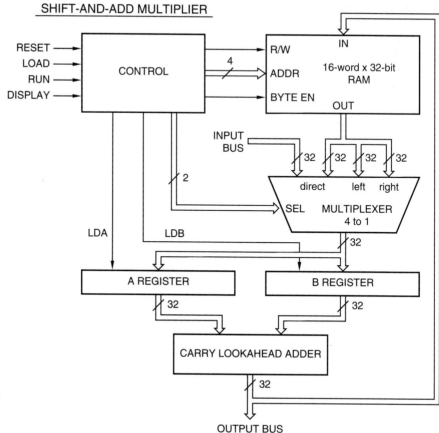

Figure 5–1
Block diagram for a digital
design project.

Figure 5–1 shows a sample block diagram. Each block is labeled with the function of the block, not the individual chips that comprise it. As another example, Figure 5–2(a) shows the block-diagram symbol for a 32-bit register. If the register is to be built using four 74x377 8-bit registers, and this information is important to someone reading the diagram (e.g., for cost reasons), then it can be conveyed as shown in (b). However, splitting the block to show individual chips as in (c) is incorrect.

bus

A *bus* is a collection of two or more related signal lines. In a block diagram, buses are drawn with a double or heavy line. A slash and a number are often used to indicate how many individual signal lines are contained in a bus. Active levels (defined later) and inversion bubbles may or may not appear in block diagrams; in most cases, they are unimportant at this level of detail. However, important control signals and buses should have names, usually the same names that appear in the more detailed schematic.

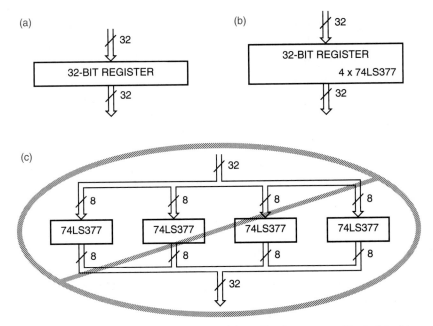

Figure 5–2   A 32-bit register block: (a) realization unspecified; (b) chips specified; (c) too much detail.

The flow of control and data in a block diagram should be clearly indicated. Logic diagrams are generally drawn with signals flowing from left to right, but in block diagrams this ideal is more difficult to achieve. Inputs and outputs may be on any side of a block, and the direction of signal flow may be arbitrary. Arrowheads are used on buses and ordinary signal lines to eliminate ambiguity.

5.1.2   Gate Symbols

The symbol shapes for AND and OR gates and buffers are shown in Figure 5–3(a). (Recall from Chapter 3 that a buffer is a circuit that simply converts "weak" logic signals into "strong" ones.) To draw logic gates with more than about four inputs, we expand the AND and OR symbols as shown in (b). A small circle, called an *inversion bubble*, denotes logical complementation and is used in the symbols   *inversion bubble* for NAND and NOR gates and inverters as shown in (c).

Using the generalized DeMorgan's theorem, we can manipulate the logic expressions for gates with complemented outputs. For example, if X and Y are the inputs of a NAND gate with output Z, then we can write

$$Z \;=\; (X \cdot Y)'$$
$$=\; X' + Y'$$

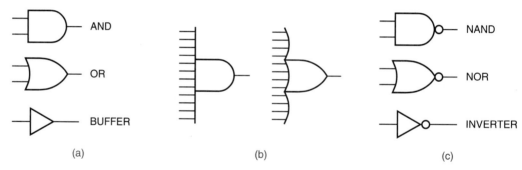

Figure 5-3 Shapes for basic logic gates: (a) AND, OR, and buffers; (b) expansion of inputs; (c) inversion bubbles.

This gives rise to two different but equally correct symbols for a NAND gate, as we demonstrated in Figure 4–3 on page 201. In fact, this sort of manipulation may be applied to gates with uncomplemented inputs, too. For example, consider the following equations for an AND gate:

$$
\begin{aligned}
Z &= X \cdot Y \\
&= ((X \cdot Y)')' \\
&= (X' + Y')'
\end{aligned}
$$

Thus, an AND gate may be symbolized as an OR gate with inversion bubbles on its inputs and output.

Equivalent symbols for standard gates that can be obtained by these manipulations are summarized in Figure 5–4. Even though both symbols in a pair represent the same logic function, the choice of one symbol or the other in a logic diagram is not arbitrary, at least not if we are adhering to good documentation standards. As we'll show in the next three subsections, proper choices of signal names and gate symbols can make logic diagrams much easier to use and understand.

IEEE STANDARD
LOGIC SYMBOLS

Together with the American National Standards Institute (ANSI), the Institute of Electrical and Electronic Engineers (IEEE) has developed a standard set of logic symbols. The most recent revision of the standard is ANSI/IEEE Std 91-1984, *IEEE Standard Graphic Symbols for Logic Functions.* The standard allows both rectangular- and distinctive-shape symbols for logic gates. We have been using and will continue to use the distinctive shape symbols throughout this book, but the rectangular-shape symbols are described in Appendix A.

Figure 5-4
Equivalent gate symbols under
the generalized DeMorgan's
theorem.

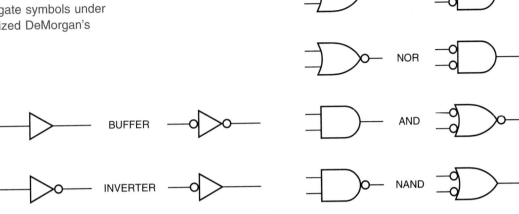

OR

NOR

BUFFER

INVERTER

AND

NAND

5.1.3 Signal Names and Active Levels

Each input and output signal in a logic circuit should have a descriptive alphanumeric label, the signal's name. Most computer-aided design systems for drawing logic circuits also allow certain special characters, such as *, _, and !, to be included in signal names. In the analysis and synthesis examples in Chapter 4, we used mostly single-character signal names (X, Y, etc.) because the circuits didn't do much. However, in a real system, well-chosen signal names convey information to someone reading the logic diagram the same way that variable names in a software program do. The names usually indicate an action that is controlled (GO, PAUSE) or a condition that is detected (READY, ERROR) by the signal.

Each signal name should have an *active level* associated with it. A signal is *active high* if it performs the named action or denotes the named condition when it is HIGH or 1. (Under the positive-logic convention, which we use throughout this book, "HIGH" and "1" are equivalent.) A signal is *active low* if it performs the named action or denotes the named condition when it is LOW or 0. A signal is said to be *asserted* when it is at its active level. A signal is said to be *negated* (or, sometimes, *deasserted*) when it is not at its active level.

active level
active high

active low
assert
negate
deassert

The active level of each signal in a circuit is normally specified as part of its

---

Although it is absolutely necessary to name only a circuit's main inputs and outputs, most logic designers find it useful to name internal signals as well. During circuit debugging, it's nice to have a name to use when pointing to an internal signal that's behaving strangely. Most computer-aided design systems automatically generate labels for unnamed signals, but a user-chosen name is preferable to a computer-generated one like XSIG1057.

NAME THAT
SIGNAL!

Table 5–1
Each line shows a different naming convention for active levels.

Active Low	Active High
READY–	READY+
ERROR.L	ERROR.H
ADDR15(L)	ADDR15(H)
RESET*	RESET
ENABLE~	ENABLE
/TRANSMIT	TRANSMIT

name, according to some convention. Examples of several different active-level naming conventions are shown in Table 5–1. The choice of one of these or other signal naming conventions is sometimes just a matter of personal preference, but more often it is constrained by the engineering environment. Since the active-level designation is part of the signal name, the naming convention must be compatible with the input requirements of any computer-aided design tools, such as schematic editors and simulators, that will process the signal names. In this text, we'll use the last convention in the table: An active-low signal name has a prefix of /, and an active-high signal has no prefix. The / prefix may be read as "not."

active-level naming convention

/

It's extremely important to understand the difference between signal names, expressions, and equations. A *signal name* is just a name—an alphanumeric label. A *logic expression* combines signal names using the operators of switching algebra—AND, OR, and NOT—as we explained in Section 4.1. A *logic equation* is an assignment of a logic expression to a signal name—it describes one signal's function in terms of other signals.

signal name
logic expression
logic equation

The distinction between signal names and logic expressions can be related to a concept used in computer programming languages: The left-hand side of an assignment statement contains a variable *name*, and the right-hand side contains an *expression* whose value will be given to the named variable [e.g., Z := -(X+Y) in Pascal]. In a programming language, you can't put an expression on the left-hand side of an assignment statement. In logic design, you shouldn't use a logic expression as a signal name.

Logic signals may have names like X, READY, and /GO. The "/" in /GO is just part of the signal's name, like an underscore in a Pascal variable name. There is *no* signal whose name is READY'—this is an expression, since ' is an operator. However, there *may* be another signal named /READY such that /READY = READY' during normal operation of the circuit. We are very careful in this book to distinguish between signal names, which are always printed in black, and logic expressions, which are always printed in color when they are written near the corresponding signal lines.

CAD NOTE

In a computer-aided design (CAD) system, you can associate *names*, not expressions, with all of the signals in a logic circuit, following conventions similar to what we describe in this section. Expressions that describe each signal in terms of other signals are a consequence of the circuit's structure. If you need to know these expressions, you can derive them by hand or, if you're fortunate, just give a command to your computer-aided design system to work them out for you.

5.1.4   Active Levels for Pins

When we draw the outline of an AND or OR symbol, or a rectangle representing a larger-scale logic element, we think of the given logic function as occurring *inside* that symbolic outline. In Figure 5–5(a), we show the logic symbols for an AND and OR gate and for a larger-scale element with an ENABLE input. The AND and OR gates have active-high inputs—they require 1s on the input to assert their outputs. Likewise, the larger-scale element has an active-high ENABLE input, which must be 1 to enable the element to do its thing. In (b), we show the same logic elements with active-low input and output pins. Exactly the same logic functions are performed *inside* the symbolic outlines, but the inversion bubbles indicate that 0s must now be applied to the input pins to activate the logic functions, and that the outputs are 0 when they are "doing their thing."

Thus, active levels may be associated with the input and output pins of gates and larger-scale logic elements. We use an inversion bubble to indicate an active-low pin and the absence of a bubble to indicate an active-high pin. For example, the AND gate in Figure 5–6(a) performs the logical AND of two active-

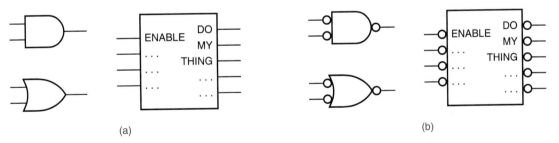

(a)   (b)

Figure 5–5 Logic symbols: (a) AND, OR, and a larger-scale logic element; (b) the same elements with active-low inputs and outputs.

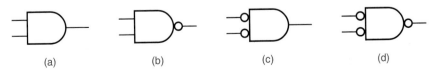

(a)   (b)   (c)   (d)

Figure 5–6 Four ways of obtaining an AND function: (a) AND gate (7408); (b) NAND gate (7400); (c) NOR gate (7402); (d) OR gate (7432).

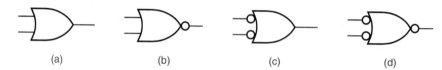

Figure 5-7 Four ways of obtaining an OR function: (a) OR gate (7432); (b) NOR gate (7402); (c) NAND gate (7400); (d) AND gate (7408).

high inputs and produces an active-high output: if both inputs are asserted (1), the output is asserted (1). The NAND gate in (b) also performs the AND function, but it produces an active-low output. Even a NOR or OR gate can be construed to perform the AND function using active-low inputs and outputs, as shown in (c) and (d). All four gates in the figure can be said to perform the same function: the output of each gate is asserted if both of its inputs are asserted. Figure 5-7 shows the same idea for the OR function: The output of each gate is asserted if either of its inputs is asserted.

Sometimes a noninverting buffer is used simply to boost the fanout of a logic signal without changing its function. Figure 5-8 shows the possible logic symbols for both inverters and noninverting buffers. In terms of active levels, all of the symbols perform exactly the same function: Each asserts its output signal if and only if its input is asserted.

Figure 5-8 Alternate logic symbols: (a, b) inverters; (c, d) noninverting buffers.

5.1.5 Bubble-to-Bubble Logic Design

Experienced logic designers formulate their circuits in terms of the logic functions performed *inside* the symbolic outlines, and select discrete gates to fit the constraints of the design. Design constraints include input and output active levels of the selected larger-scale elements, speed, cost, and the availability of discrete gate types. However, it is still possible to choose logic symbols and descriptive signal names, including active-level designators, to make the function of a logic circuit quite easy to understand. Usually, this means choosing signal names and gate types and symbols so that most of the inversion bubbles "cancel out." In this way, simplified logic expressions using AND and OR operators may be read directly from the logic diagram.

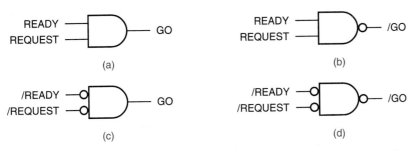

Figure 5-9 Many ways to GO: (a) active-high inputs and output; (b) active-high inputs, active-low output; (c) active-low inputs, active-high output; (d) active-low inputs and outputs.

For example, suppose we need to produce a signal that tells a device to "GO" when we are "READY" and we get a "REQUEST." Clearly, an AND function is required; in switching algebra, we can write GO = READY·REQUEST. However, we can use different gates to perform the AND function, depending on the active level required for the GO signal and the active levels of the available input signals. Figure 5–9(a) shows the simplest case, where GO must be active-high and the available input signals are also active-high; we use an AND gate. If, on the other hand, the device that we're controlling requires an active-low /GO signal, we can use a NAND gate as shown in (b). If the available input signals are active-low, we can use a NOR or OR gate as shown in (c) and (d).

The active levels of available signals don't always match the active levels of available gates. For example, suppose we are given input signals /READY (active-low) and REQUEST (active-high). Figure 5–10 shows two different ways to generate GO using an inverter to generate the appropriate active level for the AND function. The second way is generally preferred, since inverting gates like NOR are generally faster than noninverting ones like AND. We drew the inverter differently in each case to make the output's active level match its signal name.

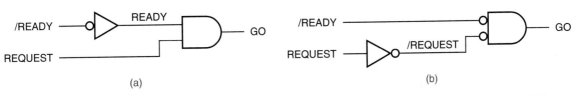

Figure 5-10 Two more ways to GO, with mixed input levels: (a) with an AND gate; (b) with a NOR gate.

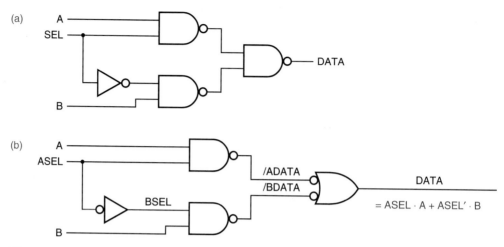

Figure 5–11   A 2-input multiplexer (you're not supposed to know what that is yet): (a) cryptic logic diagram; (b) proper logic diagram using active-level designators and alternate logic symbols.

To understand the benefits of bubble-to-bubble logic design, consider the circuit in Figure 5–11(a). What does it do? In Section 4.2 we showed several ways to analyze such a circuit, and we could certainly obtain a logic expression for the DATA output using these techniques. However, when the circuit is redrawn in Figure 5–11(b), the output function can be read directly from the logic diagram, as follows. The DATA output is asserted when either /ADATA or /BDATA is asserted. If ASEL is asserted, then /ADATA is asserted if and only if A is asserted; that is, /ADATA is a copy of A. If ASEL is negated, BSEL is asserted and /BDATA is a copy of B. In other words, DATA is a copy of A if ASEL is asserted, and DATA is a copy of B if ASEL is negated. Even though there are five inversion bubbles in the logic diagram, we mentally had to perform only one negation to understand the circuit—that BSEL is asserted if ASEL is not.

If we wish, we can write an algebraic expression for the DATA output. We use the technique of Section 4.2, simply propagating expressions through gates toward the output. In doing so, we can ignore pairs of inversion bubbles that cancel, and directly write the expression shown in color in the figure.

Another example is shown in Figure 5–12. Reading directly from the logic diagram, we see that /ENABLE is asserted if /READY and /REQUEST are asserted or if TEST is asserted. The HALT output is asserted if /READY and /REQUEST are not both asserted or if /LOCK is asserted. Once again, this example has only

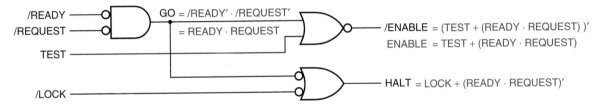

Figure 5–12   Another properly drawn logic diagram.

one place where a gate input's active level does not match the input signal level, and this is reflected in the verbal description of the circuit.

We can, if we wish, write algebraic equations for the /ENABLE and HALT outputs. As we propagate expressions through gates towards the output, we obtain expressions like /READY'·/REQUEST'. However, we can use our active-level naming convention to simplify terms like /READY'. The circuit contains no signal with the name READY; but if it did, it would satisfy the relationship READY = /READY' according to the naming convention. This allows us to write the /ENABLE and HALT equations as shown. Complementing both sides of the /ENABLE equation, we obtain an equation that describes a hypothetical active-high ENABLE output in terms of hypothetical active-high inputs.

We'll see more examples of bubble-to-bubble logic design in this and later chapters, especially as we begin to use larger-scale logic elements.

The following rules are useful for performing bubble-to-bubble logic design:

- The signal name on a device's output should have the same active level as the device's output pin, that is, active-low if the device symbol has an inversion bubble on the output pin, active-high if not.

- If the active level of an input signal is the same as that of the input pin to which it is connected, then the logic function inside the symbolic outline is activated when the signal is asserted. This should be the most common case in a logic diagram.

- If the active level of an input signal is the opposite of that of the input pin to which it is connected, then the logic function inside the symbolic outline is activated when the signal is negated. This case should be avoided whenever possible because it forces us to keep track mentally of a logical negation to understand the circuit.

BUBBLE-TO-
BUBBLE
LOGIC DESIGN
RULES

5.1.6 Drawing Layout

Logic diagrams and schematics should be drawn with gates in their "normal" orientation with inputs on the left and outputs on the right. The logic symbols for larger-scale logic elements are also normally drawn with inputs on the left and outputs on the right.

A complete schematic page should be drawn with system inputs on the left and outputs on the right, and the general flow of signals should be from left to right. If an input or output appears in the middle of the page, it should be extended to the left or right edge, respectively. In this way, a reader can find all inputs and outputs by looking at the edges of the page only. All signal paths on the page should be connected when possible; paths may be broken if the drawing gets crowded, but breaks should be flagged in both directions, as described later.

Sometimes block diagrams are drawn without crossing lines to give a neater appearance, but for logic diagrams this is not recommended. For logic diagrams it is best to allow lines to cross and to indicate connections clearly with a dot. However, some computer-aided design systems (and some designers) can't draw legible connection dots. To distinguish between crossing lines and connected lines, they adopt the convention that only "T"-type connections are allowed, as shown in Figure 5–13. This is a good convention to follow in any case.

Figure 5–13
Line crossings and connections.

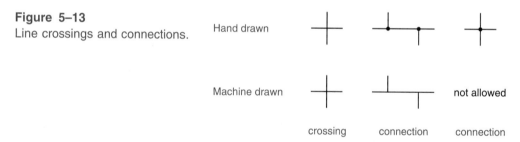

Schematics that fit on a single page are the easiest to work with. The largest practical paper size for a schematic might be E-size (34×44 inches). Although its drawing capacity is great, such a large paper size is fairly unwieldy to work with. The best compromise of drawing capacity and practicality is B-size (11×17 inches) paper. It can be easily folded for storage and quick reference in standard 3-ring notebooks, and it can be copied on most office copiers (because of all the accountants with their B-size ledger paper). Some designers draw their original schematics on D-size (22×34 inches) paper, and use a 50%-reduced B-size copy for reference. Regardless of paper size, schematics seem to come out best when the page is used in landscape format, that is, with its long dimension oriented from left to right, the direction of most signal flow.

Schematics that don't fit on a single page should be broken up into indi-

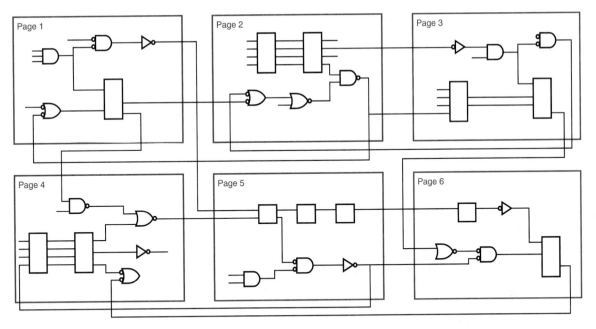

Figure 5–14 Flat schematic structure.

vidual pages in a way that minimizes the connections (and confusion) between pages. They may also use a coordinate system, like that of a road map, to flag the source and destination of signals that travel from one page to another. An outgoing signal should have flags referring to all of the destinations of that signal, while an incoming signal should have a flag referring to the source only. That is, an incoming signal should be flagged to the place where it is generated, not to a place somewhere in the middle of a chain of destinations that use the signal. *signal flags*

A multiple-page schematic usually has a "flat" structure. As shown in Figure 5–14, each page is carved out from the complete schematic and can connect to any other page as if all the pages were on one large sheet. However, much like programs, schematics can also be constructed hierarchically, as illustrated in Figure 5–15. In this approach, the "top-level" schematic is just a single page that may take the place of a block diagram. Typically, the top-level schematic contains no gates or other logic elements; it shows the major subsystems and their interconnections. The subsystems are in turn defined on lower-level pages, which may contain ordinary gate-level descriptions, or may themselves define even lower-level hierarchies. If a particular lower-level hierarchy is used more than once, it may be reused (or "called," in the programming sense) multiple times by the higher-level pages. Most computer-aided logic design systems support hierarchical schematics. *flat schematic structure*

hierarchical schematic structure

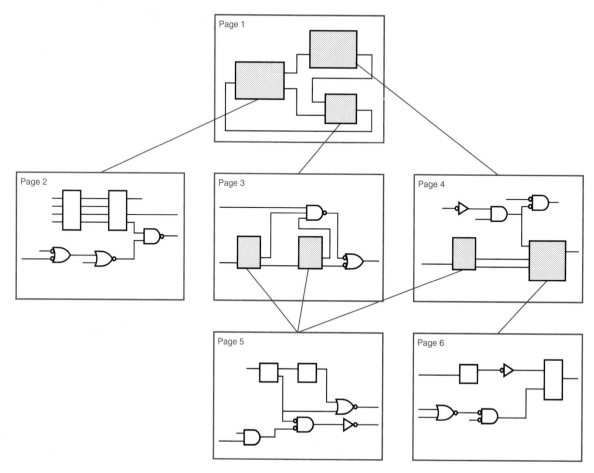

Figure 5–15 Hierarchical schematic structure.

5.1.7 Buses

As defined previously, a bus is a collection of two or more related signal lines. For example, a microprocessor system might have an address bus containing 16 address lines, ADDR0–ADDR15, and a data bus with 8 data lines, DATA0–DATA7. The signal names in a bus are not necessarily related or ordered as in these first examples. For example, a microprocessor system might have a control bus containing five signals, ALE, MIO, /RD, /WR, and RDY.

Logic diagrams use special notation for buses in order to reduce the amount of drawing and to improve readability. As shown in Figure 5–16, a bus typically has its own descriptive name, such as ADDR[15–0], DATA[7–0], or CONTROL, and is drawn with a double line or a thicker line than ordinary signals. Individual signals may be put into or pulled out of the bus by connecting an ordinary signal

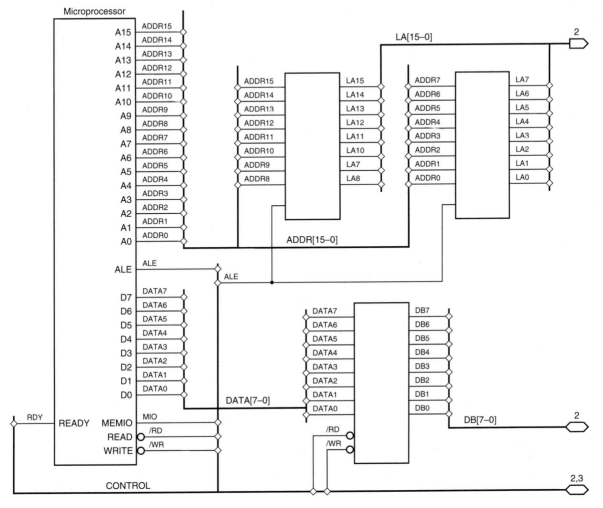

Figure 5–16 Examples of buses.

line to the bus and writing its name on the signal line. Sometimes a special connection dot is also used, as shown in the example. A computer-aided design system keeps track of the individual signals in a bus. When it actually comes time to build a circuit from the schematic, signal lines in a bus are treated just as though they had all been drawn individually.

The symbols at the right-hand edge of Figure 5–16 are interpage signal flags. They indicate that the LA bus goes out to page 2, the DB bus is bidirectional and connects to page 2, and the CONTROL bus is bidirectional and connects to pages 2 and 3.

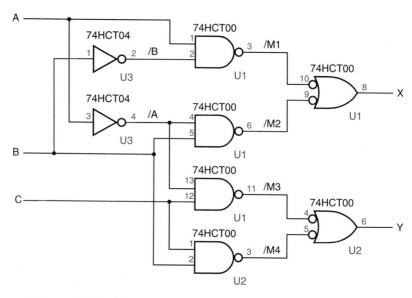

Figure 5–17 Schematic diagram for a circuit using a 74HCT00.

5.1.8 Additional Schematic Information

IC type

Complete schematic diagrams indicate IC types, reference designators, and pin numbers, as in Figure 5–17. The *IC type* is a part number identifying the integrated circuit that performs a given logic function. For example, a 2-input NAND gate might be identified as a 74LS00 or a 74LS37. Even when there is only one possible generic part number for a given gate type (e.g., 7408 for a 2-input AND gate), the IC type identifies the logic family and speed for that gate (e.g., 74HCT08, 74ACT08, 74LS08, 74ALS08, etc.).

reference designator

The *reference designator* for an IC identifies a particular instance of that IC type installed in the system. In conjunction with the system's mechanical documentation, the reference designator allows a particular IC to be located during assembly, test, and maintenance of the system. Traditionally, reference designators for ICs begin with the letter U (for "unit").

pin number

Once a particular IC is located, *pin numbers* are used to locate individual logic signals on its pins. Sometimes the manufacturers' catalogs contain pin diagrams for ICs in the form shown in Figure 5–18. However, logic diagrams and schematics should never contain these pin diagrams. Instead, the pin numbers are written near the corresponding inputs and outputs of the standard logic symbol, as shown in Figure 5–17.

Figure 5–18
Pin diagram for a 74HCT00 quad
2-input NAND gate.

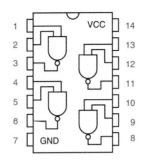

In the rest of this book, just to make you comfortable with properly drawn schematics, we'll include reference designators and pin numbers for all of our logic circuit examples.

Figure 5–19 shows the pinouts of many different SSI ICs that are used in examples throughout this book. Some special graphic elements appear in a few of the symbols:

- Symbols for the 74x14 Schmitt-trigger inverter has a special element inside the symbol to indicate hysteresis.

- Symbols for the 74x03 quad NAND and the 74x266 quad Exclusive NOR have a special element to indicate an open-drain or open-collector output.

Note that an IC's pin numbers may differ depending on package type. The figure shows the pin numbers that are used in a dual-inline package, the type of package that you would use in a digital design laboratory course or in a low-density, "thru-hole" commercial printed-circuit board.

5.2 CIRCUIT TIMING

5.2.1 Timing Diagrams

A *timing diagram* illustrates the logical behavior of signals in a digital circuit as a function of time. Timing diagrams are an important part of the documentation of any digital system. They can be used both to explain the timing relationships among signals within a system, and to define the timing requirements of external signals that are applied to the system.

timing diagram

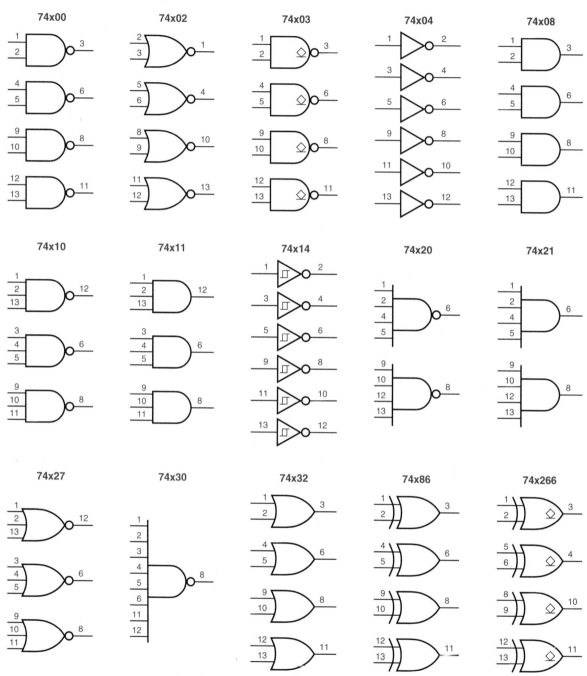

Figure 5-19 Pinouts for SSI ICs in standard dual-inline packages.

Figure 5–20(a) is the block diagram of a simple combinational circuit with two inputs and two outputs. Assuming that the ENB input is held at a constant value, (b) shows the delay of the two outputs with respect to the GO input. In each waveform, the upper line represents a logic 1, and the lower line a logic 0. Signal transitions are drawn as slanted lines to remind us that they do not occur in zero time in real circuits. (Also, slanted lines look nicer than vertical ones.)

Arrows are sometimes drawn, espccially in complex timing diagrams, to show *causality*—which input transitions cause which output transitions. In any case, the most important information provided by a timing diagram is a specification of the *delay* between transitions.

causality

delay

Different paths through a circuit may have different delays. For example, Figure 5–20(b) shows that the delay from GO to READY is shorter than the delay from GO to DAT. Similarly, the delays from the ENB input to the outputs may vary. And, as we'll discuss later, the delay through any given path may vary depending on whether the output is changing from LOW to HIGH or from HIGH to LOW (this phenomenon is not shown in the figure).

Delay in a real circuit is normally measured between the centerpoints of transitions, so the delays in a timing diagram are marked this way. A single timing diagram may contain many different delay specifications. Each different

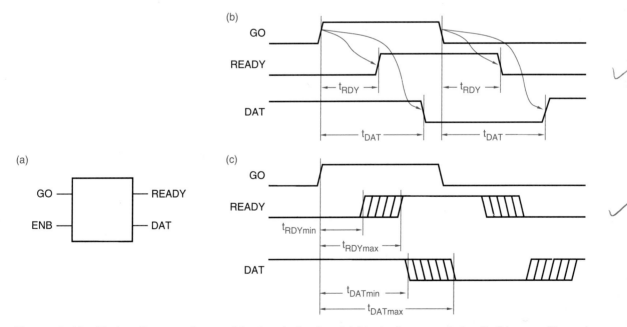

Figure 5–20   Timing diagrams for combinational circuits: (a) block diagram of circuit; (b) causality and propagation delay; (c) minimum and maximum delays.

delay is marked with a different identifier, such as t_{RDY} and t_{DAT} in the figure. In large timing diagrams, the delay identifiers are usually numbered for easier reference (e.g., $t_1, t_2, \ldots, t_{42}$). In either case, the timing diagram is normally accompanied by a *timing table* that specifies each delay amount and the conditions under which it applies.

timing table

Since the delays of real digital components can vary depending on voltage, temperature, and manufacturing parameters, delay is seldom specified as a single number. Instead, a timing table may specify a range of values by giving *minimum*, *typical*, and *maximum* values for each delay. The idea of a range of delays is sometimes carried over into the timing diagram itself by showing the transitions to occur at uncertain times, as in Figure 5–20(c).

For some signals, the timing diagram needn't show whether the signal changes from 1 to 0 or from 0 to 1 at a particular time, only that a transition occurs at that time. Any signal that carries a bit of "data" has this characteristic—the actual value of the data bit varies according to circumstances but, regardless of value, the bit is transferred, stored, or processed at a particular time relative to "control" signals in the system. Figure 5–21(a) is a timing diagram that illustrates this concept. The "data" signal is normally at a steady 0 or 1 value, and transitions occur only at the times indicated. The idea of an uncertain delay time can also be used with "data" signals, as shown for the DATAOUT signal.

Quite often in digital systems, a group of data signals in a bus is processed by identical circuits. In this case, all signals in the bus have the same timing,

Figure 5–21
Timing diagrams for "data" signals: (a) certain and uncertain transitions; (b) sequence of values on an 8-bit bus.

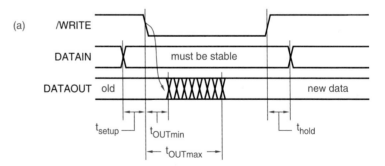

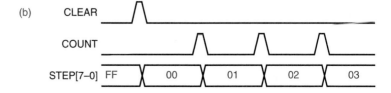

and can be represented by a single line in the timing diagram and corresponding specifications in the timing table. If the bus bits are known to take on a particular combination at a particular time, this is sometimes shown in the timing diagram using binary, octal, or hexadecimal numbers, as in Figure 5–21(b).

5.2.2 Propagation Delay

In Section 3.6.2, we formally defined the *propagation delay* of a signal path as the time that it takes for a change at the input of the path to produce a change at the output of the path. A combinational circuit with many inputs and outputs has many different paths, and each one may have a different propagation delay. Also, the propagation delay when the output changes from LOW to HIGH (t_{pLH}) may be different from the delay when it changes from HIGH to LOW (t_{pHL}). *propagation delay*

    The manufacturer of a combinational-logic IC normally specifies all of these different propagation delays, or at least the delays that would be of interest in typical applications. A logic designer who combines ICs in a larger circuit uses the individual device specifications to analyze the overall circuit timing. The delay of a path through the overall circuit is the sum of the delays through subpaths in the individual devices.

    The timing specification for a device may give minimum, typical, and maximum values for each propagation-delay path and transition direction:

- *Maximum.* This specification is the one that is most often used by experienced designers, since a path "never" has a propagation delay longer than the maximum. However, the definition of "never" varies among logic families and manufacturers. For example, "maximum" propagation delays of 74LS and 74S TTL devices are specified with $V_{CC} = 5$ V, $T_A = 25°C$, and almost no capacitive load. If the voltage or temperature is different, or if the capacitive load is more than 15 pF, the delay may be longer. On the other hand, a "maximum" propagation delay is specified for 74AC and 74ACT devices over the full operating voltage and temperature range, and with a heavier capacitive load of 50 pF. *maximum delay*

- *Typical.* This specification is the one that is most often used by designers who don't expect to be around when their product leaves the friendly environment of the engineering lab and is shipped to customers. The "typical" delay is what you see from a device that was manufactured on a good day and is operating under near-ideal conditions. *typical delay*

- *Minimum.* This is the smallest propagation delay that a path will ever exhibit. Most well-designed circuits don't depend on this number; that is, *minimum delay*

Table 5–2 Propagation delay in nanoseconds of selected TTL and CMOS SSI parts.

	74LS				74HCT		74ACT			
	Typical		Maximum		Typical	Maximum	Typical		Maximum	
Part number	t_{pLH}	t_{pHL}	t_{pLH}	t_{pHL}	t_{pLH}, t_{pHL}	t_{pLH}, t_{pHL}	t_{pLH}	t_{pHL}	t_{pLH}	t_{pHL}
'00, '04, '10, '20	9	10	15	15	11	35	5.5	4.0	9.5	8.0
'02, '27	10	10	15	15	9	29	5.5	6.0	9.0	10
'08, '11, '21	8	10	15	20	11	35	6.0	6.0	10	10
'14	15	15	22	22	16	48	7.5	6.5	12.5	11
'30	8	13	15	20	11	35				
'32	14	14	22	22	9	30	6.0	6.0	10	10
'86 (2 levels)	12	10	23	17	13	40	8.5	7.0	10	10.5
'86 (3 levels)	20	13	30	22	13	40	8.5	7.0	10	10.5

they will work properly even if the delay is zero. That's good because manufacturers don't specify minimum delay in most moderate-speed logic families, including 74LS and 74S TTL. However, in high-speed families, including ECL and 74AC and 74ACT CMOS, a nonzero minimum delay is specified to help the designer ensure that timing requirements of latches and flip-flops, discussed in Section 7.2, are met.

Table 5–2 lists the typical and maximum delays of several 74-series CMOS and TTL gates. Table 5–3 does the same thing for most of the CMOS and TTL MSI parts that are introduced later in this chapter.

All inputs of an SSI gate have the same propagation delay to the output. Note that TTL gates usually have different delays for LOW-to-HIGH and HIGH-to-LOW transitions (t_{pLH} and t_{pHL}), but CMOS gates often do not. CMOS gates have a more symmetrical output driving capability, so any difference between the two cases may not be worth noting.

Also note that two different sets of numbers are given for the 74LS86 XOR gate (discussed in Section 5.7.1). If one input is LOW, and the other is changed, the change propagates through two levels of logic internally, and we observe the first set of delays. If one input is HIGH, and the other is changed, the change propagates through *three* levels of logic internally, and we observe the second set of delays. Similar behavior is exhibited by the 74LS138 and 74LS139 in the MSI table. However, the manufacturers of the corresponding CMOS parts do not specify such differences; apparently they are small enough to be ignored.

To accurately analyze the timing of a circuit containing multiple SSI and MSI devices, a designer may have to study its logical behavior in excruciating

Table 5–3   Propagation delay in nanoseconds of selected TTL and CMOS MSI parts.

			74LS				74HCT		74FCT	
			Typical		Maximum		Typical	Maximum	Typical	Maximum
Part	From	To	t_{pLH}	t_{pHL}	t_{pLH}	t_{pHL}	t_{pLH},t_{pHL}	t_{pLH},t_{pHL}	t_{pLH},t_{pHL}	t_{pLH},t_{pHL}
'138	any select	output (2)	11	18	20	41	23	45	5	9
	any select	output (3)	21	20	27	39	23	45	5	9
	/G2A, /G2B	output	12	20	18	32	22	42	4	8
	G1	output	14	13	26	38	22	42	4	8
'139	any select	output (2)	13	22	20	33	14	43	5	9
	any select	output (3)	18	25	29	38	14	43	5	9
	enable	output	16	21	24	32	11	43	5	9
'151	any select	Y	27	18	43	30	17	51	5	9
	any select	/Y	14	20	23	32	18	54	5	9
	any data	Y	20	16	32	26	16	48	4	7
	any data	/Y	13	12	21	20	15	45	4	7
	enable	Y	26	20	42	32	12	36	4	7
	enable	/Y	15	18	24	30	15	45	4	7
'153	any select	output	19	25	29	38	14	43	5	9
	any data	output	10	17	15	26	12	43	4	7
	enable	output	16	21	24	32	11	34	4	7
'157	select	output	15	18	23	27	15	46	7	10.5
	any data	output	9	9	14	14	12	38	4	6
	enable	output	13	14	21	23	12	38	7	10.5
'182	any /Gi, /Pi	C1–3					13	41		
	any /Gi, /Pi	/G					13	41		
	any /Pi	/P					11	35		
	C0	C1–3					17	50		
'280	any input	EVEN	33	29	50	45	18	53	6	10
	any input	ODD	23	31	35	50	19	56	6	10
'283	C0	any Si	16	15	24	24	22	66		
	any Ai, Bi	any Si	15	15	24	24	21	61		
	C0	C4	11	11	17	22	19	58		
	any Ai, Bi	C4	11	12	17	17	20	60		
'381	CIN	any Fi	18	14	27	21				
	any Ai, Bi	/G	20	21	30	33				
	any Ai, Bi	/P	21	33	23	33				
	any Ai, Bi	any Fi	20	15	30	23				
	any select	any Fi	35	34	53	51				
	any select	/G, /P	31	32	47	48				
'682	any Pi	/PEQQ	13	15	25	25	26	69	7	11
	any Qi	/PEQQ	14	15	25	25	26	69	7	11
	any Pi	/PGTQ	20	15	30	30	26	69	9	14
	any Qi	/PGTQ	21	19	30	30	26	69	9	14

HOW TYPICAL
IS TYPICAL?

Most ICs, perhaps 99%, really are manufactured on "good" days and exhibit delays near the "typical" specifications. However, if you design a system that works only if all of its 100 ICs meet the "typical" timing specs, probability theory suggests that 63% $(1 - .99^{100})$ of the systems won't work.

detail. For example, when TTL inverting gates (NAND, NOR, etc.) are placed in series, a LOW-to-HIGH change at one gate's output causes a HIGH-to-LOW change at the next one's, and so the differences between t_{pLH} and t_{pHL} tend to average out. On the other hand, when noninverting gates (AND, OR, etc.) are placed in series, a transition causes all outputs to change in the same direction, and so the gap between t_{pLH} and t_{pHL} tends to widen. The analysis gets more complicated if there are MSI devices in the delay path, or if there are multiple paths from a given input signal to a given output signal.

Thus, in large circuits, analysis of all of the different delay paths and transition directions can be very complex. To permit a simplified "worst-case" *worst-case delay*   analysis, designers often use a single *worst-case delay* specification that is the

A COROLLARY OF
MURPHY'S LAW

Murphy's law states, "If something can go wrong, it will." A corollary to this is, "If you want something to go wrong, it won't."

In the example described in the previous box, you might think that you have a 63% chance of detecting the potential timing problems in the engineering lab. The problems aren't spread out evenly, though, since all ICs from a given batch tend to behave about the same. Murphy's Corollary says that *all* of the engineering prototypes will be built with ICs from the same, "good" batches. Therefore, everything works fine for a while, just long enough for the system to get into volume production and for everyone to become complacent and self-congratulatory.

Then, unbeknownst to the production department, a "slow" batch of some IC type arrives from a supplier and gets used in every system that is built, so that *nothing* works. The production engineers scurry around trying to analyze the problem (not easy, because the designer is long gone and didn't bother to write a circuit description), and in the meantime the company loses big bucks because it is unable to ship its product.

ESTIMATING
MINIMUM
DELAYS

If the minimum delay of an IC is not specified, a conservative designer assumes that it has a minimum delay of zero.

Some circuits won't work if the propagation delay actually goes to zero, but the cost of modifying a circuit to handle the zero-delay case may be unreasonable, especially since this case is expected never to occur. To obtain a design that always works under "reasonable" conditions, logic designers often estimate that ICs have minimum delays of one-fourth to one-third of their published *typical* delays.

LOGIC FAMILIES

As of this writing, 74LS and 74HCT are the most inexpensive and commonly used MSI logic families, but most logic gates and larger-scale elements are available in a variety of families. For example, the 74LS139, 74S139, 74ALS139, 74AS139, 74HCT139, 74ACT139, 74AC139, and 74FCT139 are all dual 2-to-4 decoders with the same logic function and pinouts, but in electrically different TTL and CMOS families. In addition, "macro" logic elements with the same pin names and functions as the '139 and other popular 74-series devices are available as building blocks in most ASIC design environments. Thus, throughout this text, we use "74x" as a generic prefix. And we'll sometimes omit the prefix and write, for example, '139. In a real schematic diagram for a circuit that you are going to build or simulate, you should include the full part number, since timing and loading information depend on the family.

maximum of t_{pLH} and t_{pHL} specifications. The worst-case delay through a circuit is then computed as the sum of the worst-case delays through the individual components, independent of the transition direction and other circuit conditions. This may give an overly pessimistic view of the overall circuit delay, but it saves design time and it's guaranteed to work. Of course, designers who have access to sophisticated CAD software can simply run a timing analysis program to obtain true minimum, typical, and maximum delays for any desired circuit path.

5.3 DECODERS

A *decoder* is a multiple-input, multiple-output logic circuit that converts coded inputs into coded outputs, where the input and output codes are different. The input code generally has fewer bits than the output code, and there is a one-to-one mapping from input code words into output code words. In a *one-to-one mapping*, each input code word produces a different output code word.

  The general structure of a decoder circuit is shown in Figure 5–22. The enable inputs, if present, must be asserted for the decoder to perform its normal mapping function. Otherwise, the decoder maps all input code words into a single, "disabled," output code word.

decoder ✓

one-to-one mapping

Figure 5–22
Decoder circuit structure.

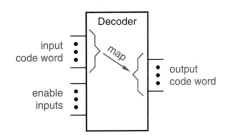

The most commonly used input code is an *n*-bit binary code, where an *n*-bit word represents one of 2^n different coded values, normally the integers from 0 through $2^n - 1$. Sometimes an *n*-bit binary code is truncated to represent fewer than 2^n values. For example, in the BCD code, the 4-bit combinations 0000 through 1001 represent the decimal digits 0–9, and combinations 1010 through 1111 are not used.

The most commonly used output code is a 1-out-of-*m* code, which contains *m* bits, where one bit is asserted at any time. Thus, in a 1-out-of-4 code with active-high outputs, the code words are 0001, 0010, 0100, and 1000. With active-low outputs, the code words are 1110, 1101, 1011, and 0111.

5.3.1 Binary Decoders

binary decoder

The most commonly used decoder circuit is an *n*-to-2^n decoder or *binary decoder*. Such a decoder has an *n*-bit binary input code and a 1-out-of-2^n output code. A binary decoder is used when you need to activate exactly one of 2^n outputs based on an *n*-bit input value.

For example, Figure 5–23(a) shows the inputs and outputs and Table 5–4 is the truth table of a 2-to-4 decoder. The input code word I1,I0 represents an integer in the range 0–3. The output code word Y3,Y2,Y1,Y0 has Y*i* equal to 1 if

Figure 5–23
A 2-to-4 decoder: (a) inputs and outputs; (b) logic diagram.

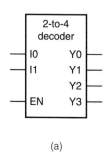

(a)

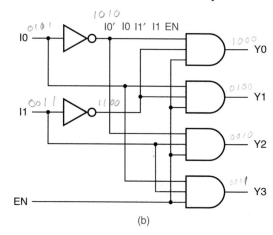

(b)

Table 5–4
Truth table for a 2-to-4 decoder.

Inputs			Outputs			
EN	I1	I0	Y3	Y2	Y1	Y0
0	X	X	0	0	0	0
1	0	0	0	0	0	1
1	0	1	0	0	1	0
1	1	0	0	1	0	0
1	1	1	1	0	0	0

and only if the input code word is the binary representation of *i* and the *enable*   *enable input*
input EN is 1. If EN is 0, then all of the outputs are 0. A gate-level circuit
for the 2-to-4 decoder is shown in Figure 5–23(b). Each AND gate *decodes* one   *decode*
combination of the input code word I1,I0.

The binary decoder's truth table introduces a "don't-care" notation for input
combinations. If one or more input values do not affect the output values for
some combination of the remaining inputs, they are marked with an "x" for that
input combination. This convention can greatly reduce the number of rows in
the truth table, as well as make the functions of the inputs more clear.

The input code of an *n*-bit binary decoder need not represent the integers
from 0 through $2^n - 1$. For example, Table 5–5 shows the 3-bit Gray-code output
of a mechanical encoding disk with eight positions. The eight disk positions
can be decoded with a 3-bit binary decoder with the appropriate assignment of
signals to the decoder outputs, as shown in Figure 5–24.

Also, it is not necessary to use all of the outputs of a decoder. For example,
a *decimal* or *BCD decoder* can be obtained from a 4-to-16 decoder by using only   *decimal decoder*
the first ten outputs, Y0–Y9. Commercially available decimal decoders, such as   *BCD decoder*
the 74x42, are just 4-to-16 decoders in which outputs Y10–Y15 have been omitted
because of pin limitations.

Table 5–5
Position encoding for a 3-bit
mechanical encoding disk.

Disk Position	I2	I1	I0	Binary Decoder Output
0°	0	0	0	Y0
45°	0	0	1	Y1
90°	0	1	1	Y3
135°	0	1	0	Y2
180°	1	1	0	Y6
225°	1	1	1	Y7
270°	1	0	1	Y5
315°	1	0	0	Y4

Figure 5–24
Using a 3-to-8 binary decoder to
decode a Gray code.

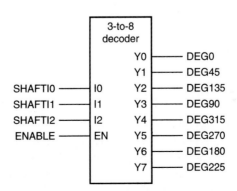

2 in 1

Figure 5–25
Logic symbol for one-half of a
74x139 dual 2-to-4 decoder:
(a) conventional symbol; (b) default signal names associated
with external pins.

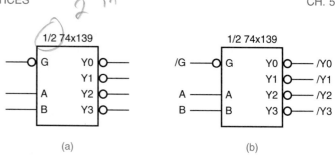

(a)   (b)

5.3.2   Logic Symbols for Larger-Scale Elements

Before describing some commercially available 74-series MSI decoders, we need
to discuss general guidelines for drawing logic symbols for larger-scale logic
elements.

The most basic rule is that logic symbols are drawn with inputs on the
left and outputs on the right. The top and bottom edges of a logic symbol are
not normally used for signal connections. However, explicit power and ground
connections are sometimes shown at the top and bottom, especially if these
connections are made on "nonstandard" pins. (Most MSI parts have power and
ground connected to the corner pins, e.g., pins 8 and 16 of a 16-pin DIP package.)

Like gate symbols, the logic symbols for larger-scale elements associate an
active level with each pin. With respect to active levels, it's important to use a
consistent convention to naming the internal signals and external pins.

Larger-scale elements almost always have their signal names defined in
terms of the functions performed *inside* their symbolic outline, as explained in
Section 5.1.4. For example, Figure 5–25(a) shows the logic symbol for one
section of a 74x139 dual 2-to-4 decoder, an MSI part that we'll fully describe in
the next subsection. When the G input is asserted, one of the outputs Y0–Y3 is
asserted, as selected by a 2-bit code applied to the A and B inputs. It is apparent
from the symbol that the G input pin and all of the output pins are active low.

When the 74x139 symbol appears in the logic diagram for a real application,
its inputs and outputs have signals connected to other devices, and each such
signal has a name that indicates its function in the application. However, when
we describe the 74x139 in isolation, we might still like to have a name for the
signal on each external pin. Figure 5–25(b) shows our naming convention in
this case. Active-high pins are given the same name as the internal signal, while
active-low pins have the internal signal name preceded by a /.

5.3.3   The 74x139 Dual 2-to-4 Decoder

74x139

Two independent and identical 2-to-4 decoders are contained in a single TTL
MSI part, the *74x139*. The gate-level circuit diagram for this IC is shown in
Figure 5–26(a). Notice that the outputs and the enable input of the '139 are

Throughout this book, we use "traditional" symbols for larger-scale logic elements. The IEEE standard uses somwhat different symbols for larger-scale logic elements. IEEE standard symbols, as well as the pros and cons of IEEE vs. traditional symbols, are discussed in Appendix A.

IEEE STANDARD
LOGIC SYMBOLS

Unfortunately, not all IC manufacturers use consistent drawing and naming conventions for internal signals, external pins, and active levels. For example, the Texas Instruments *TTL Logic Data Book* (1988) properly draws the G input in the 74LS139 symbol as active low, but improperly shows the Y0–Y3 outputs as active high. Furthermore, it includes an overbar in the external names of some active-low pins, including $\overline{G}$ in the 74LS138 and 74LS139, but fails to do so for others, such as Y0–Y7 in the 74LS138. The result is that a digital designer still must study the truth table and internal logic diagram of a larger-scale element to be absolutely sure what it does.

BAD NAMES

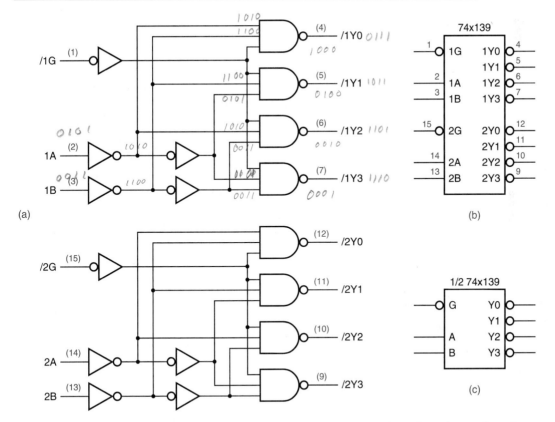

Figure 5–26 The 74x139 dual 2-to-4 decoder: (a) logic diagram, including pin numbers for a standard 16-pin dual in-line package; (b) traditional logic symbol; (c) logic symbol for one decoder.

Table 5–6
Truth table for one-half of a
74x139 dual 2-to-4 decoder.

Inputs			Outputs			
/G	B	A	/Y3	/Y2	/Y1	/Y0
1	x	x	1	1	1	1
0	0	0	1	1	1	0
0	0	1	1	1	0	1
0	1	0	1	0	1	1
0	1	1	0	1	1	1

active-low. Most commercially available decoders are designed with active-low outputs, since inverting gates are generally faster than noninverting ones. Also notice that the '139 has extra inverters on its select inputs. Without these inverters, each select input would present three unit loads instead of one, consuming much more of the fanout budget of the device that drives it.

A logic symbol for the 74x139 is shown in Figure 5–26(b). Notice that all of the signal names inside the symbol outline are active-high (no /), and that inversion bubbles indicate active-low inputs and outputs. Sometimes a schematic may contain a generic symbol for just one decoder, one-half of a '139, as shown in (c). In this case, a computer-aided design system may defer the assignment of the generic function to one half or the other of a particular '139 package until the printed-circuit board is laid out, in much the same way as the assignment of 2-input NAND gates to 74x00 packages can be deferred.

Table 5–6 is the truth table for a 74x139-type decoder. The truth tables in some manufacturers' data books use L and H to denote the input and output signal voltage levels, so there can be no ambiguity about the electrical function of the device; a truth table written this way is sometimes called a *function table*. However, since we use positive logic throughout this book, we can use 0 and 1 without ambiguity. In any case, the truth table gives the logic function in terms of the *external pins* of the device. A truth table for the function performed *inside* the symbol outline would look just like Table 5–4, except that the input signal names would be G, B, A.

function table

Some logic designers draw the symbol for 74x139s and other logic functions without inversion bubbles. Instead, they use an overbar on signal names inside the symbol outline to indicate negation, as shown in Figure 5–27(a). This notation is self-consistent, but it is inconsistent with our drawing standards for bubble-to-bubble logic design. The symbol shown in (b) is absolutely *incorrect*: according to this symbol, a logic 1, not 0, must be applied to the enable pin to enable the decoder.

Some manufacturers' data books have similar inconsistencies. For example, the Texas Instruments *TTL Logic Data Book* uses names like 1$\overline{\text{G}}$ for the '139's

Figure 5–27
More ways to symbolize a
74x139: (a) correct but to be
avoided; (b) incorrect because of
double negations.

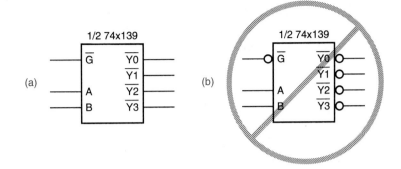

enable inputs, with the overbar indicating an active-low pin, but names like 1Y0 for active-low output pins. Worse, the same name, 1Y0, is used *inside* the symbol outline to denote the active-high internal signal.

Apparently, the TI data book really intends for 1Y0 to denote the external signal, since the columns of its '139 function table, which is written in terms of external signal levels, have headings like 1Y0. Unfortunately, the notation is not consistent among different parts, so you have to study the description of each part to know what's really going on. Be assured, however, that the signal names in this text *are* consistent.

5.3.4   The 74x138 3-to-8 Decoder

The *74x138* is a commercially available MSI 3-to-8 decoder whose gate-level   *74x138* circuit diagram and symbol are shown in Figure 5–28; its truth table is given in Table 5–7. Like the 74x139, the 74x138 has active-low outputs, and it has three enable inputs (G1, /G2A, /G2B), all of which must be asserted for the selected output to be asserted.

The logic function of the '138 is straightforward—an output is asserted if and only if the decoder is enabled and the output is selected. Thus, we can easily write logic equations for an internal output signal such as Y5 in terms of the internal input signals:

$$Y5 \;=\; \underbrace{G1 \cdot G2A \cdot G2B}_{\text{enable}} \cdot \underbrace{C \cdot B' \cdot A}_{\text{select}}$$

However, because of the inversion bubbles, we have the following relations between internal and external signals:

$$G2A \;=\; /G2A'$$
$$G2B \;=\; /G2B'$$
$$Y5 \;=\; /Y5'$$

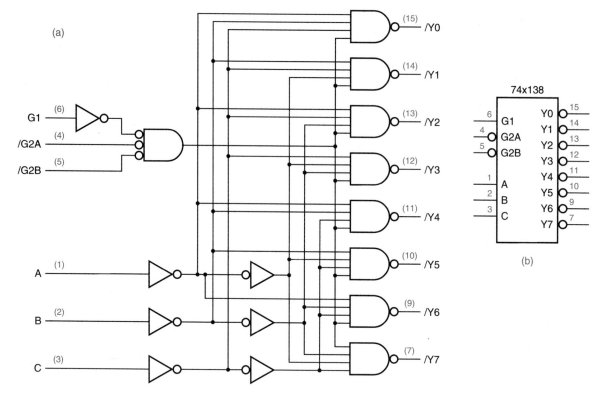

Figure 5–28 The 74x138 3-to-8 decoder: (a) logic diagram, including pin numbers for a standard 16-pin dual in-line package; (b) traditional logic symbol.

Table 5–7 Truth table for a 74x138 3-to-8 decoder.

Inputs						Outputs							
G1	/G2A	/G2B	C	B	A	/Y7	/Y6	/Y5	/Y4	/Y3	/Y2	/Y1	/Y0
0	x	x	x	x	x	1	1	1	1	1	1	1	1
x	1	x	x	x	x	1	1	1	1	1	1	1	1
x	x	1	x	x	x	1	1	1	1	1	1	1	1
1	0	0	0	0	0	1	1	1	1	1	1	1	0
1	0	0	0	0	1	1	1	1	1	1	1	0	1
1	0	0	0	1	0	1	1	1	1	1	0	1	1
1	0	0	0	1	1	1	1	1	1	0	1	1	1
1	0	0	1	0	0	1	1	1	0	1	1	1	1
1	0	0	1	0	1	1	1	0	1	1	1	1	1
1	0	0	1	1	0	1	0	1	1	1	1	1	1
1	0	0	1	1	1	0	1	1	1	1	1	1	1

Therefore, if we're interested, we can write the following equation for the external output signal /Y5 in terms of external input signals:

$$/Y5 = Y5' = (G1 \cdot /G2A' \cdot /G2B' \cdot C \cdot B' \cdot A)'$$
$$= G1' + /G2A + /G2B + C' + B + A'$$

On the surface, this equation doesn't resemble what we might expect for a decoder, since it is a logical sum rather than a product. However, if we practice bubble-to-bubble logic design, we don't have to worry about this; we just give the output signal an active-low name and remember that it's active low when we connect it to other inputs.

5.3.5 Cascading Binary Decoders

Multiple binary decoders can be used to decode larger code words. Figure 5–29 shows how two 3-to-8 decoders can be combined to make a 4-to-16 decoder. The availability of both active-high and active-low enable inputs on the 74x138 makes it possible to enable one or the other directly based on the state of the most significant input bit. The top decoder (U1) is enabled when N3 is 0, and the bottom one (U2) is enabled when N3 is 1.

Figure 5–29

Design of a 4-to-16 decoder using 74x138s.

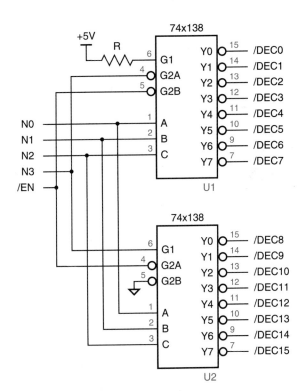

Figure 5–30
Design of a 5-to-32 decoder
using 74x138s and a 74x139.

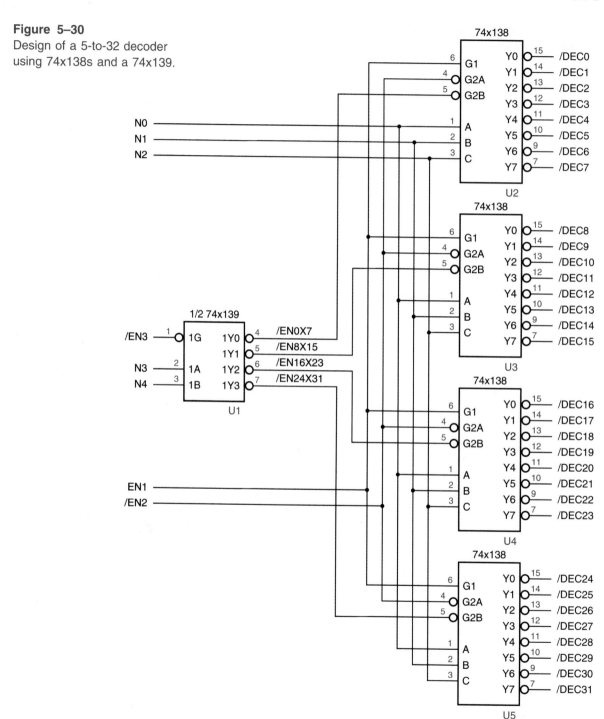

For even larger code words, binary decoders can be cascaded hierarchically. Figure 5–30 shows how to use half of a 74x139 to decode the two high-order bits of a 5-bit code word, thereby enabling one of four 74x138s that decode the three low-order bits.

* 5.3.6 Binary Decoders as Minterm Generators

The outputs of an n-to-2^n binary decoder correspond to n-variable minterms since, when the decoder is enabled, each output is asserted for exactly one row of an n-variable truth table. If an n-input logic function is specified as a minterm list, a decoder can be used to build the function directly, as a canonical sum (sum of minterms).

For example, consider the function $F = \Sigma_{X,Y,Z}(0,2,3,5)$, whose Karnaugh map is shown in Figure 5–31(a). A traditional sum-of-products circuit using NAND gates is shown in (b). Alternatively, the canonical sum can be realized with a decoder and one NAND gate as shown in (c), simply by "picking off" and summing the appropriate minterms. (Note that X, Y, and Z must be connected to the decoder inputs in the proper order.) A designer might prefer one design or

excellent!

---

*Throughout this book, optional sections are marked with an asterisk.

Figure 5–31
Designing a circuit for the logic
function $F = \Sigma_{X,Y,Z}(0,2,3,5)$:
(a) Karnaugh map; (b) NAND-
based minimal sum of products;
(c) decoder-based canonical
sum.

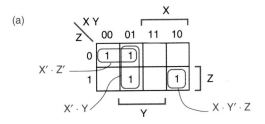

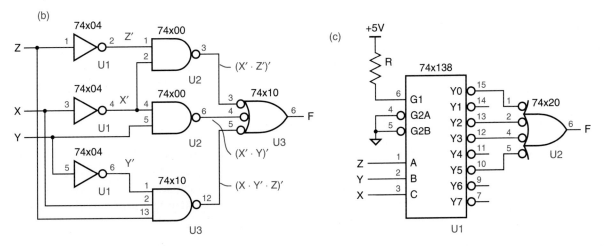

the other, depending on tradeoffs in speed, cost, and other things. The decoder solution is particularly flexible, since the designer can change it as required to any canonical sum with four or fewer minterms, simply by connecting different decoder outputs to the NAND gate (see Exercise 5.40).

Using a decoder to realize the canonical sum is impractical for logic functions with more than a few inputs, of course, since the number of minterms and decoder outputs grows exponentially with the number of inputs. On the other hand, the decoder approach makes a lot of sense for *multiple-output* logic functions with a few inputs. For example, Figure 5–32 shows how a decoder and three NAND gates can be used to realize three different canonical sums of the same input variables.

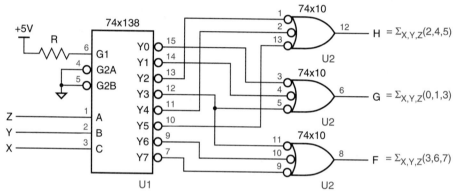

Figure 5–32   Using a decoder to realize a multiple-output function.

Probably the most attractive aspect of designing with decoders is that it's so easy—mindless, if you will. Don't knock it! The decoder approach isn't used too much anymore, but not because it's mindless. Instead, programmable logic devices (PLDs) have expanded the approach, allowing logic designers to realize not only sums of minterms, but also sums of arbitrary product terms of a large number of input variables, simply by listing the desired terms. These devices are introduced in the next chapter.

* 5.3.7   Seven-Segment Decoders

seven-segment display

Look at your wrist and you'll probably see a *seven-segment display*. This type of display, which normally uses light-emitting diodes (LEDs) or liquid-crystal display (LCD) elements, is used in watches, calculators, and instruments to display decimal data. A digit is displayed by illuminating a subset of the seven line segments shown in Figure 5–33(a).

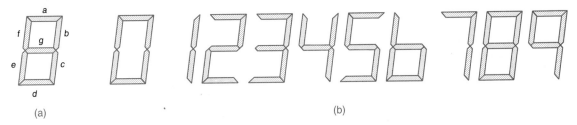

Figure 5–33 Seven-segment display: (a) segment identification; (b) decimal digits.

A *seven-segment decoder* has 4-bit BCD as its input code and the "seven-segment code," which is graphically depicted in Figure 5–33(b), as its output code. Table 5–8 and Figure 5–34 are the truth table and logic diagram for a *74x49* seven-segment decoder. Except for the strange (clever?) connection of the "blanking input" /BI, each output of the 74x49 is a minimal product-of-sums realization for the corresponding segment, assuming "don't-cares" for the non-decimal input combinations. The INVERT-OR-AND structure used for each output may seem a little strange, but it is equivalent under the generalized DeMorgan's theorem to an AND-OR-INVERT gate, which is a fairly fast and compact structure to build in CMOS or TTL.

seven-segment decoder

74x49

Table 5–8
Truth table for a 74x49 seven-segment decoder.

Inputs					Outputs						
/BI	D	C	B	A	a	b	c	d	e	f	g
0	x	x	x	x	0	0	0	0	0	0	0
1	0	0	0	0	1	1	1	1	1	1	0
1	0	0	0	1	0	1	1	0	0	0	0
1	0	0	1	0	1	1	0	1	1	0	1
1	0	0	1	1	1	1	1	1	0	0	1
1	0	1	0	0	0	1	1	0	0	1	1
1	0	1	0	1	1	0	1	1	0	1	1
1	0	1	1	0	0	0	1	1	1	1	1
1	0	1	1	1	1	1	1	0	0	0	0
1	1	0	0	0	1	1	1	1	1	1	1
1	1	0	0	1	1	1	1	0	0	1	1
1	1	0	1	0	0	0	0	1	1	0	1
1	1	0	1	1	0	0	1	1	0	0	1
1	1	1	0	0	0	1	0	0	0	1	1
1	1	1	0	1	1	0	0	1	0	1	1
1	1	1	1	0	0	0	0	1	1	1	1
1	1	1	1	1	0	0	0	0	0	0	0

Figure 5–34
The 74x49 seven-segment
decoder: (a) logic diagram,
including pin numbers; (b) traditional logic symbol.

(a)

(b)

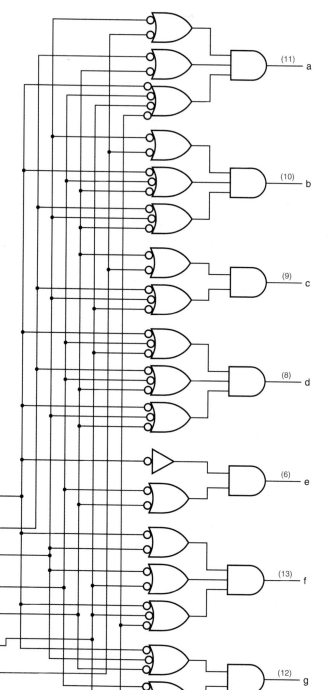

Most modern seven-segment display elements have decoders built into them, so that a 4-bit BCD word can be applied directly to the device. Many of the older, discrete seven-segment decoders have special high-voltage or high-current outputs that are well suited for driving large, high-powered display elements.

5.4 THREE-STATE BUFFERS

In Sections 3.7.2 and 3.11.4, we showed the electrical design of CMOS and TTL devices whose outputs may be in one of three states—0, 1, or Hi-Z. In this section, we'll show how to use them.

The most basic three-state device is a *three-state buffer*, often called a *three-state driver*. The logic symbols for four physically different three-state buffers are shown in Figure 5–35. The basic symbol is that of a noninverting buffer (a, b) or an inverter (c, d). The extra signal at the top of the symbol is a *three-state enable* input, which may be active high (a, c) or active low (b, d). When the enable input is asserted, the device behaves like an ordinary buffer or inverter. When the enable input is negated, the device output "floats"; that is, it goes to a high-impedance (Hi-Z), disconnected state and functionally behaves as if it weren't even there.

three-state buffer

three-state driver

three-state enable

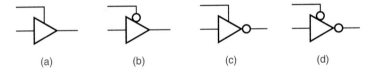

(a)   (b)   (c)   (d)

Figure 5–35   Various three-state buffers: (a) noninverting, active-high enable; (b) noninverting, active-low enable; (c) inverting, active-high enable; (d) inverting, active-low enable.

Three-state devices allow multiple sources to share a single "party line," as long as only one device "talks" on the line at a time. Figure 5–36 gives an example of how this can be done. Three input bits, SSRC2–SSRC0, select one of eight sources of data that may drive a single line, SDATA. A 3-to-8 decoder, the 74x138, ensures that only one of the eight SEL lines is asserted at a time, enabling only one three-state buffer to drive SDATA. However, if not all of the EN lines are asserted, then none of the three-state buffers is enabled. The logic value on SDATA is undefined in this case.

Typical three-state devices are designed so that they go into the Hi-Z state faster than they come out of the Hi-Z state. (In terms of the specifications in a data book, t_{pLZ} and t_{pHZ} are both less than t_{pZL} and t_{pZH}; also see Section 3.7.2.) This means that if the outputs of two three-state devices are connected to the same party line, and we simultaneously disable one and enable the other, the first

Figure 5–36
Eight sources sharing a three-state party line.

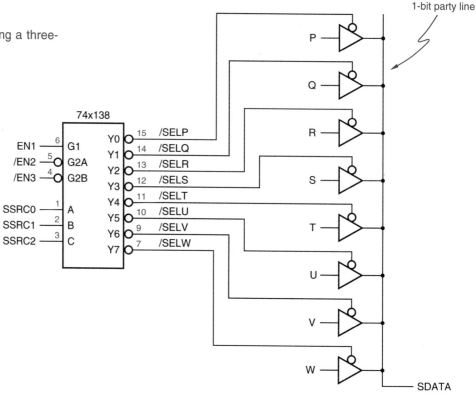

device will get off the party line before the second one gets on. This is important because, if both devices were to drive the party line at the same time, and if both were trying to maintain opposite output values (0 and 1), then excessive current would flow and create noise in the system, as discussed in Section 3.7.5. This is often called *fighting*.

fighting

Unfortunately, delays and timing skews in control circuits make it difficult to ensure that the enable inputs of different three-state devices change "simultaneously." Even when this is possible, a problem arises if three-state devices from different-speed logic families (or even different ICs manufactured on different days) are connected to the same party line. The turn-on time (t_{pZL} or t_{pZH}) of a "fast" device may be shorter than the turn-off time (t_{pLZ} or t_{pHZ}) of a "slow" one, and the outputs may still fight.

dead time

The only really safe way to use three-state devices is to design control logic that guarantees a *dead time* on the party line during which no one is driving it. The dead time must be long enough to account for the worst-case differences between turn-off and turn-on times of the devices and for skews in the three-state control signals. A timing diagram that illustrates this sort of operation for the party line of Figure 5–36 is shown in Figure 5–37. This timing diagram also illustrates a drawing convention for three-state signals—when in the Hi-Z state, they are shown at an "undefined" level halfway between 0 and 1.

Figure 5–37
Timing diagram for the three-state party line.

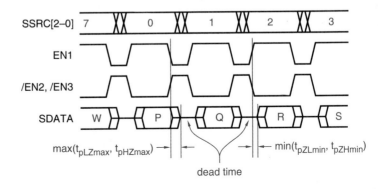

The actual voltage level of a floating signal depends on circuit details, such as resistive and capacitive load, and may vary over time. Also, the interpretation of this level by other circuits depends on the input characteristics of those circuits, so it's best not to count on a floating signal as being anything other than "undefined." Sometimes a pull-up resistor is used on three-state party lines to ensure that a floating value is pulled to a HIGH voltage and interpreted as logic 1. This is especially important on party lines that drive CMOS devices, which may consume excessive current when their input voltage is halfway between logic 0 and 1.

DEFINING
"UNDEFINED"

5.4.1 Standard SSI and MSI Three-State Buffers

Like logic gates, several independent three-state buffers may be packaged in a single SSI IC. For example, Figure 5–38 shows the pinouts of *74x125* and *74x126*, each of which contains four independent noninverting three-state buffers in a 14-pin package. The three-state enable inputs in the '125 are active low, and in the '126 they are active high.

    Most party-line applications use a bus with more than one bit of data. For example, in an 8-bit microprocessor system, the data bus is eight bits wide, and peripheral devices normally place data on the bus eight bits at a time. Thus, a peripheral device enables eight three-state drivers to drive the bus, all at the same time. Independent enable inputs, as in the '125 and '126, are not necessary.

    Thus, to reduce the package size in wide-bus applications, several commonly used MSI parts contain multiple three-state buffers with common enable inputs. For example, Figure 5–39 shows the logic diagram and symbol for a *74x541* octal noninverting three-state buffer. *Octal* means that the part contains eight individual buffers. Both enable inputs, /G1 and /G2, must be asserted to

74x125
74x126

74x541
octal

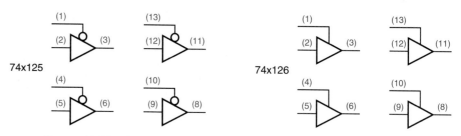

Figure 5–38 Pinouts of the 74x125 and 74x126 three-state buffers.

hysteresis

enable the device's three-state outputs. The little rectangular symbols inside the buffer symbols indicate *hysteresis*, an electrical characteristic of the inputs that improves noise immunity, as we explained in Section 3.7.1. The 74x541 inputs typically have 0.4 volts of hysteresis.

Figure 5–40 shows part of a microprocessor system with an 8-bit data bus, DB[0–7], and a 74x541 used as an input port. The microprocessor selects Input Port 1 by asserting INSEL1 and requests a read operation by asserting READ. The selected 74x541 responds by driving the microprocessor data bus with user-

(a)

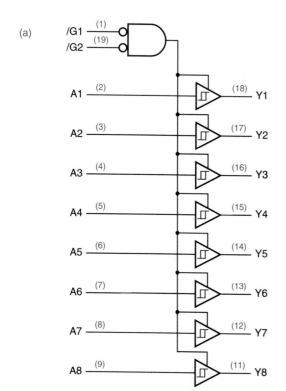

Figure 5–39
The 74x541 octal three-state buffer: (a) logic diagram, including pin numbers for a standard 20-pin dual in-line package; (b) traditional logic symbol.

(b)

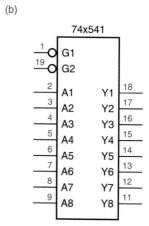

Figure 5–40
Using a 74x541 as a microprocessor input port.

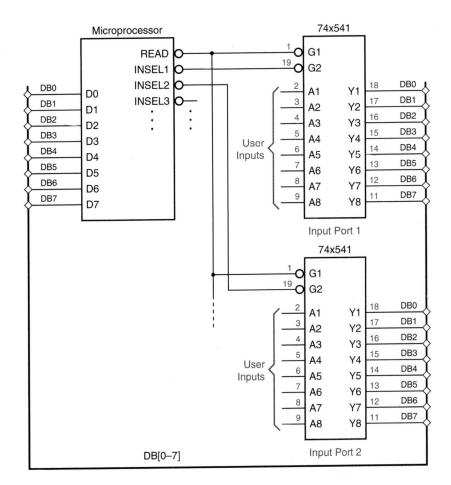

supplied input data. Other input ports may be selected when a different INSEL line is asserted along with READ.

Many other varieties of octal three-state buffers are commercially available. For example, the *74x540* is identical to the 74x541 except that it contains in- *74x540* verting buffers. The *74x240* and *74x241* are similar to the '540 and '541, except *74x240* that they are split into two 4-bit sections, each with a single enable line. *74x241*

A *bus transceiver* contains pairs of three-state buffers connected in opposite *bus transceiver* directions between each pair of pins, so that data can be transferred in either direction. For example, Figure 5–41 shows the logic diagram and symbol for a *74x245* octal three-state transceiver. The DIR input determines the direction of *74x245* transfer, from A to B (DIR=1) or from B to A (DIR=0). The three-state buffer for the selected direction is enabled only if /G is asserted.

A bus transceiver is typically used between two *bidirectional buses*, as *bidirectional bus* shown in Figure 5–42. Three different modes of operation are possible, depending on the state of /G and DIR, as shown in Table 5–9. As usual, it is the designer's responsibility to ensure that neither bus is ever driven simultaneously by two devices. However, independent transfers where both buses are driven at the same time may occur when the transceiver is disabled, as indicated in the last line of the table.

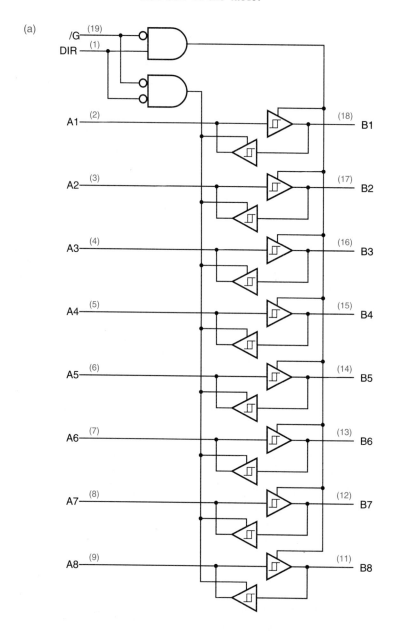

(a)

(b)

Figure 5–41
The 74x245 octal three-state transceiver: (a) logic diagram; (b) traditional logic symbol.

Figure 5–42
Bidirectional buses and
transceiver operation.

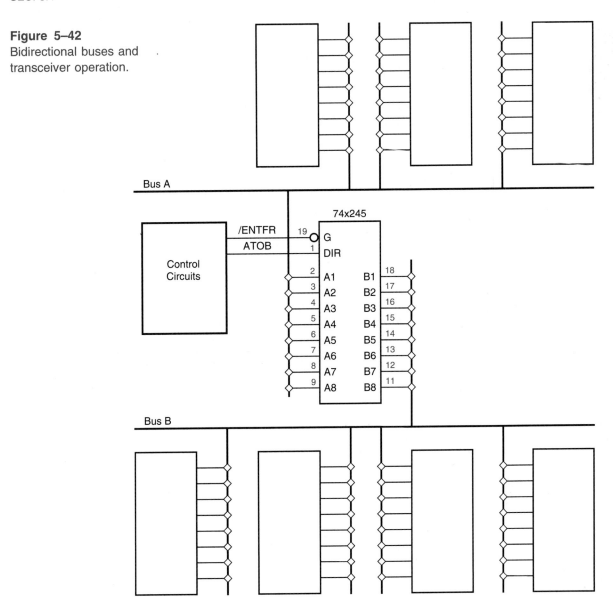

Bus A

74x245

/ENTFR 19 G

ATOB 1 DIR

Control
Circuits

	2 A1	B1 18
	3 A2	B2 17
	4 A3	B3 16
	5 A4	B4 15
	6 A5	B5 14
	7 A6	B6 13
	8 A7	B7 12
	9 A8	B8 11

Bus B

Table 5–9 Modes of operation for a pair of bidirectional buses.

/ENTFR	ATOB	Operation
0	0	Transfer data from a source on bus B to a destination on bus A.
0	1	Transfer data from a source on bus A to a destination on bus B.
1	x	Transfer data on buses A and B independently.

5.5 ENCODERS

encoder

A decoder's output code normally has more bits than its input code. If the device's output code has *fewer* bits than the input code, the device is usually called an *encoder*. For example, consider a device with eight input bits representing an unsigned binary number, and two output bits indicating whether the number is prime or divisible by 7. We might call such a device a lucky/prime encoder.

2^n-to-n encoder
binary encoder

Probably the simplest encoder to build is a 2^n-to-*n* or *binary encoder*. As shown in Figure 5–43(a), it has just the opposite function as a binary *decoder*—its input code is the 1-out-of-2^n code and its output code is *n*-bit binary. The equations for an 8-to-3 encoder with inputs I0–I7 and outputs Y0–Y2 are given below:

$$Y0 = I1 + I3 + I5 + I7$$
$$Y1 = I2 + I3 + I6 + I7$$
$$Y2 = I4 + I5 + I6 + I7$$

The corresponding logic circuit is shown in (b). In general, a 2^n-to-*n* encoder can be built from *n* 2^{n-1}-input OR gates. Bit *i* of the input code is connected to OR gate *j* if bit *j* in the binary representation of *i* is 1.

Figure 5–43
Binary encoder: (a) general structure; (b) 8-to-3 encoder.

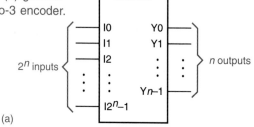

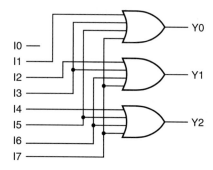

5.5.1 Priority Encoders

The 1-out-of-2^n coded outputs of an *n*-bit binary decoder are generally used to control a set of 2^n devices, where at most one device is supposed to be active at any time. Conversely, consider a system with 2^n *inputs*, each of which indicates a request for service, as in Figure 5–44. This structure is often found in microprocessor input/output subsystems, where the inputs might be interrupt requests.

In this situation, it may seem natural to use a binary encoder of the type shown in Figure 5–43 to observe the inputs and indicate which one is requesting service at any time. However, this encoder works properly only if the inputs are

Figure 5–44
A system with 2^n requestors,
and a "request encoder" that
indicates which request signal is
asserted at any time.

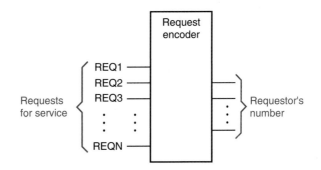

guaranteed to be asserted at most one at a time. If multiple requests can be made
simultaneously, the encoder gives undesirable results. For example, suppose that
inputs I2 and I4 of the 8-to-3 encoder are both 1; then the output is 110, the
binary encoding of 6.

Either 2 or 4, not 6, would be a useful output in the preceding example, but
how can the encoding device decide which? The solution is to assign *priority* to   *priority*
the input lines, so that when multiple requests are asserted, the encoding device
produces the number of the highest-priority requestor. Such a device is called a
priority encoder.   *priority encoder*

The logic symbol for an 8-input priority encoder is shown in Figure 5–45(a).
Input I7 has the highest priority. Outputs A2–A0 contain the number of the
highest-priority asserted input, if any. The IDLE output is asserted if no inputs
are asserted.

Figure 5–45
Logic symbol for 8-input
priority encoders: (a) generic;
(b) 74x148.

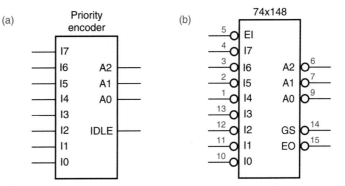

In order to write logic equations for the priority encoder's outputs, we first
define eight intermediate variables H0–H7, such that Hi is 1 if and only if Ii is
the highest priority 1 input:

$$H7 = I7$$
$$H6 = I6 \cdot I7'$$

$$H5 \quad = \quad I5 \cdot I6' \cdot I7'$$

$$\cdots$$

$$H0 \quad = \quad I0 \cdot I1' \cdot I2' \cdot I3' \cdot I4' \cdot I5' \cdot I6' \cdot I7'$$

Using these signals, the equations for the A2–A0 outputs are similar to the ones for a simple binary encoder:

$$A2 \quad = \quad H4 + H5 + H6 + H7$$
$$A1 \quad = \quad H2 + H3 + H6 + H7$$
$$A0 \quad = \quad H1 + H3 + H5 + H7$$

The IDLE output is 1 if no inputs are 1:

$$IDLE \quad = \quad (I0 + I1 + I2 + I3 + I4 + I5 + I6 + I7)'$$
$$= \quad I0' \cdot I1' \cdot I2' \cdot I3' \cdot I4' \cdot I5' \cdot I6' \cdot I7'$$

5.5.2   The 74x148 Priority Encoder

74x148

The *74x148* is a commercially available, MSI 8-input priority encoder. Its logic symbol is shown in Figure 5–45(b) and its schematic is shown in Figure 5–46. The main difference between this IC and the "generic" priority encoder of Figure 5–45(a) is that its inputs and outputs are active low. Also, it has an enable input, /EI, that must be asserted for any of its outputs to be asserted. The complete truth table is given in Table 5–10.

Table 5–10   Truth table for a 74x148 8-input priority encoder.

Inputs									Outputs				
/EI	/I0	/I1	/I2	/I3	/I4	/I5	/I6	/I7	/A2	/A1	/A0	/GS	/EO
1	x	x	x	x	x	x	x	x	1	1	1	1	1
0	x	x	x	x	x	x	x	0	0	0	0	0	1
0	x	x	x	x	x	x	0	1	0	0	1	0	1
0	x	x	x	x	x	0	1	1	0	1	0	0	1
0	x	x	x	x	0	1	1	1	0	1	1	0	1
0	x	x	x	0	1	1	1	1	1	0	0	0	1
0	x	x	0	1	1	1	1	1	1	0	1	0	1
0	x	0	1	1	1	1	1	1	1	1	0	0	1
0	0	1	1	1	1	1	1	1	1	1	1	0	1
0	1	1	1	1	1	1	1	1	1	1	1	1	0

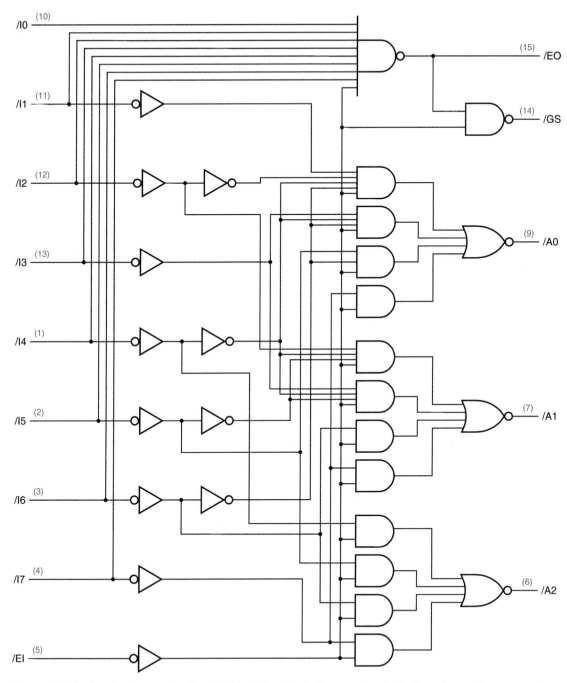

Figure 5–46   Logic diagram for the 74x148 8-input priority encoder, including pin numbers for a standard 16-pin dual in-line package.

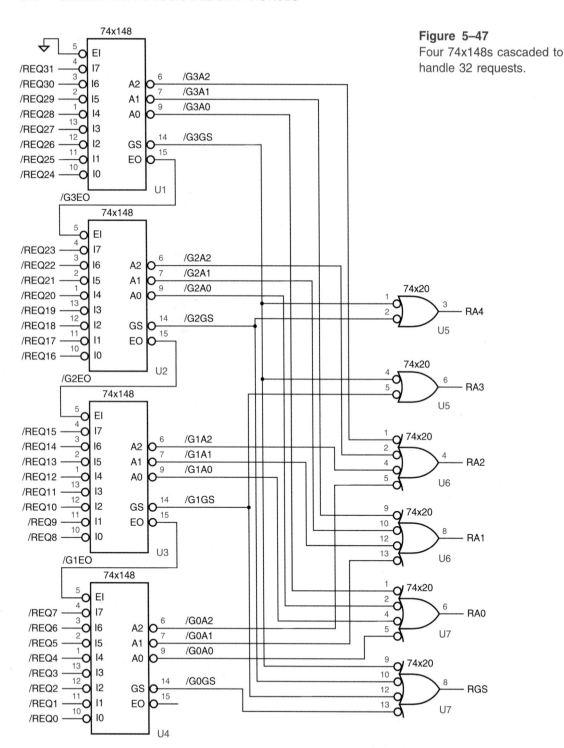

Figure 5–47
Four 74x148s cascaded to handle 32 requests.

Instead of an IDLE output, the '148 has a /GS output that is asserted when the device is enabled and one or more of the request inputs is asserted. The manufacturer calls this "Group Select," but it's easier to remember as "Got Something." The /EO signal is an enable *output* designed to be connected to the /EI input of another '148 that handles lower-priority requests. /EO is asserted if /EI is asserted but no request input is asserted; thus, a lower-priority '148 may be enabled. Figure 5–47 shows how four 74x148s can be connected in this way to accept 32 request inputs and produce a 5-bit output, RA4–RA0, indicating the highest-priority requestor. Since the A2–A0 outputs of at most one '148 will be enabled at any time, the outputs of the individual '148s can be OR'ed to produce RA2–RA0. Likewise, the individual /GS outputs can be combined in a 4-to-2 encoder to produce RA4 and RA3. The RGS output is asserted if any GS output is asserted.

5.6 MULTIPLEXERS

A *multiplexer* is a digital switch—it connects data from one of n sources to its output. Figure 5–48(a) shows the inputs and outputs of an n-input, b-bit multiplexer. There are n sources of data, each of which is b bits wide, and there are b output bits. In typical commercially available multiplexers, $n = 1, 2, 4, 8,$ or 16, and $b = 1, 2,$ or 4. There are s inputs that select among the n sources, so $s = \lceil \log_2 n \rceil$. An enable input EN allows the multiplexer to "do its thing"; when EN = 0, all of the outputs are 0. A multiplexer is often called a *mux* for short.

multiplexer

mux

Figure 5–48
Multiplexer structure: (a) inputs and outputs; (b) functional equivalent.

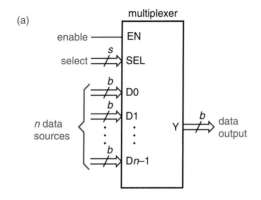

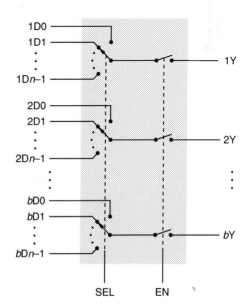

Figure 5–48(b) shows a switch circuit that is roughly equivalent to the multiplexer. However, unlike a mechanical switch, a multiplexer is a unidirectional device: information flows only from inputs (on the left) to outputs (on the right). We can write a general logic equation for a multiplexer output:

$$iY = \sum_{j=0}^{n-1} EN \cdot M_j \cdot iDj$$

Here, the summation symbol represents a logical sum of product terms. Variable iY is a particular output bit ($1 \leq i \leq b$), and variable iDj is input bit i of source j ($0 \leq j \leq n - 1$). M_j represents minterm j of the s select inputs. Thus, when the multiplexer is enabled and the value on the select inputs is j, each output iY equals the corresponding bit of the selected input, iDj.

Multiplexers are obviously useful devices in any application in which data must be switched from multiple sources to a destination. A common application in computers is the multiplexer between the processor's registers and its arithmetic logic unit (ALU). For example, consider a 16-bit processor in which each instruction has a 3-bit field that specifies one of eight registers to use. This 3-bit field is connected to the select inputs of an 8-input, 16-bit multiplexer. The multiplexer's data inputs are connected to the eight registers, and its data outputs are connected to the ALU to execute the instruction using the selected register.

5.6.1   Standard MSI Multiplexers

The sizes of commercially available MSI multiplexers are limited by the number of pins available in an inexpensive IC package. The most commonly used muxes come in 16-pin packages. At one extreme, we have the *74x151*, shown in Figure 5–49, which selects among eight 1-bit inputs. The select inputs are named

74x151

Table 5–11
Truth table for a 74x151 8-input, 1-bit multiplexer.

/EN	C	B	A	Y	/Y
1	x	x	x	0	1
0	0	0	0	D0	D0'
0	0	0	1	D1	D1'
0	0	1	0	D2	D2'
0	0	1	1	D3	D3'
0	1	0	0	D4	D4'
0	1	0	1	D5	D5'
0	1	1	0	D6	D6'
0	1	1	1	D7	D7'

with header "Inputs" spanning /EN, C, B, A and "Outputs" spanning Y, /Y.

C, B, and A; as in a decoder, C is most significant numerically. The enable input is active low; both active-high (Y) and active-low (/Y) versions of the output are provided.

The 74x151's truth table is shown in Table 5–11. Here we have once again extended our notation for truth tables. Up until now, our truth tables have specified an output of 0 or 1 for each input combination. In the 74x151's table, only a few of the inputs are listed under the "Inputs" heading. Each output is

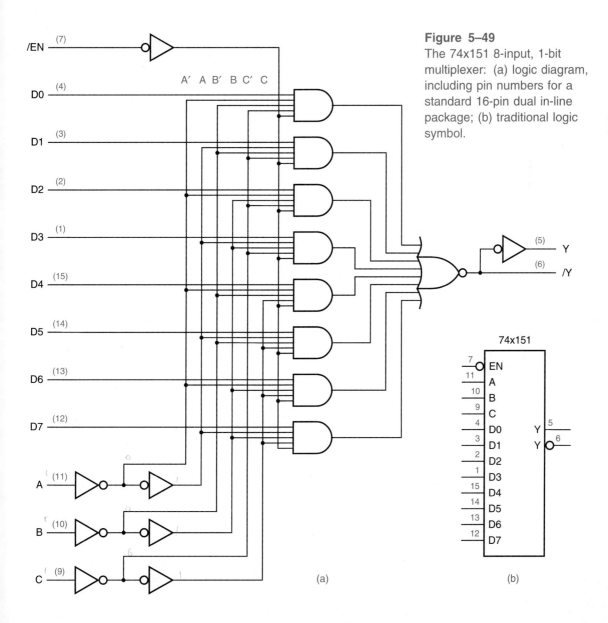

Figure 5–49
The 74x151 8-input, 1-bit multiplexer: (a) logic diagram, including pin numbers for a standard 16-pin dual in-line package; (b) traditional logic symbol.

specified as 0, 1, or a simple logic function of the remaining inputs (e.g., D0 or D0′). This notation saves eight columns and eight rows in the table, and presents the logic function more clearly than a larger table would.

74x157

At the other extreme of muxes in 16-pin packages, we have the *74x157*, shown in Figure 5–50, which selects between two 4-bit inputs. Just to confuse things, the manufacturer has named the select input S and the active-low enable

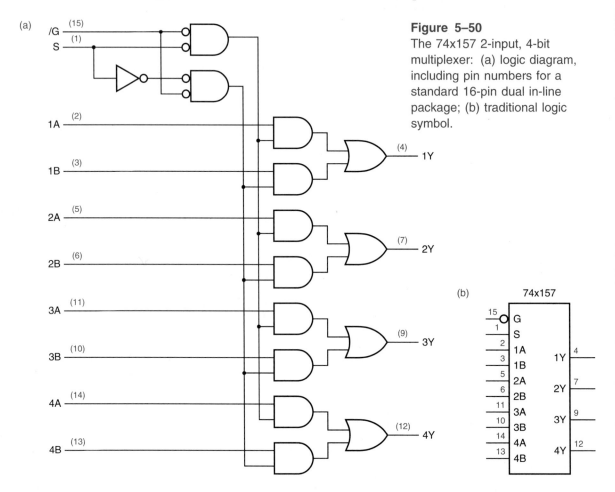

Figure 5–50
The 74x157 2-input, 4-bit multiplexer: (a) logic diagram, including pin numbers for a standard 16-pin dual in-line package; (b) traditional logic symbol.

Table 5–12
Truth table for a 74x157 2-input, 4-bit multiplexer.

Inputs		Outputs			
/G	S	1Y	2Y	3Y	4Y
1	x	0	0	0	0
0	0	1A	2A	3A	4A
0	1	1B	2B	3B	4B

input /G. Also note that the data sources are named A and B instead of D0 and D1 as in our generic example. Our extended truth-table notation makes the 74x157's description very compact, as shown in Table 5–12.

Intermediate between the 74x151 and 74x157 is the *74x153*, a 4-input, 2-bit multiplexer. As shown in Figure 5–51, this device has separate enable inputs (1G, 2G) for each bit. Its function is very straightforward.

74x153

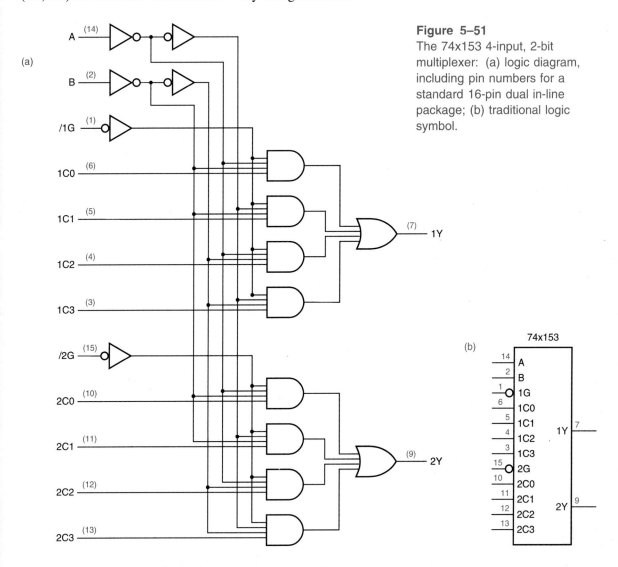

(a)

Figure 5–51
The 74x153 4-input, 2-bit multiplexer: (a) logic diagram, including pin numbers for a standard 16-pin dual in-line package; (b) traditional logic symbol.

(b)

Some multiplexers have three-state outputs. The enable input of such a multiplexer, instead of forcing the outputs to zero, forces them to the Hi-Z state. For example, the logic diagram and symbol for the 74x251 are shown in

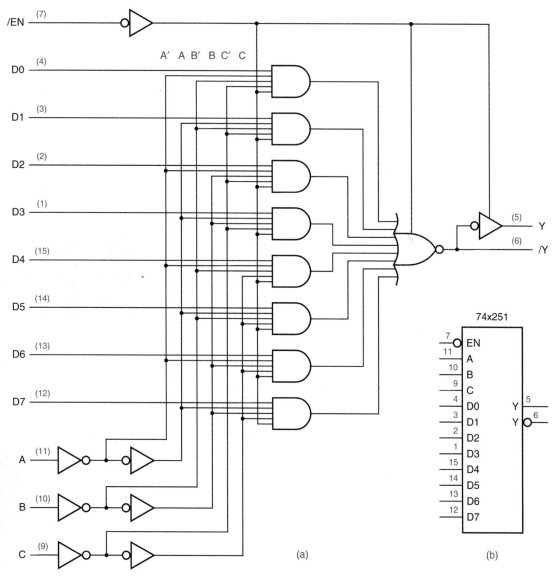

Figure 5–52 The 74x251 8-input, 1-bit multiplexer with three-state output: (a) logic diagram, including pin numbers; (b) traditional logic symbol.

74x253
74x257

Figure 5–52. The '251 is identical to the '151, except for the three-state outputs. Similarly, the *74x253* and *74x257* are three-state versions of the '153 and '157. The three-state outputs are especially useful when *n*-input muxes are combined to form larger muxes, as shown in the next subsection.

5.6.2 Expanding Multiplexers

Seldom does the size of an MSI multiplexer match the characteristics of the problem at hand. For example, we suggested earlier that an 8-input, 16-bit multiplexer might be used in the design of a computer processor. This function could be performed by 16 74x151 8-input, 1-bit multiplexers or equivalent ASIC cells, each handling one bit of all the inputs and the output. The processor's 3-bit register field would be connected to the A, B, and C inputs of all 16 muxes, so they would all select the same register source at any given time.

Of course, the device that produces the 3-bit register field in this example must have enough fanout to drive 16 loads. With 74LS-series ICs this is possible because typical devices have a fanout of 20 LS-TTL loads. However, we're lucky that the '151 was designed so that each of the A, B, and C inputs presents only one LS-TTL load to the circuit driving it. Theoretically, the '151 could have been designed without the first rank of three inverters shown on the select inputs in Figure 5–49, but then each select input would have presented five LS-TTL loads, and the drivers in our present application would need a fanout of 80.

Another dimension in which multiplexers can be expanded is the number of data sources. For example, suppose we needed a 32-input, 1-bit multiplexer. Figure 5–53 shows one way to build it. Five select bits are required. A 2-to-4 decoder (one-half of a 74x139) decodes the two high-order select bits to enable one of four 74x151 8-input multiplexers. Since only one '151 is enabled at a time, the '151 outputs can simply be OR'ed to obtain the final output.

The 32-to-1 multiplexer can also be built using 74x251s. The circuit is identical to Figure 5–53, except that the output NAND gate is eliminated. Instead, the Y (and, if desired, /Y) outputs of the four '251s are simply tied together, as shown in Figure 5–54. The '139 decoder ensures that at most one of the '251s has its three-state outputs enabled at any time. If the '139 is disabled (/XEN is

---

The use of bubble-to-bubble logic design should help your understanding of these multiplexer design examples. Since the decoder outputs and the multiplexer enable inputs are all active low, they can be hooked up directly. You can ignore the inversion bubbles when thinking about the logic function that is performed—you just say that when a particular decoder output is asserted, the corresponding multiplexer is enabled.

Bubble-to-bubble design also provides two options for the final OR function in Figure 5–53. The most obvious design would have used a 4-input OR gate connected to the Y outputs. However, for faster operation, we used an inverting gate, a 4-input NAND connected to the /Y outputs. This eliminated the delay of two inverters—the one used inside the '151 to generate Y from /Y, and the extra inverter circuit that is used to obtain an OR function from a basic NOR circuit in a CMOS or TTL OR gate.

TURN ON THE
BUBBLE MACHINE

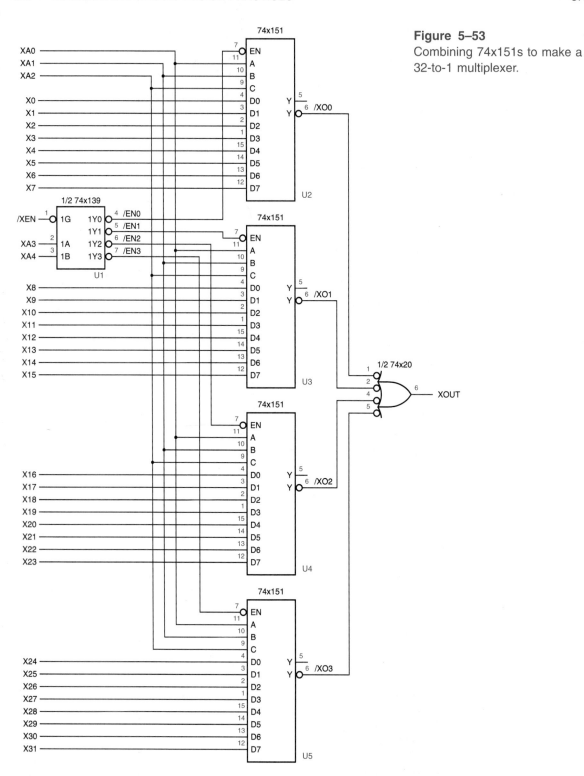

Figure 5–53
Combining 74x151s to make a 32-to-1 multiplexer.

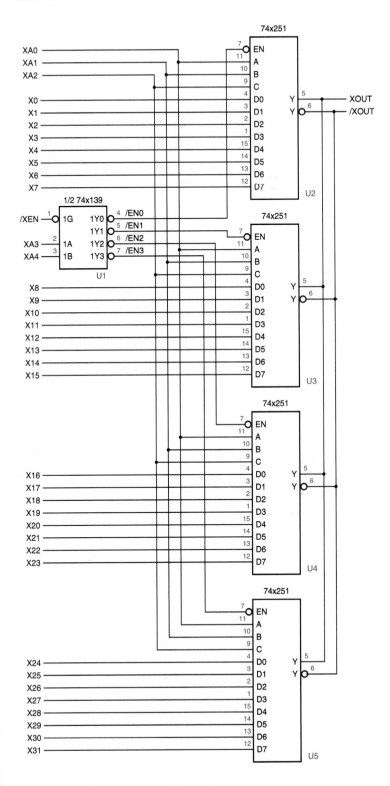

Figure 5–54
Another design for a 32-to-1 multiplexer.

negated), then all of the '251s are disabled, and the XOUT and /XOUT outputs
are undefined. However, if desired, resistors may be connected between these
signals and +5 volts to pull the output HIGH in this case.

5.6.3 Multiplexers, Demultiplexers, and Buses

demultiplexer

A multiplexer can be used to select one of *n* sources of data to transmit on a bus.
At the far end of the bus, a *demultiplexer* can be used to route the bus data to one
of *m* destinations. Such an application, using a 1-bit bus, is depicted in terms of
our switch analogy in Figure 5–55(a). In fact, block diagrams for logic circuits
often depict multiplexers and demultiplexers using the wedge-shaped symbols in
(b), to suggest visually how a selected one of multiple data sources gets directed
onto a bus and routed to a selected one of multiple destinations.

Figure 5–55
A multiplexer driving a bus,
with a demultiplexer receiving
the bus: (a) switch equivalent;
(b) block diagram symbols.

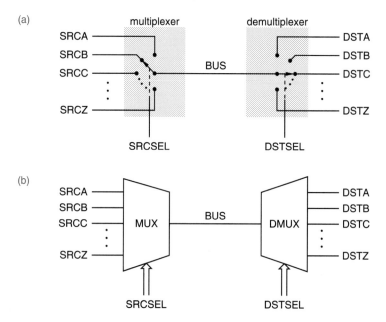

The function of a demultiplexer is just the inverse of a multiplexer's. For
example, a 1-bit, *n*-output demultiplexer has one data input and *s* inputs to select
one of $n = 2^s$ data outputs. In normal operation, all outputs except the selected
one are 0; the selected output equals the data input. This definition may be
generalized for a *b*-bit, *n*-output demultiplexer; such a device has *b* data inputs,
and its *s* select inputs choose one of $n = 2^s$ sets of *b* data outputs.

A binary decoder with an enable input can be used as a demultiplexer, as
shown in Figure 5–56. The decoder's enable input is connected to the data line,
and its select inputs determine which of its output lines is driven with the data
bit. The remaining output lines are negated. Thus, the 74x139 can be used as

Figure 5–56 Using a 2-to-4 binary decoder as a 1-bit, 4-output demultiplexer: (a) generic decoder; (b) 74x139.

a 2-bit, 4-output demultiplexer with active-low data inputs and outputs, and the 74x138 can be used as a 1-bit, 8-output demultiplexer. In fact, the manufacturer's catalog typically lists these ICs as "decoders/demultiplexers."

Another standard TTL MSI decoder/demultiplexer is the *74x155*, shown in Figure 5–57. This IC is most useful as a 2-bit, 4-output demultiplexer. It's not as flexible as the 74x139 when used as a dual 2-to-4 decoder, because both sections use the same select inputs.

74x155

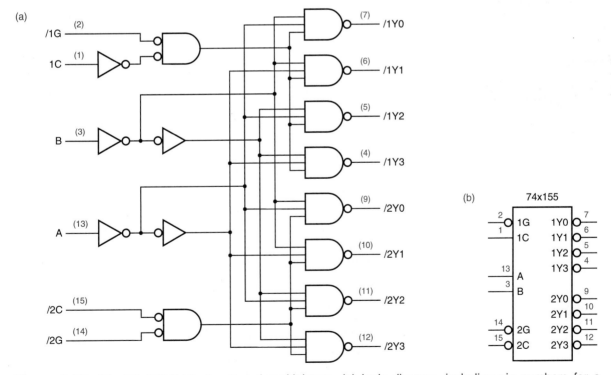

Figure 5–57 The 74x155 2-bit, 4-output demultiplexer: (a) logic diagram, including pin numbers for a standard 16-pin dual in-line package; (b) traditional logic symbol.

Table 5–13
General truth table for a
3-variable logic function, F(X,Y,Z).

Row	X	Y	Z	F
0	0	0	0	F_0
1	0	0	1	F_1
2	0	1	0	F_2
3	0	1	1	F_3
4	1	0	0	F_4
5	1	0	1	F_5
6	1	1	0	F_6
7	1	1	1	F_7

* 5.6.4 Multiplexers as Function Generators

In the heyday of MSI-based design, multiplexers were often used as a sort of
"universal logic element," not for switching data sources to outputs, but for
performing arbitrary logic functions in a nicely structured way. In recent years,
this design approach has been almost completely replaced by the widespread
use of programmable logic devices, but we describe it here nonetheless as an
interesting application of multiplexers.

A 2^s-input multiplexer can perform any logic function of s variables in
a very straightforward way. For example, Table 5–13 shows the truth table
for a general logic function of three variables. The function's value for each
input combination can be an arbitrary value, denoted by F_0, F_1, and so on.
Regardless of the values chosen in the truth table, the logic function can always
be performed by connecting logic 0s and 1s to the corresponding inputs of an
8-input multiplexer, as shown in Figure 5–58(a). A specific example, a function
in which F is 1 if and only if an even number of inputs are 1, is shown in (b).

*functional
decomposition*

Multiplexers can be used even more effectively using a simple form of
functional decomposition. Consider the general truth table for a logic function of
three variables, which we showed earlier in Table 5–13. As shown in Table 5–14,
we can "decompose" this table using only half as many rows. Each row displays
a particular combination of variables X and Y, and indicates that the output is
strictly a function of Z for that input combination. To perform a 3-variable logic
function with a 4-input multiplexer, we connect each data input of the multiplexer

Table 5–14
Decomposed truth table for a
3-variable logic function, F(X,Y,Z),
after "pulling out" variable Z.

Row	X	Y	Z	F
0,1	0	0	x	$F_{00}(Z)$
2,3	0	1	x	$F_{01}(Z)$
4,5	1	0	x	$F_{10}(Z)$
6,7	1	1	x	$F_{11}(Z)$

Figure 5–58
Realizing logic functions with
multiplexers: (a) general
scheme; (b) $F = \Sigma_{X,Y,Z}(0,3,5,6)$.

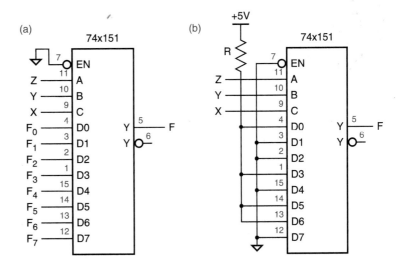

to the appropriate function of Z, according to the decomposed truth table. But
there are only four different functions of Z possible: 0, 1, Z, and Z′. Thus, we
need only the multiplexer, constant 0 and 1 logic signals, and if Z′ appears in
the decomposed truth table, an inverter for Z.

Figure 5–59 shows the decomposition for the example that we did previously using an 8-input multiplexer. Figure 5–60 shows an example where a
4-variable function is performed by an 8-input multiplexer. In general, an $s + 1$-
variable function can be performed by a 2^s-input multiplexer.

Figure 5–59
Realizing $F = \Sigma_{X,Y,Z}(0,3,5,6)$ with
a 4-input multiplexer.

Row	X	Y	Z	F	
0	0	0	0	1	Z′
1	0	0	1	0	
2	0	1	0	0	Z
3	0	1	1	1	
4	1	0	0	0	Z
5	1	0	1	1	
6	1	1	0	1	Z′
7	1	1	1	0	

(a)

(b)

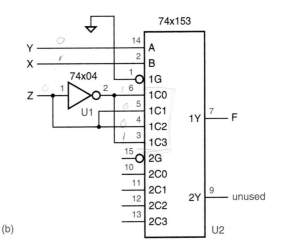

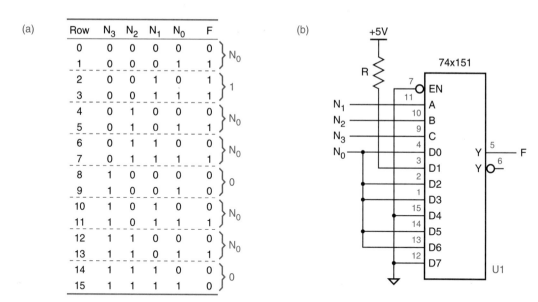

Figure 5–60   Realizing $F = \Sigma_{N_3,N_2,N_1,N_0}(1,2,3,5,7,11,13)$ with an 8-input multiplexer.

5.7 EXCLUSIVE OR GATES AND PARITY CIRCUITS

5.7.1 EXCLUSIVE OR and EXCLUSIVE NOR Gates

EXCLUSIVE OR (XOR)

EXCLUSIVE NOR (XNOR)

EQUIVALENCE

$\oplus$

An *EXCLUSIVE OR (XOR)* gate is a 2-input gate whose output is 1 if exactly one of its inputs is 1. Stated another way, an XOR gate produces a 1 output if its inputs are different. An *EXCLUSIVE NOR (XNOR)* or *EQUIVALENCE* gate is just the opposite—it produces a 1 output if its inputs are the same. A truth table for these functions is shown in Table 5–15. The XOR operation is sometimes denoted by the symbol "$\oplus$", that is,

$$X \oplus Y = X' \cdot Y + X \cdot Y'$$

Although EXCLUSIVE OR is not one of the basic functions of switching algebra, discrete XOR gates are fairly commonly used in practice. Most switching technologies cannot perform the XOR function directly; instead, they use multigate designs like the ones shown in Figure 5–61.

Table 5–15
Truth table for XOR and XNOR functions.

X	Y	X⊕Y (XOR)	(X⊕Y)' (XNOR)
0	0	0	1
0	1	1	0
1	0	1	0
1	1	0	1

Figure 5–61
Multigate designs for the 2-input XOR function: (a) AND-OR; (b) three-level NAND.

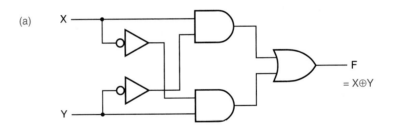

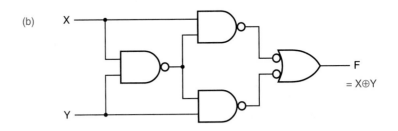

The logic symbols for XOR and XNOR functions are shown in Figure 5–62. There are four equivalent symbols for each function. All of these alternatives are a consequence of a simple rule:

- Any two signals (inputs or output) of an XOR or XNOR gate may be complemented without changing the resulting logic function.

In bubble-to-bubble logic design, we choose the symbol that is most expressive of the logic function being performed.

Four XOR gates are provided in a single 14-pin SSI IC, the *74x86* shown in Figure 5–63. Likewise, four XNOR gates are provided in the *74x266*, shown in Figure 5–64. The underlined diamond in the 74x266 symbol indicates an "open-collector" or "open-drain" output. As discussed in Sections 3.7.3 and 3.11.4, such outputs requires external pull-up resistors for proper operation. We'll show how 74x266s can be used to build comparators in Section 5.8.1.

74x86
74x266

Figure 5–62   Equivalent symbols for (a) XOR gates; (b) XNOR gates.

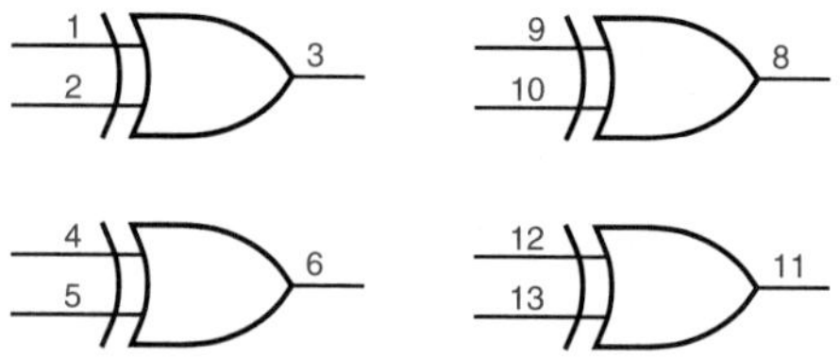

Figure 5–63
Pinouts of the 74x86 quadruple
2-input EXCLUSIVE OR gate.

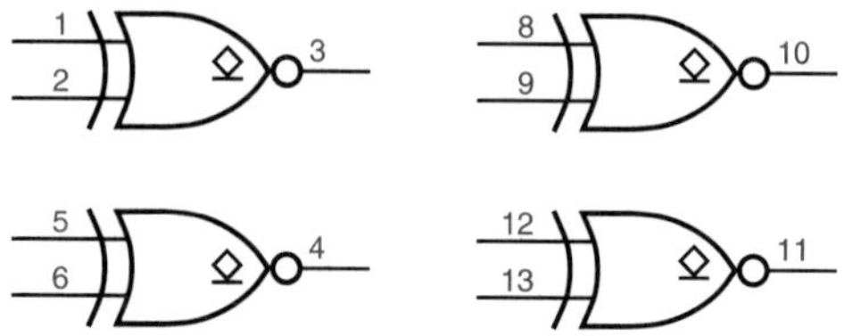

Figure 5–64
Pinouts of the 74x266 quadruple
2-input EXCLUSIVE NOR gate.

5.7.2 Parity Circuits

odd-parity circuit

even-parity circuit

As shown in Figure 5–65(a), n XOR gates may be cascaded to form a circuit with $n + 1$ inputs and a single output. This is called an *odd-parity circuit*, because its output is 1 if an odd number of its inputs are 1. The circuit in (b) is also an odd-parity circuit, but it's faster because its gates are arranged in a tree-like structure. If the output of either circuit is inverted, we get an *even-parity circuit*, whose output is 1 if an even number of its inputs are 1.

(a)

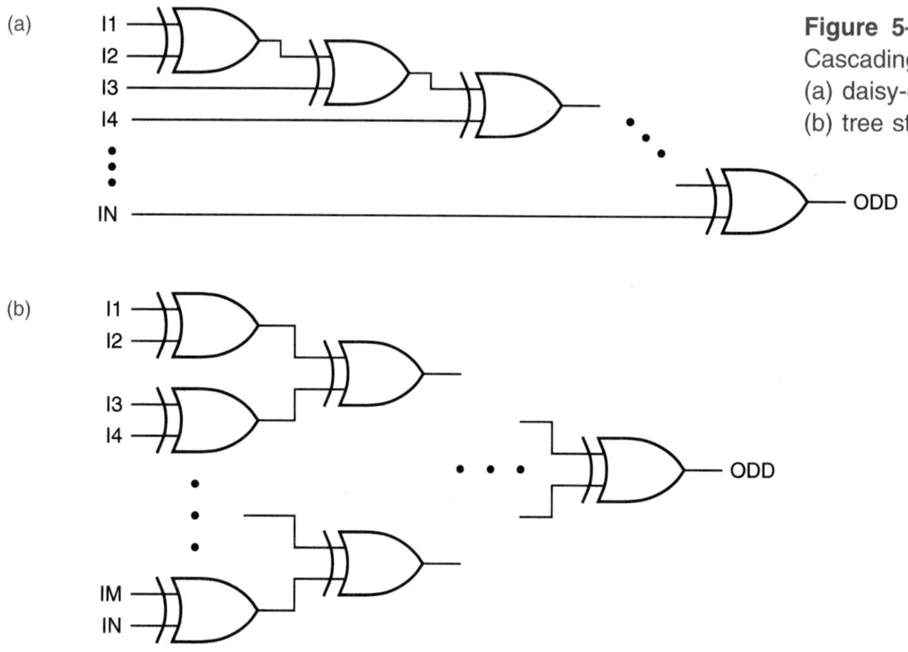

Figure 5–65
Cascading XOR gates:
(a) daisy-chain connection;
(b) tree structure.

(b)

Figure 5–66
The 74x280 9-bit odd/even parity generator: (a) logic diagram, including pin numbers for a standard 16-pin dual in-line package; (b) traditional logic symbol.

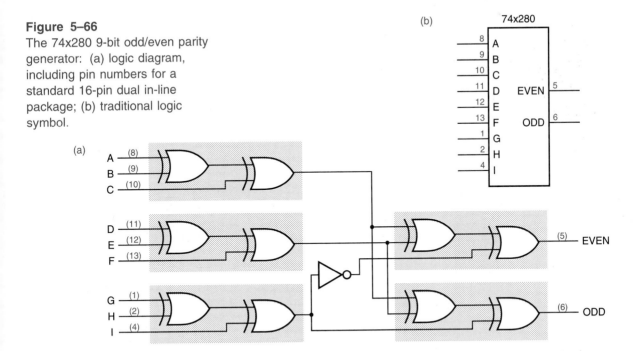

5.7.3 The 74x280 9-Bit Parity Generator

Rather than build a multibit parity circuit with discrete XOR gates, it is more economical to put all of the XORs in a single MSI package with just the primary inputs and outputs available at the external pins. The *74x280* 9-bit parity gener- *74x280* ator, shown in Figure 5–66, is such a device. It has nine inputs and two outputs that indicate whether an even or odd number of inputs are 1.

5.7.4 Parity-Checking Applications

In Section 2.15, we described error-detecting codes that use an extra bit, called a parity bit, to detect errors in the transmission and storage of data. In an even-parity code, the parity bit is chosen so that the total number of 1 bits in a code

---

If each XOR gate in Figure 5–66 were built using discrete NAND gates as in Figure 5–61(b), the 74x280 would be pretty slow, having a propagation delay equivalent to $4 \cdot 3 + 1$, or 13, NAND gates. Instead, a typical implementation of the 74x280 uses a 4-wide AND-OR-INVERT gate to perform the function of each shaded pair of XOR gates in the figure with about the same delay as a single NAND gate. The A–I inputs are buffered through two levels of inverters so that each input presents just one unit load to the circuit driving it. Thus, the total propagation delay through this implementation of the 74x280 is about the same as five inverting gates, not 13.

SPEEDING UP
THE XOR TREE

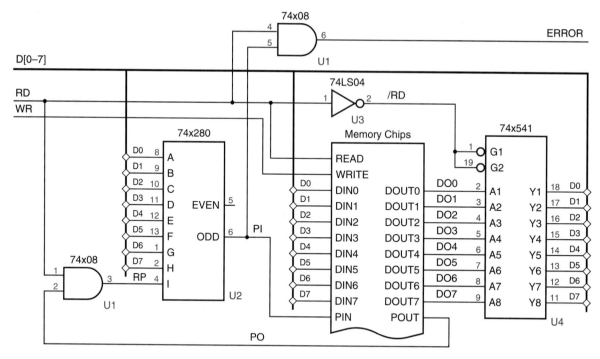

Figure 5–67 Parity generation and checking for an 8-bit-wide memory system.

word is even. Parity circuits like the 74x280 are used both to generate the correct value of the parity bit when a code word is stored or transmitted, and to check the parity bit when a code word is retrieved or received.

Figure 5–67 shows how a parity circuit might be used to detect errors in the memory of a microprocessor system. The memory stores 8-bit bytes, plus a parity bit for each byte. The microprocessor uses a bidirectional bus D[0–7] to transfer data to and from the memory. Two control lines, RD and WR, are used to indicate whether a read or write operation is desired, and an ERROR signal is asserted to indicate parity errors during read operations. Complete details of the memory chips, such as addressing inputs, are not shown; memory chips are described in detail in Chapter 11.

To store a byte into the memory chips, we specify an address (not shown), place the byte on D[0–7], generate its parity bit on PIN, and assert WR. The AND gate on the I input of the 74x280 ensures that I is 0 except during read operations, so that during writes the '280's output depends only on the parity of the D-bus data. The '280's ODD output is connected to PIN, so that the total number of 1s stored is even.

To retrieve a byte, we specify an address (not shown) and assert RD; the byte value appears on DOUT[0–7] and its parity appears on POUT. A 74x541 drives the byte onto the D bus, and the '280 checks its parity. If the parity of the 9-bit word DOUT[0–7],POUT is odd during a read, the ERROR signal is asserted.

Parity circuits are also used with error-correcting codes such as the Hamming codes described in Section 2.15.3. We showed the parity-check matrix for a 7-bit Hamming code in Figure 2–13 on page 59. We can correct errors in this code as shown in Figure 5–68. A 7-bit word, possibly containing an error, is presented on DU[1–7]. Three 74x280s compute the parity of the three bit-groups defined by the parity-check matrix. The outputs of the '280s form the syndrome,

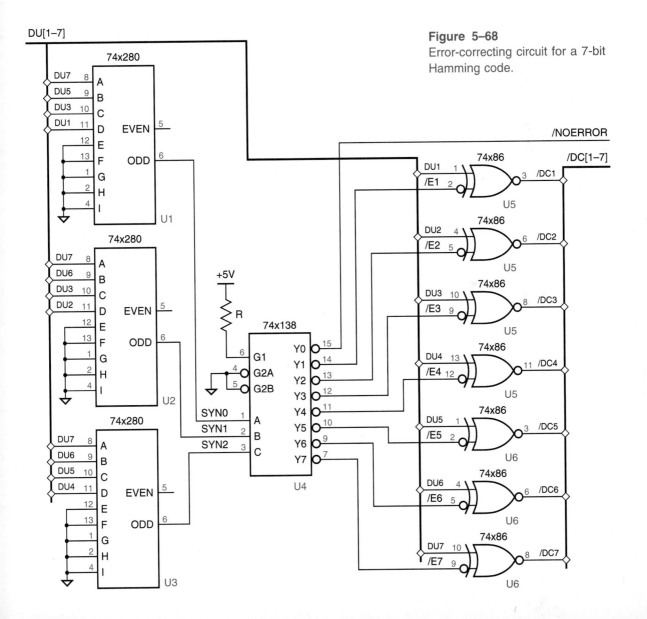

Figure 5–68

Error-correcting circuit for a 7-bit Hamming code.

which is the number of the erroneous input bit, if any. A 74x138 is used to decode the syndrome. If the syndrome is zero, the /NOERROR signal is asserted (this signal also could be named ERROR). Otherwise, the erroneous bit is corrected by complementing it. The corrected code word appears on the /DC bus.

Note that the active-low outputs of the '138 led us to use an active-low /DC bus. If we required an active-high DC bus, we could have put a discrete inverter on each XOR input or output, or used a decoder with active-high outputs, or used XNOR gates.

5.8 COMPARATORS

comparator

Comparing two binary words for equality is a commonly used operation in computer systems and device interfaces. For example, in Figure 2–7(a) on page 52, we showed a system structure in which devices are enabled by comparing a "device select" word with a predetermined "device ID." A circuit that compares two binary words and indicates whether they are equal is called a *comparator*. Some comparators interpret their input words as signed or unsigned numbers and also indicate an arithmetic relationship (greater or less than) between the words.

5.8.1 Comparator Structure

EXCLUSIVE OR and EXCLUSIVE NOR gates may be viewed as 1-bit comparators. Figure 5–69(a) shows an interpretation of the 74x86 XOR gate as a 1-bit comparator. The active-high output, DIFF, is asserted if the inputs are different. The outputs of four XOR gates are ORed to create a 4-bit comparator in (b). The DIFF output is asserted if any of the input-bit pairs are different. Given enough

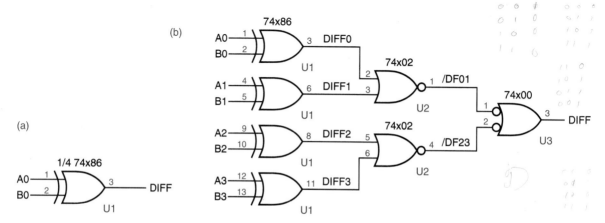

Figure 5–69 Comparators using the 74x86: (a) 1-bit comparator; (b) 4-bit comparator.

Figure 5–70
Comparators using the 74x266:
(a) 1-bit comparator; (b) 4-bit
comparator.

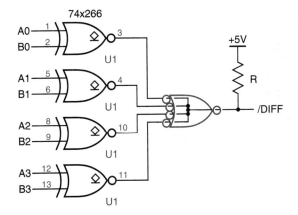

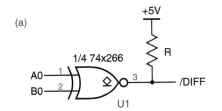

XOR gates and wide enough OR gates, comparators with any number of input
bits can be built.

Figure 5–70(a) shows an interpretation of the 74x266 XNOR gate as a 1-bit
comparator. The active-low output, /DIFF, is asserted if the inputs are different.
Recall that the '266 has an open-collector or open-drain output, so a pull-up resistor is required. Four XNOR gates can be combined to create a 4-bit comparator
as shown in (b). Taking advantage of the open-collector outputs, a wired-AND
(NOT-OR-NOT) function is used to create the /DIFF output. Using this structure, an n-bit comparator can be obtained for any value of n by wire-ANDing
the outputs of n '266s, subject to the electrical limitations of the device (mainly
HIGH-state output leakage current, I_{OH}; see Exercise 5.62).

The n-bit comparators in the preceding subsection might be called *parallel com-parators* because they look at each pair of input bits simultaneously and deliver the 1-bit comparison results in parallel to an n-input OR or AND function. It is also possible to design an "iterative comparator" that looks at its bits one at a time using a small, fixed amount of logic per bit. Before looking at the iterative comparator design, you should understand the general class of "iterative circuits" described in the next subsection.	AN ITERATIVE COMPARATOR

* 5.8.2   Iterative Circuits

An *iterative circuit* is a special type of combinational circuit, with the structure
shown in Figure 5–71. The circuit contains n identical modules, each of which
has both *primary inputs and outputs* and *cascading inputs and outputs*. The
leftmost cascading inputs are called *boundary inputs* and are connected to fixed
logic values in most iterative circuits. The rightmost cascading outputs are called
boundary outputs and usually provide important information.

iterative circuit

primary inputs and outputs

cascading inputs and outputs

boundary inputs

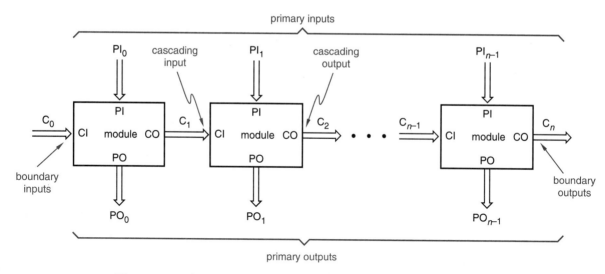

Figure 5–71   General structure of an iterative combinational circuit.

Iterative circuits are well suited to problems that can be solved by a simple iterative algorithm:

(1) Set C_0 to its initial value and set i to 0.

(2) Use C_i and PI_i to determine the values of PO_i and C_{i+1}.

(3) Increment i.

(4) If $i < n$, go to step 2.

In an iterative circuit, the loop of steps 2–4 is "unwound" by providing a separate combinational circuit that performs step 2 for each value of i.

Examples of iterative circuits are the comparator circuit in the next subsection and the ripple adder in Section 5.9.2. The 74x85 4-bit comparator and the 74x283 4-bit adder are examples of MSI circuits that can be used as the individual modules in a larger iterative circuit. In Section 9.5 we'll explore the relationship between iterative circuits and corresponding sequential circuits that execute the 4-step algorithm above in discrete time steps.

* 5.8.3   An Iterative Comparator Circuit

Two n-bit values X and Y can be compared one bit at a time using a single bit EQ_i at each step to keep track of whether all of the bit-pairs have been equal so far:

(1) Set EQ_0 to 1 and set i to 0.

(2) If EQ_i is 1 and X_i and Y_i are equal, set EQ_{i+1} to 1. Else set EQ_{i+1} to 0.

(3) Increment i.

(4) If $i < n$, go to step 2.

Figure 5–72 shows a corresponding iterative circuit. Note that this circuit has no primary outputs; the boundary output is all that interests us. Other iterative circuits, such as the ripple adder of Section 5.9.2, have primary outputs of interest.

Figure 5–72
An iterative comparator circuit:
(a) module for one bit; (b) complete circuit.

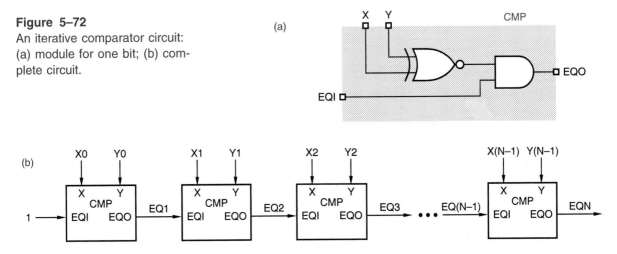

Given a choice between the iterative comparator circuit in this subsection and one of the parallel comparators shown previously, you would probably prefer the parallel comparator. The iterative comparator saves little if any cost, and it's very slow because the cascading signals need time to "ripple" from the leftmost to the rightmost module. Iterative circuits that process more than one bit at a time, using modules like the 74x85 4-bit comparator and 74x283 4-bit adder, are much more likely to be used in practical designs.

5.8.4 Standard MSI Comparators

Comparator applications are common enough that several MSI comparators have been developed commercially. The *74x85* is a 4-bit comparator with the logic symbol shown in Figure 5–73. It provides a greater-than output (AGTBOUT) and a less-than output (ALTBOUT) as well as an equal output (AEQBOUT). The '85 also has *cascading inputs* (AGTBIN, ALTBIN, AEQBIN) for combining multiple '85s to create comparators for more than four bits. Both the cascading inputs and the outputs are arranged in a 1-out-of-3 code, since in normal operation exactly one input and one output should be asserted.

74x85

cascading inputs

Figure 5–73
Traditional logic symbol for the
74x85 8-bit comparator.

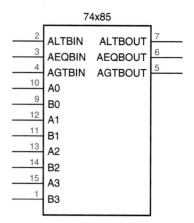

The cascading inputs are defined so that the outputs of an '85 that compares less-significant bits are connected to the inputs of an '85 that compares more-significant bits, as shown in Figure 5–74 for a 12-bit comparator. This is an iterative circuit according to the definition in Section 5.8.2. Each '85 develops its cascading outputs roughly according to the following pseudo-logic equations:

$$\text{AGTBOUT} = (A > B) + (A = B) \cdot \text{AGTBIN}$$

$$\text{AEQBOUT} = (A = B) \cdot \text{AEQBIN}$$

$$\text{ALTBOUT} = (A < B) + (A = B) \cdot \text{ALTBIN}$$

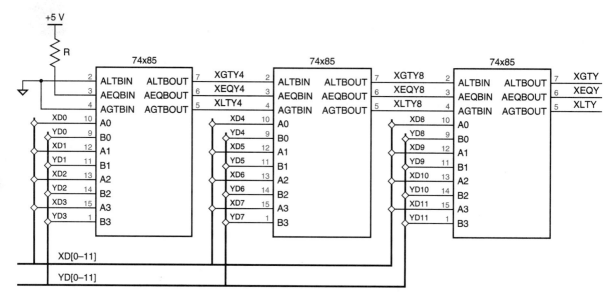

Figure 5–74 A 12-bit comparator using 74x85s.

The parenthesized subexpressions above are not normal logic expressions, but indicate an arithmetic comparison that occurs between the A3–A0 and B3–B0 inputs. In other words, AGTBOUT is asserted if A > B or if A = B and AGTBIN is asserted (if the higher-order bits are equal, we have to look at the lower-order bits for the answer). The arithmetic comparisons can be expressed using normal logic expressions, for example,

$$
\begin{aligned}
(A > B) \;=\; & A3 \cdot B3' + \\
& (A3 \oplus B3)' \cdot A2 \cdot B2' + \\
& (A3 \oplus B3)' \cdot (A2 \oplus B2)' \cdot A1 \cdot B1' + \\
& (A3 \oplus B3)' \cdot (A2 \oplus B2)' \cdot (A1 \oplus B1)' \cdot A0 \cdot B0'
\end{aligned}
$$

Such expressions must be substituted into the pseudo-logic equations above to obtain genuine logic equations for the comparator outputs.

DETAILS, DETAILS

> For some reason, the 74LS85 actually uses (AEQBIN + ALTBIN)' in place of AGTBIN in the AGTBOUT equation, and (AEQBIN + AGTBIN)' in place of ALTBIN in the ALTBOUT equation. This produces outputs different from the equations above when all of the cascading inputs are negated, but this input combination is not used in most applications.

74x682

    Several 8-bit MSI comparators are also available. The simplest of these is the *74x682*, whose internal logic diagram is shown in Figure 5–75 and whose logic symbol is shown in Figure 5–76. The top half of the circuit checks the two 8-bit input words for equality. Each XNOR-gate output is asserted if its inputs are equal, and the /PEQQ output is asserted if all eight input-bit pairs are equal. The bottom half of the circuit compares the input words arithmetically, and asserts /PGTQ if P[7–0] > Q[7–0].
    Unlike the 74x85, the 74x682 does not have cascading inputs. Also unlike the '85, the '682 does not provide a "less than" output. However, any desired condition, including ≤ and ≥, can be formulated as a function of the /PEQQ and /PGTQ outputs, as shown in Figure 5–77.

COMPARING
COMPARATORS

> The individual 1-bit comparators (XNOR gates) in the '682 are drawn in the opposite sense as the examples of the preceding subsection—outputs are asserted for *equal* inputs and then ANDed, rather than asserted for *different* inputs and then ORed. We can look at a comparator's function either way, as long as we're consistent.

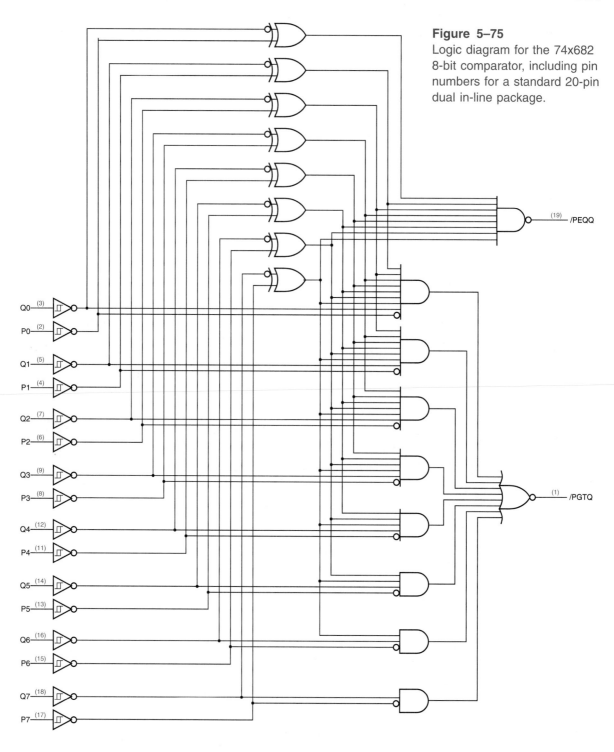

Figure 5–75
Logic diagram for the 74x682
8-bit comparator, including pin
numbers for a standard 20-pin
dual in-line package.

Figure 5–76
Traditional logic symbol for the
74x682 8-bit comparator.

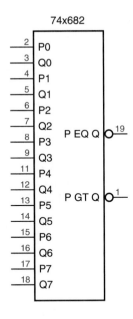

* 5.9 ADDERS, SUBTRACTERS, AND ALUS

Addition is the most commonly performed arithmetic operation in digital systems.
An *adder* combines two arithmetic operands using the addition rules described in   *adder*
Chapter 2. As we showed in Section 2.6, the same addition rules and therefore
the same adders are used for both unsigned and two's-complement numbers. An
adder can perform subtraction as the addition of the minuend and the comple-

Figure 5–77
Arithmetic conditions derived
from 74x682 outputs.

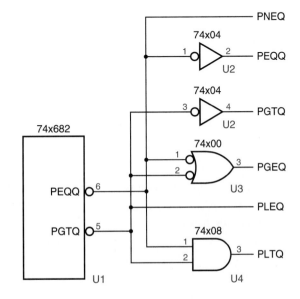

subtracter

mented (negated) subtrahend, but it is also possible to build *subtracter* circuits that perform subtraction directly. MSI devices called ALUs, described at the end of this section, perform addition, subtraction, or any of several other operations according to an operation code applied to the device's control inputs.

* 5.9.1   Half Adders and Full Adders

half adder

The simplest adder, called a *half adder*, adds two 1-bit operands X and Y, producing a 2-bit sum. The sum can range from 0 to 2, which requires two bits to express. The low-order bit of the sum may be named HS (half sum), and the high-order bit may be named CO (carry out). We can write the following equations for HS and CO:

$$HS = X \oplus Y$$
$$= X \cdot Y' + X' \cdot Y$$
$$CO = X \cdot Y$$

full adder

To add operands with more than one bit, we must provide for carries between bit positions. The basic building block for this operation is called a *full adder*. Besides the addend-bit inputs X and Y, a full adder has a carry-bit input, CIN. The sum of the three inputs can range from 0 to 3, which can still be expressed with just two output bits, S and COUT, having the following equations:

$$S = X \oplus Y \oplus CIN$$
$$= X \cdot Y' \cdot CIN' + X' \cdot Y \cdot CIN' + X' \cdot Y' \cdot CIN + X \cdot Y \cdot CIN$$
$$COUT = X \cdot Y + X \cdot CIN + Y \cdot CIN$$

Here, S is 1 if an odd number of the inputs are 1, and COUT is 1 if two or more of the inputs are 1. These equations represent the same operation that was specified by the binary addition table in Table 2–3 on page 28.

One possible circuit that performs the full-adder equations is shown in Figure 5–78(a). The corresponding logic symbol is shown in (b). Sometimes the symbol is drawn as shown in (c), so that cascaded full adders can be drawn more neatly, as in the next subsection.

* 5.9.2   Ripple Adders

ripple adder

Two binary words, each with *n* bits, can be added using a *ripple adder*—a cascade of *n* full-adder stages, each of which handles one bit. Figure 5–79 shows the circuit for a 4-bit ripple adder. The carry input to the least significant bit (c_0) is normally set to 0, and the carry output of each full adder is connected to the carry input of the next most significant full adder. The ripple adder is a classic example of an iterative circuit as defined in Section 5.8.2.

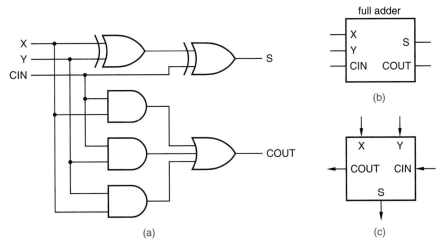

(a)   (c)

Figure 5–78   Full adder: (a) gate-level circuit diagram; (b) logic symbol; (c) alternate logic symbol suitable for cascading.

A ripple adder is slow, since in the worst case a carry must propagate from the least significant full adder to the most significant one. This occurs if one addend is $11\ldots11$ and the other is $00\ldots01$. Assuming that all of the addend bits are presented simultaneously, the total worst-case delay is

$$t_{\text{ADD}} \;=\; t_{\text{XYCout}} + (n-2) \cdot t_{\text{CinCout}} + t_{\text{CinS}}$$

where t_{XYCout} is the delay from X or Y to COUT in the least significant stage, t_{CinCout} is the delay from CIN to COUT in the middle stages, and t_{CinS} is the delay from CIN to S in the most significant stage.

A faster adder can be built by obtaining each sum output s_i with just two levels of logic. This can be accomplished by writing an equation for s_i in terms of x_0–x_i, y_0–y_i, and c_0, "multiplying out" or "adding out" to obtain a sum-of-products or product-of-sums expression, and building the corresponding AND-OR

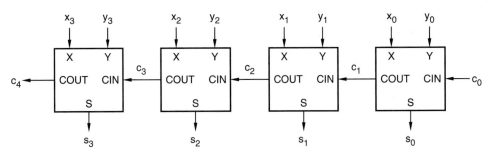

Figure 5–79   A 4-bit ripple adder.

or OR-AND circuit. Unfortunately, beyond s_2, the resulting expressions have too many terms, requiring too many first-level gates and more inputs than typically possible on the second-level gate. For example, even assuming that $c_0 = 0$, a two-level AND-OR circuit for s_2 requires fourteen 4-input ANDs, four 5-input ANDs, and an 18-input OR gate; higher-order sum bits are even worse. Nevertheless, it is possible to build adders with just a few levels of delay using a more reasonable number of gates, as we'll see in Section 5.9.4.

* 5.9.3 Subtracters

full subtracter

A binary subtraction operation analogous to binary addition was also specified in Table 2–3 on page 28. A *full subtracter* handles one bit of the binary subtraction algorithm, having input bits X (minuend), Y (subtrahend), and BIN (borrow in), and output bits D (difference) and BOUT (borrow out). We can write logic equations corresponding to the binary subtraction table as follows:

$$D = X \oplus Y \oplus BIN$$
$$BOUT = X' \cdot Y + X' \cdot BIN + Y \cdot BIN$$

These equations are very similar to equations for a full adder, which should not be surprising. We showed in Section 2.6 that a two's-complement subtraction operation, $X - Y$, can be performed by an addition operation, namely, by adding the two's complement of Y to X. The two's complement of Y is $\overline{Y} + 1$, where $\overline{Y}$ is the bit-by-bit complement of Y. We also showed in Exercise 2.25 that a binary adder can be used to perform an unsigned subtraction operation $X - Y$, by performing the operation $X + \overline{Y} + 1$. We can now confirm that these statements are true by manipulating the logic equations above:

$$
\begin{aligned}
BOUT &= X' \cdot Y + X' \cdot BIN + Y \cdot BIN \\
BOUT' &= (X + Y') \cdot (X + BIN') \cdot (Y' + BIN') \quad \text{(generalized DeMorgan's theorem)} \\
&= X \cdot Y' + X \cdot BIN' + Y' \cdot BIN' \quad \text{(multiply out)} \\
D &= X \oplus Y \oplus BIN \\
&= X \oplus Y' \oplus BIN' \quad \text{(complementing XOR inputs)}
\end{aligned}
$$

For the last manipulation, recall that we can complement the two inputs of an XOR gate without changing the function performed.

Comparing with the equations for a full adder, the above equations tell us that we can build a full subtracter from a full adder as shown in Figure 5–80. Just to keep things straight, we've given the full adder circuit in (a) a fictitious name, the "74x999." As shown in (c), we can interpret the function of this same physical circuit to be a full subtracter by giving it a new symbol with active-low borrow in, borrow out, and subtrahend signals. Thus, to build a ripple subtracter for two *n*-bit active-high operands, we can use *n* 74x999s and inverters, as

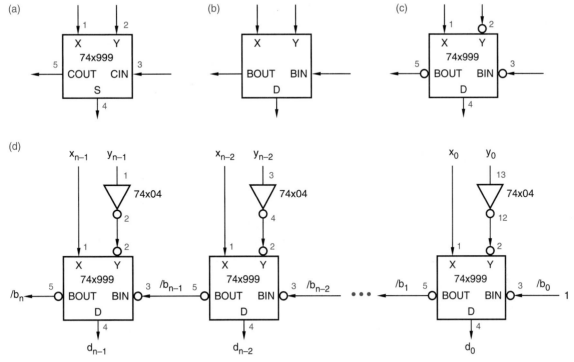

Figure 5–80   Designing subtracters using adders: (a) full adder; (b) full subtracter; (c) interpreting the device in (a) as a full subtracter; (d) ripple subtracter.

shown in (d). Note that for the subtraction operation, the borrow input of the least significant bit should be negated (no borrow), which for an active-low input means that the physical pin must be 1 or HIGH. This is just the opposite as in addition, where the same input pin is an active-high carry-in that is 0 or LOW.

By going back to the math in Chapter 2, we can show that this sort of manipulation works for all adder and subtracter circuits, not just ripple adders and subtracters. That is, any *n*-bit adder circuit can be made to function as a subtracter by complementing the subtrahend and treating the carry-in and carry-out signals as borrows with the opposite active level. The rest of this section discusses addition circuits only, with the understanding that they can easily be made to perform subtraction.

* 5.9.4   Carry Lookahead Adders

The logic equation for sum bit *i* of a binary adder can actually be written quite simply:

$$s_i = x_i \oplus y_i \oplus c_i$$

Figure 5–81
Structure of one stage of a carry
lookahead adder.

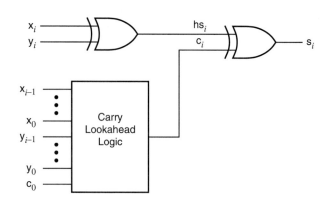

More complexity is introduced when we expand c_i above in terms of x_0–x_{i-1}, y_0–y_{i-1}, and c_0, and we get a real mess expanding the XORs. However, if we're willing to forego the XOR expansion, we can at least streamline the design of c_i

carry lookahead   logic using ideas of *carry lookahead* discussed in this subsection.

Figure 5–81 shows the basic idea. The block labeled "Carry Lookahead Logic" calculates c_i in a fixed, small number of logic levels for any reasonable value of i. Two definitions are the key to carry lookahead logic:

carry generate

- For a particular combination of inputs x_i and y_i, adder stage i is said to *generate* a carry if it produces a carry-out of 1 ($c_{i+1} = 1$) independent of the inputs on x_0–x_{i-1}, y_0–y_{i-1}, and c_0.

carry propagate

- For a particular combination of inputs x_i and y_i, adder stage i is said to *propagate* carries if it produces a carry-out of 1 ($c_{i+1} = 1$) in the presence of an input combination of x_0–x_{i-1}, y_0–y_{i-1}, and c_0 that causes a carry-in of 1 ($c_i = 1$).

Corresponding to these definitions, we can write logic equations for a carry-generate signal, g_i, and a carry-propagate signal, p_i, for each stage of a carry lookahead adder:

$$g_i = x_i \cdot y_i$$
$$p_i = x_i + y_i$$

That is, a stage unconditionally generates a carry if both of its addend bits are 1, and it propagates carries if at least one of its addend bits is 1. The carry output of a stage can now be written in terms of the generate and propagate signals:

$$c_{i+1} = g_i + p_i \cdot c_i$$

To eliminate carry ripple, we recursively expand the c_i term for each stage, and multiply out to obtain a 2-level AND-OR expression. Using this technique, we can obtain the following carry equations for the first four adder stages:

$$c_1 = g_0 + p_0 \cdot c_0$$
$$c_2 = g_1 + p_1 \cdot c_1$$
$$= g_1 + p_1 \cdot (g_0 + p_0 \cdot c_0)$$
$$= g_1 + p_1 \cdot g_0 + p_1 \cdot p_0 \cdot c_0$$
$$c_3 = g_2 + p_2 \cdot c_2$$
$$= g_2 + p_2 \cdot (g_1 + p_1 \cdot g_0 + p_1 \cdot p_0 \cdot c_0)$$
$$= g_2 + p_2 \cdot g_1 + p_2 \cdot p_1 \cdot g_0 + p_2 \cdot p_1 \cdot p_0 \cdot c_0$$
$$c_4 = g_3 + p_3 \cdot c_3$$
$$= g_3 + p_3 \cdot (g_2 + p_2 \cdot g_1 + p_2 \cdot p_1 \cdot g_0 + p_2 \cdot p_1 \cdot p_0 \cdot c_0)$$
$$= g_3 + p_3 \cdot g_2 + p_3 \cdot p_2 \cdot g_1 + p_3 \cdot p_2 \cdot p_1 \cdot g_0 + p_3 \cdot p_2 \cdot p_1 \cdot p_0 \cdot c_0$$

Each equation corresponds to a circuit with just three levels of delay—one for the generate and propagate signals, and two for the sum-of-products shown. A *carry lookahead adder* uses three-level equations such as these in each adder stage for the block labeled "carry lookahead" in Figure 5–81. The sum output for a stage is produced by combining the carry bit with the two addend bits for the stage as we showed in the figure. In the next subsection, we'll study some commercial MSI adders and ALUs that use carry lookahead.

carry lookahead adder

* 5.9.5 MSI Adders

The *74x283* is a 4-bit binary adder that forms its sum and carry outputs with just a few levels of logic, using the carry lookahead technique. Figure 5–82 is a logic symbol for the 74x283. The older *74x83* is identical except for its pinout, which has nonstandard locations for power and ground.

74x283

74x83

Figure 5–82
Traditional logic symbol for the
74x283 4-bit binary adder.

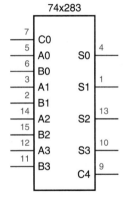

Figure 5–83
Logic diagram for the 74x283
4-bit binary adder, including pin
numbers for a standard 16-pin
dual in-line package.

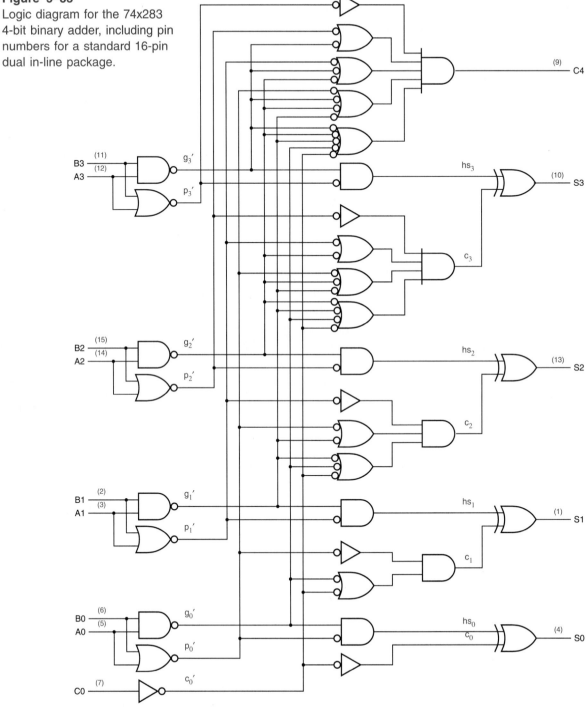

The logic diagram for the '283, shown in Figure 5–83, has a few differences from the general carry-lookahead design that we described in the preceding subsection. First of all, its addends are named A and B instead of X and Y; no big deal. Second, it produces active-low versions of the carry-generate (g_i') and carry-propagate (p_i') signals, since inverting gates are faster than noninverting ones. Third, it takes advantage of the fact that we can algebraically manipulate the half-sum equation as follows:

$$
\begin{aligned}
hs_i &= x_i \oplus y_i \\
&= x_i \cdot y_i' + x_i' \cdot y_i \\
&= x_i \cdot y_i' + x_i \cdot x_i' + x_i' \cdot y_i + y_i \cdot y_i' \\
&= (x_i + y_i) \cdot (x_i' + y_i') \\
&= (x_i + y_i) \cdot (x_i \cdot y_i)' \\
&= p_i \cdot g_i'
\end{aligned}
$$

Thus, an AND gate with an inverted input can be used instead of an XOR gate to create each half-sum bit.

Finally, the '283 creates the carry signals using an INVERT-OR-AND structure (the DeMorgan equivalent of an AND-OR-INVERT), which has about the same delay as a single CMOS or TTL inverting gate. This requires some explaining, since the carry equations that we derived in the preceding subsection are used in a slightly modified form. In particular, the c_{i+1} equation uses the term $p_i \cdot g_i$ instead of g_i. This has no effect on the output, since p_i is always 1 when g_i is 1. However, it allows the equation to be factored as follows:

$$
\begin{aligned}
c_{i+1} &= p_i \cdot g_i + p_i \cdot c_i \\
&= p_i \cdot (g_i + c_i)
\end{aligned}
$$

This leads to the following carry equations, which are used by the circuit:

$$
\begin{aligned}
c_1 &= p_0 \cdot (g_0 + c_0) \\
c_2 &= p_1 \cdot (g_1 + c_1) \\
&= p_1 \cdot (g_1 + p_0 \cdot (g_0 + c_0)) \\
&= p_1 \cdot (g_1 + p_0) \cdot (g_1 + g_0 + c_0) \\
c_3 &= p_2 \cdot (g_2 + c_2) \\
&= p_2 \cdot (g_2 + p_1 \cdot (g_1 + p_0) \cdot (g_1 + g_0 + c_0)) \\
&= p_2 \cdot (g_2 + p_1) \cdot (g_2 + g_1 + p_0) \cdot (g_2 + g_1 + g_0 + c_0) \\
c_4 &= p_3 \cdot (g_3 + c_3) \\
&= p_3 \cdot (g_3 + p_2 \cdot (g_2 + p_1) \cdot (g_2 + g_1 + p_0) \cdot (g_2 + g_1 + g_0 + c_0)) \\
&= p_3 \cdot (g_3 + p_2) \cdot (g_3 + g_2 + p_1) \cdot (g_3 + g_2 + g_1 + p_0) \cdot (g_3 + g_2 + g_1 + g_0 + c_0)
\end{aligned}
$$

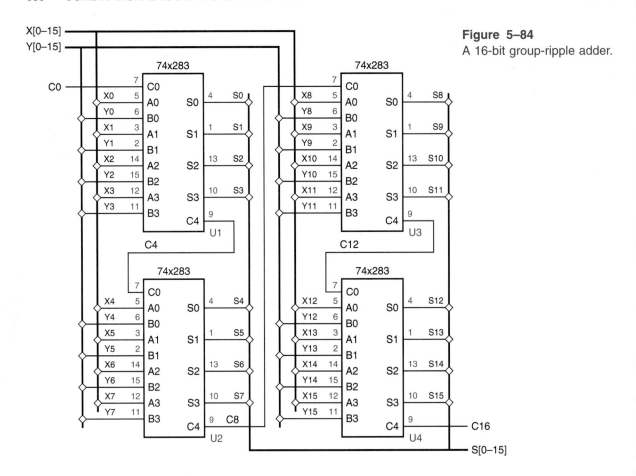

Figure 5–84

A 16-bit group-ripple adder.

If you've followed the derivation of these equations and can obtain the same ones by reading the '283 logic diagram, then congratulations, you're up to speed on switching algebra! If not, you may want to review Sections 4.1 and 4.2.

group-ripple adder   The propagation delay from the C0 input to the C4 output of the '283 is very short, about the same as two inverting gates. As a result, fairly fast *group-ripple adders* with more than four bits can be made simply by cascading the carry outputs and inputs of '283s, as shown in Figure 5–84 for a 16-bit adder. The total propagation delay from C0 to C16 in this circuit is about the same as that of eight inverting gates.

* 5.9.6 MSI Arithmetic and Logic Units

arithmetic and logic   An *arithmetic and logic unit (ALU)* is a combinational circuit that can perform
unit (ALU)   any of a number of different arithmetic and logical operations on a pair of *b*-bit operands. The operation to be performed is specified by a set of function-select

Figure 5–85
Logic symbol for the 74x181
4-bit ALU.

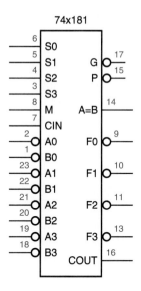

inputs. Typical MSI ALUs have 4-bit operands and three to five function select inputs, allowing up to 32 different functions to be performed.

Figure 5–85 is a logic symbol for the *74x181* 4-bit ALU. The operation performed by the '181 is selected by the M and S3–S0 inputs, as detailed in Table 5–16. Note that the identifiers A, B, and F in the table refer to the 4-bit

74x181

Table 5–16
Functions performed by the
74x181 4-bit ALU.

Inputs				Function	
S3	S2	S1	S0	M = 0 (arithmetic)	M = 1 (logic)
0	0	0	0	F = A minus 1 plus CIN	F = A$'$
0	0	0	1	F = A·B minus 1 plus CIN	F = A$'$ + B$'$
0	0	1	0	F = A·B$'$ minus 1 plus CIN	F = A$'$ + B
0	0	1	1	F = 1111 plus CIN	F = 1111
0	1	0	0	F = A plus (A + B$'$) plus CIN	F = A$'$·B$'$
0	1	0	1	F = A·B plus (A + B$'$) plus CIN	F = B$'$
0	1	1	0	F = A minus B minus 1 plus CIN	F = A⊕B$'$
0	1	1	1	F = A + B$'$ plus CIN	F = A + B$'$
1	0	0	0	F = A plus (A + B) plus CIN	F = A$'$·B
1	0	0	1	F = A plus B plus CIN	F = A⊕B
1	0	1	0	F = A·B$'$ plus (A + B) plus CIN	F = B
1	0	1	1	F = A + B plus CIN	F = A + B
1	1	0	0	F = A plus A plus CIN	F = 0000
1	1	0	1	F = A·B plus A plus CIN	F = A·B$'$
1	1	1	0	F = A·B$'$ plus A plus CIN	F = A·B
1	1	1	1	F = A plus CIN	F = A

words A3–A0, B3–B0, and F3–F0; and the symbols · and + refer to logical AND and OR operations.

The 181's M input selects between arithmetic and logical operations. When M = 1, logical operations are selected, and each output Fi is a function only of the corresponding data inputs, Ai and Bi. No carries propagate between stages, and the CIN input is ignored. The S3–S0 inputs select a particular logical operation; any of the 16 different combinational logic functions on two variables may be selected.

When M = 0, arithmetic operations are selected, carries propagate between the stages, and CIN is used as a carry input to the least significant stage. For operations larger than four bits, multiple '181 ALUs may be cascaded like the group-ripple adder in the Figure 5–84, with the carry-out (COUT) of each ALU connected to the carry-in (CIN) of the next most significant stage. The same function-select signals (M, S3–S0) are applied to all the '181s in the cascade.

To perform two's-complement addition, we use S3–S0 to select the operation "A plus B plus CIN." The CIN input of the least-significant ALU is normally set to 0 during addition operations. To perform two's-complement subtraction, we use S3–S0 to select the operation A minus B minus 1 plus CIN. In this case, the CIN input of the least significant ALU is normally set to 1, since CIN acts as the complement of the borrow during subtraction.

The '181 provides other arithmetic operations, such as "A minus 1 plus CIN," that are useful in some applications (e.g., decrement by 1). It also provides a bunch of weird arithmetic operations, such as "A·B′ plus (A + B) plus CIN," that are almost never used in practice, but that "fall out" of the circuit for free.

Notice that the operand inputs /A3–/A0 and /B3–/B0 and the function outputs /F3–/F0 of the '181 are active low. The '181 can also be used with active-

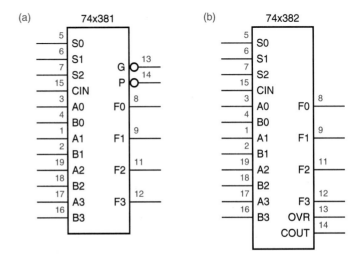

Figure 5–86
Logic symbols for 4-bit ALUs:
(a) 74x381; (b) 74x382.

Table 5–17
Functions performed by the
74x381 and 74x382 4-bit ALUs.

Inputs			
S2	S1	S0	Function
0	0	0	F = 0000
0	0	1	F = B minus A minus 1 plus CIN
0	1	0	F = A minus B minus 1 plus CIN
0	1	1	F = A plus B plus CIN
1	0	0	F = A⊕B
1	0	1	F = A + B
1	1	0	F = A·B
1	1	1	F = 1111

high operand inputs and function outputs. In this case, a different version of the function table must be constructed. When M = 1, logical operations are still performed, but for a given input combination on S3–S0, the function obtained is precisely the dual of the one listed in Table 5–16. When M = 0, arithmetic operations are performed, but the function table is once again different. Refer to a '181 data sheet for more details.

Two other MSI ALUs, the *74x381* and *74x382* shown in Figure 5–86, en- *74x381* code their select inputs more compactly, and provide only eight different but *74x382* useful functions, as detailed in Table 5–17. The only difference between the '381 and '382 is that one provides group-carry lookahead outputs (which we explain next), while the other provides ripple carry and overflow outputs.

* 5.9.7 Group-Carry Lookahead

The '181 and '381 provide *group-carry lookahead* outputs that allow multiple *group-carry* ALUs to be cascaded without rippling carries between 4-bit groups. Like the *lookahead* 74x283, the ALUs use carry lookahead to produce carries internally. However, they also provide /G and /P outputs that are carry lookahead signals for the entire 4-bit group. The /G output is asserted if the ALU generates a carry, that is, if it will produce a carry-out (COUT = 1) independent of whether or not there is a carry-in (CIN = 1):

$$\text{/G} = (g_3 + p_3 \cdot g_2 + p_3 \cdot p_2 \cdot g_1 + p_3 \cdot p_2 \cdot p_1 \cdot g_0)'$$

The /P output is asserted if the ALU propagates a carry, that is, if it will produce a carry-out if there is a carry-in:

$$\text{/P} = (p_3 \cdot p_2 \cdot p_1 \cdot p_0)'$$

When ALUs are cascaded, the group-carry lookahead outputs may be combined in just two levels of logic to produce the carry input to each ALU. A *lookahead carry circuit*, the *74x182* shown in Figure 5–87, performs this operation. The '182 inputs are C0, the carry input to the least significant ALU ("ALU 0"), and G0–G3 and P0–P3, the generate and propagate outputs of ALUs 0–3. Using these inputs, the '182 produces the carry inputs C1–C3 for ALUs 1–3. Figure 5–88 shows the connections for a 16-bit ALU using four '381s and a '182.

Figure 5–87
Logic symbol for the 74x182 carry lookahead circuit.

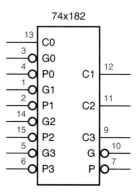

The 182's carry equations are obtained by "adding out" the basic carry lookahead equation of Section 5.9.4:

$$c_{i+1} = g_i + p_i \cdot c_i$$
$$= (g_i + p_i) \cdot (g_i + c_i)$$

Expanding for the first three values of i, we obtain the following equations:

$$C1 = (G0 + P0) \cdot (G0 + C0)$$
$$C2 = (G1 + P1) \cdot (G1 + G0 + P0) \cdot (G1 + G0 + C0)$$
$$C3 = (G2 + P2) \cdot (G2 + G1 + P1) \cdot (G2 + G1 + G0 + P0) \cdot (G2 + G1 + G0 + C0)$$

The '182 realizes each of these equations with just one level of delay—an INVERT-OR-AND gate.

When more than four ALUs are cascaded, they may be partitioned into "supergroups," each with its own '182. For example, a 64-bit adder would have four supergroups, each containing four ALUs and a '182. The /G and /P outputs of each '182 can be combined in a next-level '182, since they indicate whether the supergroup generates or propagates carries:

$$/G = ((G3 + P3) \cdot (G3 + G2 + P2) \cdot (G3 + G2 + G1 + P1) \cdot (G3 + G2 + G1 + G0))'$$
$$/P = (P0 \cdot P1 \cdot P2 \cdot P3)'$$

Figure 5–88
A 16-bit ALU using group-carry
lookahead.

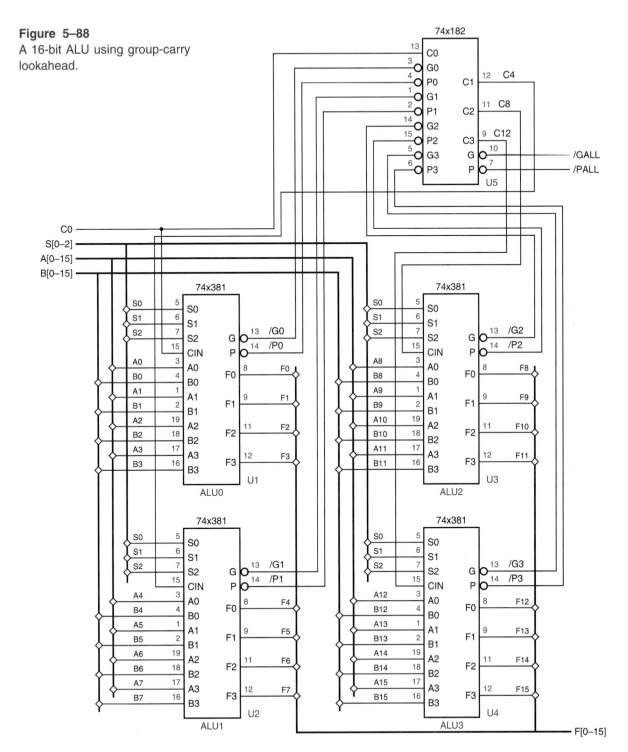

* 5.10   COMBINATIONAL MULTIPLIERS

In Section 2.8, we outlined an algorithm that uses n shifts and adds to multiply n-bit binary numbers. Although the shift-and-add algorithm emulates the way that we do paper-and-pencil multiplication of decimal numbers, there is nothing inherently "sequential" or "time dependent" about multiplication. That is, given two n-bit input words A and B, it is possible to write a truth table that expresses the $2n$-bit product $P = A \cdot B$ as a *combinational* function of A and B. A *combinational multiplier* is a logic circuit with such a truth table.

*combinational
multiplier*
74284
74285

The *74284* and *74285* are commercially available MSI 4×4 combinational multipliers. The two chips are used together to obtain the 8-bit product of two 4-bit unsigned inputs. As shown in Figure 5–89, the '284 produces the four high-order bits of the product, and the '285 produces the four low-order bits. Pull-up resistors are shown in the figure because the '284 and '285 have open-collector outputs (Section 3.11.4).

Figure 5–89
A 4×4 multiplication element using a 74284 and a 74285, each in a standard 16-pin package.

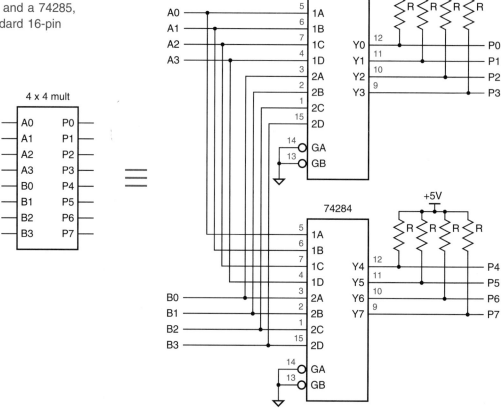

Figure 5–90
Producing an 8×8 product as a
sum of 4×4 products.

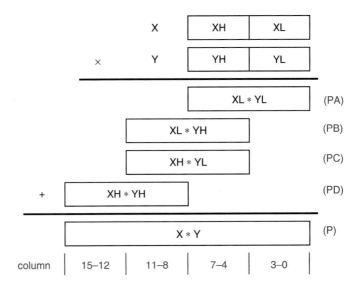

There is no inherent reason for a combinational multiplier to use two separate chips for the low-order and high-order product bits. The '284 and '285 were designed that way because of the pin, power, and gate limitations in TTL technology in 1972. More recently designed combinational multipliers, such as the Cypress Semiconductor 16×16-bit 7C516, use CMOS LSI technology and larger packages to perform larger multiplications.

All by itself, a 4×4 multiplier using the '284 and '285 isn't too useful—you could almost do a 4×4 multiplication in your head in the 60 ns that the '284 and '285 take to do it (well, maybe not quite that fast...). However, most applications require larger multiplications, such as 16×16 or 32×32, that can be performed hierarchically using adders and 4×4 multipliers as building blocks.

Figure 5–90 shows the basic idea for a hierarchical 8×8 multiplier. The 8-bit multiplier and multiplicand X and Y are each broken into two 4-bit components. Each component of X is multiplied by each component of Y, and the partial products are added, with the appropriate shifts. Figure 5–91 is a combinational circuit corresponding to this scheme. This circuit has three levels:

(1) The multipliers (U1–U4) form all of the 4×4 products in parallel.

Each 4×4 multiplier in Figure 5–91 contains a '284 and a '285, designated U1 and U2 in Figure 5–89. When preparing the mechanical documentation for the system, a computer-aided design system must "flatten" the hierarchical schematic, in which case it might give the individual '284s and '285s reference designators like U1.1, U1.2, ..., U4.1, U4.2.

HOW TO GET
FLATTENED

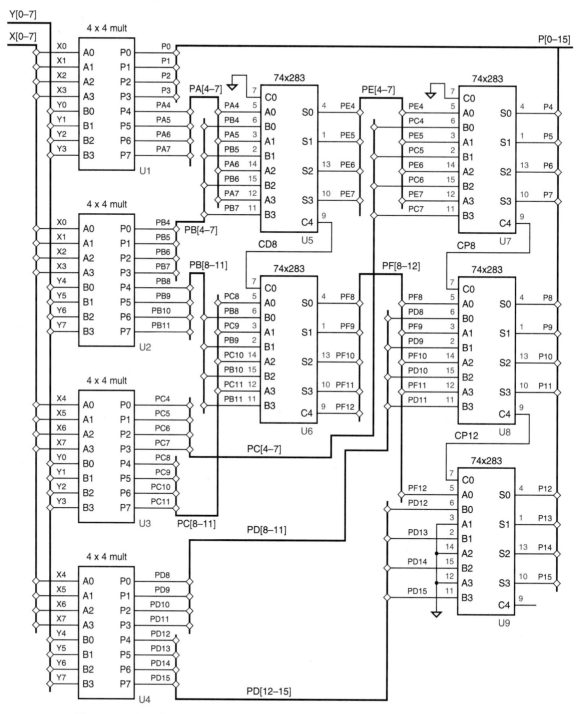

Figure 5–91 Structure of an 8×8 multiplier using 4×4 multipliers and adders.

(2) The second-level adders form a partial product combining the first two operands in columns 11–4 of Figure 5–90.

(3) The third-level adders combine the result of the preceding step with the last operand in columns 15–4 of the figure. Notice that even though columns 15–12 have only one operand, an adder is required to accommodate the high-order carries from the preceding step.

It is instructive to analyze the total worst-case delay of the 8×8 multiplier circuit. Assuming that LS-TTL is used, the individual circuit components have the following worst-case delays:

$t_{4\times4}$ 60 ns. Input to output delay of a 4×4 multiplier.

t_{c0c4} 16 ns. Delay from the C0 input to the C4 output in a 74LS283.

t_{abc4} 16 ns. Delay from an Ai or Bi input to the C4 output in the 74LS283.

t_{c0s} 21 ns. Delay from the C0 input to any Si output in a 74LS283.

t_{abs} 24 ns. Delay from an Ai or Bi input to any Si output in a 74LS283.

We must determine the worst-case delay from any Xi or Yi input to any Pi output along any path through the circuit. A worst-case delay can be determined by adding the delays of each level in the circuit:

(1) 60 ns. All of the 4×4 multipliers produce their outputs at the same time, t_{4x4} after the inputs change.

(2) 37 ns. The worst-case delay for the second-level (8-bit) adder to have a valid output is t_{abc4} (through U5) plus $\max(t_{c0s}, t_{c0c4})$ (through U6).

(3) 56 ns. The worst-case delay for the third-level (12-bit) adder to have a valid output is t_{abc4} (through U7) plus t_{c0c4} (through U8) plus $\max(t_{c0s}, t_{abs})$ (through U9).

The sum of the three levels' delay is 153 ns. However, this bound is overly pessimistic, since U7 does not need all of the second-level outputs to be valid in order to produce a valid output. Several different paths involving U5–U8 must be analyzed to calculate the actual worst-case delay in Exercise 5.75.

5.11 COMBINATIONAL LOGIC DESIGN WITH MSI FUNCTIONS

We began this chapter by pointing out that practical combinational circuits cannot be designed by brute-force application of theoretical techniques. The key to practical logic design, we said, is structured thinking—decomposing a complex problem into smaller chunks that are easier to swallow. At this point, you should be familiar with the most commonly used combinational MSI chunks, so we can now offer some practical examples of combinational logic design with MSI.

As we discussed in Section 1.9, an important goal of practical logic design is to minimize cost, including design time and debugging time as well as component cost. The examples in this section try to attain that goal. In Chapter 6, we show how programmable logic devices can be used to attain that goal even more effectively.

5.11.1 Using MSI Functions for "Random Logic"

random logic

In Section 4.3 we gave many examples of combinational logic functions and showed how to build them using discrete gates. Most designers would classify these functions as *random logic*—functions that do not correspond to a common, structured operation such as decoding, multiplexing, comparing, or adding. Yet in Sections 5.3.6 and 5.6.4 we showed two structured methods using MSI decoders and multiplexers that can economically perform "random" combinational logic functions with a small number of inputs. For example, compare two designs for the pair of functions $F = \Sigma_{X,Y,Z}(3,6,7)$ and $G = \Sigma_{X,Y,Z}(0,1,3)$:

- The minimal multiple-output sum-of-products (NAND-NAND) circuit based on Figure 4–43(b) on page 237 requires two inverters (2/6 74x04), four 2-input NANDs (4/4 74x00), and one 3-input NAND (1/3 74x10), a total of 1.67 IC packages.

- A structured solution based on Figure 5–32 uses a 74x138 3-to-8 decoder and two 3-input NANDs (2/3 74x10), also a total of 1.67 IC packages. At the board level, the MSI solution costs about the same as the SSI solution, but it's a "no-brainer" to design.

Next, compare two designs for a prime-number detector:

- A minimal sum-of-products (NAND-NAND) circuit based on Figure 4–29 on page 226 requires three inverters (3/6 74x04), one 2-input NAND (1/4 74x00), three 3-input NANDs (3/3 74x10), and one 4-input NAND (1/2 74x20), a total of 2.25 IC packages.

- The structured solution given in Figure 5–60 uses a single 74x151 8-input multiplexer, just one IC package. Not only is it easier to design, but also it costs substantially less at the PCB level.

Of course, the decoder and multiplexer approaches are uneconomical for circuits with more than a few inputs. However, in Chapter 6 we'll describe programmable logic devices that can handle random-logic functions with lots of inputs.

Table 5–18
Function table for an almost-
binary decoder.

/CS	/RD	A2	A1	A0	Output to Assert
1	x	x	x	x	none
x	1	x	x	x	none
0	0	0	0	x	/BILL
0	0	0	x	0	/MARY
0	0	0	1	0	/JOAN
0	0	0	1	1	/PAUL
0	0	1	0	0	/ANNA
0	0	1	0	1	/FRED
0	0	1	1	0	/DAVE
0	0	1	1	1	/KATE

5.11.2   Almost-MSI Functions

Sometimes a required function almost matches what is provided by a standard
MSI IC, but not quite. A common example of such a function is an *almost-binary*   *almost-binary*
decoder. Suppose you needed to generate a set of enable signals according to   *decoder*
Table 5–18. Each output is asserted for one row of the table in which /CS and
/RD are asserted, and negated for all other input combinations. This almost
matches the function of a 3-to-8 decoder with two enable inputs, except that
/BILL and /MARY are each asserted for two combinations of the A2–A0 inputs.
Rather than design a custom decoder from scratch, in a board-level design it is
more economical to adapt an MSI decoder as shown in Figure 5–92.

Figure 5–92
Almost-binary decoder circuit.

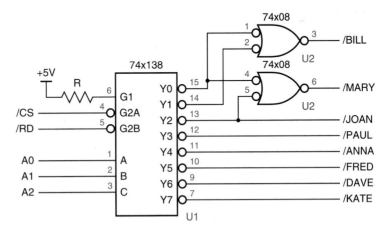

Another example is an *almost-binary-select multiplexer.* Suppose that you   *almost-binary-select*
needed a circuit that selects one of four 16-bit input buses, A, B, C, or D, to drive   *multiplexer*
a 16-bit output bus F, as specified in Table 5–19 by three control bits (ignore the

Table 5–19
Function table for an almost-
binary-select multiplexer.

S2	S1	S0	Input to Select	MS1	MS0
0	0	0	A	0	0
0	0	1	B	0	1
0	1	0	A	0	0
0	1	1	C	1	0
1	0	0	A	0	0
1	0	1	D	1	1
1	1	0	A	0	0
1	1	1	B	0	1

last two columns for now). One way to design this circuit is to use 16 74x151
8-input multiplexers, one for each bit connected as shown in Figure 5–93(a).
However, this solution is wasteful because it uses 8-input multiplexers to select
among only four distinct sources. As shown in (b), a more economical solution
uses only eight 74x153 4-input multiplexers, each of which handles two bits,
and a "code converter" circuit. The code converter produces the select inputs
for the '153s according to the last two columns of Table 5–19. Only one copy
of the code converter is needed to drive all of the '153s, so the net savings with
this approach is eight multiplexer packages minus whatever it takes to build the
code converter (just one package, according to Exercise 5.78).

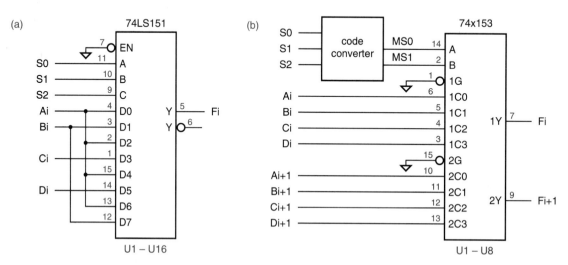

Figure 5–93 Design of the almost-binary-select multiplexer: (a) duplicating the data inputs; (b) re-
encoding the select inputs.

A final example is a *mode-dependent comparator.* This example is taken   *mode-dependent comparator*
from a system that compares two *n*-bit words under normal circumstances, but
that must sometimes ignore one or two low-order bits of the input words, as
specified by two mode-control bits. One way to obtain this function is to use
three comparators—an *n*-bit, an $(n-1)$-bit, and an $(n-2)$-bit one—and combine
their outputs with a 3-input multiplexer controlled by the mode bits. However, a
more economical solution is shown in Figure 5–94 for an 8-bit mode-dependent
comparator. Instead of combining multiple comparator outputs, this design uses
gates on a single comparator's low-order input bits, forcing low-order bits to be
equal in the appropriate modes. The design can be further simplified using a
different encoding for the mode-control bits (see Exercise 5.81).

Figure 5–94

Design of an 8-bit mode-
dependent comparator:
(a) mode-control encoding;
(b) logic diagram.

M1	M0	Comparison
0	0	8-bit
0	1	7-bit
1	0	6-bit
1	1	not used

(a)

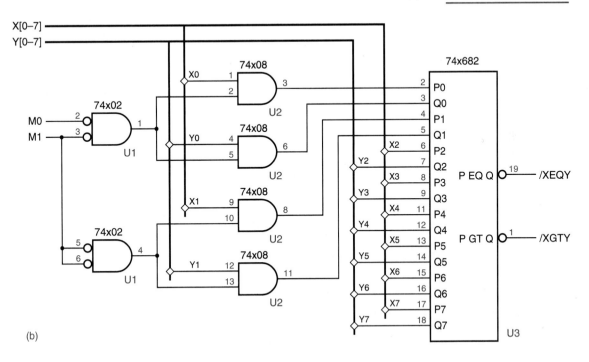

(b)

5.11.3 Finding the MSI Functions in Larger Problems

Quite often, the requirements for a digital-circuit application are specified in a way that makes an MSI solution obvious. Consider the following problem:

- Design a combinational circuit whose inputs are two 8-bit unsigned binary integers, X and Y, and a control signal MIN/MAX. The output of the circuit is an 8-bit unsigned binary integer Z such that Z = min(X, Y) if MIN/MAX = 1, and Z = max(X, Y) if MIN/MAX = 0.

This circuit is fairly easy to visualize in terms of MSI functions. Clearly, we can use a comparator to determine whether X > Y. We can use the comparator's output to control multiplexers that produce min(X, Y) and max(X, Y), and we can use another multiplexer to select one of these results depending on MIN/MAX. Figure 5–95(a) is the block diagram of a circuit corresponding to this approach.

Our first solution approach works, but it's more expensive than it needs to be. Although it has three two-input multiplexers, there are only two input words, X and Y, that may ultimately be selected and produced at the output of the circuit. Therefore, we should be able to use just a single two-input mux, and use some other logic to figure out which input to tell it to select. This approach is shown in Figure 5–95(b) and (c). The "other logic" is very simple indeed, just a single XOR gate.

Now let's look at a design problem whose MSI solution is not quite so obvious, a "fixed-point to floating-point encoder." An unsigned binary integer B in the range $0 \leq B < 2^{11}$ can be represented by 11 bits in "fixed-point" format, $B = b_{10}b_9 \ldots b_1 b_0$. We can represent numbers in the same range with less precision using only 7 bits in a floating-point notation, $F = M \cdot 2^E$, where M is a 4-bit mantissa $m_3 m_2 m_1 m_0$ and E is a 3-bit exponent $e_2 e_1 e_0$. The smallest integer in this format is $0 \cdot 2^0$ and the largest is $(2^4 - 1) \cdot 2^7$.

USING MSI
EFFECTIVELY

The wastefulness of our original design approach in Figure 5–95(a) may have been obvious to you from the beginning, but it demonstrates an important approach to designing with MSI:

- Use standard MSI parts to handle data, and look for ways that a single MSI part can perform different functions at different times or in different modes. Design control circuits, possibly using SSI, to select the appropriate functions as needed, to reduce the total parts count of the design.

As Figure 5–95(c) dramatically shows, this approach can save a lot of chips. When designing with IC chips, you should *not* heed the slogan, "Have all you want, we'll make more"!

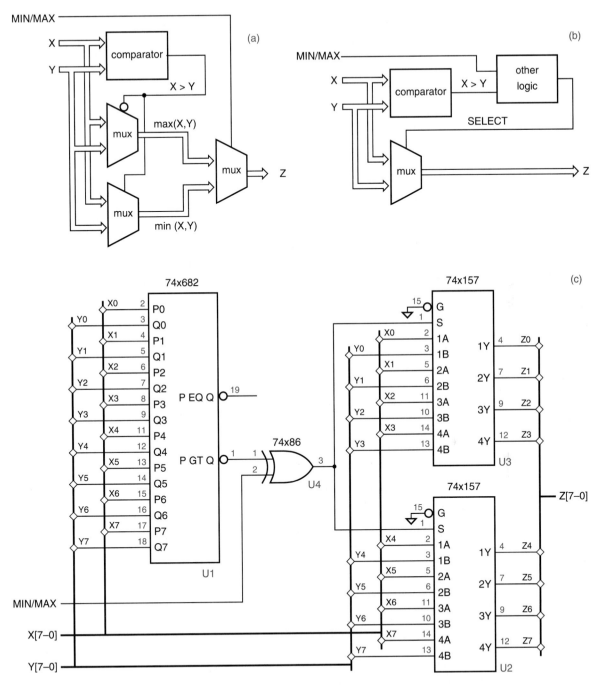

Figure 5–95 Min/max circuit: (a) block diagram of a "first-cut" solution; (b) block diagram of a more cost-effective solution; (c) logic diagram for the second solution.

Given an 11-bit fixed-point integer B, we can convert it to our 7-bit floating-point notation by "picking off" four high-order bits beginning with the most significant 1, for example,

$$11010110100 = 1101 \cdot 2^7 + 0110100$$
$$00100101111 = 1001 \cdot 2^5 + 01111$$
$$00000111110 = 1111 \cdot 2^2 + 10$$
$$00000001011 = 1011 \cdot 2^0 + 0$$
$$00000000010 = 0010 \cdot 2^0 + 0$$

The last term in each equation is a truncation error that results from the loss of precision in the conversion. Corresponding to this conversion operation, we can write the specification for a fixed-point to floating-point encoder circuit:

- A combinational circuit is to convert an 11-bit unsigned binary integer B into a 7-bit floating-point number M, E, where M and E have 4 and 3 bits, respectively. The numbers have the relationship $B = M \cdot 2^E + T$, where T is the truncation error, $0 \leq T < 2^E$.

Starting with a problem statement like the one above, it takes some creativity to come up with an efficient circuit design—the specification gives no clue. However, we can get some ideas by looking at how we converted numbers by hand earlier. We basically scanned each input number from left to right to find the first position containing a 1, stopping at the b_3 position if no 1 was found. We picked off four bits starting at that position to use as the mantissa, and the starting position number determined the exponent. These operations are beginning to sound like MSI functions.

"Scanning for the first 1" is what a generic priority encoder does. The output of the priority encoder is a number that tells us the position of the first 1. The position number determines the exponent; first-1 positions of b_{10}–b_3 imply exponents of 7–0, and positions of b_2–b_0 or no-1-found imply an exponent of 0. Therefore, we can scan for the first 1 with an 8-input priority encoder with inputs I7 (highest priority) through I0 connected to b_{10}-b_3. We can use the priority encoder's A2–A0 outputs directly as the exponent, as long as the no-1-found case produces A2–A0=000.

"Picking off four bits" sounds like a selecting or multiplexing operation. The 3-bit exponent determines which four bits of B we pick off, so we can use the exponent bits to control an 8-input, 4-bit multiplexer that selects the appropriate four bits of B to form M.

A circuit that results from these ideas is shown in Figure 5–96. It contains several optimizations:

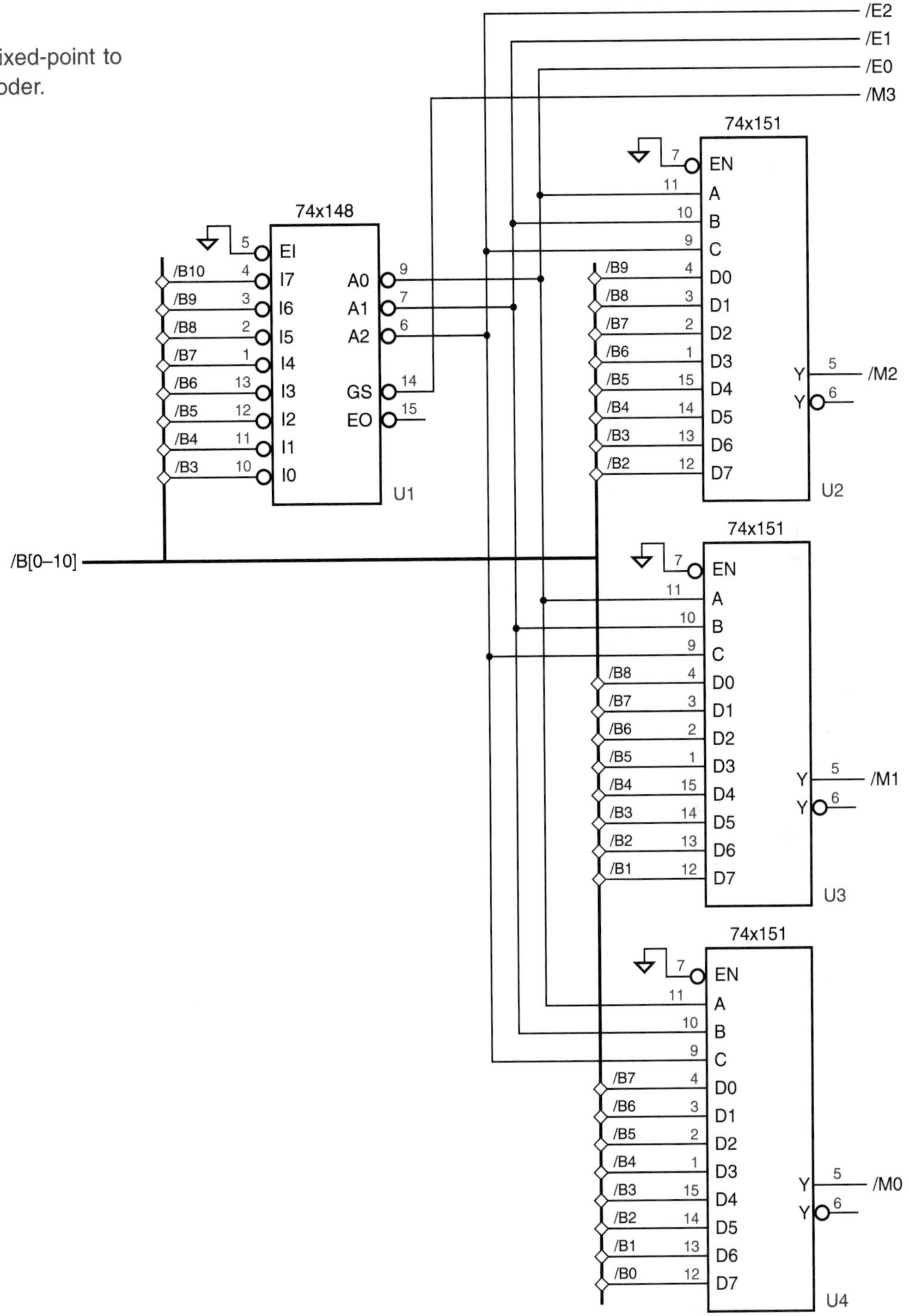

Figure 5–96
A combinational fixed-point to
floating-point encoder.

- Since the available MSI priority encoder, the 74x148, has active-low inputs, the input number B is assumed to be available on an active-low bus /B[0–10]. If only an active-high version of B is available, then eight inverters can be used to obtain the active-low version.

- If you think about the conversion operation a while, you'll realize that the most significant bit of the mantissa, m_3, is always 1, except in the no-1-found case. The '148 has a /GS output that indicates this case, allowing us to eliminate the multiplexer for m_3.

- The '148 has active-low outputs, so the exponent bits (/E0–/E2) are produced in active-low form. Naturally, three inverters could be used to produce an active-high version.

- Since everything else is active-low, active-low mantissa bits are used too. Active-high bits are also readily available on the '148 /EO and the '151 /Y outputs.

Strictly speaking, the multiplexers in Figure 5–96 are drawn incorrectly. The 74x151 has an alternate symbol shown in Figure 5–97. In words, if the multiplexer's data inputs are active low, then the data outputs have an active level opposite that shown in the original symbol. The "active-low-data" symbol should be preferred in Figure 5–96, since the active levels of the '151 inputs and outputs would then match their signal names. However, in data transfer and storage applications, designers (and the book) don't always "go by the book." It is usually clear from the context that a multiplexer (or a multibit register, in Section 9.2.4) does not alter the active level of its data.

Figure 5–97
Alternate logic symbol for the
74x151 8-input multiplexer.

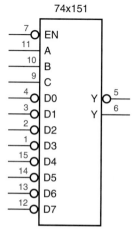

R E F E R E N C E S

Digital designers who want to write better should start by reading the classic *Elements of Style*, 3rd ed., by William Strunk, Jr. and E. B. White (Macmillan, 1979). Another book on writing style, especially directed at nerds, is *Effective Writing for Engineers, Managers, and Scientists*, 2nd ed., by H. J. Tichy (Wiley, 1988).

The ANSI/IEEE standard for logic symbols, Std 91-1984, *IEEE Standard Graphic Symbols for Logic Functions*, is available from the IEEE (New York, NY 10017). Another standard of interest to logic designers is ANSI/IEEE 991-1986, *Logic Circuit Diagrams*. These two standards and others, including standard symbols for 10-inch gongs and maid's-signal plugs, can be found in one handy, five-pound reference published by the IEEE, *Electrical and Electronics Graphics Symbols and Reference Designations*, 2nd ed. (Wiley, 1987).

Real logic devices are described in hundreds of data book published by different manufacturers. Updated editions of these books are published every few years. For a given logic family such as 74ALS, all manufacturers list generally equivalent specifications, so you can get by with just one data book per family. Some specifications, especially timing, may vary slightly between manufacturers, so when timing is tight it's best to check a couple of different sources and use the worst case. That's a *lot* easier than convincing your manufacturing department to buy a component only from a single supplier.

Data books for the 74FCT family of high-speed CMOS MSI devices include the *1993 High-Performance Logic Data Book* from Integrated Device Technology, Inc. (Santa Clara, CA 95054) and the *1993 Databook* from Quality Semiconductor, Inc. (Santa Clara, CA 95050). Texas Instruments publishes a series of data books, listed in Table 5–20, that are standard references for digital design with SSI and MSI parts, both TTL and CMOS. Other manufacturers that publish fairly comprehensive data books include Motorola, National Semiconductor, and Fairchild Semiconductor.

On the technical side of digital design, lots of textbooks cover digital design

Table 5–20 Texas Instruments logic data books.

Order number	Topics	Title	Year
SDLD001A	74, 74S, 74LS TTL	*TTL Logic Data Book*	1988
SDAD001B	74AS, 74ALS TTL	*ALS/AS Logic Data Book*	1988
SDFD001	74F TTL	*F Logic Data Book*	1987
SCLD001B	74HC, 74HCT CMOS	*High-Speed CMOS Logic Data Book*	1988
SCAD001A	74AC, 74ACT CMOS	*Advanced CMOS Logic Data Book*	1987

principles, but only a few cover practical aspects of design. A classic is *Digital Design with Standard MSI and LSI* by Thomas R. Blakeslee, 2nd ed. (Wiley, 1979), which includes chapters on everything from structured combinational design with MSI and LSI to "the social consequences of engineering." A more recent, excellent short book focusing on digital design practices is *The Well-Tempered Digital Design* by Robert B. Seidensticker (Addison-Wesley, 1986). It contains hundreds of readily accessible digital-design "proverbs" in areas ranging from high-level design philosophy to manufacturability.

DRILL PROBLEMS

5.1   Give three examples of combinational logic circuits that require *billions and billions* of rows to describe in a truth table. For each circuit, describe the circuit's inputs and output(s), and indicate exactly how many rows the truth table contains; you need not write out the truth table. (*Hint:* You can find several such circuits right in this chapter.)

5.2   Draw the DeMorgan equivalent symbol for a 74x30 8-input NAND gate.

5.3   Draw the DeMorgan equivalent symbol for a 74x27 3-input NOR gate.

5.4   What's wrong with the signal name "READY'"?

5.5   You may find it annoying to have to keep track of the active levels of all the signals in a logic circuit. Why not use only noninverting gates, so all signals are active high?

5.6   True or false: In bubble-to-bubble logic design, outputs with a bubble can be connected only to inputs with a bubble.

5.7   A digital communication system is being designed with twelve identical network ports. Which type of schematic structure is probably most appropriate for the design?

5.8   Determine the exact maximum propagation delay from IN to OUT of the circuit in Figure X5.8 for both LOW-to-HIGH and HIGH-to-LOW transitions, using the timing information given in Table 5–2. Repeat, using a single worst-case delay number for each gate and compare and comment on your results.

5.9   Repeat Drill 5.8, substituting 74HCT00s for the 74LS00s.

5.10 Repeat Drill 5.8, substituting 74ACT00s for the 74LS00s.

5.11 Repeat Drill 5.8, substituting 74LS08s for the 74LS00s.

5.12 Repeat Drill 5.8, substituting 74ACT08s for the 74LS00s.

Figure X5.8

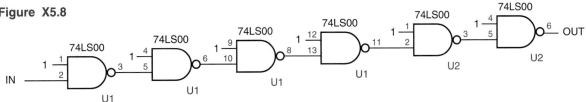

5.13 Estimate the minimum propagation delay from IN to OUT for the circuit shown in Figure X5.13. Justify your answer.

5.14 Determine the exact maximum propagation delay from IN to OUT of the circuit in Figure X5.13 for both LOW-to-HIGH and HIGH-to-LOW transitions, using the timing information given in Table 5–2. Repeat, using a single worst-case delay number for each gate and compare and comment on your results.

5.15 Repeat Drill 5.14, substituting 74ACT86s for the 74LS86s.

5.16 Which would expect to be faster, a decoder with active-high outputs or one with active-low outputs?

5.17 Using the information in Table 5–3, determine the maximum propagation delay from any input to any output in the 5-to-32 decoder circuit of Figure 5–30. You may use the "worst-case" analysis method.

5.18 Repeat Drill 5.17, performing a detailed analysis for each transition direction, and compare your results.

5.19 Show how to build each of the following single- or multiple-output logic functions using one or more 74x138 or 74x139 binary decoders and NAND gates:

(a) $F = \Sigma_{X,Y,Z}(2,4,7)$ (b) $F = \Pi_{A,B,C}(3,4,5,6,7)$

(c) $F = \Sigma_{A,B,C,D}(2,4,6,14)$ (d) $F = \Sigma_{W,X,Y,Z}(0,1,2,3,5,7,11,13)$

(e) $F = \Sigma_{W,X,Y}(1,3,5,6)$ (f) $F = \Sigma_{A,B,C}(0,4,6)$
$\quad\ \ G = \Sigma_{W,X,Y}(2,3,4,7)$ $\quad\ \ G = \Sigma_{C,D,E}(1,2)$

5.20 Draw the digits created by a 74x49 seven-segment decoder for the nondecimal inputs 1010 through 1111.

5.21 Starting with the logic diagram for the 74x148 priority encoder, write logic equations for its /A2, /A1, and /A0 outputs. How do they differ from the "generic" equations given in Section 5.5.1?

Figure X5.13

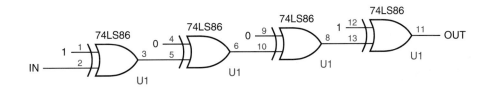

5.22 What's terribly wrong with the circuit in Figure X5.22? Suggest a change that eliminates the terrible problem.

Figure X5.22

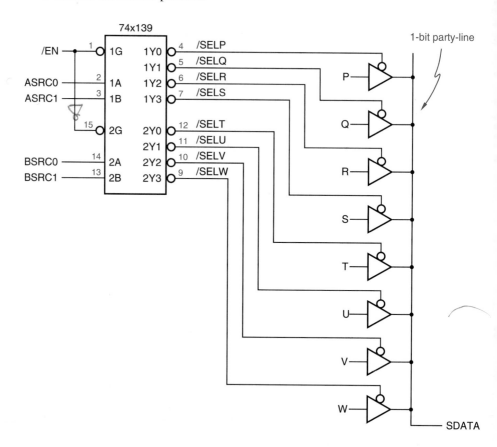

5.23 Using the information in Tables 5–2 and 5–3, determine the maximum propagation delay from any input to any output in the 32-to-1 multiplexer circuit of Figure 5–53. You may use the "worst-case" analysis method.

5.24 Write a truth table for the 74x153 4-input, 2-bit multiplexer.

5.25 An *n*-input parity tree can be built with XOR gates in the style of Figure 5–65(a). Under what circumstances does a similar *n*-input parity tree built using XNOR gates perform exactly the same function?

5.26 Using the information in Tables 5–2 and 5–3, determine the maximum propagation delay from the DU bus to the DC bus in the error-correction circuit of Figure 5–68. You may use the "worst-case" analysis method.

5.27 Starting with the equations given in Section 5.8.4, write a complete logic expression for the ALTBOUT output of the 74x85.

5.28 Starting with the logic diagram for the 74x682, write a logic expression for the /PGTQ output in terms of the inputs.

5.29 Write an algebraic expression for s_2, the third sum bit of a binary adder, as a function of inputs x_0, x_1, x_2, y_0, y_1, and y_2. Assume that $c_0 = 0$, and do not attempt to "multiply out" or minimize the expression.

5.30 Using the information in Table 5–3, determine the maximum propagation delay from any input to any output of the 16-bit group ripple adder of Figure 5–84. You may use the "worst-case" analysis method.

E X E R C I S E S

5.31 A possible definition of a BUT gate (Exercise 4.44) is "Y1 is 1 if A1 and B1 are 1 *but* either A2 or B2 is 0; Y2 is defined symmetrically." Write the truth table and find minimal sum-of-products expressions for the BUT-gate outputs. Draw the logic diagram for a NAND-NAND circuit for the expressions, assuming that only uncomplemented inputs are available. You may use 74x00, '04, '10, '20, and '30 gates.

5.32 Find a minimum-SSI-package design for the BUT gate defined in Exercise 5.31. You are restricted to using 74x00, '02, '04, '10, '20, and '30 gates and no more than two packages total. Write the output expressions (which need not be two-level sums-of-products), and draw the logic diagram.

5.33 For each circuit in the two preceding exercises, compute the worst-case delay from input to output, using the delay numbers for LS-TTL in Table 5–2. Compare the cost (package count), speed, and input loading of the two designs. Which is better?

5.34 Butify the function $F = \Sigma_{W,X,Y,Z}(3,7,11,12,13,14)$. That is, show how to perform   *butification* F with a single BUT gate as defined in Exercise 5.31 and a single 2-input OR gate.

5.35 Design a 1-out-of-4 checker with four inputs, A, B, C, D, and a single output ERR. The output should be 1 if two or more of the inputs are 1, and 0 if no input or one input is 1. Use SSI parts from Figure 5–19, and try to minimize the number of packages required. (*Hint:* It can be done with three packages.)

5.36 Suppose that a 74LS138 decoder is connected so that all enable inputs are asserted and C B A = 101. Using the information in Table 5–3 and the '138 internal logic diagram, determine the propagation delay from input to all relevant outputs for each possible single-input change. (*Hint:* There are a total of nine delay numbers, since a change on A, B, or C affects two outputs, and a change on any of the three enable inputs affects one output.)

5.37 Suppose that you are asked to design a new MSI part, a decimal decoder that is optimized for applications in which only decimal input combinations are expected to occur. How can the cost of such a decoder be minimized compared to one that is simply a 4-to-16 decoder with six outputs removed? Write the logic equations for all ten outputs of the minimized decoder, assuming active-high inputs and outputs and no enable inputs.

5.38 How many Karnaugh maps would be required to work Exercise 5.37 using the formal multiple-output minimization procedure described in Section 4.3.8?

5.39 Suppose that a system requires a 5-to-32 binary decoder with a single active-low enable input, a design similar to Figure 5–30. With the EN1 input pulled HIGH, either the /EN2 or the /EN3 input in the figure could be used as the enable, with the other input grounded. Discuss the pros and cons of using /EN2 versus /EN3.

5.40 A logic designer built a circuit to perform the function shown in Figure 5–31, only to find out during debugging that a different function was required—minterm 6 should have been included instead of minterm 2. Compare the designer's difficulty in changing the circuit depending on whether Figure 5–31(b) or (c) was built.

5.41 Determine whether the circuits driving the a, b, and c outputs of the 74x49 seven-segment decoder correspond to minimal product-of-sums expressions for these segments, assuming that the nondecimal input combinations are "don't cares" and /BI = 1.

5.42 Redesign the 74x49 seven-segment decoder so that the digits 6 and 9 have tails as shown in Figure X5.42. Are any of the digit patterns for nondecimal inputs 1010 through 1111 affected by your redesign?

Figure X5.42

5.43 A famous logic designer decided to quit teaching and make a fortune by fabricating huge quantities of the MSI circuit shown in Figure X5.43.

   (a) Label the inputs and outputs of the circuit with appropriate signal names, including active-level indications.

   (b) What does the circuit do? Be specific and account for all inputs and outputs.

   (c) Draw the MSI logic symbol that would go on the data sheet of this wonderful device.

   (d) With what standard MSI parts does the new part compete? Do you think it would be successful in the marketplace?

Figure X5.43

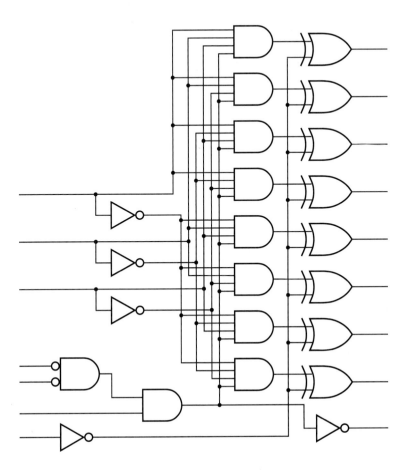

5.44 An FCT three-state buffer drives ten FCT inputs and a 4.7-KΩ pull-up resistor to 5.0 V. When the output changes from LOW to Hi-Z, estimate how long it takes for the FCT inputs to see the output as HIGH. State any assumptions that you make.

5.45 On a three-state bus, ten FCT three-state buffers drive ten FCT inputs and a 4.7-KΩ pull-up resistor to 5.0 V. Assuming that no other devices are driving the bus, estimate how long the bus signal remains at a valid logic level when an active output enters the Hi-Z state. State any assumptions that you make.

5.46 Design a 10-to-4 encoder with inputs in the 1-out-of-10 code and outputs in BCD.

5.47 Draw the logic diagram for a 16-to-4 encoder using just four 8-input NAND gates. What are the active levels of the inputs and outputs in your design?

5.48 Draw the logic diagram for a circuit that uses the 74x148 to resolve priority among eight active-high inputs, I0–I7, where I7 has the highest priority. The circuit should produce active-high address outputs A2–A0 to indicate the number of the highest-priority asserted input. If no input is asserted, then A2–A0 should be 111 and an

IDLE output should be asserted. You may use discrete SSI gates in addition to the '148. Be sure to name all signals with the proper active levels.

5.49 Draw the logic diagram for a circuit that resolves priority among eight active-low inputs, /I0–/I7, where /I0 has the highest priority. The circuit should produce active-high address outputs A2–A0 to indicate the number of the highest-priority asserted input. If at least one input is asserted, then an AVALID output should be asserted. Be sure to name all signals with the proper active levels. This circuit can be built with a single 74x148 and no other gates.

5.50 A purpose of Exercise 5.49 was to demonstrate that it is not always possible to maintain consistency in active-level notation unless you are willing to define alternate logic symbols for MSI parts that can be used in different ways. Define an alternate symbol for the 74x148 that provides this consistency in Exercise 5.49.

5.51 Design a combinational circuit with eight active-low request inputs, /R0–/R7, and eight outputs, A2–A0, AVALID, B2–B0, and BVALID. The /R0–/R7 inputs and A2–A0 and AVALID outputs are defined as in Exercise 5.49. The B2–B0 and BVALID outputs identify the second-highest priority request input that is asserted. You should be able to design this circuit with no more than six SSI and MSI packages, but don't use more than 10 in any case.

5.52 Design a 3-input, 5-bit multiplexer that fits in a 24-pin IC package. Write the truth table and draw a logic diagram and logic symbol for your multiplexer.

5.53 Show how, using external connections, the 74x155 2-bit, 4-output demultiplexer can be made to function as a 1-bit, 8-output demultiplexer.

5.54 Write the truth table and a logic diagram for the logic function performed by the CMOS circuit in Figure X5.54. (The circuit contains transmission gates, which were introduced in Section 3.7.8.)

Figure X5.54

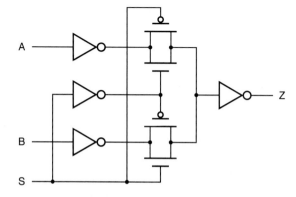

5.55 Design a circuit for the BUT gate defined in Exercise 5.31, using a minimum number of MSI packages. You may use any of the 74x MSI parts described in this chapter, and 74x04 inverters if necessary. Draw the logic diagram and compare the package count of your circuit with the solutions to Exercises 5.31 and 5.32.

5.56 A famous logic designer decided to quit teaching and make a fortune by fabricating huge quantities of the MSI circuit shown in Figure X5.56.

   (a) Label the inputs and outputs of the circuit with appropriate signal names, including active-level indications.

   (b) What does the circuit do? Be specific and account for all inputs and outputs.

   (c) Draw the MSI logic symbol that would go on the data sheet of this wonderful device.

   (d) With what standard MSI parts does the new part compete? Do you think it would be successful in the marketplace?

Figure X5.56

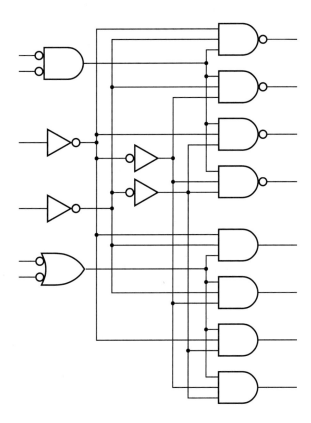

5.57 A 16-bit *barrel shifter* is a combinational logic circuit with 16 data inputs, 16 data outputs, and 4 control inputs. The output word equals the input word, rotated by a number of bit positions specified by the control inputs. For example, if the input word equals ABCDEFGHIJKLMNOP (each letter represents one bit), and the control

barrel shifter

inputs are 0101 (5), then the output word is FGHIJKLMNOPABCDE. Design a 16-bit barrel shifter using combinational MSI parts discussed in this chapter. Your design should contain 20 or fewer ICs. Do not draw a complete schematic, but sketch and describe your design in general terms and indicate the types and total number of ICs required.

5.58 A digital designer who built the circuit in Figure 5–67 accidentally used 74x00s instead of '08s in the circuit, and found that the circuit still worked, except for a change in the active level of the ERROR signal. How was this possible?

5.59 What logic function is performed by the CMOS circuit shown in Figure X5.59?

Figure X5.59

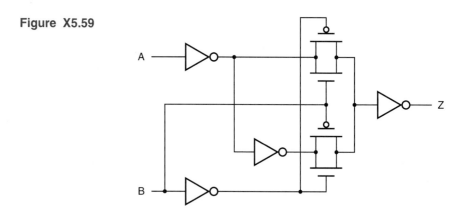

5.60 Write the truth table and a logic diagram for the logic function performed by the CMOS circuit in Figure X5.60.

5.61 An odd-parity circuit with 2^n inputs can be built with $2^n - 1$ XOR gates. Describe two different structures for this circuit, one of which gives a minimum worst-case input to output propagation delay and the other of which gives a maximum. For each structure, state the worst-case number of XOR-gate delays, and describe a situation where that structure might be preferred over the other.

5.62 Using an n-bit comparator structure similar to Figure 5–70(b), what is the maximum value of n that can be obtained, given the electrical limitations of the 74LS266? Assume that the /DIFF output drives a single 74LS04 input and that normal 74LS TTL noise margins are to be maintained.

5.63 Write a 4-step iterative algorithm corresponding to the iterative comparator circuit of Figure 5–72.

5.64 Design a 16-bit comparator using five 74x85s in a tree-like structure, such that the maximum delay for a comparison equals twice the delay of one 74x85.

Figure X5.60

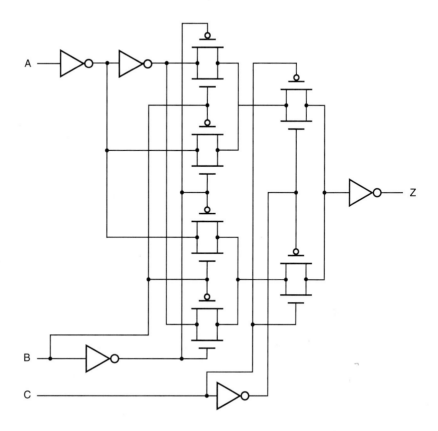

5.65 Starting with a manufacturer's logic diagram for the 74x85, write a logic expression for the ALTBOUT output, and prove that it algebraically equals the expression derived in Exercise 5.64.

5.66 Design a comparator similar to the 74x85 that uses the opposite cascading order. That is, to perform a 12-bit comparison, the cascading outputs of the high-order comparator would drive the cascading inputs of the mid-order comparator, and the mid-order outputs would drive the low-order inputs. You needn't do a complete logic design and schematic; a truth table and an application note showing the interconnection for a 12-bit comparison are sufficient.

5.67 Design a 24-bit comparator using three 74x682s and additional SSI/MSI components as required. Your circuit should compare three 24-bit unsigned numbers P and Q and produce two output bits that indicate whether $P = Q$ or $P > Q$.

5.68 Draw a 6-variable Karnaugh map for the s_2 function of Exercise 5.67, and find all of its prime implicants. Using the 6-variable map format of Exercise 4.64, label the variables in the order $x_0, y_0, x_2, y_2, x_1, y_1$ instead of U, V, W, X, Y, Z. You need not write out the algebraic product corresponding to each prime implicant; simply identify each one with a number (1, 2, 3, ...) on the map. Then make a list that shows for each prime implicant whether or not it is essential and how many inputs are needed on the corresponding AND gate.

5.69   Starting with the logic diagram for the 74x283 in Figure 5–83, write a logic expression for the S2 output in terms of the inputs, and prove that it does indeed equal the third sum bit in a binary addition as advertised. You may assume that $c_0 = 0$ (i.e., ignore c_0).

5.70   Using the information in Table 5–3, determine the maximum propagation delay from any A or B bus input to any F bus output of the 16-bit carry lookahead adder of Figure 5–88. You may use the "worst-case" analysis method.

5.71   Referring to the data sheet of a 74S182 carry lookahead circuit, determine whether or not its outputs match the equations given in Section 5.9.7.

5.72   Estimate the number of product terms in a minimal sum-of-products expression for the c_{32} output of a 32-bit binary adder. Be more specific than "billions and billions," and justify your answer.

5.73   Draw the logic diagram for a 64-bit ALU using sixteen 74x181s and five 74S182s for full carry lookahead (two levels of '182s). For the '181s, you need only show the CIN inputs and /G and /P outputs.

5.74   Sketch the structure of a 16×16 multiplication using addition and 4×4 multiplication in the format of Figure 5–90. How many 4-bit adders and 4×4 multipliers are needed to build a 16×16 combinational multiplier?

5.75   Calculate the true worst-case delay of the 8×8 multiplier in Figure 5–91 using the delay numbers given in the text. Describe the signal path over which the worst-case delay occurs.

5.76   Show how to build all four of the following functions using no more than one SSI package and one of the following MSI parts: 74x138, 74x139, 74x151, 74x153, 74x157.

$$F1 = X' \cdot Y' \cdot Z' + X \cdot Y \cdot Z \qquad F2 = X' \cdot Y' \cdot Z + X \cdot Y \cdot Z'$$
$$F3 = X' \cdot Y \cdot Z' + X \cdot Y' \cdot Z \qquad F4 = X \cdot Y' \cdot Z' + X' \cdot Y \cdot Z$$

5.77   Design an almost-binary decoder with the function table in Table X5.77. Minimize the number of MSI and SSI IC packages in your design.

5.78   Design a 3-input, 2-output combinational circuit that performs the code conversion specified by Table 5–19, using a single MSI IC.

5.79   Design an almost binary-select multiplexer with four 4-bit input buses P, Q, R, and S, selecting one of the buses to drive a 4-bit output bus T according to Table X5.79. Use the basic structure of Figure 5–93(b), but select a new ordering for data buses on the '153 data inputs that minimizes the size and propagation delay of the code converter.

Table X5.77

/CS	A2	A1	A0	Output to assert
1	x	x	x	none
0	0	0	x	/BILL
0	0	x	0	/MARY
0	0	1	x	/JOAN
0	0	x	1	/PAUL
0	1	0	x	/ANNA
0	1	x	0	/FRED
0	1	1	x	/DAVE
0	1	x	1	/KATE

5.80 Design an almost binary-select multiplexer with five 4-bit input buses A, B, C, D, and E, selecting one of the buses to drive a 4-bit output bus T according to Table X5.80. You may use no more than three ICs.

5.81 Redesign the mode-dependent comparator of Figure 5–94 to minimize the size and delay of the SSI mode-control logic. You may use a different 2-bit code for mode control.

Table X5.79

S2	S1	S0	Input to select
0	0	0	P
0	0	1	P
0	1	0	P
0	1	1	Q
1	0	0	P
1	0	1	P
1	1	0	R
1	1	1	S

Table X5.80

S2	S1	S0	Input to select
0	0	0	A
0	0	1	B
0	1	0	A
0	1	1	C
1	0	0	A
1	0	1	D
1	1	0	A
1	1	1	E

5.82 Replace the SSI mode-control logic of Figure 5–94 with two units of an MSI chip from this chapter, and show that the resulting design can be used with *any* mode-control encoding.

5.83 Repeat Exercise 5.82 using a single MSI chip. (*Hint:* Think!)

5.84 Design a 3-bit equality checker with six inputs, SLOT[2–0] and GRANT[2–0], and one active-low output, /MATCH. The SLOT inputs are connected to fixed values when the circuit installed in the system, but the GRANT values are changed on a cycle-by-cycle basis during normal operation of the system. Using only SSI and MSI parts that appear in Tables 5–2 and 5–3, design a comparator with the shortest possible maximum propagation delay from GRANT[2–0] to /MATCH. (*Note:* The author recently had to solve this problem "in real life" to shave 2 ns off the critical-path delay in a 25-MHz system design.)

5.85 Design a combinational circuit whose inputs are two 8-bit unsigned binary integers, X and Y, and a control signal MIN/MAX. The output of the circuit is an 8-bit unsigned binary integer Z such that $Z = 0$ if $X = Y$; otherwise, $Z = \text{min}(X,Y)$ if MIN/MAX $= 1$, and $Z = \text{max}(X,Y)$ if MIN/MAX $= 0$.

5.86 Design a combinational circuit whose inputs are two 8-bit unsigned binary integers, X and Y, and whose output is an 8-bit unsigned binary integer $Z = \text{max}(X,Y)$. Oh, by the way, a U.S. law limiting the import of foreign-made ICs has just raised the price of MSI comparators to over $10, but all of the other SSI and MSI parts in this chapter are still available for about $1. Do a minimum-cost design, of course.

5.87 Repeat Exercise 5.86, assuming that tinkering by Congress has also raised the price of MSI multiplexers to over $10.

5.88 Most MSI parts are designed so that each of their inputs presents only one unit-load to the circuit that is driving it. However, the internal logic diagrams of some MSI parts, including the 74x148, 74x157, and 74x283, show multiple loads on certain inputs. Referring to the electrical specifications of the 74LS versions of these parts in a TTL data book, list the inputs, if any, that present multiple unit loads, and determine whether or not the load specifications match the requirements implied by the internal logic diagrams.

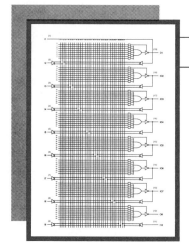

COMBINATIONAL LOGIC DESIGN WITH PLDS

Discrete SSI and MSI chips in board-level designs have been largely supplant-ed by LSI chips that can be programmed to perform the same functions. A *programmable logic device (PLD)* is an LSI chip that contains a "regular" circuit structure, but that allows the designer to customize it for a specific application. When a typical PLD leaves the IC manufacturer's factory, it is not yet customized with a specific function. Instead, it is *programmed* by the purchaser to perform a function required by a particular application.

programmable logic device (PLD)

program a PLD

     PLD-based board-level designs often cost less than SSI/MSI designs for a number of reasons. Since PLDs provide more functionality per chip, the total chip count and printed-circuit-board (PCB) area are less. Manufacturing costs are reduced in other ways, too. A PLD-based board manufacturer needs to keep supplies of only a few, "generic" PLD types, instead of many different MSI part types. This reduces overall inventory requirements and simplifies handling.

     PLD-type structures also appear as logic elements embedded in LSI chips, where chip count and board area are not issues. Despite the fact that a PLD may "waste" a certain number of gates, a PLD structure can actually reduce circuit cost because its "regular" physical structure may use less chip area than a "ran-dom logic" circuit. More importantly, the logic function performed by the PLD

PLD-BASED
DESIGN IS EASY!

> Board-level designs using PLDs are usually smaller and faster than their SSI/MSI
> equivalents, and they often cost less, too. Just as important, PLDs usually simplify
> the logic designer's job. A PLD equivalent to a larger SSI/MSI circuit can often
> be specified with a simple program, and can easily be changed to fix bugs or to
> improve functionality. In fact, every time you see a PLD program in this chapter,
> you should think to yourself, "Would I rather write this program or design the
> corresponding SSI/MSI circuit?"

structure can often be "tweaked" in successive chip revisions by changing just
one or a few metal mask layers that define signal connections in the array, instead of requiring the wholesale addition of gates and gate inputs and subsequent
re-layout of a "random logic" design.

PLDs are "structured" logic devices not only in their internal circuit design
but also in the software tools that are used to specify their functions. Later in
this chapter, we'll describe a programming language for specifying the functionality of a PLD. A software tool, the PLD compiler, can convert this functional
description into a set of interconnections that can be programmed into the PLD
to make it perform the desired function.

Many modern PLDs are "erasable"—they can be reprogrammed to perform
different functions. This capability is especially useful during circuit debugging;
it makes digital design and debugging a lot more like software programming and
debugging, where you can try different ideas without a huge investment.

combinational PLD
logic block

The first PLDs that we'll study are combinational chips. A *combinational
PLD* typically contains an array of identical cells or *logic blocks*, each of which
contains a certain number of gates and programmable interconnections, but no
sequential elements such as flip-flops or latches. Each logic block can be programmed to perform any combinational logic function that "fits" within the logic
block's number of inputs and outputs and its type of gates and interconnections.

Read-only memory (ROM) is sometimes considered to be a programmable
logic device, since a ROM can be programmed to perform any combinational
logic function with a corresponding number of inputs and outputs. We'll explore this possibility later, in Section 11.1. Combinational PLDs are sometimes
programmed to have feedback loops and thereby perform sequential logic functions, but we'll also defer discussion of these applications, until the chapter on
sequential PLDs.

6.1 PROGRAMMABLE LOGIC ARRAYS

programmable logic
array (PLA)

Historically, the first PLDs were *programmable logic arrays (PLAs)*. A PLA is a
combinational, two-level AND-OR device that can be programmed to realize any

sum-of-products logic expression, subject to the size limitations of the device. Limitations are

- the number of inputs (n),   *inputs*

- the number of outputs (m), and   *outputs*

- the number of product terms (p).   *product terms*

We might describe such a device as "an $n \times m$ PLA with p product terms." In general, p is far less than the number of n-variable minterms (2^n). Thus, a PLA cannot perform arbitrary n-input, m-output logic functions; its usefulness is limited to functions that can be expressed in sum-of-products form using p or fewer product terms.

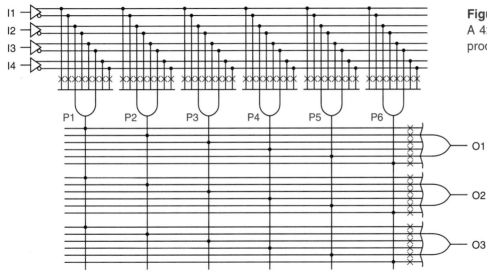

Figure 6–1
A 4×3 PLA with six product terms.

An $n \times m$ PLA with p product terms contains p 2n-input AND gates and m p-input OR gates. Figure 6–1 shows a small PLA with four inputs, six AND gates, and three OR gates and outputs. Each input is connected to a buffer that produces both a true and a complemented version of the signal for use within the array. Potential connections in the array are indicated by X's; the device is programmed by establishing only the connections that are actually needed. Thus, each AND gate can be connected to any subset of the primary input signals and their complements. Similarly, each OR gate can be connected to any subset of the AND-gate outputs.

As shown in Figure 6–2, a more compact diagram can be used to represent a PLA. Moreover, the layout of this diagram more closely resembles the actual   *PLA diagram* internal layout of a PLA chip (e.g., Figure 6–5 on page 396).

Figure 6–2
Compact representation of a 4×3
PLA with six product terms.

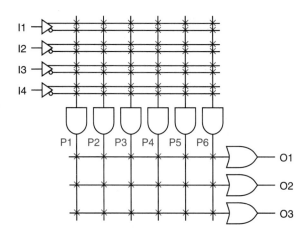

The PLA in Figure 6–2 can perform any three 4-input combinational logic functions that can be written as sums of products using a total of six or fewer distinct product terms, for example:

$$O1 = I1 \cdot I2 + I1' \cdot I2' \cdot I3' \cdot I4'$$
$$O2 = I1 \cdot I3' + I1' \cdot I3 \cdot I4 + I2$$
$$O3 = I1 \cdot I2 + I1 \cdot I3' + I1' \cdot I2' \cdot I4'$$

These equations have a total of eight product terms, but the first two terms in the O3 equation are the same as the first terms in the O1 and O2 equations. The programmed connection pattern in Figure 6–3 matches these logic equations.

PLA constant outputs

Sometimes a PLA output must be programmed to be a constant 1 or a constant 0. That's no problem, as shown in Figure 6–4. Product term P1 is always 1 because its product line is connected to no inputs and is therefore always pulled HIGH; this constant-1 term drives the O1 output. No product term

Figure 6–3
A 4×3 PLA programmed with a
set of three logic equations.

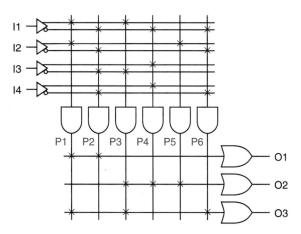

Figure 6–4
A 4×3 PLA programmed to
produce constant 0 and 1
outputs.

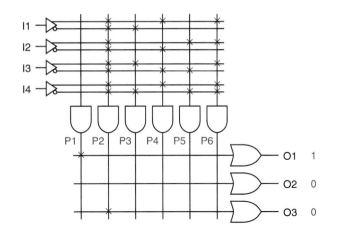

drives the O2 output, which is therefore always 0. Another method of obtaining
a constant-0 output is shown for O3. Product term P2 is connected to each input
variable and its complement; therefore, it's always 0 ($X \cdot X' = 0$).

Our example PLA has too few inputs, outputs, and AND gates (product
terms) to be very useful. An n-input PLA could conceivably use as many as 2^n
product terms, to realize all possible n-variable minterms. The actual number of
product terms in typical commercial PLAs is far fewer, on the order of 4 to 16
per output, regardless of the value of n.

The Signetics 82S100 is a typical example of the PLAs that were introduced
in the mid-1970s. It has 16 inputs, 48 AND gates, and 8 outputs. Thus, it has
$2 \times 16 \times 48 = 1536$ fusible links in the AND array and $8 \times 48 = 384$ in the OR array.

Theoretically, if *all* of the input variables in Figure 6–4 change *simultaneously*, the
output of product term P2 could have a brief 0-1-0 glitch. This is highly unlikely
in typical applications, and is impossible if one input happens to be unused and is
connected to a constant logic signal.

AN UNLIKELY
GLITCH

Although the 82S100 had somewhat fewer product terms per output (6) than most of
today's PLDs (7–8 or more), it was still more effective in typical designs, because
any product term could be applied to any output, and a single product term could
be shared among multiple outputs containing that product term. PAL devices, as
we'll see in Section 6.3, limit the number of product terms that can be applied to
any single output, and do not allow product terms to be shared.

NEWER ISN'T
NECESSARILY
BETTER

6.2 PLD CIRCUIT AND PROGRAMMING TECHNOLOGIES

There are several different circuit technologies for building and physically programming a PLD.

6.2.1 Bipolar PLDs

Early commercial PLAs, as well as some of today's fastest devices, use bipolar circuits. For example, Figure 6–5 shows how the example 4×3 PLA circuit of the preceding section might be built in a bipolar, TTL-like technology. Each potential connection is made by a diode in series with a metal link that may be present or absent. If the link is present, then the diode connects its input into a diode-AND function. If the link is missing, then the corresponding input has no effect on that AND function.

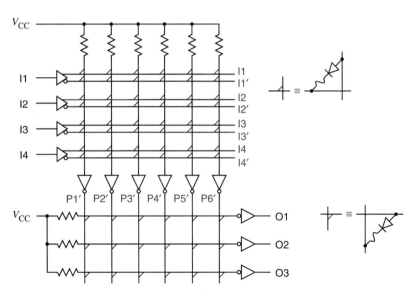

Figure 6–5
A 4×3 PLA built using TTL-like open-collector gates and diode logic.

A diode-AND function is performed because each and every horizontal "input" line that is connected via a diode to a particular vertical "AND" line must be HIGH in order for that AND line to be HIGH. If an input line is LOW, it pulls LOW all of the AND lines to which it is connected. This first matrix of circuit elements that perform the AND function is called the *AND plane*.

AND plane

Each AND line is followed by an inverting buffer, so overall a NAND function is obtained. The outputs of the first-level NAND functions are combined by another set of programmable diode AND functions, once again followed by inverters. The result is a two-level NAND-NAND structure that is functionally equivalent to the AND-OR PLA structure described in the preceding section. The matrix of circuit elements that perform the OR function (or the second NAND function, depending on how you look at it) is called the *OR plane*.

OR plane

A bipolar PLD chip is manufactured with all of its diodes present, but with a tiny *fusible link* in series with each one (the little squiggles in Figure 6–5). It is then possible, by applying special input patterns to the device, to select individual links, apply a high voltage (10–30 V), and thereby vaporize selected links.

fusible link

Early bipolar PLDs had reliability problems. Sometimes the stored patterns changed because of incompletely vaporized links that would "grow back," and sometimes intermittent failures occurred because of floating shrapnel inside the IC package. However, these problems have been largely worked out, and reliable fusible-link technology is used in today's bipolar PLDs.

If a customer requires a large volume of PLDs, all with the same program, a bipolar PLD manufacturer can make a special metal mask that preprograms the device, which would now be called a *mask-programmed PLD*.

MASK-
PROGRAMMED
PLDS

Mask-programmed PLDs potentially save money for customers by eliminating the piece-by-piece programming step, and by improving reliability. However, they are missing a key benefit of today's erasable PLDs—*re*programmability. If a customer orders 10,000 mask-programmed parts, installs the first 1,000, and then finds a bug requiring a program change, that's just too bad; 9,000 mask-programmed parts have to be scrapped!

So mask-programmed PLDs are used only when high volumes reduce costs significantly, and the designers are *very* sure that they finally got the program right.

6.2.2 CMOS PLDs

Despite the improved reliability of bipolar PLDs, in recent years they have been largely supplanted by CMOS PLDs with a number of advantages, including reduced power consumption and reprogrammability. Figure 6–6 shows a CMOS design for the 4×3 PLA circuit of Section 6.1.

Instead of a diode, an *n*-channel transistor with a programmable connection is placed at each intersection between an input line and a word line. If the input is LOW, then the transistor is "off," but if the input is HIGH, then the transistor is "on," which pulls the AND line LOW. Overall, an inverted-input AND (i.e., NOR) function is obtained. This is similar in structure and function to a normal CMOS *k*-input NOR gate, except that the usual series connection of *k* *p*-channel pull-up transistors has been replaced with a passive pull-up resistor (in practice, the pull-up is a single *p*-channel transistor with a constant bias).

As shown in color on Figure 6–6, the effects of using an *inverted*-input AND gate are canceled by using the opposite (complemented) input lines for

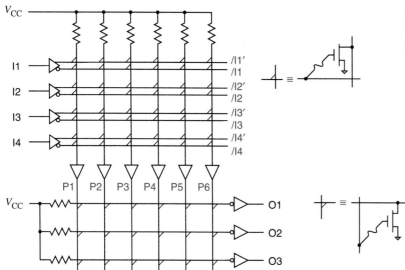

Figure 6–6
A 4×3 PLA built using CMOS logic.

each input, compared with Figure 6–5. Also notice that the connection between the AND plane and the OR plane is noninverting, so the AND plane performs a true AND function.

The outputs of the first-level AND functions are combined in the OR plane by another set of NOR functions with programmable connections. The output of each NOR function is followed by an inverter, so a true OR function is realized, and overall the PLA performs an AND-OR function as desired.

In CMOS PLD technologies, the programmable links shown in Figure 6–6 are not normally fuses. In non-field-programmable devices, such as custom VLSI chips, the presence or absence of each link is simply established as part of the metal mask pattern for the manufacture of the device. This is also the case for CMOS "mask-programmed" parts, similar to the bipolar mask-programmed parts discussed previously. By far the most common programming technology, however, is used in CMOS EPLDs, as discussed next.

*erasable
programmable
logic device
(EPLD)*

An *erasable programmable logic device (EPLD)* can be programmed with any desired link configuration, as well as "erased" to its original state, either electronically or by exposing it to ultraviolet light. No, erasing does not cause links to suddenly appear or disappear! Rather, EPLDs use a different technology, called "floating-gate MOS."

*floating-gate MOS
transistor*

As shown in Figure 6–7, an EPLD uses *floating-gate MOS transistors.* Such a transistor has two gates. The "floating" gate is unconnected and is surrounded by extremely high-impedance insulating material. In the original, manufactured state, the floating gate has no charge on it and has no effect on circuit operation.

Figure 6–7
AND plane of an EPLD using
floating-gate MOS transistors.

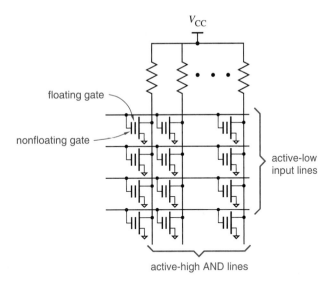

In this state, all transistors are effectively "connected"; that is, there is a logical link present at every crosspoint in the AND and OR planes.

To program an EPLD, the programmer applies a high voltage to the nonfloating gate at each location where a logical link is not wanted. This causes a temporary breakdown in the insulating material and allows a negative charge to accumulate on the floating gate. When the high voltage is removed, the negative charge remains on the floating gate. During subsequent operations, the negative charge prevents the transistor from turning "on" when a HIGH signal is applied to the nonfloating gate; the transistor is effectively disconnected from the circuit.

EPLD manufacturers claim that a properly programmed bit will retain 70% of its charge for at least 10 years, even if the part is stored at 125°C, so for most applications the programming can be considered to be permanent. However, EPLDs can also be erased.

At least one EPLD manufacturer (Intel) uses light for erasing. The insulating material surrounding the floating gate becomes slightly conductive if it is exposed to ultraviolet light with a certain wavelength. Thus, an EPLD can be erased by exposing the chip to ultraviolet light, typically for 5–20 minutes. Such an EPLD chip must be housed in an (expensive) package with a quartz lid through which the chip may be exposed to the erasing light. If the chip is housed in a normal, opaque plastic package, then it becomes a *one-time programmable (OTP)* part.

EPLD erasing

one-time programmable (OTP)

More popular are *electrically erasable PLDs*; these are made by several different manufacturers, including AMD, Cypress Semiconductor, and Lattice Semiconductor. The floating gates in an electrically erasable PLD are surrounded by an extremely thin insulating layer, and can be erased by applying a voltage

electrically erasable PLD

of the opposite polarity as the charging voltage to the nonfloating gate. Thus, the same piece of equipment that is normally used to program a PLD can also be used to erase an EPLD before programming it.

field-programmable
gate array
(FPGA)

Larger-scale PLDs, often called *field-programmable gate arrays (FPGAs)*, can be programmed in a variety of different ways, depending on the manufacturer. Some use the floating-gate MOS transistor technology discussed previously. Others use read/write memory cells to control the state of each connection. The read/write memory cells are volatile—they do not retain their state when power is removed. Therefore, when power is first applied to the FPGA, all of its read/write memory must be initialized to a state specified by a separate, external nonvolatile memory, such as a programmable read-only memory (PROM) chip.

CHANGING
HARDWARE
ON THE FLY

PROMs are normally used to supply the connection pattern for a read/write-memory-based FPGA, but there are also applications where the pattern is actually read from a floppy disk. You just received a floppy with a new software version? Guess what, you just got a new hardware version too!

This leads us to think of the intriguing possibility, already being studied for some applications, of "reconfigurable hardware," where a hardware subsystem is redefined, on the fly, to optimize its performance for the particular task at hand.

6.2.3 Programming and Testing

A special piece of equipment is used to vaporize fuses, charge up floating-gate transistors, or do whatever else is required to program a PLD. This piece of equipment, found nowadays in almost all digital design labs and production facilities, is called a *PLD programmer* or a *PROM programmer*. (It can be used for programmable read-only memories, "PROMs," as well as for PLDs.) A typical PLD programmer is a self-contained unit that includes a socket or sockets that physically accept the devices to be programmed, and a way to "download" desired programming patterns into the programmer, typically by means of an RS-232 serial-communication port or a floppy-disk drive.

PLD programmers typically place a PLD into a special mode of operation in order to program it. For example, a PLD programmer typically programs the PLDs described in this chapter eight fuses at a time as follows:

(1) Raise a certain pin to a predetermined, high voltage (such as 14 V) to put the device into programming mode.

(2) Select a group of eight fuses by applying a binary "address" to certain inputs of the device. (For example, the 82S100 has 1920 fuses, and would therefore require 8 inputs to select one of 240 groups of 8 fuses.)

(3) Apply an 8-bit value to the *outputs* of the device to specify the desired programming for each fuse (the outputs are used as inputs in programming mode).

(4) Raise a second predetermined pin to the high voltage for a predetermined length of time (such as 100 microseconds) to program the eight fuses.

(5) Lower the second predetermined pin to a low voltage (such as 0 V) to read out and verify the programming of the eight fuses.

(6) Repeat steps 1–5 for each group of eight fuses.

As noted in step 5 above, fuse patterns are verified as they are programmed into a device. If a fuse fails to program properly the first time, the operation can be retried; if it fails to program properly after a few tries, the device is discarded (often with great predjudice and malice aforethought).

While verifying the fuse pattern of a programmed device proves that its fuses are programmed properly, it does not prove that the device will perform the logic function specified by those fuses. This is true because the device may have unrelated internal defects such as missing connections between the fuses and elements of the AND-OR array. The only way to test for these defects is to put the device into its normal operational mode, apply a set of normal logic inputs, and observe the outputs. The required input and output patterns, called *test vectors*, can be specified by the designer, or can be generated automatically by a special test-vector-generation program. Regardless of how the test vectors are generated, most PLD programmers have the ability to apply test-vector inputs to a PLD and to check its outputs against the expected results.

Most PLDs have a *security fuse* which, when programmed, disables the device's verify capability. Manufacturers can program this fuse to prevent others from reading out the PLD fuse patterns to copy the product design. Even if the security fuse is programmed, test vectors still work, so the PLD can be checked.

security fuse

6.3 PROGRAMMABLE ARRAY LOGIC DEVICES

A special case of a PLA, and today's most commonly used type of PLD, is the *programmable array logic (PAL) device*. Unlike a PLA, in which both the AND and OR arrays are programmable, a PAL device has a *fixed* OR array.

programmable array logic (PAL) device

PAL devices were introduced by Monolithic Memories, now part of AMD, in the late 1970s. Key innovations in the first PAL devices, besides the introduction of a catchy acronym, were the use of a fixed OR array and bidirectional input/output pins (but see note on page 404).

These ideas are well illustrated by the *PAL16L8*, shown in Figures 6–8 and 6–9 and probably today's most commonly used combinational PLD structure.

PAL16L8

Figure 6–8
Logic diagram of
the PAL16L8.

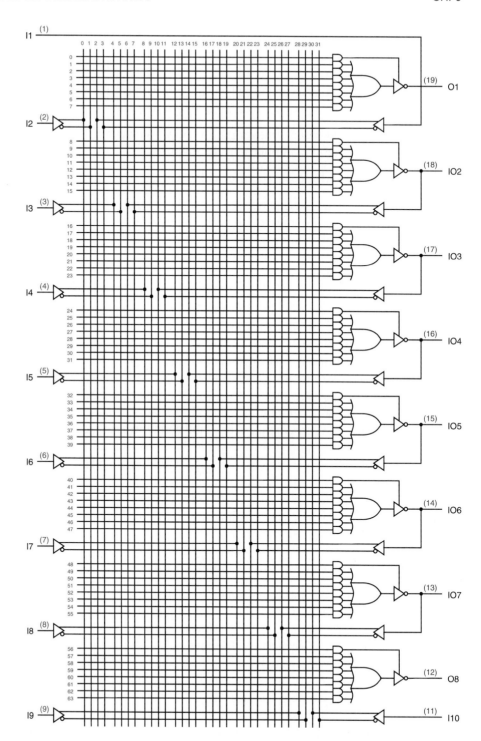

Figure 6–9
Traditional logic symbol for the
PAL16L8.

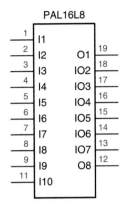

Its programmable AND array has 64 rows and 32 columns, identified for programming purposes by the small numbers in the logic diagram, and 64×32 = 2048 fusible links. Each of the 64 AND gates in the array has 32 inputs, accommodating 16 variables and their complements (hence, the "16" in "PAL16L8").

Eight AND gates are associated with each output pin of the PAL16L8. Seven of them provide inputs to a fixed 7-input OR gate. The eighth, which we call the *output-enable gate*, is connected to the three-state enable input of the output buffer; the buffer is enabled only when the output-enable gate has a 1 output. Thus, an output of the PAL16L8 can perform only logic functions that can be written as sums of seven or fewer product terms. Each product term can be a function of any or all 16 inputs, but only seven such product terms are available.

output-enable gate

The worst-case logic function for two-level AND-OR design is an n-input XOR (parity) function, which requires 2^{n-1} product terms. However, less perverse functions with more than seven product terms of a PAL16L8 can often be built by decomposing them into a 4-level structure (AND-OR-AND-OR) that can be realized with two passes through the AND-OR array. Unfortunately, besides using up PLD outputs for the first-pass terms, this doubles the delay, since a first-pass input must pass through the PLD twice to propagate to the output. Still, we'll discuss two-pass logic further in Section 6.6.

HOW USEFUL
ARE SEVEN
PRODUCT
TERMS?

Although the PAL16L8 has up to 16 inputs and up to 8 outputs, it is housed in a package with only 20 pins, including two for power and ground (the corner pins, 10 and 20). This magic is achieved through the use of six bidirectional pins (13–18) that may be used as inputs or outputs or both. This and other differences between the PAL16L8 and a PLA structure are summarized next:

- The PAL16L8 has a fixed OR array, with seven AND gates permanently

> A step *backwards* in MMI's introduction of PAL devices was their popularization of the word "combinatorial" to describe combinational circuits. *Combinational* circuits have no memory—their output at any time depends on the current input *combination*. For well-rounded computer engineers, the word "combinatorial" should conjure up vivid images of binomial coefficients, problem-solving complexity, and computer-science-great Donald Knuth.

connected to each OR gate. AND-gate outputs cannot be shared; if a product term is needed by two OR gates, it must be generated twice.

- Each output of the PAL16L8 has an individual three-state output enable signal, controlled by a dedicated AND gate (the output-enable gate). Thus, outputs may be programmed as always enabled, always disabled, or enabled by a product term involving the device inputs.

- There is an inverter between the output of each OR gate and the external pin of the device.

I/O pin

- Six of the output pins, called *I/O pins*, may also be used as inputs. This provides many possibilities for using each I/O pin, depending on how the device is programmed:

  (1) If an I/O pin's output-control gate produces a constant 0, then the output is always disabled and the pin is used strictly as an input.

  (2) If the input signal on an I/O pin is not used by any gates in the AND array, then the pin may be used strictly as an output. Depending on the programming of the output-enable gate, the output may always be enabled, or it may be enabled only for certain input conditions.

  (3) If an I/O pin's output-control gate produces a constant 1, then the output is always enabled, but the pin may still be used as an input too. In this way, outputs can be used to generate first-pass "helper terms" for logic functions that cannot be performed in a single pass with the limited number of AND terms available for a single output. We'll discuss this case in Section 6.6.

FRIENDS
AND FOES

> PAL is a registered trademark of Advanced Micro Devices, Inc. Like other trademarks, it should be used only as an adjective. Use it as a noun or without a trademark notice at your own peril (as I learned in a letter from AMD's lawyers in February 1989).
>
> To get around AMD's trademark, I suggest that you use a descriptive name that is more indicative of the device's internal structure: a *fixed-OR element (FOE)*.

The speed of combinational PLDs is usually stated as a single number that gives the propagation delay from any input to any output for either direction of transition. PLDs are available in a variety of speed grades; for example, a PAL16L8-25 would have a delay of 25 ns. In 1993, the fastest available combinational PLDs included the bipolar PAL16L8-5 and the CMOS GAL16V8-5 and PALCE16V8-5. (The GAL and PALCE devices are introduced in Section 6.12.)

COMBINATIONAL PLD SPEED

---

(4) In another case with an I/O pin always output-enabled, the output may be used as an input to AND gates that affect the very same output. That is, we can embed a feedback sequential circuit in a PAL16L8. We'll discuss this case in Section 10.4.4.

6.4 THE ABEL PLD PROGRAMMING LANGUAGE

A PLD is ultimately programmed by specifying a diode or fuse pattern. However, few designers like to hack away at fuse patterns directly, or even indirectly in a hexadecimal text file. Instead, most designers use a *PLD programming language* to specify logic functions symbolically.

PLD programming language

In this section, we'll describe the basic features of an industry-standard PLD programming language called *ABEL*. Other PLD programming languages have similar features, but vary somewhat in their syntax.

ABEL

A PLD programming language is supported by a *PLD language processor*, which we'll simply call a *compiler*. The compiler's job is to translate a text file written in the language into a fuse pattern for a physical PLD. Even though most PLDs can be physically programmed only with sum-of-products expressions, languages like ABEL allow PLD equations to be written in almost any format. The compiler algebraically manipulates and minimizes the equations to fit, if possible, into the available PLD structure. As we'll discuss in Section 10.7, some compilers for sequential PLDs allow state machines to be defined in a high-level language and automatically select an appropriate PLD, perform state assignments, and develop logic equations.

PLD language processor
compiler

6.4.1 ABEL Program Structure

Table 6–1 shows the typical structure of a PLD program in the ABEL language, and Table 6–2 shows an actual program. This example exhibits the following language features:

ABEL (Advanced Boolean Equation Language) is a trademark of the Data I/O Corporation (Redmond, WA 97046).

LEGAL NOTICE

Table 6–1
Typical structure of an ABEL
program.

```
module module name
title string
deviceID device deviceType;
pin declarations
other declarations
equations
equations
end module name
```

identifier
- *Identifiers* must begin with a letter or underscore, may contain up to 31 letters, digits, and underscores, and are case sensitive.

module
- A program file begins with a `module` statement, which associates an identifier (e.g., `Memory_Decoder`) with the program module. Large programs can have multiple modules, each with its own local title, declarations, and equations.

title
- The `title` statement specifies a title string that will be inserted into the documentation files that are created by the compiler.

string
- A *string* is a series of characters enclosed by single quotes.

device
- The `device` declaration includes a device identifier (e.g., `MEMDEC`) and a string that denotes the device type (e.g., `'P16L8'` for a PAL16L8). The compiler uses the device identifier in the names of documentation files that it generates, and it uses the device type to determine whether the device can really perform the logic functions requested in the program.

@ALTERNATE
- The `@ALTERNATE` directive tells the compiler to recognize an alternate set of symbols to denote logic operations:

*
      * AND.

+
      + OR.

/
      / NOT (used as a prefix).

:+:
      :+: XOR.

:*:
      :*: XNOR.

      The default symbols in ABEL for these operations are &, #, !, $, and !$.

comment
- *Comments* begin with a double quote and end with another double quote or the end of the line, whichever comes first.

pin declarations
- *Pin declarations* tell the compiler about symbolic names associated with

Table 6–2   An ABEL program for a memory-decoder PLD.

```
module Memory_Decoder
title 'Memory Decoder PLD
J. Wakerly, Micro Systems Engineering'
MEMDEC device 'P16L8';
@ALTERNATE


" Input pins
LARGE                        pin 1;
A16, A17, A18, A19           pin 2, 3, 4, 5;
A20, A21, A22, A23           pin 6, 7, 8, 9;
/RTEST, IOSEL                pin 11, 18;

" Output pins
/ROMCS                       pin 12;
LOCAL                        pin 13;

equations

ROMCS = LARGE * A23 * A22 * A21 * A20 * A19 * A18 * A17 * A16
      + /LARGE * A19 * A18 * A17 * A16
      + RTEST;

LOCAL = LARGE * A23 * A22 * A21 * A20 * A19 * A18 * A17 * A16
      + /LARGE * A19 * A18 * A17 * A16
      + RTEST
      + IOSEL;

end Memory_Decoder
```

the device's external pins. Signals whose names have the NOT prefix (/) are active low at the external pin; others are active high.

- *Other declarations* allow the designer to define constants and expressions   *other declarations* to simplify program design and improve program readability.

- The equations statement indicates that logic equations defining output   equations signals as functions of input signals will follow.

- *Equations* are written like assignment statements in a programming lan-   *equations* guage. Each equation is terminated by a semicolon.

- The end statement marks the end of the module.   end

Equations for combinational PLD outputs use the *unclocked assignment*   *unclocked* *operator,* =. The left-hand side of an equation normally contains a signal name.   *assignment* The right-hand side is a logic expression, not necessarily in sum-of-products   *operator,* =

form. The signal name on the left-hand side of an equation may be optionally preceded by the NOT operator /; this is equivalent to complementing the right-hand side. The compiler's job is to generate a fuse pattern such that the signal named on the left-hand side realizes the logic expression on the right-hand side.

6.5   ABEL DOCUMENTATION

6.5.1   Program Listings

The example ABEL program in Table 6–2 defines the logic function to be performed by a PAL16L8 device, but the logic diagram for a system using this device does not list the ABEL program. Instead, the logic diagram contains a generic symbol for the PAL16L8 and uses signal names that correspond to the names in the program, as shown in Figure 6–10. Obviously, the assignment of pins to logic signals in the logic diagram and in the program must match, or the circuit will not work as desired. PLD program listings are normally kept as a separate but very important part of the documentation package for a digital system.

6.5.2   Fuse Patterns

The ABEL compiler generates the complete fuse pattern for a PLD from its program. This fuse pattern is stored in a special file (whose name typically has the suffix ".JED") that can be "downloaded" into a PROM programmer to customize an unprogrammed PLD for the desired functions. The compiler also produces a documentation file (suffix ".DOC") that shows the final equations resulting from Boolean manipulations on the original ABEL equations.

Figure 6–10
Logic diagram for memory
decoder using a PAL16L8.

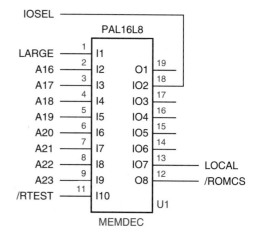

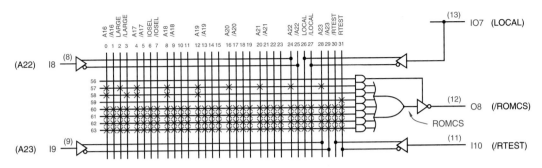

Figure 6–11   Fuse configuration for the /ROMCS output.

It is instructive to study part of the fuse pattern for the ABEL program in Table 6–2. The internal fuse configuration for the /ROMCS output that results from the program is shown in Figure 6–11. Several aspects of this figure must be explained.

Previously in this book, / has been used as an ordinary character in alphanumeric signal names, but in ABEL it is used as a prefix operator indicating logical negation. However, this usage is consistent with our signal-naming convention, in which / is used as a prefix to indicate an active-low signal. Under our naming convention, if *signal* and */signal* are the names of two different signals, then in normal operation

$$signal \;=\; (/signal)'$$

In ABEL, */signal* is an expression, not a signal name, but the equation above simply expresses a double negation, and is still true.

Let us consider input pin 11 in Figure 6–11, which was declared to be active low (/RTEST). This signal is buffered and driven onto column line 30 (/RTEST), and its complement is driven onto column line 31 (RTEST). If pin 11 was declared to be active high (RTEST), the internal signals would be available

---

An ABEL program doesn't necessarily define a PLD fuse pattern unambiguously. When you run the ABEL compiler, you can specify the level of optimization ("reduction") that it should apply when deriving the fuse pattern. Depending on the algorithm, it may find a reduced equation with more or less product terms. Also, with some PLDs, such as the GAL16V8 discussed in Section 6.12, the compiler may be able to choose between sum-of-products and product-of-sums solutions. Thus, only the .JED and .DOC files contain an unambiguous specification of a PLD's fuse pattern.

   In the example in this section, the reduced equations happen to be identical to the ones given in the original ABEL program.

ONE PROGRAM, MANY RESULTS

on the opposite column lines. Both signals are available to the AND-OR array in either case; the compiler just has to keep track of where to find them. Thus, the active level of an input signal does not affect whether or not a given logic equation can be realized, only the details of the fuse pattern.

On the other hand, consider the /ROMCS output signal on pin 12. The output signal is active low, but there is an inverter between the pin and the OR gate that generates the sum-of-products expression. Therefore, the OR gate must produce the signal /ROMCS', or ROMCS. We happened to specify the pin-12 output by writing ROMCS as a sum of three product terms. Therefore, the compiler can use these three terms directly. The effect would be the same if we wrote the equation

```
/ROMCS = /( LARGE * A23 * A22 * A21 * A20
                 * A19 * A18 * A17 * A16
         + /LARGE * A19 * A18 * A17 * A16
         + RTEST );
```

Either way, we get the fuse pattern shown in rows 57–59 of Figure 6–11. (An X represents a connection.)

All of the fuses in row 56 are blown so that the output-enable gate produces a constant-1 output and the three-state buffer is always enabled. All of the fuses in rows 60–63 are left intact, so the corresponding, unused AND gates produce constant-0 outputs.

TILDE (˜)
IN .DOC FILES

> Documentation files generated by ABEL use ˜ as a *prefix character* in the names of signals on external pins that have been declared to be active low, while continuing to use ! or / as a *negation operator*. Thus, when you see something like "!˜ROMCS" in an ABEL DOC file, it means "the complement of the signal on the external pin named ˜ROMCS." To be consistent, ABEL should really use the ˜ prefix instead of ! or / in a program's pin declarations to denote active-low inputs. In any case, "!˜ROMCS" contains a double negation and may be read simply as "ROMCS."

6.5.3 Signal Polarity

Consistent use of active-level notation requires the choice of "true" signal names (without a /) that describe conditions detected and actions performed in a positive sense. In full signal names, a consistent suffix or prefix (/ in this book) is used to denote the active level.

When active levels are used consistently in the external signal names in a circuit, PLD equations involving those signals can be written in a quite natural and consistent manner. Equations are written in terms of the "true" signal names, and a complement operator (/, !, or whatever) is used to perform logical negation as required to express the desired logic function. The correspondence between each internal "true" signal and an external signal that has a particular active level is established by the pin declarations or equivalent section in the PLD program. The active levels are easily changed without rewriting any of the logic equations.

As mentioned earlier, the choice of active level for input signals has no effect on whether a particular logic equation can be realized in a PLD. Since both true and complemented versions of each input signal are available at all the AND gates, the compiler can always select the appropriate one.

On the other hand, things aren't quite so simple for output signals. In the example in the preceding subsection, we happened to write the ROMCS in just the right form for a sum-of-products realization. Let us now consider the LOCAL equation:

```
LOCAL = LARGE * A23 * A22 * A21 * A20 * A19 * A18 * A17 * A16
      + /LARGE * A19 * A18 * A17 * A16
      + RTEST
      + IOSEL;
```

At first glance, this looks like no problem, since it is a sum of four product terms, and we have seven available. However, since the output pin is declared to be active high (LOCAL), the AND-OR logic must produce an internal signal that is active low (/LOCAL). To obtain an equation for the internal signal, we (or the compiler) can use DeMorgan's theorem to complement both sides of the preceding equation:

```
/LOCAL = (/LARGE + /A23 + /A22 + /A21 + /A20
                 + /A19 + /A18 + /A17 + /A16)
       * (LARGE + /A19 + /A18 + /A17 + /A16)
       * /RTEST
       * /IOSEL;
```

We (or the compiler) can multiply out and simplify this equation to obtain the required form for the PLD structure, a sum of products:

```
/LOCAL = /RTEST * /IOSEL * LARGE * /A23
       + /RTEST * /IOSEL * LARGE * /A22
       + /RTEST * /IOSEL * LARGE * /A21
       + /RTEST * /IOSEL * LARGE * /A20
       + /RTEST * /IOSEL * /A19
       + /RTEST * /IOSEL * /A18
       + /RTEST * /IOSEL * /A17
       + /RTEST * /IOSEL * /A16;
```

This equation illustrates a common problem that occurs when you try to fit expressions into a PLD structure. When a product of n variables is complemented, it may give rise to a sum of n product terms, exceeding the number of product terms available in the PLD. In particular, the equation above is a sum of *eight* product terms, and the PAL16L8 has only seven per output. A way around this problem is described in the next section.

6.6 TWO-PASS LOGIC

A combinational PLD can handle decoder-like functions very well, since it has many inputs and can easily generate "wide" products of many variables. However, if the output polarity is such that a wide product term must be complemented, we may run out of AND terms in the array. The simplest solution to this problem is to redefine the output to have the opposite polarity, usually requiring an external inverter to generate the polarity that we wanted in the first place. Sometimes this approach is undesirable, or we may be stuck with an expression that has too many product terms in either polarity. Another solution is discussed in this section.

The number of product terms in the /LOCAL equation in the previous section can't be reduced, so we can't generate /LOCAL in one pass through a PAL16L8. However, we could split it into two parts by generating a "helper" term, using one of the six output pins that are also available as inputs:

```
/LCLHELP                          pin 14;
...
LCLHELP = /RTEST * /IOSEL * LARGE * /A23
        + /RTEST * /IOSEL * LARGE * /A22
        + /RTEST * /IOSEL * LARGE * /A21
        + /RTEST * /IOSEL * LARGE * /A20;

/LOCAL = LCLHELP
        + /RTEST * /IOSEL * /A19
        + /RTEST * /IOSEL * /A18
        + /RTEST * /IOSEL * /A17
        + /RTEST * /IOSEL * /A16;
```

TWO-PASS
TIMING

> A good first estimate of the propagation delay for two passes through a combinational PLD is simply to double the delay for one pass. For example, two passes through a 15-ns PAL16L8 take 30 ns.
>
> Some larger-scale PLDs have special internal bypass paths that allow a signal to be fed back for the second pass without going off the chip and back on again. These devices avoid the delay of the off-chip driver and input receiver between the first and second pass, thereby providing a two-pass delay as fast as 150% of the one-pass delay.

Better yet, by observing the close relationship between the ROMCS and LOCAL signals, we could take advantage of the existing logic and simply rewrite the LOCAL equation in terms of ROMCS and IOSEL:

```
LOCAL = ROMCS + IOSEL;
```

Applying DeMorgan's theorem to this equation, we find that /LOCAL can be expressed as a single product term:

```
/LOCAL = /ROMCS * /IOSEL;
```

However, both of the solutions presented above require signals to take two passes through the PLD. That is, a transition on a signal such as A23 must propagate through the PLD twice before affecting the LOCAL output.

In two-pass logic, helper terms (LCLHELP in our first example) must be assigned to I/O pins (IO2–IO7 in a 16L8), not to output-only pins (O1 and O8 in a 16L8). The I/O pin always output-enabled, and is simultaneously used as an input and an output.

In our second two-pass example (/LOCAL = /ROMCS * /IOSEL), we are trying to use an output signal (ROMCS) as an input signal as well. Therefore, ROMCS must be assigned to an I/O pin; it would have to be moved from pin 12 in the original design to, say, pin 15.

HELPER PIN
ASSIGNMENTS

6.7 COMBINATIONAL PLD APPLICATIONS

Combinational PLDs can easily duplicate the functions of most standard combinational MSI devices, such as decoders, encoders, multiplexers, and the like. On one hand, if you need precisely the function that is performed by a standard MSI device, the standard device will probably be the smallest, fastest, and cheapest solution. On the other hand, if you need a special-purpose version of one of these functions, or if you feel you may need to "tweak" the design at a later time, a PLD is usually the best choice. A few examples of PLD implementations of such functions are given in this section.

6.7.1 Decoders

Nothing in logic design is much easier than writing the PLD equations for a decoder. For example, Table 6–3 is an ABEL program that implements the same functionality as a 74x138 3-to-8 binary decoder in a PAL16L8. The corresponding logic symbol is shown in Figure 6–12.

Note that this example defines a *constant expression* for ENB in the "other   *constant expression*

Table 6–3   An ABEL program for 74x138-like 3-to-8 binary decoder.

```
module Binary_Decoder
title '74x138 Decoder PLD
J. Wakerly, Stanford University'
Z74X138 device 'P16L8';
@ALTERNATE

" Input pins
A, B, C, /G2A, /G2B, G1        pin 1, 2, 3, 4, 5, 6;

" Output pins
/Y0, /Y1, /Y2, /Y3             pin 19, 18, 17, 16;
/Y4, /Y5, /Y6, /Y7             pin 15, 14, 13, 12;

" Constant expression

ENB = G1 * G2A * G2B;

equations

Y0 = ENB * /C * /B * /A;
Y1 = ENB * /C * /B *  A;
Y2 = ENB * /C *  B * /A;
Y3 = ENB * /C *  B *  A;
Y4 = ENB *  C * /B * /A;
Y5 = ENB *  C * /B *  A;
Y6 = ENB *  C *  B * /A;
Y7 = ENB *  C *  B *  A;

end Binary_Decoder
```

Figure 6–12
Logic diagram for the PAL16L8
used as a 74x138 decoder.

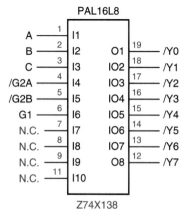

declarations" section of the ABEL program. Here, ENB is not an input or output signal, but merely a user-defined name. In the `equations` section, the compiler substitutes the expression (G1 * G2A * G2B) everywhere that "ENB" appears. Assigning the constant expression to a user-defined name improves the readability and maintainability of the ABEL program.

As we said in the introduction to this section, if all you needed was a '138, you'd be better off using a real '138 than a PLD. However, if you need nonstandard functionality, then the PLD can usually achieve it much more cheaply and easily than an MSI/SSI-based solution. For example, if you need the functionality of a '138 but with active-high outputs, you need only to change two lines in the pin declarations of Table 6–3:

```
Y0, Y1, Y2, Y3              pin 19, 18, 17, 16;
Y4, Y5, Y6, Y7             pin 15, 14, 13, 12;
```

Since each of the equations required a single product of six variables (including the three in the ENB expression), each complemented equation requires a *sum* of six variables, less than the seven available in a PAL16L8.

Another easy change is to provide alternate enable inputs that are OR'ed with the main enable inputs. To do this, you need only define additional pins and modify the definition of ENB:

```
EN1, /EN2                 pin 7, 8;
...
ENB = G1 * G2A * G2B + EN1 + EN2;
```

If you make this change in conjunction with the original version of the program with active-low outputs, after reduction you get three product terms for each output, each having a form similar to

```
Y0 = G1 * G2A * G2B * /C * /B */A
   + EN1 * /C * /B */A
   + EN2 * /C * /B */A;
```

(Remember that while the AND-OR array generates an expression for an active-high Y0 signal as shown above, there is an inversion between the AND-OR array and the output of the PLD, so the actual output is active low as desired.)

If you add the extra enables to the version of the program with active-high outputs, then the PLD must realize the complement of the sum-of-products expression above. It's not immediately obvious how many product terms this expression will have, and whether it will fit in a PAL16L8, but we can use the ABEL compiler to get the answer for us:

```
/Y0 = C + B + A
    + /G2B * /EN1 * /EN2
    + /G2A * /EN1 * /EN2
    + /G1  * /EN1 * /EN2;
```

The expression has a total of six product terms, so it fits in a PAL16L8.

As a final tweak on our 74x138-like decoder design, we can add an input to dynamically control whether the output is active high or active low, and modify all of the equations as follows:

```
POL                     pin 9;
...
Y0 = POL :+: ENB * /C * /B * /A;
Y1 = POL :+: ENB * /C * /B *  A;
...
Y7 = POL :+: ENB *  C *  B *  A;
```

Because of the XOR operation, it is hardly apparent whether the reduced equations will fit within the seven product terms per output available in a PAL16L8; you're asked to use a PLD compiler to find out the answer in Exercise 6.25.

To begin another example, consider the almost-binary-select decoder of Figure 5–92 on page 369. It is easily built using a single PAL16L8; Table 6–4 is an ABEL program for the device, and Figure 6–13 is a corresponding logic diagram. Each of the last six equations uses a single AND gate in the PLD. The ABEL compiler can also minimize the BILL and MARY equations to use just one AND gate each. Once again, active-high output signals could be obtained just by changing two lines in the declaration section:

```
BILL, MARY, JOAN, PAUL      pin 19, 18, 17, 16;
ANNA, FRED, DAVE, KATE      pin 15, 14, 13, 12;
```

Since each of the (minimized) equations required a single product of four or five variables, each complemented equation requires a *sum* of four or five variables, which fits in a PAL16L8.

Figure 6–13
Logic diagram for the PAL16L8
used as an almost-binary-select
decoder.

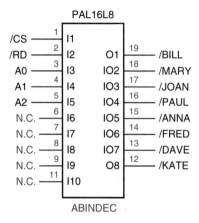

Table 6–4 An ABEL program for an almost-binary-select decoder.

```
module Almost_Binary_Decoder
title 'Almost-Binary-Select Decoder PLD
J. Wakerly, Stanford University'
ABINDEC device 'P16L8';
@ALTERNATE

" Input pins
/CS, /RD, A0, A1, A2            pin 1, 2, 3, 4, 5;

" Output pins
/BILL, /MARY, /JOAN, /PAUL    pin 19, 18, 17, 16;
/ANNA, /FRED, /DAVE, /KATE    pin 15, 14, 13, 12;

equations

BILL = CS * RD * (/A2 * /A1 * /A0 + /A2 * /A1 *  A0);
MARY = CS * RD * (/A2 * /A1 * /A0 + /A2 *  A1 * /A0);
JOAN = CS * RD * (/A2 *  A1 * /A0);
PAUL = CS * RD * (/A2 *  A1 *  A0);
ANNA = CS * RD * ( A2 * /A1 * /A0);
FRED = CS * RD * ( A2 * /A1 *  A0);
DAVE = CS * RD * ( A2 *  A1 * /A0);
KATE = CS * RD * ( A2 *  A1 *  A0);

end Almost_Binary_Decoder
```

6.7.2 Multiplexers

Multiplexers are also easy to implement in combinational PLDs. For example, the function of a 74x153 4-input, 2-bit multiplexer can be duplicated easily in a PAL16L8, as shown in Figure 6–14 and Table 6–5. Several characteristics of the PLD-based design and program are worth noting:

- Signal names in the ABEL program are changed slightly from the signal names shown for a 74x153 on page 327, since ABEL does not allow a number to be used as the first character of a signal name.

> As you see in these examples, it's very easy to program a PLD to performer decoder and multiplexer functions. Still, if you need the logic function of a standard decoder or multiplexer, it's usually less costly to use a standard MSI chip than it is to use a PLD. The PLD-based approach is best if the multiplexer has some nonstandard functional requirements, or if you think you may have to change its function as a result of debugging.

EASIEST, BUT
NOT CHEAPEST

Figure 6–14
Logic diagram for the PAL16L8
used as a 74x153 multiplexer.

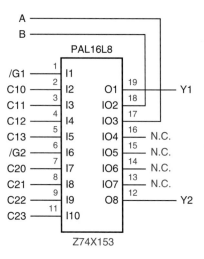

Table 6–5
ABEL program for a 74x153-like
4-input, 2-bit multiplexer.

```
module Four_Input_Multiplexer
title '74x153 Multiplexer PLD
J. Wakerly, Stanford University'
Z74X153 device 'P16L8';
@ALTERNATE

" Input pins
A, B, /G1, /G2                  pin 17, 18, 1, 6;
C10, C11, C12, C13              pin 2, 3, 4, 5;
C20, C21, C22, C23              pin 7, 8, 9, 11;

" Output pins
Y1, Y2                          pin 19, 12;

equations

Y1 = G1 * ( /B * /A * C10
          + /B *  A * C11
          +  B * /A * C12
          +  B *  A * C13);

Y2 = G2 * ( /B * /A * C20
          + /B *  A * C21
          +  B * /A * C22
          +  B *  A * C23);

end Four_Input_Multiplexer
```

Table 6–6
Inverted, reduced equations
for 74x153-like 4-input, 2-bit
multiplexer.

```
/Y1 = /A * /B * /C10
    +  A * /B * /C11
    + /A *  B * /C12
    +  A *  B * /C13
    + /G1;

/Y2 = /A * /B * /C20
    +  A * /B * /C21
    + /A *  B * /C22
    +  A *  B * /C23
    + /G2;
```

- A 74x153 has twelve inputs, while a PAL16L8 has only ten inputs. There-fore, two of the '153 inputs are assigned to 16L8 I/O pins, which are no longer usable as outputs.

- The '153 outputs (1Y and 2Y) are assigned to pins 19 and 12 on the 16L8, which are usable *only* as outputs. This is preferable to assigning them to I/O pins; given a choice, it's better to leave I/O pins than output-only pins as spares.

- Although the multiplexer equations in the table are written quite naturally in sum-of-products form, they don't map directly onto the 16L8's structure because of the inverter between the AND-OR array and the actual output pins. Therefore, the ABEL compiler must complement the equations in the table and then reduce the result to sum-of-products form. On first thought, it's not at all obvious that the final equations will fit into the seven product terms per output available in a 16L8. However, if you stare for a while at the resulting equations in Table 6–6, you should see that not only do they fit, but they make good sense in terms of the '153 multiplexer's function.

An example of a nonstandard multiplexer function is the almost-binary-select multiplexer of Table 5–19 on page 370. A 4-input, 3-bit multiplexer with the required behavior can be built in a single PAL16L8 with the logic diagram shown in Figure 6–15. An ABEL program that creates the required behavior is shown in Table 6–7. Since this function uses all of the available pins on the PAL16L8, we had to make the pin assignment carefully. In particular, we were forced to assign two output signals to the two output-only pins (O1 and O8), to maximize the number of input pins available.

An 18-bit almost-binary-select multiplexer can be built with six PAL16L8 chips programmed and wired according to Table 6–7 and Figure 6–15. Fur-thermore, its select coding can be easily changed with a few keystrokes. This compares favorably with the MSI approach of Figure 5–93(b) on page 370, which required nine chips for a 16-bit mux.

Figure 6–15
Logic diagram for the PAL16L8
used as an almost-binary-select
multiplexer.

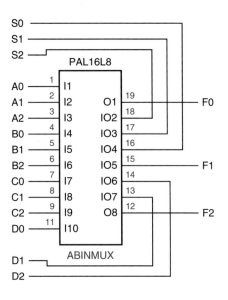

Table 6–7
An ABEL program for
an almost-binary-select
multiplexer.

```
module Almost_Binary_Multiplexer
title 'Almost-Binary-Select Multiplexer PLD'
ABINMUX device 'P16L8';
@ALTERNATE

" Input pins
S0, S1, S2                          pin 16, 17, 18;
A0, A1, A2, B0, B1, B2              pin 1, 2, 3, 4, 5, 6;
C0, C1, C2, D0, D1, D2              pin 7, 8, 9, 11, 13, 14;

" Output pins
F0, F1, F2                          pin 19, 15, 12;

equations

F0 = /S0 * A0
   + (/S2 * /S1 * S0 + S2 * S1 * S0) * B0
   + (/S2 *  S1 * S0) * C0
   + ( S2 * /S1 * S0) * D0;
F1 = /S0 * A1
   + (/S2 * /S1 * S0 + S2 * S1 * S0) * B1
   + (/S2 *  S1 * S0) * C1
   + ( S2 * /S1 * S0) * D1;
F2 = /S0 * A2
   + (/S2 * /S1 * S0 + S2 * S1 * S0) * B2
   + (/S2 *  S1 * S0) * C2
   + ( S2 * /S1 * S0) * D2;

end Almost_Binary_Multiplexer
```

6.8 ABEL SETS AND RELATIONS

We can take this opportunity to introduce another useful capability of PLD programming languages. Most digital systems include buses, registers, and other circuits that handle a group of two or more signals in an identical fashion. In the multiplexer example of Table 6–7, the three output equations were identical except for the bit numbers 0, 1, 2. The ABEL language includes the concept of "sets" for writing such equations more compactly (and therefore with less chance of errors or inconsistencies).

A *set* is simply a defined collection of signals that is handled as a unit.   *set*
When an operation is applied to a set, it is applied to each element of the set. Table 6–8 shows how sets can be used to write the ABEL equations for the almost-binary-select multiplexer. Each set is defined at the beginning of the program by associating a set name (e.g., ABUS) with a bracketed list of the set

Table 6–8
Almost-binary-select-
multiplexer program using
set notation.

```
module Almost_Binary_Multiplexer
title 'Almost-Binary-Select Multiplexer PLD
J. Wakerly, Kalvij Telecom, Inc.'
ABINMXS device 'P16L8';
@ALTERNATE

" Input pins
S0, S1, S2                    pin 16, 17, 18;
A0, A1, A2, B0, B1, B2        pin 1, 2, 3, 4, 5, 6;
C0, C1, C2, D0, D1, D2        pin 7, 8, 9, 11, 13, 14;

" Output pins
F0, F1, F2                    pin 19, 15, 12;

" Set definitions
ABUS = [A2,A1,A0];
BBUS = [B2,B1,B0];
CBUS = [C2,C1,C0];
DBUS = [D2,D1,D0];
FBUS = [F2,F1,F0];

equations

FBUS = /S0 * ABUS
    + (/S2 * /S1 * S0 + S2 * S1 * S0) * BBUS
    + (/S2 *  S1 * S0) * CBUS
    + ( S2 * /S1 * S0) * DBUS;

end Almost_Binary_Multiplexer
```

elements (e.g., [A2,A1,A0]). The number and order of elements in a set are significant, as we'll see.

Most of the operators in ABEL, including all of the ones in this book, can be used with sets. When an operation is applied to two or more sets, all of the sets must have the same number of elements, and the operation is applied individually to set elements in corresponding positions. For example, the equation

```
FBUS = ABUS + BBUS + (CBUS * DBUS);
```

is equivalent to three equations,

```
F2 = A2 + B2 + (C2 * D2);
F1 = A1 + B1 + (C1 * D1);
F0 = A0 + B0 + (C0 * D0);
```

When an operation includes both set and nonset variables, the nonset variables are combined individually with set elements in each position. Thus, the single FBUS equation in Table 6–8 is equivalent to the F0, F1, and F2 equations in Table 6–7.

relational
expression

relational operator

Another interesting feature of ABEL is its ability to convert "relational expressions" into logic expressions. A *relational expression* is a pair of operands combined with one of the *relational operators* listed in Table 6–9. The compiler converts a relational expression into a logic expression that is 1 if and only if the relation is true. Either operand in a relation may be an integer or a set. If the operand is a set, it is treated as an unsigned binary integer with the leftmost variable representing the most significant bit. Thus, we could write the output equations for a 74x682 8-bit comparator in two lines in ABEL:

```
PEQQ = [P7,P6,P5,P4,P3,P2,P1,P0] == [Q7,Q6,Q5,Q4,Q3,Q2,Q1,Q0];
PGTQ = [P7,P6,P5,P4,P3,P2,P1,P0] > [Q7,Q6,Q5,Q4,Q3,Q2,Q1,Q0];
```

Of course, the fact we can write the equations easily doesn't mean that they'll fit in a PLD. The PEQQ expression above expands into 16 product terms, and PGTQ expands into 373 product terms (no, I didn't figure that one out by hand!).

Table 6–9
Relational operators in ABEL.

Symbol	Relation
==	equal
!=	not equal
<	less than
<=	less than or equal
>	greater than
>=	greater than or equal

6.9 ADDITIONAL COMBINATIONAL PLD APPLICATIONS

We introduced the 74x49 seven-segment decoder in Section 5.3.7 on page 308. A PLD-based version of this function appears in the ABEL program in Table 6–10. This version differs slightly from a standard 74x49; the displays for digits 6 and 9 have "tails," as in Exercise 5.42, and the decoded outputs for hexadecimal input combinations A–F yield reasonable visual approximations of the letters A, b, C, d, E, and F.

The 74x49 program uses both sets and relations to create a straightforward description of the decoder's function. Four input bits representing a BCD digit are declared to be a set, DIGITIN = [D,C,B,A], with the bits arranged from most significant to least significant. In the equations section of the program, a relational expression such as DIGITIN == 5 returns "1" if and only if the four bits of DIGITIN are the binary representation of 5.

Similarly, the seven decoded segment-output bits are defined to be a set, SEGOUT = [SEGA,SEGB,SEGC,SEGD,SEGE,SEGF,SEGG]. The segment pattern for each displayed digit is then established with a single constant expression, for example, DIG2 = [0,1,1,0,0,0,0]. A single output equation is then used to specify the values on all segment outputs for all input combinations. This equation is easy to understand but would be tedious to evaluate or simplify by hand. Nevertheless, the ABEL compiler is able to obtain the reduced equations shown in Table 6–11, which easily fit in a PAL16L8. Note that ABEL generates a sum-of-products expression for the *complement* of each desired output signal, taking into account the inverter between the AND-OR array and the output pin of a PAL16L8.

Table 6–10
An ABEL program for a 74x49-like seven-segment decoder.

```
module Seven_Segment_Decoder
title 'Seven-Segment_Decoder
J. Wakerly, Micro Design Resources, Inc.'
Z74X49H device 'P16L8';
@ALTERNATE

" Input pins
A, B, C, D                          pin 1, 2, 3, 4;
/BI                                 pin 5;

" Output pins
SEGA, SEGB, SEGC, SEGD              pin 19, 18, 17, 16;
SEGE, SEGF, SEGG                    pin 15, 14, 13;

" (continued on next page!)
```

Table 6–10
(continued)

```
" Set definitions
DIGITIN = [D,C,B,A];
SEGOUT = [SEGA,SEGB,SEGC,SEGD,SEGE,SEGF,SEGG];

" Segment encodings for digits
DIG0 = [1,1,1,1,1,1,0];   " 0
DIG1 = [0,1,1,0,0,0,0];   " 1
DIG2 = [1,1,0,1,1,0,1];   " 2
DIG3 = [1,1,1,1,0,0,1];   " 3
DIG4 = [0,1,1,0,0,1,1];   " 4
DIG5 = [1,0,1,1,0,1,1];   " 5
DIG6 = [1,0,1,1,1,1,1];   " 6   "tail" included
DIG7 = [1,1,1,0,0,0,0];   " 7
DIG8 = [1,1,1,1,1,1,1];   " 8
DIG9 = [1,1,1,1,0,1,1];   " 9   "tail" included
DIGA = [1,1,1,0,1,1,1];   " A
DIGB = [0,0,1,1,1,1,1];   " b
DIGC = [1,0,0,1,1,1,0];   " C
DIGD = [0,1,1,1,1,0,1];   " d
DIGE = [1,0,0,1,1,1,1];   " E
DIGF = [1,0,0,0,1,1,1];   " F

equations

SEGOUT = /BI * (
            (DIGITIN == 0) * DIG0
          + (DIGITIN == 1) * DIG1
          + (DIGITIN == 2) * DIG2
          + (DIGITIN == 3) * DIG3
          + (DIGITIN == 4) * DIG4
          + (DIGITIN == 5) * DIG5
          + (DIGITIN == 6) * DIG6
          + (DIGITIN == 7) * DIG7
          + (DIGITIN == 8) * DIG8
          + (DIGITIN == 9) * DIG9
          + (DIGITIN == 10) * DIGA
          + (DIGITIN == 11) * DIGB
          + (DIGITIN == 12) * DIGC
          + (DIGITIN == 13) * DIGD
          + (DIGITIN == 14) * DIGE
          + (DIGITIN == 15) * DIGF);

end Seven_Segment_Decoder
```

Table 6–11   Reduced equations for 74x49-like seven-segment decoder.

```
/SEGA = (  /A * /B *  C * /D        /SEGE = (        /B *  C * /D
         +  A *  B * /C *  D               + A * /B * /C
         +  A * /B * /C * /D               + A            * /D
         +  A * /B *  C *  D               + BI );
         +  BI );

/SEGB = (   A * /B *  C * /D        /SEGF = (   A * /B *  C *  D
         +  A *  B      *  D               +       B * /C * /D
         + /A      *  C *  D               + A         * /C * /D
         + /A *  B *  C                    + A *  B *         /D
         +  BI );                          + BI);

/SEGC = (  /A *  B * /C * /D        /SEGG = (   A *  B *  C * /D
         +       B *  C *  D               + /A * /B *  C *  D
         + /A      *  C *  D               +       /B * /C * /D
         +  BI);                           + BI);

/SEGD = (  /A *  B * /C *  D
         +  A * /B * /C * /D
         + /A * /B *  C * /D
         +  A *  B *  C
         +  BI);
```

The next application we consider is based on the mode-dependent comparator design of Figure 5–94 on page 371. Comparing, adding, and other "iterative" operations are usually poor candidates for PLD-based design, because an equivalent two-level sum-of-products expression has far too many product terms. Based on the results in the preceding subsection, we certainly wouldn't be able to replace the 74x682 in Figure 5–94 with a PLD; the 74x682 8-bit comparator is just about the most efficient possible single chip we can use to perform an 8-bit comparison. However, a PLD-based design is quite reasonable for the "random" mode-control logic.

Taking this idea one step further, suppose we needed to design a 32-bit mode-dependent comparator. Once again, the 74x682 is the most efficient building block. However, the '682 does not have cascading inputs, so the outputs of four '682s must somehow be combined to perform a 32-bit comparison. The combining logic and the mode-control logic can be put together in a single PAL16L8. Figure 6–16 shows the complete circuit design, and Table 6–12 is the ABEL program for the MODECOMP PLD. The strategy for this design is to use the '682s to compare most of the bits, and to use a PAL16L8 to combine the '682 outputs and to handle the two low-order bits as a function of the mode.

Notice that since the outputs of the MODECOMP PLD are active low, and

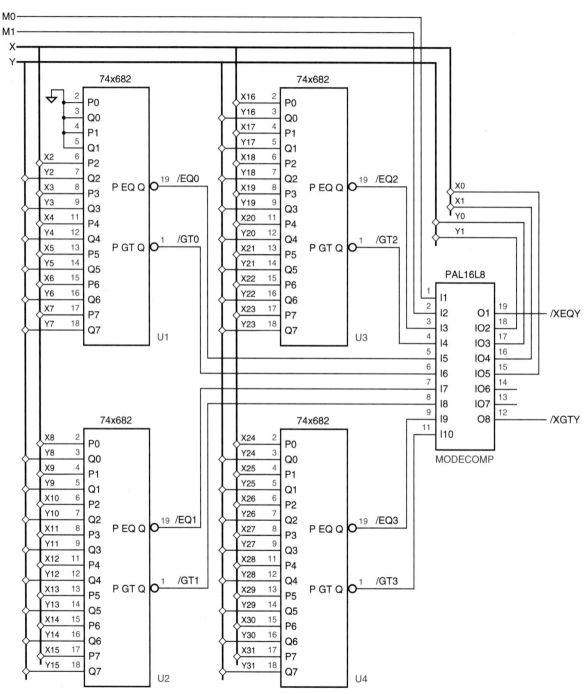

Figure 6–16   A 32-bit mode-dependent comparator.

Table 6-12   An ABEL program for the MODECOMP PLD.

```
module Comparator_Control
title 'Control PLD for Mode-Dependent Comparator
J. Wakerly, Micro Design Resources, Inc.'
MODECOMP device 'P16L8';
@ALTERNATE


" Input pins
M0, M1                      pin 1, 2;
/EQ0, /GT0, /EQ1, /GT1      pin 5, 6, 7, 8;
/EQ2, /GT2, /EQ3, /GT3      pin 3, 4, 9, 11;
X0, X1, Y0, Y1              pin 15, 16, 17, 18;

" Output pins
/XEQY, /XGTY                pin 19, 12;

equations

XEQY = EQ3 * EQ2 * EQ1 * EQ0 * /M1 * /M0 * (          " Mode 00 (32-bit)
          /X1 * /Y1 * /X0 * /Y0 +
          /X1 * /Y1 *  X0 *  Y0 +
           X1 *  Y1 * /X0 * /Y0 +
           X1 *  Y1 *  X0 *  Y0 )
     + EQ3 * EQ2 * EQ1 * EQ0 * /M1 *  M0 * (          " Mode 01 (31-bit)
          /X1 * /Y1 + X1 * Y1 )
     + EQ3 * EQ2 * EQ1 * EQ0 *  M1 * /M0;             " Mode 10 (30-bit)

XGTY = GT3 + (EQ3 * GT2) + (EQ3 * EQ2 * GT1)          " Easy cases
     + (EQ3 * EQ2 * EQ1 * GT0)                        " Covers all modes
     + EQ3 * EQ2 * EQ1 * EQ0 * /M1 *  M0 * X1 * /Y1   " Mode 01 (31-bit)
     + EQ3 * EQ2 * EQ1 * EQ0 * /M1 * /M0 * (          " Mode 00 (32-bit)
          X1 * /Y1 +
         /X1 * /Y1 *  X0 * /Y0 +                      " No special case
          X1 *  Y1 *  X0 * /Y0 );                     "  needed for 30-bit mode

end Comparator_Control
```

the equations are written in terms of active-high output signals, the right-hand side of each equation can be multiplied out to obtain directly an expression suitable for programming into the PLD's AND-OR structure. Both outputs use seven product terms, the maximum available in the PAL16L8. (When expanded into a sum of products, the XGTY equation yields eight product terms. However, two terms can be combined; can you see which ones?) If the output pins were declared to be active high, the resulting equations would "blow up," requiring more product terms than are available in the PAL16L8.

* 6.10 OTHER COMBINATIONAL PLDS AND APPLICATIONS

PAL20L8

The *PAL20L8* is another combinational PLD similar to the PAL16L8, except that it comes in a 24-pin package and has four more input-only pins. An application that pushes the PAL20L8 just about to its limit is a 15-input priority encoder, similar in structure to the MSI 74x148.

A logic diagram for the priority encoder is given in Figure 6–17. Inputs P0–P14 are asserted to indicate requests, with P14 having the highest priority. If EN (Enable) is asserted, then the /Y3–/Y0 outputs give the number (active low) of the highest-priority request, and GS (Got Something) is asserted if any request is present. If EN is negated, then the /Y3–/Y0 outputs are negated and GS is negated. /ENOUT is asserted if EN is asserted and no request is present.

Figure 6–17

Logic diagram for a PLD-based 15-input priority encoder.

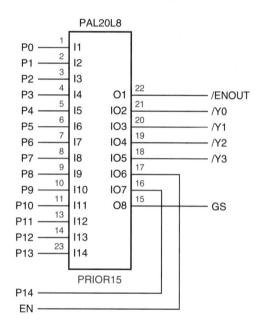

An ABEL program for the priority encoder is given in Table 6–13. Recall that ABEL allows constant expressions to be defined; such expressions are useful when the same product terms are used in many equations. For the priority encoder, we wrote a series of expressions for constants H0–H14, where Hi is 1 if input Pi is the highest-priority asserted request input. These expressions allow us to write the equations for Y3–Y0 as simple sums, each with seven product terms.

The active levels for the /Y3–/Y0, GS, and /ENOUT outputs were not chosen at random. If the opposite polarity were specified, more product terms would be needed, and the function wouldn't fit in a PAL20L8.

---

*Throughout this book, optional sections are marked with an asterisk.

Table 6–13   An ABEL program for a 15-input priority encoder.

```
module Priority_Encoder_15
title '15-Input Priority Encoder
J. Wakerly, DAVID Systems, Inc.'
PRIOR15 device 'P20L8';
@ALTERNATE


" Input pins

P0, P1, P2, P3, P4, P5, P6, P7       pin 1, 2, 3, 4, 5, 6, 7, 8;
P8, P9, P10, P11, P12, P13, P14      pin 9, 10, 11, 13, 14, 23, 16;
EN                                   pin 17;

" Output pins

/Y3, /Y2, /Y1, /Y0                   pin 18, 19, 20, 21;
GS, /ENOUT                           pin 15, 22;

" Constant expressions

H14 = EN*P14;
H13 = EN*/P14*P13;
H12 = EN*/P14*/P13*P12;
H11 = EN*/P14*/P13*/P12*P11;
H10 = EN*/P14*/P13*/P12*/P11*P10;
H9  = EN*/P14*/P13*/P12*/P11*/P10*P9;
H8  = EN*/P14*/P13*/P12*/P11*/P10*/P9*P8;
H7  = EN*/P14*/P13*/P12*/P11*/P10*/P9*/P8*P7;
H6  = EN*/P14*/P13*/P12*/P11*/P10*/P9*/P8*/P7*P6;
H5  = EN*/P14*/P13*/P12*/P11*/P10*/P9*/P8*/P7*/P6*P5;
H4  = EN*/P14*/P13*/P12*/P11*/P10*/P9*/P8*/P7*/P6*/P5*P4;
H3  = EN*/P14*/P13*/P12*/P11*/P10*/P9*/P8*/P7*/P6*/P5*/P4*P3;
H2  = EN*/P14*/P13*/P12*/P11*/P10*/P9*/P8*/P7*/P6*/P5*/P4*/P3*P2;
H1  = EN*/P14*/P13*/P12*/P11*/P10*/P9*/P8*/P7*/P6*/P5*/P4*/P3*/P2*P1;
H0  = EN*/P14*/P13*/P12*/P11*/P10*/P9*/P8*/P7*/P6*/P5*/P4*/P3*/P2*/P1*P0;

equations

Y3 = H8 + H9 + H10 + H11 + H12 + H13 + H14;
Y2 = H4 + H5 + H6 + H7 + H12 + H13 + H14;
Y1 = H2 + H3 + H6 + H7 + H10 + H11 + H14;
Y0 = H1 + H3 + H5 + H7 + H9 + H11 + H13;

GS    = EN*(P14+P13+P12+P11+P10+P9+P8+P7+P6+P5+P4+P3+P2+P1+P0);
ENOUT = EN*/P14*/P13*/P12*/P11*/P10*/P9*/P8*/P7*/P6*/P5*/P4*/P3*/P2*/P1*/P0;

end Priority_Encoder_15
```

THANKS FOR THE
COMPLEMENT

> The equation for GS in Table 6–13 could also be written in terms of the comple-
> mented output variable:
>
> ```
> /GS   = /EN + (/P14*/P13*/P12*/P11*/P10*/P9*/P8*/P7
>                 */P6*/P5*/P4*/P3*/P2*/P1*/P0);
> ```
>
> To some people, the equations may be easier to understand and explain in that form:
> "We haven't got something if we're not enabled or all request inputs are negated."
> This form also makes it evident that the /GS equation can be implemented as a
> sum of two product terms, one of which has 15 inputs.

Another priority example uses a single PLD to replace the four MSI parts
in the fixed-point to floating-point encoder of Figure 5–96 on page 375. A logic
diagram is shown in Figure 6–18 and an ABEL program is given in Table 6–14.
The constant terms S0–S7 select exponent values of 0–7 depending on the location of the most significant 1 bit in the fixed-point input number. The exponent
E2–E0 is simply a binary encoding of the select terms, and the mantissa bits
M3–M0 are generated by four equations that serve as a 4-bit, 8-input multiplexer.
Unfortunately, the PAL20L8 has available only seven product terms per output,
and these equations require eight product terms each. There's only one spare
output, not enough to generate all four required outputs using two-pass logic.

We're stuck—the fixed-point to floating-point encoder cannot be built with
a single PAL20L8. However, the equations in Table 6–14 could be used in
another PLD that has at least eight product terms available on four of its outputs,
such as the PAL22V10 discussed in Section 10.2.

Figure 6–18
Fixed-point to floating-point
encoder using a single PLD.

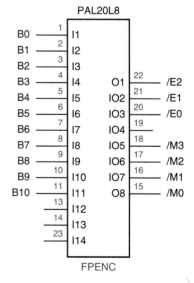

Table 6–14 An ABEL program for the fixed-point to floating-point PLD.

```
module Fixed_to_Floating_Encoder
title 'Fixed-point to Floating-point Encoder
J. Wakerly, Xanthrax Systems'
FPENC device 'P20L8';
@ALTERNATE


" Input pins
B0, B1, B2, B3, B4, B5        pin 1, 2, 3, 4, 5, 6;
B6, B7, B8, B9, B10           pin 7, 8, 9, 10, 11;

" Output pins
/E0, /E1, /E2                 pin 20, 21, 22;
/M0, /M1, /M2, /M3            pin 15, 16, 17, 18;

" Constant expressions
S7 = B10;
S6 = /B10 * B9;
S5 = /B10 * /B9 * B8;
S4 = /B10 * /B9 * /B8 * B7;
S3 = /B10 * /B9 * /B8 * /B7 * B6;
S2 = /B10 * /B9 * /B8 * /B7 * /B6 * B5;
S1 = /B10 * /B9 * /B8 * /B7 * /B6 * /B5 * B4;
S0 = /B10 * /B9 * /B8 * /B7 * /B6 * /B5 * /B4;

equations

E2 = S7 + S6 + S5 + S4;
E1 = S7 + S6 + S3 + S2;
E0 = S7 + S5 + S3 + S1;


M3 = S7*B10 + S6*B9 + S5*B8 + S4*B7 + S3*B6 + S2*B5 + S1*B4 + S0*B3;
M2 = S7*B9  + S6*B8 + S5*B7 + S4*B6 + S3*B5 + S2*B4 + S1*B3 + S0*B2;
M1 = S7*B8  + S6*B7 + S5*B6 + S4*B5 + S3*B4 + S2*B3 + S1*B2 + S0*B1;
M0 = S7*B7  + S6*B6 + S5*B5 + S4*B4 + S3*B3 + S2*B2 + S1*B1 + S0*B0;

end Fixed_to_Floating_Encoder
```

* 6.11 THREE-STATE OUTPUT CONTROL

The combinational-PLD applications in previous sections have used the bidirectional I/O pins (IO2–IO7 on a PAL16L8) statically, that is, always output-enabled or always output-disabled. In such applications, the compiler can take care of programming the output-enable gates appropriately—all fuses blown, or all fuses intact. By default in ABEL, a three-state output pin is programmed to be always enabled if its signal name appears on the left-hand side of an equation, and always disabled otherwise.

ENABLE

Three-state output pins can also be controlled dynamically, by a single input, by a product term, or, using two-pass logic, by a more complex logic expression. In ABEL, the keyword ENABLE indicates that the following equation applies to the output-enable for the indicated pin. In a PAL16L8, the output-enable is controlled by a single AND gate, so the right-hand side of the ENABLE equation must reduce to a single product term.

* 6.11.1 Three-State Output Example

Table 6–15 shows a simple PLD program example with three-state control. Derived from the program for a 74x153-like multiplexer on page 418, this program provides the function of a 74x253-like multiplexer by using the /EN input as a three-state control for the outputs.

Table 6–15
ABEL program for a 74x253-like 4-input, 2-bit multiplexer with three-state output control.

```
module Four_Input_Mux_With_OE
title '74x253 Multiplexer PLD
J. Wakerly, Stanford University'
Z74X253 device 'P16L8';
@ALTERNATE

" Input pins
A, B, /G1, /G2             pin 17, 18, 1, 6;
C10, C11, C12, C13         pin 2, 3, 4, 5;
C20, C21, C22, C23         pin 7, 8, 9, 11;

" Output pins
Y1, Y2                     pin 19, 12;

equations

ENABLE Y1 = G1;
Y1 = /B * /A * C10
   + /B *  A * C11
   +  B * /A * C12
   +  B *  A * C13);

ENABLE Y2 = G2;
Y2 = /B * /A * C20
   + /B *  A * C21
   +  B * /A * C22
   +  B *  A * C23);

end Four_Input_Mux_With_OE
```

* 6.11.2 Bidirectional I/O Pins

In the preceding example, the output pins Y1 and Y2 are always either enabled or floating, and are used strictly as "output pins." I/O pins (IO2–IO7 in a 16L8) can be used as "bidirectional pins"; that is, they can be used dynamically as inputs or outputs depending on whether the output-enable gate is producing a 0 or a 1.

An example application of I/O pins is a four-way, 2-bit bus transceiver with the following specifications:

- The transceiver handles four 2-bit bidirectional buses, A[1-2], B[1-2], C[1-2], and D[1-2].

- Each bus has its own output-enable signal, /AOE, /BOE, /COE, or /DOE. There is also a "master" output-enable signal, /MOE. The transceiver drives a particular bus if and only if /MOE and the output-enable signal for that bus are both asserted.

- The source of data to drive the buses is selected by three select inputs, S[2–0], according to Table 6–16. If S2 is 0, the buses are driven with a constant value, otherwise they are driven with one of the other buses. However, when the selected source is a bus, that bus is driven with 00.

Figure 6–19 is a logic diagram for a PAL16L8 with the required inputs and outputs. Since the device has only six bidirectional pins and the specification requires eight, the A bus uses one pair of pins for input and another for output. Table 6–17 is an ABEL program that performs the transceiver function. According to the ENABLE equations, each bus is output-enabled if MOE and its own OE are asserted. Each bus is driven with S1 and S0 if S2 is 0, and with the selected bus if a different bus is selected. If the bus itself is selected, the output equation evaluates to 0, and the bus is driven with 00 as required.

In the examples in this section, an I/O pin is output-disabled when it is used as input. However, this is not always the case. For example, in two-pass

Table 6–16
Bus selection codes for a four-way bus transceiver.

S2	S1	S0	Source selected
0	0	0	00
0	0	1	01
0	1	0	10
0	1	1	11
1	0	0	A bus
1	0	1	B bus
1	1	0	C bus
1	1	1	D bus

Table 6–17   An ABEL program for four-way, 2-bit bus transceiver.

```
module Four_Way_Transceiver
title 'Four-way 2-bit Bus Transceiver
J. Wakerly, BNR, Inc.'
XCVR4X2 device 'P16L8';
@ALTERNATE

" Input pins
A1I, A2I                        pin 1, 11;
/AOE, /BOE, /COE, /DOE, /MOE    pin 2, 3, 4, 5, 6;
S0, S1, S2                      pin 7, 8, 9;

" Output and bidirectional pins
A1O, A2O                        pin 19, 12;
B1, B2, C1, C2, D1, D2          pin 18, 17, 16, 15, 14, 13;

" Set definitions
ABUSO = [A1O,A2O];
ABUSI = [A1I,A2I];
BBUS  = [B1,B2];
CBUS  = [C1,C2];
DBUS  = [D1,D2];
SEL   = [S2,S1,S0];
CONST = [S1,S0];

" Constants
SELA = [1,0,0];
SELB = [1,0,1];
SELC = [1,1,0];
SELD = [1,1,1];

equations

ENABLE ABUSO = AOE * MOE;
ABUSO = /S2*CONST + (SEL==SELB)*BBUS + (SEL==SELC)*CBUS + (SEL==SELD)*DBUS;

ENABLE BBUS = BOE * MOE;
BBUS = /S2*CONST + (SEL==SELA)*ABUSI + (SEL==SELC)*CBUS + (SEL==SELD)*DBUS;

ENABLE CBUS = COE * MOE;
CBUS = /S2*CONST + (SEL==SELA)*ABUSI + (SEL==SELB)*BBUS + (SEL==SELD)*DBUS;

ENABLE DBUS = DOE * MOE;
DBUS = /S2*CONST + (SEL==SELA)*ABUSI + (SEL==SELB)*BBUS + (SEL==SELC)*CBUS;

end Four_Way_Transceiver
```

Figure 6–19
PLD inputs and outputs for a
four-way, 2-bit bus transceiver.

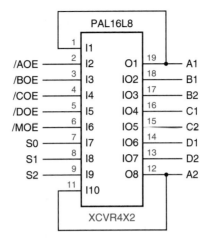

logic (Section 6.6), the I/O pin that provides the output of the first pass and the input to the second must necessarily be output-enabled. And in Section 10.4.4, we'll show examples of how the signal on an I/O pin can be fed back onto itself to create a sequential circuit inside a combinational PLD.

6.12 GENERIC ARRAY LOGIC DEVICES

In Chapter 10 we'll introduce sequential PLDs, programmable logic devices that provide flip-flops at some or all OR-gate outputs. These devices can be programmed to perform a variety of useful sequential-circuit functions.

One type of sequential PLD, originally introduced by Lattice Semiconductor, is called *generic array logic* or a *GAL device*, is particularly popular. A single GAL device type, the *GAL16V8*, can be configured (via programming and a resulting fuse pattern) to emulate the AND-OR, flip-flip, and output structure of any of a variety of combinational and sequential PAL devices, including the PAL16L8 introduced in this chapter. What's more, the GAL device can be erased electrically and reprogrammed and, as we'll see in Section 10.2, can be configured and programmed for functions that are a superset of what can be achieved with standard PAL devices.

Figure 6–20 shows the logic diagram for a GAL16V8 when it has been configured as a strictly combinational device similar to the PAL16L8. This configuration is achieved by programming two "architecture-control" fuses, not shown. In this configuration, the device is called a *GAL16V8C*.

generic array logic
GAL device
GAL16V8

GAL16V8C

Figure 6–20
Logic diagram of the
GAL16V8C.

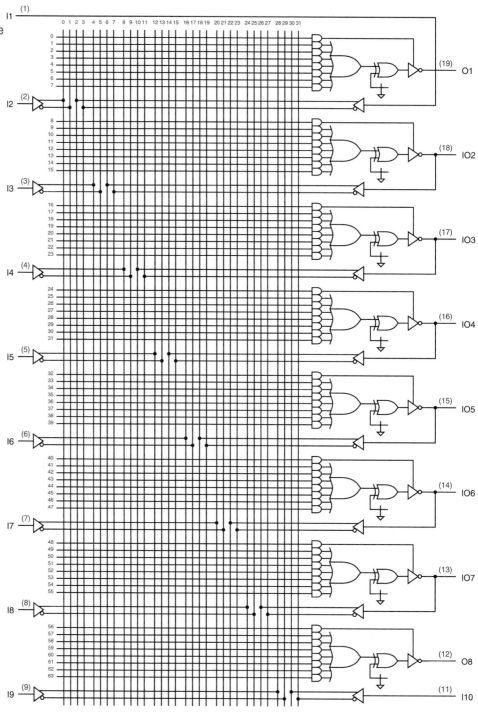

The most important thing to note about the GAL16V8C logic diagram, compared to that of a PAL16L8 on page 402, is that an XOR gate has been inserted between each OR output and the three-state output driver. One input of the XOR gate is "pulled up" to a logic 1 value but connected to ground (0) via a fuse. If this fuse is intact, the XOR gate simply passes the OR-gate's output unchanged, but if the fuse is blown the XOR gate inverts the OR-gate's output. Thus, this fuse is said to control the *output polarity* of the corresponding output pin.   *output polarity*

Output-polarity control is a very important feature of recent generations of PLDs, including the GAL16V8. By giving the logic designer the opportunity to invert the output of the AND-OR array, it allows functions to be realized that could not otherwise be obtained. As a simple example, suppose we needed to obtain the following output as a function of 10 inputs:

$$O1 = I1 \cdot I2 \cdot I3 \cdot I4 \cdot I5 \cdot I6 \cdot I7 \cdot I8 \cdot I9 \cdot I10$$

This is a trivial sum-of-products expression (it has only one product term) and should map directly onto the output structure of a PAL16L8, except for that nasty inversion between AND-OR array and the output pin. Because of this, the AND-OR array must realize a sum-of-products expression for the *complement* of the desired output function, which can be written using DeMorgan's theorem:

$$O1' = I1' + I2' + I3' + I4' + I5' + I6' + I7' + I8' + I9' + I10'$$

The complements of the 10 inputs are readily available but, unfortunately, the new expression is a sum of 10 product terms, which won't fit in a PAL16L8. Well, 10 product terms won't fit in a GAL16V8 either, but with the GAL16V8 we have the option of programming the output-polarity fuse to cancel inversion in the output driver, and use our original sum-of-products expression directly.

In general, you have two ways to realize a function $O_i = F(I_1, I_2, \ldots, I_n)$:

(1) You can find a minimal sum-of-products expression for the function. If it fits in the available product terms, you can program the output-polarity fuse to cancel the inversion in the output driver and use the expression directly.

(2) You can find a minimal sum-of-products expression for the *complement* of the given function:

$$G(I_1, I_2, \ldots, I_n) = [F(I_1, I_2, \ldots, I_n)]'$$

If it fits, you can program the output-polarity fuse to preserve the inversion in the output driver and realize the desired function as follows:

$$
\begin{aligned}
O_i &= [G(I_1, I_2, \ldots, I_n)]' \\
&= [[F(I_1, I_2, \ldots, I_n)]']' \\
&= F(I_1, I_2, \ldots, I_n)
\end{aligned}
$$

To illustrate the usefulness of the 16V8's output-polarity control, let's look at an enhanced version of the seven-segment decoder program that we described in Table 6–10 on page 423. The modified program is shown in Table 6–18 and adds the following features:

- The outputs are all active low.

- Two new inputs, ENHEX and ERRDET, control the decoding of the segment outputs.

- If ENHEX = 0, the outputs match the behavior of a 74x49.

- If ENHEX = 1, then the outputs for digits 6 and 9 have tails, and the outputs for digits A–F are controlled by ERRDET.

- If ENHEX = 1 and ERRDET = 0, then the outputs for digits A–F look like letters A–F, as in the program of Table 6–10.

- If ENHEX = 1 and ERRDET = 1, then the output for digits A–F looks like the letter S.

Table 6–18
ABEL program for an enhanced seven-segment decoder.

```
module Seven_Segment_Decoder
title 'Seven-Segment_Decoder
J. Wakerly, Micro Design Resources, Inc.'
Z74X49HV device 'P16V8C';
@ALTERNATE

" Input pins
A, B, C, D                pin 1, 2, 3, 4;
/BI, ENHEX, ERRDET        pin 5, 6, 7;

" Output pins
SEGA, SEGB, SEGC, SEGD    pin 19, 18, 17, 16;
SEGE, SEGF, SEGG          pin 15, 14, 13;

" Set definitions
DIGITIN = [D,C,B,A];
SEGOUT = [SEGA,SEGB,SEGC,SEGD,SEGE,SEGF,SEGG];

" Segment encodings for digits
DIG0  = [1,1,1,1,1,1,0];   " 0
DIG1  = [0,1,1,0,0,0,0];   " 1
DIG2  = [1,1,0,1,1,0,1];   " 2
DIG3  = [1,1,1,1,0,0,1];   " 3
DIG4  = [0,1,1,0,0,1,1];   " 4

" Continued on next page
```

Table 6–18
(continued)

```
DIG5   = [1,0,1,1,0,1,1];    " 5
DIG6   = [0,0,1,1,1,1,1];    " 6  no tail
DIG6T  = [1,0,1,1,1,1,1];    " 6  tail
DIG7   = [1,1,1,0,0,0,0];    " 7
DIG8   = [1,1,1,1,1,1,1];    " 8
DIG9   = [1,1,1,0,0,1,1];    " 9  no tail
DIG9T  = [1,1,1,1,0,1,1];    " 9  tail
DIGA   = [0,0,0,1,1,0,1];    " 74x49-style
DIGAH  = [1,1,1,0,1,1,1];    " A
DIGB   = [0,0,1,1,0,0,1];    " 74x49-style
DIGBH  = [0,0,1,1,1,1,1];    " b
DIGC   = [0,1,0,0,0,1,1];    " 74x49-style
DIGCH  = [1,0,0,1,1,1,0];    " C
DIGD   = [1,0,0,1,0,1,1];    " 74x49-style
DIGDH  = [0,1,1,1,1,0,1];    " d
DIGE   = [0,0,0,1,1,1,1];    " 74x49-style
DIGEH  = [1,0,0,1,1,1,1];    " E
DIGF   = [0,0,0,0,0,0,0];    " 74x49-style
DIGFH  = [1,0,0,0,1,1,1];    " F
DIGS   = [1,0,1,1,0,1,1];    " S

" Constant expressions
STD = /ENHEX;
TAILS = ENHEX;
HEX = ENHEX * /ERRDET;
ERR = ENHEX * ERRDET * DIGS;

equations

SEGOUT = /BI * (
              (DIGITIN == 0) * DIG0
            + (DIGITIN == 1) * DIG1
            + (DIGITIN == 2) * DIG2
            + (DIGITIN == 3) * DIG3
            + (DIGITIN == 4) * DIG4
            + (DIGITIN == 5) * DIG5
            + (DIGITIN == 6) * (STD * DIG6 + TAILS * DIG6T)
            + (DIGITIN == 7) * DIG7
            + (DIGITIN == 8) * DIG8
            + (DIGITIN == 9) * (STD * DIG9 + TAILS * DIG9T)
            + (DIGITIN == 10) * (STD * DIGA + HEX * DIGAH + ERR)
            + (DIGITIN == 11) * (STD * DIGB + HEX * DIGBH + ERR)
            + (DIGITIN == 12) * (STD * DIGC + HEX * DIGCH + ERR)
            + (DIGITIN == 13) * (STD * DIGD + HEX * DIGDH + ERR)
            + (DIGITIN == 14) * (STD * DIGE + HEX * DIGEH + ERR)
            + (DIGITIN == 15) * (STD * DIGF + HEX * DIGFH + ERR) );

end Seven_Segment_Decoder
```

Table 6–19
Product-term requirements for
the enhanced seven-segment
decoder.

Output	True	Complement
SEGA	8	7^*
SEGB	7^*	9
SEGC	6	6^*
SEGD	7^*	9
SEGE	5^*	8
SEGF	7	7^*
SEGG	7^*	9

The ABEL compiler provides an option to reduce both the true and the complemented form of the equation for each output, and to select the one with the smallest number of product terms. For the equations in Table 6–18, the compiler obtains the results shown in Table 6–19, which are explained below:

- The "True" column shows the number of product terms in the minimal sum-of-products expression for each signal specified by the "Equations" section of the PLD program. If the output-polarity fuse is programmed not to invert, then these expressions yield active-high signals at the PLD output pins.

- The "Complement" column shows the number of product terms in the minimal sum-of-products expression for the complement of each signal. If the output-polarity fuse is programmed not to invert, then these expressions yield active-low signals at the PLD output pins.

The results show that the required equations don't fit in a PAL16L8 if the outputs are specified to be either all active low or all active high, since at least one equation requires 8 or more product terms in each case. However, with a 16V8, we can realize any output signal with either polarity, because each signal has at least one expression with 7 or fewer product terms:

SUMS OF
PRODUCTS
AND PRODUCTS
OF SUMS
(SAY THAT
5 TIMES FAST)

You may recall from Section 4.3.6 that the minimal sum-of-products expression for the complement of a function can be manipulated through DeMorgan's theorem to obtain a minimal product-of-sums expression for the original function. You may also recall that the number of product terms in the minimal sum of products may differ from the number of sum terms in the minimal product of sums. The discussion in this section shows that if the number of terms in *either* minimal form is less than or equal to the number of product terms available in a 16V8's AND-OR array, then the function can be made to fit.

In the design examples in previous sections, we were forced to choose a particular polarity, active high or active low, for some outputs in order to obtain reduced equations that would fit in a PAL16L8. When a GAL16V8 is used, no such restriction exists. If an equation *or its complement* can be reduced to the number of product terms available, then the corresponding output can be made active high or active low by programming the output-polarity fuse appropriately.

HAVE IT
YOUR WAY

- For active-high outputs, if the minimal expression is in the column labeled "True," program the output-polarity control not to invert; otherwise program it to invert.

- For active-low outputs, if the minimal expression is in the column labeled "True," program the output-polarity control to invert; otherwise program it not to invert.

For example, we can have all active-high outputs by selecting the form of the equation for each output marked with an asterisk in the table, and programming the output-polarity control not to invert if the selected form is in the column labeled "True." And we can have all active-low outputs by programming the polarity control in the opposite way.

Advanced Micro Devices (AMD) makes a part that is equivalent to the GAL16V8, called the *PALCE16V8*. Both Lattice and AMD make a 24-pin GAL device, the *GAL20V8* or *PALCE20V8*, that can be configured to emulate the structure of the PAL20L8 or any of a variety of sequential PLDs, as described in Section 10.2.

PALCE16V8
GAL20V8
PALCE20V8

REFERENCES

Programmable logic devices have been around for a while, but it was only in the mid- to late 1980s that their cost, reliability, and ease of use finally caused an explosion in PLD-based designs.

The best resources for learning about PLD-based design still seem to be the PLD manufacturers. For example, AMD publishes a comprehensive *PAL Device Data Book and Design Guide* (Sunnyvale, CA 94088, 1993). In addition to data sheets for all of AMD's programmable logic devices, it contains over 100 pages of application notes on PLD-based design methods and "real-world" issues like testability, ground bounce, and latch-up.

A much more detailed discussion of the internal operation of structured LSI and VLSI devices, including ROMs, RAMs, and PLDs, can be found in electronics texts such as *Microelectronics*, 2nd ed., by J. Millman and A. Grabel (McGraw-Hill, 1987) and *VLSI Engineering* by Thomas E. Dillinger (Prentice Hall, 1988). Additional PLD references are cited at the end of Chapter 10.

DRILL PROBLEMS

6.1   What is the maximum number of product terms that a PLD might require to realize an *n*-input combinational logic function? How many times larger is this than the number of product terms available to a single output of an 82S100 PLA? Of a PAL16L8?

6.2   Write logic expressions for product terms P3–P6 in Figure 6–4.

6.3   How many fuses does it take to program a PAL16L8? How many does it take to program a PAL20L8?

6.4   To what do the "16" and the "8" refer in "PAL16L8"?

6.5   Explain the difference between the use of "/" previously in this book and its use in ABEL programs.

6.6   Indicate how many product terms are used by each output of the PAL16L8 defined by Table 6–3.

6.7   Suppose that one line is changed in the ABEL program in Table 6–5:

```
/Y1, /Y2                        pin 19, 12;
```

How many product terms does each of the outputs use in this case? (You should be able to figure this out in a snap without using an ABEL compiler.)

6.8   Rewrite the ABEL program in Table 6–12 using sets and relations to handle the X and Y inputs.

6.9   Using a single PAL16L8, design a circuit with the same inputs, outputs, and function as the 74x139 dual 2-to-4 decoder. Create a pin assignment that is similar to the 139's. Draw a logic diagram showing the assignment of signals to device pins, and write a complete ABEL program for the function.

6.10 Using a single PAL16L8, design a circuit with the same inputs, outputs, and function as the 74x157 4-bit 2-input multiplexer. Create a pin assignment that is similar to the 157's. Draw a logic diagram showing the assignment of signals to device pins, and write a complete ABEL program for the function.

6.11 Repeat Drill 6.10 for the 74x257.

6.12 Using a single PAL16L8, design a circuit with the same inputs, outputs, and function as the 74x541 octal three-state buffer. Create a pin assignment that is similar to the 541's. Draw a logic diagram showing the assignment of signals to device pins, and write a complete ABEL program for the function. Use sets to minimize the size of your ABEL program.

6.13 Using a single PAL16L8, design a circuit with the same inputs, outputs, and function as the 74x280 9-input parity generator. Draw a logic diagram showing the assignment of signals to device pins, and write a complete ABEL program for the function. (*Hint:* You'll have to use two-pass logic. Why?)

6.14 Using a single PAL16L8, design an 8-bit one/zero/true/complement element. This circuit should have two select inputs S[1–0], eight data inputs X[1–8], and eight data outputs Z[1–8]. Depending on the value of the select inputs, Z[1–8] should be all 0s, all 1s, X[1–8], or the complement of X[1–8]. Draw a logic diagram showing the assignment of signals to device pins, and write a complete ABEL program for the function. Use sets to minimize the size of the program. (A 4-bit version of this function was once sold, the TTL 74H87.)

6.15 Modify the ABEL program for the 74x49-like seven-segment decoder in Table 6–10 to use a GAL16V8. Without using a PLD compiler, determine the maximum number of product terms our required for each output.

6.16 Modify the ABEL program for the 74x49-like seven-segment decoder in Table 6–10 so no segments are activated when the input is a hexadecimal digit A–F.

6.17 Write a single ABEL program that you can compile with reduction to find a minimal sum-of-products expression for all eight of the following logic functions:

(a) $F = X' \cdot Y' \cdot Z' + X \cdot Y \cdot Z + X \cdot Y' \cdot Z$ (b) $F = M' \cdot N' + M \cdot P + N' \cdot P$

(c) $F = A \cdot B + A \cdot B' \cdot C' + A' \cdot B \cdot C$ (d) $F = A' \cdot B \cdot (C \cdot B \cdot A' + B \cdot C')$

(e) $F = X \cdot Y \cdot (X' \cdot Y \cdot Z + X \cdot Y' \cdot Z + X \cdot Y \cdot Z' + X' \cdot Y' \cdot Z)$ (f) $F = M \cdot N + M' \cdot N' \cdot P'$

(g) $F = (A + A') \cdot B + B \cdot A \cdot C' + C \cdot (A + B') \cdot (A' + B)$ (h) $F = X \cdot Y' + Y \cdot Z + Z' \cdot X$

6.18 Explain how you would modify the program in Drill 6.17 and its results to obtain a minimal product-of-sums expression for each logic function.

6.19 State the rule used by an ABEL compiler to program an output-enable gate if no output-enable equation has been written for the corresponding output.

6.20 What is the most important difference between a PAL16L8 and a GAL16V8C?

6.21 A device called the GAL18V8 is available in the same package as a GAL16V8. Just based on the name, what enhancement would you expect the GAL18V8 to contain? (*Hint:* Look at the logic diagram of the GAL16V8C.)

6.22 Discuss the impact of active-level selection for inputs and outputs on whether logic functions "fit" in PAL16L8s and GAL16V8s (a total of four cases to consider).

E X E R C I S E S

6.23 Indicate how many product terms are used by each output of the PAL16L8 defined by Table 6–3 if three inputs are added and the ENB expression is modified as follows:

```
EN1, EN2, EN3           pin 7, 8, 9;
...
ENB = (G1 * G2A * G2B) + (EN1 * EN2) + (/EN1 * EN3) + (EN2 * EN3);
```

6.24 Describe circumstances under which programs for an *n*-output PLD benefit from multiple-output minimization, compared to *n* individual single-output minimizations.

6.25 Using an ABEL compiler or other PLD compiler, determine how many product terms are required per output for the 74x138-like decoder with auxiliary enable inputs (EN1, /EN2) and output polarity control (POL) described in Section 6.7.1. Does this fit in a PAL16L8? How many product terms are required if the auxiliary enable inputs are omitted?

6.26 Using a single PAL16L8, design a circuit similar to the 74x139 dual 2-to-4 decoder, including an active-low output-enable and a polarity-control input for each section. Draw a logic diagram showing the assignment of signals to device pins, and write a complete ABEL program for the function. (Create a pin assignment that is similar to the 139's.)

6.27 In the style of Figure 6–14 and Table 6–5, draw a logic diagram and write two ABEL programs for a PAL16L8 that performs the function of a 74x151 8-input, 1-bit multiplexer. One version of the program should derive the Y output from the /Y output using two-pass logic, and the other should produce both outputs in a single pass. Determine if either version fits in a PAL16L8.

6.28 Using a single PAL16L8, design a circuit with the same inputs, outputs, and function as the 74x283 4-bit adder. Draw a logic diagram showing the assignment of signals to device pins, and write a complete ABEL program for the function. (*Hint:* Use all the outputs.) Assuming that the propagation delay from a PAL16L8 input to output is 10 ns, compare the speed of your solution with that of a 74LS283.

6.29 Think of an interesting and useful function to provide using the spare input pin in the solution to Exercise 6.28. Modify the ABEL program to include your new capability, and convince yourself that it still fits in the available product terms.

6.30 Using only as many PAL16L8s as absolutely necessary, design the fastest possible 4-bit adder with the same inputs and outputs as the 74x283. You need not write the equations, but draw a logic diagram for your solution, including signal names and an explanation of their functions. Assuming that the propagation delay from a PAL16L8 input to output is 10 ns, how fast is your adder?

6.31 Complete Exercise 6.30 by writing the ABEL equations for each PLD and compiling them to ensure that they fit in the available product terms.

6.32 Without using an ABEL compiler, prove that the logic function of a 74x182 carry-lookahead circuit can be realized in one pass through a PAL16L8.

6.33 Design a 3-input, 5-bit multiplexer in a single PAL20L8. The select inputs are S1 and S0, the data inputs are A1–A5, B1–B5, and C1–C5, and the data outputs are Y1–Y5. The Y outputs should equal the A, B, or C for select-input combinations 00, 01, and 10, respectively, and should be zero when the select inputs are 11. Draw a logic diagram showing the assignment of signals to device pins, and write a complete ABEL program for the function. Use sets and relations in the program. Does the PAL20L8 have enough product terms for the function? If not, what changes could you make to the design specification to make it fit?

6.34 Indicate how the ABEL program in Exercise 6.33 must be modified if the Y outputs are to be floating when the select inputs are 11.

6.35 Using a single PAL16L8, design a circuit with the same inputs, outputs, and function as the 74x85 4-bit magnitude comparator. Draw a logic diagram showing the assignment of signals to device pins, and write a complete ABEL program for the function.

6.36 Use an ABEL compiler to determine whether your solution to Exercise 6.35 fits in the available product terms. If it doesn't, use two-pass logic to create a design that does. (*Hint:* If that doesn't work, you're off track; try again.)

6.37 Write an equation for a /XGEY output for the MODECOMP PLD of Table 6–12 that is asserted when $X \geq Y$. Does your equation fit in a PAL16L8?

6.38 How many product terms are required for each output in Table 6–13 if each is assigned an active level opposite the one shown?

6.39 Modify the logic diagram and the ABEL program for the four-way, 2-bit transceiver of Figure 6–19 so that the data bits appear on output pins 19–12 in the order A1, A2, ..., D1, D2. (*Hint:* It takes more than a change in the declarations.)

6.40 Modify the ABEL program for the 74x49-like seven-segment decoder in Table 6–10 to have active-low outputs. Using ABEL or another PLD compiler, determine whether the reduced equations fit in a PAL16L8.

6.41 Add a "POL" input and modify the ABEL program for the seven-segment decoder in Table 6–18 so the segment outputs are active high or active low depending on the value of POL. Without using a PLD compiler, estimate the number of product terms that each reduced equation will have in the modified design. Do you think that the design will still fit in a PAL16V8? *Optional:* Using ABEL or another PLD compiler, determine exactly how many product terms are used per output.

6.42 In the MODECOMP PLD of Figure 6–16, we assigned outputs to device pins so as to maximize the possible uses of the unused pins (compare with the situation if we used pins 13–14 for the outputs). Experience dictates this approach, since after a PCB is designed and built, you can often fix bugs or add last-minute features without too much pain if there are spare PLD resources available on the board. Find three other instances in this chapter where we assigned outputs to maximize flexibility in this way, and list the possible uses of each spare pin.

6.43   Determine whether any of the following PLD examples in this chapter could benefit from the product-term sharing available in a PLA (as opposed to a PAL device): MEMDEC, Z74X138, ABINDEC, Z74X153, Z74X49H, PRIOR15. Briefly explain or justify each answer.

6.44   Building on the results in Exercise 6.43, give your opinion on whether it was a good idea to the inventors of PAL devices to use a fixed OR structure.

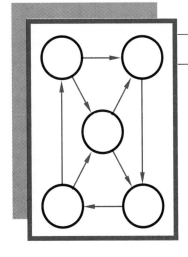

7

SEQUENTIAL LOGIC DESIGN PRINCIPLES

Logic circuits are classified into two types, "combinational" and "sequential." A *combinational* logic circuit is one whose outputs depend only on its current inputs. The rotary channel selector knob on an inexpensive TV is like a combinational circuit—its "output" selects a channel based only on the current position of the knob ("input").

combinational circuit

A *sequential* logic circuit is one whose outputs depend not only on its current inputs, but also on the past sequence of inputs, possibly arbitrarily far back in time. The circuit controlled by the channel-up and channel-down pushbuttons on a TV or VCR is a sequential circuit—the channel selection depends on the past sequence of up/down pushes, at least since when you started viewing 10 hours before, and perhaps as far back as when you first powered-up the device.

sequential circuit

So it is inconvenient, often impossible, to describe the behavior of a sequential circuit by means of a table that lists outputs as a function of the input sequence that has been received up until the current time. With the TV channel selector, it is impossible to determine what channel is currently selected by looking only at the preceding sequence of presses on the up and down pushbuttons, whether we look at the preceding 10 presses or the preceding 1,000. More information, the current "state" of the channel selector, is needed.

state

state variable

The *state* of a sequential circuit is a collection of *state variables* whose values at any one time contain all the information about the past necessary to account for the circuit's future behavior.[1]

In the channel-selector example, the current channel number is the current state. Inside the TV, this state might be stored as seven binary state variables representing a decimal number between 0 and 127. Given the current state (channel number), we can always predict the next state as a function of the inputs (presses of the up/down pushbuttons). In this example, one highly visible output of the sequential circuit is an encoding of the state itself—the channel-number display. Other outputs, internal to the TV, may be combinational functions of the state alone (e.g., VHF/UHF tuner selection) or of both state and input (e.g., turning off the TV if the current state is 0 and the "down" button is pressed).

State variables need not have direct physical significance, and there are often many ways to choose them to describe a given sequential circuit. For example, in the TV channel selector, the state might be stored as three BCD digits or 12 bits, with many of the bit combinations (4,096 possible) being unused.

In a digital logic circuit, state variables are binary values, corresponding to certain logic signals in the circuit, as we'll see in later sections. A circuit with n binary state variables has 2^n possible states. As large as it might be, 2^n is always finite, never infinite, so sequential circuits are sometimes called

finite-state machine

finite-state machines.

NON-FINITE-
STATE MACHINES

A group of mathematicians recently proposed a non-finite-state machine, but they're still working on the state table.... Sorry, that's just a joke. There *are* mathematical models for infinite-state machines, such as Turing machines. They typically contain a small finite-state-machine control unit, and an infinite amount of auxiliary memory, such as an endless tape.

clock

clock period

clock frequency

The state changes of most sequential circuits occur at times specified by a free-running *clock* signal. Figure 7–1 gives timing diagrams and nomenclature for typical clock signals. By convention, a clock signal is active high if state changes occur at the clock's rising edge or when the clock is HIGH, and active low in the complementary case. The *clock period* is the time between successive transitions in the same direction, and the *clock frequency* is the reciprocal of the

[1]Herbert Hellerman, *Digital Computer Systems Principles*, McGraw-Hill, 1967, p. 237.

period. The first edge or pulse in a clock period or sometimes the period itself is called a *clock tick*. The *duty cycle* is the percentage of time that the clock signal is at its asserted level. Typical digital systems, from digital watches to supercomputers, use a quartz-crystal oscillator to generate a free-running clock signal. Clock frequencies range from 32.768 kHz (for a watch) to 400 MHz (for a CMOS RISC microprocessor with a cycle time of 2.5 ns); "typical" systems using TTL and CMOS parts have clock frequencies in the 5–66 MHz range.

clock tick
duty cycle

Figure 7–1
Clock signals: (a) active high;
(b) active low.

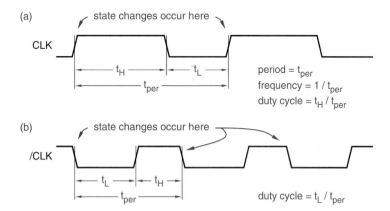

In this chapter we'll discuss two types of sequential circuits that account for the majority of practical discrete designs. A *feedback sequential circuit* uses ordinary gates and feedback loops to obtain memory in a logic circuit, thereby creating sequential-circuit building blocks such as flip-flops and latches that are used in higher-level designs. A *clocked synchronous state machine* uses these building blocks, in particular edge-triggered D flip-flops, to create circuits whose inputs are examined and whose outputs change with respect to a controlling clock signal. Other sequential circuit types, such as general fundamental mode, multiple pulse mode, and multiphase, are sometimes useful in high-performance systems and VLSI, and are discussed in advanced texts.

*feedback sequential
circuit*

*clocked synchronous
state machine*

7.1   BISTABLE ELEMENTS

The simplest sequential circuit consists of a pair of inverters forming a feedback loop, as shown in Figure 7–2. It has *no* inputs and two outputs, Q and /Q.

7.1.1   Digital Analysis

The circuit of Figure 7–2 is often called a *bistable*, since a strictly digital analysis shows that it has two stable states. If Q is HIGH, then the bottom inverter has a HIGH input and a LOW output, which forces the top inverter's output HIGH as

bistable

Figure 7–2
A pair of inverters forming a
bistable element.

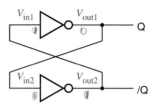

we assumed in the first place. But if Q is LOW, then the bottom inverter has
a LOW input and a HIGH output, which forces Q LOW, another stable situation.
We could use a single state variable, the state of signal Q, to describe the state
of the circuit; there are two possible states, $Q = 0$ and $Q = 1$.

The bistable element is so simple that it has no inputs and therefore no
way of controlling or changing its state. When power is first applied to the
circuit, it randomly comes up in one state or the other and stays there forever.
Nevertheless, it serves our illustrative purposes very well, and we *will* actually
conjure up an application for it in Section 9.2.3.

7.1.2   Analog Analysis

The analysis of the bistable has more to reveal if we consider its operation from
an analog point of view. The dark line in Figure 7–3 shows the steady-state
(DC) transfer function T for a single inverter; the output voltage is a function
of input voltage, $V_{out} = T(V_{in})$. With two inverters connected in a feedback
loop as in Figure 7–2, we know that $V_{in1} = V_{out2}$ and $V_{in2} = V_{out1}$; therefore,
we can plot the transfer functions for both inverters on the same graph with an
appropriate labeling of the axes. Thus, the dark line is the transfer function for
the top inverter in Figure 7–2, and the colored line is the transfer function for
the bottom one.

Considering only the steady-state behavior of the bistable's feedback loop,
and not dynamic effects, the loop is in equilibrium if the input and output voltages
of both inverters are constant DC values consistent with the loop connection and
the inverters' DC transfer function. That is, we must have

Figure 7–3
Transfer functions for inverters in
a bistable feedback loop.

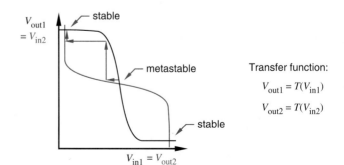

Transfer function:

$$V_{out1} = T(V_{in1})$$
$$V_{out2} = T(V_{in2})$$

$$V_{in1} = V_{out2}$$
$$= T(V_{in2})$$
$$= T(V_{out1})$$
$$= T(T(V_{in1}))$$

Likewise, we must have

$$V_{in2} = T(T(V_{in2}))$$

We can find these equilibrium points graphically from Figure 7–3; they are the points at which the two transfer curves meet. Surprisingly, we find that there are not two but *three* equilibrium points. Two of them, labeled *stable*, correspond to the two states that our "strictly digital" analysis identified earlier, with Q either 0 (LOW) or 1 (HIGH).

stable

The third equilibrium point, labeled *metastable*, occurs with V_{out1} and V_{out2} about halfway between a valid logic 1 voltage and a valid logic 0 voltage; so Q and /Q are not valid logic signals at this point. Yet the loop equations are satisfied; if we can get the circuit to operate at the metastable point, it could theoretically stay there indefinitely.

metastable

7.1.3 Metastable Behavior

Closer analysis of the situation at the metastable point shows that it is aptly named. It is not truly stable, because random noise will tend to drive a circuit that is operating at the metastable point toward one of the stable operating points as we'll now demonstrate. Suppose the bistable is operating precisely at the metastable point in Figure 7–3. Now let us assume that a small amount of circuit noise reduces V_{in1} by a tiny amount. This tiny change causes V_{out1} to *increase* by a small amount. But since V_{out1} produces V_{in2}, we can follow the first horizontal arrow from near the metastable point to the second transfer characteristic, which now demands a lower voltage for V_{out2}, which is V_{in1}. Now we're back where we started, except we have a much larger change in voltage at V_{in1} than the original noise produced, and the operating point is still changing. This "regenerative" process continues until we reach the stable operating point at the upper left-hand corner of Figure 7–3. However, if we perform a "noise" analysis for either of the stable operating points, we find that feedback brings the circuit back toward the stable operating point, rather than away from it.

Metastable behavior of a bistable can be compared to the behavior of a ball dropped onto a hill, as shown in Figure 7–4. If we drop a ball from overhead, it will probably immediately roll down to one side of the hill or the other. But if it lands right at the top, it may precariously sit there for a while before random forces (wind, rodents, earthquakes) start it rolling down the hill. Like

Figure 7–4
Ball and hill analogy for
metastable behavior.

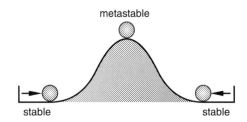

the ball at the top of the hill, the bistable may stay in the metastable state for an unpredictable length of time before nondeterministically settling into one stable state or the other.

If the *simplest* sequential circuit is susceptible to metastable behavior, you can be sure that *all* sequential circuits are susceptible. And this behavior is not something that only occurs at power-up.

Returning to the ball-and-hill analogy, consider what happens if we try to kick the ball from one side of the hill to the other. Apply a strong force (Arnold Schwarzenegger), and the ball goes right over the top and lands in a stable resting place on the other side. Apply a weak force (Mr. Rogers), and the ball falls back to its original starting place. But apply a wishy-washy force (Charlie Brown), and the ball goes to the top of the hill, teeters, and eventually falls back to one side or the other.

This behavior is completely analogous to what happens to flip-flops under marginal triggering conditions. For example, we'll soon study S-R flip-flops, where a pulse on the S input forces the flip-flop from the 0 state to the 1 state. A minimum pulse width is specified for the S input. Apply a pulse of this width or longer, and the flip-flop immediately goes to the 1 state. Apply a very short pulse, and the flip-flop stays in the 0 state. Apply a pulse just under the minimum width, and the flip-flop may go into the metastable state. Once the flip-flop is in the metastable state, its operation depends on "the shape of its hill." Flip-flops built from high-gain, fast technologies tend to come out of metastability faster than ones built from low-performance technologies.

We'll say more about metastability in the next section in connection with specific flip-flop types, and in Section 9.8 with respect to synchronous design methodology and synchronizer failure.

7.2   LATCHES AND FLIP-FLOPS

Latches and flip-flops are the basic building blocks of most sequential circuits. Typical digital systems use latches and flip-flops that are prepackaged, functionally specified devices in a standard integrated circuit. In ASIC design environments, latches and flip-flops are typically predefined cells specified by the ASIC vendor. However, within a standard IC or an ASIC, each latch or flip-flop is typically designed as a feedback sequential circuit using individual logic gates and feedback loops. We'll study these discrete designs for two reasons—to better understand the behavior of the prepackaged elements, and to gain the capability of building a latch or flip-flop "from scratch" as is required occasionally in digital-design practice and often in digital-design exams.

All digital designers use the name *flip-flop* for a sequential device that normally samples its inputs and changes its outputs only at times determined by a clocking signal. On the other hand, most digital designers use the name *latch* for a sequential device that watches all of its inputs continuously and changes its outputs at any time, independent of a clocking signal. We follow this convention in this text. However, some textbooks and digital designers may use the name "flip-flop" for a device that we call a "latch."

flip-flop

latch

In any case, because the functional behaviors of latches and flip-flops are quite different, it is important for the logic designer to know which type is being used in a design, either from the device's part number (e.g., 74x374 vs. 74x373) or other contextual information. We discuss the most commonly used types of latches and flip-flops in the following subsections.

7.2.1   S-R Latch

An *S-R (set-reset) latch* based on NOR gates is shown in Figure 7–5(a). The circuit has two inputs, S and R, and two outputs, Q and /Q, where /Q ("not Q") is normally the complement of Q.

S-R latch

Figure 7–5
S-R latch: (a) circuit design using NOR gates; (b) function table.

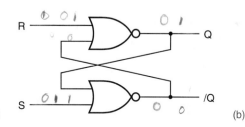

S	R	Q	/Q
0	0	last Q	last /Q
0	1	0	1
1	0	1	0
1	1	0	0

(a)   (b)

If S and R are both 0, the circuit behaves like the bistable element—we have a feedback loop that retains one of two logic states, Q=0 or Q=1. As shown in Figure 7–5(b), either S or R may be asserted to force the feedback loop to a desired state. S *sets* or *presets* the Q output to 1; R *resets* or *clears* the Q output to 0. After the S or R input is negated, the latch remains in the state that it was forced into. Figure 7–6(a) shows the functional behavior of an S-R latch for a typical sequence of inputs. Colored arrows indicate causality, that is, which input transitions cause which output transitions.

set
preset
reset
clear

Three different logic symbols for the same S-R latch circuit are shown in Figure 7–7. The symbols differ in the treatment of the active-low /Q output.

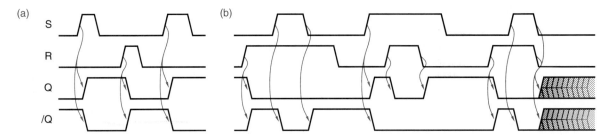

Figure 7–6   Typical operation of an S-R latch: (a) "normal" inputs; (b) S and R asserted simultaneously.

Historically, the first symbol was used, showing the active-low or complemented signal name inside the function rectangle. However, in bubble-to-bubble logic design the second form of the symbol is preferred, showing an inversion bubble outside the function rectangle. The last form of the symbol is obviously wrong.

propagation delay   Figure 7–8 defines timing parameters for an S-R latch. The *propagation delay* is the time it takes for a transition on an input signal to produce a transition on an output signal. A given latch or flip-flop may have several different propagation delay specifications, one for each pair of input and output signals. Also, the propagation delay may be different depending on whether the output makes a LOW-to-HIGH or HIGH-to-LOW transition. With an S-R latch, a LOW-to-HIGH transition on S can cause a LOW-to-HIGH transition on Q, so a propagation delay $t_{pLH(SQ)}$ occurs as shown in transition 1 in the figure. Similarly, a LOW-to-HIGH transition on R can cause a HIGH-to-LOW transition on Q, with propagation delay $t_{pHL(RQ)}$ as shown in transition 2. Not shown in the figure are the corresponding transitions on /Q, which would have propagation delays $t_{pHL(S/Q)}$ and $t_{pLH(R/Q)}$.

Q VERSUS /Q

> In most applications of an S-R latch, the /Q output is always the complement of the Q output. However, /Q is not quite correctly named, because there is one case where /Q is *not* the complement of Q. If both S and R are 1, as they are in several places in Figure 7–6(b), then both outputs are forced to 0. Once we negate either input, the outputs return to complementary operation as usual. However, if we negate both inputs simultaneously, the latch goes to an unpredictable next state, and it may in fact oscillate or enter the metastable state. Metastability may also occur if a 1 pulse that is too short is applied to S or R.

Figure 7–7
Symbols for an S-R latch: (a) old
style; (b) preferred; (c) incorrect
because of double negation.

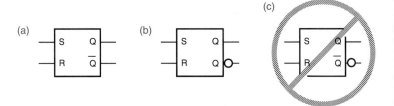

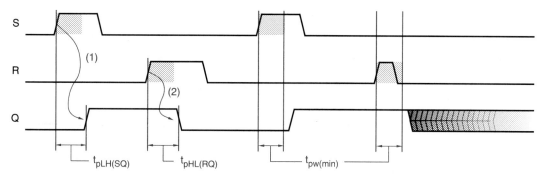

Figure 7–8 Timing parameters for an S-R latch.

Minimum pulse width specifications are usually given for the S and R   *minimum pulse*
inputs. As shown in Figure 7–8, the latch may go into the metastable state and   *width*
remain there for a random length of time if a pulse shorter than the minimum
width $t_{pw(min)}$ is applied to S or R. The latch can be deterministically brought out
of the metastable state only by applying a pulse to S or R that meets or exceeds
the minimum pulse width requirement.

7.2.2 /S-/R Latch

A */S-/R latch* with active-low set and reset inputs may be built from NAND gates   */S-/R latch*
as shown in Figure 7–9(a). In TTL and CMOS logic families, /S-/R latches are
used much more often than S-R latches because NAND gates are preferred over
NOR gates.

As shown by the function table, Figure 7–9(b), operation of the /S-/R latch
is similar to that of the S-R, with two major differences. First, /S and /R are
active low, so the latch remembers its previous state when /S = /R = 1; the
active-low inputs are clearly indicated in the symbols in (c). Second, when both
/S and /R are asserted simultaneously, both latch outputs go to 1, not 0 as in the
S-R latch. Except for these differences, operation of the /S-/R is the same as the
S-R, including timing and metastability considerations.

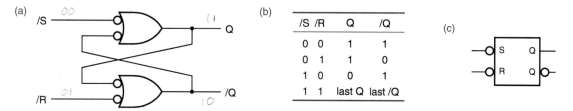

Figure 7–9 /S-/R latch: (a) circuit design using NAND gates; (b) function table; (c) logic symbol.

> As mentioned in the previous note, an S-R latch may go into the metastable state if S and R are negated simultaneously. Often, but not always, a commercial latch's specifications define "simultaneously" (e.g., S and R negated within 20 ns of each other). In any case, the minimum delay between negating S and R for them to be considered nonsimultaneous is closely related to the mimimum pulse width specification. Both specifications are measures of how long it takes for the latch's feedback loop to stabilize during a change of state.

7.2.3   S-R Latch with Enable

*S-R latch with
enable*

An S-R or /S-/R latch is sensitive to its S and R inputs at all times. However, it may easily be modified to create a device that is sensitive to these inputs only when an enabling input C is asserted. Such an *S-R latch with enable* is shown in Figure 7–10. As shown by the function table, the circuit behaves like an S-R latch when C is 1, and retains its previous state when C is 0. The latch's behavior for a typical set of inputs is shown in Figure 7–11. If both S and R are 1 when C changes from 1 to 0, the circuit behaves like an S-R latch in which S and R are negated simultaneously—the next state is unpredictable and the output may become metastable.

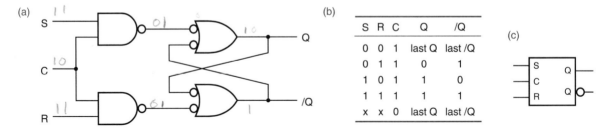

(a)   (b)

S	R	C	Q	/Q
0	0	1	last Q	last /Q
0	1	1	0	1
1	0	1	1	0
1	1	1	1	1
x	x	0	last Q	last /Q

(c)

Figure 7–10   S-R latch with enable: (a) circuit using NAND gates; (b) function table; (c) logic symbol.

7.2.4   D Latch

D latch

S-R latches are useful in control applications, where we often think in terms of setting a flag in response to some condition, and resetting it when conditions change; so we control the set and reset inputs somewhat independently. However, we often need latches simply to store bits of information—each bit of information is presented on a signal line, and we'd like to store it somewhere. A *D latch* may be used in such an application.

Figure 7–12 shows a D latch. Its logic diagram is recognizable as that of an S-R latch with enable, with an inverter added to generate S and R inputs from the single D (data) input. This eliminates the troublesome situation in S-R

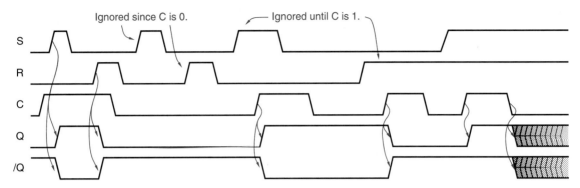

Figure 7–11   Typical operation of an S-R latch with enable.

latches, where S and R may be asserted simultaneously. The control input of a
D latch, labeled C in (c), is sometimes named ENABLE, CLK, or G, and is active
low in some D-latch designs.

An example of a D latch's functional behavior is given in Figure 7–13.
When the C input is asserted, the Q output follows the D input. In this situation,
the latch is said to be "open" and the path from D input to Q output is "trans-
parent"; the circuit is often called a *transparent latch* for this reason. When the   *transparent latch*
C input is negated, the latch "closes"; the Q output retains its last value and no
longer changes in response to D, as long as C remains negated.

More detailed timing behavior of the D latch is shown in Figure 7–14. Four
different delay parameters are shown for signals that propagate from the C or D

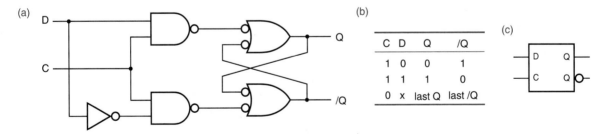

Figure 7–12   D latch: (a) circuit design using NAND gates; (b) function table; (c) logic symbol.

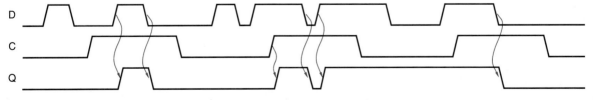

Figure 7–13   Functional behavior of a D latch for various inputs.

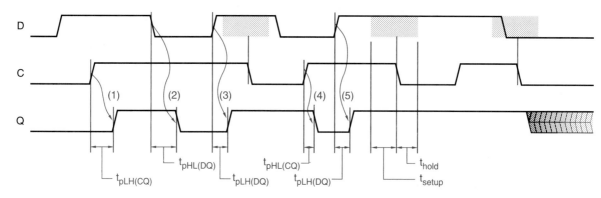

Figure 7–14   Timing parameters for a D latch.

input to the Q output. For example, at transitions 1 and 4 the latch is initially "closed" and the D input is the opposite of Q output, so that when C goes to 1 the latch "opens up" and the Q output changes after delay $t_{pLH(CQ)}$ or $t_{pHL(CQ)}$. At transitions 2 and 3 the C input is already 1 and the latch is already open, so that Q transparently follows the transitions on D with delay $t_{pHL(DQ)}$ and $t_{pLH(DQ)}$. Another four parameters exist for signals propagating to the /Q output, not shown.

    Although the D latch eliminates the S = R = 1 problem of the S-R latch, it does not eliminate the metastability problem. As shown in Figure 7–14, there is a (shaded) window of time around the falling edge of C when the D input must not change. This window begins at time t_{setup} before the falling (latching) edge

setup time   of C; t_{setup} is called the *setup time*. The window ends at time t_{hold} afterward; t_{hold}
hold time   is called the *hold time*. If D changes at any time during the setup- and hold-time window, the output of the latch is unpredictable and may become metastable, as shown for the last latching edge in the figure.

7.2.5   Edge-Triggered D Flip-Flop

positive-edge-   A *positive-edge-triggered D flip-flop* combines a pair of D latches, as shown in
triggered   Figure 7–15, to create a circuit that samples its D input and changes its Q and
D flip-flop   /Q outputs only at the rising edge of a controlling CLK signal. The first latch is
master   called the *master*; it is open and follows the input when CLK is 0. When CLK goes to 1, the master latch is closed and its output is transferred to the second
slave   latch, called the *slave*. The slave latch is open all the while that CLK is 1, but changes only at the beginning of this interval, because the master is closed and unchanging during the rest of the interval.

dynamic-input   The triangle on the D flip-flop's CLK input indicates edge-triggered behavior,
indicator   and is called a *dynamic-input indicator*. Examples of the flip-flop's functional behavior for several input transitions are presented in Figure 7–16. The QM

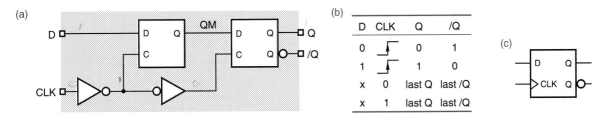

Figure 7–15   Positive-edge-triggered D flip-flop: (a) circuit design using D latches; (b) function table; (c) logic symbol.

signal shown is the output of the master latch. Notice that QM changes only when CLK is 0. When CLK goes to 1, the current value of QM is transferred to Q, and QM is prevented from changing until CLK goes to 0 again.

Figure 7–17 shows more detailed timing behavior for the D flip-flop. All propagation delays are measured from the rising edge of CLK, since that's the only event that causes an output change. Different delays may be specified for LOW-to-HIGH and HIGH-to-LOW output changes.

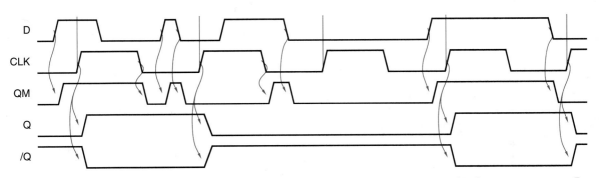

Figure 7–16   Functional behavior of a positive-edge-triggered D flip-flop.

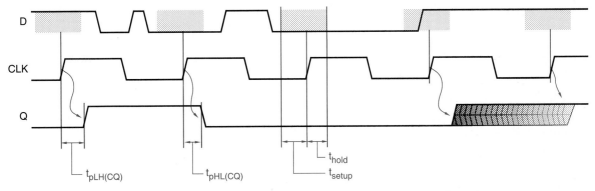

Figure 7–17   Timing behavior of a positive-edge-triggered D flip-flop.

Like a D latch, the edge-triggered D flip-flop has a setup and hold time window during which the D inputs must not change. This window occurs around the triggering edge of CLK, and is indicated by shaded color in Figure 7–17. If the setup and hold times are not met, the flip-flop output will usually go to a stable, though unpredictable, 0 or 1 state. In some cases, however, the output will oscillate or go to a metastable state halfway between 0 and 1, as shown at the second-to-last clock tick in the figure. If the flip-flop goes into the metastable state, it will return to a stable state on its own only after a probabilistic delay, as explained in Section 9.8. It can also be forced into a stable state by applying another triggering clock edge with a D input that meets the setup- and hold-time requirements, as shown at the last clock tick in the figure.

negative-edge-triggered D flip-flop

A *negative-edge-triggered D flip-flop* simply inverts the clock input, so that all the action takes place on the falling edge of /CLK; its function table and logic symbol are shown in Figure 7–18.

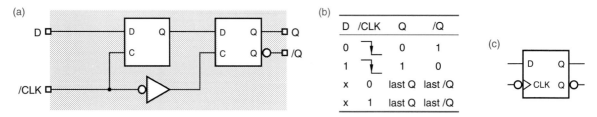

Figure 7–18   Negative-edge triggered D flip-flop: (a) circuit design using D latches; (b) function table; (c) logic symbol.

asynchronous inputs

Some D flip-flops have *asynchronous inputs* that may be used to force the flip-flop to a particular state independent of the CLK and D inputs. These inputs, typically labeled PR (preset) and CLR (clear), behave like the set and reset inputs on an S-R latch. The logic symbol and NAND circuit for an edge-triggered D flip-flop with these inputs is shown in Figure 7–19. Although asynchronous inputs are used by some logic designers to perform tricky sequential functions, they are best reserved for initialization and testing purposes, to force a sequential circuit into a known starting state; more on this when we discuss synchronous design methodology in Section 8.1.

preset

clear

TIME FOR A
COMMERCIAL

Commercial TTL positive-edge-triggered D flip-flops do not use the master-slave latch design of Figure 7–15 or Figure 7–19. Instead, flip-flops like the 74LS74 use the six-gate design of Figure 7–20, which is smaller and faster. We'll show how to formally analyze the next-state behavior of both designs in Section 7.5.

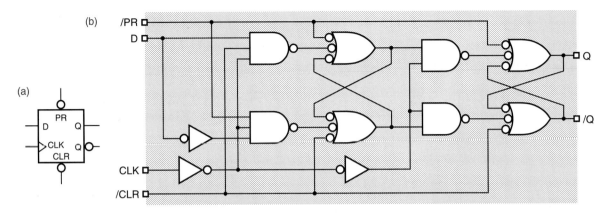

Figure 7–19   Positive-edge-triggered D flip-flop with preset and clear: (a) logic symbol; (b) circuit design using NAND gates.

* 7.2.6 Master/Slave S-R Flip-Flop

We indicated earlier that S-R latches are useful in "control" applications, where we may have independent conditions for setting and resetting a control bit. If the control bit is supposed to be changed only at certain times with respect to a clock signal, then we need an S-R flip-flop that, like a D flip-flop, changes its outputs only on a certain edge of the clock signal. This subsection and the next two describe flip-flops that are useful for such applications.

[*]Throughout this book, optional sections are marked with an asterisk.

Figure 7–20
Commercial circuit for a positive-edge-triggered D flip-flop such as 74LS74.

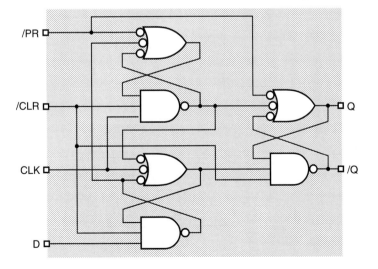

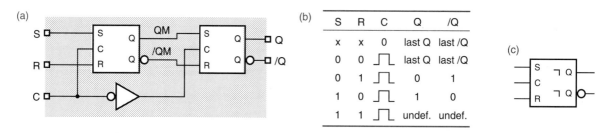

(a)

(b)

S	R	C	Q	/Q
x	x	0	last Q	last /Q
0	0	⎍	last Q	last /Q
0	1	⎍	0	1
1	0	⎍	1	0
1	1	⎍	undef.	undef.

(c)

Figure 7–21 Master/slave S-R flip-flop: (a) circuit using S-R latches; (b) function table; (c) logic symbol.

master/slave S-R flip-flop

If we substitute S-R latches for the D latches in the negative-edge-triggered D flip-flop of Figure 7–18(a), we get a *master/slave S-R flip-flop*, shown in Figure 7–21. Like a D flip-flop, the S-R flip-flop changes its outputs only at the falling edge of a control signal C. However, the new output value depends on input values not just at the falling edge, but during the entire interval in which C is 1 prior to the falling edge. As shown in Figure 7–22, a short pulse on S any time during this interval can set the master latch; likewise, a pulse on R can reset it. The value transferred to the flip-flop output on the falling edge of C depends on whether the master latch was last set or cleared while C was 1.

Shown in Figure 7–21(c), the logic symbol for the master/slave S-R flip-flop does not use a dynamic-input indicator, because the flip-flop is not truly edge triggered. It is more like a latch that follows its input during the entire interval that C is 1, but that changes its output to reflect the final latched value only when C goes to 0. In the symbol, a *postponed-output indicator* indicates

postponed-output indicator

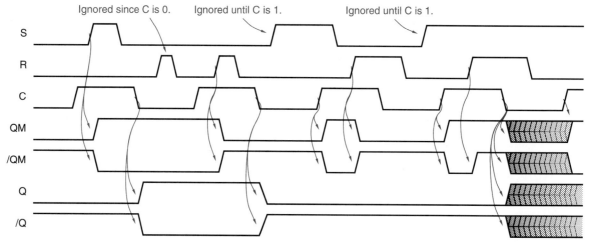

Figure 7–22 Internal and functional behavior of a master/slave S-R flip-flop.

that the output signal does not change until enable input C is negated. Flip-flops with this kind of behavior are sometimes called *pulse-triggered flip-flops*.

The operation of the master/slave S-R flip-flop is unpredictable if both S and R are asserted at the falling edge of C. In this case, just before the falling edge, both the Q and /Q outputs of the master latch are 1. When C goes to 0, the master latch's outputs change unpredictably and may even become metastable. At the same time, the slave latch opens up and propagates this garbage to the flip-flop output.

pulse-triggered
flip-flop

* 7.2.7 Master/Slave J-K Flip-Flop

The problem of what to do when S and R are asserted simultaneously is solved in a *master/slave J-K flip-flop*. The J and K inputs are analogous to S and R. However, as shown in Figure 7–23, asserting J asserts the master's S input only if the flip-flop's /Q output is currently 1 (i.e., Q is 0), and asserting K asserts the master's R input only if Q is currently 1. Thus, if J and K are asserted simultaneously, the flip-flop goes to the opposite of its current state.

master/slave J-K
flip-flop

Figure 7–24 shows the functional behavior of a J-K master/slave flip-flop for a typical set of inputs. Note that the J and K inputs need not be asserted at the end of the triggering pulse for the flip-flop output to change at that time. In fact, because of the gating on the master latch's S and R inputs, it is possible for the flip-flop output to change to 1 even though K and not J is asserted at the end of the triggering pulse. This behavior, known as *1s catching*, is illustrated in the second-to-last triggering pulse in the figure. An analogous behavior known as *0s catching* is illustrated in the last triggering pulse. Because of this behavior, the J and K inputs of a J-K master/slave flip-flop must be held valid during the entire interval that C is 1.

1s catching

0s catching

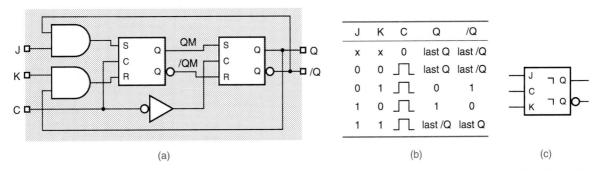

(a)   (b)   (c)

Figure 7–23   Master/slave J-K flip-flop: (a) circuit design using S-R latches; (b) function table; (c) logic symbol.

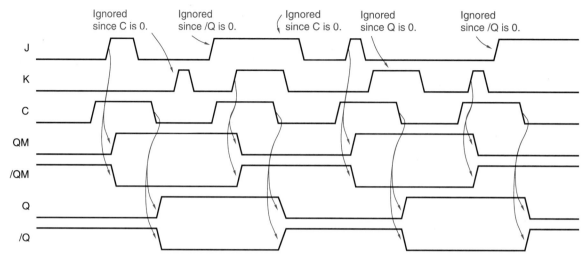

Figure 7–24   Internal and functional behavior of a master/slave J-K flip-flop.

* 7.2.8   Edge-Triggered J-K Flip-Flop

edge-triggered J-K
flip-flop

The problem of 1s and 0s catching is solved in an *edge-triggered J-K flip-flop*, whose functional equivalent is shown in Figure 7–25. Using an edge-triggered D flip-flop internally, the edge-triggered J-K flip-flop samples its inputs at the rising edge of the clock and produces its next output according to the "characteristic equation" $Q^* = J \cdot Q' + K' \cdot Q$ (see Section 7.3.2).

Typical functional behavior of an edge-triggered J-K flip-flop is shown in Figure 7–26. Like the D input of an edge-triggered D flip-flop, the J and K inputs of a J-K flip-flop must meet published setup- and hold-time specifications with respect to the triggering clock edge for proper operation.

Because they eliminate the problems of 1s and 0s catching and of simultaneously asserting both control inputs, edge-triggered J-K flip-flops are used much

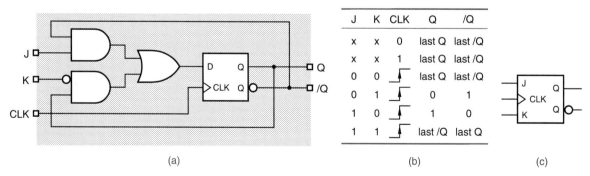

(a)   (b)   (c)

Figure 7–25   Edge-triggered J-K flip-flop: (a) equivalent function using an edge-triggered D flip-flop; (b) function table;   (c) logic symbol.

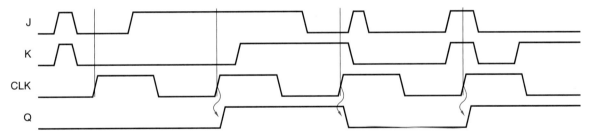

Figure 7–26   Functional behavior of a positive-edge-triggered J-K flip-flop.

more commonly nowadays than are the older pulse-triggered types of the preceding two subsections. The *74x109* is a TTL positive-edge-triggered J-/K flip-flop   *74x109*
with an active-low K input (named /K or $\overline{K}$).

The most common application of J-K flip-flops is in clocked synchronous state machines. As we'll explain in Section 7.4.5, the next-state logic for J-K flip-flops is sometimes simpler than for D flip-flops. However, most state machines are still designed using D flip-flops because the design methodology is a bit simpler and because most sequential programmable logic devices contain D, not J-K, flip-flops. Therefore, we'll give most of our attention to D flip-flops.

Figure 7–27
Internal logic diagram for
a 74LS109 positive-edge-
triggered J-/K flip-flop.

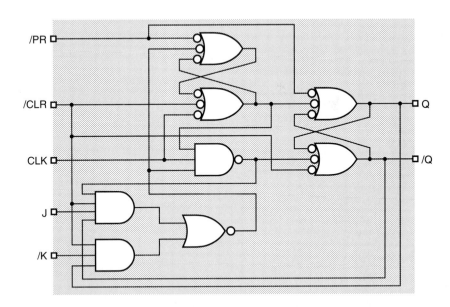

The internal design of the 74LS109 is very similar to that of the 74LS74, which we showed in Figure 7–20. As shown in Figure 7–27, the '109 simply replaces the bottom-left gate of the '74, which realizes the characteristic equation $Q^* = D$, with an AND-OR structure that realizes the J-/K characteristic equation, $Q^* = J \cdot Q' + /K \cdot Q$.

ANOTHER
COMMERCIAL
(FLIP-FLOP,
THAT IS)

7.2.9 T Flip-Flop

T flip-flop

A *T (toggle) flip-flop* changes state on every tick of the clock. Figure 7–28 shows the symbol and illustrates the behavior of a positive-edge-triggered T flip-flop. Notice that the signal on the flip-flop's Q output has precisely half the frequency of the T input. Figure 7–29 shows how to obtain a T flip-flop from a D or J-K flip-flop. T flip-flops are most often used in counters and frequency dividers, as we'll show in Section 9.3.

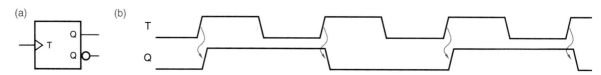

Figure 7–28   Positive-edge-triggered T flip-flop: (a) logic symbol; (b) functional behavior.

Figure 7–29
Possible circuit designs for a
T flip-flop: (a) using a D flip-flop;
(b) using a J-K flip-flop.

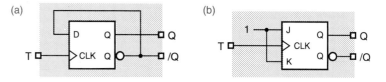

T flip-flop with enable

In many applications of T flip-flops, the flip-flop need not be toggled on every clock tick. Such applications can use a *T flip-flop with enable*. As shown in Figure 7–30, the flip-flop changes state at the triggering edge of the clock only if the enable signal E is asserted. Like the D, J, and K inputs on other edge-triggered flip-flops, the E input must meet specified setup and hold times with respect to the triggering clock edge. The circuits of Figure 7–29 are easily modified to provide an E input, as shown in Figure 7–31.

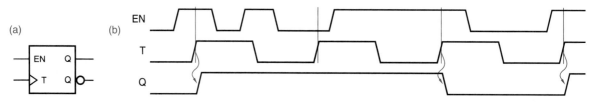

Figure 7–30   Positive-edge-triggered T flip-flop with enable: (a) logic symbol; (b) functional behavior.

Figure 7–31
Possible circuits for a
T flip-flop with enable:
(a) using a D flip-flop;
(b) using a J-K flip-flop.

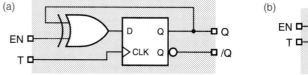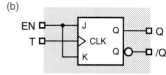

7.3 CLOCKED SYNCHRONOUS STATE-MACHINE ANALYSIS

Although latches and flip-flops, the basic building blocks of sequential circuits, are themselves feedback sequential circuits that can be formally analyzed, we'll first study the operation of *clocked synchronous state machines*, since they are the easiest to understand. "State machine" is a generic name given to these sequential circuits; "clocked" refers to the fact that their storage elements (flip-flops) employ a clock input; and "synchronous" means that all of the flip-flops use the same clock signal. Such a state machine changes state only when a triggering edge or "tick" occurs on the clock signal.   *clocked synchronous state machine*

7.3.1 State-Machine Structure

Figure 7–32 shows the general structure of a clocked synchronous state machine. The *state memory* is a set of *n* flip-flops that store the current state of the machine, and has 2^n distinct states. The flip-flops are all connected to a common clock signal that causes the flip-flops to change state at each *tick* of the clock. What constitutes a tick depends on the flip-flop type (edge triggered, pulse triggered, etc.). For the positive-edge-triggered D and J-K flip-flops considered in this section, a tick is the rising edge of the clock signal.   *state memory*   *tick*

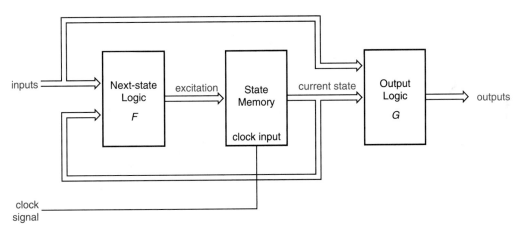

Figure 7–32 Clocked synchronous state-machine structure (Mealy machine).

The next state of the state machine in Figure 7–32 is determined by the *next-state logic F* as a function of the current state and input. The *output logic G* determines the output as a function of the current state and input. Both *F* and *G* are strictly combinational logic circuits. We can write   *next-state logic*   *output logic*

$$\text{Next state} = F(\text{current state}, \text{input})$$
$$\text{Output} = G(\text{current state}, \text{input})$$

MEALY AND
MOORE
MACHINES

A sequential circuit whose output depends on both state and input as shown in Figure 7–32 is called a *Mealy machine*. In some sequential circuits, the output depends on the state alone:

$$\text{Output} \;=\; G(\text{current state})$$

Such a circuit is called a *Moore machine*, and its general structure is shown in Figure 7–33.

Obviously, the only difference between the two state-machine models is in how outputs are generated. In practice, most state machines must be categorized as Mealy machines, because they have one or more *Mealy-type outputs* that depend on input as well as state. However, many of these same machines also have one or more *Moore-type outputs* that depend only on state. We'll find the distinction between output types to be important when we study the design of PLD-based state machines in Section 10.7.

State machines may use positive-edge-triggered D flip-flops for their state memory, in which case a tick occurs at each rising edge of the clock signal. It is also possible for the state memory to use negative-edge-triggered D flip-flops, D latches, or J-K flip-flops. However, inasmuch as most state machines are designed nowadays using programmable logic devices with positive-edge-triggered D flip-flops, that's what we'll concentrate on.

7.3.2 Characteristic Equations

characteristic equation

* suffix

The functional behavior of a latch or flip-flop can be described formally by a *characteristic equation* that specifies the flip-flop's next state as a function of its current state and inputs. The characteristic equations of the flip-flops in Section 7.2 are listed in Table 7–1. By convention, the $*$ suffix in $Q*$ means "the next value of Q." Notice that the characteristic equation does not describe

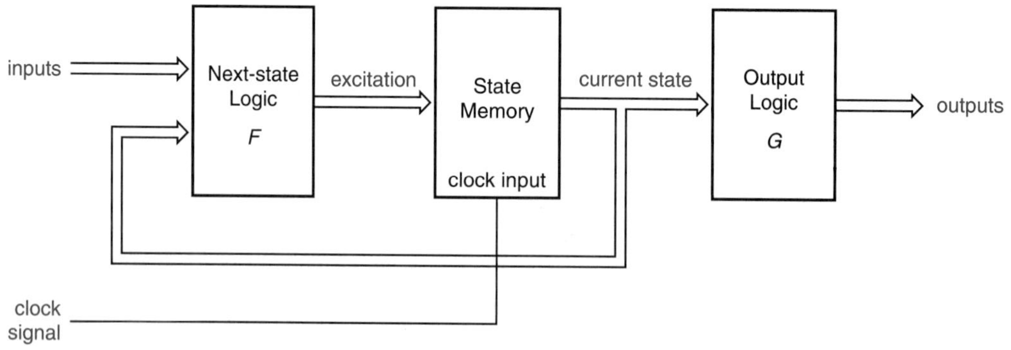

Figure 7–33 Clocked synchronous state-machine structure (Moore machine).

detailed timing behavior of the device (latching vs. edge-triggered, etc.), only the functional response to the control inputs. This simplified description is useful in the analysis of state machines, as we'll soon show.

Table 7-1
Latch and flip-flop characteristic equations.

Device Type	Characteristic Equation
S-R latch	$Q* = S + R' \cdot Q$
D latch	$Q* = D$
Edge-triggered D flip-flop	$Q* = D$
Master/slave S-R flip-flop	$Q* = S + R' \cdot Q$
Master/slave J-K flip-flop	$Q* = J \cdot Q' + K' \cdot Q$
Edge-triggered J-K flip-flop	$Q* = J \cdot Q' + K' \cdot Q$
T flip-flop	$Q* = Q'$
T flip-flop with enable	$Q* = EN \cdot Q' + EN' \cdot Q$

7.3.3 Analysis of State Machines with D Flip-Flops

Consider the formal definition of a state machine that we gave previously:

$$\text{Next state} = F(\text{current state, input})$$
$$\text{Output} = G(\text{current state, input})$$

Recalling our notion that "state" embodies all we need to know about the past history of the circuit, the first equation tells us that what we next need to know can be determined from what we currently know and the current input. The second equation tells us that the current output can be determined from the same information. The goal of sequential circuit analysis is to determine the next-state and output functions so that the behavior of a circuit can be predicted.

The analysis of a clocked synchronous state machine has three basic steps:

(1) Determine the next-state and output functions F and G.

(2) Use F and G to construct a *state/output table* that completely specifies the   *state/output table*
next state and output of the circuit for every possible combination of current state and input.

(3) (Optional) Draw a *state diagram* that presents the information from the   *state diagram*
previous step in graphical form.

Figure 7–34 shows a simple state machine with two positive-edge-triggered D flip-flops. To determine the next-state function F, we must first consider the behavior of the state memory. At the rising edge of the clock signal, each D flip-flop samples its D input and transfers this value to its Q output; the characteristic

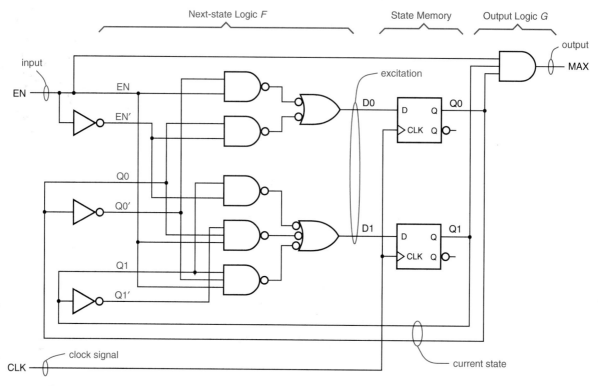

Figure 7–34   Clocked synchronous state machine using positive-edge-triggered D flip-flops.

equation of a D flip-flop is Q* = D. Therefore, in order to determine the next value of Q (i.e., Q*), we must first determine the current value of D.

In Figure 7–34 there are two D flip-flops, and we have named the signals on their outputs Q0 and Q1. These two outputs are the state variables; their value is the current state of the machine. We have named the signals on the corresponding D inputs D0 and D1. These signals provide the *excitation* for the D flip-flops at each clock tick. Logic equations that express the excitation signals as functions of the current state and input are called *excitation equations* and can be derived from the circuit diagram:

excitation

excitation equation

$$D0 = Q0 \cdot EN' + Q0' \cdot EN$$
$$D1 = Q1 \cdot EN' + Q1' \cdot Q0 \cdot EN + Q1 \cdot Q0' \cdot EN$$

By convention, the next value of a state variable after a clock tick is denoted by appending a star to the state-variable name, for example, Q0* or Q1*. Using the characteristic equation of D flip-flops, Q* = D, we can describe the next-state

* suffix

Table 7–2
Transition table for the state
machine in Figure 7–34.

Q1 Q0	EN	
	0	1
00	00	01
01	01	10
10	10	11
11	11	00
	Q1* Q0*	

function of the example machine with equations for the next value of the state variables:

$$Q0* = D0$$
$$Q1* = D1$$

Substituting the excitation equations for D0 and D1, we can write

$$Q0* = Q0 \cdot EN' + Q0' \cdot EN$$
$$Q1* = Q1 \cdot EN' + Q1' \cdot Q0 \cdot EN + Q1 \cdot Q0' \cdot EN$$

These equations, which express the next value of the state variables as a function of current state and input, are called *transition equations*.   *transition equation*

For each combination of current state and input value, the transition equations predict the next state. Each state is described by two bits, the current values of Q0 and Q1: (Q1 Q0) = 00, 01, 10, or 11. (The reason for "arbitrarily" picking the order (Q1 Q0) instead of (Q0 Q1) will become apparent shortly.) For each state, our example machine has only two possible input values, EN = 0 or EN = 1, so there are a total of 8 state/input combinations. (In general, a machine with s state bits and i inputs has 2^{s+i} state/input combinations.)

Table 7–2 shows a *transition table* that is created by evaluating the transition   *transition table*
equations for every possible state/input combination. Traditionally, a transition table lists the states along the left and the input combinations along the top of the table, as shown in the example.

The function of our example machine is apparent from its transition table—it is a 2-bit binary counter with an enable input EN. When EN=0 the machine maintains its current count, but when EN=1 the count advances by 1 at each clock tick, rolling over to 00 when it reaches a maximum value of 11.

If we wish, we may assign alphanumeric *state names* to each state. The   *state names*
simplest naming is 00 = A, 01 = B, 10 = C, and 11 = D. Substituting the state names for combinations of Q1 and Q0 (and Q1* and Q0*) in Table 7–2 produces the *state table* in Table 7–3. Here "S" denotes the current state, and "S*" denotes   *state table*
the next state of the machine. A state table is usually easier to understand than a

Table 7–3
State table for the state machine in Figure 7–34.

	EN	
S	0	1
A	A	B
B	B	C
C	C	D
D	D	A
	S*	

transition table because in complex machines we can use state names that have meaning. However, a state table contains less information than a transition table because it does not indicate the binary values assumed by the state variables in each named state.

Once a state table is produced, we have only the output logic of the machine left to analyze. In the example machine, there is only a single output signal, and it is a function of both current state and input (this is a Mealy machine). So we can write a single *output equation*:

output equation

$$\text{MAX} = \text{Q1} \cdot \text{Q0} \cdot \text{EN}$$

state/output table

The output behavior predicted by this equation can be combined with the next-state information to produce a *state/output table* as shown in Table 7–4.

Table 7–4
State/output table for the machine in Figure 7–34.

	EN	
S	0	1
A	A, 0	B, 0
B	B, 0	C, 0
C	C, 0	D, 0
D	D, 0	A, 1
	S*, MAX	

State/output tables for Moore machines are slightly simpler. For example, in the circuit of Figure 7–34 suppose we removed the EN signal from the AND gate that produces the MAX output, producing a Moore-type output MAXS. Then MAXS is a function of the state only, and the state/output table can list MAXS in a single column, independent of the input values. This is shown in Table 7–5.

state diagram
node
directed arc

A *state diagram* presents the information from the state/output table in a graphical format. It has one circle (or *node*) for each state, and an arrow (or *directed arc*) for each transition. Figure 7–35 shows the state diagram for our example state machine. The letter inside each circle is a state name. Each arrow

Table 7–5
State/output table for a Moore
machine.

	EN		
S	0	1	MAXS
A	A	B	0
B	B	C	0
C	C	D	0
D	D	A	1
		S*	

leaving a given state points to the next state for a given input combination. The
arrow also shows the output value produced in the given state for the given input
combination.

The state diagram for a Moore machine can be somewhat simpler. In

Figure 7–35
State diagram corresponding to
the state machine of Table 7–4.

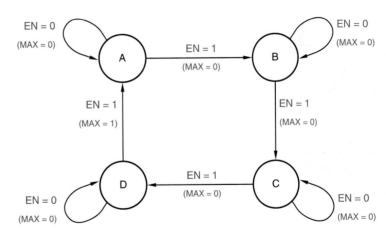

The state-diagram notation for output values in Mealy machines is a little mislead-   A CLARIFICATION
ing. You should remember that the listed output value is produced continuously
when the machine is in the indicated state and has the indicated input, not just
during the transition to the next state.

Since there is only one input in our example machine, there are only two possible   LITTLE ARROWS,
input combinations, and two arrows leaving each state. In a machine with n inputs,   LITTLE ARROWS
we would have 2^n arrows leaving each state. This is messy if n is large. Later, in   EVERYWHERE
Figure 7–40, we'll describe a convention whereby a state needn't have one arrow
leaving it for each input combination, only one arrow for each different next state.

Figure 7–36
State diagram corresponding to
the state machine of Table 7–5.

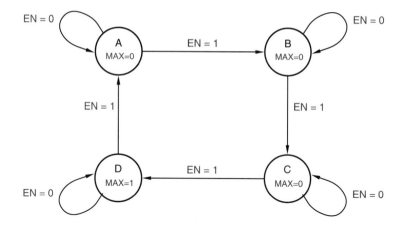

TIMING
DIAGRAMS

Using the transition, state, and output tables, we can construct a timing diagram
that show the behavior of a state machine for any desired starting state and input
sequence. For example, Figure 7–37 shows the behavior of our example machine
with a starting state of 00 (A) and a particular pattern on the EN input.

Notice that the value of the EN input affects the next state only at the rising
edge of the CLOCK input; that is, the counter counts only if EN=1 at the rising
edge of CLOCK. On the other hand, since MAX is a Mealy-type output, its value
is affected by EN at all times. If we also provide a Moore-type output MAXS as
suggested in the text, its value depends only on state as shown in the figure.

The timing diagram is drawn in a way that shows changes in the MAX and
MAXS outputs occurring slightly later than the state and input changes that cause
them, reflecting the combinational-logic delay of the output circuits. Naturally, the
drawings are merely suggestive; precise timing is normally indicated by a timing
table of the type suggested in Section 5.2.1.

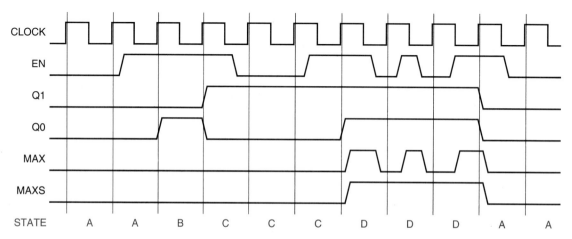

Figure 7–37 Timing diagram for example state machine.

this case, the output values can be shown inside each state circle, since they are functions of state only. The state diagram for a Moore machine using this convention is shown in Figure 7–36.

The original logic diagram of our example state machine, Figure 7–34, was laid out to match our conceptual model of a Mealy machine. However, nothing requires us to group the next-state logic, state memory, and output logic in this way. Figure 7–38 shows another logic diagram for the same state machine. To analyze this circuit, the designer (or analyzer, in this case) can still extract the required information from the diagram as drawn. The only circuit difference in the new diagram is that we have used the flip-flops' /Q outputs (which are normally the complement of Q) to save a couple of inverters.

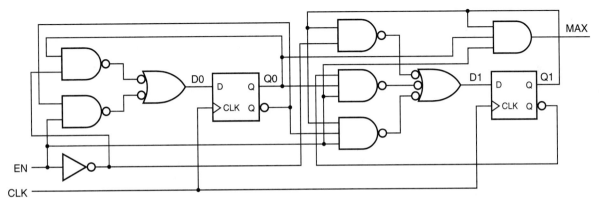

Figure 7–38 Redrawn logic diagram for a clocked synchronous state machine.

In summary, the detailed steps for analyzing a clocked synchronous state machine are as follows:

(1) Determine the excitation equations for the flip-flop control inputs.   *excitation equations*

(2) Substitute the excitation equations into the flip-flop characteristic equations   *transition equations*
    to obtain transition equations.

(3) Use the transition equations to construct a transition table.   *transition table*

(4) Determine the output equations.   *output equations*

(5) Add output values to the transition table for each state (Moore) or state/input   *transition/output*
    combination (Mealy) to create a transition/output table.   *table*

(6) Name the states and substitute state names for state-variable combinations   *state names*
    in the transition/output table to obtain a state/output table.   *state/output table*

(7) (Optional) Draw a state diagram corresponding to the state/output table.   *state diagram*

We'll go through this complete sequence of steps to analyze another clocked synchronous state machine, shown in Figure 7–39. Reading the logic diagram, we find that the excitation equations are as follows:

$$D0 = Q1' \cdot X + Q0 \cdot X' + Q2$$
$$D1 = Q2' \cdot Q0 \cdot X + Q1 \cdot X' + Q2 \cdot Q1$$
$$D2 = Q2 \cdot Q0' + Q0' \cdot X' \cdot Y$$

Substituting into the characteristic equation for D flip-flops, we obtain the transition equations:

$$Q0* = Q1' \cdot X + Q0 \cdot X' + Q2$$
$$Q1* = Q2' \cdot Q0 \cdot X + Q1 \cdot X' + Q2 \cdot Q1$$
$$Q2* = Q2 \cdot Q0' + Q0' \cdot X' \cdot Y$$

A transition (and output) table based on these equations is shown in Table 7–6.

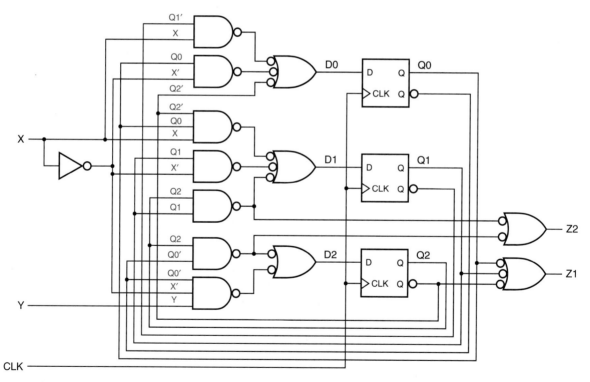

Figure 7–39 A clocked synchronous state machine with three flip-flops and eight states.

Table 7–6
Transition and output table
for the state machine in
Figure 7–39.

	X Y				
Q2 Q1 Q0	00	01	10	11	Z1 Z2
000	000	100	001	001	10
001	001	001	011	011	10
010	010	110	000	000	10
011	011	011	010	010	00
100	101	101	101	101	11
101	001	001	001	001	10
110	111	111	111	111	11
111	011	011	011	011	11
	Q2* Q1* Q0*				

Reading the logic diagram, we can write two output equations:

$$Z1 = Q2 + Q1' + Q0'$$
$$Z2 = Q2 \cdot Q1 + Q2 \cdot Q0'$$

The resulting output values are shown in the last column of Table 7–6. Assigning state names A–H, we obtain the state/output table shown in Table 7–7.

A state diagram for the example machine is shown in Figure 7–40. Since our example is a Moore machine, the output values are written with each state. Each arc is labeled with a *transition expression*; a transition is taken for input combinations for which the transition expression is 1. Transitions labeled "1" are always taken.

transition expression

Table 7–7
State/output table for the state
machine in Figure 7–39.

	X Y				
S	00	01	10	11	Z1 Z2
A	A	E	B	B	10
B	B	B	D	D	10
C	C	G	A	A	10
D	D	D	C	C	00
E	F	F	F	F	11
F	B	B	B	B	10
G	H	H	H	H	11
H	D	D	D	D	11
	S*				

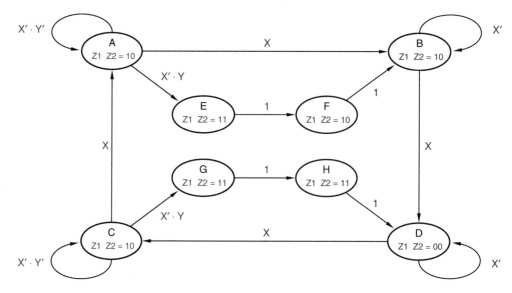

Figure 7–40   State diagram corresponding to Table 7–7.

The transition expressions on arcs leaving a particular state must be mutually exclusive and all inclusive, for the following reasons:

mutual exclusion

- No two transition expressions evaluate to 1 for the same input combination, since a machine cannot have two next states for one input combination.

all inclusion

- For every possible input combination, some transition expression is 1, so that all next states are defined.

Starting with the state table, a transition expression for a particular current state and next state can be written as a sum of minterms for the input combinations that cause that transition. If desired, the expression can then be minimized to give the information in a more compact form. Transition expressions are most useful in the *design* of state machines, where the expressions may be developed from the word description of the problem, as we'll show in Section 8.3.

* 7.3.4   Analysis of State Machines with J-K Flip-Flops

Clocked synchronous state machines built from J-K flip-flops can also be analyzed by the basic procedure in the preceding subsection. The only difference is that there are two excitation equations for each flip-flop—one for J and the other for K. To obtain the transition equations, both of these must be substituted into the J-K's characteristic equation, $Q* = J \cdot Q' + K' \cdot Q$.

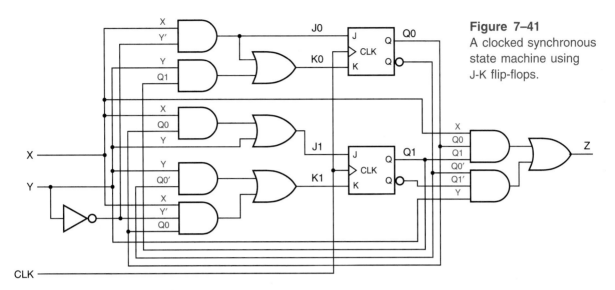

Figure 7–41
A clocked synchronous
state machine using
J-K flip-flops.

Figure 7–41 is an example state machine using J-K flip-flops. Reading the
logic diagram, we can derive the following excitation equations:

$$J0 = X \cdot Y'$$
$$K0 = X \cdot Y' + Y \cdot Q1$$
$$J1 = X \cdot Q0 + Y$$
$$K1 = Y \cdot Q0' + X \cdot Y' \cdot Q0$$

Substituting into the characteristic equation for J-K flip-flops, we obtain the transition equations:

$$
\begin{aligned}
Q0* &= J0 \cdot Q0' + K0' \cdot Q0 \\
&= X \cdot Y' \cdot Q0' + (X \cdot Y' + Y \cdot Q1)' \cdot Q0 \\
&= X \cdot Y' \cdot Q0' + X' \cdot Y' \cdot Q0 + X' \cdot Q1' \cdot Q0 + Y \cdot Q1' \cdot Q0 \\
Q1* &= J1 \cdot Q1' + K1' \cdot Q1 \\
&= (X \cdot Q0 + Y) \cdot Q1' + (Y \cdot Q0' + X \cdot Y' \cdot Q0)' \cdot Q1 \\
&= X \cdot Q1' \cdot Q0 + Y \cdot Q1' + X' \cdot Y' \cdot Q1 + Y' \cdot Q1 \cdot Q0' + X' \cdot Q1 \cdot Q0 + Y \cdot Q1 \cdot Q0
\end{aligned}
$$

A transition (and output) table based on these equations is shown in Table 7–8.
Reading the logic diagram, we can write the output equation:

$$Z = X \cdot Q1 \cdot Q0 + Y \cdot Q1' \cdot Q0'$$

The resulting output values are shown in each column of Table 7–8 along with
the next state. Assigning state names A–D, we obtain the state/output table shown
in Table 7–9. A state diagram is shown in Figure 7–42.

Table 7–8
Transition and output table
for the state machine in
Figure 7–41.

	X Y			
Q1 Q0	00	01	10	11
00	00, 0	10, 1	01, 0	10, 1
01	01, 0	11, 0	10, 0	11, 0
10	10, 0	00, 0	11, 0	00, 0
11	11, 0	10, 0	00, 1	10, 1
		Q1* Q0*, Z		

Table 7–9
State/output table for the state
machine in Figure 7–41.

	X Y			
S	00	01	10	11
A	A, 0	C, 1	B, 0	C, 1
B	B, 0	D, 0	C, 0	D, 0
C	C, 0	A, 0	D, 0	A, 0
D	D, 0	C, 0	A, 1	C, 1
		S*, Z		

Figure 7–42
State diagram corre-
sponding to the state
machine of Table 7–9.

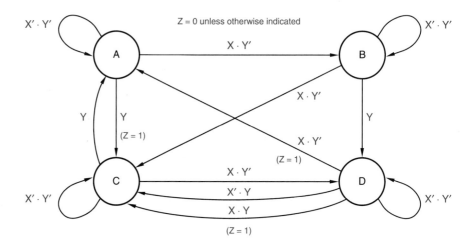

ANOTHER WAY
TO ANALYZE J-K

> There's another way to obtain the transition table for a state machine based on J-K flip-flops. Instead of substituting the excitation equations algebraically into the characteristic equation, you can construct an excitation table that shows the excitation values (J0, K0, J1, K1) for each state/input combination. You can then apply the characteristic equation individually to each state/excitation entry to obtain the transition table.

7.4 CLOCKED SYNCHRONOUS STATE-MACHINE DESIGN

The steps for designing a clocked synchronous state machine, starting from a word description or specification, are just about the reverse of the analysis steps that we used in the preceding section:

(1) Construct a state/output table corresponding to the word description or spec- *state/output table*
    ification, using mnemonic names for the states. (It's also possible to start
    with a state diagram; this method is discussed in Section 8.3.)

(2) (Optional) Minimize the number of states in the state/output table. *state minimization*

(3) Choose a set of state variables and assign state-variable combinations to *state assignment*
    the named states.

(4) Substitute the state-variable combinations into the state/output table to cre- *transition/output*
    ate a transition/output table that shows the desired next state-variable com- *table*
    bination and output for each state/input combination.

(5) Choose a flip-flop type (e.g., D or J-K) for the state memory. In most cases,
    you'll already have a choice in mind at the outset of the design, but this
    step is your last chance to change your mind.

(6) Construct an excitation table that shows the excitation values required to *excitation table*
    obtain the desired next state for each state/input combination.

(7) Derive excitation equations from the excitation table. *excitation equations*

(8) Derive output equations from the transition/output table. *output equations*

(9) Draw a logic diagram that shows the state-variable storage elements and re- *logic diagram*
    alizes the required excitation and output equations. (Or realize the equations
    directly in a programmable logic device.)

   In this section, we'll describe each of these basic steps in state-machine design. Step 1 is the most important, since it is here that the designer really *designs*, going through the creative process of translating a (perhaps ambiguous) *design*
English-language description of the state machine into a formal tabular description. Step 2 is hardly ever performed by experienced digital designers, but designers bring much of their experience to bear in step 3.
   Once the first three steps are completed, all of the remaining steps can be completed by "turning the crank," that is, by following a well-defined synthesis procedure. We'll discuss all of the steps in the rest of this section.

LET YOUR
COMPUTER DO
THE CRANKING

All nine steps of the state-machine design procedure are discussed in the remainder of this section. Steps 4–9 are the most tedious, but they are easily automated. For example, when you design a state machine that will be realized in a programmable logic device, you can use the PLD compiler to do the cranking, as shown in Section 10.7. Still, it's important for you to understand the details of the synthesis procedure, both to give you an appreciation of the compiler's function and to give you a chance of figuring out what's going on when the compiler yields unexpected results.

7.4.1 State-Table Design Example

There are several different ways to describe a state machine's state table. In later chapters, we'll see tools such as ASM charts and state-machine description language that specify a state table indirectly. In this chapter, however, we deal only with state tables that are specified directly, in the same tabular format that we used in the previous section for analysis.

We'll present the state-table design process, as well as the synthesis procedure in later subsections, using the simple design problem below:

Design a clocked synchronous state machine with two inputs, A and B, and a single output Z that is 1 if:

- A had the same value at each of the two previous clock ticks, *or*

- B has been 1 since the last time that the first condition was true.

Otherwise, the output should be 0.

If the meaning of this specification isn't crystal clear to you at this point, don't worry. Part of your job as a designer is to convert such a specification into a state table that is absolutely unambiguous; even if it doesn't match what was originally intended, it at least forms a basis for further discussion and refinement.

As an additional "hint" or requirement, state-table design problems often include timing diagrams that show the state machine's expected behavior for one or more sequences of inputs. Such a timing diagram is unlikely to specify unambiguously the machine's behavior for all possible sequences of inputs but, again, it's a good starting point for discussion and a benchmark against which proposed designs can be checked. Figure 7–43 is such a timing diagram for our example state-table design problem.

The first step in the state-table design is to construct a template. From the word description, we know that our example is a Moore machine—its output depends only on the current state, that is, what happened in previous clock periods. Thus, as shown in Figure 7–44(a), we provide one next-state column for each possible input combination and a single column for the output values.

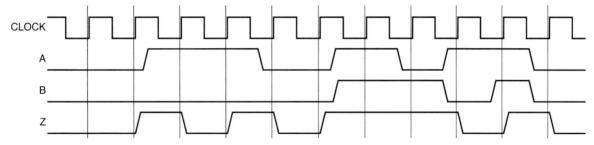

Figure 7–43   Timing diagram for example state machine.

The order in which the input combinations are written doesn't affect this part of the process, but we've written them in Karnaugh-map order to simplify the derivation of excitation equations later. In a Mealy machine we would omit the output column and write the output values along with the next-state values under each input combination. The leftmost column is simply an English-language reminder of the meaning of each state or the "history" associated with it.

   The word description isn't specific about what happens when this machine is first started, so we'll just have to improvise. We'll assume that when power is first applied to the system, the machine enters an *initial state*, called INIT in this example. We write the name of the initial state (INIT) in the first row, and leave room for enough rows (states) to complete the design. We can also fill in the value of Z for the INIT state; common sense says it should be 0 because there were *no* inputs beforehand.

initial state

For proper system operation, the hardware design of a state machine should ensure that it enters a known initial state on power-up, such as the INIT state in our design example. Most systems have a RESET signal that is asserted during power-up.

   The RESET signal is typically generated by an analog circuit. Such a reset circuit typically detects a voltage (say, 4.5 V) close to the power supply's full voltage, and follows that with a delay (say, 100 mS) to ensure that all components (including oscillators) have had time to stabilize before it "unresets" the system. The Texas Instruments TL7705 is such an analog reset IC; it has an internal 4.5-V reference for the detector and uses an external resistor and capacitor to determine the "unreset" time constant.

   If a state machine is built using discrete flip-flops with asynchronous preset and clear inputs, the RESET signal can be applied to these inputs to force the machine into the desired initial state. If preset and clear inputs are not available, or if reset must be synchronous (as in systems using high-speed microprocessors such as the Intel 80386 and 80486), then the RESET signal may be used as another input to the state machine, with all of the next-state entries going to the desired initial state when RESET is asserted.

REALIZING
RELIABLE
RESET

(a)

Meaning	S	00	01	11	10	Z
Initial state	INIT					0
. . .						
. . .						
. . .						

A B (columns 00 01 11 10)

S*

(b)

Meaning	S	00	01	11	10	Z
Initial state	INIT	A0	A0	A1	A1	0
Got a 0 on A	A0					0
Got a 1 on A	A1					0

A B (columns 00 01 11 10)

S*

(c)

Meaning	S	00	01	11	10	Z
Initial state	INIT	A0	A0	A1	A1	0
Got a 0 on A	A0	OK	OK	A1	A1	0
Got a 1 on A	A1					0
Got two equal A inputs	OK					1

A B (columns 00 01 11 10)

S*

(d)

Meaning	S	00	01	11	10	Z
Initial state	INIT	A0	A0	A1	A1	0
Got a 0 on A	A0	OK	OK	A1	A1	0
Got a 1 on A	A1	A0	A0	OK	OK	0
Got two equal A inputs	OK					1

A B (columns 00 01 11 10)

S*

Figure 7–44   Evolution of a state table.

Next, we must fill in the next-state entries for the INIT row. The Z output can't be 1 until we've seen at least two inputs on A, so we'll provide two states, A0 and A1, that "remember" the value of A on the previous clock tick, as shown in Figure 7–44(b). In both of these states, Z is 0, since we haven't satisified the conditions for a 1 output yet. The precise meaning of state A0 is "Got $A = 0$ on the previous tick, $A \neq 0$ on the tick before that, and $B \neq 1$ at some time since the previous pair of equal A inputs." State A1 is defined similarly.

At this point we know that our state machine has at least three states, and we have created two more blank rows to fill in. Hmmmm, this isn't such a good trend! In order to fill in the next-state entries for *one* state (INIT), we had to create *two* new states A0 and A1. If we kept going this way, we could end up with 65,535 states by bedtime! Instead, we should be on the lookout for existing states that have the same meaning as new ones that we might otherwise create. Let's see how it goes.

In state A0, we know that input A was 0 at the previous clock tick. Therefore, if A is 0 again, we go to a new state OK with $Z = 1$, as shown in Figure 7–44(c). If A is 1, then we don't have two equal inputs in a row, so we go to state A1 to remember that we just got a 1. Likewise in state A1, shown in (d), we go to OK if we get a second 1 input in a row, or to A0 if we get a 0.

Once we get into the OK state, the machine description tells us we can stay there as long as $B = 1$, irrespective of the A input, as shown in Figure 7–45(a). If $B = 0$, we have to look for two 1s or two 0s in a row on A again. However, we've got a little problem in this case. The current A input may or may not be

(a)

Meaning	S	A B 00	01	11	10	Z
Initial state	INIT	A0	A0	A1	A1	0
Got a 0 on A	A0	OK	OK	A1	A1	0
Got a 1 on A	A1	A0	A0	OK	OK	0
Got two equal A inputs	OK	?	OK	OK	?	1
			S*			

(b)

Meaning	S	A B 00	01	11	10	Z
Initial state	INIT	A0	A0	A1	A1	0
Got a 0 on A	A0	OK0	OK0	A1	A1	0
Got a 1 on A	A1	A0	A0	OK1	OK1	0
Two equal, A=0 last	OK0					1
Two equal, A=1 last	OK1					1
			S*			

(c)

Meaning	S	A B 00	01	11	10	Z
Initial state	INIT	A0	A0	A1	A1	0
Got a 0 on A	A0	OK0	OK0	A1	A1	0
Got a 1 on A	A1	A0	A0	OK1	OK1	0
Two equal, A=0 last	OK0	OK0	OK0	OK1	A1	1
Two equal, A=1 last	OK1					1
			S*			

(d)

Meaning	S	A B 00	01	11	10	Z
Initial state	INIT	A0	A0	A1	A1	0
Got a 0 on A	A0	OK0	OK0	A1	A1	0
Got a 1 on A	A1	A0	A0	OK1	OK1	0
Two equal, A=0 last	OK0	OK0	OK0	OK1	A1	1
Two equal, A=1 last	OK1	A0	OK0	OK1	OK1	1
			S*			

Figure 7–45 Continued evolution of a state table.

the second equal input in a row, so we may still be "OK" or we may have to go back to A0 or A1. We defined the OK state too broadly—it doesn't "remember" enough to tell us which way to go.

The problem is solved in Figure 7–45(b) by splitting OK into two states, OK0 and OK1, that "remember" the previous A input. All of the next states for OK0 and OK1 can be selected from existing states, as shown in (c) and (d). For example, if we get A = 0 in OK0, we can just stay in OK0; we don't have to create a new state that "remembers" three 0s in a row, because the machine's description doesn't require us to distinguish that case. Thus, we have achieved "closure" of the state table, which now describes a *finite*-state machine. As a sanity check, Figure 7–46 repeats the timing diagram of Figure 7–43, listing the states that should be visited according to our final state table.

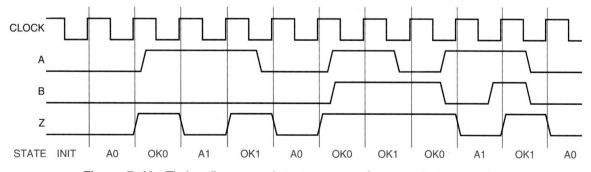

Figure 7–46 Timing diagram and state sequence for example state machine.

STATE-TABLE
DESIGN AS
A KIND OF
PROGRAMMING

Designing a state table (or equivalently, a state diagram) is a creative process that
is like writing a computer program in many ways:

- You start with a fairly precise description of inputs and outputs, but a possibly
  ambiguous description of the desired relationship between them, and usually
  no clue about how to actually obtain the desired outputs from the inputs.

- During the design, you may have to identify and choose among different ways
  of doing things, sometimes using common sense, and sometimes arbitrarily.

- You may have to identify and handle special cases that weren't included in
  the original description.

- You will probably have to keep track of several ideas in your head during
  the design process.

- Since the design process is not an algorithm, there's no guarantee that you
  can complete the state table or program using a finite number of states or
  lines of code. However, unless you work for the government, you must try
  to do so.

- When you finally run the state machine or program, it will do exactly what
  you told it to do—no more, no less.

- There's no guarantee that the thing will work the first time; you may have
  to debug and iterate on the whole process.

Although state-table design is a challenge, there's no need to be intimidated. If
you've made it this far, then you've probably written a few programs that worked,
and you can become just as good at designing state tables.

7.4.2   State Minimization

Figure 7–45(d) is a "minimal" state table for our original word description, in the
sense that it contains the fewest possible states. However, Figure 7–47 shows
other state tables, with more states, that also do the job. Formal procedures
can be used to minimize the number of states in such tables. The basic idea of
equivalent states   formal minimization procedures is to identify *equivalent states*, where two states
are equivalent if it is impossible to distinguish the states by observing only the
current and future *outputs* of the machine (and *not* the internal state variables).
A pair of equivalent states can be replaced by a single state.

It can be proved that two states S1 and S2 are equivalent if two conditions
are true. First, S1 and S2 must produce the same values at the state-machine
output(s); in a Mealy machine, this must be true for all input combinations.
Second, for each input combination, S1 and S2 must have the same next state or
equivalent next states.

Thus, a formal state-minimization procedure shows that states OK00 and

(a)

Meaning	S	AB 00	01	11	10	Z
Initial state	INIT	A0	A0	A1	A1	0
Got a 0 on A	A0	OK00	OK00	A1	A1	0
Got a 1 on A	A1	A0	A0	OK11	OK11	0
Got 00 on A	OK00	OK00	OK00	OKA1	A1	1
Got 11 on A	OK11	A0	OKA0	OK11	OK11	1
OK, got a 0 on A	OKA0	OK00	OK00	OKA1	A1	1
OK, got a 1 on A	OKA1	A0	OKA0	OK11	OK11	1
		S*				

(b)

Meaning	S	AB 00	01	11	10	Z
Initial state	INIT	A0	A0	A1	A1	0
Got a 0 on A	A0	OK00	OK00	A1	A1	0
Got a 1 on A	A1	A0	A0	OK11	OK11	0
Got 00 on A	OK00	OK00	OK00	A001	A1	1
Got 11 on A	OK11	A0	A110	OK11	OK11	1
Got 001 on A, B=1	A001	A0	AE10	OK11	OK11	1
Got 110 on A, B=1	A110	OK00	OK00	AE01	A1	1
Got bb...10 on A, B=1	AE10	OK00	OK00	AE01	A1	1
Got bb...01 on A, B=1	AE01	A0	AE10	OK11	OK11	1
		S*				

Figure 7–47 Nonminimal state tables equivalent to Figure 7–45(d).

OKA0 in Figure 7–47(a) are equivalent because they produce the same output and their next-state entries are identical. Since the states are equivalent, state OK00 may be eliminated and its occurrences in the table replaced by OKA0, or vice versa. Likewise, states OK11 and OKA1 are equivalent.

To minimize the state table in Figure 7–47(b), a formal procedure must use a bit of circular reasoning. States OK00, A110, and AE10 all produce the same output and have *almost* identical next-state entries, so they *might* be equivalent. They are equivalent only if A001 and AE01 are equivalent. Similarly, OK11, A001, and AE01 are equivalent only if A110 and AE10 are equivalent. In other words, the states in the first set are equivalent if the states in the second set are, and vice versa. So, let's just go ahead and say they're equivalent.

IS THIS REALLY ALL NECESSARY?

Details of formal state-minimization procedures are discussed in advanced textbooks, cited in the References. However, these procedures are seldom used by most digital designers. By carefully matching state meanings to the requirements of the problem, experienced digital designers produce state tables for small problems with a minimal or near-minimal number of states, without using a formal minimization procedure. Also, there are situations where *increasing* the number of states may simplify the design, so even an automated state-minimization procedure doesn't necessarily help. A designer can do more to simplify a state machine during the state-assignment phase of the design, discussed in the next subsection.

7.4.3 State Assignment

The next step in the design process is to determine how many binary variables are required to represent the states in the state table, and to assign a specific combination to each named state. We'll call the binary combination assigned to a particular state a *coded state*. The *total number of states* in a machine with *n*

coded state

total number of states

Table 7–10
State and output table for example problem.

	AB				
S	00	01	11	10	Z
INIT	A0	A0	A1	A1	0
A0	OK0	OK0	A1	A1	0
A1	A0	A0	OK1	OK1	0
OK0	OK0	OK0	OK1	A1	1
OK1	A0	OK0	OK1	OK1	1
	S*				

flip-flops is 2^n, so the number of flip-flops needed to code s states is $\lceil \log_2 s \rceil$, the smallest integer greater than or equal to $\log_2 s$.

unused states

For reference, the state/output table of our example machine is repeated in Table 7–10. It has five states, so it requires three flip-flops. Of course, three flip-flops provide a total of eight states, so there will be $8-5 = 3$ *unused states*. We'll discuss alternatives for handling the unused states at the end of this subsection. Right now, we have to deal with lots of choices for the five coded states.

CAUTION: MATH

The number of different ways to choose m coded states out of a set of n possible states is given by a *binomial coefficient*, denoted $\binom{n}{m}$, whose value is $\frac{n!}{m!\cdot(n-m)!}$. (We used binomial coefficients previously in Section 2.10, in the context of decimal coding.) In our example, there are $\binom{8}{5}$ different ways to choose five coded states out of eight possible states, and 5! ways to assign the five named states to each different choice. So there are $\frac{8!}{5!\cdot 3!} \cdot 5!$ or 6,720 different ways to assign the five states of our example machine to combinations of three binary state variables. We don't have time to look at all of them.

The simplest assignment of s coded states to 2^n possible states is to use the first s binary integers in binary counting order, as shown in the first assignment column of Table 7–11. However, the simplest state assignment does not always lead to the simplest excitation equations, output equations, and resulting logic circuit. In fact, the state assignment often has a major effect on circuit cost, and may interact with other factors, such as the choice of storage elements (e.g., D vs. J-K flip-flops) and the realization approach for excitation and output logic (e.g., sum-of-products, product-of-sums, or ad hoc).

So, how do we choose the best state assignment for a given problem? In general, the only formal way to find the *best* assignment is to try *all* the assignments. That's too much work, even for students. Instead, most digital designers rely on experience and several practical guidelines for making reasonable state assignments:

Table 7-11
Possible state assignments for
the state machine in Table 7-10.

State name	Assignment			
	Simplest Q1–Q3	Decomposed Q1–Q3	One-hot Q1–Q5	Almost one-hot Q1–Q4
INIT	000	000	00001	0000
A0	001	100	00010	0001
A1	010	101	00100	0010
OK0	011	110	01000	0100
OK1	100	111	10000	1000

- Choose an initial coded state into which the machine can easily be forced at reset (00...00 or 11...11 in typical circuits).

- Minimize the number of state variables that change on each transition.

- Maximize the number of state variables that don't change in a group of related states (i.e., a group of states in which most of transitions stay in the group).

- Exploit symmetries in the problem specification and the corresponding symmetries in the state table. That is, suppose that one state or group of states means almost the same thing as another. Once an assignment has been established for the first, a similar assignment, differing only in one bit, should be used for the second.

- If there are unused states (i.e., if $s < 2^n$ where $n = \lceil \log_2 s \rceil$), then choose the "best" of the available state-variable combinations to achieve the foregoing goals. That is, don't limit the choice of coded states to the first s n-bit integers.

- Decompose the set of state variables into individual bits or fields where each bit or field has a well-defined meaning with respect to the input effects or output behavior of the machine.

- Consider using more than the minimum number of state variables to make a decomposed assignment possible.

Some of these ideas are incorporated in the "decomposed" state assignment in Table 7-11. As before, the initial state is 000, which is easy to force either asynchronously (applying the RESET signal to the flip-flop CLR inputs) or synchronously (by AND'ing RESET' with all of the D flip-flop inputs). After this point, the assignment takes advantage of the fact that there are only four states in addition to INIT, which is a fairly "special" state that is never re-entered once

INITIAL VERSUS
IDLE STATES

> The example state machine in this subsection visits its initial state only during reset. Many machines are designed instead with an "idle" state that is entered both at reset and whenever the machine has nothing in particular to do.

the machine gets going. Therefore, Q1 can be used to indicate whether or not the machine is in the INIT state, and Q2 and Q3 can be used to distinguish among the four non-INIT states.

The non-INIT states in the "decomposed" column of Table 7–11 appear to have been assigned in binary counting order, but that's just a coincidence. State bits Q2 and Q3 actually have individual meanings in the context of the state machine's inputs and output. Q3 gives the previous value of A, and Q2 indicates that the conditions for a 1 output are satisfied in the current state. By decomposing the state-bit meanings in this way, we can expect the next-state and output logic to be simpler than in a "random" assignment of Q2,Q3 combinations to the non-INIT states. We'll continue the state-machine design based on this assignment in later subsections.

one-hot assignment

Another useful state assignment, one that can be adapted to any state machine, is the *one-hot assignment* shown in Table 7–11. This assignment uses more than the minimum number of state variables—it uses one bit per state. In addition to being simple, a one-hot assignment has the advantage of usually leading to small excitation equations, since each flip-flop must be set to 1 for transitions into only one state. An obvious disadvantage of a one-hot assignment, especially for machines with many states, is that it requires (many) more than the minimum number of flip-flops. However, the one-hot encoding is ideal for a machine with s states that is required to have a set of 1-out-of-s coded outputs indicating its current state. The one-hot-coded flip-flop outputs can be used directly for this purpose, with no additional combinational output logic.

The last column of Table 7–11 is an "almost one-hot assignment" that uses the "no-hot" combination for the initial state. This makes a lot of sense for two reasons: It's easy to initialize most storage devices to the all-0s state, and the initial state in this machine is never revisited once the machine gets going. Completing the state-machine design using this state assignment is considered in Exercises 7.31 and 7.34.

unused states

We promised earlier to consider the disposition of *unused states* when the number of states available with n flip-flops, 2^n, is greater than the number of states required, s. There are two approaches that make sense, depending on the application requirements:

- *Minimal risk.* This approach assumes that it is possible for the state machine somehow to get into one of the unused (or "illegal") states, perhaps because of a hardware failure, an unexpected input, or a design error. Therefore, all of the unused state-variable combinations are identified, and explicit next-state entries are made so that, for any input combination, the unused states go to the "initial" state, the "idle" state, or some other "safe" state. This is an automatic consequence of some design methodologies if the initial state is coded 00...00.

- *Minimal cost.* This approach assumes that the machine will never enter an unused state. Therefore, in the transition and excitation tables, the next-state entries of the unused states can be marked as "don't-cares." In most cases, this simplifies the excitation logic. However, the machine's behavior if it ever does enter an unused state may be pretty weird.

We'll look at both of these approaches as we complete the design of our example state machine.

7.4.4 Synthesis Using D Flip-Flops

Once we've assigned coded states to the named states of a machine, the rest of the design process is pretty much "turning the crank." In fact, in Section 10.7 we'll describe software tools that can turn the crank for you. Just so that you'll appreciate those tools, however, we'll go through the process by hand in this subsection.

Coded states are substituted for named states in the (possibly minimized) state table to obtain a *transition table*. The transition table shows the next coded state for each combination of current coded state and input. Table 7–12 shows the transition and output table that is obtained from the example state machine of Table 7–10 on page 488 using the "decomposed" assignment of Table 7–11 on page 489.   *transition table*

The next step is to write an *excitation table* that shows, for each combination   *excitation table* of coded state and input, the flip-flop excitation input values needed to make the machine go to the desired next coded state. This structure and content of this table depend on the type of flip-flops that are used (D, J-K, T, etc.). We *usually*

Table 7–12
Transition and output table for example problem.

Q1 Q2 Q3	A B 00	01	11	10	Z
000	100	100	101	101	0
100	110	110	101	101	0
101	100	100	111	111	0
110	110	110	111	101	1
111	100	110	111	111	1
	Q1* Q2* Q3*				

Table 7–13
Excitation and output table for
Table 7–12, using D flip-flops.

	A B				
Q1 Q2 Q3	00	01	11	10	Z
000	100	100	101	101	0
100	110	110	101	101	0
101	100	100	111	111	0
110	110	110	111	101	1
111	100	110	111	111	1
	D1 D2 D3				

have a particular flip-flop type in mind at the beginning of a design—and we
certainly do in this subsection, given its title. In fact, most state-machine designs
nowadays use D flip-flops, because of their availability in both discrete packages
and programmable logic devices, and because of their ease of use (compare with
J-K flip-flops in the next subsection).

Of all flip-flop types, a D flip-flop has the simplest characteristic equation,
$Q* = D$. Each D flip-flop in a state machine has a single excitation input, D,
and the excitation table must show the value required at each flip-flop's D input
for each coded-state/input combination. Table 7–13 shows the excitation table
for our example problem. Since $D = Q*$, the excitation table is identical to the
transition table, except for labeling of its entries. Thus, with D flip-flops, you
don't really need to write a separate excitation table; you can just call the first
table a *transition/excitation table*.

transition/excitation table

The excitation table is like a truth table for three combinational logic functions (D1, D2, D3) of five variables (A, B, Q1, Q2, Q3). Accordingly, we can
design circuits to realize these functions using any of the combinational design
methods at our disposal. In particular, we can transfer the information in the
excitation table to Karnaugh maps, which we may call *excitation maps*, and find
a minimal sum-of-products or product-of-sums expression for each function.

excitation maps

Excitation maps for our example state machine are shown in Figure 7–48.
Each function, such as D1, has five variables and therefore uses a *5-variable
Karnaugh map*. A 5-variable map is drawn as a pair of 4-variable maps, where
cells in the same position in the two maps are considered to be adjacent. These
maps are a bit unwieldy, but if you want to design by hand any but the most
trivial state machines, you're going to get stuck with 5-variable maps and worse.
At least we had the foresight to label the input combinations of the original state
table in Karnaugh-map order, which makes it easier to transfer information to the
maps in this step. However, note that the *states* were not assigned in Karnaugh-
map order; in particular, the rows for states 110 and 111 are in the opposite order
in the map as in the excitation table.

5-variable Karnaugh map

It is in this step, transferring the excitation table to excitation maps, that

Figure 7–48

Excitation maps for D1, D2, and
D3 assuming that unused states
go to state 000.

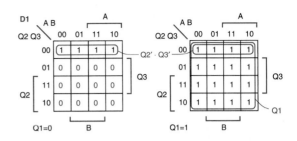

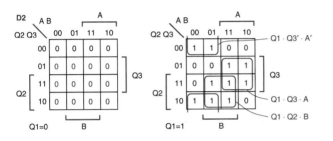

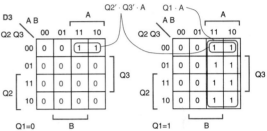

we discover why the excitation table is not quite a truth table—it does not
specify functional values for *all* input combinations. In particular, the next-state
information for the unused states, 001, 010, and 011, is not specified. Here we
must make a choice, discussed in the preceding subsection, between a minimal-
risk and a minimal-cost strategy for handling the unused states. Figure 7–48 has
taken the minimal-risk approach: The next state for each unused state and input
combination is 000, the INIT state. The three rows of colored 0s in each Karnaugh
map are the result of this choice. With the maps completely filled in, we can now
obtain minimal sum-of-products expressions for the flip-flop excitation inputs:

$$D1 = Q1 + Q2' \cdot Q3'$$
$$D2 = Q1 \cdot Q3' \cdot A' + Q1 \cdot Q3 \cdot A + Q1 \cdot Q2 \cdot B$$
$$D3 = Q1 \cdot A + Q2' \cdot Q3' \cdot A$$

An output equation can easily be developed directly from the information
in Table 7–13. The output equation is simpler than the excitation equations,
because the output is a function of state only. We could use a Karnaugh map,
but it's easy to find a minimal-risk output function algebraically, by writing it as
the sum of the two coded states (110 and 111) in which Z is 1:

$$Z = Q1 \cdot Q2 \cdot Q3' + Q1 \cdot Q2 \cdot Q3$$
$$= Q1 \cdot Q2$$

At this point, we're just about done with the state-machine design. If
the state machine is going to be built with discrete flip-flops and gates, then

the final step is to draw a logic diagram. On the other hand, if we are using a programmable logic device, then we only have to enter the excitation and output equations into a computer file that specifies how to program the device, as discussed in Section 10.7. Or, if we're lucky, we specified the machine using a high-level state-machine description language (Section 10.8) in the first place, and the computer did all the work in this subsection for us!

MINIMAL-COST
SOLUTION

If we choose in our example to derive minimal-cost excitation equations, we write "don't-cares" in the next-state entries for the unused states. The colored d's in Figure 7–49 are the result of this choice. The excitation equations obtained from this map are somewhat simpler than before:

$$D1 = 1$$
$$D2 = Q1 \cdot Q3' \cdot A' + Q3 \cdot A + Q2 \cdot B$$
$$D3 = A$$

For a minimal-cost output function, the value of Z is a "don't-care" for the unused states. This leads to an even simpler output function, $Z = Q2$. The logic diagram for the minimal-cost solution is shown in Figure 7–50.

Figure 7–49
Excitation maps for D1, D2, and D3 assuming that next states of unused states are "don't-cares."

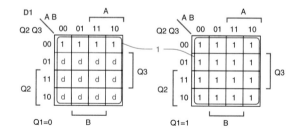

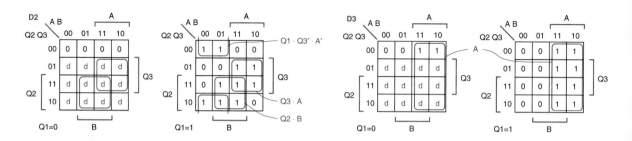

* 7.4.5 Synthesis Using J-K Flip-Flops

At one time, J-K flip-flops were popular for discrete SSI state-machine designs, since a J-K flip-flop embeds more functionality than a D flip-flop in the same

Figure 7–50
Logic diagram for example state
machine using D flip-flops and
minimal-cost excitation logic.

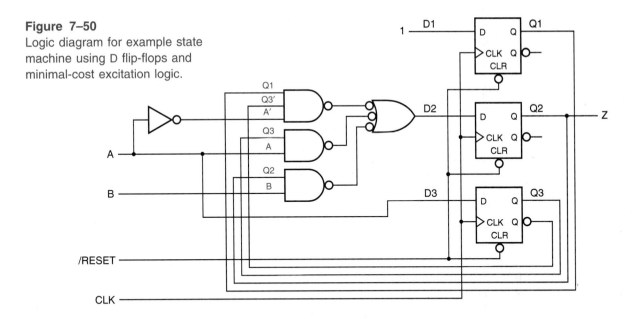

size SSI package. By "more functionality" we mean that the combination of J
and K inputs yields more possibilities for controlling the flip-flop than a single D
input does. As a result, a state machine's excitation logic may be simpler using
J-K flip-flops than using D flip-flops, reducing package count when SSI gates are
used for the excitation logic.

Up through the state-assignment step, the design procedure with J-K flip-
flops is basically the same as with D flip-flops. The only difference is that a
designer might select a slightly different state assignment, knowing the sort of
behavior that can easily be obtained from J-K flip-flops (e.g., "toggling" by setting
J and K to 1).

While minimizing excitation logic was a big deal in the days of SSI-based design,
the name of the game has changed with PLDs and ASICs. As you might guess
from your knowledge of the AND-OR structure of combinational PLDs, the need
to provide separate AND-OR arrays for the J and K inputs of a J-K flip-flop would
be a distinct disadvantage in a sequential PLD.

In ASIC technologies, a J-K flip-flop might require over 25% more chip area
than a D flip-flop. For example, in LSI Logic Corp.'s LCA10000 series of CMOS
gate arrays, an FJK1 J-K flip-flop macrocell uses 9 "gate cells," while an FD1 D
flip-flop macrocell uses only 7 gate cells. Therefore, a more cost-effective design
usually results from sticking with D flip-flops and using the extra chip area for
more complex excitation logic in just the cases where it's really needed.

Still, this subsection describes the J-K synthesis process "just for fun."

JUST FOR FUN

Table 7–14
Application table for J-K flip-flops.

Q	Q*	J	K
0	0	0	d
0	1	1	d
1	0	d	1
1	1	d	0

The big difference occurs in the derivation of an excitation table from the transition table. With D flip-flops, the two tables are identical; using the D's characteristic equation, $Q* = D$, we simply substitute $D = Q*$ for each entry. With J-K flip-flops, each entry in the excitation table has twice as many bits as in the transition table, since there are two excitation inputs per flip-flop.

A J-K flip-flop's characteristic equation, $Q* = J \cdot Q' + K' \cdot Q$, cannot be rearranged to obtain independent equations for J and K. Instead, the required values for J and K are expressed as functions of Q and Q* in a *J-K application table*, Table 7–14. According to the first row, if Q is currently 0, all that is required to obtain 0 as the next value of Q is to set J to 0; the value of K doesn't matter. Similarly, according to the third row, if Q is currently 1, the next value of Q will be 0 if K is 1, regardless of J's value. Each desired transition can be obtained by either of two different combinations on the J and K inputs, so we get a "don't-care" entry in each row of the application table.

J-K application table

To obtain a J-K excitation table, the designer must look at both the current and desired next value of each state bit in the transition table and substitute the corresponding pair of J and K values from the application table. For the transition table in Table 7–12 on page 491, these substitutions produce the excitation table in Table 7–15. For example, in state 100 under input combination 00, Q1 is 1 and the required Q1* is 1; therefore, "d0" is entered for J1 K1. For the same state/input combination, Q2 is 0 and Q2* is 1, so "1d" is entered for J2 K2. Obviously, it takes quite a bit of patience and care to fill in the entire excitation table (a job best left to a computer).

Table 7–15
Excitation and output table for Table 7–12, using J-K flip-flops.

Q1 Q2 Q3	A B 00	01	11	10	Z
	00	**01**	**11**	**10**	**Z**
000	1d, 0d, 0d	1d, 0d, 0d	1d, 0d, 1d	1d, 0d, 1d	0
100	d0, 1d, 0d	d0, 1d, 0d	d0, 0d, 1d	d0, 0d, 1d	0
101	d0, 0d, d1	d0, 0d, d1	d0, 1d, d0	d0, 1d, d0	0
110	d0, d0, 0d	d0, d0, 0d	d0, d0, 1d	d0, d1, 1d	1
111	d0, d1, d1	d0, d0, d1	d0, d0, d0	d0, d0, d0	1

J1 K1, J2 K2, J3 K3

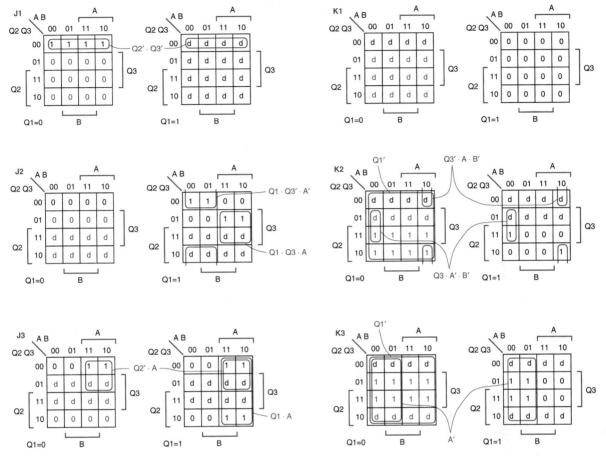

Figure 7–51   Excitation maps for J1, K1, J2, K2, J3, and K3, assuming that unused states go to state 000.

As in the D synthesis example of the preceding subsection, the excitation table is *almost* a truth table for the excitation functions. This information is transferred to Karnaugh maps in Figure 7–51.

The excitation table does not specify next states for the unused states, so once again we must choose between the minimal-risk and minimal-cost approaches. The colored entries in the Karnaugh maps result from taking the minimal-risk approach.

Note that even though the "safe" next state for unused states is 000, we didn't just put 0s in the corresponding map cells, as we were able to do in the D case. Instead, we still had to work with the application table to determine the proper combination of J and K needed to get Q* = 0 for each unused state entry, once again a tedious and error-prone process.

Using the maps in Figure 7–51, we can derive sum-of-products excitation equations:

$$
\begin{aligned}
J1 &= Q2' \cdot Q3' & K1 &= 0 \\
J2 &= Q1 \cdot Q3' \cdot A' + Q1 \cdot Q3 \cdot A & K2 &= Q1' + Q3' \cdot A \cdot B' + Q3 \cdot A' \cdot B' \\
J3 &= Q2' \cdot A + Q1 \cdot A & K3 &= Q1' + A'
\end{aligned}
$$

These equations take two more gates to realize than do the preceding subsection's minimal-risk equations using D flip-flops, so J-K flip-flops didn't save us anything in this example, least of all design time.

MINIMAL-COST
SOLUTION

In the preceding design example, excitation maps for the minimal-cost approach would have been somewhat easier to construct, since we could have just put d's in all of the unused state entries. Sum-of-products excitation equations obtained from the minimal-cost maps (not shown) are as follows:

$$
\begin{aligned}
J1 &= 1 & K1 &= 0 \\
J2 &= Q1 \cdot Q3' \cdot A' + Q3 \cdot A & K2 &= Q3' \cdot A \cdot B' + Q3 \cdot A' \cdot B' \\
J3 &= A & K3 &= A'
\end{aligned}
$$

The state encoding for the J-K circuit is the same as in the D circuit, so the output equation is the same, $Z = Q1 \cdot Q2$ for minimal risk, $Z = Q2$ for minimal cost.

A logic diagram corresponding to the minimal-cost equations is shown in Figure 7–52. This circuit has two more gates than the minimal-cost D circuit in Figure 7–50, so J-K flip-flops still didn't save us anything.

Figure 7–52
Logic diagram for example state machine using J-K flip-flops and minimal-cost excitation logic.

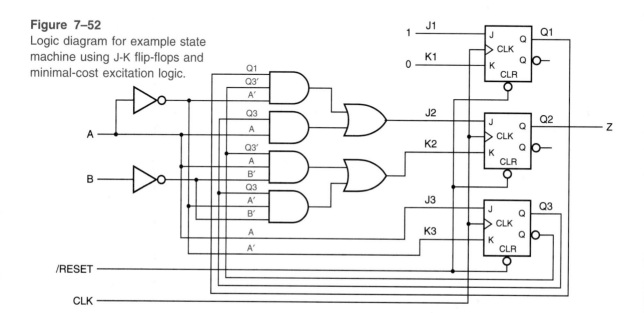

7.4.6 More Design Examples Using D Flip-Flops

We'll conclude this section with two more state-machine design examples using D flip-flops. The first example is a "1s-counting machine":

> Design a clocked synchronous state machine with two inputs, X and Y, and one output, Z. The output should be 1 if the number of 1 inputs on X and Y since reset is a multiple of 4, and 0 otherwise.

At first glance, you might think the machine needs an infinite number of states, to count 1 inputs over an arbitrarily long time. However, since the output indicates the number of inputs received *modulo 4*, four states are sufficient. We'll name the states S0–S3, where the total number of 1s received in state S*i* is *i* modulo 4, and S0 is the initial state. Table 7–16 is the resulting state and output table.

The 1s-counting machine can use two state variables to code its four states, with no unused states. In this case, there are only 4! possible assignments of coded states to named states. Still, we'll try only one of them. We'll assign coded states to the named states in Karnaugh-map order (00, 01, 11, 10) for two reasons: In this state table, it minimizes the number of state variables that change for most transitions, potentially simplifying the excitation equations; and it simplifies the mechanical transfer of information to excitation maps.

A transition/excitation table based on our chosen state assignment is shown in Table 7–17. Since we're using D flip-flops, the transition and excitation tables are the same. Corresponding Karnaugh maps for D1 and D2 are shown in Figure 7–53. Since there are no unused states, all of the information we need is

Table 7–16
State and output table for 1s-counting machine.

		XY				
Meaning	S	00	01	11	10	Z
Got zero 1s (modulo 4)	S0	S0	S1	S2	S1	1
Got one 1 (modulo 4)	S1	S1	S2	S3	S2	0
Got two 1s (modulo 4)	S2	S2	S3	S0	S3	0
Got three 1s (modulo 4)	S3	S3	S0	S1	S0	0
		S*				

Table 7–17
Transition/excitation and output table for 1s-counting machine.

	XY				
Q1 Q2	00	01	11	10	Z
00	00	01	11	01	1
01	01	11	10	11	0
11	11	10	00	10	0
10	10	00	01	00	0
	Q1* Q2* or D1 D2				

Figure 7–53
Excitation maps for D1 and D2 in
1s-counting machine.

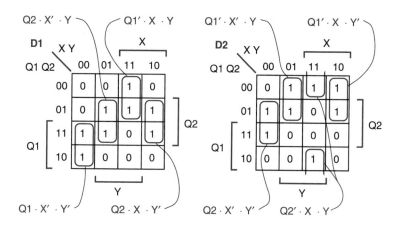

in the excitation table; no choice is required between minimal-risk and minimal-cost approaches. The excitation equations can be read from the maps, and the output equation can be read directly from the transition/excitation table:

$$D1 = Q2 \cdot X' \cdot Y + Q1' \cdot X \cdot Y + Q1 \cdot X' \cdot Y' + Q2 \cdot X \cdot Y'$$
$$D2 = Q1' \cdot X' \cdot Y + Q1' \cdot X \cdot Y' + Q2 \cdot X' \cdot Y' + Q2' \cdot X \cdot Y$$
$$Z = Q1' \cdot Q2'$$

A logic diagram using D flip-flops and AND-OR or NAND-NAND excitation logic can be drawn from these equations.

The second example is a "combination lock" state machine that activates an "unlock" output when a certain binary input sequence is received:

Design a clocked synchronous state machine with one input, X, and two outputs, UNLK and HINT. The UNLK output should be 1 if and only if X is 0 and the sequence of inputs received on X at the preceding seven clock ticks was 0110111. The HINT output should be 1 if and only if the current value of X is the correct one to move the machine closer to being in the "unlocked" state (with UNLK = 1).

It should be apparent from word description that this is a Mealy machine. The UNLK output depends on both the past history of inputs and X's current value, and HINT depends on both the state and the current X (indeed, if the current X produces HINT = 0, the user may want to change X before the clock tick).

A state and output table for the combination lock is presented in Table 7–18. In the initial state, A, we assume that we have received no inputs in the required sequence; we're looking for the first 0 in the sequence. Therefore, as long as we get 1 inputs, we stay in state A, and we move to state B when we receive a 0. In state B, we're looking for a 1. If we get it, we move on to C; if we

Table 7–18
State and output table for
combination-lock machine.

| | | X | |
Meaning	S	0	1
Got zip	A	B, 01	A, 00
Got 0	B	B, 00	C, 01
Got 01	C	B, 00	D, 01
Got 011	D	E, 01	A, 00
Got 0110	E	B, 00	F, 01
Got 01101	F	B, 00	G, 01
Got 011011	G	E, 00	H, 01
Got 0110111	H	B, 11	A, 00
		S∗, UNLK HINT	

don't, we can stay in B, since the 0 we just received might still turn out to be the first 0 in the required sequence. In each successive state, we move on to the next state if we get the correct input, and we go back to A or B if we get the wrong one. An exception occurs in state G; if we get the wrong input (a 0) there, the previous three inputs might still turn out to be the first three inputs of the required sequence, so we go back to state E instead of B. In state H, we've received the required sequence, so we set UNLK to 1 if X is 0. In each state, we set HINT to 1 for the value of X that moves us closer to state H.

The combination lock's eight states can be coded with three state variables, leaving no unused states. There are 8! state assignments to choose from. To keep things simple, we'll use the simplest, and assign the states in binary counting order, yielding the transition/excitation table in Table 7–19. Corresponding Karnaugh maps for D1, D2, and D3 are shown in Figure 7–54. The excitation equations can be read from the maps:

Table 7–19
Transition/excitation table for
combination-lock machine.

| | X | |
Q1 Q2 Q3	0	1
000	001, 01	000, 00
001	001, 00	010, 01
010	001, 00	011, 01
011	100, 01	000, 00
100	001, 00	101, 01
101	001, 00	110, 01
110	100, 00	111, 01
111	001, 11	000, 00
	Q1∗ Q2∗ Q3∗, UNLK HINT	

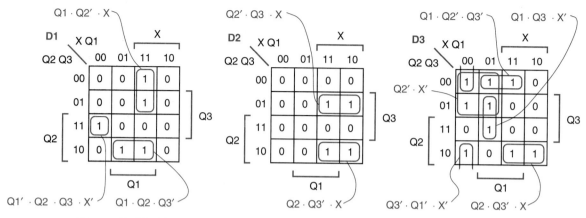

Figure 7–54   Excitation maps for D1, D2, and D3 in combination-lock machine.

$$D1 = Q1 \cdot Q2' \cdot X + Q1' \cdot Q2 \cdot Q3 \cdot X' + Q1 \cdot Q2 \cdot Q3'$$
$$D2 = Q2' \cdot Q3 \cdot X + Q2 \cdot Q3' \cdot X$$
$$D3 = Q1 \cdot Q2' \cdot Q3' + Q1 \cdot Q3 \cdot X' + Q2' \cdot X' + Q3' \cdot Q1' \cdot X' + Q2 \cdot Q3' \cdot X$$

The output values are transferred from the transition/excitation and output table to another set of maps in Figure 7–55. The corresponding output equations are:

$$UNLK = Q1 \cdot Q2 \cdot Q3 \cdot X'$$
$$HINT = Q1' \cdot Q2' \cdot Q3' \cdot X' + Q1 \cdot Q2' \cdot X + Q2' \cdot Q3 \cdot X + Q2 \cdot Q3 \cdot X' + Q2 \cdot Q3' \cdot X$$

Note that some product terms are repeated in the excitation and output equations, yielding a slight savings in the cost of the AND-OR realization. If we went to the trouble of performing a formal multiple-output minimization of all five excitation and output functions, we could save two more gates (see Exercise 7.47).

Figure 7–55
Karnaugh maps for output
functions UNLK and HINT in
combination-lock machine.

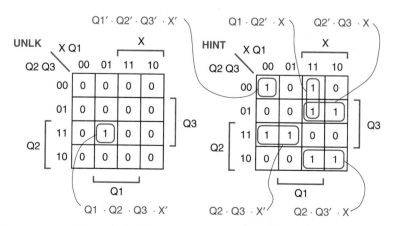

* 7.5   FEEDBACK SEQUENTIAL CIRCUITS

The simple bistable and the various latches and flip-flops that we studied earlier in this chapter are all feedback sequential circuits. Each has one or more feedback loops that, ignoring their behavior during state transitions, store a 0 or a 1 at all times. The feedback loops are memory elements, and the circuits' behavior depends on both the current inputs and the values stored in the loops.

Only rarely does a logic designer encounter a situation where a feedback sequential circuit must be analyzed or designed. The most commonly used feedback sequential circuits are the flip-flops and latches that are used as the building blocks in larger sequential-circuit designs. Their internal design and operating specifications are supplied by an IC manufacturer.

    Even an ASIC designer typically does not design gate-level flip-flop or latch circuits, since these elements are supplied in a "library" of commonly used functions in the particular ASIC technology.

    Still, you may be curious about how off-the-shelf flip-flops and latches "do their thing"; this section will show you how to analyze any such circuit built from conventional gates.

KEEP YOUR
FEEDBACK
TO YOURSELF

CMOS flip-flops typically use transmission gates to form their feedback loops, requiring a slightly different analysis procedure than what is described here; see the References.

CMOS
FLIP-FLOPS

---

* 7.5.1   Analysis

Feedback sequential circuits are the most common example of *fundamental-mode circuits*. In such circuits, inputs are not normally allowed to change simultaneously. The analysis procedure assumes that inputs change one at a time, allowing enough time between successive changes for the circuit to settle into a stable internal state. This differs from clocked circuits, in which multiple inputs can change at almost arbitrary times without affecting the state, and all input values are sampled and state changes occur with respect to a clock signal.

    Like clocked synchronous state machines, feedback sequential circuits may be structured as Mealy or Moore circuits, as shown in Figures 7–56 and 7–57. A circuit with n feedback loops has n binary state variables and 2^n states.

    To analyze a feedback sequential circuit, we must break the feedback loops in Figure 7–56 or 7–57 so that the next value stored in each loop can be predicted

fundamental-mode circuit

---

*This section and all of its subsections are optional.

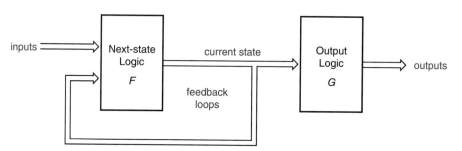

Figure 7–56   Feedback sequential circuit structure (Mealy model).

as a function of the circuit inputs and the current value stored in all loops. Figure 7–58 shows how to do this for the NAND circuit for a D latch, which has only one feedback loop. We conceptually break the loop by inserting a fictional buffer in the loop as shown. The output of the buffer, named Y, is the single state variable for this example.

Let us assume that the propagation delay of the fictional buffer is 10 ns (but any nonzero number will do), and that all of the other circuit components have zero delay. If we know the circuit's current state (Y) and inputs (D and C), then we can predict the value Y will have in 10 ns. The next value of Y, denoted

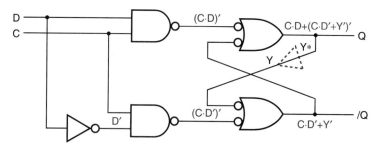

Figure 7–57   Feedback sequential circuit structure (Moore model).

Figure 7–58
Feedback analysis of a D latch.

The way the circuit in Figure 7–58 is drawn, it may look like there are two feedback loops. However, once we make one break as shown, there are no more loops. That is, each signal can be written as a combinational function of the other signals, not including itself.

JUST ONE LOOP

Y∗, is a combinational function of the current state and inputs. Thus, reading the circuit diagram, we can write an *excitation equation* for Y∗:

excitation equation

$$Y∗ = (C·D) + (C·D' + Y')'$$
$$= C·D + C'·Y + D·Y$$

Now the state of the feedback loop (and the circuit) can be written as a function of the current state and input, and enumerated by a *transition table* as shown in Table 7–20. Each cell in the transition table shows the fictional-buffer output value that will occur 10 ns (or whatever delay you've assumed) after the corresponding state and input combination occurs.

transition table

A transition table has one row for each possible combination of the state variables, so a circuit with n feedback loops has 2^n rows in its transition table. The table has one column for each possible input combination, so a circuit with m inputs has 2^m columns in its transition table.

By definition, a fundamental-mode circuit such as a feedback sequential circuit does not have a clock to tell it when to sample its inputs. Instead, we can imagine that the circuit is evaluating its current state and input *continuously* (or every 10 ns, if you prefer). As the result of each evaluation, it goes to a next state predicted by the transition table. Most of the time, the next state is the same as the current state; this is the essence of fundamental-mode operation. We make some definitions below that will help us study this behavior in more detail.

In a fundamental-mode circuit, a *total state* is a particular combination of *internal state* (the values stored in the feedback loops) and *input state* (the current value of the circuit inputs). A *stable total state* is a combination of internal state and input state such that the next internal state predicted by the transition table is the same as the current internal state. If the next internal state is different,

total state
internal state
input state
stable total state

Table 7–20
Transition table for the D latch in Figure 7–58.

	C D			
Y	00	01	11	10
0	0	0	1	0
1	1	1	1	0
		Y∗		

Table 7–21
State table for the D latch in
Figure 7–58, showing stable total
states.

S	00	01	11	10
		C D		
S0	(S0)	(S0)	S1	(S0)
S1	(S1)	(S1)	(S1)	S0
		S*		

unstable total state
state table

then the combination is an *unstable total state*. We have rewritten the transition
table for the D latch in Table 7–21 as a *state table*, giving the names S0 and S1
to the states and drawing a circle around the stable total states.

To complete the analysis of the circuit, we must also determine how the
outputs behave as functions of the internal state and inputs. There are two
outputs, and hence two *output equations*:

output equation

$$Q = C \cdot D + C' \cdot Y + D \cdot Y$$
$$/Q = C \cdot D' + Y'$$

Note that Q and /Q are *outputs*, not state variables. Even though the circuit has
two output variables, which can theoretically take on four combinations, it has
only one state variable Y, and hence only two states.

The output values predicted by the Q and /Q equations can be incorporated
in a combined state and output table that completely describes the operation of
the circuit, as shown in Table 7–22. In spite of their names, it is possible for Q
and /Q to have the same value (1) momentarily, during the transition from S0 to
S1 under the CD = 11 column of the table.

We can now predict the behavior of the circuit from the transition and
output table. First of all, notice that we have written the column labels in our
state tables in "Karnaugh map" order, so that only a single input bit changes
between adjacent columns of the table. This layout helps our analysis because
we assume that only one input changes at a time, and that the circuit always
reaches a stable total state before another input changes.

Table 7–22
State and output table for the
D latch in Figure 7–58.

S	00	01	11	10
		C D		
S0	(S0), 01	(S0), 01	S1 , 11	(S0), 01
S1	(S1), 10	(S1), 10	(S1), 10	S0 , 01
		S*, Q /Q		

Table 7–23

Analysis of the D latch in
Figure 7–58 for a few transitions.

S		C D		
	00	01	11	10
S0	(S0), 01	(S0), 01	S1 , 11	(S0), 01
S1	(S1), 10	(S1), 10	(S1), 10	S0 , 01
			S*, Q /Q	

At any time, the circuit is in a particular internal state and a particular input is applied to it; we called this combination the total state of the circuit. Let us start at the stable total state "S0/00" (S = S0, C D = 00), as shown in Table 7–23. Now suppose that we change D to 1. The total state moves to one cell to the right; we have a new stable total state, S0/01. The D input is different, but the internal state and output are the same as before. Next, let us change C to 1. The total state moves one cell to the right to S0/11, which is unstable. The next-state entry in this cell sends the circuit to internal state S1, so the total state moves down one cell, to S1/11. Examining the next-state entry in the new cell, we find that we have reached a stable total state. We can trace the behavior of the circuit for any desired sequence of single input changes in this way.

Now we can revisit the question of simultaneous input changes. Even though "almost simultaneous" input changes may occur in practice, we must assume that nothing happens simultaneously in order to analyze the behavior of sequential circuits. The impossibility of simultaneous events is supported by the varying delays of circuit components themselves, which depend on voltage, temperature, and fabrication parameters. What this tells us is that a set of n inputs that appear to us to change "simultaneously" may actually change in any of $n!$ different orders from the point of view of the circuit operation.

For example, consider the operation of the D latch as shown in Table 7–24. Let us assume that it starts in stable total state S1/11. Now suppose that C and D are both "simultaneously" set to 0. In reality, the circuit behaves as if one or the other input went to 0 first. Suppose that C changes first. Then the sequence of

Table 7–24

Multiple input changes for a
D latch.

S		C D		
	00	01	11	10
S0	(S0), 01	(S0), 01	S1 , 11	(S0), 01
S1	(S1), 10	(S1), 10	(S1), 10	S0 , 01
			S*, Q /Q	

two left-pointing arrows in the table tells us that the circuit goes to stable total state S1/00. However, if D changes first, then the other sequence of arrows tells us that the circuit goes to stable total state S0/00. So the final state of the circuit is unpredictable, a clue that the feedback loop may actually become metastable if we set C and D to 0 simultaneously. The time span over which this view of simultaneity is relevant is the setup- and hold-time window of the D latch.

Simultaneous input changes don't always cause unpredictable behavior. However, we must analyze the effects of all possible orderings of signal changes to determine this; if all orderings give the same result, then the circuit output is predictable. For example, consider the behavior of the D latch starting in total state S0/00 with C and D simultaneously changing from 0 to 1; it always ends up in total state S1/11.

* 7.5.2 Analyzing Circuits with Multiple Feedback Loops

cut set
minimal cut set

In circuits with multiple feedback loops, we must break all of the loops, creating one fictional buffer and state variable for each loop that we break. There are many possible ways, which mathematicians call *cut sets*, to break the loops in a given circuit, so how do we know which one is best? The answer is that any *minimal cut set*—a cut set with a minimum number of cuts—is fine. Mathematicians can give you an algorithm for finding a minimal cut set, but as a digital designer working on small circuits, you can just eyeball the circuit to find one.

Different cut sets for a circuit lead to different excitation equations, transition tables, and state/output tables. However, the stable total states derived from one minimal cut set correspond one-to-one to the stable total states derived from any other minimal cut set for the same circuit. That is, state/output tables derived from different minimal cut sets display the same input/output behavior, with only the names and coding of the states changed (to protect the innocent?).

If you use more than the minimal number of cuts to analyze a feedback sequential circuit, the resulting state/output table will still describe the circuit correctly. However, it will use 2^m times as many states as necessary, where m is the number of extra cuts. Formal state-minimization procedures can be used to reduce this larger table to the proper size, but it's a much better idea to select a minimal cut set in the first place.

A good example of a sequential circuit with multiple feedback loops is the commercial circuit design for a positive edge-triggered D flip-flop that we showed in Figure 7–20. The circuit is redrawn in simplified form in Figure 7–59, assuming that the original circuit's /PR and /CLR inputs are never asserted, and also showing fictional buffers to break the three feedback loops. Interestingly, the three loops give rise to eight states, compared with the minimum of four states used by the two-loop design in Figure 7–19. We'll readdress this curious difference later.

Figure 7–59
Simplified positive edge-triggered
D flip-flop for analysis.

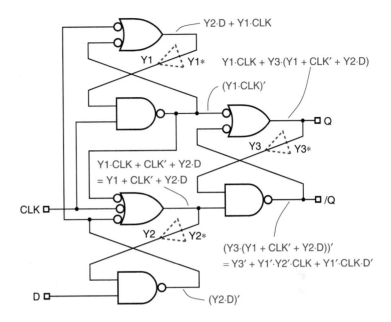

The following excitation and output equations can be derived from the logic diagram in Figure 7–59:

$$Y1* \;=\; Y2 \cdot D + Y1 \cdot CLK$$
$$Y2* \;=\; Y1 + CLK' + Y2 \cdot D$$
$$Y3* \;=\; Y1 \cdot CLK + Y1 \cdot Y3 + Y3 \cdot CLK' + Y2 \cdot Y3 \cdot D$$
$$Q \;=\; Y1 \cdot CLK + Y1 \cdot Y3 + Y3 \cdot CLK' + Y2 \cdot Y3 \cdot D$$
$$/Q \;=\; Y3' + Y1' \cdot Y2' \cdot CLK + Y1' \cdot CLK \cdot D'$$

The corresponding transition table is shown in Table 7–25, with the stable total states circled. Before we analyze this table further, we must introduce the concept of "races."

* 7.5.3   Races

In a feedback sequential circuit, a *race* is said to occur when multiple internal *race* variables change state as a result of a single input changing state. In the example of Table 7–25, a race occurs in stable total state 011/00 when CLK is changed from 0 to 1. The table indicates that the next internal state is 000, a 2-variable change from 011.

As we've discussed previously, logic signals never really change "simultaneously." Thus, the internal state may make the change $011 \rightarrow 000$ as either $011 \rightarrow 001 \rightarrow 000$ or $011 \rightarrow 010 \rightarrow 000$. Table 7–26 indicates that the example circuit, starting in total state 011/00, should go to total state 000/10 when CLK

Table 7–25
Transition table for the D flip-flop
in Figure 7–59.

Y1 Y2 Y3	CLK D			
	00	01	11	10
000	010	010	(000)	(000)
001	011	011	000	000
010	(010)	110	110	000
011	(011)	111	111	000
100	010	010	111	111
101	011	011	111	111
110	010	(110)	111	111
111	011	(111)	(111)	(111)
	Y1* Y2* Y3*			

changes from 0 to 1. However, it may temporarily visit total state 001/10 or 010/10 along the way. That's OK, because the next internal state for both of these temporary states is 000; therefore, even in the temporary states, the excitation logic continues to drive the feedback loops toward the same stable total state, 000/10. Since the final state does not depend on the order in which the state variables change, this is called a *noncritical race*.

noncritical race

Now suppose that the next-state entry for total state 010/10 is changed to 110, as shown in Table 7–27, and consider the case that we just analyzed. Starting in stable total state 011/00 and changing CLK to 1, the circuit may end up in internal state 000 or 111 depending on the order and speed of the internal variable changes. This is called a *critical race*.

critical race

Table 7–26
Portion of a transition table
showing a noncritical race.

Y1 Y2 Y3	CLK D			
	00	01	11	10
000	010	010	(000)	(000)
001	011	011	000	000
010	(010)	110	110	000
011	(011)	111	111	000
	Y1* Y2* Y3*			

Table 7–27
A transition table containing a
critical race.

Y1 Y2 Y3	CLK D			
	00	01	11	10
000	010	010	(000)	(000)
001	011	011	000	000
010	(010)	110	110	110
011	(011)	111	111	000
100	010	010	111	111
101	011	011	111	111
110	010	(110)	111	111
111	011	(111)	(111)	(111)
	Y1∗ Y2∗ Y3∗			

* 7.5.4 State Tables and Flow Tables

Analysis of the real transition table for our example circuit, Table 7–25, shows
that it does not have any critical races; in fact, it has no races except the noncritical one we identified earlier. Once we've determined this fact, we no longer
need to refer to state variables. Instead, we can name the state-variable combinations and determine the output values for each state/input combination to obtain
a state/output table such as Table 7–28.

The state table shows that for some single input changes, the circuit takes
multiple "hops" to get to a new stable total state. For example, in state S0/11,
an input change to 01 sends the circuit first to state S2 and then to stable total
state S6/01. A *flow table* eliminates multiple hops and shows only the ultimate   *flow table*
destination for each transition. The flow table also eliminates the rows for unused
internal states—ones that are stable for no input combination—and eliminates
the next-state entries for total states that cannot be reached from a stable total
state as the result of a single input change. Using these rules, Table 7–29 shows
the flow table for our positive edge-triggered flip-flop example.

When you design a feedback-based sequential circuit, you must ensure that its
transition table does not contain any critical races. Otherwise, the circuit may
operate unpredictably, with the next state for racy transitions depending on factors
like temperature, voltage, and the phase of the moon.

WATCH OUT FOR
CRITICAL RACES!

Table 7–28
State and output table for the
D flip-flop in Figure 7–59.

S	00	01	11	10
		CLK D		
S0	S2 , 01	S2 , 01	(S0), 01	(S0), 01
S1	S3 , 10	S3 , 10	S0 , 01	S0 , 01
S2	(S2), 01	S6 , 01	S6 , 01	S0 , 01
S3	(S3), 10	S7 , 10	S7 , 10	S0 , 01
S4	S2 , 01	S2 , 01	S7 , 11	S7 , 11
S5	S3 , 10	S3 , 10	S7 , 10	S7 , 10
S6	S2 , 01	(S6), 01	S7 , 11	S7 , 11
S7	S3 , 10	(S7), 10	(S7), 10	(S7), 10
		S*, Q /Q		

The flip-flop's edge-triggered behavior can be observed in the series of state transitions shown in Table 7–30. Let us assume that the flip-flop starts in internal state S0/10. That is, the flip-flop is storing a 0 (since $Q = 0$), CLK is 1, and D is 0. Now suppose that D changes to 1; the flow table shows that we move one cell to the left, still a stable total state with the same output value. We can change D between 0 and 1 as much as we want, and just bounce back and forth between these two cells. However, once we change CLK to 0, we move to internal state S2 or S6, depending on whether D was 0 or 1 at the time; but still the output is unchanged. Once again, we can change D between 0 and 1 as much as we want, this time bouncing between S2 and S6 without changing the output.

Table 7–29
Flow and output table for the
D flip-flop in Figure 7–59.

S	00	01	11	10
		CLK D		
S0	S2 , 01	S6 , 01	(S0), 01	(S0), 01
S2	(S2), 01	S6 , 01	— , —	S0 , 01
S3	(S3), 10	S7 , 10	— , —	S0 , 01
S6	S2 , 01	(S6), 01	S7 , 11	— , —
S7	S3 , 10	(S7), 10	(S7), 10	(S7), 10
		S*, Q /Q		

Table 7–30
Flow and output table showing
D flip-flop's edge-triggered
behavior.

	CLK D			
S	00	01	11	10
S0	S2 , 01	S6 , 01	(S0), 01	(S0), 01
S2	(S2), 01	S6 , 01	— , –	S0 , 01
S3	(S3), 10	S7 , 10	— , –	S0 , 01
S6	S2 , 01	(S6), 01	S7 , 11	— , –
S7	S3 , 10	(S7), 10	(S7), 10	(S7), 10
		S*, Q /Q		

The moment of truth finally comes when CLK changes to 1. Depending on whether we are in S2 or S6, we go back to S0 (leaving Q at 0) or to S7 (setting Q to 1). Similar behavior involving S3 and S7 can be observed on a rising clock edge that causes Q to change from 1 to 0.

* 7.5.5 Comments

In Figure 7–19, we showed a circuit for a positive edge-triggered D flip-flop with only two feedback loops and hence four states. The circuit that we just analyzed has three loops and eight states. Even after eliminating unused states, the flow table has five states. However, a formal state-minimization procedure can be used to show that states S0 and S2 are "compatible" so that they can be merged into a single state SB that handles the transitions for both, as shown in Table 7–31. Thus, the job really could have been done by a four-state circuit. In fact, in Exercise 7.48 you'll show that the circuit in Figure 7–19 does the job specified by the reduced flow table.

Table 7–31
Reduced flow and output table
for a positive edge-triggered
D flip-flop.

	CLK D			
S	00	01	11	10
SB	(SB), 01	S6 , 01	(SB), 01	(SB), 01
S3	(S3), 10	S7 , 10	— , –	SB , 01
S6	SB , 01	(S6), 01	S7 , 11	— , –
S7	S3 , 10	(S7), 10	(S7), 10	(S7), 10
		S*, Q /Q		

The feedback sequential circuits that we've analyzed in this section exhibit quite reasonable behavior since, after all, they are latch and flip-flop circuits that have been used for years. However, if we throw together a "random" collection of gates and feedback loops, we won't necessarily get "reasonable" sequential circuit behavior. In a few rare cases, we may not get a sequential circuit at all (see Exercise 7.49), and in many cases, the circuit may be unstable for some or all input combinations (see Exercise 7.54).

* 7.6   FEEDBACK SEQUENTIAL CIRCUIT DESIGN

It's sometimes useful to design a small feedback sequential circuit, such as a specialized latch or a pulse catcher; this section will show you how. It's even possible that you might go on to be an IC designer, and be responsible for designing high-performance latches and flip-flops from scratch. This section will serve as an introduction to the basic concepts you'll need, but you'll still need considerably more study, experience, and finesse to do it right.

7.6.1   Latches

Although the design of feedback sequential circuits is generally a hard problem, some circuits can be designed pretty easily. Any circuit with one feedback loop is just a variation of an S-R or D latch. It has the general structure shown in Figure 7–60, and an excitation equation with the following format:

$$Q* \;=\; \text{(forcing term)} + \text{(holding term)} \cdot Q$$

For example, the excitation equations for S-R and D latches are

$$Q* \;=\; S + R' \cdot Q$$
$$Q* \;=\; C \cdot D + C' \cdot Q$$

Corresponding circuits are shown in Figure 7–61(a) and (b).

hazard-free
excitation logic   In general, *the excitation logic in a feedback sequential circuit must be hazard free;* we'll demonstrate this fact by way of an example. Figure 7–62(a) is a Karnaugh map for the D-latch excitation circuit of Figure 7–61(b). The map

Figure 7–60
General structure of a latch.

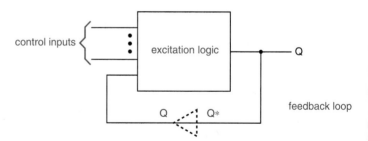

feedback loop

Figure 7–61
Latch circuits: (a) S-R latch;
(b) unreliable D latch; (c) hazard-
free D latch.

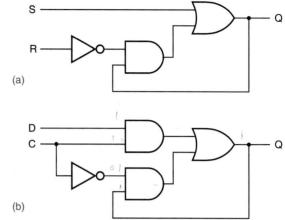

(a)

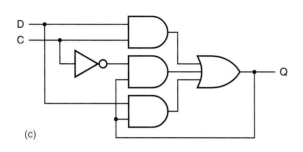

(c)

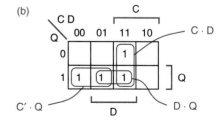

(b)

exhibits a static-1 hazard when D and Q are 1 and C is changing. Unfortunately, the latch's feedback loop may not hold its value if a hazard-induced glitch occurs. For example, consider what happens in Figure 7–61(b) when D and Q are 1 and C changes from 1 to 0; the circuit should latch a 1. However, unless the inverter is very fast, the output of the top AND gate goes to 0 before the output of the bottom one goes to 1, the OR-gate output goes to 0, and the feedback loop stores a 0.

Hazards can be eliminated using the methods described in Section 4.5. In the D latch, we simply include the consensus term in the excitation equation:

$$Q* \ = \ C{\cdot}D + C'{\cdot}Q + D{\cdot}Q$$

Figure 7–61(c) shows the corresponding hazard-free, correct D-latch circuit.

Now, suppose we need a specialized "D" latch with three data inputs, D1–D3, that stores a 1 only if D1–D3 = 010. We can convert this word description into an excitation equation that mimics the equation for a simple D latch:

$$Q* \ = \ C{\cdot}(D1'{\cdot}D2{\cdot}D3') + C'{\cdot}Q$$

Eliminating hazards, we get

$$Q* \ = \ C{\cdot}D1'{\cdot}D2{\cdot}D3' + C'{\cdot}Q + D1'{\cdot}D2{\cdot}D3'{\cdot}Q$$

Figure 7–62
Karnaugh maps for D-
latch excitation functions:
(a) original, containing
a static-1 hazard;
(b) hazard eliminated.

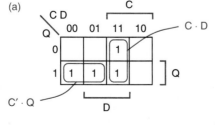

$$Q* = C \cdot D + C' \cdot Q$$

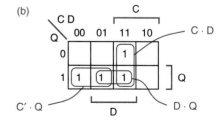

$$Q* = C \cdot D + C' \cdot Q + D \cdot Q$$

PRODUCT-TERM
EXPLOSION

In some cases, the need to cover hazards can cause an explosion in the number of product terms in a two-level realization of the excitation logic. For example, suppose we need a specialized latch with two control inputs, C1 and C2, and three data inputs as before. The latch is to be "open" only if both control inputs are 1, and is to store a 1 if any data input is 1. The minimal excitation equation is

$$Q* = C1 \cdot C2 \cdot (D1 + D2 + D3) + (C1 \cdot C2)' \cdot Q$$
$$= C1 \cdot C2 \cdot D1 + C1 \cdot C2 \cdot D2 + C1 \cdot C2 \cdot D3 + C1' \cdot Q + C2' \cdot Q$$

However, it takes six consensus terms to eliminate hazards (see Exercise 7.56).

The hazard-free excitation equation can be realized with discrete gates or in a PLD, as we'll show in Section 10.4.4.

7.6.2 Designing Fundamental-Mode Flow Tables

To design feedback sequential circuits more complex than latches, we must first convert the word description into a flow table. Once we have a flow table, we can turn the crank (with some effort) to obtain a circuit.

When we construct the flow table for a feedback sequential circuit, we give each state a meaning in the context of the problem, much like we did in the design of clocked state machines. However, it's easier to get confused when constructing the flow table for a feedback sequential circuit. Therefore, *primitive flow table* the recommended procedure is to construct a *primitive flow table*—one that has only one stable total state in each row. Since there is only one stable state per row, the output may be shown as a function of state only.

In a primitive flow table, each state has a more precise "meaning" than it might otherwise have, and the table's structure clearly displays the underlying fundamental-mode assumption: Inputs change one at a time, with enough time between changes for the circuit to settle into a new stable state. A primitive flow table usually has extra states, but we can "turn the crank" later to minimize the number of states once we have a flow table that we believe to be correct.

We'll use the following problem, a "pulse-catching" circuit, to demonstrate flow-table design:

Design a feedback sequential circuit with two inputs, P (pulse) and R (reset), and a single output Z that is normally 0. The output should be set to 1 whenever a 0-to-1 transition occurs on P, and should be reset to 0 whenever R is 1. Typical functional behavior is shown in Figure 7–63.

Table 7–32 is a primitive flow table for the pulse catcher. Let's walk through how this table was developed.

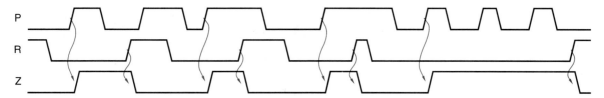

Figure 7–63 Typical functional behavior of a pulse-catching circuit.

We assume that the pulse catcher is initially idle, with P and R both 0; this is the IDLE state, with Z = 0. In this state, if reset occurs (R = 1), we could probably stay in the same state, but since this is supposed to be a *primitive* flow table, we create a new state, RES1, so as not to have two stable total states in the same row. On the other hand, if a pulse occurs (P = 1) in the IDLE state, we definitely want to go to a different state, which we've named PLS1, since we've caught a pulse and we must set the output to 1. Input combination 11 is not considered in the IDLE state, because of the fundamental-mode assumption that only one input changes at a time; we assume the circuit always makes it to another stable state before input combination 11 can occur.

Next, we fill in the next-state entries for the newly created RES1 state. If reset goes away, we can go back to the IDLE state. If a pulse occurs, we must remain in a "reset" state since, according to the timing diagram, a 0-to-1 transition that occurs while R is 1 is ignored. Again, to keep the flow table in primitive form, we must create a new state for this case, RES2.

Now that we have one stable total state in each column, we may be able to go to existing states for more transitions, instead of always defining new states. Sure enough, starting in stable total state PLS1/10, if R = 1 we can go to RES2,

Table 7–32 Primitive flow table for pulse-catching circuit.

Meaning	S	00	01	11	10	Z
Idle, waiting for pulse	IDLE	(IDLE)	RES1	—	PLS1	0
Reset, no pulse	RES1	IDLE	(RES1)	RES2	—	0
Got pulse, output on	PLS1	PLS2	—	RES2	(PLS1)	1
Reset, got pulse	RES2	—	RES1	(RES2)	PLSN	0
Pulse gone, output on	PLS2	(PLS2)	RES1	—	PLS1	1
Got pulse, but output off	PLSN	IDLE	—	RES2	(PLSN)	0

P R

S*

which fits the requirement of producing a 0 output. On the other hand, where should we go if P = 0? IDLE is a stable total state in the 00 column, but it produces the wrong output value. In PLS1, we've gotten a pulse and haven't seen a reset, so if the pulse goes away, we should go to a state that still has Z = 1. Thus, we must create a new state PLS2 for this case.

In RES2, we can safely go to RES1 if the pulse goes away. However, we've got to be careful if reset goes away, as shown in the timing diagram. Since we've already passed the pulse's 0-to-1 transition, we can't go to the PLS1 state, since that would give us a 1 output. Instead, we create a new state PLSN with a 0 output.

Finally, we can complete the next-state entries for PLS2 and PLSN without creating any new states. Notice that starting in PLS2, we bounce back and forth between PLS2 and PLS1 and maintain a continuous 1 output if we get a series of pulses without an intervening reset input.

7.6.3 Flow-Table Minimization

As we mentioned earlier, a primitive flow table usually has more states than required. However, there exists a formal procedure, discussed in the References, for minimizing the number of states in a flow table. This procedure is often complicated by the existence of don't-care entries in the flow table.

Fortunately, our example flow table is small and simple enough to minimize by inspection. States IDLE and RES1 produce the same output, and they have the same next-state entry for input combinations where they are both specified. Therefore, they are compatible and may be replaced by a single state (IDLE) in a reduced flow table. The same can be said for states PLS1 and PLS2 (replaced by PLS) and RES2 and PLSN (replaced by RES). The resulting reduced flow table, which has only three states, is shown in Table 7–33.

Table 7–33
Reduced flow table for pulse-catching circuit.

S	00	01	11	10	Z
		P R			
IDLE	(IDLE)	(IDLE)	RES	PLS	0
PLS	(PLS)	IDLE	RES	(PLS)	1
RES	IDLE	IDLE	(RES)	(RES)	0
			S*		

Table 7–34
Example flow table for state
assignment problem.

S	\[XY\] 00	01	11	10
A	(A)	B	(A)	B
B	(B)	(B)	D	(B)
C	(C)	A	A	(C)
D	(D)	B	(D)	C
		S*		

7.6.4 Race-Free State Assignment

The next somewhat creative (read "difficult") step in feedback sequential circuit design is to find a race-free assignment of coded states to the named states in the reduced flow table. Recall from Section 7.5.3 that a race occurs when multiple internal variables change state as a result of a single input change. A feedback-based sequential circuit must not contain any critical races; otherwise, the circuit may operate unpredictably. As we'll see, eliminating races often necessitates *increasing* the number of states in the circuit.

A circuit's potential for having races in its transition table can be analyzed by means of a *state adjacency diagram* for its flow table. The adjacency diagram is a simplified state diagram that omits self-loops and does not show the direction of other transitions (A→B looks the same as B→A), or the input combinations that cause them. Table 7–34 is an example fundamental-mode flow table and Figure 7–64(a) is the corresponding adjacency diagram.

state adjacency diagram

Two states are *adjacent* if there is an arc between them in the state adjacency diagram. For race-free transitions, adjacent coded states must differ in only one bit. If two states A and B are adjacent, it doesn't matter whether the orginal flow table had transitions from A to B, from B to A, or both. Any one of these

adjacent states

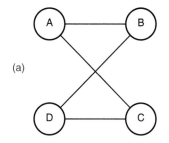
(a)

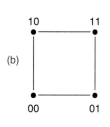
(b)

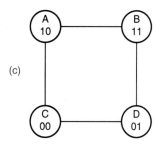
(c)

Figure 7–64 State-assignment example: (a) adjacency diagram; (b) a 2-cube; (c) one of eight possible race-free state assignments.

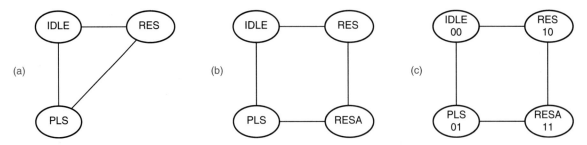

Figure 7–65 Adjacency diagrams for the pulse catcher: (a) using original flow table; (b) after adding a state; (c) showing one of eight possible race-free state assignments.

transitions is a race if A and B differ in more than one variable. That's why we don't bother to show the direction of transitions in an adjacency diagram.

The problem of finding a race-free assignment of states to *n* state variables is equivalent to the problem of mapping the nodes and arcs of the adjacency diagram onto the nodes and arcs of an *n*-cube. In Figure 7–64, the problem is to map the adjacency diagram (a) onto a 2-cube (b). You can visually identify eight ways to do this (four rotations times two flips), one of which produces the state assignment shown in (c).

Figure 7–65(a) is an adjacency diagram for our pulse-catching circuit, based on the reduced flow table in Table 7–33. Clearly, there's no way to map this "triangle" of states onto a 2-cube. At this point, we have to go back and modify the original flow table. In particular, the flow table tells us the destination state that we *eventually* must reach for each transition, but it doesn't prevent us from going through other states on the way. As shown in Table 7–35, we can create a new state RESA and make the transition from PLS to RES by going through RESA. The modified state table has the new adjacency diagram shown in Figure 7–65(b), which has many race-free assignments possible. A transition table based on the assignment in (c) is shown in Table 7–36. Note

Table 7–35
State table allowing a race-free assignment for pulse-catching circuit.

	P R				
S	00	01	11	10	Z
IDLE	(IDLE)	(IDLE)	RES	PLS	0
PLS	(PLS)	IDLE	RESA	(PLS)	1
RESA	—	—	RES	—	—
RES	IDLE	IDLE	(RES)	(RES)	0
		S*			

Table 7–36
Race-free transition table for
pulse-catching circuit.

		P R			
Y1 Y2	00	01	11	10	Z
00	(00)	(00)	10	01	0
01	(01)	00	11	(01)	1
11	—	—	10	—	—
10	00	00	(10)	(10)	0
		Y1* Y2*			

that the PLS→RESA→RES transition will be slower than the other transitions
in the original flow table because it requires two internal state changes, with two
propagation delays through the feedback loops.

   Even though we added a state in the previous example, we were still able to
get by with just two state variables. However, we may sometimes have to add one
or more state variables to make a race-free assignment. Figure 7–66(a) shows the
worst possible adjacency diagram for four states—every state is adjacent to every
other state. Clearly, this adjacency diagram cannot be mapped onto a 2-cube.
However, there is a race-free assignment of states to a 3-cube, shown in (b),
where each state in the original flow table is represented by two equivalent states
in the final state table. Both states in a pair, such as A1 and A2, are equivalent
and produce the same output. Each state is adjacent to one of the states in every
other pair, so a race-free transition may be selected for each next-state entry.

Figure 7–66
A worst-case scenario:
(a) 4-state adjacency diagram;
(b) assignment using pairs of
equivalent states.

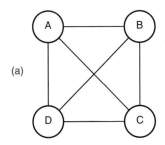

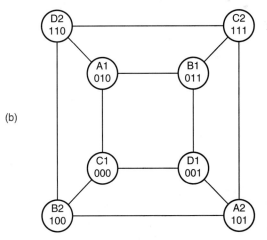

HANDLING THE
GENERAL CASE

In the general case of a flow table with 2^n rows, it can be shown that a race-free assignment can be obtained using $2n - 1$ state variables (see the References). However, there aren't many applications for fundamental-mode circuits with more than a few states, so the general case is of little more than academic interest.

7.6.5 Excitation Equations

Once we have a race-free transition table for a circuit, we can just "turn the crank" to obtain excitation equations for the feedback loops. Figure 7–67 shows Karnaugh maps derived from Table 7–36, the pulse catcher's transition table. Notice that "don't-care" next-state and output entries give rise to corresponding entries in the maps, simplifying the excitation and output logic. The resulting minimal sum-of-products excitation and output equations are as follows:

$$Y1* \ = \ P{\cdot}R + P{\cdot}Y1$$
$$Y2* \ = \ Y2{\cdot}R' + Y1'{\cdot}Y2{\cdot}P + Y1'{\cdot}P{\cdot}R'$$
$$Z \ = \ Y2$$

Recall that the excitation logic in a feedback sequential circuit must be hazard free. The sum-of-products expressions above happen to be hazard free as well as minimal. The logic diagram of Figure 7–68 uses these expressions to build the pulse-catching circuit.

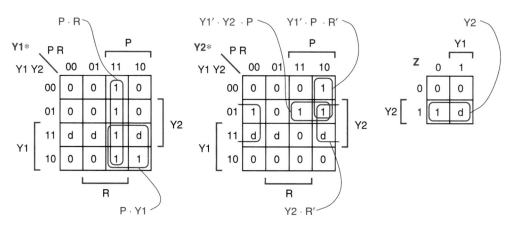

Figure 7–67 Karnaugh maps for pulse-catcher excitation and output logic.

7.6.6 Essential Hazards

After all this effort, you'd think that we'd have a pulse-catching circuit that would work reliably all of the time. Well, we're not quite there yet. A fundamental-mode circuit must generally satisfy five requirements for proper operation:

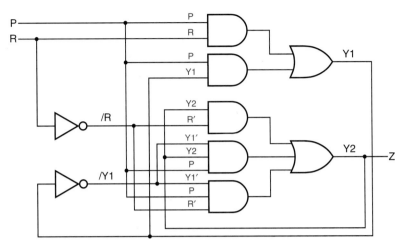

Figure 7–68   Pulse-catching circuit.

all important

(1) Only one input signal changes at a time, with a minimum bound between successive input changes.

(2) There is a maximum propagation delay through excitation logic and feedback paths; this maximum is less than the time between successive input changes.

(3) The state assignment (transition table) is free of critical races.

(4) The excitation logic is hazard free.

(5) The minimum propagation delay through excitation logic and feedback paths is greater than the maximum timing skew through "input logic."

Without the first requirement, it would be impossible to satisfy the major premise of fundamental-mode operation—that the circuit has time to settle into a stable total state between successive input changes. The second requirement says that the excitation logic is fast enough to do just that. The third requirement ensures that the proper state changes are made even if the excitation circuits for different state variables have different delays. The fourth requirement ensures that state variables that aren't supposed to change on a particular transition don't.

The last requirement deals with subtle timing-dependent errors that can occur in fundamental-mode circuits, even ones that satisfy the first four requirements. An *essential hazard* is the possibility of a circuit going to an erroneous *essential hazard* next state as the result of a single input change; the error occurs if the input change is not seen by all of the excitation circuits before the resulting state-variable transition(s) propagate back to the inputs of the excitation circuits. In a

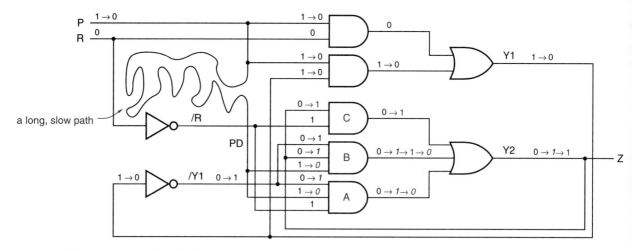

Figure 7–69 Physical conditions in pulse-catching circuit for exhibiting an essential hazard.

world where "faster is better" is the usual rule, a designer may sometimes have to *slow down* excitation logic to mask these hazards.

Essential hazards are best explained in terms of an example, our pulse-catching circuit. Suppose we built our circuit on a PCB or a chip, and we (or, more likely, our CAD system) inadvertently connected input signal P through a long, slow path at the point shown in Figure 7–69. Let's assume that this delay is longer than the propagation delay of the AND-OR excitation logic.

Now consider what can happen if P R = 10, the circuit is in internal state 10, and P changes from 1 to 0. According to the transition table, which is repeated in Table 7–37, the circuit should go to internal state 00, and that's that. However, let's look at the actual operation of the circuit as traced in Figure 7–69:

- (Changes shown in nonitalic black.) The first thing that happens after P changes is that Y1 changes from 1 to 0, and /Y1 changes from 0 to 1 shortly thereafter. Now the circuit is in internal state 00.

- (Changes shown in italic black.) The change in /Y1 at AND gate A causes its output to go to 1, which in turn forces Y2 to 1. Whoops, now the circuit is in internal state 01.

- (Changes shown in nonitalic color.) The change in Y2 at AND gates B and C causes their outputs to go to 1, reinforcing the 1 output at Y2. All this time, we've been waiting for the 1-to-0 change in P to appear at point PD.

- (Changes shown in italic color.) Finally, PD changes from 1 to 0, forcing the outputs of AND gates A and B to 0. However, AND gate C still has a 1 output, and the circuit remains in state 01—*the wrong state.*

Table 7–37
Transition table for pulse-
catching circuit, exhibiting an
essential hazard.

	P R				
Y1 Y2	00	01	11	10	Z
00	(00)	(00)	10	01	0
01	(01)	00	11	(01)	1
11	—	—	10	—	–
10	00	00	(10)	(10)	0
			Y1* Y2*		

The only way to avoid this erroneous behavior in general is to ensure that changes in P arrive at the inputs of all the excitation circuits before any changes in state variables do. Thus, the inevitable difference in input arrival times, called *timing skew*, must be less than the propagation delay of the excitation circuits and feedback loops. This timing requirement can generally be satisfied only by careful design *at the electrical circuit level*.   *timing skew*

In the example circuit, it would appear that the hazard is easily masked, even by non-electrical engineers, since the designer need only ensure that a straight wire has shorter propagation delay than an AND-OR structure, easy in most technologies. However, other circuits, such as the edge-triggered D flip-flop in Figure 7–19, have essential hazards in which the input skew paths include inverters. In such cases, the input inverters must be guaranteed to be faster than the excitation logic; that's not so trivial in either board-level or IC design. (If the excitation circuit in Figure 7–69 were physically built using AND-OR-INVERT gates, the delay from input changes to /Y1 could be very short indeed, as short as the delay through a single inverter.)

Essential hazards can be found in most but not all fundamental-mode circuits. There's an easy rule for detecting them; in fact, this is the definition of "essential hazard" in some texts:

- A fundamental-mode flow table contains an essential hazard for a stable total state S and an input variable X if, starting in state S, the stable total state reached after three successive transitions in X is different from the stable total state reached after one transition in X.

Thus, the essential hazard in the pulse catcher is detected by the arrows in Table 7–37 starting in internal state 10 with P R = 10.

A fundamental-mode circuit must have at least three states to have an essential hazard, so latches don't have them. On the other hand, all flip-flops (circuits that sample inputs on a clock edge) do.

THESE HAZARDS
ARE, WELL,
ESSENTIAL!

> Essential hazards are called "essential" because they are inherent in the flow table for a particular sequential function, and will appear in any circuit realization of that function. They can be masked only by controlling the delays in the circuit. Compare with static hazards in combinational logic, where we could eliminate hazards by adding consensus terms to a logic expression.

7.6.7 Summary

In summary, you use the following steps to design a feedback sequential circuit:

(1) Construct a primitive flow table from the circuit's word description.

(2) Minimize the number of states in the flow table.

(3) Find a race-free assignment of coded states to named states, adding auxiliary states or splitting states as required.

(4) Construct the transition table.

(5) Construct excitation maps and find a hazard-free realization of the excitation equations.

(6) Draw the logic diagram.

(7) Check for essential hazards. Modify the circuit if necessary to ensure that minimum excitation and feedback delays are greater than maximum inverter or other input-logic delays.

Note that some circuits routinely violate the basic fundamental-mode assumption that inputs change one at a time. For example, in a positive-edge-triggered D flip-flop, the D input may change at the same time that CLK changes from 1 to 0, and the flip-flop still operates properly. The same thing certainly cannot be said at the 0-to-1 transition of CLK. Such situations require analysis of the transition table and circuit on a case-by-case basis if proper operation in "special cases" is to be guaranteed.

A FINAL
QUESTION

> Given the difficulty of designing fundamental-mode circuits that work properly, let alone ones that are fast or compact, how did anyone ever come up with the 6-gate, 8-state, commercial D flip-flop design in Figure 7–20? Don't ask me, I don't know!

REFERENCES

The problem of metastability has been around for a long time. Greek philosophers wrote about the problem of indecision thousands of years ago. A group of modern philosophers named Devo sang about metastability in the title song of their *Freedom of Choice* album in 1980. Congress *still* can't decide whether high taxes are a good thing or a bad thing.

Most ASICs and MSI-based designs use the sequential-circuit types described in this chapter. However, there are other types that are used in both older discrete designs and in modern, custom VLSI designs.

For example, clocked synchronous state machines are a special case of a more general class of *pulse-mode circuits*. Such circuits have one or more *pulse inputs* such that (a) only one pulse occurs at a time; (b) nonpulse inputs are stable when a pulse occurs; (c) only pulses can cause state changes; and (d) a pulse causes at most one state change. In clocked synchronous state machines, the clock is the single pulse input, and a "pulse" is the triggering edge of the clock. However, it is also possible to build circuits with multiple pulse inputs, and it is possible to use storage elements other than the familiar edge-triggered flip-flops. These possibilities are discussed thoroughly by Edward J. McCluskey in *Logic Design Principles* (Prentice Hall, 1986).

pulse-mode circuit
pulse input

A particularly important type of pulse-mode circuit discussed by McCluskey and others is the *two-phase latch machine*. The rationale for a two-phase clocking approach in VLSI circuits is discussed by Carver Mead and Lynn Conway in *Introduction to VLSI Systems* (Addison-Wesley, 1980).

two-phase latch machine

Fundamental-mode circuits need not use feedback loops as the memory elements. For example, McCluskey's *Introduction to the Theory of Switching Circuits* (McGraw-Hill, 1965) gives several examples of fundamental-mode circuits built from S-R flip-flops. His 1986 book shows how transmission gates are used to create memory in CMOS latches and flip-flops, and such circuits are analyzed.

Methods for reducing both completely and incompletely-specified state tables are described in advanced logic design texts, including McCluskey's 1986 book. A more mathematical discussion of these methods and other "theoretical" topics in sequential machine design appears in *Switching and Finite Automata Theory*, 2nd ed., by Zvi Kohavi (McGraw-Hill, 1978).

DRILL PROBLEMS

7.1   Give three examples of metastability that occur in everyday life, other than ones discussed in this chapter.

7.2   Sketch the outputs of an S-R latch of the type shown in Figure 7–5 for the input waveforms shown in Figure X7.2. Assume that input and output rise and fall times are zero, that the propagation delay of a NOR gate is 10 ns, and that each time division below is 10 ns.

Figure X7.2

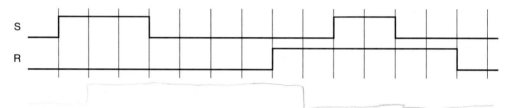

7.3   Repeat Drill 7.2 using the input waveforms shown in Figure X7.3. Although you may find the result unbelievable, this behavior can actually occur in real devices whose transition times are short compared to their propagation delay.

Figure X7.3

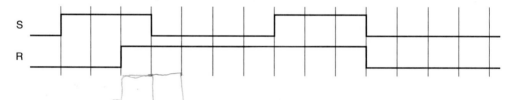

7.4   Figure 7–31 showed how to build a T flip-flop with enable using a D flip-flop and combinational logic. Show how to build a D flip-flop using a T flip-flop with enable and combinational logic.

7.5   Show how to build a J-K flip-flop using a T flip-flop with enable and combinational logic.

7.6   Show how to build an S-R latch using a single 74x74 positive-edge-triggered D flip-flop and *no* other components.

7.7   Show how to build a flip-flop equivalent to the 74x109 positive-edge-triggered J-/K flip-flop using a 74x74 positive-edge-triggered D flip-flop and one or more gates from a 74x00 package.

7.8   Show how to build a flip-flop equivalent to the 74x74 positive-edge-triggered D flip-flop using a 74x109 positive-edge-triggered J-/K flip-flop and *no* other components.

7.9 Analyze the clocked synchronous state machine in Figure X7.9. Write excitation equations, excitation/transition table, and state/output table (use state names A–D for Q1 Q2 = 00–11).

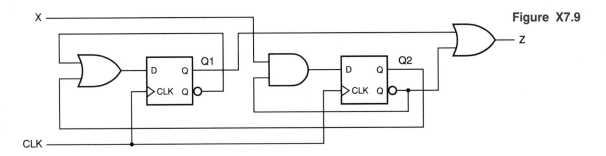

Figure X7.9

7.10 Repeat Drill 7.9, swapping AND and OR gates in the logic diagram. Is the new state/output table the "dual" of the original one? Explain.

7.11 Draw a state diagram for the state machine described by Table 7–10.

7.12 Draw a state diagram for the state machine described by Table 7–16.

7.13 Draw a state diagram for the state machine described by Table 7–18.

7.14 Construct a state and output table equivalent to the state diagram in Figure X7.14. Note that the diagram is drawn with the convention that the state does not change except for input conditions that are explicitly shown.

Figure X7.14

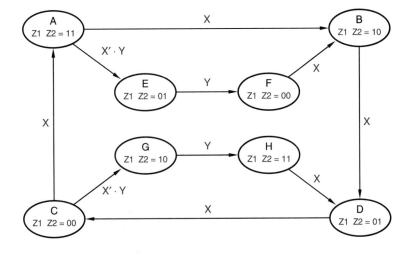

7.15 Analyze the clocked synchronous state machine in Figure X7.15. Write excitation equations, excitation/transition table, and state table (use state names A–H for Q2 Q1 Q0 = 000–111).

Figure X7.15

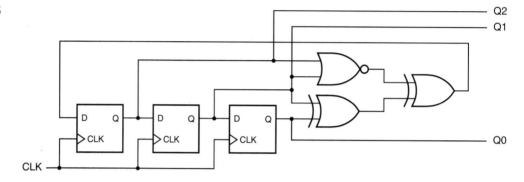

7.16 Analyze the clocked synchronous state machine in Figure X7.16. Write excitation equations, excitation/transition table, and state/output table (use state names A–H for Q1 Q2 Q3 = 000–111).

Figure X7.16

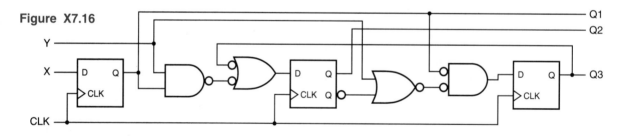

7.17 Analyze the clocked synchronous state machine in Figure X7.17. Write excitation equations, transition equations, transition table, and state/output table (use state names A–D for Q1 Q2 = 00–11). Draw a state diagram, and draw a timing diagram

Figure X7.17

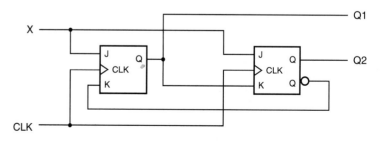

for CLK, EN, Q1, and Q2 for 10 clock ticks, assuming that the machine starts in state 00 and EN is continuously 1.

7.18 Analyze the clocked synchronous state machine in Figure X7.18. Write excitation equations, transition equations, transition table, and state/output table (use state names A–D for Q1 Q0 = 00–11). Draw a state diagram, and draw a timing diagram for CLK, X, Q1, and Q0 for 10 clock ticks, assuming that the machine starts in state 00 and X is continuously 1.

Figure X7.18

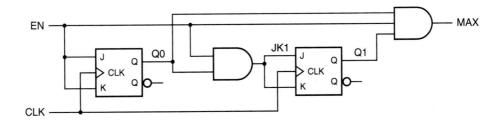

7.19 Analyze the clocked synchronous state machine in Figure X7.19. Write excitation equations, excitation/transition table, and state/output table (use state names A–D for Q1 Q2 = 00–11).

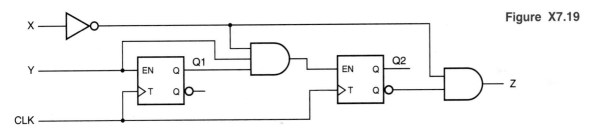

Figure X7.19

EXERCISES

7.20 Explain how metastability occurs in a D latch when the setup and hold times are not met, analyzing the behavior of the feedback loop inside the latch.

7.21 What is the minimum setup time of a pulse-triggered flip-flop such as a master/slave J-K or S-R flip-flop? (*Hint:* It depends on certain characteristics of the clock.)

7.22 Describe a situation, other than the metastable state, in which the Q and /Q outputs of a 74x74 edge-triggered D flip-flop may be noncomplementary for an arbitrarily long time.

7.23 Compare the circuit in Figure X7.23 with the D latch in Figure 7–12. Prove that the circuits function identically. In what way is Figure X7.23, which is used in some commercial D latches, better?

Figure X7.23

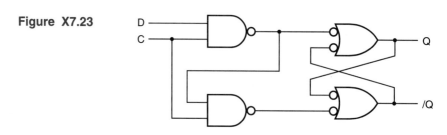

7.24 Suppose that a clocked synchronous state machine with the structure of Figure 7–32 is designed using D latches with active-high C inputs as storage elements. For proper next-state operation, what relationships must be satisfied among the following timing parameters?

t_{Fmin}, t_{Fmax} Minimum and maximum propagation delay of the next-state logic.

t_{CQmin}, t_{CQmax} Minimum and maximum clock-to-output delay of a D latch.

t_{DQmin}, t_{DQmax} Minimum and maximum data-to-output delay of a D latch.

t_{setup}, t_{hold} Setup and hold times of a D latch.

t_H, t_L Clock HIGH and LOW times.

7.25 Redesign the state machine in Drill 7.9 using just three inverting gates—NAND or NOR—and no inverters.

7.26 Draw a state diagram for a clocked synchronous state machine with two inputs, INIT and X, and one Moore-type output Z. As long as INIT is asserted, Z is continuously 0. Once INIT is negated, Z should remain 0 until X has been 0 for two successive ticks and 1 for two successive ticks, regardless of the order of occurrence. Then Z should go to 1 and remain 1 until INIT is asserted again. Your state diagram should be neatly drawn and planar (no crossed lines). (*Hint:* No more than ten states are required).

7.27 Design a clocked synchronous state machine that checks a serial data line for even parity. The circuit should have two inputs, SYNC and DATA, in addition to CLOCK, and one Moore-type output, ERROR. Devise a state/output table that does the job using just four states, and include a description of each state's meaning in the table. Choose a 2-bit state assignment, write transition and excitation equations, and draw the logic diagram. Your circuit may use D flip-flops, J-K flip-flops, or one of each.

7.28 Design a clocked synchronous state machine with the state/output table shown in Table X7.28, using D flip-flops. Use two state variables, Q1 Q2, with the state assignment A = 00, B = 01, C = 11, D = 10.

Table X7.28

	X		
S	0	1	Z
A	B	D	0
B	C	B	0
C	B	A	1
D	B	C	0
	S*		

7.29 Repeat Exercise 7.28 using J-K flip-flops.

7.30 Write a new transition table and derive minimal-cost excitation and output equations for the state table in Table 7–10 using the "simplest" state assignment in Table 7–11 and D flip-flops. Compare the cost of your excitation and output logic (when realized with a two-level AND-OR circuit) with the circuit in Figure 7–50.

7.31 Repeat Exercise 7.30 using the "almost one-hot" state assignment in Table 7–11.

7.32 Suppose that the state machine in Figure 7–50 is to be built using 74LS74 D flip-flops. What signals should be applied to the flip-flop preset and clear inputs?

7.33 Write new transition and excitation tables and derive minimal-cost excitation and output equations for the state table in Table 7–10 using the "simplest" state assignment in Table 7–11 and J-K flip-flops. Compare the cost of your excitation and output logic (when realized with a two-level AND-OR circuit) with the circuit in Figure 7–52.

7.34 Repeat Exercise 7.33 using the "almost one-hot" state assignment in Table 7–11.

7.35 Construct a new excitation table and derive minimal-cost excitation and output equations for the state machine of Table 7–12 using T flip-flops with enable inputs (Figure 7–30). Compare the cost of your excitation and output logic (when realized with a two-level AND-OR circuit) with the circuit in Figure 7–50.

7.36 Determine the full 8-state table of the circuit in Figure 7–50. Use the names U1, U2, and U3 for the unused states (001, 010, and 011). Draw a state diagram and explain the behavior of the unused states.

7.37 Repeat Exercise 7.36 for the circuit of Figure 7–52.

7.38 Write a transition table for the nonminimal state table in Figure 7–47(a) that results from assigning the states in binary counting order, INIT–OKA1 = 000–110. Write corresponding excitation equations for D flip-flops, assuming a minimal-cost disposition of the unused state 111. Compare the cost of your equations with the minimal-cost equations for the minimal state table presented in the text.

7.39 Write the application table for a T flip-flop with enable.

7.40   In many applications, the outputs produced by a state machine during or shortly after reset are irrelevant, as long as the machine begins to behave correctly a short time after the reset signal is removed. If this idea is applied to Table 7–10, the INIT state can be removed and only two state variables are needed to code the remaining four states. Redesign the state machine using this idea. Write a new state table, transition table, excitation table for D flip-flops, minimal-cost excitation and output equations, and logic diagram. Compare the cost of the new circuit with that of Figure 7–50.

7.41   Repeat Exercise 7.40 using J-K flip-flops, and use Figure 7–52 to compare cost.

7.42   Redesign the 1s-counting machine of Table 7–16, assigning the states in binary counting order (S0–S3 = 00, 01, 10, 11). Compare the cost of the resulting sum-of-products excitation equations with the ones derived in the text.

7.43   Repeat Exercise 7.42 using J-K flip-flops.

7.44   Repeat Exercise 7.42 using T flip-flops with enable.

7.45   Redesign the combination-lock machine of Table 7–18, assigning coded states in Gray-code order (A–H = 000, 001, 011, 010, 110, 111, 101, 100). Compare the cost of the resulting sum-of-products excitation equations with the ones derived in the text.

7.46   Find a 3-bit state assignment for the combination-lock machine of Table 7–18 that results in less costly excitation equations than the ones derived in the text. (*Hint:* Use the fact that inputs 1–3 are the same as inputs 4–6 in the required input sequence.)

7.47   What changes would be made to the excitation and output equations for the combination-lock machine in Section 7.4.6 as the result of performing a formal multiple-output minimization procedure (Section 4.3.8) on the five functions? You need not construct 31 product maps and go through the whole procedure; you should be able to "eyeball" the excitation and output maps in Section 7.4.6 to see what savings are possible.

7.48   Analyze the feedback sequential circuit in Figure 7–19, assuming that the /PR and /CLR inputs are always 1. Derive excitation equations, construct a transition table, and analyze the transition table for critical and noncritical races. Name the states, and write a state/output table and a flow/output table. Show that the flow table performs the same function as Table 7–31.

7.49   Draw the logic diagram for a circuit that has one feedback loop, but that is *not* a sequential circuit. That is, the circuit's output should be a function of its current input only. In order to prove your case, break the loop and analyze the circuit as if it were a feedback sequential circuit, and demonstrate that the outputs for each input combination do not depend on the "state."

7.50 A *BUT flop* may be constructed from a single NBUT gate as shown in Figure X7.50.   *BUT flop*
(An *NBUT gate* is simply a BUT gate with inverted outputs; see Exercise 5.31 for   *NBUT gate*
the definition of a BUT gate.) Analyze the BUT flop as a feedback sequential circuit
and obtain excitation equations, transition table, and flow table. Is this circuit good
for anything, or is it a flop?

Figure X7.50

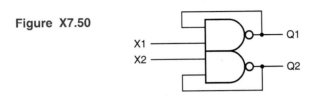

7.51 Repeat Exercise 7.50 for the BUT flop in Figure X7.51.

Figure X7.51

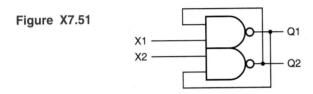

7.52 A clever student designed the circuit in Figure X7.52 as a solution to Exercise 5.55.
However, the circuit didn't work correctly 100% of the time. Analyze the circuit
and explain why.

Figure X7.52

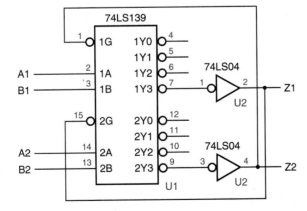

7.53 Show that a 4-bit ones'-complement adder with end-around carry is a feedback
sequential circuit.

7.54 Analyze the feedback sequential circuit in Figure X7.54. Break the feedback loops, write excitation equations, and construct a transition and output table, showing the stable total states. What application might this circuit have?

Figure X7.54

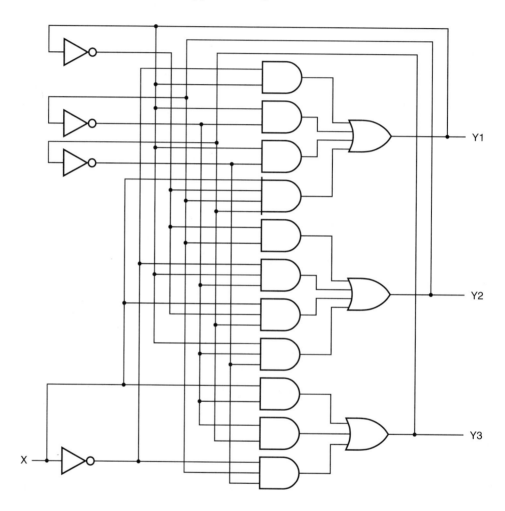

7.55 We claimed in Section 7.6.1 that all single-loop feedback sequential circuits have an excitation equation of the form

$$Q* = \text{(forcing term)} + \text{(holding term)} \cdot Q$$

Why aren't there any practical circuits whose excitation equation substitutes Q' for Q above?

7.56 Design a latch with two control inputs, C1 and C2, and three data inputs, D1, D2, and D3. The latch is to be "open" only if both control inputs are 1, and it is to store a 1 if any of the data inputs is 1. Use hazard-free two-level sum-of-products circuits for the excitation functions.

7.57 Repeat Exercise 7.56, but minimize the number of gates required; the excitation circuits may have multiple levels of logic.

7.58 Redraw the timing diagram in Figure 7–63, showing the internal state variables of the pulse-catching circuit of Figure 7–68, assuming that it starts in state 00.

7.59 The general solution for obtaining a race-free state assignment of 2^n states using $2n - 1$ state variables yields the adjacency diagram shown in Figure X7.59 for the $n = 2$ case. Compare this diagram with Figure 7–66. Which is better, and why?

Figure X7.59

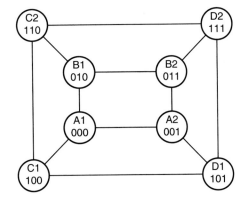

7.60 Design a fundamental-mode flow table for a pulse-catching circuit similar to the one described in Section 7.6.2, except that the circuit should detect both 0-to-1 and 1-to-0 transitions on P.

7.61 Design a fundamental-mode flow table for a double-edge-triggered D flip-flop, one that samples its inputs and changes its outputs on both edges of the clock signal.

7.62 Design a fundamental-mode flow table for a circuit with two inputs, EN and CLKIN, and a single output, CLKOUT, with the following behavior. A clock period is defined to be the interval between successive rising edges of CLKIN. If EN is asserted during an entire given clock period, then CLKOUT should be "on" during the next clock period; that is, it should be identical to CLKIN. If EN is negated during an entire given clock period, then CLKOUT should be "off" (constant 1) during the next clock period. If EN is both asserted and negated during a given clock period, then CLKOUT should be on in the next period if it had been off, and off if it had been on. After writing the fundamental-mode flow table, reduce it by combining "compatible" states if possible.

7.63 Design a circuit that meets the specifications of Exercise 7.62 using edge-triggered D flip-flops (74LS74) or J-K flip-flops (74LS109) and NAND and NOR gates without

8.1 SYNCHRONOUS DESIGN METHODOLOGY

synchronous system   In a *synchronous system*, all flip-flops are clocked by the same, common clock signal, and preset and clear inputs are not used, except for system initialization. Although it's true that all the world does not march to the tick of a common clock, within the confines of a digital system or subsystem we can make it so. When we interconnect digital systems or subsystems that use different clocks, we can usually identify a limited number of asynchronous signals that need special treatment, as we'll show later, in Section 9.7.3.

Races and hazards are not a problem in synchronous systems, for two reasons. First, the only fundamental-mode circuits that might be subject to races or essential hazards are predesigned elements (e.g., discrete flip-flops or ASIC cells), guaranteed by the manufacturer to work properly. Second, even though the combinational circuits that drive flip-flop control inputs may contain static or dynamic or function hazards, these hazards have no effect, since the control inputs are sampled only *after* the hazard-induced glitches have had a chance to settle out.

Aside from designing the functional behavior of each state machine, the designer of a practical synchronous system or subsystem must perform just three well-defined tasks to ensure reliable system operation:

(1) Minimize and determine the amount of clock skew in the system, as discussed in Section 9.7.1.

(2) Ensure that flip-flops have positive setup- and hold-time margins, including an allowance for clock skew, as described in Section 9.1.4.

(3) Identify asynchronous inputs, synchronize them with the clock, and ensure that the synchronizers have an adequately low probability of failure, as described in Sections 9.7.3 and 9.8.

These practical issues are discussed in the next chapter. In this chapter we'll show ways of designing the required state-machine functionality. We begin by looking at a general model for synchronous system structure and an example.

8.1.1 Synchronous System Structure

The sequential-circuit design examples that we gave in Chapter 7 were mostly individual state machines with a small number of states. If a sequential circuit has more than a few flip-flops, then it's not desirable (and often not possible) to treat the circuit as a single, monolithic state machine, because the number of states would be too large to handle.

Fortunately, most digital systems or subsystems can be partitioned into two or more parts. Whether the system processes numbers, digitized voice signals, or

Figure 8–1
Synchronous system structure.

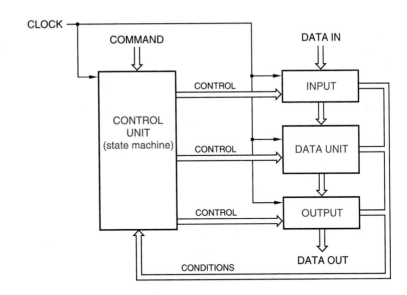

a stream of spark-plug pulses, a certain part of the system, which we'll call the *data unit*, can be viewed as storing, routing, combining, and generally processing "data." Another part, which we'll call the *control unit*, can be viewed as starting and stopping actions in the data unit, testing conditions, and deciding what to do next according to circumstances. In general, only the control unit must be designed as a state machine. The data unit and its components are typically handled at a higher level of abstraction, such as:

data unit
control unit

- Registers. A collection of flip-flops is loaded in parallel with many bits of "data," which can then be used or retrieved together.

- Specialized functions. These include multibit counters and shift registers, which increment or shift their contents on command.

- Read/write memory. Individual latches or flip-flops in a collection of the same can be written or read out.

These topics are discussed in later chapters.

Figure 8–1 is a general block diagram of a system with a control unit and a data unit. We have also included explicit blocks for input and output, but we could have just as easily absorbed these functions into the data unit itself. The control unit is a state machine whose inputs include *command inputs* that indicate how the machine is to function, and *condition inputs* provided by the data unit. The command inputs may be supplied by another subsystem or by a user to set the general operating mode of the control machine (RUN/HALT, NORMAL/TURBO, etc.), while the condition inputs allow the control unit to change its behavior as required by circumstances in the data unit (ZERO DETECT, MEMORY FULL, etc.).

command input
condition input

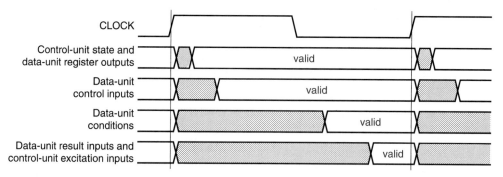

Figure 8–2   Operations during one clock cycle in a synchronous system.

A key characteristic of the structure in Figure 8–1 is that the control unit and the data unit both use the same common clock. Figure 8–2 illustrates the operations of both units during a typical clock cycle:

(1) Shortly after the beginning of the clock period, the control-unit state and the data-unit register outputs are valid.

(2) Next, Moore-type outputs of the control-unit state machine become valid. These signals are control *inputs* to the data unit. They determine what data-unit functions are performed in the rest of the clock period, for example, selecting memory addresses, multiplexer paths, and arithmetic operations.

(3) Near the end of the clock period, data-unit condition outputs such as zero- or overflow-detect are valid, and are made available to the control unit.

(4) At the end of the clock period, just before the setup-time window begins, the next-state logic of the control-unit state machine has determined the next state based on the current state and command and condition inputs, and computational results in the data unit are available to be loaded into data-unit registers.

(5) After the clock edge, the whole cycle may repeat.

Control-unit state-machine outputs may be of the Mealy or Moore type; the latter was assumed in Figure 8–2. Moore-type outputs control the data unit's actions strictly according to the current state, which depends on *past* conditions in the data unit. Mealy-type outputs may select different actions in the data unit according to *current* conditions in the data unit. This increases flexibility, but may also increase delay and thereby lengthen the minimum clock period for correct system operation. Also, Mealy-type outputs must not create feedback loops. For example, a signal that adds 1 to an adder's input if the adder output is not 0 causes an oscillation if the adder output is −1.

8.1.2   A Synchronous System Design Example

To give you a preview of several elements of synchronous system design, this
subsection presents a representative example of a synchronous system.   The
example is a *shift-and-add multiplier* for unsigned integers using the algorithm   *shift-and-add*
of Section 2.8. Its data unit uses several off-the-shelf components, including shift   *multiplier*
registers that are introduced later, in Section 9.4. Its control unit is described by
an ASM chart of the type introduced in Section 8.4. You should come back and
review the description below after you have studied both of these topics.

      Figure 8–3 illustrates data-unit registers and functions that are used to per-
form an 8-bit multiplication:

MPY/LPROD   A shift register that initially stores the multiplier, and accumulates
             the low-order bits of the product as the algorithm is executed.

    HPROD   Initially cleared, and accumulates the high-order bits of the product
             as the algorithm is executed.

     MCND   Stores the multiplicand throughout the algorithm.

        F   A combinational function equal to the 9-bit sum of HPROD and
             MCND if the low-order bit of MPY/LPROD is 1, and equal to HPROD
             (extended to 9 bits) otherwise.

The MPY/LPROD shift register serves a dual purpose, holding both untested multiplier bits (on the right) and unchanging product bits (on the left) as the algorithm
is executed.   At each step it shifts right one bit, discarding the multiplier bit
that was just tested, moving the next multiplier bit to be tested to the rightmost
position, and loading into the leftmost position one more product bit that will
not change for the rest of the algorithm.

Figure 8–3
Registers and functions used by
the shift-and-add multiplication
algorithm.

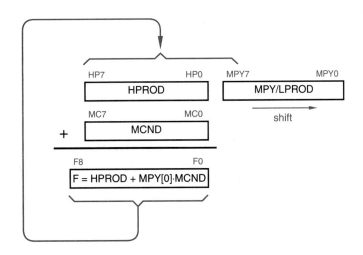

Figure 8–4 is an MSI design for the data unit. An 8-bit bidirectional data bus, D[0–7] is used to load the multiplier and multiplicand into two registers before a multiplication begins. When the multiplication is completed, the same bus is used to read the 16-bit result in two steps. The following signals are used for these operations:

/LDMCND   When asserted, enables the multiplicand register U1 to be loaded at the next clock tick.

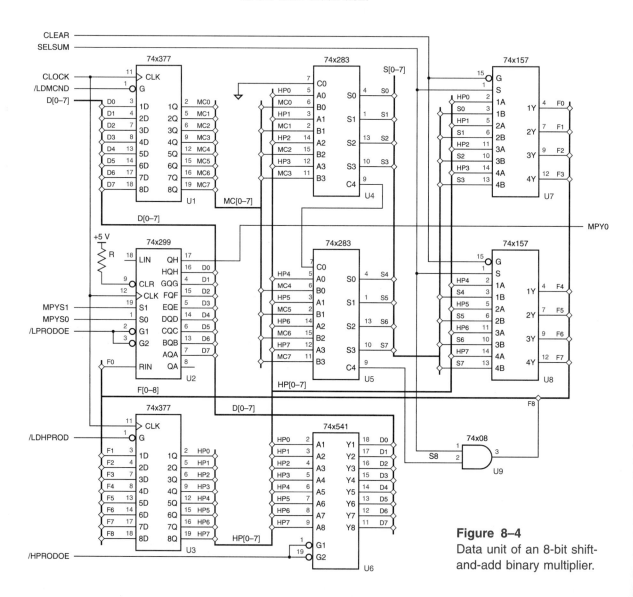

Figure 8–4

Data unit of an 8-bit shift-and-add binary multiplier.

MPYS1 MPYS0   When 11, these signals enable the multiplier/low-product register U2 to be loaded at the next clock tick. They are set to 01 during the multiplication operation to enable the register to shift right, and are 00 at other times to preserve the register's contents.

/LPRODOE   When asserted, enables the multiplier/low-product register U2 to drive the D[0–7] bus; used to read the low-order result byte.

/HPRODOE   When asserted, enables the high-product register U3 to be driven through U6 onto the D[0–7] bus; used to read the high-order result byte.

Among these signals, only MPYS0 is used by the control-unit state machine; the rest are assumed to be generated elsewhere and should be negated while the multiplication algorithm is being executed.

An ASM chart for the multiplier's control-unit state machine appears in Figure 8–5. The state machine has the following inputs and outputs:

START   An external command input that starts a multiplication.

MPY0   A condition input from the data unit, the next multiplier bit to test.

MPYS0   A control output that enables the MPY/LPROD register to shift right.

LDHPROD   A control output that enables the HPROD register to be loaded.

SELSUM   A control output that selects between the shifted adder output or shifted HPROD to be loaded back into HPROD.

DONE   A status output indicating completion of the multiplication algorithm.

Note that the state machine's MPYS0 output must be combined with an external signal, not shown, that loads the MPY/LPROD register before a multiplication begins; also see Exercise 9.42.

| The M1–M8 states in the ASM chart are very similar and repetitive. This is a good indication that the state machine is a good candidate for decomposition using a counter, a possibility explored in Exercise 9.43. | ANOTHER APPROACH |

8.2 AD HOC STATE-MACHINE DESIGN

Small state machines can often be designed by directly translating the word description into a circuit. The circuit cost and design effort resulting from this approach may be better or worse than what is obtained with the formal state-table design and synthesis approach described in Section 7.4.

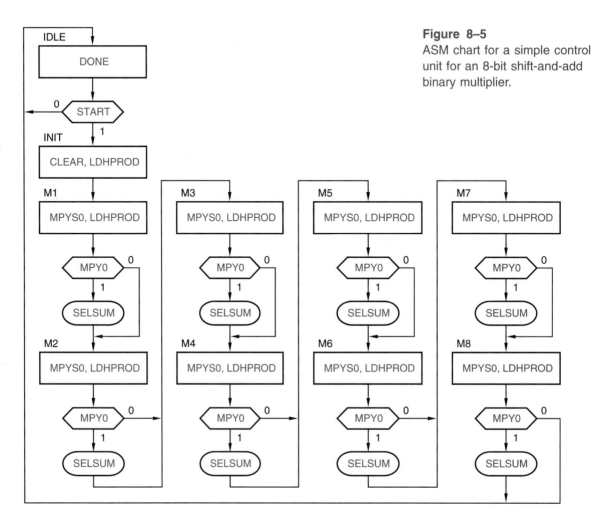

Figure 8–5
ASM chart for a simple control unit for an 8-bit shift-and-add binary multiplier.

8.2.1 An Ad Hoc Example

As an example, let's reconsider a design problem from Section 7.4:

> Design a clocked synchronous state machine with two inputs, A and B, and a single output Z that is 1 if:
>
> - A had the same value at each of the two previous clock ticks, *or*
> - B has been 1 since the last time that the first condition was true.
>
> Otherwise, the output should be 0.

If the second condition is ignored, the machine is very easy to design. Two flip-flops can be used to store the value of the A input for the previous two clock

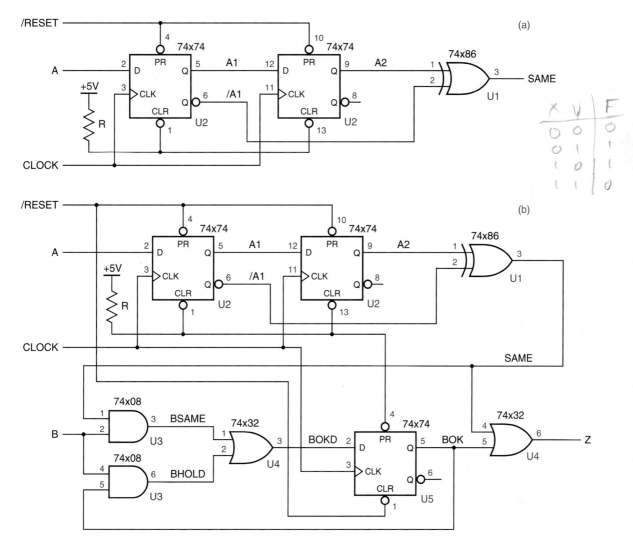

Figure 8–6 Ad hoc design of the example state machine: (a) ignoring the second condition in the design specification; (b) complete design.

ticks. An XOR gate can then be connected to the flip-flop outputs to indicate whether A has had the same value for two clock ticks. Figure 8–6(a) is a circuit resulting from these ideas.

The next step in the ad hoc design is to handle the second condition in the design specification. This can be accomplished by adding another flip-flop, one that gets set to 1 if the first condition is satisfied, and remains 1 as long as B is 1. The Z output is then the OR of the new flip-flop's output and the first signal, as shown in Figure 8–6(b). The figure also shows connections for an active-low

/RESET input. Comparing with Figure 7–50 on page 495, the ad hoc design has the same number of flip-flops and one more gate. Depending on how your brain works, you may or may not be more confident that the circuit obtained with the ad hoc approach satisfies the design specification (see Exercise 8.1).

8.2.2 Finite-Memory Design

finite-memory machine

An entire class of machines, called *finite-memory machines*, are quite appropriate for ad-hoc design. Such a machine's outputs are completely determined by its current input and its inputs and outputs during the previous n clock ticks, where n is a finite, bounded integer. Any machine that can be realized as shown in Figure 8–7 is a finite-memory machine. Note that a finite-*state* machine need not be a finite-*memory* machine; for example, a modulo-n counter with an enable input and a "MAX" output has only n states, but its output may depend on the value of the enable input at every clock tick since initialization.

The "combination-lock" state machine of Section 7.4.6 on page 501 is a finite-memory machine; its design specification is repeated below:

> Design a clocked synchronous state machine with one input, X, and two outputs, UNLK and HINT. The UNLK output should be 1 if and only if X is 0 and the sequence of inputs received on X at the preceding seven clock ticks was 0110111. The HINT output should be 1 if and only if the current value of X is the correct one to move the machine closer to being in the "unlocked" state (with UNLK = 1).

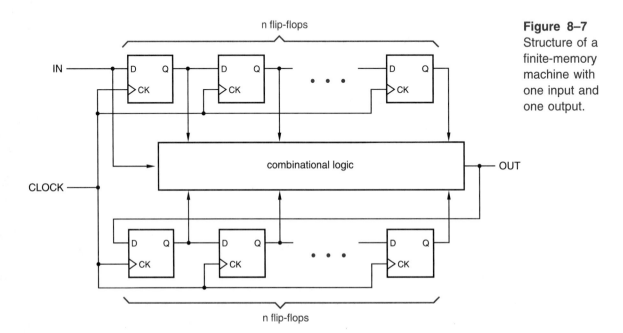

Figure 8–7
Structure of a finite-memory machine with one input and one output.

If we ignore the HINT output, the combination lock is quite obviously a finite-memory machine: the UNLK output can be obtained by looking at the last seven inputs and determining whether they were 0110111. A shift register can be used to store the last seven inputs, and two gates decode the required pattern, as shown in Figure 8–8.

Compared to the standard state-table design in Section 7.4.6, the circuit in Figure 8–8 is a "no-brainer" to design. It might be preferred for that reason, even though it has more flip-flops. On the other hand, providing the HINT output is a bit tougher. The machine is still a finite-memory machine, but the HINT output is a real mess to design (think about it, or try working Exercise 8.15). For that reason, we may elect to go back to the state-table approach.

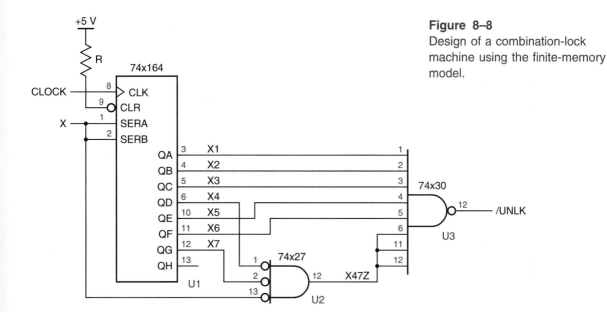

Figure 8–8
Design of a combination-lock machine using the finite-memory model.

In summary, an ad hoc design approach may be appropriate for small machines whose word descriptions translate directly into flip-flops that "remember things" and combinational logic that decodes the things remembered. However, we need better structured approaches to solve more complex problems.

Since the word description or design specification for a state machine is often fuzzy and sometimes nonexistent, a useful design approach must make the designer confident that the machine will really solve the problem at hand. That is, the design approach should highlight, rather than hide, any ambiguities in the design specification, allowing the ambiguities to be resolved. The next two sections describe two approaches to state-machine design that can provide a high degree of confidence.

AD HOC
SUMMARY

8.3   DESIGNING STATE MACHINES USING STATE DIAGRAMS

Most people like to take a graphical approach to design—you've probably solved many problems just by doodling. For that reason, state diagrams are often used to design small- to medium-sized state machines. In this section, we'll give examples of state-diagram design, and describe a simple procedure for synthesizing circuits from the state diagrams. Even when a state-machine description language for PLD-based state machines is available, as in Section 10.7, some designers like to sketch a state diagram before writing a detailed program.

Designing a state diagram is much like designing a state table, which, as we showed in Section 7.4.1, is much like writing a program. However, there is one fundamental difference between a state diagram and a state table, a difference that makes state-diagram design simpler but also more error prone:

- A state table is an exhaustive listing of the next states for each state/input combination. No ambiguity is possible.

- A state diagram contains a set of arcs labeled with transition expressions. Even when there are many inputs, only one transition expression is required per arc. However, when a state diagram is constructed, there is no guarantee that the transition expressions written on the arcs leaving a particular state cover all the input combinations exactly once.

*ambiguous state
diagram*

In an improperly constructed (*ambiguous*) state diagram, the next state for some input combinations may be unspecified, which is generally undesirable, while multiple next states may be specified for others, which is just plain wrong.

Figure 8–9
T-bird tail lights.

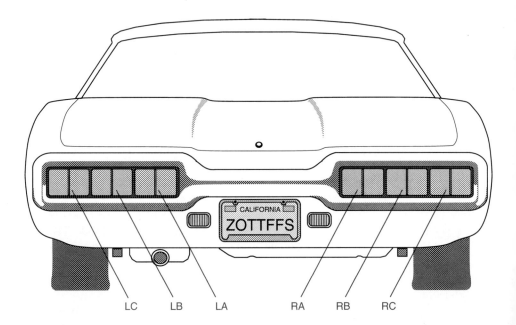

LC   LB   LA   RA   RB   RC

Thus, considerable care must be taken in the design of state diagrams; we'll give several examples.

Our first example is a state machine that controls the tail lights of a 1965 Ford Thunderbird, shown in Figure 8–9. There are three lights on each side, and for turns they operate in sequence to show the turning direction, as illustrated in Figure 8–10. The state machine has two input signals, LEFT and RIGHT, that indicate the driver's request for a left turn or a right turn. It also has an emergency-flasher input, HAZ, that requests the tail lights to be operated in hazard mode—all six lights flashing on and off in unison. We also assume the existence of a free-running clock signal whose frequency equals the desired flashing rate for the lights.

Given the foregoing requirements, we can design a clocked synchronous state machine to control the T-bird tail lights. We will design a Moore machine, so that the state alone determines which lights are on and which are off. For a left turn, the machine should cycle through four states, in which the right-hand lights are off and 0, 1, 2, or 3 of the left-hand lights are on. Likewise, for a right turn, it should cycle through four states in which the left-hand lights are off and 0, 1, 2, or 3 of the right-hand lights are on. In hazard mode, only two states are required—all lights on and all lights off.

Figure 8–11 shows our first cut at a state diagram for the machine. A common IDLE state is defined in which all of the lights are off. When a left turn is requested, the machine goes through three states in which 1, 2, and 3 of the left-hand lights are on, and then back to IDLE; right turns work similarly. In

---

Actually, Figure 8–9 looks more like the rear end of a Mercury Capri, which also had sequential tail lights.

WHOSE
REAR END?

Figure 8–10

Flashing sequence for T-bird tail lights: (a) left turn; (b) right turn.

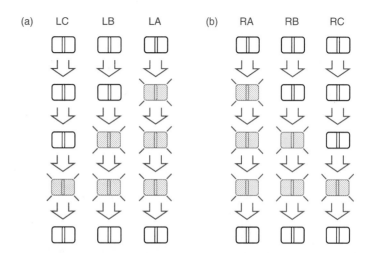

Figure 8–11
Initial state diagram and output
table for T-bird tail lights.

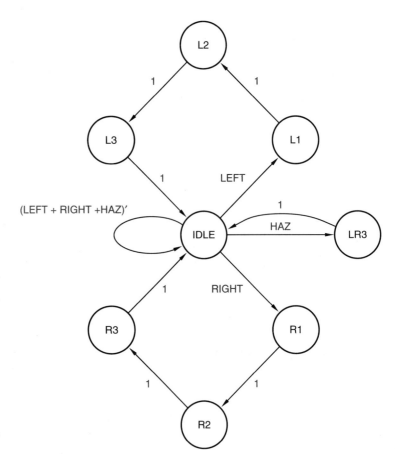

Output Table

State	LC	LB	LA	RA	RB	RC
IDLE	0	0	0	0	0	0
L1	0	0	1	0	0	0
L2	0	1	1	0	0	0
L3	1	1	1	0	0	0
R1	0	0	0	1	0	0
R2	0	0	0	1	1	0
R3	0	0	0	1	1	1
LR3	1	1	1	1	1	1

the hazard mode, the machine cycles back and forth between the IDLE state and a state in which all six lights are on. Since there are so many outputs, we've included a separate output table rather than writing output values on the state diagram. Even without assigning coded states to the named states, we can write output equations from the output table, if we let each state name represent a logic expression that is 1 only in that state:

$$LA = L1 + L2 + L3 + LR3 \qquad RA = R1 + R2 + R3 + LR3$$
$$LB = L2 + L3 + LR3 \qquad RB = R2 + R3 + LR3$$
$$LC = L3 + LR3 \qquad RC = R3 + LR3$$

There's one big problem with the state diagram of Figure 8–11—it doesn't properly handle the case of multiple inputs asserted simultaneously. For example, what happens in the IDLE state if both LEFT and HAZ are asserted? According to the state diagram, the machine goes to two states, L1 and LR3, which is impossible. In reality, the machine would have only one next state, which could

Figure 8–12
Corrected state diagram for
T-bird tail lights.

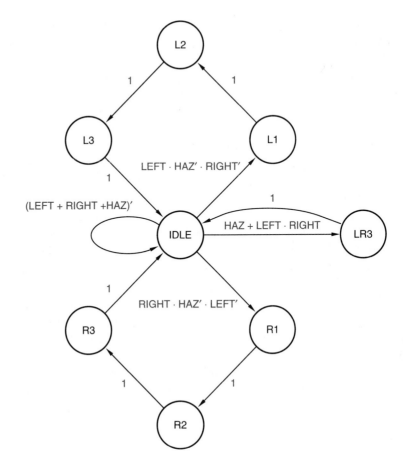

be L1, LR3, or a totally unrelated (and possibly unused) third state, depending on details of the state machine's realization (e.g., see Exercise 8.18).

The problem is fixed in Figure 8–12, where we have given the HAZ input priority. Also, we treat LEFT and RIGHT asserted simultaneously as a hazard request, since the driver is clearly confused and needs help.

The new state diagram is unambiguous because the transition expressions on the arcs leaving each state are mutually exclusive and all-inclusive. That is, for each state, no two expressions are 1 for the same input combination, and some expression is 1 for every input combination. This can be confirmed algebraically for this or any other state diagram by performing two steps:

(1) *Mutual exclusion.* For each state, show that the logical product of each possible pair of transition expressions on arcs leaving that state is 0. If there are n arcs, then there are $n(n-1)/2$ logical products to evaluate.   *mutual exclusion*

(2) *All inclusion.* For each state, show that the logical sum of the transition expressions on all arcs leaving that state is 1.   *all inclusion*

Figure 8–13
Enhanced state diagram for
T-bird tail lights.

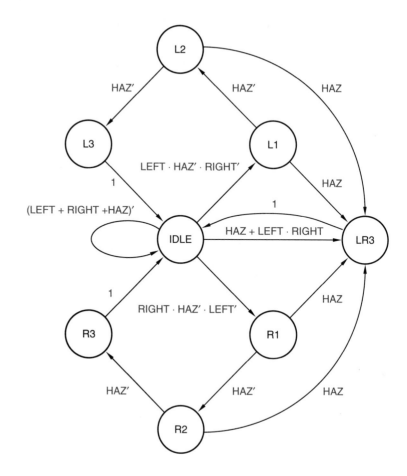

If there are many transitions leaving each state, these steps, especially the first one, are very difficult to perform. However, typical state machines, even ones with lots of states and inputs, don't have many transitions leaving each state, since most designers can't dream up such complex machines in the first place. This is where the trade-off between state-table and state-diagram design occurs. In state-table design, the foregoing steps are not required, because the structure of a state table guarantees mutual exclusion and all inclusion. But if there are a lot of inputs, the state table has *lots* of columns.

Verifying that a state diagram is unambiguous may be difficult in principle, but it's not too bad in practice for small state diagrams. In Figure 8–12, most of the states have a single arc with a transition expression of 1, so verification is trivial. Real work is needed only to verify the IDLE state, which has four transitions leaving it. This can be done on a sheet of scratch paper by listing the eight combinations of the three inputs, and checking off the combinations covered by each transition expression. Each combination should have exactly

Table 8–1
State assignment for T-bird tail
lights machine.

State	Q2	Q1	Q0
IDLE	0	0	0
L1	0	0	1
L2	0	1	1
L3	0	1	0
R1	1	0	1
R2	1	1	1
R3	1	1	0
LR3	1	0	0

one check. As another example, consider the state diagrams in Figures 7–40 and 7–42 on pages 478 and 480; both can be verified mentally.

Returning to the T-bird tail lights machine, we can now synthesize a circuit from the state diagram if we wish. However, if we want to make any changes in the machine's behavior, now is the time to do it, before we've gone to the trouble of synthesizing a circuit. In particular, notice that once a left- or right-turn cycle has begun, the state diagram in Figure 8–12 allows the cycle to run to completion, even if HAZ is asserted. While this may have a certain aesthetic appeal, it would be safer for the car's occupants to have the machine go into hazard mode as soon as possible. The state diagram is modified to provide this behavior in Figure 8–13.

Now we're finally ready to synthesize a circuit for the T-bird machine. The state diagram has eight states, so we'll need a minimum of three flip-flops to code the states. Obviously, there are many state assignments possible (8! to be exact); we'll use the one in Table 8–1 for the following reasons:

(1) An initial (idle) state of 000 is compatible with most flip-flops and registers, which are easily initialized to the 0 state.

(2) Two state variables, Q1 and Q0, are used to "count" in Gray-code sequence for the left-turn cycle (IDLE→L1→L2→L3→IDLE). This minimizes the number of state-variable changes per state transition, which should simplify the excitation logic.

(3) Because of the symmetry in the state diagram, the same sequence on Q1 and Q0 is used to "count" during a right-turn cycle, while Q2 is used to distinguish between left and right.

(4) The remaining state-variable combination is used for the LR3 state.

The next step is to write a sort of transition table. However, we must use a format different from the transition tables of Section 7.3.3, because the

Table 8–2   Transition list for T-bird tail lights machine.

S	Q2	Q1	Q0	Transition expression	S*	Q2*	Q1*	Q0*
IDLE	0	0	0	(LEFT + RIGHT + HAZ)′	IDLE	0	0	0
IDLE	0	0	0	LEFT·HAZ′·RIGHT′	L1	0	0	1
IDLE	0	0	0	HAZ + LEFT·RIGHT	LR3	1	0	0
IDLE	0	0	0	RIGHT·HAZ′·LEFT′	R1	1	0	1
L1	0	0	1	HAZ′	L2	0	1	1
L1	0	0	1	HAZ	LR3	1	0	0
L2	0	1	1	HAZ′	L3	0	1	0
L2	0	1	1	HAZ	LR3	1	0	0
L3	0	1	0	1	IDLE	0	0	0
R1	1	0	1	HAZ′	R2	1	1	1
R1	1	0	1	HAZ	LR3	1	0	0
R2	1	1	1	HAZ′	R3	1	1	0
R2	1	1	1	HAZ	LR3	1	0	0
R3	1	1	0	1	IDLE	0	0	0
LR3	1	0	0	1	IDLE	0	0	0

transition list

transitions in a state diagram are specified by expressions rather than by an exhaustive tabulation of next states. We'll call the new format a *transition list* because it has one row for each transition or arc in the state diagram.

Table 8–2 is the transition list for the state diagram of Figure 8–13 and the state assignment of Table 8–1. Each row contains the current state, next state, and transition expression for one arc in the state diagram. Both the named and coded versions of the current state and next state are shown. The named states are useful for reference purposes, while the coded states are used to develop transition equations.

Once we have a transition list, the rest of the synthesis steps are pretty much "turning the crank," especially in gate-level or PLD-based design. If you're interested, synthesis procedures are described in Section 8.5. Although these procedures are sometimes applied manually, more often they are embedded in a CAD software package.

We also encountered one "turn-the-crank" step in this section—finding the ambiguities in state diagrams. Even though the procedure we discussed can be easily automated and doesn't require much computer time for most practical-size designs, few CAD programs perform this step. Thus, in most design environments, the designer is responsible for providing a state-machine description that is unambiguous. The algorithmic-state-machine approach described in the next section is one way to do this.

Not
covered

8.4 DESIGNING STATE MACHINES USING ASM CHARTS

Back in Section 7.4, we said that designing a state table or diagram is much
like writing a computer program; perhaps by now you believe it. The *synthesis*
procedures that we'll cover in Section 8.5 may remind you of what compilers do
for programs; that is, they convert a high-level description of a state machine into
a low-level realization. This subsection is still concerned with ideas in *design*,
and shows another way to create a correct, high-level state-machine description
to feed into the CAD software.

An *algorithmic-state-machine (ASM) chart* is a graphical specification of   *algorithmic-state-*
state-machine behavior that is more like a programmer's flowchart than a state   *machine (ASM)*
diagram. Figure 8–14 shows the basic elements that appear in ASM charts: *chart*

- *State box.* An ASM chart has one *state box* per state, showing both the state   *state box*
  name and the state code, and containing a list of Moore-type outputs (if
  any) that are asserted in that state. The main difference between a state box
  and a state diagram's bubble is that the state box has just a *single exit point*
  representing the next-state transition, shown by a single transition arrow
  leaving the box. The arrow leads to a single state box or decision box.

- *Decision box.* A single transition arrow is split into two alternative transitions by a *decision box*, which contains a *condition expression*—a logic   *decision box*
  expression involving the machine's inputs. For input combinations such   *condition expression*
  that the condition expression is 1, the exit path labeled 1 is taken; other-

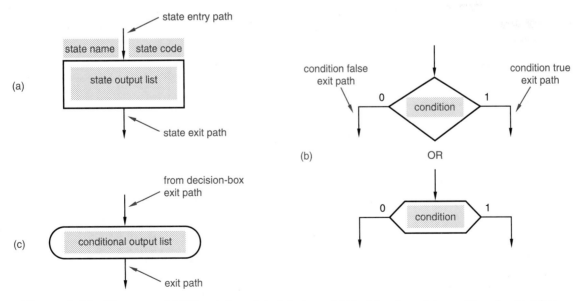

Figure 8–14   Elements of ASM notation: (a) state box; (b) decision box; (c) conditional output box.

wise, the exit path labeled 0 is taken. Each exit path leads to a state box or to another decision box. Multiple decision boxes with different condition expressions are placed in series when a state has more than two different next states.

conditional output box
- *Conditional output box.* This element is placed on top of the exit path of a decision box to specify Mealy-type outputs. It lists outputs that are asserted in the current state (as determined by looking back along the path to a state box), given an input combination that would cause that path to be taken at the clock tick. Like the state-diagram notation for Mealy machines, this is a little misleading, since the outputs are normally asserted for the entire clock period when the conditions are satisfied, not just at the clock tick when the transition is taken.

Figure 8–15
ASM charts: (a) free-running modulo-4 counter; (b) modulo-4 counter with enable; (c) another modulo-4 counter with enable; (d) modulo-4 counter with a Mealy output.

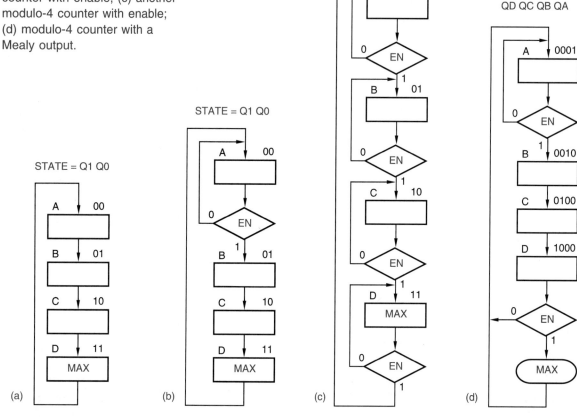

8.4.1 ASM-Chart Examples

A few simple ASM charts are shown in Figure 8–15. The first example, in (a), is a free-running divide-by-4 counter with a MAX output. A note at the top of the chart indicates the signal names of the state variables, Q1 and Q0 in this example. By looking at the coded states, we can see that the designer has decided to realize the state machine as a binary counter (as opposed to a ring counter, Johnson counter, or some other type).

The ASM chart in Figure 8–15(b) uses a decision box to define an enable input. Note that the enable input is only looked at in state A so that, once started, the machine can only be stopped in state A. A more usual enabling behavior is shown in (c); EN is checked in every state. The machine in (d) uses a different state coding. Also, the MAX output is shown as a Mealy output using a conditional output box; MAX is asserted in state D only if EN = 1.

Another example is a "1s-counting machine," which we previously designed using the state-table approach in Table 7–16 on page 499. The machine's behavior is defined as follows:

> Design a clocked synchronous state machine with two inputs, X and Y, and a single output, Z. The output should be 1 if the number of 1 inputs on X and Y, since the machine started is a multiple of 4, and 0 otherwise.

An ASM chart that produces this behavior is shown in Figure 8–16. Note that the conditions $X \cdot Y$ and $X + Y$ in each pair of decision boxes are not mutually exclusive. However, the structure of the ASM chart guarantees that $X \cdot Y = 0$ before the $X + Y$ test is made, so the chart is not ambiguous. A state diagram would have to use the transition expression $X \cdot Y' + X' \cdot Y$ instead of $X + Y$ to prevent ambiguity.

Figure 8–16
ASM chart for the
1s-counting machine.

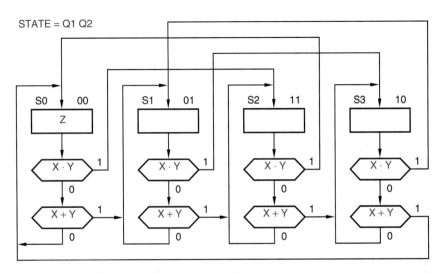

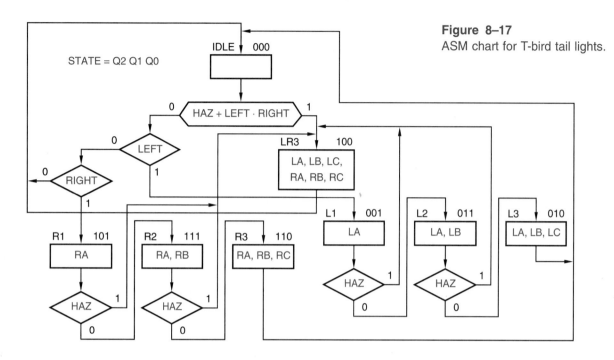

Figure 8–17
ASM chart for T-bird tail lights.

A final example, shown in Figure 8–17, is an ASM chart for the T-bird tail-lights machine of Figure 8–13. The IDLE state in the original state diagram has several transitions leaving it, so the ASM chart requires several decision boxes in series to define the transitions. Each state box contains a list of the outputs that are asserted in that state. The drawing conventions are somewhat relaxed; for example, exit paths leave the most convenient side of a decision box. You are encouraged to study the T-bird tail-lights chart well enough to convince yourself that it does the job.

A big advantage of designing with ASM charts is that a properly constructed chart is guaranteed to provide an unambiguous description of next-state behavior. That is, in each state, every input combination leads to exactly one next state in the chart. This is true because every input combination yields a definite outcome in each decision box and, in a properly constructed chart, all exit paths go to a single next state or to another decision box. Thus, no verification is needed; the mutual-exclusion and all-inclusion properties that we defined for state diagrams are automatic.

AMBIGUOUS
ASM CHARTS

> Some authors allow ASM-chart decision boxes to be put in parallel; that is, they allow a single exit path to go to multiple decision boxes. In this case, an ambiguous ASM chart can be constructed.

ASM charts can be viewed as the graphical equivalent of state-machine description languages, such as ABEL's "state diagram" language that we'll introduce in Section 10.8. An important aspect of both is that they avoid the ambiguity in next-state transitions that can occur when transition expressions are simply written on a state diagram:

- In an ASM chart, a sequence of decision boxes precisely defines the order in which conditions are tested.

- In ABEL, this job is done by nested IF-THEN-ELSE statements.

Extending the comparison between state-machine design and programming, many computer scientists think it is a mistake to write flowcharts and then convert them into code; a program should have only one description, the commented code itself. Likewise, some digital designers (the author included) seldom draw a state diagram or an ASM chart, preferring to go directly to ABEL or another state-machine description that can be used by a CAD program to realize a state-machine directly.

ASM CHARTS
VERSUS
STATE-MACHINE
DESCRIPTION
LANGUAGE

8.4.2 ASM Transition Lists and Expressions

Transition lists can be developed directly from an ASM chart, but the procedure is a bit more involved than the one for state diagrams. In a state diagram, each possible transition from a state has its own arc leaving the state bubble. In an ASM chart, each state box has only one exit path; the possible transitions correspond to all possible paths to next states going through zero or more decision boxes. Each path yields one transition expression (and row in the transition list). In the T-bird machine, there are four possible paths leaving the IDLE state, as shown in Figure 8–18.

Figure 8–18
Possible paths leaving the IDLE
state in T-bird machine.

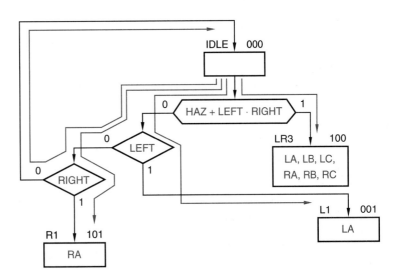

Table 8–3   Transition expressions for the IDLE state in Figure 8–17.

Condition Expressions				
HAZ + LEFT·RIGHT	LEFT	RIGHT	Transition Expression	S*
1	–	–	$Q2' \cdot Q1' \cdot Q0' \cdot (HAZ + LEFT \cdot RIGHT)$	LR3
0	1	–	$Q2' \cdot Q1' \cdot Q0' \cdot (HAZ + LEFT \cdot RIGHT)' \cdot (LEFT)$	L1
0	0	0	$Q2' \cdot Q1' \cdot Q0' \cdot (HAZ + LEFT \cdot RIGHT)' \cdot (LEFT)' \cdot (RIGHT)'$	IDLE
0	0	1	$Q2' \cdot Q1' \cdot Q0' \cdot (HAZ + LEFT \cdot RIGHT)' \cdot (LEFT)' \cdot (RIGHT)$	R1

The transition expression corresponding to an ASM-chart path is the logical product of the current state's minterm and the condition expressions that appear in the decision boxes along the chosen path. Each condition expression is complemented if the chosen path goes through the "false" or "0" exit path of the decision box; otherwise, it appears as is. Thus, the IDLE state in Figure 8–17 has the transition expressions shown in Table 8–3. The L1, L2, R1, and R2 states have a single decision box and therefore two transition expressions each, and the remaining states have no decision boxes and one transition expression each. Using these expressions, you can construct exactly the same transition list as Table 8–2.

WHO USES
ASM CHARTS?

> Many digital designers consider ASM charts to be the most convenient way to specify state-machine behavior. However, in much the same way that most programmers have given up flowcharts in favor of block-structured high-level languages and self-documenting code, experienced logic designers have given up ASM charts in favor of state-machine description languages. Therefore, if you're just getting started in this business, don't worry too much about ASM charts; most modern state machines are designed with ABEL and other state-machine description languages.

* 8.5 SYNTHESIZING STATE MACHINES FROM TRANSITION LISTS

Once a machine's state diagram has been designed and a state assignment has been made, the creative part of the design process is pretty much over. The rest of the synthesis procedure can be carried out by CAD programs.

As we showed previously, a transition list can be constructed from a machine's state diagram and state assignment, or from an ASM chart. This section shows how to synthesize a state machine from its transition list. It also delves into some of the options and nuances of state-machine design using transition

*This section and all of its subsections are optional.

lists. This material is useful not only for synthesizing machines by hand, but also for understanding the internal operation and the external quirks of CAD programs and languages that deal with state machines.

* 8.5.1   Transition Equations

The first step in synthesizing a state machine from a transition list is to develop a set of transition equations that define each next-state variable V* in terms of the current state and input. The transition list can be viewed as a sort of hybrid truth table in which the state-variable combinations for current-state are listed explicitly and input combinations are listed algebraically. Reading down a V* column in a transition list, we find a sequence of 0s and 1s, indicating the value of V* for various (if we've done it right, all) state/input combinations. For reference, the transition list for the T-bird tail-lights machine of Figure 8–13 is repeated in Table 8–4.

A transition equation for a next-state variable V* can be written using a sort of hybrid canonical sum:

$$V* = \sum_{\text{transition-list rows where } V* = 1} (\text{transition p-term})$$

That is, the transition equation has one "transition p-term" for each row of the transition list that contains a 1 in the V* column. A row's *transition p-term*   *transition p-term*
is the product of the current state's minterm and the transition expression. If

Table 8–4   Transition list for T-bird tail-lights machine.

S	Q2	Q1	Q0	Transition Expression	S*	Q2*	Q1*	Q0*
IDLE	0	0	0	(LEFT + RIGHT + HAZ)'	IDLE	0	0	0
IDLE	0	0	0	LEFT·HAZ'·RIGHT'	L1	0	0	1
IDLE	0	0	0	HAZ + LEFT·RIGHT	LR3	1	0	0
IDLE	0	0	0	RIGHT·HAZ'·LEFT'	R1	1	0	1
L1	0	0	1	HAZ'	L2	0	1	1
L1	0	0	1	HAZ	LR3	1	0	0
L2	0	1	1	HAZ'	L3	0	1	0
L2	0	1	1	HAZ	LR3	1	0	0
L3	0	1	0	1	IDLE	0	0	0
R1	1	0	1	HAZ'	R2	1	1	1
R1	1	0	1	HAZ	LR3	1	0	0
R2	1	1	1	HAZ'	R3	1	1	0
R2	1	1	1	HAZ	LR3	1	0	0
R3	1	1	0	1	IDLE	0	0	0
LR3	1	0	0	1	IDLE	0	0	0

the transition expression itself happens to be a simple product term, then the transition equation expresses V* in standard sum-of-products form; we'll find this to be important for PLD-based state-machine designs in Section 10.7.

The transition equation for Q2* in the T-bird machine can be written as the sum of eight p-terms:

$$
\begin{aligned}
Q2* \;=\; & Q2'\cdot Q1'\cdot Q0'\cdot(\text{HAZ} + \text{LEFT}\cdot\text{RIGHT}) \\
& + Q2'\cdot Q1'\cdot Q0'\cdot(\text{RIGHT}\cdot\text{HAZ}'\cdot\text{LEFT}') \\
& + Q2'\cdot Q1'\cdot Q0\cdot(\text{HAZ}) \\
& + Q2'\cdot Q1\cdot Q0\cdot(\text{HAZ}) \\
& + Q2\cdot Q1'\cdot Q0\cdot(\text{HAZ}') \\
& + Q2\cdot Q1'\cdot Q0\cdot(\text{HAZ}) \\
& + Q2\cdot Q1\cdot Q0\cdot(\text{HAZ}') \\
& + Q2\cdot Q1\cdot Q0\cdot(\text{HAZ})
\end{aligned}
$$

Some straightforward algebraic manipulations lead to a simplified transition equation that combines the first two, second two, and last four p-terms above:

$$
\begin{aligned}
Q2* \;=\; & Q2'\cdot Q1'\cdot Q0'\cdot(\text{HAZ} + \text{RIGHT}) \\
& + Q2'\cdot Q0\cdot(\text{HAZ}) \\
& + Q2\cdot Q0
\end{aligned}
$$

Transition equations for Q1* and Q0* may be obtained in a similar manner:

$$
\begin{aligned}
Q1* \;=\; & Q2'\cdot Q1'\cdot Q0\cdot(\text{HAZ}') \\
& + Q2'\cdot Q1\cdot Q0\cdot(\text{HAZ}') \\
& + Q2\cdot Q1'\cdot Q0\cdot(\text{HAZ}') \\
& + Q2\cdot Q1\cdot Q0\cdot(\text{HAZ}') \\
\;=\; & Q0\cdot\text{HAZ}' \\[4pt]
Q0* \;=\; & Q2'\cdot Q1'\cdot Q0'\cdot(\text{LEFT}\cdot\text{HAZ}'\cdot\text{RIGHT}') \\
& + Q2'\cdot Q1'\cdot Q0'\cdot(\text{RIGHT}\cdot\text{HAZ}'\cdot\text{LEFT}') \\
& + Q2'\cdot Q1'\cdot Q0\cdot(\text{HAZ}') \\
& + Q2\cdot Q1'\cdot Q0\cdot(\text{HAZ}') \\
\;=\; & Q2'\cdot Q1'\cdot Q0'\cdot\text{HAZ}'\cdot(\text{LEFT}\oplus\text{RIGHT}) + Q1'\cdot Q0\cdot\text{HAZ}'
\end{aligned}
$$

Except for Q1*, there's no guarantee that the transition equations above are in any sense minimal—in fact, the expressions for Q2* and Q0* aren't even in standard sum-of-products or product-of-sums form. The simplified equations, or the original unsimplified ones, merely provide an unambiguous starting point for whatever combinational design method you might choose to synthesize the excitation logic for the state machine—ad hoc, NAND-NAND, MSI-based, or whatever. As you'll see in Section 10.7, in a PLD-based design you could simply plug the equations into a program that calculates minimal sum-of-products expressions for realization in the PLD.

* 8.5.2 Excitation Equations

While we're on the subject of excitation logic, note that so far we have derived only *transition equations*, not *excitation equations*. However, if we use D flip-flops as the memory elements in our state machines, then the excitation equations are trivial to derive from the transition equations, since the characteristic equation of a D flip-flop is $Q* = D$. Therefore, if the transition equation for a state variable $Qi*$ is

$$Qi* \ = \ \text{expression}$$

then the excitation equation for the corresponding D flip-flop input is

$$Di \ = \ \text{expression}$$

Efficient excitation equations for other flip-flop types, especially J-K, are not so easy to derive (see Exercise 8.24). For that reason, the vast majority of discrete and PLD-based state-machine designs employ D flip-flops.

* 8.5.3 Variations on the Scheme

There are other ways to obtain transition and excitation equations from a transition list. If the column for a particular next-state variable contains fewer 0s than 1s, it may be advantageous to write that variable's transition equation in terms of the 0s in its column. That is, we write

$$V*' = \sum_{\text{transition-list rows where } V* = 0} (\text{transition p-term})$$

That is, $V*'$ is 1 for all of the p-terms for which $V*$ is 0. Thus, a transition equation for $Q2*'$ may be written as the sum of seven p-terms:

$$
\begin{aligned}
Q2*' \ = \ & Q2' \cdot Q1' \cdot Q0' \cdot ((\text{LEFT} + \text{RIGHT} + \text{HAZ})') \\
& + Q2' \cdot Q1' \cdot Q0' \cdot (\text{LEFT} \cdot \text{HAZ}' \cdot \text{RIGHT}') \\
& + Q2' \cdot Q1' \cdot Q0 \cdot (\text{HAZ}') \\
& + Q2' \cdot Q1 \cdot Q0 \cdot (\text{HAZ}') \\
& + Q2' \cdot Q1 \cdot Q0' \cdot (1) \\
& + Q2 \cdot Q1 \cdot Q0' \cdot (1) \\
& + Q2 \cdot Q1' \cdot Q0' \cdot (1) \\
= \ & Q2' \cdot Q1' \cdot Q0' \cdot \text{HAZ}' \cdot \text{RIGHT}' + Q2' \cdot Q0 \cdot \text{HAZ}' + Q1 \cdot Q0' + Q2 \cdot Q0'
\end{aligned}
$$

Both sides of the simplified equation can be complemented to obtain an equation for Q2∗.

To obtain an expression for a next-state variable V∗ directly using the 0s in the transition list, we can complement the right-hand side of the general V∗′ equation using DeMorgan's theorem, obtaining a sort of hybrid canonical product:

$$V* = \prod_{\text{transition-list rows where } V* = 0} (\text{transition s-term})$$

transition s-term

Here, a row's *transition s-term* is the sum of the maxterm for the current state and the complement of the transition expression. If the transition expression is a simple product term, then its complement is a sum, and the transition equation expresses V∗ in product-of-sums form.

* 8.5.4   Realizing the State Machine

Once you have the excitation equations for for a state machine, all you're left with is a multiple-output combinational logic design problem. Of course, there are many ways to realize combinational logic from equations, but the easiest way is just to type them into a PLD program.

Combinational PLDs such as the PAL16L8 that we introduced in Section 6.3 can be used to realize excitation equations up to a certain number of inputs, outputs, and product terms.

Better yet, in Chapter 10 we'll introduce sequential PLDs that include D flip-flops on the same chip with the combinational AND-OR array. For a given number of PLD input and output pins, these sequential PLDs can actually realize larger state machines than their combinational counterparts, because the excitation signals never have to go off the chip. In Section 10.7.3, we'll show how to realize the T-bird tail-lights excitation equations in a sequential PLD.

* 8.6   ANOTHER EXAMPLE STATE MACHINE

This section gives one more example of state-machine design using a state diagram. The example provides a basis for further discussion of a few topics: unused states, output-coded state assignments, and "don't-care" state codings.

* 8.6.1   The Guessing Game

Our final state-machine example is a "guessing game" that can be built as an amusing lab project:

Design a clocked synchronous state machine with four inputs, G1–G4, that are connected to pushbuttons. The machine has four outputs, L1–L4, connected to lamps or LEDs located near the like-numbered pushbuttons. There is also an ERR output connected to a red lamp. In normal operation, the L1–L4 outputs display a 1-out-of-4 pattern. At each clock tick, the pattern is rotated by one position, as in a ring counter; the clock frequency is about 4 Hz.

Guesses are made by pressing a pushbutton, which asserts an input Gi. When any Gi input is asserted, the ERR output is asserted if the "wrong" pushbutton was pressed, that is, if the Gi input detected at the clock tick does not have the same number as the lamp output that was asserted before the clock tick. Once a guess has been made, play stops and the ERR output maintains the same value for one or more clock ticks until the Gi input is negated, then play resumes.

Clearly, we will have to provide four states, one for each position of the rotating pattern, and we'll need at least one state to indicate that play has stopped. A possible state diagram is shown in Figure 8–19. The machine cycles through states S1–S4 as long as no Gi input is asserted, and it goes to the STOP state when a guess is made. Each Li output is asserted in the like-numbered state.

Figure 8–19
First try at a state diagram
for the guessing game.

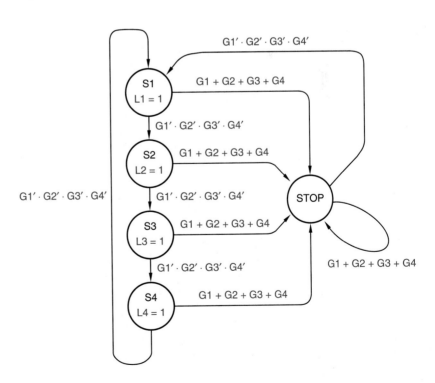

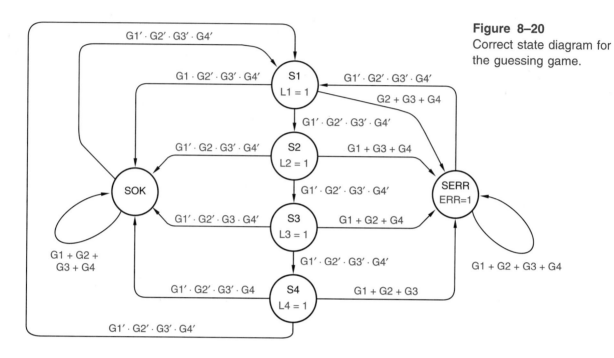

Figure 8–20
Correct state diagram for the guessing game.

Table 8–5   Transition list for guessing-game machine.

Current State				Next State				Output				
S	Q2	Q1	Q0	Transition Expression	S*	Q2*	Q1*	Q0*	L1	L2	L3	L4	ERR
S1	0	0	0	$G1' \cdot G2' \cdot G3' \cdot G4'$	S2	0	0	1	1	0	0	0	0
S1	0	0	0	$G1 \cdot G2' \cdot G3' \cdot G4'$	SOK	1	0	0	1	0	0	0	0
S1	0	0	0	$G2 + G3 + G4$	SERR	1	0	1	1	0	0	0	0
S2	0	0	1	$G1' \cdot G2' \cdot G3' \cdot G4'$	S3	0	1	1	0	1	0	0	0
S2	0	0	1	$G1' \cdot G2 \cdot G3' \cdot G4'$	SOK	1	0	0	0	1	0	0	0
S2	0	0	1	$G1 + G3 + G4$	SERR	1	0	1	0	1	0	0	0
S3	0	1	1	$G1' \cdot G2' \cdot G3' \cdot G4'$	S4	0	1	0	0	0	1	0	0
S3	0	1	1	$G1' \cdot G2' \cdot G3 \cdot G4'$	SOK	1	0	0	0	0	1	0	0
S3	0	1	1	$G1 + G2 + G4$	SERR	1	0	1	0	0	1	0	0
S4	0	1	0	$G1' \cdot G2' \cdot G3' \cdot G4'$	S1	0	0	0	0	0	0	1	0
S4	0	1	0	$G1' \cdot G2' \cdot G3' \cdot G4$	SOK	1	0	0	0	0	0	1	0
S4	0	1	0	$G1 + G2 + G3$	SERR	1	0	1	0	0	0	1	0
SOK	1	0	0	$G1 + G2 + G3 + G4$	SOK	1	0	0	0	0	0	0	0
SOK	1	0	0	$G1' \cdot G2' \cdot G3' \cdot G4'$	S1	0	0	0	0	0	0	0	0
SERR	1	0	1	$G1 + G2 + G3 + G4$	SERR	1	0	1	0	0	0	0	1
SERR	1	0	1	$G1' \cdot G2' \cdot G3' \cdot G4'$	S1	0	0	0	0	0	0	0	1

The only problem with the preceding state diagram is that it doesn't "remember" in the STOP state whether the guess was correct, so it has no way to control the ERR output. This problem is fixed in Figure 8–20, which has two "stopped" states, SOK and SERR. On an incorrect guess, the machine goes to SERR, where ERR is asserted; otherwise, it goes to SOK. Although the machine's word description doesn't require it, the state diagram is designed to go to SERR even if the user tries to fool it by pressing two or more pushbuttons simultaneously, or by changing guesses while stopped.

A transition list corresponding to the state diagram is shown in Table 8–5, using a simple 3-bit binary state encoding with Gray-code order for the S1–S4 cycle. Transition equations for Q1* and Q0* can be obtained from the table as follows:

$$
\begin{aligned}
Q1* \; &= \; Q2' \cdot Q1' \cdot Q0 \cdot (G1' \cdot G2' \cdot G3' \cdot G4') \\
&\quad + Q2' \cdot Q1 \cdot Q0 \cdot (G1' \cdot G2' \cdot G3' \cdot G4') \\
&= \; Q2' \cdot Q0 \cdot G1' \cdot G2' \cdot G3' \cdot G4' \\
Q0* \; &= \; Q2' \cdot Q1' \cdot Q0' \cdot (G1' \cdot G2' \cdot G3' \cdot G4') \\
&\quad + Q2' \cdot Q1' \cdot Q0' \cdot (G2 + G3 + G4) \\
&\quad + Q2' \cdot Q1' \cdot Q0 \cdot (G1' \cdot G2' \cdot G3' \cdot G4') \\
&\quad + Q2' \cdot Q1' \cdot Q0 \cdot (G1 + G3 + G4) \\
&\quad + Q2' \cdot Q1 \cdot Q0 \cdot (G1 + G2 + G4) \\
&\quad + Q2' \cdot Q1 \cdot Q0' \cdot (G1 + G2 + G3) \\
&\quad + Q2 \cdot Q1' \cdot Q0 \cdot (G1 + G2 + G3 + G4)
\end{aligned}
$$

Using a logic minimization program, the Q0* expression can be reduced to 11 product terms in two-level sum-of-products form. An expression for Q2* is best formulated in terms of the 0s in the Q2* column of the transition list:

$$
\begin{aligned}
Q2*' \; &= \; Q2' \cdot Q1' \cdot Q0' \cdot (G1' \cdot G2' \cdot G3' \cdot G4') \\
&\quad + Q2' \cdot Q1' \cdot Q0 \cdot (G1' \cdot G2' \cdot G3' \cdot G4') \\
&\quad + Q2' \cdot Q1 \cdot Q0 \cdot (G1' \cdot G2' \cdot G3' \cdot G4') \\
&\quad + Q2' \cdot Q1 \cdot Q0' \cdot (G1' \cdot G2' \cdot G3' \cdot G4') \\
&\quad + Q2 \cdot Q1' \cdot Q0' \cdot (G1' \cdot G2' \cdot G3' \cdot G4') \\
&\quad + Q2 \cdot Q1' \cdot Q0 \cdot (G1' \cdot G2' \cdot G3' \cdot G4') \\
&= \; (Q2' + Q1') \cdot (G1' \cdot G2' \cdot G3' \cdot G4')
\end{aligned}
$$

The last five columns of Table 8–5 show output values. Thus, output equations can be developed in much the same way as transition equations. However, since this example is a Moore machine, outputs are independent of the transition expressions; only one row of the transition list must be considered for each current state. The output equations are

$$L1 = Q2' \cdot Q1' \cdot Q0' \qquad L3 = Q2' \cdot Q1 \cdot Q0 \qquad ERR = Q2 \cdot Q1' \cdot Q0$$
$$L2 = Q2' \cdot Q1' \cdot Q0 \qquad L4 = Q2' \cdot Q1 \cdot Q0'$$

* 8.6.2   Unused States

Our state diagram for the guessing game has six states, but the actual state machine, built from three flip-flops, has eight. By omitting the unused states from the transition list, we treated them as "don't-cares" in a very limited sense:

- When we wrote equations for Q1* and Q0*, we formed a sum of transition p-terms for state/input combinations that had an explicit 1 in the corresponding columns of the transition list. Although we didn't consider the unused states, our procedure implicitly treated them as if they had 0s in the Q1* and Q0* columns.

- Conversely, we wrote the Q2*′ equation as a sum of transition p-terms for state/input combinations that had an explicit 0 in the corresponding columns of the transition list. Unused states were implicitly treated as if they had 1s in the Q2* column.

As a consequence of these choices, all of the unused states in the guessing-game machine have a coded next state of 100 for all input combinations. That's acceptable or "safe" behavior should the machine stray into an unused state, since 100 is the coding for one of the normal states (SOK).

To treat the unused states as true "don't-cares," we would have to allow them to go to any next state under any input combination. This is simple in principle but may be difficult in practice.

In Section 7.4.4, we showed how to handle unused states as "don't-cares" in the Karnaugh-map method for developing transition/excitation equations. Unfortunately, for all but the smallest problems, Karnaugh maps are unwieldy. On the other hand, commercially available logic minimization programs can easily handle larger problems, but many of them don't handle "don't-cares." Perhaps over the coming years, we'll see more CAD tools that automatically determine don't-care conditions and synthesize minimum-cost excitation logic directly from a transition list or equivalent state-machine description.

* 8.6.3   Output-Coded State Assignment

Let's look at another realization of the guessing-game machine. The machine's outputs are a function of state only; furthermore, a *different* output combination is produced in each named state. Therefore, we can use the outputs as state

Table 8–6 Transition list for guessing-game machine using outputs as state variables.

| | Current State | | | | | | Next State | | | | | |
S	L1	L2	L3	L4	ERR	Transition Expression	S*	L1*	L2*	L3*	L4*	ERR*
S1	1	0	0	0	0	$G1'·G2'·G3'·G4'$	S2	0	1	0	0	0
S1	1	0	0	0	0	$G1·G2'·G3'·G4'$	SOK	0	0	0	0	0
S1	1	0	0	0	0	$G2 + G3 + G4$	SERR	0	0	0	0	1
S2	0	1	0	0	0	$G1'·G2'·G3'·G4'$	S3	0	0	1	0	0
S2	0	1	0	0	0	$G1'·G2·G3'·G4'$	SOK	0	0	0	0	0
S2	0	1	0	0	0	$G1 + G3 + G4$	SERR	0	0	0	0	1
S3	0	0	1	0	0	$G1'·G2'·G3'·G4'$	S4	0	0	0	1	0
S3	0	0	1	0	0	$G1'·G2'·G3·G4'$	SOK	0	0	0	0	0
S3	0	0	1	0	0	$G1 + G2 + G4$	SERR	0	0	0	0	1
S4	0	0	0	1	0	$G1'·G2'·G3'·G4'$	S1	1	0	0	0	0
S4	0	0	0	1	0	$G1'·G2'·G3'·G4$	SOK	0	0	0	0	0
S4	0	0	0	1	0	$G1 + G2 + G3$	SERR	0	0	0	0	1
SOK	0	0	0	0	0	$G1 + G2 + G3 + G4$	SOK	0	0	0	0	0
SOK	0	0	0	0	0	$G1'·G2'·G3'·G4'$	S1	1	0	0	0	0
SERR	0	0	0	0	1	$G1 + G2 + G3 + G4$	SERR	0	0	0	0	1
SERR	0	0	0	0	1	$G1'·G2'·G3'·G4'$	S1	1	0	0	0	0

variables and assign each named state to the required output combination. This sort of *output-coded state assignment* can sometimes result in excitation equations that are simpler than the set of excitation and output equations obtained with a state assignment using a minimum number of state variables.

output-coded state assignment

Table 8–6 is a guessing-game transition list resulting from an output-coded state assignment. Each transition/excitation equation has very few transition p-terms because the state diagram has so few 1s in the next-state columns:

$$L1* = L1'·L2'·L3'·L4·ERR'·(G1'·G2'·G3'·G4')$$
$$+ L1'·L2'·L3'·L4'·ERR'·(G1'·G2'·G3'·G4')$$
$$+ L1'·L2'·L3'·L4'·ERR·(G1'·G2'·G3'·G4')$$
$$L2* = L1·L2'·L3'·L4'·ERR'·(G1'·G2'·G3'·G4')$$
$$L3* = L1'·L2·L3'·L4'·ERR'·(G1'·G2'·G3'·G4')$$
$$L4* = L1'·L2'·L3·L4'·ERR'·(G1'·G2'·G3'·G4')$$
$$ERR* = L1·L2'·L3'·L4'·ERR'·(G2 + G3 + G4)$$
$$+ L1'·L2·L3'·L4'·ERR'·(G1 + G3 + G4)$$
$$+ L1'·L2'·L3·L4'·ERR'·(G1 + G2 + G4)$$
$$+ L1'·L2'·L3'·L4·ERR'·(G1 + G2 + G3)$$
$$+ L1'·L2'·L3'·L4'·ERR·(G1 + G2 + G3 + G4)$$

There are no output equations, of course. The ERR* equation above is the worst in the group, requiring 16 terms to express in minimal sum-of-products or product-of-sums form.

As a group, the equations developed above have about the same complexity as the transition and output equations that we developed from Table 8–5. Even though the output-coded assignment does not produce a simpler set of equations in this example, it can still save cost in a PLD-based design, as we'll show in Section 10.7.

* 8.6.4 "Don't-Care" State Codings

Out of the 32 possible coded states using five variables, only six are used in Table 8–6. The rest of the states are unused and have a next state of 00000 if the machine is built using the equations in the preceding subsection. Another possible disposition for unused states, one that we haven't discussed before, is obtained by careful use of "don't-cares" in the assignment of coded states to current states.

Table 8–7 shows one such state assignment for the guessing-game machine, derived from the output-coded state assignment of the preceding subsection. In this example, every possible combination of current-state variables corresponds to one of the coded states (e.g., 10111 = S1, 00101 = S3). However, *next states* are coded using the same unique combinations as in the preceding subsection. Table 8–8 shows the resulting transition list.

In this approach, each unused current state behaves like a nearby "normal" state; Figure 8–21 illustrates this concept. The machine is well-behaved and goes to a "normal state" if it inadvertently enters an unused state. Yet the approach still allows some simplification of the excitation and output logic by introducing don't-cares in the transition list. When a row's transition p-term is written, current-state variables that are don't-cares in that row are omitted, for example,

$$
\begin{aligned}
\text{ERR*} \;=\; & \text{L1} \cdot (\text{G2} + \text{G3} + \text{G4}) \\
& + \text{L1}' \cdot \text{L2} \cdot (\text{G1} + \text{G3} + \text{G4}) \\
& + \text{L1}' \cdot \text{L2}' \cdot \text{L3} \cdot (\text{G1} + \text{G2} + \text{G4}) \\
& + \text{L1}' \cdot \text{L2}' \cdot \text{L3}' \cdot \text{L4} \cdot (\text{G1} + \text{G2} + \text{G3}) \\
& + \text{L1}' \cdot \text{L2}' \cdot \text{L3}' \cdot \text{L4}' \cdot \text{ERR} \cdot (\text{G1} + \text{G2} + \text{G3} + \text{G4})
\end{aligned}
$$

Compared with the ERR* equation in the preceding subsection, the one above still requires 16 terms to express as a sum of products. However, it requires only five terms in minimal product-of-sums form, which makes its complement more suitable for realization in a PLD.

Table 8–7
Current-state assignment for guessing-game machine using don't-cares.

State	L1	L2	L3	L4	ERR
S1	1	x	x	x	x
S2	0	1	x	x	x
S3	0	0	1	x	x
S4	0	0	0	1	x
SOK	0	0	0	0	0
SERR	0	0	0	0	1

Table 8–8   Transition list for guessing-game machine using don't-care state codings.

	Current State						Next State					
S	L1	L2	L3	L4	ERR	Transition Expression	S*	L1*	L2*	L3*	L4*	ERR*
S1	1	x	x	x	x	$G1' \cdot G2' \cdot G3' \cdot G4'$	S2	0	1	0	0	0
S1	1	x	x	x	x	$G1 \cdot G2' \cdot G3' \cdot G4'$	SOK	0	0	0	0	0
S1	1	x	x	x	x	$G2 + G3 + G4$	SERR	0	0	0	0	1
S2	0	1	x	x	x	$G1' \cdot G2' \cdot G3' \cdot G4'$	S3	0	0	1	0	0
S2	0	1	x	x	x	$G1' \cdot G2 \cdot G3' \cdot G4'$	SOK	0	0	0	0	0
S2	0	1	x	x	x	$G1 + G3 + G4$	SERR	0	0	0	0	1
S3	0	0	1	x	x	$G1' \cdot G2' \cdot G3' \cdot G4'$	S4	0	0	0	1	0
S3	0	0	1	x	x	$G1' \cdot G2' \cdot G3 \cdot G4'$	SOK	0	0	0	0	0
S3	0	0	1	x	x	$G1 + G2 + G4$	SERR	0	0	0	0	1
S4	0	0	0	1	x	$G1' \cdot G2' \cdot G3' \cdot G4'$	S1	1	0	0	0	0
S4	0	0	0	1	x	$G1' \cdot G2' \cdot G3' \cdot G4$	SOK	0	0	0	0	0
S4	0	0	0	1	x	$G1 + G2 + G3$	SERR	0	0	0	0	1
SOK	0	0	0	0	0	$G1 + G2 + G3 + G4$	SOK	0	0	0	0	0
SOK	0	0	0	0	0	$G1' \cdot G2' \cdot G3' \cdot G4'$	S1	1	0	0	0	0
SERR	0	0	0	0	1	$G1 + G2 + G3 + G4$	SERR	0	0	0	0	1
SERR	0	0	0	0	1	$G1' \cdot G2' \cdot G3' \cdot G4'$	S1	1	0	0	0	0

Figure 8–21
State assignment using don't-cares for current states.

Current coded states   Next coded states

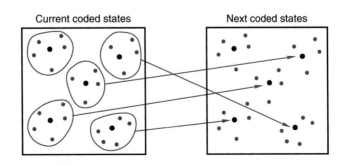

8.7 DECOMPOSING STATE MACHINES

Just like large procedures or subroutines in a programming language, large state machines are difficult to conceptualize, design, and debug. Therefore, when faced with a large state-machine problem, digital designers often look for opportunities to solve it with a collection of smaller state machines.

state-machine
decomposition

There's a well-developed theory of *state-machine decomposition* that you can use to analyze any given, monolithic state machine to determine whether it can be realized as a collection of smaller ones. However, decomposition theory is not too useful for designers who want to avoid designing large state machines in the first place. Rather, a practical designer tries to cast the original design problem into a natural, hierarchical structure, so that the uses and functions of submachines are obvious, making it unnecessary ever to write a state table for the equivalent monolithic machine.

main machine
submachine

The simplest and most commonly used type of decomposition is illustrated in Figure 8–22. A *main machine* provides the primary inputs and outputs and executes the top-level control algorithm. *Submachines* perform low-level steps under the control of the main machine, and may optionally handle some of the primary inputs and outputs.

Perhaps the most commonly used submachine is a counter. The main machine starts the counter when it wishes to stay in a particular main state for *n* clock ticks, and the counter asserts a DONE signal when *n* ticks have occurred. The main machine is designed to wait in the same state until DONE is asserted. This adds an extra output and input to the main machine (START and DONE), but it saves $n - 1$ states.

A decomposed state machine designed along these lines is based on the guessing game of Section 8.6. The original guessing game is easy to win after a minute of practice because the lamps cycle at a very consistent rate of 4 Hz. To make the game more challenging, we can double or triple the clock speed,

Figure 8–22
A typical, hierarchical
state-machine structure.

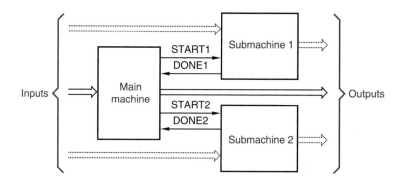

A REALLY
BAD JOKE

Note that the title of this section has nothing to do with the "buried flip-flops" found in some PLDs.

Figure 8–23
Block diagram of guessing game
with random delay.

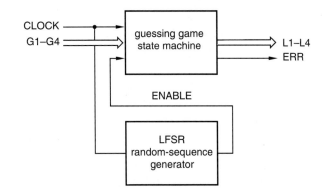

but allow the lamps to stay in each state for a random length of time. Then the user truly must guess whether a given lamp will stay on long enough for the corresponding pushbutton to be pressed.

A block diagram for the enhanced guessing game is shown in Figure 8–23. The main machine is basically the same as before, except that it only advances from one lamp state to the next if the enable input EN is asserted, as shown by the ASM chart in Figure 8–24. The enable input is driven by the output of a pseudo-random sequence generator, a linear feedback shift register (LFSR).

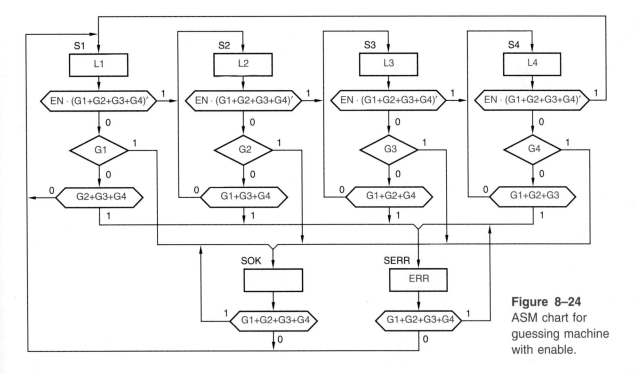

Figure 8–24
ASM chart for
guessing machine
with enable.

A SHIFTY CIRCUIT

LFSR circuits are described in Section 9.4.8. In Figure 8–23, the low-order bit of an n-bit LFSR counter is used as the enable signal. Thus, the length of time that a particular lamp stays on depends on the counting sequence of the LFSR.

In the best case for the user, the LFSR contains $10\ldots00$; in this case the lamp is on for $n-1$ clock ticks because it takes that long for the single 1 to shift into the low-order bit position. In the worst case, the LFSR contains $11\ldots11$ and shifting occurs for n consecutive clock ticks. At other times, shifting stops for a time determined by the number of consecutive 0s starting with the low-order bit of the LFSR.

All of these cases are unpredictable unless the user has memorized the shifting cycle ($2^n - 1$ states) or is very fast at Galois-field arithmetic. Obviously, a large value of n (≥ 16) provides the most fun.

Another obvious candidate for decomposition is a state machine that performs binary multiplication using the shift-and-add algorithm, or binary division using the shift-and-subtract algorithm. To perform an n-bit operation, these algorithms require an initialization step, n computation steps, and possible cleanup steps. The main machine for such an algorithm contains states for initialization, generic computation, and cleanup steps, and a modulo-n counter can be used as a submachine to control the number of generic computation steps executed. In Section 8.1.2, we showed an ASM chart for a monolithic state machine for 8-bit shift-and-add multiplication, while the design of a corresponding decomposed machine for n-bit multiplication was left to the reader as Exercise 9.43.

R E F E R E N C E S

Algorithmic-state-machine notation was pioneered at Hewlett-Packard Laboratories by Thomas E. Osborne and was further developed by Osborne's colleague Christopher R. Clare in a book, *Designing Logic Systems Using State Machines* (McGraw-Hill, 1973). It has subsequently found a home in many digital design texts, including *The Art of Digital Design* by F. P. Prosser and D. E. Winkel (Prentice-Hall, 1987, 2nd ed.) and *Digital Design* by M. Morris Mano (Prentice-Hall, 1984). Another notation for describing state machines, an extension of "traditional" state-diagram notation, is the mnemonic documented state (MDS) diagram developed by William I. Fletcher in *An Engineering Approach to Digital Design* (Prentice-Hall, 1980).

DRILL PROBLEMS

8.1   Analyze the state machine in Figure 8–6(b), producing a state and output table.

8.2   Show that the state and output table derived in Drill 8.1 is equivalent to Table 7–10 on page 488.

8.3   Demonstrate that the state machine in Figure 8–6(b) can be realized as a finite-memory machine. Draw a logic diagram for a realization of the machine that matches the general structure shown in Figure 8–7.

8.4   Synthesize a circuit from a transition list for the ASM chart of Exercise 8.13. Write a transition equation for each state variable as a sum of p-terms, and write simplified transition/excitation equations for a realization using D flip-flops. Draw a circuit diagram using TTL SSI and MSI components.

8.5   All but one of the state diagrams in Figure X8.5 are ambiguous or incorrect or both. List all of the ambiguities and errors in these state diagrams. (*Hint:* Use Karnaugh maps where necessary to find uncovered and double-covered input combinations.)

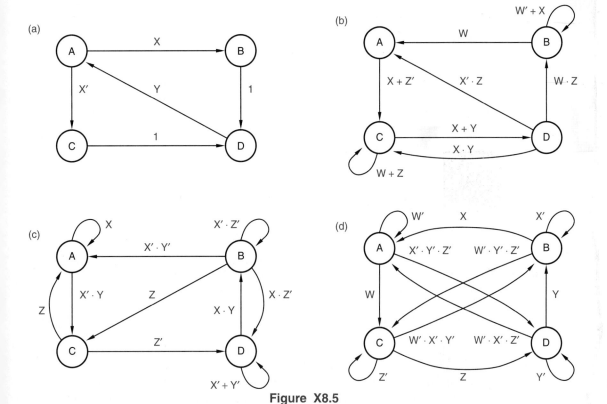

Figure X8.5

8.6   Synthesize a circuit for the state diagram of Figure 8–13 using six variables to encode the state, where the LA–LC and RA–RC outputs equal the state variables themselves. Write a transition list, a transition equation for each state variable as a sum of p-terms, and simplified transition/excitation equations for a realization using D flip-flops. Draw a circuit diagram using TTL SSI and MSI components.

8.7   Starting with the transition list in Table 8–5, find a mimimal sum-of-products expression for Q2*, assuming that the next states for the unused states are true don't-cares.

8.8   Write a new transition list and transition/excitation equations for the T-bird tail-lights machine of Figure 8–13 using an output-coded state assignment.

8.9   Modify the state diagram of Figure 8–13 so that the machine goes into hazard mode immediately if LEFT and RIGHT are asserted simultaneously during a turn. Write the corresponding transition list.

8.10   Repeat Drill 8.9 using the ASM chart of Figure 8–17.

8.11   Construct an ASM chart for the state machine described by Table 7–10. Don't use more than seven decision boxes.

8.12   Construct an ASM chart for the state machine described by Table 7–18. Don't use more than eight decision boxes.

8.13   Construct an ASM chart for the combination-lock machine of Section 7.4.6.

8.14   Construct a state diagram for the 1s-counting machine of Table 7–16 and compare with the ASM chart in Figure 8–16.

E X E R C I S E S

8.15   Write an equation for a HINT output of the finite-memory machine in Figure 8–8. You need not multiply out or minimize your result. (*Hint:* Consider all of the possible initial sequences of 0110111 that might be stored in the finite memory.)

8.16   Design a circuit with two inputs, CLOCK and /SYNC, that produces the output waveforms shown in Figure X8.16. The /SYNC pulse occurs only once every 256 clock cycles. Your design should use at most one SSI package and two MSI parts that were described in this chapter. The output signals should be glitch-free and must change synchronously on the rising edge of CLOCK.

8.17   Synthesize a circuit for the ambiguous state diagram in Figure 8–11. Use the state assignment in Table 8–1. Write a transition list, a transition equation for each state variable as a sum of p-terms, and simplified transition/excitation equations for a realization using D flip-flops. Determine the actual next state of the circuit, starting from the IDLE state, for each of the following input combinations on (LEFT, RIGHT, HAZ): (1,0,1), (0,1,1), (1,1,0), (1,1,1). Comment on the machine's behavior in these cases.

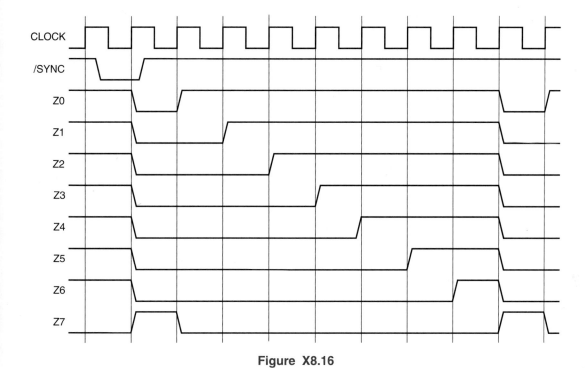

Figure X8.16

8.18 Suppose that for a state SA and an input combination I, an ambiguous state diagram indicates that there are two next states, SB and SC. The actual next state
SD for this transition depends on the state machine's realization. If the state machine is synthesized using the $V* = \Sigma$(p-terms where $V* = 1$) method to obtain
transition/excitation equations for D flip-flops, what is the relationship between the
coded states for SB, SC, and SD? Explain.

8.19 Repeat Exercise 8.18, assuming that the machine is synthesized using the $V*' = \Sigma$(p-terms where $V* = 0$) method.

8.20 Suppose that for a state SA and an input combination I, an ambiguous state diagram
does not define a next state. The actual next state SD for this transition depends on
the state machine's realization. Suppose that the state machine is synthesized using
the $V* = \Sigma$(p-terms where $V* = 1$) method to obtain transition/excitation equations
for D flip-flops. What coded state is SD? Explain.

8.21 Repeat Exercise 8.20, assuming that the machine is synthesized using the $V*' = \Sigma$(p-terms where $V* = 0$) method.

8.22 Synthesize a circuit from the ASM chart of Figure 8–16. Compare your results
with the results of Section 7.4.6, and explain why they are the same or different.

8.23 Given the transition equations for a clocked synchronous state machine that is to be built using master/slave S-R flip-flops, how can the excitation equations for the S and R inputs be derived? (*Hint:* Show that any transition equation, $Qi* = expr$, can be written in the form $Qi* = Qi \cdot expr1 + Qi' \cdot expr2$, and see where that leads.)

8.24 Repeat Exercise 8.23 for J-K flip-flops. How can the "don't-cares" that are possible in a J-K design be specified?

8.25 Draw a logic diagram for the output logic of the guessing-game machine in Table 8–5 using a single 74x139 dual 2-to-4 decoder. (*Hint:* Use active-low outputs.)

8.26 What does the personalized license plate in Figure 8–9 stand for? (*Hint:* It's a computer engineer's version of OTTFFSS.)

8.27 Redesign the excitation logic for flip-flop U5 in Figure 8–6 so that it uses only a single AND gate.

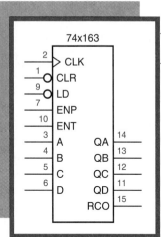

74x163

2	CLK
1	CLR
9	LD
7	ENP
10	ENT

3	A	QA	14
4	B	QB	13
5	C	QC	12
6	D	QD	11
		RCO	15

SEQUENTIAL LOGIC DESIGN PRACTICES

9

The purpose of this chapter is to familiarize you with the most commonly used and dependable sequential-circuit design methods. Therefore, our emphasis continues to be on *synchronous systems*, that is, systems in which all flip-flops are clocked by the same, common clock signal. Although it's true that all the world does not march to the tick of a common clock, within the confines of a digital system or subsystem we can make it so. When we interconnect digital systems or subsystems that use different clocks, we can usually identify a limited number of asynchronous signals that need special treatment, as we'll show later.

synchronous system

We begin this chapter with a quick summary of sequential circuit documentation standards. Then we introduce the most basic building blocks of sequential-circuit design—latches, flip-flops, counters, and shift registers—and show applications of some of the most common MSI device types. Finally, we show how all these elements come together in synchronous systems and how the inevitable asynchronous inputs are handled. This all sets the stage for our discussion of sequential PLDs in Chapter 10.

9.1 SEQUENTIAL CIRCUIT DOCUMENTATION STANDARDS

9.1.1 General Requirements

Basic documentation standards in areas like signal naming, logic symbols, and schematic layout, which we introduced in Chapter 5, apply to digital systems as a whole and therefore to sequential circuits in particular. However, there are several ideas to highlight for system elements that are specifically "sequential":

- *State-machine layout.* Within a logic diagram, a collection of flip-flops and combinational logic that forms a state machine should be drawn together in a logical format on the same page, so the fact that it is a state machine is obvious. (That is, you shouldn't have to flip pages to find the feedback path!)

- *Cascaded elements.* In a similar way, registers, counters, and shift registers that use multiple ICs should have the ICs grouped together in the schematic so that the cascading structure is obvious.

- *Flip-flops.* The symbols for individual sequential-circuit elements, especially flip-flops, should rigorously follow the appropriate drawing standards, so that the type, function, and clocking behavior of the elements are clear.

- *State-machine descriptions.* State machines should be described by state tables, state diagrams, transition lists, ASM charts, or text files in a state-machine description language.

- *Timing diagrams.* The documentation package for sequential circuits should include timing diagrams that show the general timing assumptions and timing behavior of the circuit.

- *Timing specifications.* A sequential circuit should be accompanied by a specification of the timing requirements for proper internal operation (e.g., maximum clock frequency), as well as the requirements for any externally supplied inputs (e.g., setup- and hold-time requirements with respect to the system clock, minimum pulse widths, etc.).

9.1.2 Logic Symbols

We introduced traditional symbols for flip-flops in Section 7.2. Flip-flops are always drawn as rectangular-shaped symbols, and follow the same general guidelines as other rectangular-shaped symbols—inputs on the left, outputs on the right, bubbles for active levels, and so on. In addition, some specific guidelines apply to flip-flop symbols:

- A dynamic indicator is placed on edge-triggered clock inputs.

- A postponed-output indicator is placed on master/slave outputs that change at the end interval during which the clock is asserted.

- Asynchronous preset and clear inputs may be shown at the top and bottom of a flip-flop symbol—preset at the top and clear at the bottom.

The logic symbols for larger-scale sequential elements, such as the counters and shift register described later in this chapter, are generally drawn with all inputs, including presets and clears, on the left, and all outputs on the right. Bidirectional signals may be drawn on the left or the right, whichever is convenient.

Like individual flip-flops, larger-scale sequential elements use a dynamic indicator to indicate edge-triggered clock inputs. In "traditional" symbols, the names of the inputs and outputs give a clue of their function, but they are sometimes ambiguous. Comparing two elements described later in this chapter, the 74x163 counter has a CLR input that is sampled only at the clock edge, while the 74x161, which has exactly the same traditional symbol, has an asynchronous CLR input.

IEEE standard symbols, which we show in Appendix A for all of the sequential elements in this chapter, have a rich set of notation that can provide an unambiguous definition of every signal's function.

IEEE STANDARD
SYMBOLS

9.1.3 State-Machine Descriptions

So far we have dealt with five different representations of state machines:

- Word descriptions

- State tables

- State diagrams

- Transition lists

- ASM charts

And in the next chapter, we'll introduce yet another way, state-machine description language.

You might think that having all these different ways to represent state machines is a problem—too much for you to learn! Well, they're *not* all that different or difficult to learn, but there *is* a subtle problem here. Consider a similar problem in assembly-language programming, where high-level "pseudo-code" might be used to document how a program works. The pseudo-code may express the programmer's intentions very well, but errors, misinterpretations, and typos can occur when the pseudo-code is translated into real code. In any creative process, inconsistencies can occur when there are multiple representations of how things work.

inconsistent state-machine representations

The same kind of inconsistencies can occur in state-machine design. A logic designer may document a machine's desired behavior with a 100%-correct hand-drawn state diagram, but there are still plenty of opportunities to mess up while "turning the crank" to translate the state diagram into a state table, transition table, excitation equations, and logic diagram.

The solution to this problem is similar to the one adopted by programmers who write self-documenting code using a high-level language. The key is to select a representation that is both expressive of the designer's intentions *and* that can be translated into a physical realization using an error-free, automated process. (You don't hear many programmers screaming "Compiler bug!" when their programs don't work the first time.) The transition-list and ASM-chart notations discussed in Chapter 8 are a step in the right direction for state-machine designers. However, the ultimate solution (for now, at least) is to write state-machine "programs" in a high-level state-machine description language, allowing automatic translation into a PLD-based design. We'll study this solution in detail in Section 10.7.

9.1.4   Timing Diagrams and Specifications

We showed many examples of timing diagrams in Chapters 5 and 7. In the design of synchronous systems, most timing diagrams show the relationship between the clock and various input, output, and internal signals.

Figure 9–1 shows a fairly typical timing diagram that specifies the requirements and characteristics of input and output signals in a synchronous circuit. The first line shows the system clock and its nominal timing parameters. The remaining lines show a range of delays for other signals.

For example, the second line shows that flip-flops and other clocked devices change their outputs at some time between the rising edge of CLOCK and time t_{ffpd} afterward. External circuits that sample these signals should not do so while they are changing. The timing diagram is drawn as if the minimum value of t_{ffpd} is zero; a complete documentation package would include a timing table indicating the actual minimum, typical, and maximum values of t_{ffpd} and all other timing parameters.

Figure 9–1
A detailed timing diagram
showing propagation delays
and setup and hold times
with respect to the clock.

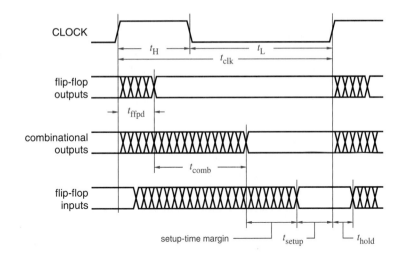

The third line of the timing diagram shows the additional time, t_{comb}, required for the flip-flop output changes to propagate through combinational logic elements, such as flip-flop excitation logic. The excitation inputs of flip-flops and other clocked devices require a setup time of t_{setup}, as shown in the fourth line. For proper circuit operation we must have $t_{clk} - t_{ffpd} - t_{comb} > t_{setup}$.

timing margin

Timing margins indicate how much "worse than worst-case" the individual components of a circuit can be without causing the circuit to fail. Well-designed systems have positive, nonzero timing margins to allow for unexpected circumstances (marginal components, brown-outs, engineering errors, etc.) and clock skew (Section 9.7.1).

setup-time margin

The difference $t_{clk} - t_{ffpd(max)} - t_{comb(max)} - t_{setup}$ is called the *setup-time margin*; if this number is negative, the circuit won't work. Note that *maximum* propagation delays are used to calculate setup-time margin. Another timing margin involves the hold-time requirement t_{hold}; the sum of the *minimum* values of t_{ffpd} and t_{comb} must be greater than t_{hold}, and the *hold-time margin* is $t_{ffpd(min)} + t_{comb(min)} - t_{hold}$.

hold-time margin

The timing diagram in Figure 9–1 does not show the timing differences between different flip-flop inputs or combinational-logic signals, even though such differences exist in most circuits. For example, one flip-flop's Q output may be connected directly to another flip-flop's D input, so that t_{comb} for that path is zero, while another's may go the ripple-carry path of a 32-bit adder before reaching a flip-flop input. When proper synchronous design methodology is used, these relative timings are not critical, since none of these signals affect the state of the circuit until a clock edge occurs. You merely have to find the longest delay path in one clock period to determine whether the circuit will work.

However, you may have to analyze several different paths in order to find the worst-case one.

Another, perhaps more common, type of timing diagram shows only functional behavior and is not concerned with actual delay amounts; an example is shown in Figure 9–2. Here, the clock is "perfect." Whether to show signal changes as vertical or slanted lines is strictly a matter of personal taste in this and all other timing diagrams, unless rise and fall times must be explicitly indicated. Clock transitions are shown as vertical lines in this and other figures in keeping with the idea that the clock is a "perfect" reference signal.

The other signals in Figure 9–2 may be flip-flop outputs, combinational outputs, or flip-flop inputs. Shading is used to indicate "don't-care" signal values; cross-hatching as in Figure 9–1 on the preceding page could be used instead. All of the signals are shown to change immediately after the clock edge. In reality, the outputs change sometime later, and inputs may change just barely before the next clock edge. However, "lining up" everything on the clock edge allows the timing diagram to display more clearly which functions are performed during each clock period. Signals that are lined up with the clock are simply understood to change sometime *after* the clock edge, with timing that meets the setup- and hold-time requirements of the circuit. Many timing diagrams of this type appear in this chapter.

Figure 9–2
Functional timing of a
synchronous circuit.

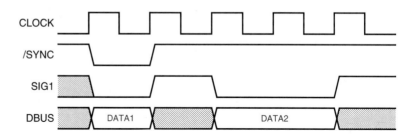

9.2 LATCHES AND FLIP-FLOPS

9.2.1 SSI Latches and Flip-Flops

Several different types of discrete latches and flip-flops are available as SSI parts. These devices are most commonly used in the design of state machines and "unstructured" sequential circuits that don't fall into the categories of shift registers, counters, and other sequential MSI functions that you'll study later in this chapter. However, SSI latches and flip-flops are being eliminated to a large extent in modern designs as their functions are embedded in larger, structured logic devices such as PLDs, as we'll show in Chapter 10. Nevertheless, a handful

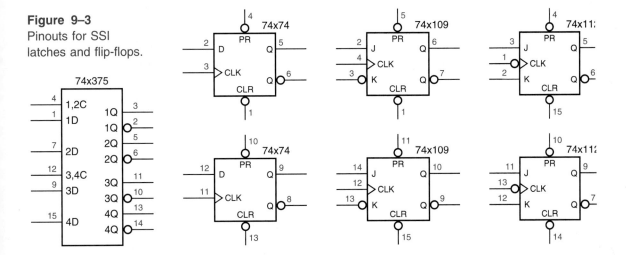

Figure 9–3
Pinouts for SSI
latches and flip-flops.

of these discrete building blocks still appear in most digital systems, so it's important to be familiar with them.

Figure 9–3 shows the pinouts for several SSI sequential devices. The only latch in the figure is the *74x375*, which contains four D latches, similar in function to the "generic" D latches described in Section 7.2.4. Because of pin limitations, the latches are arranged in pairs with a common C control line for each pair.

Among the devices in Figure 9–3, the most important is the *74x74*, which contains two independent positive-edge-triggered D flip-flops with preset and clear inputs. We described the functional operation, timing, and internal structure of edge-triggered D flip-flops in general, and the 74x74 in particular, in Section 7.2.5. Besides the 74x74's use in "random" sequential circuits, fast versions of the part, such as the *74F74* and *74ACT74*, find application in synchronizers for asynchronous input signals, as discussed in Section 9.8.

The *74x109* is a positive-edge-triggered J-/K flip-flop with an active-low K input (named /K or $\overline{K}$). We discussed the internal structure of the '109 in Section 7.2.8. Another J-K flip-flop is the *74x112*, which has an active-low clock input.

74x375

74x74

74F74
74ACT74
74x109

74x112

* 9.2.2 Switch Debouncing

A very common application of simple bistables and latches is in debouncing switches. We're all familiar with electrical switches from experience with lights, garbage disposals, and other appliances. Switches connected to sources of constant logic 0 and 1 are often used in digital systems to supply "user inputs."

---

*Throughout this book, optional sections are marked with an asterisk.

However, in digital logic applications we must consider another aspect of switch operation, the time dimension. A simple make or break operation, which occurs instantly as far as we slow-moving humans are concerned, actually has several phases that are discernible by high-speed digital logic.

Figure 9–4(a) shows how a single-pole, double-throw (SPDT) switch might be used to generate a single logic input. As shown in (b), the switch contacts and wiper have a "break before make" behavior, so the wiper terminal is "floating" at some time halfway through the switch depression. The figure assumes that the gate interprets this floating input as a 1, but in a real CMOS or TTL circuit we must provide a pull-up resistor on the /SW signal to guarantee a valid logic signal when the input is floating.

Figure 9–4
Switch input without debouncing.

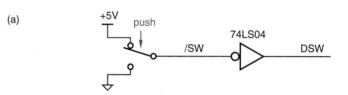

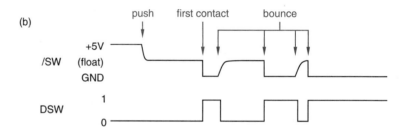

In any case, once the wiper hits the bottom contact, it doesn't stay there for long; it bounces a few times before finally settling. It bounces far enough to lose its connection to the bottom contact, though not far enough to touch the top contact again. The result is that several transitions are seen on the /SW and /DSW

contact bounce

logic signals for each single switch depression. This behavior is called *contact bounce*. Typical switches bounce for 10–20 ms, a very long time compared to the switching speeds of logic gates.

Contact bounce may or may not be a problem, depending on the switch application. For example, many computers have configuration information spec-

DIP switch

ified by small switches, called *DIP switches* because they are the same size as a dual in-line package (DIP). Since DIP switches are normally changed only when the computer is inactive, there's no problem. Contact bounce *is* a problem if a switch is being used to count or signal some event (e.g., laps in a race).

debounce

Then we must provide a circuit (or, in microprocessor-based systems, software) to *debounce* the switch—to provide just one signal change or pulse for each external event.

Figure 9–5
Switch input using a bistable for debouncing.

(a)

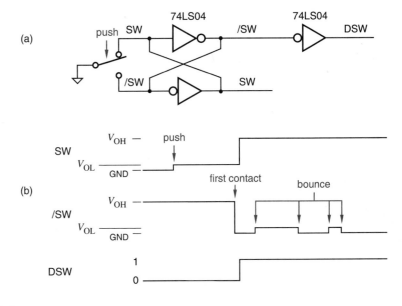

(b)

* 9.2.3 The Simplest Switch Debouncer

Switch debouncing is a good application for the simplest sequential circuit, the bistable element of Section 7.1, which can be used as shown in Figure 9–5. Before the switch is pushed, the top contact holds SW at 0 V, a valid logic 0, and the top inverter produces a logic 1 on /SW and on the bottom contact. When the button is pushed and contact is broken, feedback in the bistable continues to hold SW at V_{OL} (≤ 0.5 V for LS-TTL), still a valid logic 0.

When the wiper hits the bottom contact, the circuit operates unconventionally for a moment. The top inverter in the bistable is trying to maintain a logic 1 on the /SW signal; the top transistor in its totem-pole output is "on" and connecting /SW through a small resistance to +5 V. Suddenly, the switch contact makes a metallic connection of /SW to ground, 0 V. Not surprisingly, the switch contact wins.

A short time later (30 ns for LS-TTL), the forced logic 0 on /SW propagates through the two inverters of the bistable, so that the top inverter gives up its vain attempt to drive a 1, and instead drives a logic 0 onto /SW. At this point, the

---

The circuit in Figure 9–5, while elegant, should not be used with high-speed CMOS devices, like the 74ACT04, whose outputs are capable of sourcing large amounts of current in the HIGH state. While shorting such outputs to ground momentarily may not cause any damage, it will generate a noise pulse on power and ground signals that may trigger improper operation of the circuit elsewhere. The debouncing circuit in the figure works best with wimpy logic families like LS-TTL.

DON'T DO THIS
WITH CMOS!

top inverter output is no longer shorted to ground, and feedback in the bistable maintains the logic 0 on /SW even if the wiper bounces off the bottom contact, as it does.

Advantages of this circuit compared to other debouncing approaches are that it has a low chip count (one-third of a 74LS04), no pull-up resistors are required, and both polarities of the input signal (active-high and active-low) are produced. In situations where momentarily shorting gate outputs must be avoided, a similar circuit can be designed using a /S-/R latch and pull-up resistors, as suggested in Figure 9–6.

Figure 9–6
Switch input using a /S-/R latch for debouncing.

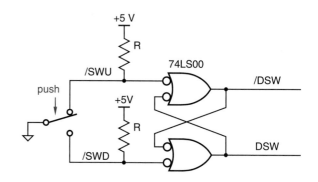

9.2.4 Multibit Registers and Latches

register

A collection of two or more D flip-flops with a common clock input is called a *register*. Registers are often used to store a collection of related bits, such as a byte of data in a computer. However, a single register can also be used to store unrelated bits of data or control information; the only real constraint is that all of the bits are stored using the same clock signal.

74x175

Figure 9–7(a) and (b) shows the logic diagram and logic symbol for a commonly used MSI register, the *74x175*. The 74x175 contains four edge-triggered D flip-flops and provides both active-high and active-low outputs at the

74x174

external pins of the device. The *74x174*, shown in (c), is similar, except that it eliminates the active-low outputs and provides two more flip-flops instead.

The individual flip-flops in a '175 are negative-edge triggered, as indicated by the inversion bubbles on their CLK inputs. However, the circuit also contains an inverter that makes the flip-flops positive-edge triggered with respect to the device's external CLK input pin. A common, active-low, clear signal (/CLR) is connected to the asynchronous clear inputs of all four flip-flops. Both CLK and /CLR are buffered before fanning out to the four flip-flops, so that a device driving one of these inputs sees only one unit load instead of four. This is especially important if a common clock or clear signal must drive many such registers.

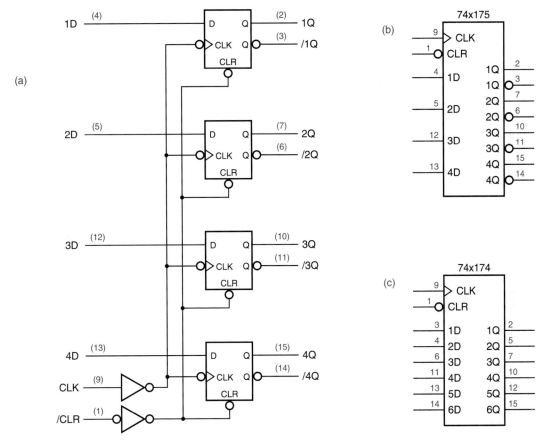

Figure 9–7 MSI registers: (a) logic diagram for the 74x175 4-bit register, including pin numbers for a standard 16-pin dual in-line package; (b) traditional logic symbol for the 74x175; (c) traditional logic symbol for the 74x174 6-bit register.

Many digital systems, including computers, telecommunications devices, and stereo equipment, process information 8 or 16 bits at a time; as a result, ICs that handle eight bits are very popular. One such MSI IC is the *74x374* octal edge-triggered D flip-flop, also known simply as an 8-bit register. (Once again, "octal" means that the device has eight sections.) As shown in Figure 9–8(a) and (b), the 74x374 contains eight edge-triggered D flip-flops that all sample their inputs and change their outputs on the rising edge of a common CLK input. Each flip-flop output drives a three-state buffer that in turn drives an active-high output. All of the three-state buffers are enabled by a common /OE (output enable) input. As in the other registers that we've studied, the control inputs (CLK and /OE) are buffered so that they present only one unit load to a device that drives them.

74x374

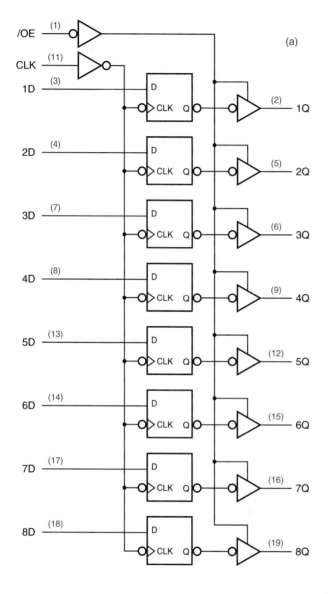

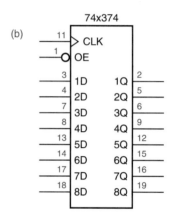

Figure 9–8
MSI registers and latches: (a) logic diagram for the 74x374 8-bit register, including pin numbers for a standard 20-pin dual in-line package; (b) traditional logic symbol for the 74x374; (c) traditional logic symbol for the 74x373 8-bit latch.

74x373

One variation of the 74x374 is the *74x373*, whose symbol is shown in Figure 9–8(c). The '373 uses D latches instead of edge-triggered flip-flops. Therefore, its outputs follow the corresponding inputs whenever C is asserted, and latch the last input values when C is negated.

74x377

The *74x377*, whose symbol is shown in Figure 9–9(a), is an edge-triggered register like the '374, but it does not have three-state outputs. Instead, pin 1 is used as an active-low clock enable input /G. If /G is asserted (LOW) at the rising edge of the clock, then the flip-flops are loaded from the data inputs; otherwise, they retain their present values, as shown logically in (b).

Figure 9–9
The 74x377 8-bit register with
gated clock: (a) logic symbol;
(b) logical behavior of one bit.

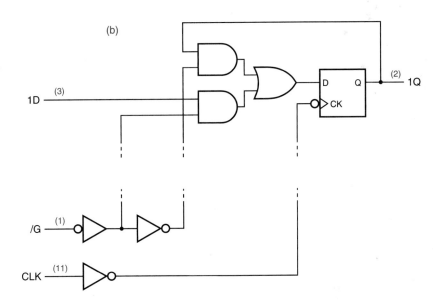

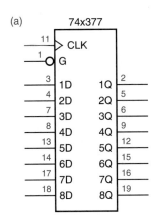

9.3 COUNTERS

The name *counter* is generally used for any clocked sequential circuit whose state   *counter*
diagram contains a single cycle, as in Figure 9–10. The *modulus* of a counter is   *modulus*
the number of states in the cycle. A counter with *m* states is called a *modulo-m*   *modulo-m counter*
counter or, sometimes, a *divide-by-m counter*. A counter whose modulus is not   *divide-by-m counter*
a power of 2 will, of necessity, contain extra states that are not used in normal
operation.

Probably the most commonly used counter type is an *n-bit binary counter*.   *n-bit binary counter*
Such a counter has n flip-flops and has 2^n states, which are visited in the sequence
$0, 1, 2, \ldots, 2^n - 1, 0, 1, \ldots$. Each of the foregoing states is encoded as the
corresponding n-bit binary integer.

Figure 9–10
General structure of a counter's
state diagram—a single cycle.

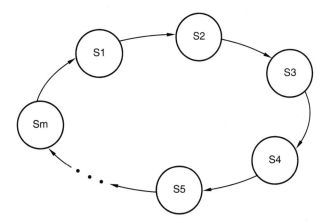

Figure 9–11
A 4-bit binary ripple counter.

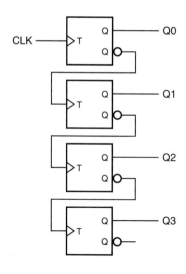

9.3.1 Ripple Counters

An *n*-bit binary counter can be constructed with just *n* flip-flops and no other
components, for any value of *n*. Figure 9–11 shows such a counter for $n = 4$.
Recall that a T flip-flop changes state (toggles) on every rising edge of its clock
input. Thus, each bit of the counter toggles if and only if the immediately
preceding bit changes from 1 to 0. This corresponds to a normal binary counting
sequence—when a particular bit changes from 1 to 0, it generates a carry to the

ripple counter   next most significant bit. The counter is called a *ripple counter* because the carry
information ripples from the less significant bits to the more significant bits one
bit at a time.

9.3.2 Synchronous Counters

Although a ripple counter requires fewer components than any other type of
binary counter, it does so at a price—it is slower than any other type of binary
counter. In the worst case, when the most significant bit must change, the output
is not valid until time $n \cdot t_{pTQ}$ after the rising edge of CLK, where t_{pTQ} is the
propagation delay from input to output of a T flip-flop.

synchronous counter   A *synchronous counter* connects all of its flip-flop clock inputs to the same
common CLK signal, so that all of the flip-flop outputs change at the same time,
after only t_{pTQ} ns of delay. As shown in Figure 9–12, this requires the use of
T flip-flops with enable inputs; the output toggles on the rising edge of T if and
only if EN is asserted. Combinational logic on the EN inputs determines which,
if any, flip-flops toggle on each rising edge of T. As shown in the figure, it is
also possible to provide a master count-enable signal CNTEN. Each T flip-flop
toggles if and only if CNTEN is asserted and all of the lower-order counter bits

Figure 9–12
A synchronous 4-bit binary
counter with serial enable logic.

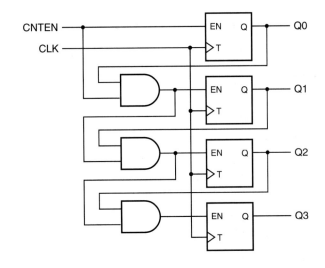

are 1. Like the binary ripple counter, a synchronous n-bit binary counter can be
built with a fixed amount of logic per bit—in this case, a T flip-flop with enable
and a 2-input AND gate.

The counter structure in Figure 9–12 is sometimes called a *synchronous serial counter* because the combinational enable signals propagate serially from the least significant to the most significant bits. If the clock period is too short, there may not be enough time for a change in the counter's LSB to propagate to the MSB. This problem is eliminated in Figure 9–13 by driving each EN input with a dedicated AND gate, just a single level of logic. Called a *synchronous parallel counter*, this is the fastest binary counter structure.

synchronous serial counter

synchronous parallel counter

Figure 9–13
A synchronous 4-bit binary
counter with parallel enable logic.

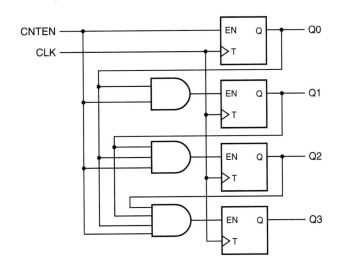

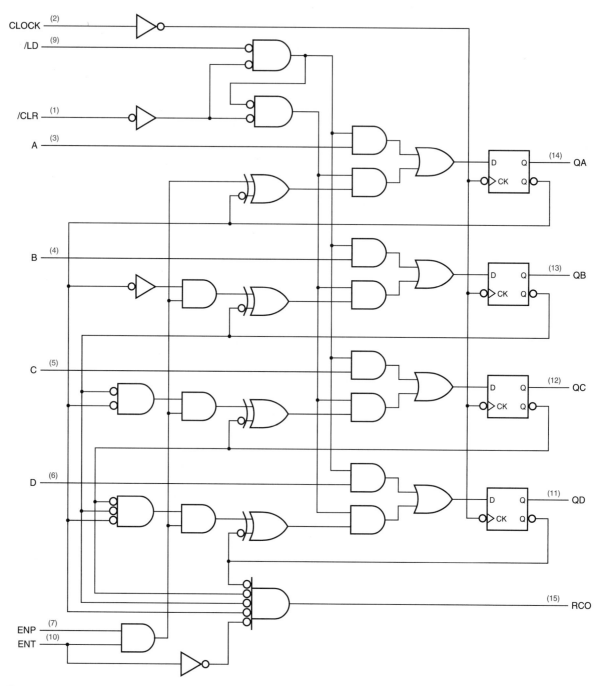

Figure 9–14   Logic diagram for the 74x163 synchronous 4-bit binary counter, including pin numbers for a standard 16-pin dual in-line package.

9.3.3   MSI Counters and Applications

The most popular MSI counter is the *74x163*, a synchronous 4-bit binary counter   *74x163*
with load and clear inputs, whose internal logic diagram is shown in Figure 9–14.
Its function is summarized by the state table in Table 9–1, and its traditional logic
symbol is shown in Figure 9–15.

The '163 uses D flip-flops rather than T flip-flops internally to facilitate
the load and clear functions. Each D input is driven by a 2-input multiplexer
consisting of an OR gate and two AND gates. The multiplexer output is 0 if the
/CLR input is asserted. Otherwise, the top AND gate passes the data input (A, B,
C, or D) to the output if /LD is asserted. If neither /CLR nor /LD is asserted,
then the bottom AND gate passes the output of an XNOR gate to the multiplexer
output.

The XNOR gates perform the counting function in the '163. One input of
each XNOR is the corresponding count bit (QA, QB, QC, or QD); the other input
is 1, which complements the count bit, if and only if both enables ENP and ENT

Table 9–1   State table for a 74x163 4-bit binary counter.

Inputs				Current State				Next State			
/CLR	/LD	ENT	ENP	QD	QC	QB	QA	QD*	QC*	QB*	QA*
0	X	X	X	X	X	X	X	0	0	0	0
1	0	X	X	X	X	X	X	D	C	B	A
1	1	0	X	X	X	X	X	QD	QC	QB	QA
1	1	X	0	X	X	X	X	QD	QC	QB	QA
1	1	1	1	0	0	0	0	0	0	0	1
1	1	1	1	0	0	0	1	0	0	1	0
1	1	1	1	0	0	1	0	0	0	1	1
1	1	1	1	0	0	1	1	0	1	0	0
1	1	1	1	0	1	0	0	0	1	0	1
1	1	1	1	0	1	0	1	0	1	1	0
1	1	1	1	0	1	1	0	0	1	1	1
1	1	1	1	0	1	1	1	1	0	0	0
1	1	1	1	1	0	0	0	1	0	0	1
1	1	1	1	1	0	0	1	1	0	1	0
1	1	1	1	1	0	1	0	1	0	1	1
1	1	1	1	1	0	1	1	1	1	0	0
1	1	1	1	1	1	0	0	1	1	0	1
1	1	1	1	1	1	0	1	1	1	1	0
1	1	1	1	1	1	1	0	1	1	1	1
1	1	1	1	1	1	1	1	0	0	0	0

Figure 9–15
Traditional logic symbol for the
74x163.

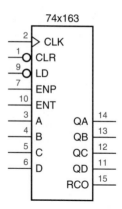

are asserted and all of the lower-order count bits are 1. The RCO ("ripple carry out") signal indicates a carry from the most significant bit position, and is 1 when all of the count bits are 1 and ENT is asserted.

free-running counter

Even though most MSI counters have enable inputs, they are often used in a *free-running* mode in which they are enabled continuously. Figure 9–16 shows the connections to make a '163 operate in this way, and Figure 9–17 shows the resulting output waveforms. Notice that starting with QA, each signal has half the frequency of the preceding one. Thus, a free-running '163 can be used as a divide-by-2, -4, -8, or -16 counter, by ignoring any unnecessary high-order output bits.

74x161

Note that the '163 is fully synchronous; that is, its outputs change only on the rising edge of CLK. A few applications require an asynchronous clear function, which is provided in the *74x161*. The '161 has the same pinout as the '163, but its /CLR input is connected to the asynchronous clear inputs of the internal flip-flops.

Figure 9–16
Connections for the 74x163 to
operate in a free-running mode.

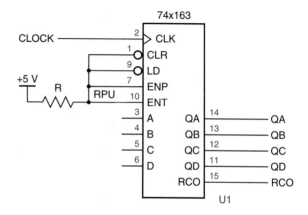

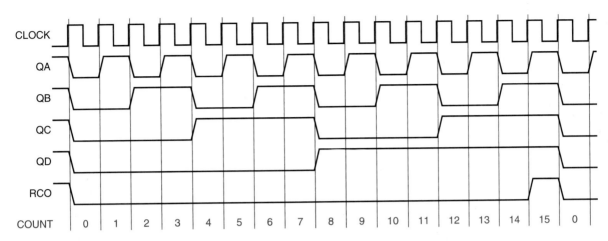

Figure 9–17 Clock and output waveforms for a free-running divide-by-16 counter.

The *74x160* and *74x162* are more variations with the same pinouts and gen-   *74x160*
eral functions as the '161 and '163, except that the counting sequence is modified   *74x162*
to go to state 0 after state 9. In other words, these are modulo-10 counters, sometimes called *decade counters*. Figure 9–18 shows the output waveforms for a   *decade counter*
free-running '160 or '162. Notice that although the QD and QC outputs have
one-tenth of the CLK frequency, they do not have a 50% duty cycle, and the
QC output, with one-fifth of the input frequency, does not have a constant duty
cycle. We'll show the design of a divide-by-10 counter with a 50% duty-cycle
output later in this subsection.

Although the '163 is a modulo-16 counter, it can be made to count in a

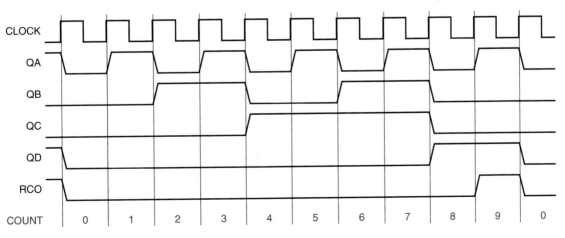

Figure 9–18 Clock and output waveforms for a free-running divide-by-10 counter.

Figure 9–19
Using the 74x163 as a
modulo-11 counter with
the counting sequence
5, 6, ..., 15, 5, 6, ...

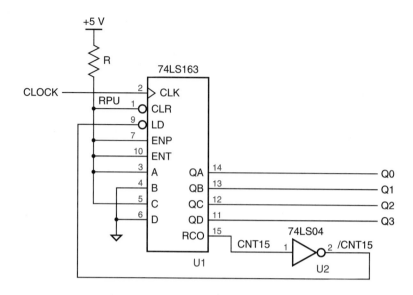

modulus less than 16 by using the /CLR or /LD input to shorten the normal
counting sequence. For example, Figure 9–19 shows one way of using the '163
as a modulo-11 counter. The RCO output, which detects state 15, is used to force
the next state to 5, so that the circuit will count from 5 to 15 and then start at 5
again, for a total of 11 states per counting cycle.

A different approach for modulo-11 counting with the '163 is shown in
Figure 9–20. This circuit uses a NAND gate to detect state 10 and force the next
state to 0. Notice that only a 2-input gate is used to detect state 10 (binary
1010). Although a 4-input gate would normally be used to detect the condition

Figure 9–20
Using the 74x163 as a
modulo-11 counter with
the counting sequence
0, 1, 2, ..., 10, 0, 1, ...

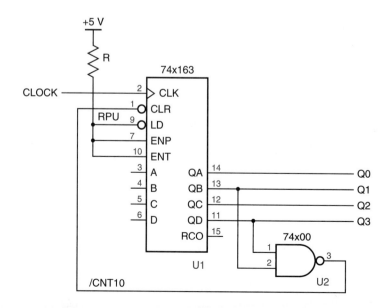

Figure 9–21
A 74x163 used as an excess-3
decimal counter.

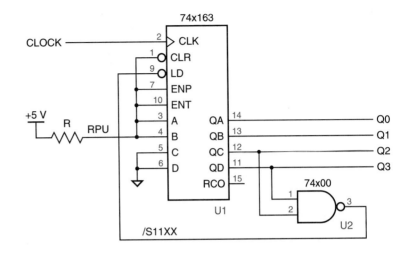

CNT10 = $Q3 \cdot Q2' \cdot Q1 \cdot Q0'$, the 2-input gate takes advantage of the fact that no
other state in the normal counting sequence of 0–10 has Q3 = 1 and Q1 = 1. In
general, to detect state N in a binary counter that counts from 0 to N, we need to
AND only the state bits that are 1 in the binary encoding of N.

There are many other ways to make a modulo-11 counter using a '163.
The choice of approach—one of the preceding or a combination of them (as in
Exercise 9.16)—depends on the application. As another example, in Section 2.10
we promised to show you how to build a circuit that counts in the excess-3 decimal code, shown in Table 2–9 on page 45. Figure 9–21 shows the connections
for a '163 to count in the excess-3 sequence. A NAND gate detects state 1100
and forces 0011 to be loaded as the next state. Figure 9–22 shows the resulting
timing waveforms. Notice that the Q3 output has a 50% duty cycle, which may
be desirable for some applications.

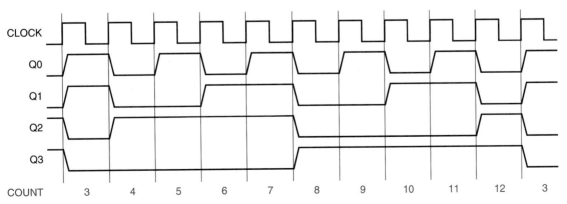

Figure 9–22   Timing waveforms for the '163 used as an excess-3 decimal counter.

A binary counter with a modulus greater than 16 can be built by cascading 74x163s. Figure 9–23 shows the general connections for such a counter. The CLK, /CLR, and /LD inputs of all the '163s are connected in parallel, so that all of them count or are cleared or loaded at the same time. A master count-enable (CNTEN) signal is connected to the low-order '163. The RCO4 output is asserted if and only if the low-order '163 is in state 15 *and* CNTEN is asserted; RCO4 is connected to the enable inputs of the high-order '163. Thus, both the carry information and the master count-enable ripple from the output of one 4-bit counter stage to the next. Like the synchronous serial counter of Figure 9–12, this scheme can be extended to build a counter with any desired number of bits, although the maximum counting speed is limited by the propagation delay of the ripple carry signal through all of the stages (but see Exercise 9.18).

Even experienced digital designers are sometimes confused about the difference between the ENP and ENT enable inputs of the '163 and similar counters,

Figure 9–23
General cascading connections for 74x163-based counters.

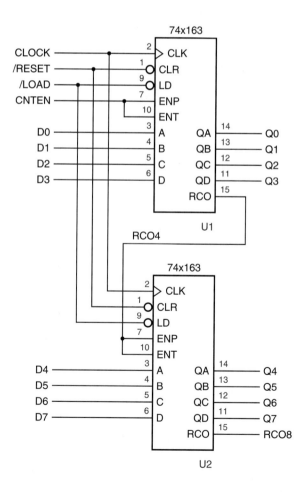

> One way to distinguish the functions of ENT and ENP in a 74x163 is to remember the "T" in ENT, which goes *to* the carry or *through* the counter.

REMEMBER
THE "T"

since both must be asserted for the counter to count. However, a glance at the 163's internal logic diagram, Figure 9–14 on page 596, shows the difference quite clearly—ENT goes to the ripple carry output as well. In many applications, this distinction is important.

For example, Figure 9–24 shows an application that uses two '163s as a modulo-193 counter that counts from 63 to 255. The MAXCNT output detects state 255 and stops the counter until /GO is asserted. When /GO is asserted, the counter is reloaded with 63 and counts up to 255 again. (Note that the value of /GO is relevant only when the counter is in state 255.) To keep the counter stopped, MAXCNT must be asserted in state 255 even while the counter is stopped.

Figure 9–24
Using 74x163s as a
modulo-193 counter with
the counting sequence
63, 64, . . . , 255, 63, 64,
. . . .

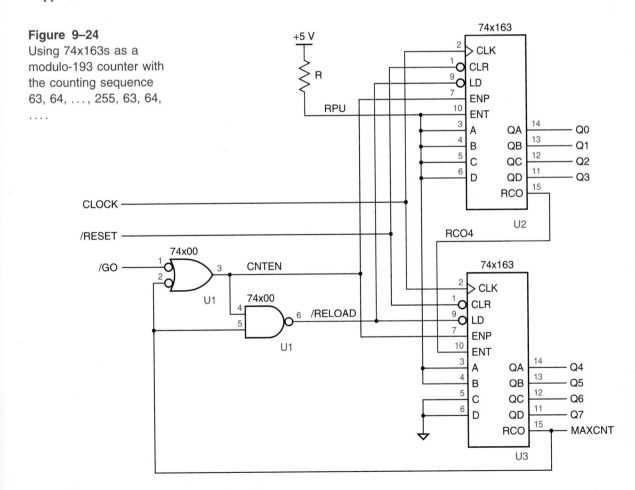

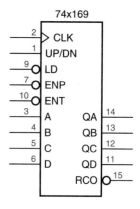

Figure 9–25
Traditional logic symbol for the
74x169 synchronous 4-bit binary
up/down counter.

Therefore, the low-order counter's ENT input is always asserted, its RCO output is connected to the high-order ENT input, and MAXCNT detects state 255 even if CNTEN is not asserted (compare with the behavior of RCO8 in Figure 9–23). To enable counting, CNTEN is connected to the ENP inputs in parallel. A NAND gate asserts /RELOAD to go back to state 63 only if /GO is asserted and the counter is in state 255.

74x169

up/down counter

Another counter with functions similar to 74x163's is the *74x169*, whose logic symbol is shown in Figure 9–25. One difference in the '169 is that its carry output and enable inputs are active low. More importantly, the '169 is an *up/down counter*; it counts in ascending or descending binary order depending on the value of an input signal, UP/DN. With a signal name like UP/DN, figuring out which value is UP and which is DN is easy if you realize that, using our active-level conventions, the signal could have been named UP or /DN. Thus, the '169 counts up when UP/DN is 1 and down when UP/DN is 0. Just to be sure, some people use the name UP/$\overline{\text{DN}}$ for this signal.

9.3.4 Decoding Binary-Counter States

A binary counter may be combined with a decoder to obtain a set of 1-out-of-m-coded signals, where one signal is asserted in each counter state. This is useful when counters are used to control a set of devices where a different device is enabled in each counter state. In this approach, each output of the decoder enables a different device.

Figure 9–26 shows how a 74x163 wired as a modulo-8 counter can be combined with a 74x138 3-to-8 decoder to provide eight signals, each one representing a counter state. Typical timing for this circuit is shown in Figure 9–27. Each decoder output is asserted during a corresponding clock period.

decoding glitches

Notice that the decoder outputs may contain "glitches" on state transitions where two or more counter bits change, even though the '163 outputs are glitch free and the '138 does not have any static hazards. In a synchronous counter like

Figure 9–26
A modulo-8 binary counter and
decoder.

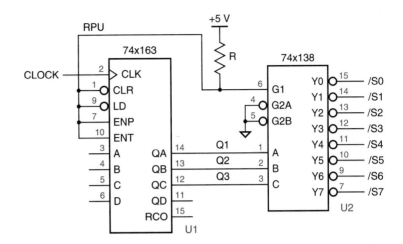

the '163, the outputs don't change at exactly the same time. More important,
multiple signal paths in a decoder like the '138 have different delays; for example,
the path from B to /Y1 is faster than the path from A to /Y1. Therefore, even if
the input changes simultaneously from 011 to 100, the decoder may behave as
if the input were temporarily 001, and the /Y1 output may have a glitch. In the
present example, it can be shown that the glitches can occur in *any* realization
of the binary decoder function; this problem is called a *function hazard*.   *function hazard*

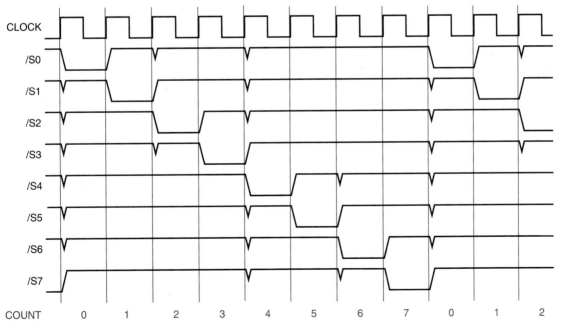

Figure 9–27 Timing diagram for a modulo-8 binary counter and decoder, showing decoding glitches.

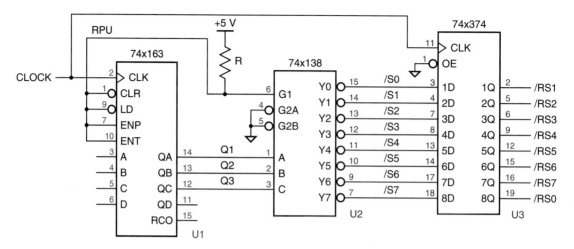

Figure 9–28   A modulo-8 binary counter and decoder with glitch-free outputs.

In most applications, the decoder output signals portrayed in Figure 9–27 would be used as control inputs to registers, counters, and other edge-triggered devices (e.g., /G in a 74x377, /LD in a 74x163, or /ENP in a 74x169). In such a case, the decoding glitches in the figure are not a problem, since they occur *after* the clock tick. They are long gone before the next tick comes along, when the decoder outputs are sampled by other edge-triggered devices. However, the glitches *would* be a problem if they were applied to something like the /S or /R inputs of a /S-/R latch. Likewise, using such potentially glitchy signals as clocks for edge-triggered devices is a definite no-no.

If necessary, one way to "clean up" the glitches in Figure 9–27 is to connect the '138 outputs to another register that samples the stable decoded outputs on the next clock tick, as shown in Figure 9–28. Notice that the decoded outputs have been renamed to account for the 1-tick delay through the register. However, once you decide to pay for an 8-bit register, a less costly solution is to use an 8-bit "ring counter," which provides glitch-free decoded outputs directly, as we'll show in Section 9.4.6.

9.4   SHIFT REGISTERS

9.4.1   Shift-Register Structure

shift register

serial input

serial output

A *shift register* is an *n*-bit register with a provision for shifting its stored data by one bit position at each tick of the clock. Figure 9–29 shows the structure of a serial-in, serial-out shift register. The *serial input*, SERIN, specifies a new bit to be shifted into one end at each clock tick. This bit appears at the *serial output*,

Figure 9–29
Structure of a serial-in, serial-out
shift register.

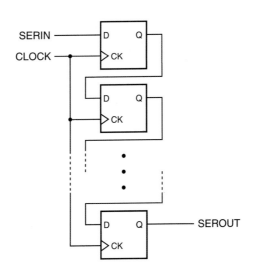

SEROUT, after *n* clock ticks, and is lost one tick later. Thus, an *n*-bit serial-in, serial-out shift register can be used to delay a signal by *n* clock ticks.

A *serial-in, parallel-out shift register*, shown in Figure 9–30, has outputs for all of its stored bits, making them available to other circuits. Such a shift register can be used to perform *serial-to-parallel conversion*, as explained later in this section.

Conversely, it is possible to build a *parallel-in, serial-out shift register*. Figure 9–31 shows the general structure of such a device. At each clock tick, the register either loads new data from inputs 1D–ND, or it shifts its current contents, depending on the value of the LOAD/SHIFT control input. Internally, the device uses a 2-input multiplexer on each flip-flop's D input to select between

*serial-in,
   parallel-out shift
   register*
*serial-to-parallel
   conversion*
*parallel-in,
   serial-out shift
   register*

Figure 9–30
Structure of a serial-in, parallel-
out shift register.

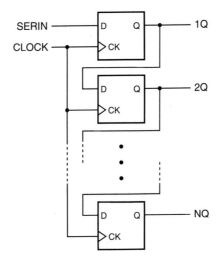

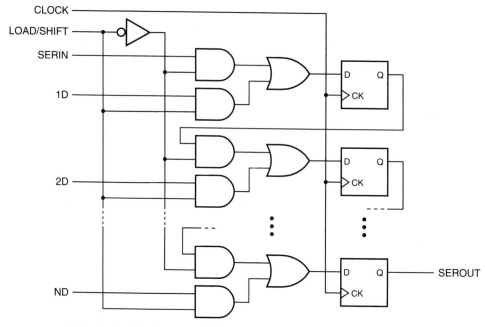

Figure 9–31   Structure of a parallel-in, serial-out shift register.

parallel-to-serial conversion

the two cases. A parallel-in, serial-out shift register can be used to perform *parallel-to-serial conversion*, as explained later in this section.

By providing outputs for all of the stored bits in a parallel-in shift register, we obtain the *parallel-in, parallel-out shift register* shown in Figure 9–32. Such a device is general enough to be used in any of the applications of the previous shift registers.

parallel-in, parallel-out shift register

9.4.2   MSI Shift Registers

Figure 9–33(a) and (b) shows logic symbols for two popular MSI 8-bit shift registers. The *74x164* is a serial-in, parallel-out device with an asynchronous clear input (/CLR). It has two serial inputs that are ANDed internally. That is, both SERA and SERB must be 1 for a 1 to be shifted into the first bit of the register.

74x164

74x166

The *74x166* is a parallel-in, serial-out shift register, also with an asynchronous clear input. This device shifts when SH/LD is 1, and loads new data otherwise. Its clocking is a little unusual, since it has two clock inputs that are connected to the internal flip-flops as shown in Figure 9–33(c). The designers of this device intended for CLK to be connected to a free-running system clock, and for CLKINH to be asserted to inhibit CLK, so that neither shifting nor loading occurs on the next clock tick, and the current register contents are held. How-

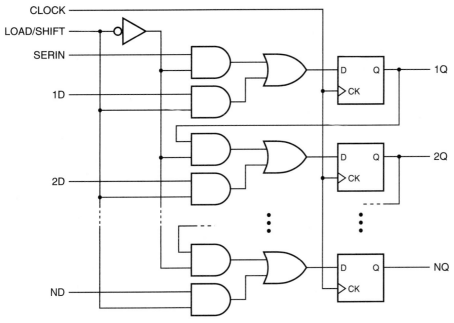

Figure 9–32 Structure of a parallel-in, parallel-out shift register.

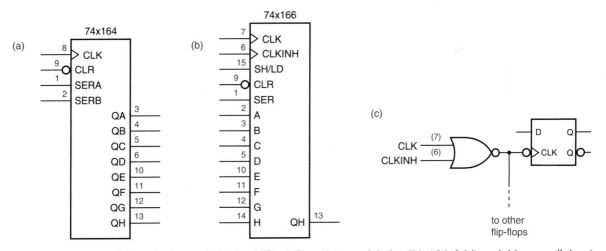

Figure 9–33 Traditional logic symbols for MSI shift registers: (a) the 74x164 8-bit serial-in, parallel-out shift register; (b) the 74x166 8-bit parallel-in, serial-out shift register; (c) equivalent circuit for 74x166 clock inputs.

Figure 9–34
Traditional logic symbol for the 74x194.

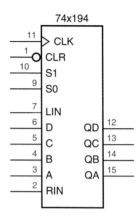

ever, for this to work, CLKINH must be changed only when CLK is 1; otherwise, undesired clock edges occur on the internal flip-flops. A much better way of obtaining a "hold" function is employed in the next device that we discuss.

74x194

unidirectional shift register
bidirectional shift register

left
right

The *74x194* is an MSI 4-bit bidirectional, parallel-in, parallel-out shift register whose logic symbol and logic diagram are shown in Figures 9–34 and 9–35. The shift registers that we've studied previously are called *unidirectional shift registers* because they shift in only one direction. The '194 is a *bidirectional shift register* because its contents may be shifted in either of two directions, depending on a control input. The two directions are called "left" and "right," even though the logic diagram and the logic symbol aren't necessarily drawn that way. In the '194, *left* means "in the direction from QD to QA," and *right* means "in the direction from QA to QD." Our logic diagram and symbol for the '194 are consistent with these names if you rotate them 90° clockwise.

Table 9–2 is the state table for the 74x194. This state table is highly compressed, since it does not contain columns for most of the inputs (A–D, RIN, LIN) or the current state QA–QD. Still, by expressing each next-state value as a function of these implicit variables, it completely defines the operation of the

Table 9–2
State table for a 74x194 4-bit universal shift register.

Function	Inputs		Next state			
	S1	S0	QA*	QB*	QC*	QD*
Hold	0	0	QA	QB	QC	QD
Shift right	0	1	RIN	QA	QB	QC
Shift left	1	0	QB	QC	QD	LIN
Load	1	1	A	B	C	D

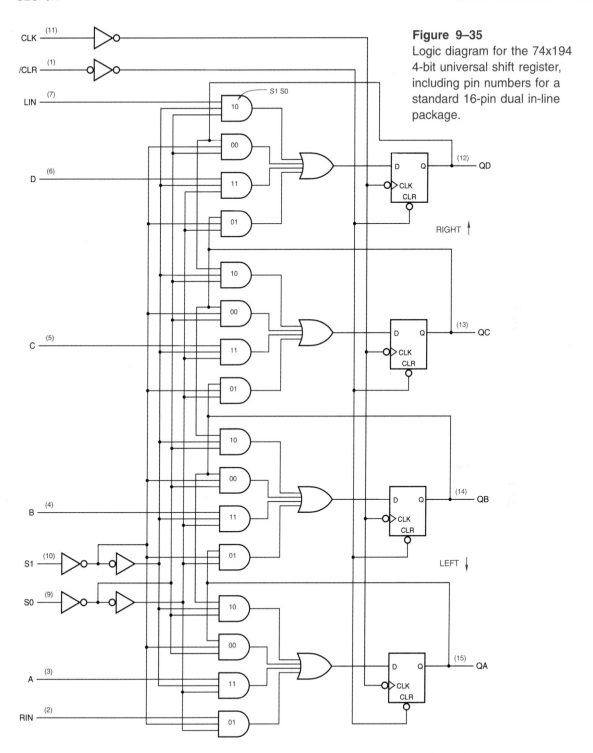

Figure 9–35
Logic diagram for the 74x194
4-bit universal shift register,
including pin numbers for a
standard 16-pin dual in-line
package.

Table 9-3 Function table for a 74x299 8-bit universal shift register.

Function	Inputs S1	S0	Next state QA*	QB*	QC*	QD*	QE*	QF*	QG*	QH*
Hold	0	0	QA	QB	QC	QD	QE	QF	QG	QH
Shift right	0	1	RIN	QA	QB	QC	QD	QE	QF	QG
Shift left	1	0	QB	QC	QD	QE	QF	QG	QH	LIN
Load	1	1	AQA	BQB	CQC	DQD	EQE	FQF	GQG	HQH

'194 for all 2^{12} possible combinations of current state and input, and it sure beats a 4,096-row table!

Note that the '194's LIN (left-in) input is conceptually located on the "right-hand" side of the chip, but it is the serial input for *left* shifts. Likewise, RIN is the serial input for right shifts.

universal shift register

The '194 is sometimes called a *universal* shift register because it can be made to function like any of the less general shift register types that we've discussed (e.g., unidirectional; serial-in, parallel-out; parallel-in, serial-out). In fact, many of our design examples in the next few subsections contain '194s configured to use just a subset of their available functions.

74x299

The *74x299* is an 8-bit universal shift register in a 20-pin package; its symbol and logic diagram are given in Figures 9–36 and 9–37. The '299's functions and state table are similar to the '194's, as shown in Table 9–3. To save pins, the '299 uses bidirectional three-state lines for input and output, as shown in the logic diagram. During load operations (S1 S0 = 11), the three-state drivers are disabled and data is loaded through the AQA–HQH pins. At other times, the stored bits are driven onto these same pins if /G1 and /G2 are asserted. The leftmost and rightmost stored bits are available at all times on separate output-only pins, QA and QH.

Figure 9-36
Traditional logic symbol for the 74x299.

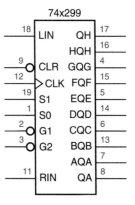

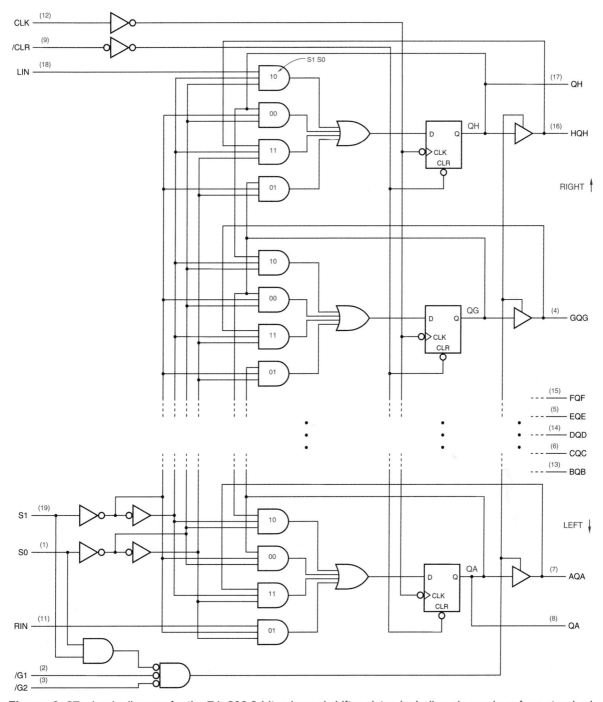

Figure 9–37 Logic diagram for the 74x299 8-bit universal shift register, including pin numbers for a standard 20-pin dual in-line package.

9.4.3 The World's Biggest Shift-Register Application

digital telephony

CO

serial channel

multiplex

space/time trade-off

The most common application of shift registers is to convert parallel data into serial format for transmission or storage, and to convert serial data back to parallel format for processing or display (see Section 2.16.1). The most common example of serial data transmission, one that you almost certainly take part in every day, is in *digital telephony*.

For years, TPC (The Phone Company) has been installing digital switching equipment in its central offices (*COs*). Most home phones have a two-wire analog connection to the central office. However, modern CO equipment samples the analog voice signal 8,000 times per second (once every 125 μs) when it enters the CO, and produces a corresponding sequence of 8,000 8-bit bytes representing the sign and amplitude of the analog signal at each sampling point. Subsequently, your voice may be transmitted digitally on 64 kbps *serial channels* throughout the phone network, until it is converted back to an analog signal at the far-end CO. Recent advances, such as ISDN (Integrated Services Digital Network), promise to extend a serial digital channel all the way to your home phone.

The 64 kbps bandwidth required by a single digital voice signal is far *less* than can be obtained on a single digital signal line or switched by digital ICs. Therefore most digital telephone equipment *multiplexes* multiple 64-kbps channels onto a single wire, saving both wires and digital ICs for switching. In the example in the next subsection, 32 channels are processed by a handful of chips. This is a classic example of a *space/time trade-off* in digital design—by running the chips faster, you can accomplish a given task with fewer chips. Indeed, this is the main reason that the telephone network has "gone digital."

9.4.4 Serial/Parallel Conversion

A typical application of serial data transfer between two modules (possibly in a piece of CO switching equipment) is shown in Figure 9–38. Three signals are normally connected between a source module and a destination module to accomplish the transfer:

I STILL
DON'T KNOW

In the first edition of this book, we noted that delays in ISDN deployment had led some people in the industry to rename it "Imaginary Services Delivered Nowhere." Although ISDN has now gained some acceptance in Europe and Japan, the jibe is still generally valid in North America.

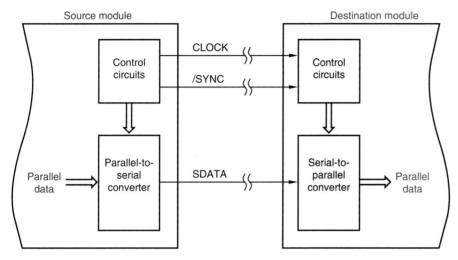

Figure 9–38   A system that transmits data serially between modules.

- *Clock.* The clock signal provides the timing reference for transfers, defining the time to transfer one bit. In systems with just two modules, the clock may be part of control circuits located on the source module as shown. In larger systems, the clock may be generated at a common point and distributed to all of the modules.

- *Synchronization.* A *synchronization pulse* (or *sync pulse*) provides a reference point for defining the data format, such as the beginning of a byte or word in the serial data stream.

  synchronization pulse
  sync pulse

- *Serial data.* The data itself is transmitted on a single line.

The general timing characteristics of these signals in a typical digital telephony application are shown in Figure 9–39(a). The **CLOCK** signal has a frequency of 2.048 MHz to allow the transmission of $32 \times 8{,}000$ 8-bit bytes per second. The 1-bit-wide pulse on the **/SYNC** signal identifies the beginning of a 125-μs interval called a *frame*. A total of 256 bits are transmitted on **SDATA** during this interval, which is divided into 32 *timeslots* containing eight bits each.

frame
timeslot

| Believe it or not, in the phone network, a 2.048 MHz clock is generated in St. Louis and distributed throughout the U.S.! The **CLOCK** signal that is distributed in a particular piece of local CO equipment is normally derived from the national clock. | THE NATION'S CLOCK |

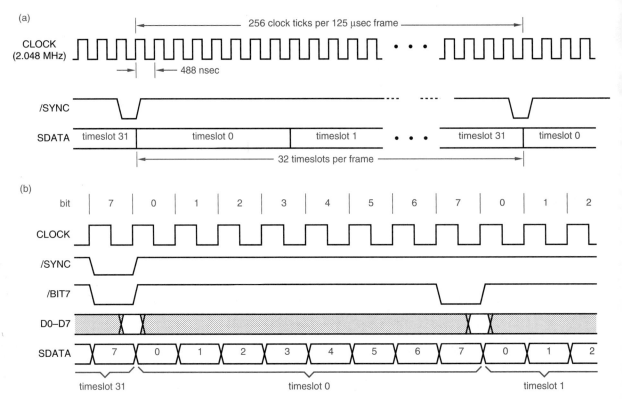

Figure 9–39 Timing diagram for parallel-to-serial conversion: (a) a complete frame; (b) one byte at the beginning of the frame.

Each timeslot carries one digitally encoded voice signal. Both timeslot numbers and bit positions within a timeslot are located relative to the SYNC pulse.

Figure 9–40 shows a circuit that converts parallel data to the serial format of Figure 9–39(a), with detailed timing shown in (b). Two 74x163 counters are wired as a free-running modulo-256 counter to define the frame. The five high-order and three low-order counter bits are the timeslot number and bit number, respectively.

A 74x166 parallel-in shift register performs the parallel-to-serial conversion. Bit 0 of the parallel data (D0–D7) is connected to the '166 input closest to the SDATA output, so bits are transmitted serially in the order 0 through 7.

WHICH BIT FIRST?

> Most real serial links for digitized voice actually transmit bit 7 first, because this is the first bit generated by the analog-to-digital converter that digitizes the voice signal. However, to simplify our examples, we transmit bit 0 first so that counter state equals bit number.

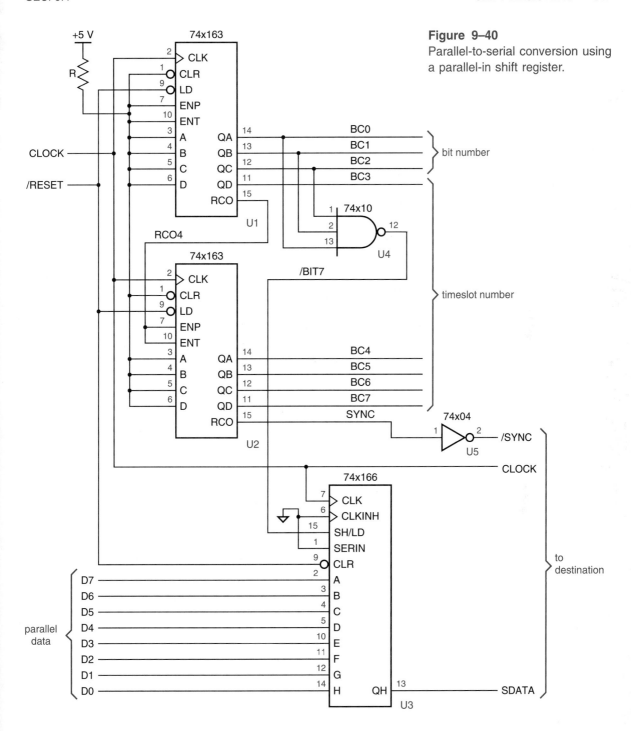

Figure 9–40
Parallel-to-serial conversion using a parallel-in shift register.

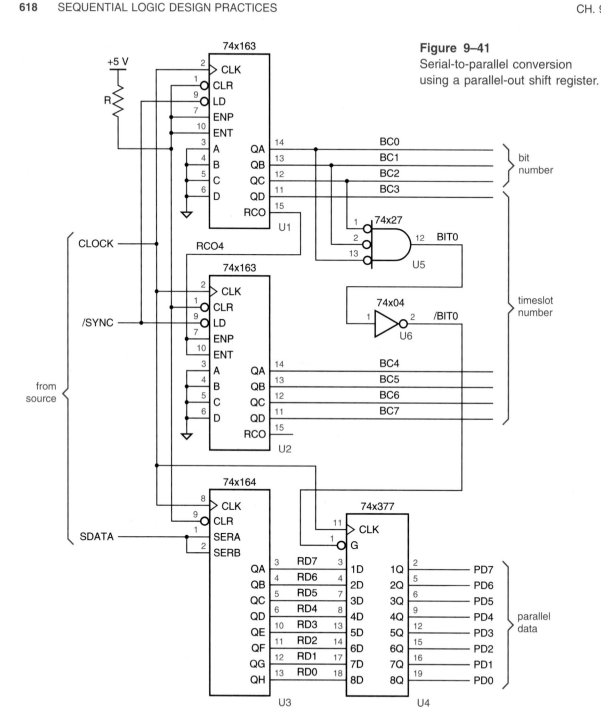

Figure 9–41
Serial-to-parallel conversion using a parallel-out shift register.

During bit 7 of each timeslot, the /BIT7 signal is asserted, which causes the '166 to be loaded with D0–D7. The value of D0–D7 is irrelevant except during the setup- and hold-time window around the clock edge on which the '166 is loaded, as shown by shading in the timing diagram. This leaves open the possibility that the parallel data bus could be used for other things at other times (see Exercise 9.24).

A destination module can convert the serial data back into parallel format using the circuit of Figure 9–41. A modulo-256 counter built from a pair of '163s is used to reconstruct the timeslot and bit numbers. Although /SYNC is asserted during state 255 of the counter on the source module, /SYNC loads the destination module's counter with 0 so that both counters go to 0 on the same clock edge. The counter's high-order bits (timeslot number) are not used in the figure, but they may be used by other circuits in the destination module to identify the byte from a particular timeslot on the parallel data bus (PD0–PD7).

Figure 9–42 shows detailed timing for the serial-to-parallel conversion circuit. A complete received byte is available at parallel output of the 74x164 shift register during the clock period following the reception of the last bit (7) of the byte. The parallel data in this example is *double-buffered*—once it is fully received, it is transferred into a 74x377 register, where it is available on PD0–PD7 for eight full clock periods until the next byte is fully received. The /BIT0 signal enables the '377 to be loaded at the proper time. Additional registers and decoding could be provided to load the byte from each timeslot into a different register, making each byte available for 125 μs (see Exercise 9.26).

double-buffered data

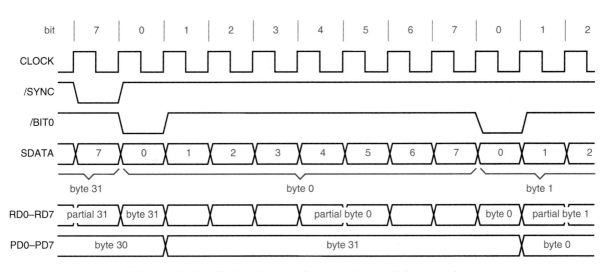

Figure 9–42   Timing diagram for serial-to-parallel conversion.

Once the received data is in parallel format, it can easily be stored or modified by other digital circuits; we'll give examples in Section 11.1.6. In digital telephony, the received parallel data is converted back into an analog voltage that is filtered and transmitted to an earpiece or speaker for 125 μs, until the next voice sample arrives.

LITTLE ENDIANS
AND BIG ENDIANS

At one point in the evolution of digital systems, the choice of which bit or byte to transmit first was a religious issue. In a famous article on the subject, "On Holy Wars and a Plea for Peace" (*Computer*, October 1981, pp. 48–54), Danny Cohen describes the differences in bit- and byte-ordering conventions and the havoc that could be (and now has been) wrought as a result.

A firm standard was never established, so that today there are some popular computer systems (such as IBM-compatible PCs) that transmit or number the low-order byte of a 32-bit word first, and others (such as the Apple Macintosh computer) that transmit or number the high-order byte first. Following Cohen's nomenclature, people refer to these conventions as "Little Endian" and "Big Endian," respectively, and talk about "endianness" as if it were actually a word.

9.4.5 Shift-Register Counters

Serial/parallel conversion is a "data" application, but shift registers have "non-data" applications as well. A shift register can be combined with combinational logic to form a state machine whose state diagram is cyclic. Such a circuit is called a *shift-register counter*. Unlike a binary counter, a shift-register counter does not count in an ascending or descending binary sequence, but it is useful in many "control" applications nonetheless.

*shift-register
counter*

9.4.6 Ring Counters

ring counter

The simplest shift-register counter uses an n-bit shift register to obtain a counter with n states, and is called a *ring counter*. Figure 9–43 is the logic diagram for a 4-bit ring counter. The 74x194 universal shift register is wired so that it normally performs a left shift. However, when RESET is asserted, it loads 0001 (refer to the '194's function table, Table 9–2 on page 610). Once RESET is negated, the '194 shifts left on each clock tick. The LIN serial input is connected to the "leftmost" output, so the next states are 0010, 0100, 1000, 0001, 0010, ... Thus, the counter visits four unique states before repeating. A timing diagram is shown in Figure 9–44. In general, an n-bit ring counter visits n states in a cycle.

Figure 9–43
Simplest design for a four-bit,
four-state ring counters with a
single circulating 1.

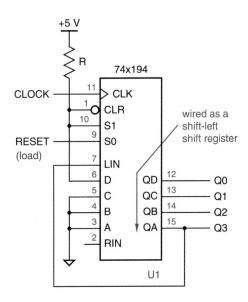

The ring counter in Figure 9–43 has one major problem—it is not robust. If its single 1 output is lost due to a temporary hardware problem (e.g., noise), the counter goes to state 0000 and stays there forever. Likewise, if an extra 1 output is set (i.e., state 0101 is created), the counter will go through an incorrect cycle of states and stay in that cycle forever. These problems are quite evident if we draw the *complete* state diagram for the counter circuit, which has 16 states. As shown in Figure 9–45, there are 12 states that are not part of the normal counting cycle. If the counter somehow gets off the normal cycle, it stays off.

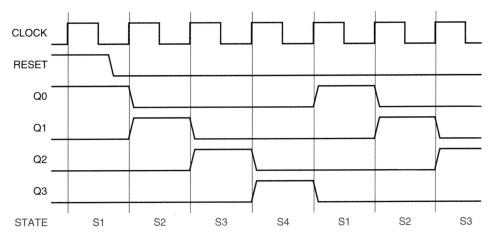

Figure 9–44 Timing diagram for a 4-bit ring counter.

Figure 9–45
State diagram for a simple ring
counter.

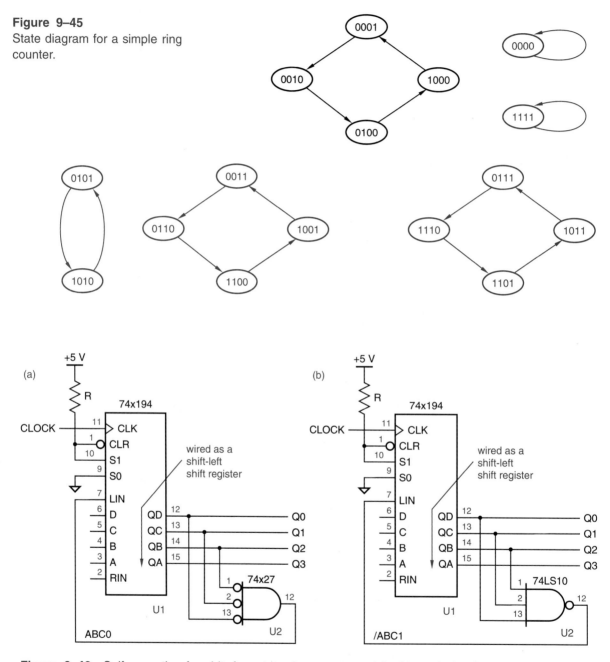

Figure 9–46   Self-correcting four-bit, four-state ring counters: (a) with a single circulating 1; (b) with a
single circulating 0.

Figure 9–47
State diagram for a self-
correcting ring counter.

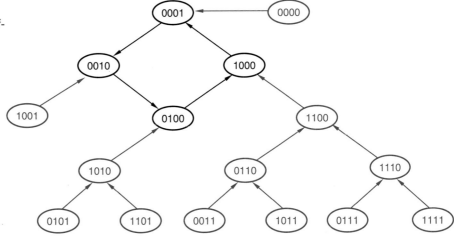

A *self-correcting counter* is designed so that all abnormal states have transitions leading to normal states. Self-correcting counters are desirable for the same
reason that we use a minimal-risk approach to state assignment in Section 7.4.3:
If something unexpected happens, a counter or state machine should go to a
"safe" state.

*self-correcting
counter*

A *self-correcting ring counter* circuit is shown in Figure 9–46(a). The
circuit uses a NOR gate to shift a 1 into LIN only when the three least significant
bits are 0. This results in the state diagram in Figure 9–47; all abnormal states
lead back into the normal cycle. Notice that, in this circuit, an explicit RESET
signal is not necessarily required. Regardless of the initial state of the shift
register on power-up, it reaches state 0001 within four clock ticks. Therefore,
an explicit reset signal is required only if it is necessary to ensure that that the
counter starts up synchronously with other devices in the system.

*self-correcting ring
counter*

For the general case, an *n*-bit self-correcting ring counter uses an *n*−1-input
NOR gate, and corrects an abnormal state within *n* − 1 clock ticks.

In TTL and CMOS logic families, NAND gates are generally easier to come
by than NORs, so it may be more convenient to design a self-correcting ring
counter as shown in Figure 9–46(b). States in this counter's normal cycle have
a single circulating 0.

The major appeal of a ring counter for control applications is that its states
appear in 1-out-of-*n* decoded form directly on the flip-flop outputs. That is,
exactly one flip-flop output is asserted in each state. Furthermore, these outputs
are "glitch free"; compare with the binary counter and decoder approach of
Figure 9–26 on page 605.

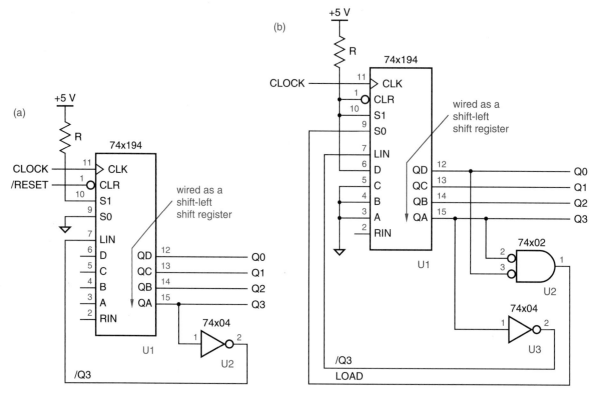

Figure 9–48   Four-bit, eight-state Johnson counters: (a) basic; (b) self-correcting.

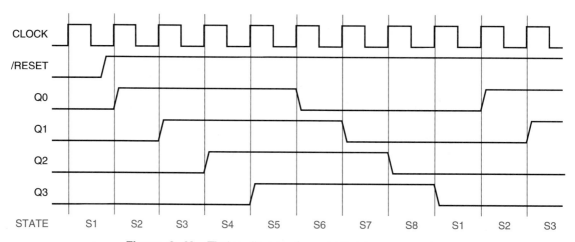

Figure 9–49   Timing diagram for a 4-bit Johnson counter.

Table 9–4
States of a 4-bit Johnson
counter.

State Name	Q3	Q2	Q1	Q0	Decoding
S1	0	0	0	0	$Q3' \cdot Q0'$
S2	0	0	0	1	$Q1' \cdot Q0$
S3	0	0	1	1	$Q2' \cdot Q1$
S4	0	1	1	1	$Q3' \cdot Q2$
S5	1	1	1	1	$Q3 \cdot Q0$
S6	1	1	1	0	$Q1 \cdot Q0'$
S7	1	1	0	0	$Q2 \cdot Q1'$
S8	1	0	0	0	$Q3 \cdot Q2'$

* 9.4.7 Johnson Counters

An *n*-bit shift register with the complement of the serial output fed back into the
serial input is a counter with $2n$ states and is called a *twisted-ring, Moebius,* or
Johnson counter. Figure 9–48(a) is the basic circuit for a Johnson counter, and
Figure 9–49 is its timing diagram. The normal states of this counter are listed
in Table 9–4. If both the true and complemented outputs of each flip-flop are
available, each normal state of the counter can be decoded with a 2-input AND
or NAND gate, as shown in the table. The decoded outputs are glitch free.

     An *n*-bit Johnson counter has $2^n - 2n$ abnormal states, and is therefore
subject to the same robustness problems as a ring counter. A 4-bit *self-correcting
Johnson counter* can be designed as shown in Figure 9–48(b). This circuit loads
0001 as the next state whenever the current state is 0xx0. A similar circuit using
a single 2-input NOR gate can perform correction for a Johnson counter with
any number of bits. The correction circuit must load 00...01 as the next state
whenever the current state is 0x...x0.

twisted-ring counter
Moebius counter
Johnson counter

*self-correcting
Johnson counter*

> We can prove that the Johnson-counter self-correction circuit corrects any abnormal
> state as follows. An abnormal state can always be written in the form x...x10x...x,
> since the only states that can't be written in this form are normal states (00...00,
> 11...11, 01...1, 0...01...1, and 0...01). Therefore, within $n - 2$ clock ticks, the
> shift register will contain 10x...x. One tick later it will contain 0x...x0, and one
> tick after that the normal state 00...01 will be loaded.

THE SELF-
CORRECTION
CIRCUIT IS
CORRECT

* 9.4.8 Linear Feedback Shift Register Counters

The *n*-bit shift register counters that we've shown so far have far less than
the maximum of 2^n normal states. An *n*-bit *linear feedback shift-register (LFSR)
counter* can have $2^n - 1$ states, almost the maximum. Such a counter is sometimes
called a *maximum-length sequence generator.*

*maximum-length
sequence
generator*

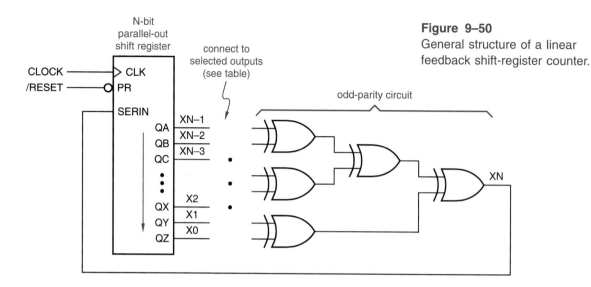

Figure 9–50
General structure of a linear feedback shift-register counter.

finite fields

The design of LFSR counters is based on the theory of *finite fields*, which was developed by French mathematician Évariste Galois (1811–1832) shortly before he was killed in a duel with a political opponent. The operation of an LFSR counter corresponds to operations in a finite field with 2^n elements.

Figure 9–50 shows the structure of an *n*-bit LFSR counter. The shift register's serial input is connected to the sum modulo 2 of a certain set of output bits. These feedback connections determine the state sequence of the counter. By convention, outputs are always numbered and shifted in the direction shown.

Using finite field theory, it can be shown that for any value of *n*, there exists

Table 9–5
Feedback equations for linear feedback shift-register counters.

n	Feedback Equation
2	$X2 = X1 \oplus X0$
3	$X3 = X1 \oplus X0$
4	$X4 = X1 \oplus X0$
5	$X5 = X2 \oplus X0$
6	$X6 = X1 \oplus X0$
7	$X7 = X3 \oplus X0$
8	$X8 = X4 \oplus X3 \oplus X2 \oplus X0$
12	$X12 = X6 \oplus X4 \oplus X1 \oplus X0$
16	$X16 = X5 \oplus X4 \oplus X3 \oplus X0$
20	$X20 = X3 \oplus X0$
24	$X24 = X7 \oplus X2 \oplus X1 \oplus X0$
28	$X28 = X3 \oplus X0$
32	$X32 = X22 \oplus X2 \oplus X1 \oplus X0$

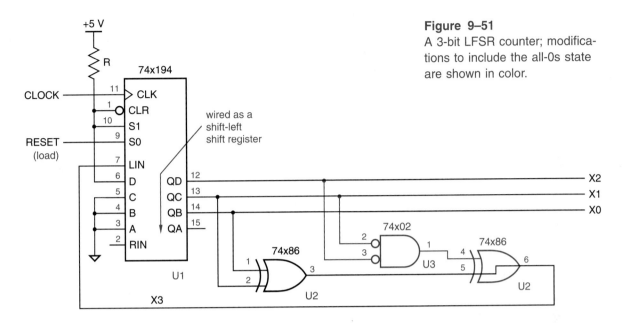

Figure 9–51
A 3-bit LFSR counter; modifications to include the all-0s state are shown in color.

at least one feedback equation such that the counter cycles through all $2^n - 1$ nonzero states before repeating. This is called a *maximum-length sequence*. (Regardless of the connection pattern, the next state for the all-0s state is all 0s.) Table 9–5 lists feedback equations that result in maximum-length sequences for selected values of n. For each value of n greater than 3, there are many other feedback equations that result in maximum-length sequences, all different.

    maximum-length sequence

    The logic diagram for a 3-bit LFSR counter is shown in Figure 9–51. The state sequence for this counter is shown in the first three columns of Table 9–6. Starting in any nonzero state, 100 in the table, the counter visits seven states before returning to the starting state.

    An LFSR counter can be modified to have 2^n states, including the all-0s state, as shown in color for the 3-bit counter in Figure 9–51. The resulting state sequence is given in the last three columns of Table 9–6. In an n-bit LFSR counter, an extra XOR gate and an $n - 1$ input NOR gate connected to all shift-register outputs except X0 accomplishes the same thing.

    The states of an LFSR counter are not visited in binary counting order. However, LFSR counters are typically used in applications where this characteristic is an advantage. A major application of LFSR counters is in generating test inputs for logic circuits. In most cases, the "pseudo-random" counting sequence of an LFSR counter is more likely than a binary counting sequence to detect errors. LFSRs can also be used in the encoding and decoding circuits for certain error-detecting and error-correcting codes, and are often used to "scramble" the data patterns transmitted by high-speed modems.

Table 9–6
State sequences for the 3-bit
LFSR counter in Figure 9–51.

Original Sequence			Modified Sequence		
X2	X1	X0	X2	X1	X0
1	0	0	1	0	0
0	1	0	0	1	0
1	0	1	1	0	1
1	1	0	1	1	0
1	1	1	1	1	1
0	1	1	0	1	1
0	0	1	0	0	1
1	0	0	0	0	0
.	.	.	1	0	0
.	.	.	.	.	.

* 9.5 ITERATIVE VERSUS SEQUENTIAL CIRCUITS

We introduced iterative circuits in Section 5.8.2. The function of an n-module
iterative circuit can be performed by a sequential circuit that uses just one copy
of the module but requires n steps (clock ticks) to obtain the result. This is an
excellent example of a space/time tradeoff in digital design.

As shown in Figure 9–52, flip-flops are used in the sequential-circuit version
to store the cascading outputs at the end of each step; the flip-flop outputs are
used as the cascading inputs at the beginning of the next step. The flip-flops
must be initialized to the boundary-input values before the first clock tick, and
they contain the boundary-output values after the nth tick.

Since an iterative circuit is a combinational circuit, all of its primary and
boundary inputs may be applied simultaneously, and its primary and boundary

Figure 9–52
General structure of the
sequential-circuit version of
an iterative circuit.

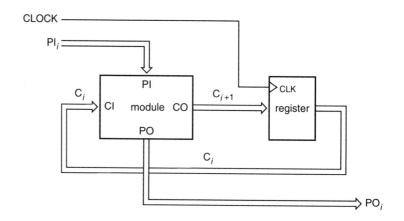

Figure 9–53
Simplified serial comparator circuit.

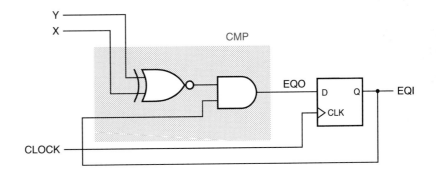

outputs are all available after a combinational delay. In the sequential-circuit version, the primary inputs must be delivered sequentially, one per clock tick, and the primary outputs are produced with similar timing. Therefore, serial-out shift registers are often used to provide the inputs, and serial-in shift registers are used to collect the outputs. For this reason, the sequential-circuit version of an "iterative widget" is often called a "serial widget."

For example, Figure 9–53 shows the basic design for a *serial comparator*   *serial comparator* circuit. The shaded block is identical to the module used in the iterative comparator of Figure 5–72 on page 345. The circuit is drawn in more detail using SSI chips in Figure 9–54. In addition, we have provided a synchronous reset input that, when asserted, forces the initial value of the cascading flip-flop to 1 at the next clock tick. (The initial value of the cascading flip-flop corresponds to the boundary input in the iterative comparator.)

With the serial comparator, an *n*-bit comparison requires $n + 1$ clock ticks. /RESET is asserted at the first clock tick. /RESET is negated and data bits are applied at the next *n* ticks. The EQI output gives the comparison result during the clock period following the last tick. Timing for two successive 4-bit comparisons is shown in Figure 9–55.

Figure 9–54
Detailed serial comparator circuit.

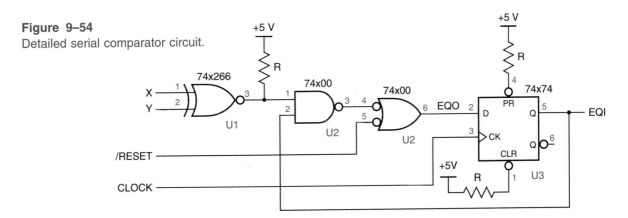

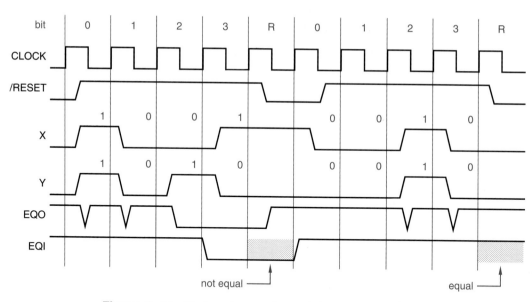

Figure 9–55 Timing diagram for serial comparator circuit.

serial binary adder   A *serial binary adder* circuit for addends of any length can be constructed from a full adder and a D flip-flop, as shown in Figure 9–56. The flip-flop, which stores the carry between successive bits of the addition, is cleared to 0 at reset. Addend bits are presented serially on the A and B inputs, starting with the LSB, and sum bits appear on S in the same order.

Because of the large size and high cost of digital logic circuits in the early days, many computers and calculators used serial adders and other serial versions of iterative circuits to perform arithmetic operations. Even though these arithmetic circuits aren't used much today, they provide a powerful reminder of the space/time trade-offs that are possible in digital design.

Figure 9–56
Serial binary adder circuit.

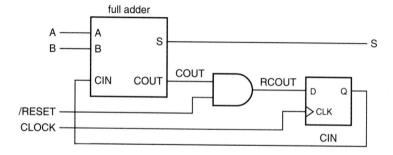

* 9.6   MSI DEVICES AS STATE MACHINES

Sometimes a state machine's word description or state diagram almost matches the behavior of an off-the-shelf MSI part. This is especially common for state machines that have an almost cyclic behavior. For these machines, it is often possible to use counters or shift registers to create the basic cycle and add combinational logic to force the noncyclic transitions.

Figure 9–57 is an example of a state diagram with an almost cyclic behavior. The main cycle is a modulo-16 counter with an enable input. In certain states, two other inputs are tested, and the counter may be forced into one of two other states, A and G. This behavior is somewhat suggestive of the 74x163 4-bit binary counter. In addition to counting, a '163 can be synchronously forced into either of two states—0000 and whatever state is defined by the A–D inputs. Thus, we can try to realize the state diagram with a '163, by using 0000 to encode state A in Figure 9–57 and assigning the rest of states in sequence as shown in Table 9–7.

Figure 9–57
State diagram with an almost
cyclic behavior.

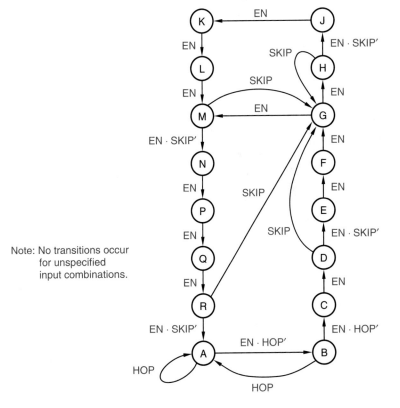

Note: No transitions occur
    for unspecified
    input combinations.

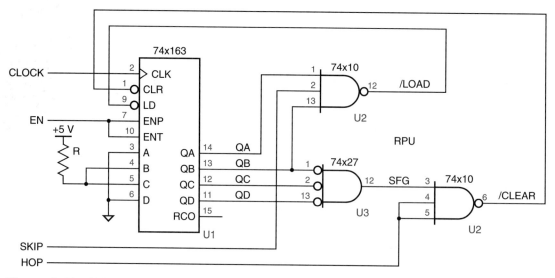

Figure 9–58 Using a 74x163 binary counter and combinational logic to realize an almost cyclic state diagram.

Figure 9–58 shows the circuit resulting from these ideas. The state machine's EN input enables the '163 in Figure 9–58 to count. State A is coded 0000, and the circuit uses the 163's /CLR input to force state A as required by the HOP input. Similarly, state G is coded 0110, and is forced using /LD as required by SKIP.

Sometimes a designer might be willing to bend a design specification a bit for convenience in an MSI design. An example is the guessing-game machine of Figure 8–20. It can be built as a Mealy machine using a 74x194 shift register and a few gates, as shown in Figure 9–59. Since the ERR output is a combinational function of the inputs, players can cheat—once the machine is stopped, the guess

Table 9–7
Counter states corresponding to state-machine named states.

State	Count	QD	QC	QB	QA	State	Count	QD	QC	QB	QA
A	0	0	0	0	0	J	8	1	0	0	0
B	1	0	0	0	1	K	9	1	0	0	1
C	2	0	0	1	0	L	10	1	0	1	0
D	3	0	0	1	1	M	11	1	0	1	1
E	4	0	1	0	0	N	12	1	1	0	0
F	5	0	1	0	1	P	13	1	1	0	1
G	6	0	1	1	0	Q	14	1	1	1	0
H	7	0	1	1	1	R	15	1	1	1	1

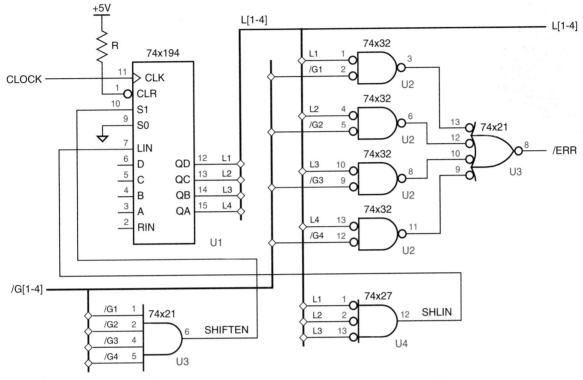

Figure 9–59   MSI realization of the guessing-game machine.

can be changed to extinguish the ERR light. (I suppose that if the designer can cheat, it's only fair for the players to cheat, too!)

When the name of the game is to minimize the number of SSI and MSI packages, designers often come up with tricky realizations that coerce an off-the-shelf part into the desired state-machine behavior. An especially tricky design example is shown in Figure 9–60, which realizes the T-bird tail-lights state diagram of Figure 8–12 (without output logic) with one 74x163 and less than one SSI package (3/4 of a 74x32 plus 1/6 of a 74x04). The state encoding for this

How do you come up with tricky realizations like the ones in this section? It's not always easy; you often must stare at the problem for a long time and experiment with different state encodings and excitation circuits. You might occasionally find this sort of approach absolutely necessary if you're in a tight situation and don't have access to PLDs, but the main use of such tricky state-machine realizations by experienced (insecure?) designers seems to be in showing off how clever they are.

FOR SHOW-OFFS
ONLY

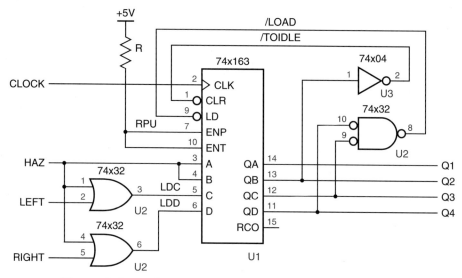

Figure 9-60   MSI realization of the T-bird tail-lights machine.

realization is shown in Table 9–8. The basic idea is that the counter is almost always loading; asserting LEFT, RIGHT, or HAZ loads a non-IDLE state and allows the counter to count until QB = 1. The ENP and ENT inputs can be always asserted, since loading and clearing override counting in the '163. The design of the output logic is left to the reader as Exercise 9.41.

Table 9–8
State assignment for T-bird tail-lights machine using a 74x163.

State	QD	QC	QB	QA
IDLE	0	0	0	0
L1	0	1	0	0
L2	0	1	0	1
L3	0	1	1	0
R1	1	0	0	0
R2	1	0	0	1
R3	1	0	1	0
LR3	1	1	1	1

9.7 IMPEDIMENTS TO SYNCHRONOUS DESIGN

While the synchronous approach is the most straightforward and reliable method of digital system design, a few nasty realities can get in the way. We'll discuss them in this section.

9.7.1 Clock Skew

Synchronous systems using edge-triggered flip-flops work properly only if all flip-flops see the triggering clock edge at the same time. Figure 9–61 shows what can happen otherwise. Here, two flip-flops are theoretically clocked by the same signal, but the clock signal seen by FF2 is delayed by a significant amount relative to FF1's clock. This difference between arrival times of the clock at different devices is called *clock skew*.   *clock skew*

We've named the delayed clock in Figure 9–61(a) "CLOCKD." If FF1's propagation delay from CLOCK to Q1 is short, and if the physical connection of Q1 to FF2 is short, then the change in Q1 caused by a CLOCK edge may actually reach FF2 *before* the corresponding CLOCKD edge. In this case, FF2 may go to an incorrect next state determined by the *next* state of FF1 instead of the current state, as shown in (b). If the change in Q1 arrives at FF2 only slightly early relative to CLOCKD, then FF2's hold-time specification may be violated, in which case FF2 may become metastable and produce an unpredictable output.

If Figure 9–61 reminds you of our illustration of an essential hazard in Section 7.6.6, you're on to something. The clock-skew problem may be viewed simply as a manifestation of the essential hazards that exist in all edge-triggered devices.

We can determine quantitatively whether clock skew is a problem in a given system by defining t_{skew} to be the amount of clock skew and using the other timing parameters defined in Figure 9–1. For proper operation, we must have

$$t_{ffpd(min)} + t_{comb(min)} - t_{hold} - t_{skew(max)} > 0$$

In other words, clock skew subtracts from the hold-time margin that we defined in Section 9.1.4.

Figure 9–61
A simple example of clock skew.

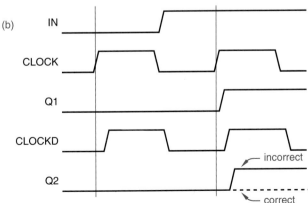

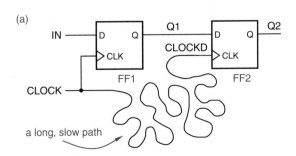

Viewed in isolation, the example in Figure 9–61 may seem a bit extreme. After all, why would a designer provide a short connection path for data and a long one for the clock, when they could just run side by side? There are several ways this can happen; some are mistakes, while others are unavoidable.

In a large system, a single clock signal may not have adequate fanout to drive all of the devices with clock inputs, so it may be necessary to provide two or more copies of the clock signal. The buffering method of Figure 9–62(a) obviously produces excessive clock skew, since CLOCK1 and CLOCK2 are delayed through an extra buffer compared to CLOCK.

A recommended buffering method is shown in Figure 9–62(b). All of the clock signals go through identical buffers, and have roughly equal delays. Ideally, all the buffers should be part of the same IC package, so that they all have similar delay characteristics and are operating at identical temperature and power-supply voltage. Some manufacturers build special buffers for just this sort of application and specify the worst-case delay variation between buffers in the same package.

Even the method in Figure 9–62(b) may produce excessive clock skew if one clock signal is loaded much more heavily than the other; transitions on the more heavily loaded clock appear to occur later because of increases in output-transistor switching delay and signal rise and fall times. Therefore, a careful designer tries to balance the loads on multiple clocks, looking at both DC load (fanout) and AC load (wiring and input capacitance).

Another bad situation can occur when signals on a PCB or in an ASIC are routed automatically by a CAD program. Figure 9–63 shows a PCB or ASIC with many flip-flops and larger-scale elements, all clocked with a common CLOCK signal. The CAD software has laid out CLOCK in a serpentine path that winds its way past all the devices that need it. Most of the other signals are routed point-to-point between an output and a small number of inputs, and so

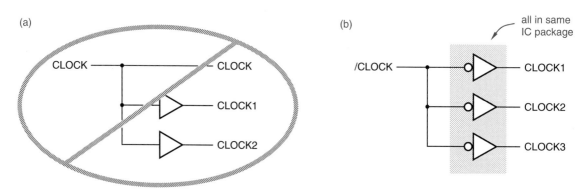

Figure 9–62   Buffering the clock: (a) excessive clock skew; (b) controllable clock skew.

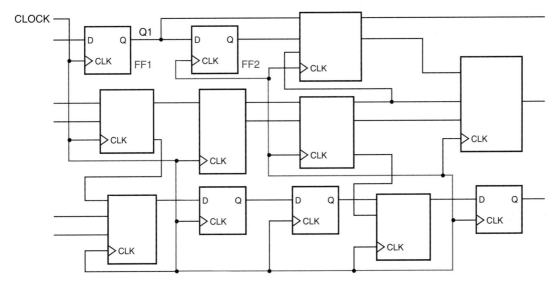

Figure 9–63   A clock-signal path leading to excessive skew in a complex PCB or ASIC.

their paths are shorter. To make matters worse, in an ASIC some types of "wire" may be slower than others (polysilicon vs. metal in CMOS technology). As a result, a CLOCK edge may indeed arrive at FF2 quite a bit later than the data change that it produces on Q1.

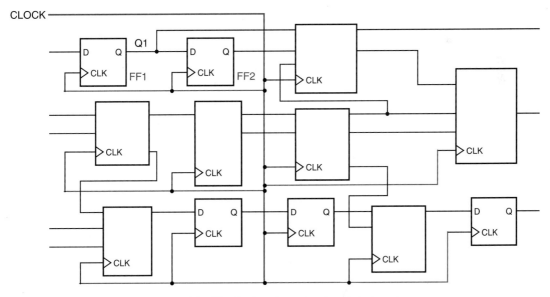

Figure 9–64   Clock-signal routing to minimize skew.

One way to minimize this sort of problem is to force CLOCK to be distributed in a tree-like structure using the fastest type of wire, as illustrated in Figure 9–64. However, in a complex design with many direct connections from one flip-flop's Q output to another's D input, it still may not be possible to guarantee that clock edges arrive everywhere before the earliest data change. A CAD timing analysis program is typically used to detect these problems, which generally can be remedied only by inserting extra delay (e.g., pairs of inverters) in the too-fast data paths.

two-phase latch design

Although synchronous design methodology simplifies the conceptual operation of large systems, we see that clock skew can be a major problem when edge-triggered flip-flops are used as the storage elements. To control this problem, many high-performance systems and VLSI chips use a *two-phase latch design*, discussed in the References. Such designs effectively split each edge-triggered D flip-flop into its two component latches, and control them with two nonoverlapping clock phases. The nonoverlap between the phases accommodates clock skew.

9.7.2   Gating the Clock

Most of the sequential MSI parts that we introduced in this chapter have synchronous function-enable inputs. The first example that we showed was the 74x377 register with synchronous load-enable input (Figure 9–9 on page 593); other parts included the 74x163 counter and 74x194 shift register with synchronous load-enable, count-enable, and shift-enable inputs. Nevertheless, many MSI parts do not have synchronous function-enable inputs; for example, the 74x374 8-bit register has three-state outputs but no load-enable input.

So, what can a designer do if an application requires an 8-bit register with both a load-enable input *and* three-state outputs? One solution is to use a 74x377 to get the load-enable, and follow it with a 74x241 three-state buffer. However, this increases both cost and delay. Another, somewhat risky alternative is to use a '374, but to suppress its clock input when it's not supposed to be loaded. This

gating the clock

is called *gating the clock*.

Figure 9–65 illustrates an obvious but wrong approach to gating the clock. A signal CLKEN is asserted to enable the clock, and is simply ANDed with the clock to produce the gated clock GCLK. This approach has two problems:

(1) If CLKEN is a state-machine output or other signal produced by a register clocked by CLOCK, then CLKEN changes some time *after* CLOCK has already gone HIGH. This produces glitches on GCLK, and false clocking of the registers controlled by GCLK.

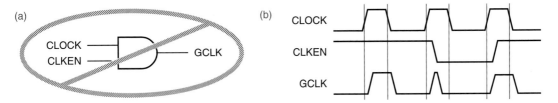

Figure 9–65 How not to gate the clock: (a) simple-minded circuit; (b) timing diagram.

(2) Even if CLKEN is somehow produced well in advance of CLOCK's rising edge (e.g., using a register clocked with the *falling* edge of CLOCK, an especially nasty kludge), the AND-gate delay gives GCLK excessive clock skew, which causes more problems all around.

A method of gating the clock that generates only minimal clock skew is shown in Figure 9–66. Here, both an ungated clock and several gated clocks are generated from the same active-low master clock signal. Gates in the same IC package are used to minimize the possible differences in their delays. The CLKEN signal may change arbitrarily whenever /CLOCK is LOW, which is when CLOCK is HIGH. That's just fine; a CLKEN signal is typically produced by a state machine whose outputs change right after CLOCK goes HIGH.

The approach of Figure 9–66 is acceptable in a particular application only if the clock skew that it creates is acceptable. Furthermore, note that CLKEN must be stable during the entire time that /CLOCK is HIGH (CLOCK is LOW). Thus, the timing margins in this approach are sensitive to the clock's duty cycle, especially if CLKEN suffers a lot of combinational-logic delay (t_{comb}) from the triggering clock edge. A truly synchronous function-enable input, such as the

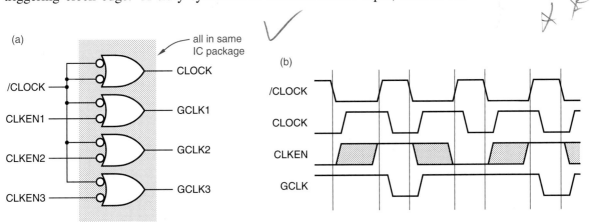

Figure 9–66 An acceptable way to gate the clock: (a) circuit; (b) timing diagram.

74x377's load-enable input in Figure 9–9, can be changed at almost any time during the entire clock period, up until a setup time before the triggering edge.

9.7.3 Asynchronous Inputs

Even though it is theoretically possible to build a computer system that is fully synchronous, you couldn't do much with it, unless you can synchronize your keystrokes with a 16 MHz clock. Digital systems of all types inevitably must deal with *asynchronous input signals* that are not synchronized with the system clock.

asynchronous input signal

Asynchronous inputs are typically requests for service (e.g., interrupts in a computer) or status flags (e.g., a resource has become available). Such inputs normally change slowly compared to the system clock frequency, and need not be recognized at a particular clock tick. If a transition is missed at one clock tick, it can always be detected at the next one. The transition rates of asynchronous signals may range from less than one per second (the keystrokes of a slow typist) to 1 MHz or more (access requests in a multiprocessor system's shared memory).

synchronizer

Ignoring the problem of metastability, it is very easy to build a *synchronizer*, a circuit that samples an asynchronous input and produces an output that meets the setup and hold times required in a synchronous system. As shown in Figure 9–67, a D flip-flop samples the asynchronous input at each tick of the system clock and produces a synchronous output that is valid during the subsequent clock period.

It is essential for asynchronous inputs to be synchronized at only *one place* in a system; Figure 9–68 shows what can happen otherwise. Because of physical delays in the circuit, the two flip-flops will not see the clock and input signals at precisely the same time. Therefore, when asynchronous input transitions occur

Figure 9–67
A single, simple synchronizer:
(a) logic diagram; (b) timing.

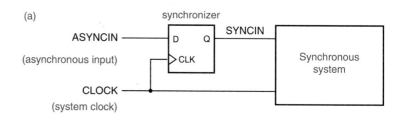

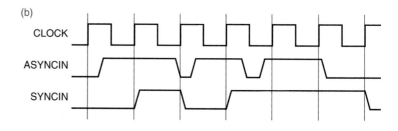

Figure 9–68
Two synchronizers for the same
asynchronous input: (a) logic
diagram; (b) possible timing.

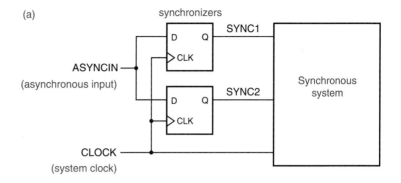

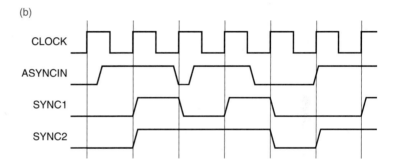

near the clock edge, there is a small window of time during which one flip-flop
may sample the input as 1 and the other may sample it as 0. This inconsistent
result may cause improper system operation, as one part of the system responds
as if the input were 1, and another part responds as if it were 0.

Sometimes combinational logic may hide the fact that there are two synchronizers, as shown in Figure 9–69. Since different paths through the combinational
logic typically have different delays, the likelihood of an inconsistent result is

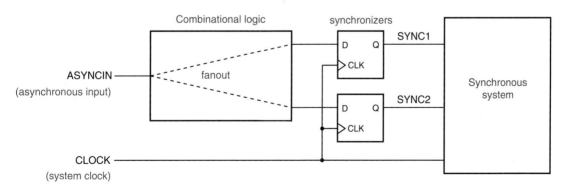

Figure 9–69 An asynchronous input driving two synchronizers through combinational logic.

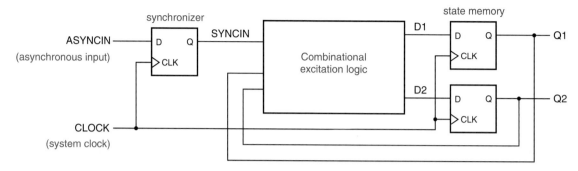

Figure 9–70 An asynchronous state-machine input coupled through a single synchronizer.

even greater. This situation is especially common when asynchronous signals are used as inputs to state machines, since the excitation logic for two or more state variables may depend on the asynchronous input. The proper way to use an asynchronous signal as a state-machine input is shown in Figure 9–70. All of the excitation logic sees the same synchronized input signal, SYNCIN.

WHO CARES?

> As you probably know, even the synchronizers in Figures 9–67 and 9–70 sometimes fail. The reason they fail is that the setup and hold times of the synchronizing flip-flop are sometimes violated because the asynchronous input can change at any time. "Well, who cares?" you may say. "If the D input changes near the clock edge, then the flip-flop will either see the change this time or miss it and pick it up next time; either way is good enough for me!" The problem is, there is a third possibility, discussed in the next section.

9.8 SYNCHRONIZER FAILURE AND METASTABILITY ESTIMATION

We showed in Section 7.1 that when the setup and hold times of a flip-flop are not met, the flip-flop may go into a third, *metastable* state halfway between 0 and 1. Worse, the length of time it may stay in this state before falling back into a legitimate 0 or 1 state is theoretically unbounded. When other gates and flip-flops are presented with a metastable input signal, some may interpret it as a 0 and others as a 1, creating the sort of inconsistent behavior that we showed in Figure 9–68. Or the other gates and flip-flops may produce metastable outputs themselves (after all, they are now operating in the *linear* part of their operating range). Luckily, the probability of a flip-flop output remaining in the metastable state decreases exponentially with time, though never all the way to zero.

9.8.1 Synchronizer Failure

Synchronizer failure is said to occur if the system uses a synchronizer output   *synchronizer failure*
while the output is still in the metastable state. The way to avoid synchronizer
failure is to ensure that the system waits "long enough" before using a synchronizer's output, "long enough" so that the mean time between synchronizer
failures is several orders of magnitude longer than the designer's expected length
of employment.

Metastability is more than an academic problem. More than a few experienced designers of high-speed digital systems have built (and released to
production) circuits that suffer from intermittent synchronizer failures. In fact,
the initial versions of several commercial ICs are said to have suffered from
metastability problems, for example, the AMD 9513 system timing controller,
the AMD 9519 interrupt controller, the Zilog Z-80 SIO, the Intel 8048 single-chip
microcomputer, and the AMD 29000 RISC microprocessor.

There are two ways to get a flip-flop out of the metastable state:

(1) Force the flip-flop into a valid logic state using input signals that meet the
published specifications for minimum pulse width, setup time, and so on.

(2) Wait "long enough," until the flip-flop comes out of metastability on its
own.

Inexperienced designers often attempt to get around metastability in other ways,
and they are usually unsuccessful. Figure 9–71 shows an attempt by a designer
who thinks that since metastability is an "analog" problem, it must have an
"analog" solution. After all, Schmitt trigger inputs and capacitors can normally
be used to clean up noisy signals. However, rather than eliminate metastability,
this circuit enhances it—with the "right" components, the circuit will oscillate
forever once it is excited by negating /S and /R simultaneously. Exercise 9.50
gives an example of a valiant but also failed attempt to eliminate metastability.
These examples should give you the sense that synchronizer problems can be
very subtle, so you must be careful. The only way to make synchronizers reliable

Figure 9–71
A failed attempt to build a
metastable-proof /S-/R flip-flop.

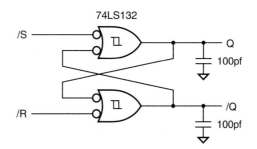

is to wait long enough for metastable outputs to resolve. We answer the question "How long is 'long enough'?" later in this section.

9.8.2   Metastability Resolution Time

If the setup and hold times of a D flip-flop are met, the flip-flop output settles to a new value within time t_{pd} after the clock edge. If they are violated, the flip-flop output may be metastable for an arbitrary length of time. In a particular system design, we use the parameter t_r, called the *metastability resolution time*, to denote the maximum time that the output can remain metastable without causing synchronizer (and system) failure.

t_r
metastability
resolution time

For example, consider the state machine in Figure 9–70 on page 642. The available metastability resolution time is

$$t_r = t_{clk} - t_{comb} - t_{setup}$$

t_{clk}
t_{comb}
t_{setup}

where t_{clk} is the clock period, t_{comb} is the propagation delay of the combinational excitation logic, and t_{setup} is the setup time of the flip-flops used in the state memory.

9.8.3   Reliable Synchronizer Design

The most reliable synchronizer is one that allows the maximum amount of time for metastability resolution. However, in the design of a digital system, we seldom have the luxury of *slowing down* the clock to make the system work more reliably. Instead, we are usually asked to *speed up* the clock to get higher performance from the system. As a result, we often need synchronizers that work reliably with very short clock periods. We'll show several such designs, and show how to predict their reliability.

We showed previously that a state machine with an asynchronous input, built as shown in Figure 9–70, has $t_r = t_{clk} - t_{comb} - t_{setup}$. In order to maximize t_r for a given clock period, we should minimize t_{comb} and t_{setup}. The value of t_{setup} depends on the type of flip-flops used in the state memory; in general, faster flip-flops have shorter setup times. The minimum value of t_{comb} is zero, and is achieved by the synchronizer design of Figure 9–72, whose operation we explain next.

Inputs to flip-flop FF1 are asynchronous with the clock, and may violate the flip-flop's setup and hold times. When this happens, the META output may become metastable and remain in that state for an arbitrary time. However, we assume that the maximum duration of metastability after the clock edge is t_r. (We show how to calculate the probability that our assumption is correct in the next subsection.) As long as the clock period is greater than t_r plus the FF2's

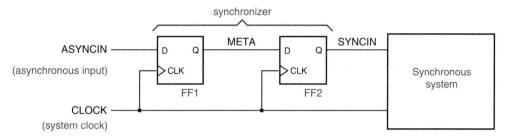

Figure 9–72 Recommended synchronizer design.

setup time, SYNCIN becomes a synchronized copy of the asynchronous input on the next clock tick without ever becoming metastable itself. The SYNCIN signal is distributed as required to the rest of the system.

9.8.4 Analysis of Metastable Timing

Figure 9–73 shows the flip-flop timing parameters that are relevant to our analysis of metastability timing. The published setup and hold times of a flip-flop with respect to its clock edge are denoted by t_s and t_h, and bracket an interval called the *decision window*, when the flip-flop samples its input and decides to change its output if necessary. As long as the D input changes outside the decision window, as in (a), the manufacturer guarantees that the output will change and settle to a valid logic state before time t_{pd}. If D changes inside the decision window, as in (b), metastability may occur and persist until time t_r.

    Theoretical research has suggested and experimental research has confirmed

decision window

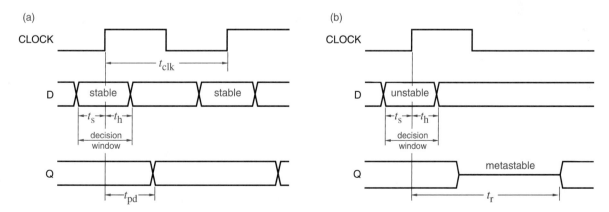

Figure 9–73 Timing parameters for metastability analysis.

that when asynchronous inputs change during the decision window, the duration of metastable outputs is governed by an exponential formula:

$$\text{MTBF}(t_r) = \frac{\exp(t_r / \tau)}{T_o \cdot f \cdot a}$$

Here $\text{MTBF}(t_r)$ is the mean time between synchronizer failures, where a failure occurs if metastability persists beyond time t_r after a clock edge, where $t_r \geq t_{pd}$. This MTBF depends on f, the frequency of the flip-flop clock; a, the number of asynchronous input changes per second applied to the flip-flop; and T_o and τ, constants that depend on the electrical characteristics of the flip-flop. For a typical 74LS74, $T_o \approx 0.4$ s and $\tau \approx 1.5$ ns.

f
a
T_o
τ

Now suppose that we build a microprocessor system with a 10 MHz clock and use the circuit of Figure 9–72 to synchronize an asynchronous input. If the asynchronous input changes during the decision window of FF1, its output META may become metastable until time t_r. If META is still metastable at the beginning of the decision window for FF2, then the synchronizer fails because FF2 may have a metastable output; system operation is unpredictable in that case.

Let us assume that the D flip-flops in Figure 9–72 are 74LS74s. The setup time t_s of a 74LS74 is 20 ns, and the clock period in our example microprocessor system is 100 ns, so the t_r required for synchronizer failure is 80 ns. If the asynchronous input changes 100,000 times per second, then the synchronizer MTBF is

$$\text{MTBF}(80\,\text{ns}) = \frac{\exp(80 / 1.5)}{0.4 \cdot 10^7 \cdot 10^5} = 3.6 \cdot 10^{11}\,\text{s}$$

That's not bad, about 100 centuries between failures! Of course, if we're lucky enough to sell 10,000 copies of our system, one of them will fail in this way every year. But, no matter, let us consider a more serious problem.

UNDERSTANDING
a AND *f*

Although a flip-flop output can go metastable *only* if D changes during the decision window, the MTBF formula does not explicitly specify how many such input changes occur. Instead, it specifies the total number of asynchronous input changes per second, *a*, and assumes that asynchronous input changes are uniformly distributed over the clock period. Therefore, the fraction of input changes that actually occur during the decision window is "built in" to the clock-frequency parameter *f*—as *f* increases, the fraction goes up.

If the system design is such that input changes might be clustered in the decision window rather than being uniformly distributed (as when synchronizing a slow input with a fixed but unknown phase difference from the system clock), then a frequency equal to the reciprocal of the decision window (based on published setup and hold times), times a safety margin of 10, should be used.

> In our analysis of metastable timing, we do not allow metastability, even briefly, on the output of FF2, because we assume that the system has been designed with zero timing margins. If the system can in fact tolerate some increase in the propagation delay of FF2, the MTBF will be somewhat better than predicted by this analysis.

DETAILS, DETAILS

Suppose we upgrade our system to use a brand-new microprocessor chip with a clock speed of 16 MHz. We may have to replace some components in our system to operate at the higher speed, but 74LS74s are still perfectly good at 16 MHz. Or are they? With a clock period of 62.5 ns, the new synchronizer MTBF is

$$\text{MTBF}(42.5\,\text{ns}) = \frac{\exp(42.5/1.5)}{0.4 \cdot 1.6 \cdot 10^7 \cdot 10^5} = 3.1\,\text{s!}$$

The only saving grace of this synchronizer at 16 MHz is that it's so lousy, we're likely to discover the problem in the engineering lab before the product ships! Thank goodness the MTBF wasn't one year.

9.8.5 Better Synchronizers

Given the poor performance of the 74LS74 as a synchronizer at moderate clock speeds, we have a couple of alternatives for building more reliable synchronizers. The simplest solution, which works for most design requirements, is simply to use a flip-flop from a faster technology. Based on published data discussed in the References, Table 9–9 lists the metastability parameters for a number of common flip-flop types and technologies. For reference purposes, note that $10^5\,\text{s} \approx 1$ day, $10^8\,\text{s} \approx 3$ years, and $10^{10}\,\text{s} \approx 3$ centuries.

Table 9–9
Metastability parameters for common flip-flops, with f = 25 MHz and a = 100 KHz.

Reference	Device	τ (ns)	T_o	MTBF(t_r) (s) t_r=20 ns	t_r=40 ns
TI	74LSxx	1.28	6.7 s	$t_r < t_{pd}$	$2 \cdot 10^0$
TI	74ALSxx	0.87	12.5 ms	$3 \cdot 10^{-1}$	$7 \cdot 10^{10}$
TI	74ASxx	0.34	6.5 μs	$2 \cdot 10^{18}$	$7 \cdot 10^{43}$
Chaney	74S74	1.70	1.0 μs	$5 \cdot 10^{-2}$	$7 \cdot 10^3$
Chaney	74S174	1.20	5.0 μs	$1 \cdot 10^0$	$2 \cdot 10^7$
Chaney	74S373	0.91	60.0 μs	$2 \cdot 10^1$	$8 \cdot 10^{10}$
Chaney	74S374	0.91	0.4 ms	$4 \cdot 10^0$	$1 \cdot 10^{10}$
Chaney	74LS74	1.5	0.4 s	$t_r < t_{pd}$	$4 \cdot 10^{-1}$
Chaney	74F74	0.4	0.2 ms	$1 \cdot 10^{13}$	$5 \cdot 10^{34}$
Chaney	74F373	0.7	4.0 μs	$3 \cdot 10^5$	$7 \cdot 10^{17}$
Chaney	74F374	0.4	0.1 ms	$2 \cdot 10^{13}$	$1 \cdot 10^{35}$

As you can see, the 74LS74 is the worst device in the table. If in the 16 MHz microprocessor system of the preceding section we replace FF1 with a 74ALS74, we get

$$\text{MTBF}(42.5\,\text{ns}) = \frac{\exp(42.5/0.87)}{12.5 \cdot 10^{-3} \cdot 1.6 \cdot 10^7 \cdot 10^5} = 8.2 \cdot 10^{10}\,\text{s}$$

If you're satisfied with a synchronizer MTBF of about 25 centuries per system shipped, you can stop here. However, if FF2 is also replaced with a 74ALS74, the MTBF gets even better, since the 'ALS74 has a shorter setup time than the 'LS74, only 10 ns. With the 'ALS74, the MTBF is about 100,000 times better:

$$\text{MTBF}(52.5\,\text{ns}) = \frac{\exp(52.5/0.87)}{12.5 \cdot 10^{-3} \cdot 1.6 \cdot 10^7 \cdot 10^5} = 8.1 \cdot 10^{15}\,\text{s}$$

Even if we ship a million systems containing this circuit, we (or our heirs) will see a synchronizer failure only once in 240 years. Now that's job security!

Actually, the margins above aren't as large as they might seem. (How large does 240 years seem to *you*?) Most of the numbers given in Table 9–9 are *averages*, and are seldom specified, let alone guaranteed, by the device manufacturer. Furthermore, calculated MTBFs are extremely sensitive to the value of τ, which in turn may depend on temperature, voltage, and the phase of the moon. So the operation of a given flip-flop in an actual system may be much worse (or much better) than predicted by our table.

9.8.6   Other Synchronizer Designs

We promised to describe a couple of ways to build more reliable synchronizers. The first way was to use faster flip-flops, that is, to reduce the value of τ in the MTBF equation. Having said that, the second way is obvious—to *increase* the value of t_r in the MTBF equation.

For a given system clock, the best value we can obtain for t_r using the circuit of Figure 9–72 is t_{clk}, if FF2 has a setup time of 0. However, we can get values of t_r on the order of $n \cdot t_{\text{clk}}$ by using the *multiple-cycle synchronizer* circuit of Figure 9–74. Here we divide the system clock by n to obtain a slower synchronizer clock and longer $t_r = (n \cdot t_{\text{clk}}) - t_{\text{setup}}$. Usually a value of $n = 2$ or $n = 3$ gives adequate synchronizer reliability. If necessary, the output of the synchronizer, which changes on the rising edge of the divide-by-n clock, may be reclocked by the system clock to eliminate skew with the rest of the system.

multiple-cycle synchronizer

The larger the value of n, the longer it takes for an asynchronous input change to be seen by the synchronous system. This is simply a price that must be paid for reliable system operation. In typical microprocessor systems, most asynchronous inputs are for external events—interrupts, DMA requests, and so on—that need not be recognized very quickly, relative to synchronizer delays.

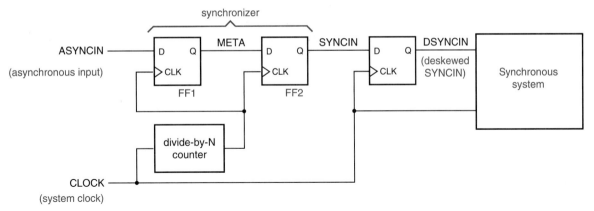

Figure 9–74 Multiple-cycle synchronizer.

In the time-critical area of main memory access, experienced designers use the processor clock to run the memory subsystem too, if possible. This eliminates the need for synchronizers and provides the fastest possible system operation.

At higher frequencies, the feasibility of the multiple-cycle synchronizer design shown in Figure 9–74 tends to be limited by clock skew. For this reason, rather than use a divide-by-n synchronizer clock, some designers use *cascaded synchronizers*. This design approach simply replaces FF1 in the synchronizer of Figure 9–72 with a cascade (shift register) of n flip-flops, all clocked with the high-speed system clock.

cascaded synchronizers

With cascaded synchronizers, the idea is that metastability will be resolved with some probability by the first flip-flop, and failing that, with an equal probability by each successive flip-flop in the cascade. So the overall probability of failure is on the order of the nth power of the failure probability of a single-flip-flop synchronizer at the system clock frequency. While this is true, the MTBF of the cascade is poorer than the MTBF of a multiple-cycle synchronizer with the same delay ($n \cdot t_{\text{clk}}$). With the cascade, the flip-flop setup time t_{setup} must be subtracted from t_r, the available metastability resolution time, n times, but in a multiple-cycle design, it is subtracted only once.

PLDs that contain internal flip-flops can be used in synchronizer designs, where the two flip-flops in Figure 9–72 are simply included in the PLD. This is very convenient in most applications, because it eliminates the need for external, discrete flip-flops. However, a PLD-based synchronizer typically has a poorer MTBF than a corresponding discrete circuit. This happens because each flip-flop in a PLD has combinational logic circuits on its D input that increase its setup time and thereby reduce the amount of time t_r available for metastability resolution during a given system clock period t_{clk}. To maximize t_r, FF2 in Figure 9–72 should be a short-setup-time discrete flip-flop.

9.8.7   Metastable-Hardened Flip-Flops

74AS4374

Going one step further, Texas Instruments has developed a series of SSI and MSI flip-flops that are specifically designed for synchronizer applications. For example, the *74AS4374* is similar to the 74AS374, except that each individual flip-flop is replaced with a pair of flip-flops, as shown in Figure 9–75. Each pair of flip-flops can be used as a synchronizer of the type shown in Figure 9–72, so eight asynchronous inputs can be synchronized with one 74AS4374.

The 'AS4374 is specified with a slightly poorer τ than the average τ that we showed in Table 9–9 for other 74AS flip-flops. The internal design of the 'AS4374 actually has been improved to reduce τ compared to other 74AS flip-flops, but TI used this improvement to give better design margins rather than a better specification. The new flip-flops have a *worst-case* $\tau = 0.42$ ns.

The 'AS4374 flip-flops have a T_o parameter of only 4 ns, compared to 6.5 μs for other 74AS flip-flops. Since T_o appears in the denominator of the MTBF equation, this improves the MTBF by a factor of 1625.

The biggest improvement in the 'AS4374 is a greatly reduced t_{setup}. Because the entire synchronizer of Figure 9–72 is built on a single chip, there are no input or output buffers between FF1 and FF2, and t_{setup} for FF2 is only 0.5 ns. Compared to a conventional 74AS flip-flop with a 5 ns t_{setup}, this improves the MTBF by a factor of $\exp(4.5/.42)$, or about 45,000.

Because of the improvements in T_o and t_{setup}, synchronizers using the 'AS4374 have MTBFs that are about $7 \cdot 10^7$ times better than ones built from discrete 74AS flip-flops. Although they cost more than discrete flip-flops, at very high system clock speeds (40 MHz and up) the 'AS4374 and related TI parts are the only 74AS-TTL parts that can reliably synchronize asynchronous inputs using only one clock period of delay, as in Figure 9–72. On the other hand, applications that use slower clocks (20 MHz or less) or that can tolerate the long latency of a multiple-cycle synchronizer can be designed more economically using conventional AS or ALS flip-flops.

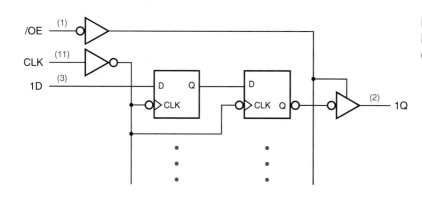

Figure 9–75
Logic diagram for the 74AS4374 octal dual-rank D flip-flop.

REFERENCES

Probably the first comprehensive set of examples of sequential MSI parts and applications appeared in *The TTL Applications Handbook*, edited by Peter Alfke and Ib Larsen (Fairchild Semiconductor, 1973). This highly practical and informative book was invaluable to this author and many others who developed digital design curricula in the 1970s.

Another book that emphasizes design with larger-scale combinational and sequential logic functions is Thomas R. Blakeslee's *Digital Design with Standard MSI and LSI*, 2nd ed., (Wiley, 1979). His coverage of the concept of space/time trade-offs is excellent, and he also introduces the microprocessor as "a universal logic circuit."

Even though this is a "practices" chapter, we introduced a few useful theoretical topics. Function hazards are discussed by Edward J. McCluskey in *Logic Design Principles* (Prentice Hall, 1986). Galois fields and their applications to error-correcting codes, as well as to the LFSR counters of this chapter, are described in introductory books on coding theory, including *Error-Control Techniques for Digital Communication* by A. M. Michelson and A. H. Levesque (Wiley-Interscience, 1985).

The general topics of clock skew and multiphase clocking are discussed in McCluskey's *Logic Design Principles*, while an illuminating discussion of these topics as applied to VLSI circuits can be found in *Introduction to VLSI Systems* by Carver Mead and Lynn Conway (Addison-Wesley, 1980). Mead and Conway's book also provides an introduction to the important topic of *self-timed systems* that eliminate the system clock, allowing each logic element to proceed at its own rate. To give credit where credit is due, we note that all of the Mead and Conway material on system timing, including an outstanding discussion of metastability, appears in their Chapter 7 written by Charles L. Seitz.

self-timed systems

Thomas J. Chaney has spent decades studying and reporting on the metastability problem. One of his more important works, "Measured Flip-Flop Responses to Marginal Triggering" (*IEEE Trans. Comput.*, Vol. C-32, No. 12, December 1983, pp. 1207–1209), reports results that we used in Table 9–9.

Another important reference is "Metastable Characteristics of Texas Instruments Advanced Bipolar Logic Families" by Robert Brueninger and Kevin Frank (TI pub. SDAA004). Besides describing the results for ALS, AS, and LS circuits that we used in Table 9–9, it contains a description of a measurement circuit that you can duplicate (with some care) in your own lab to measure the metastable characteristics of your own devices.

For the mathematically inclined, Lindsay Kleeman and Antonio Cantoni have written "On the Unavoidability of Metastable Behavior in Digital Systems" (*IEEE Trans. Comput.*, Vol. C-36, No. 1, January 1987, pp. 109–112); the title says it all. The same authors posed the question, "Can Redundancy and Masking Improve the Performance of Synchronizers?" (*IEEE Trans. Comput.*, Vol. C-35, No. 7, July 1986, pp. 643–646). Their answer in that paper was "no," but a response from a reviewer caused them to change their minds to "maybe." Obviously, they've gone metastable themselves! (Having two authors and a reviewer hasn't improved their performance, so the obvious answer to their original question is "no!") In any case, Kleeman and Antonio's papers provide a good set of pointers to mainstream scholarly references on metastability.

The most comprehensive set of references on metastability (not including Greek philosophers or Devo) is Martin Bolton's "A Guided Tour of 35 Years of Metastability Research" (*Proc. Wescon 1987*, Session 16, "Everything You Might Be Afraid to Know about Metastability," Wescon Session Records, 8110 Airport Blvd., Los Angeles, CA 90045).

DRILL PROBLEMS

9.1   What would happen if you replaced the edge-triggered D flip-flops in Figure 7–34 with D latches?

9.2   What is the counting sequence of the circuit shown in Figure X9.2?

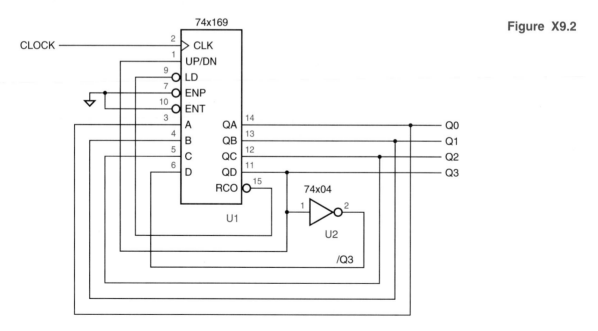

Figure X9.2

9.3   What is the behavior of the counter circuit of Figure 9–20 if it is built using a 74x161 instead of a 74x163?

9.4   A 74x163 counter is hooked up with inputs ENP, ENT, A, and D always HIGH, inputs B and C always LOW, input /LD = (QB·QC)', and input /CLR = (QC·QD)'. The CLK input is hooked up to a free-running clock signal. Draw a logic diagram for this circuit. Assuming that the counter starts in state 0000, write the output sequence on QD QC QB QA for the next 15 clock ticks.

9.5   Determine the widths of the glitches shown in Figure 9–27 on the /Y2 output of a 74x138 decoder, assuming that the '138 is internally structured as shown in Figure 5–28(a) on page 304, and that each internal gate has a delay of 10 ns.

9.6   Calculate the minimum clock period of the data unit in Figure 8–4. Use the maximum propagation delays given in Table 5–3 for LS-TTL combinational parts. Unless you can get the real numbers from a TTL data book, assume that all registers require a 10 ns minimum setup time on inputs and have a 20 ns propagation delay from clock to outputs. Indicate any assumptions you've made about delays in the control unit.

9.7   Calculate the MTBF of a synchronizer built according to Figure 9–72 using 74F74s, assuming a clock frequency of 25 MHz and an asynchronous transition rate of 1 MHz. Assume that the setup time of an 'F74 is 5 ns and the hold time is zero.

9.8   Calculate the MTBF of the synchronizer shown in Figure X9.8, assuming a clock frequency of 25 MHz and an asynchronous transition rate of 1 MHz. Assume that the setup time t_{setup} and the propagation delay t_{pd} from clock to Q or /Q in a 74ALS74 are both 10 ns.

Figure X9.8

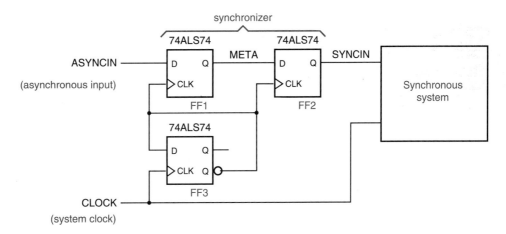

E X E R C I S E S

9.9   What does the TTL Data Book have to say about momentarily shorting the outputs of a gate to ground as we do in the switch debounce circuit of Figure 9–5?

9.10   Investigate the behavior of the switch debounce circuit of Figure 9–5 if 74HCT04 inverters are used; repeat for 74AC04 inverters.

9.11   Suppose you are asked to design a circuit that produces a debounced logic input from an SPST (single-pole, *single*-throw) switch. What inherent problem are you faced with?

9.12   Write a formula for the maximum clock frequency of the synchronous serial binary counter circuit in Figure 9–12. In your formula, let t_{pTQ} denote the propagation delay from T to Q in a T flip-flop, t_{setup} the setup time of the EN input to the rising edge of T, and t_{AND} the delay of an AND gate.

9.13   Repeat Exercise 9.12 for the synchronous parallel binary counter circuit shown in Figure 9–13, and compare results.

9.14   Repeat Exercise 9.12 for an *n*-bit synchronous serial binary counter.

9.15   Repeat Exercise 9.12 for an *n*-bit synchronous parallel binary counter. Beyond what value of *n* is your formula no longer valid?

9.16   Using a 74x163 4-bit binary counter, design a modulo-11 counter circuit with the counting sequence 3, 4, 5, . . . , 12, 13, 3, 4, ...

9.17   Look up the internal logic diagram for a 74x162 synchronous decade counter in a data book, and write its state table in the style of Table 9–1, including its counting behavior in the normally unused states 10–15.

9.18   Devise a cascading scheme for 74x163s, analogous to the synchronous parallel counter structure of Figure 9–13, such that the maximum counting speed is the same for any counter with up to 36 bits (nine '163s). Determine the maximum counting speed using worst-case delays from a manufacturer's data book for the '163s and any SSI components used for cascading.

9.19   Design a modulo-129 counter using two 74x163s and a single inverter.

9.20   Design a clocked synchronous circuit with four inputs, N3, N2, N1, and N0, that represent an integer *N* in the range 0–15. The circuit has a single output Z that is asserted for exactly *N* clock ticks during any 16-tick interval (assuming that *N* is held constant during the interval of observation). (*Hints:* Use combinational logic with a 74x163 set up as a free-running divide-by-16 counter. The ticks in which Z is asserted should be spaced as evenly as possible, that is, every second tick when $N = 8$, every fourth when $N = 4$, and so on.)

binary rate multiplier

9.21   Modify the circuit of Exercise 9.20 so that Z produces *N transitions* in each 16-tick interval. The resulting circuit is called a *binary rate multiplier*, and was once sold as a TTL MSI part, the 7497. (*Hint:* Gate the clock with the level output of the previous circuit.)

9.22 A digital designer (the author!) was asked at the last minute to add new functionality to a PCB that had room for just one more 16-pin IC. The PCB already had a 16-MHz clock signal, MCLK, and a spare microprocessor-controlled select signal, SEL. The designer was asked to provide a new clock signal, UCLK, whose frequency would be 8 MHz or 4 MHz depending on the value of SEL. To make things worse, the PCB had no spare SSI gates, and UCLK was required to have a 50% duty cycle at both frequencies. It took the designer about five minutes to come up with a circuit. Now it's your turn to do the same. (*Hint:* The designer had long considered the 74x163 to be the fundamental building block of tricky sequential-circuit design.)

9.23 Design a modulo-16 counter, using one 74x169 and at most one SSI package, with the following counting sequence: 7, 6, 5, 4, 3, 2, 1, 0, 8, 9, 10, 11, 12, 13, 14, 15, 7, ...

9.24 Design a parallel-to-serial conversion circuit with eight 2.048 Mbps, 32-channel serial links and a single 2.048 MHz, 8-bit, parallel data bus that carries 256 bytes per frame. Each serial link should have the frame format defined in Figure 9–39. Each serial data line SDATAi should have its own sync signal /SYNCi; the sync pulses should be staggered so that /SYNCi+1 has a pulse one clock tick after /SYNCi.

      Show the timing of the parallel bus and the serial links, and write a table or formula that shows which parallel-bus timeslots are transmitted on which serial links and timeslots. Draw a logic diagram for the circuit using MSI parts from this chapter; you may abbreviate repeated elements (e.g., shift registers), showing only the unique connections to each one.

9.25 Repeat Exercise 9.24, assuming that all serial data lines must reference their data to a single, common SYNC signal. How many more chips does this design require?

9.26 Show how to enhance the serial-to-parallel circuit of Figure 9–41 so that the byte received in each timeslot is stored in its own register for 125 μs, until the next byte from that timeslot is received. Draw the counter and decoding logic for 32 timeslots in detail, as well as the parallel data registers and connections for timeslots 31, 0, and 1. Also draw a timing diagram in the style of Figure 9–42 that shows the decoding and data signals associated with timeslots 31, 0, and 1.

9.27 Suppose you are asked to design a serial computer, one that moves and processes data one bit at a time. One of the first decisions you must make is which bit to transmit and process first, the LSB or the MSB. Which would you choose, and why?

9.28 Design an 8-bit self-correcting ring counter whose states are 11111110, 11111101, . . ., 01111111, using only two SSI/MSI packages.

9.29 Design two different 2-bit, 4-state counters, where each design uses just a single 74x74 package (two edge-triggered D flip-flops) and no other gates.

9.30 Design a 4-bit Johnson counter and decoding for all eight states using just three SSI/MSI packages. Your counter need not be self-correcting.

9.31 Starting with state 0001, write the sequence of states for a 4-bit LFSR counter designed according to Figure 9–50 and Table 9–5.

9.32 Prove that an even number of shift-register outputs must be connected to the odd-parity circuit in an n-bit LFSR counter if it generates a maximum-length sequence. (Note that this is a necessary but not a sufficient requirement. Also, although Table 9–5 is consistent with what you're supposed to prove, simply quoting the table is not a proof!)

9.33 Prove that X0 must appear on the right-hand side of any LFSR feedback equation that generates a maximum-length sequence. (*Note:* Assume the LFSR bit ordering and shift direction are as given in the text; that is, the LFSR counter shifts right, toward the X0 stage.)

9.34 Suppose that an n-bit LFSR counter is designed according to Figure 9–50 and Table 9–5. Prove that if the odd-parity circuit is changed to an even-parity circuit, the resulting circuit is a counter that visits $2^n - 1$ states, including all of the states except $11\ldots 11$.

9.35 Find a feedback equation for a 3-bit LFSR counter, other than the one given in Table 9–5, that produces a maximum-length sequence.

9.36 Prove that if an n-bit LFSR counter generates a maximum-length sequence ($2^n - 1$ states), then an extra XOR gate and an $n-1$ input NOR gate connected as suggested in Figure 9–51 produces a counter with 2^n states.

9.37 Prove that a sequence of 2^n states is still obtained if a NAND gate is substituted for a NOR above, but that the state sequence is different.

9.38 Design an iterative circuit with one input B_i per stage and two boundary outputs X and Y such that X = 1 if at least two B_i inputs are 1 and Y = 1 if at least two *consecutive* B_i inputs are 1.

9.39 Draw a complete state diagram for the ad hoc realization of the guessing game in Figure 9–59, including the normally unused states.

9.40 Design a combination-lock machine according to the state table of Table 7–18 on page 501 using a single 74x163 counter and combinational logic for the /LD, /CLR, and A–D inputs of the '163. Use counter values 0–7 for states A–H.

9.41 Design the output logic for the 74x163-based T-bird tail-lights machine shown in Figure 9–60. You may use any of the SSI and MSI parts introduced in Chapters 3 and 5, but try to minimize the total number of IC packages.

9.42 Modify the multiplier state machine of Figure 8–5 so that it also controls the loading of operands and the reading of results. During the first two clock ticks after START is asserted, the machine should read the multiplier and multiplicand from the data bus. Then it should perform the multiplication and drive the two result bytes onto the data bus for one clock cycle each before returning to the IDLE state. In addition to an ASM chart or state diagram, draw a timing diagram that shows what operations are performed during each clock cycle.

9.43 Decompose the state machine for the 8-bit shift-and-add multiplier of Figure 8–5, using a 74x163 to count the addition steps. Draw a block diagram showing the inputs and outputs of the main machine, a state diagram or ASM chart for the main machine, and a logic diagram for the 74x163-based submachine. Explain how the state-machine design must be changed for an n-bit multiplication.

9.44 Design a data unit and a control-unit state machine for multiplying 8-bit two's-complement numbers using the algorithm discussed in Section 2.8.

9.45 Design a data unit and control-unit state machine for dividing 8-bit unsigned numbers using the shift-and-subtract algorithm discussed in Section 2.9.

9.46 Suppose that the SYNCIN signal in Exercise 9.45 is connected to a combinational circuit in the synchronous system, which in turn drives the D inputs of 74ALS74 flip-flops that are clocked by CLOCK. What is the maximum allowable propagation delay of the combinational logic?

9.47 The circuit in Figure X9.47 includes a deskewing flip-flop so that the synchronized output from the multiple-cycle synchronizer is available as soon as possible after the edge of CLOCK. Ignoring metastability considerations, what is the maximum frequency of CLOCK? Assume that for a 74F74, $t_{setup} = 5$ ns and $t_{pd} = 7$ ns.

Figure X9.47

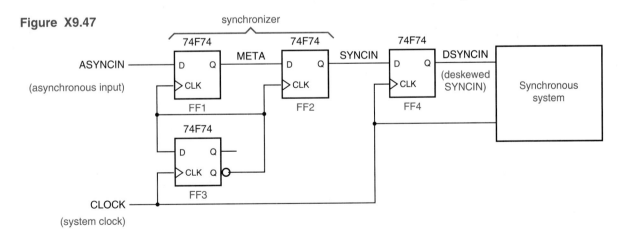

9.48 Using the maximum clock frequency determined in Exercise 9.47, and assuming an asynchronous transition rate of 4 MHz, determine the synchronizer's MTBF.

9.49 Determine the MTBF of the synchronizer in Figure X9.47, assuming an asynchronous transition rate of 4 MHz and a clock frequency of 40 MHz, which is less than the maximum determined in Exercise 9.47. In this situation, "synchronizer failure" really occurs only if DSYNCIN is metastable. In other words, SYNCIN may be allowed to be metastable for a short time, as long as it doesn't affect DSYNCIN. This yields a better MTBF.

9.50 A famous digital designer devised the circuit shown in Figure X9.50(a), which is supposed to eliminate metastability within one period of a system clock. Circuit M is a memoryless analog voltage detector whose output is 1 if Q is in the metastable state, 0 otherwise. The circuit designer's idea was that if line Q is detected to be in the metastable state when CLOCK goes low, the NAND gate will clear the D flip-flop, which in turn eliminates the metastable output, causing a 0 output from circuit M and thus negating the CLR input of the flip-flop. The circuits are all fast enough that this all happens well before CLOCK goes high again; the expected waveforms are shown in Figure X9.50(b).

Unfortunately, the synchronizer still failed occasionally, and the famous digital designer is now designing pockets for blue jeans. Explain, in detail, how it failed, including a timing diagram.

Figure X9.50

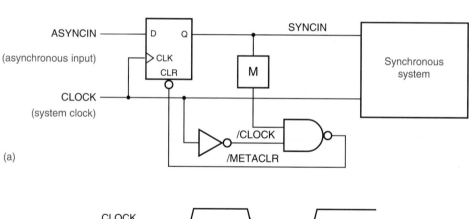

(a)

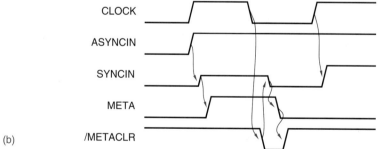

(b)

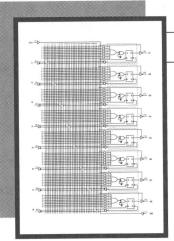

10

SEQUENTIAL LOGIC DESIGN WITH PLDS

The first commercially available programmable logic devices were PLAs. As described in Section 6.1, PLAs are combinational logic devices containing a programmable AND-OR array. Some early PLAs included a flip-flop on each output of the AND-OR array and were thus the first *sequential PLDs*, sometimes called *microsequencers*.

    In the late 1970s, first-generation PAL devices were offered in several fixed configurations, some with flip-flops and some without. The devices with flip-flops are often called *registered PLDs*; the word "register" is used because, like the MSI registers in Section 9.2.4, most sequential PLDs apply a single, common clock signal to all of their flip-flops.

    Second-generation PLDs, such as the GAL16V8 introduced in Section 6.12, allow the designer to program each output individually to have a flip-flop or not. Because of their tremendous flexibility, such devices have almost completely replaced fixed-configuration parts. Still-newer devices, so-called *field programmable gate arrays (FPGAs)*, provide functionality equivalent to several

sequential PLD
microsequencer

registered PLD

field programmable gate arrays (FPGA)

659

individual PAL or GAL devices coupled with a programmable set of on-chip interconnections. However, we concentrate on standard PLDs, because the design methods for FPGAs are equivalent to standard PLD-based design coupled with system-level partitioning considerations that are beyond the scope of this text.

We begin this chapter by showing the structure of first-generation, fixed-configuration sequential PLDs. Then we introduce the current programmable workhorses, the 16V8, 20V8, and 22V10. Timing considerations, discussed next, apply to PLDs of all types, including FPGAs.

To "ease into" design with sequential PLDs, we'll show how to obtain the functionality of several different sequential MSI parts using PLDs, and how to combine functions within a single PLD. Compared with MSI, such PLD realizations are much more easily tailored and modified for special-purpose requirements, and they are usually less expensive than MSI in both parts cost and PCB area. An optional section shows how to design counters using the specialized "X-series" of registered PLDs. Finally, we cover the most important application of sequential PLDs, namely, state-machine design. We'll describe several approaches for designing PLD-based state-machines, the most useful being state-machine description language.

10.1 BIPOLAR SEQUENTIAL PLDS

PAL16R8

The *PAL16R8*, shown in Figure 10–1, is representative of the first generation of sequential PLDs, which used bipolar (TTL) technology. This device has eight primary inputs, eight outputs, and common clock and output-enable inputs, and fits in a 20-pin package.

The PAL16R8's AND-OR array is exactly the same as the one found in the PAL16L8 combinational PLD. However, the PAL16R8 has edge-triggered D flip-flops between the AND-OR array and its eight outputs, O1–O8. All of the flip-flops are connected to a common clock input, CLK, and change state on the rising edge of the clock. Each flip-flop drives an output pin through a 3-state buffer; the buffers have a common output-enable signal, /OE. Notice that, like the combinational output pins of a PAL16L8, the registered output pins of the PAL16R8 contain the complement of the signal produced by the AND-OR array.

The possible inputs to the PAL16R8's AND-OR array are eight primary inputs (I1–I8) and the eight D flip-flop outputs. The connection from the D flip-flop outputs into the AND-OR array makes it easy to design shift registers, counters, and general state machines. Unlike the PAL16L8's combinational outputs, the PAL16R8's D-flip-flop outputs are available to the AND-OR array whether or not the O1–O8 three-state drivers are enabled. Thus, the internal flip-flops can go to a next state that is a function of the current state even when the O1–O8 outputs are disabled.

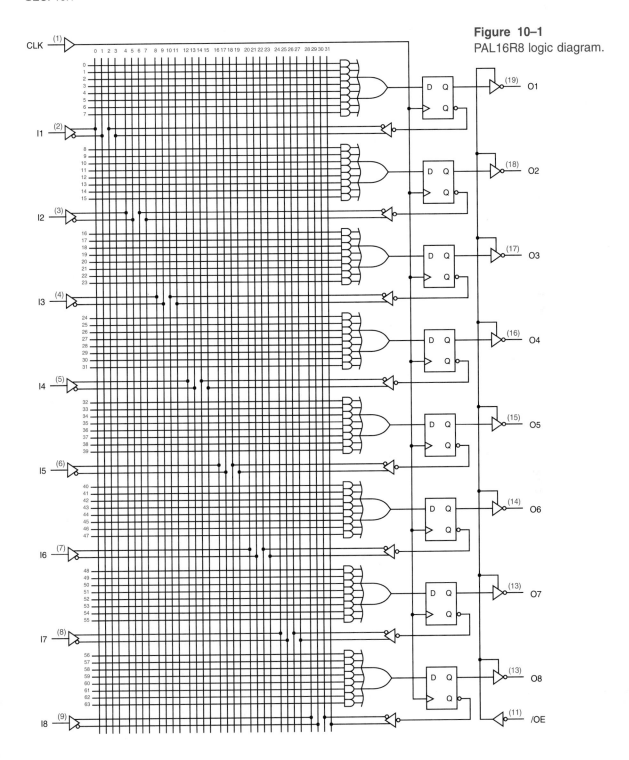

Figure 10–1
PAL16R8 logic diagram.

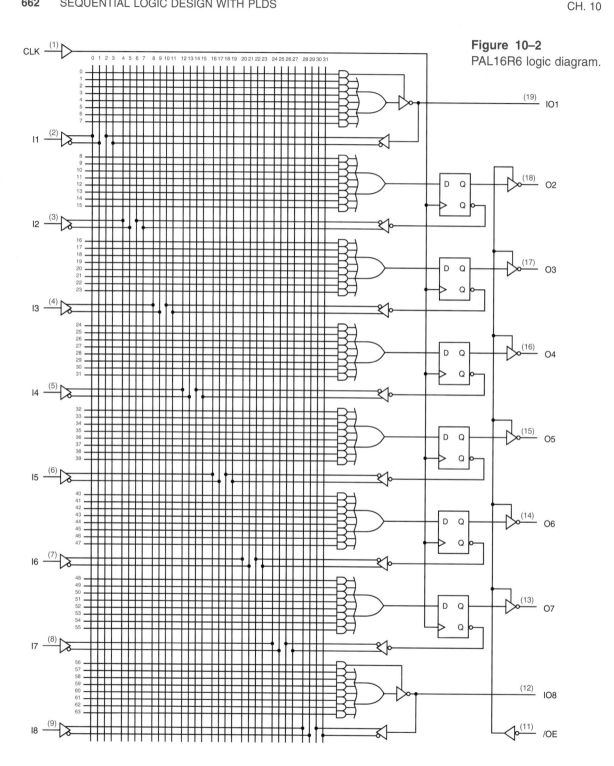

Figure 10–2
PAL16R6 logic diagram.

Table 10–1   Characteristics of standard bipolar PLDs.

Part number	Package pins	AND-gate inputs	Inputs to AND array			
			Primary inputs	Bidirectional comb. outputs	Registered outputs	Combinational outputs
PAL16L8	20	16	10	6	0	2
PAL16R4	20	16	8	4	4	0
PAL16R6	20	16	8	2	6	0
PAL16R8	20	16	8	0	8	0
PAL20L8	24	20	14	6	0	2
PAL20R4	24	20	12	4	4	0
PAL20R6	24	20	12	2	6	0
PAL20R8	24	20	12	0	8	0

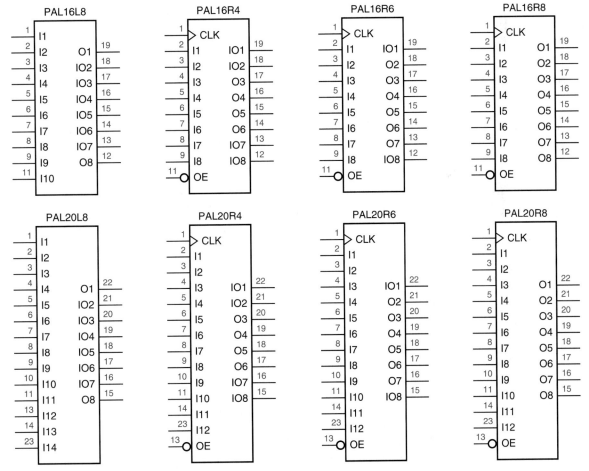

Figure 10–3   Logic symbols for popular bipolar combinational and sequential PLDs.

PAL16R6

Many applications require combinational as well as sequential PLD outputs. The manufacturers of bipolar PLDs addressed this need by providing a few variants of the PAL16R8 that omitted the D flip-flops on some output pins, and instead provided input and output capability identical to that of the PAL16L8's bidirectional pins. For example, Figure 10–2 is the logic diagram of the *PAL16R6*, which has only six registered outputs. Two pins, IO1 and IO8, are bidirectional, serving both as inputs and as combinational outputs with separate 3-state enables, just like the PAL16L8's bidirectional pins. Thus, the possible inputs to the PAL16R6's AND-OR array are the eight primary inputs (I1–I8), the six D-flip-flop outputs, and the two bidirectional pins (IO1, IO8).

PAL16L8
PAL16R4
PAL16R8
PAL20L8
PAL20R4
PAL20R6
PAL20R8

Table 10–1 on the preceding page shows eight standard bipolar PLDs with differing numbers and types of inputs and outputs. All of the PAL16xx parts in the table use the same AND-OR array, where each output has eight AND gates, each with 16 variables and their complements as possible inputs. The PAL20xx parts use a similar AND-OR array with 20 variables and their complements as possible inputs. Figure 10–3 shows logic symbols for all of the PLDs in the table.

10.2 SEQUENTIAL GALS

The GAL16V8 electrically erasable PLD was introduced in Section 6.12. Two "architecture-control" fuses are used to select among three basic configurations of this device. Section 6.12 described the *16V8C* ("complex") configuration, shown in Figure 6–20 on page 436, a structure similar to that of a bipolar combinational PAL device, the PAL16L8. The *16V8S* ("simple") configuration provides a slightly different combinational logic capability (see box on page 666).

16V8C

16V8S

16V8R

The third configuration, called the *16V8R*, allows a flip-flop to be provided on any or all of the outputs. Figure 10–4 shows the structure of the device when flip-flops are provided on all outputs. Notice that all of the flip-flops are controlled by a common clock signal on pin 1, as in the bipolar devices of the preceding subsection. Likewise, all of the output buffers are controlled by a common output-enable signal on pin 11.

PALS? GALS?

> Lattice Semiconductor introduced GAL devices such as the GAL16V8 in the mid-1980s. Advanced Micro Devices later followed up with a pin-compatible device which they call the PALCE16V8 ("C" is for CMOS, "E" is for erasable). Several other manufacturers make differently numbered but compatible devices as well. Rather than get caught up in the details of different manufacturers' names, in the rest of this chapter we'll usually refer to commonly used GAL devices with their generic names, 16V8, 20V8, and 22V10.

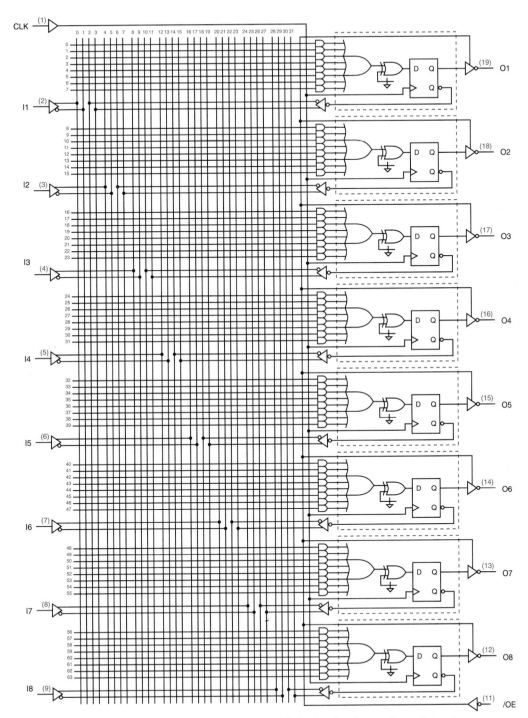

Figure 10–4 Logic diagram for the 16V8 in the "registered" configuration.

THE "SIMPLE"
16V8S

The "simple" 16V8S configuration of the GAL16V8 is not often used, because its capabilities are mostly a subset of the 16V8C's. Instead of an AND term, the 16V8S uses one fuse per output to control whether the output buffers are enabled. That is, each output pin may be programmed either to be always enabled or to be always disabled (except pins 15 and 16, which are always enabled). All of the output pins (except 15 and 16) are available as inputs to the AND array regardless of whether the output buffer is enabled.

The only advantage of a 16V8S compared to a 16V8C is that it has eight, not seven, AND terms as inputs to the OR gate on each output. The 16V8S architecture was designed mainly for emulation of certain now-obsolete bipolar PAL devices, some of which either had eight product terms per output or had inputs on pins 12 and 19, which are not inputs in the 16V8C configuration. With appropriate programming, the 16V8S can be used as a pin-for-pin compatible replacement for these devices, which included the PAL10H8, PAL12H6, PAL14H4, PAL16H2, PAL10L8, PAL12L6, PAL14L4, and PAL16L2.

output logic macrocell

The circuitry inside each dotted box in Figure 10–4 is called an *output logic macrocell*. The 16V8R is much more flexible than a PAL16R8 because each macrocell may be individually configured to bypass the flip-flop, that is, to produce a combinational output. Figure 10–5 shows the two macrocell configurations that are possible in the 16V8R; (a) is registered and (b) is combinational. Thus, it is possible to program the device to have any set of registered and combinational outputs, up to eight total.

20V8

The *20V8* is similar to the 16V8, but comes in a 24-pin package with four extra input-only pins (like the PAL20L8 discussed in Section 6.10 on page 428). Each AND gate in the 20V8 has 20 inputs, hence the "20" in "20V8."

22V10

The *22V10*, whose basic structure is shown in Figure 10–6, also comes in a 24-pin package, but is somewhat more flexible than the 20V8. The 22V10 does not have "architecture control" bits like the 16V8's and 20V8's, but it can realize any function that is realizable with a 20V8, and more. It has more product

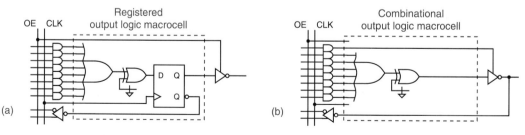

Figure 10–5 Output logic macrocells for the 16V8R: (a) registered; (b) combinational.

Figure 10–6
Logic diagram for
the 22V10.

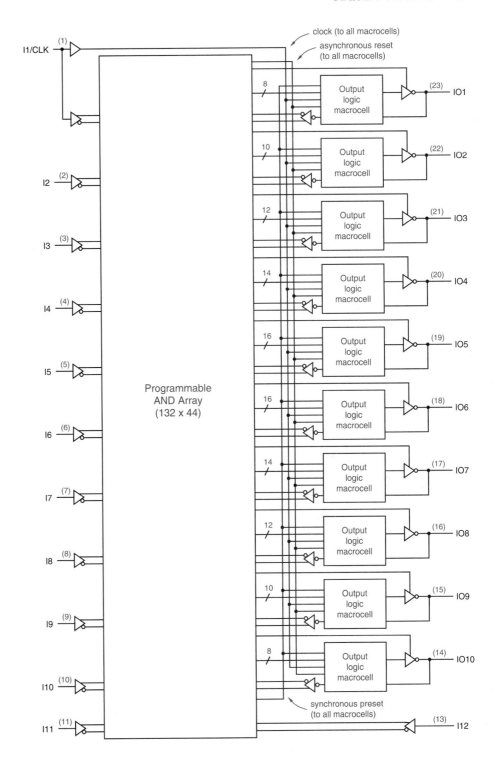

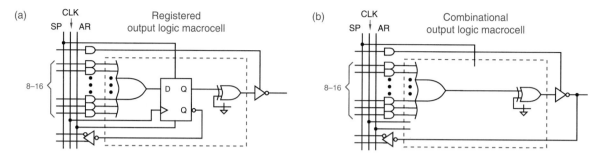

Figure 10–7 Output logic macrocells for the 22V10: (a) registered; (b) combinational.

terms, two more general-purpose inputs, and better output-enable control than the 20V8. Key differences are summarized below:

- Each output logic macrocell is configurable to have a register or not, as in the 20V8R architecture. However, the macrocells are different from the 16V8's and 20V8's, as shown in Figure 10–7.

- A single product term controls the output buffer, regardless of whether the registered or the combinational configuration is selected for a macrocell.

- Every output has at least eight product terms available, regardless of whether the registered or the combinational configuration is selected. Even more product terms are available on the inner pins, with 16 available on each of the two innermost pins. ("Innermost" is with respect to the right-hand side of the Figure 10–6, which also matches the arrangement of these pins on a 24-pin DIP package.)

- The clock signal on pin 1 is also available as a combinational input to any product term.

- A single product term is available to generate a global, asynchronous reset signal that resets all internal flip-flops to 0.

- A single product term is available to generate a global, synchronous preset signal that sets all internal flip-flops to 1 on the rising edge of the clock.

- Like the 16V8 and 20V8, the 22V10 has programmable output polarity. However, in the registered configuration, the polarity change is made at the output, rather than the input, of the D flip-flop. This affects the details of programming when the polarity is changed but does not affect the overall capability of the 22V10 (i.e., whether a given function can be realized). In fact, the difference in polarity-change location is transparent when you use a PLD programming language such as ABEL.

In 1993, the 16V8, 20V8, and 22V10 were the most popular and cost-effective PLDs (but see box below). Figure 10–8 shows generic logic symbols for these three devices. Most of the examples in the rest of this chapter use the most cost-effective of the three devices, the 16V8.

Once you understand the capabilities of different PLDs, you might ask, "Why not just always use the most capable PLD available?" For example, even if a circuit fits in a 20-pin 16V8, why not specify the slight larger, 24-pin 20V8 so that spare inputs are available in case of trouble? And, once you've specified a 20V8, why not use the somewhat more capable 22V10 which comes in the same 24-pin package?

    In the real world of product design and engineering, the constraint is cost. As in many other fields, digital devices such as PLDs are not always priced proportionally to their capabilities. In the high-volume PLD-based products most recently designed by the author, the PLD manufacturer's price for a 20V8 was 2.5 times the price of a 16V8! (If the price was proportional to capability, such as number of inputs and product terms, the ratio should have been only 1.25.) And the price of a 22V10 was almost double that of a 20V8!

    The author's most recent and successful approach to PLD-based design has been to partition designs to use only 16V8s and 20V8s, allowing 22V10s to replace 20V8s only in a pinch (usually during debugging, when a requirement surfaces for more product terms or spare outputs, and further tweaking of the board layout is physically impossible).

    PLD manufacturers also offer higher-density FPGAs that can replace a few to dozens of 22V10-class devices. Unfortunately, FPGAs usually carry the same hefty price premium as 22V10s, especially in the faster speed grades (e.g., 10–15 ns), and were not available at all in the fastest speed grades (5–7 ns) in 1993.

    For example, an FPGA that replaces four 10 ns 22V10s may cost three to five times the price of a 10 ns 22V10, or about the same as 15–25 16V8s! Under these circumstances, if the design can be partitioned to use 4, 6, or even 10 16V8s, the multichip solution is more economical than a single FPGA, unless physical space constraints or other factors necessitate the FPGA's higher level of integration.

    Of course, higher-density, fast FPGAs can be expected eventually to achieve the same low cost per function that medium-density, fast 16V8 and 20V8 devices offer today. But for now, the warning is "Let the buyer beware"!

    All of this goes to show that cost must always be considered along with design elegance and convenience to create a successful (i.e., profitable) product. And minimizing the cost of a product usually involves a plethora of common-sense economic and engineering considerations that are far removed from the turn-the-crank, mathematical gate minimization methods of Chapter 4.

HOW MUCH
DOES IT COST?

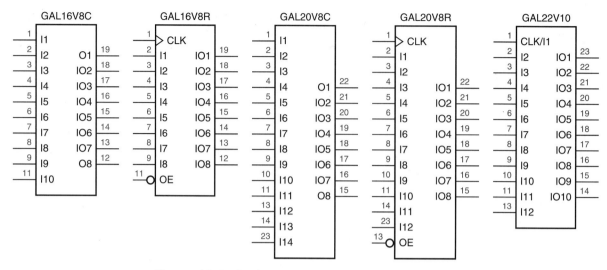

Figure 10-8   Logic symbols for popular GAL devices.

10.3   PLD TIMING SPECIFICATIONS

Several timing parameters are specified for combinational and sequential PLDs. The most important ones are illustrated in Figure 10–9 and are explained below:

t_{PD} t_{PD} This parameter applies to combinational outputs. It is the propagation delay from a primary, bidirectional, or "feedback" input to the combinational

feedback input   output. A *feedback input* an internal input of the AND-OR array that is driven by the output of an internal flip-flop.

t_{CO} t_{CO} This parameter applies to registered outputs. It is the propagation delay from the rising edge of CLK to a primary output or a feedback input.

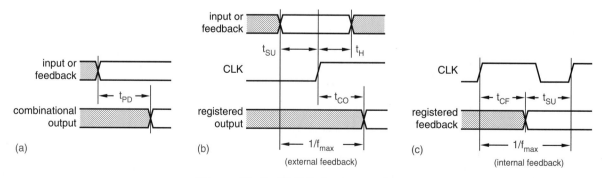

Figure 10-9   PLD timing parameters.

t_{CF} This parameter also applies to registered outputs. It is the propagation delay from the rising edge of CLK to an internal registered output that goes back to a feedback input. If specified, t_{CF} is normally less than t_{CO}. However, some manufacturers do not specify t_{CF}, in which case you must assume that $t_{CF} = t_{CO}$. t_{CF}

t_{SU} This parameter applies to primary, bidirectional, and feedback inputs that affect the D inputs of flip-flops. It is the setup-time requirement for the input signal to be stable before the rising edge of CLK. t_{SU}

t_H This parameter also applies to signals that affect the D inputs of flip-flops. It is the hold-time requirement for the input signal to be stable after the rising edge of CLK. t_H

f_{max} This parameter applies to clocked operation. It is the maximum frequency at which the PLD can operate reliably, and is the reciprocal of the minimum clock period. Two versions of this parameter can be derived from the previous specifications, depending on whether the device is operating with external or internal feedback. f_{max}

External feedback refers to a circuit in which a registered PLD output is connected to the input of another registered PLD with similar timing; for proper operation, the sum of t_{CO} for the first PLD and t_{SU} for the second must not exceed the clock period. *external feedback*

Internal feedback refers to a circuit in which a registered PLD output is fed back to a register in the same PLD; in this case, the sum of t_{CF} and t_{SU} must not exceed the clock period. *internal feedback*

Each of the PLDs that we described in previous sections is available in several different speed grades. The speed grade is usually indicated by a suffix on the part number, such as "16V8-10"; the suffix usually refers to the t_{PD} specification, in nanoseconds. Table 10–2 shows the timing of several popular bipolar and CMOS PLDs. Note that only the t_{PD} column applies to combinational outputs of a device, while the last four columns apply to registered outputs. All of the timing specifications are worst-case numbers over the commercial operating range.

When sequential PLDs are used in applications with critical timing, it's important to remember that they have longer setup times than discrete edge-triggered registers in the same technology, owing to the delay of the AND-OR array on each D input. Conversely, under typical conditions, a PLD actually has a negative hold-time requirement because of the delay through AND-OR array. However, you can't count on having a negative hold time—the worst-case specification is zero.

Table 10–2 Timing specifications, in nanoseconds, of popular bipolar and CMOS PLDs (as listed in Advanced Micro Devices and Lattice Semiconductor data books for parts available in 1992 and 1993).

Part numbers	Suffix	t_{PD}	t_{CO}	t_{CF}	t_{SU}	t_H
PAL16L8, PAL16Rx, PAL20L8, PAL20Rx	-5	5	4	–	4.5	0
PAL16L8, PAL16Rx, PAL20L8, PAL20Rx	-7	7.5	6.5	–	7	0
PAL16L8, PAL16Rx, PAL20L8, PAL20Rx	-10	10	8	–	10	0
PAL16L8, PAL16Rx, PAL20L8, PAL20Rx	B	15	12	–	15	0
PAL16L8, PAL16Rx, PAL20L8, PAL20Rx	B-2	25	15	–	25	0
PAL16L8, PAL16Rx, PAL20L8, PAL20Rx	A	25	15	–	25	0
PALCE16V8, PALCE20V8	-5	5	4	–	3	0
GAL16V8, GAL20V8	-7	7.5	5	3	5	0
GAL16V8, GAL20V8	-10	10	7.5	6	7.5	0
GAL16V8, GAL20V8	-15	15	10	8	12	0
GAL16V8, GAL20V8	-25	25	12	10	15	0
PALCE22V10	-5	5	4	–	3	0
PALCE22V10	-7	7.5	4.5	–	4.5	0
GAL22V10	-10	10	7	2.5	7	0
GAL22V10	-15	15	8	2.5	10	0
GAL22V10	-25	25	15	13	15	0

10.4 PLD REALIZATIONS OF SEQUENTIAL MSI FUNCTIONS

Most of the sequential MSI devices that we introduced in Chapter 9 can be easily realized in sequential PLDs. We'll give several examples in this section. Although TTL or CMOS MSI realization of any of these functions would probably be more cost-effective than a PLD, the examples serve as a starting point for you to design sequential PLDs for more complex applications.

clocked assignment operator, :=

Instead of an = sign, ABEL equations for registered PLD outputs use the *clocked assignment operator,* :=. In the case of bipolar PAL devices, the ABEL compiler already knows from the `device` declaration which outputs are registered and which are not, so the information provided by ":=" is redundant. However, for 16V8, 20V8, and 22V10 devices, the compiler needs this information to select the appropriate configuration of the output logic macrocell.

10.4.1 Edge-Triggered Registers

Just about the simplest application of a registered PLD is an edge-triggered register similar to the MSI 74x374 8-bit register (Figure 9–8 on page 592). A GAL16V8 can be used to perform this function as shown by the program in Table 10–3. Notice that the registered configuration of the 16V8 is specified in

Table 10–3 ABEL program for an 8-bit register.

```
module Eight_Bit_Reg
title '8-bit Edge-Triggered Register'
Z74X374 device 'P16V8R';
@ALTERNATE

" Input pins
CLK, /OE                        pin 1, 11;
D1, D2, D3, D4, D5, D6, D7, D8  pin 2, 3, 4, 5, 6, 7, 8, 9;

" Output pins
Q1, Q2, Q3, Q4, Q5, Q6, Q7, Q8  pin 19, 18, 17, 16, 15, 14, 13, 12;

" Set definitions
D = [D1,D2,D3,D4,D5,D6,D7,D8];
Q = [Q1,Q2,Q3,Q4,Q5,Q6,Q7,Q8];

equations

Q := D;

end Eight_Bit_Reg
```

the device declaration. The same function can be provided by a PAL16R8 if "P16V8R" is changed to "P16R8."

Figure 10–10 shows the correspondence between 74x374 inputs and the PLD realization. The clock and output-enable inputs of the 74x374 are provided by the corresponding inputs of the 16V8. The data inputs and Q outputs of the '374 correspond to the primary inputs and outputs of the 16V8.

The ABEL program uses sets to describe the 8-bit inputs and outputs,

Figure 10–10
PLD realization of the 74x374
MSI register.

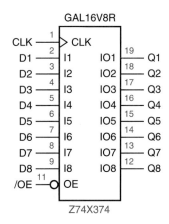

which are treated identically by the almost-trivial equation, Q := D. This equation generates just one trivial product term per output.

Another MSI 8-bit register is the 74x377. Instead of the 74x374's output-enable input, the '377 has a clock-enable input. To realize this function in a PLD, you need 9 inputs and 8 outputs, more than what's provided in a 16V8 (the /OE input is not available to the AND-OR array). So, a 20V8 must be used; Table 10–4 shows the ABEL program. In the equations, Q is set equal to D if G is asserted, and Q is set equal to the old value of Q if G is negated. Once again, the function could be realized with a PAL device (the 20R8) with a simple declaration change. Figure 10–11(a) shows the correspondence between 74x377 inputs and the PLD realization. Notice that the 20V8's /OE input has been grounded to permanently output-enable the device, since the '377 has no output-enable function.

The previous example showed a case where the PLD version of a function required more pins than the MSI version because the input/output capabilities of the devices did not quite match. Table 10–5 shows how to specify a 10-bit version of a 74x377 that fully utilizes a GAL22V10. Figure 10–11(b) shows the correspondence between 74x377 inputs and the PLD realization. Don't forget, though, that a 22V10 is more expensive than a 20V8, which is *much* more expensive than an MSI 74x377 with comparable speed.

Table 10–4   ABEL program for an 8-bit register with clock-enable.

```
module Eight_Bit_Reg_Enb
title '8-bit Edge-Triggered Register With Enable'
Z74X377 device 'P20V8R';
@ALTERNATE

" Input pins
CLK, /G                          pin 1, 10;
D1, D2, D3, D4, D5, D6, D7, D8   pin 2, 3, 4, 5, 6, 7, 8, 9;

" Output pins
Q1, Q2, Q3, Q4, Q5, Q6, Q7, Q8   pin 22, 21, 20, 19, 18, 17, 16, 15;

" Set definitions
D = [D1, D2, D3, D4, D5, D6, D7, D8];
Q = [Q1, Q2, Q3, Q4, Q5, Q6, Q7, Q8];

equations

Q := G * D + /G * Q;

end Eight_Bit_Reg_Enb
```

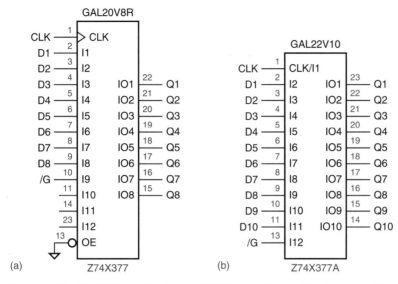

Figure 10–11 PLD realizations of MSI registers with clock enable:
(a) 74x377; (b) 10-bit version of 74x377.

Table 10–5 ABEL program for a 10-bit register with enable.

```
module Ten_Bit_Reg_Enb
title '10-bit Edge-Triggered Register With Enable'
Z74X377A device 'P22V10';
@ALTERNATE

" Input pins
CLK, /G                         pin 1, 13;
D1, D2, D3, D4, D5              pin 2, 3, 4, 5, 6;
D6, D7, D8, D9, D10            pin 7, 8, 9, 10, 11;

" Output pins
Q1, Q2, Q3, Q4, Q5             pin 23, 22, 21, 20, 19;
Q6, Q7, Q8, Q9, Q10           pin 18, 17, 16, 15, 14;

" Set definitions
D = [D1, D2, D3, D4, D5, D6, D7, D8, D9, D10];
Q = [Q1, Q2, Q3, Q4, Q5, Q6, Q7, Q8, Q9, Q10];

equations

Q := G * D + /G * Q;

end Ten_Bit_Reg_Enb
```

10.4.2 Shift Registers

Shift registers use substantially more of a PLD's capability than do the simple registers of the previous subsection. For example, Table 10–6 and Figure 10–12(a) show how to realize a function similar to that of a 74x194 universal shift register using a 16V8. Notice that pin 12 of the 16V8 is used as an input. The 16V8 realization differs from the real '194 in just one way—in the function of the /CLR input. In the real '194, /CLR is an asynchronous input, while in the 16V8 it is sampled along with other inputs at the rising edge of CLK.

If you really need to provide an asynchronous clear input, you can use the 22V10, which provides a single product line to control the reset inputs of all of its flip-flops. Table 10–7 is a program that produces the required behavior.

ABEL recognizes the ".RE" suffix to indicate control of the asynchronous flip-flop reset signal in a 22V10; similarly, a ".PR" suffix indicates control of

Table 10–6   ABEL program for a 4-bit universal shift register.

```
module Four_Bit_Shift_Reg
title '4-bit Universal Shift Register'
Z74X194 device 'P16V8R';
@ALTERNATE

" Input pins
CLK, /OE                          pin 1, 11;
RIN, A, B, C, D, LIN              pin 2, 3, 4, 5, 6, 7;
S1, S0, /CLR                      pin 8, 9, 12;

" Output pins
QA, QB, QC, QD                    pin 19, 18, 17, 16;

" Set definitions
INPUT   = [A,    B,   C,   D  ];
LEFTIN  = [B,    C,   D,   LIN];
RIGHTIN = [RIN,  A,   B,   C  ];
OUT     = [ QA,  QB,  QC,  QD ];

CTRL  = [S1,S0];
HOLD  = (CTRL == [0,0]);
RIGHT = (CTRL == [0,1]);
LEFT  = (CTRL == [1,0]);
LOAD  = (CTRL == [1,1]);

equations

OUT := /CLR * (HOLD * OUT + RIGHT * RIGHTIN + LEFT * LEFTIN + LOAD * INPUT);

end Four_Bit_Shift_Reg
```

Figure 10–12
PLD realizations of a 74x194
universal shift register:
(a) with synchronous clear;
(b) with asynchronous clear.

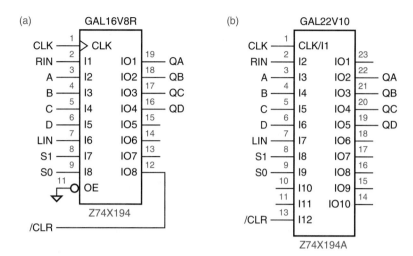

Table 10–7 Program for a 4-bit universal shift register with asynchronous clear.

```
module Four_Bit_Shift_Reg_A
title '4-bit Universal Shift Register With Asynchronous Clear'
Z74X194A device 'P22V10';
@ALTERNATE

" Input pins
CLK, RIN, A, B, C, D, LIN           pin 1, 2, 3, 4, 5, 6, 7;
S1, S0, /CLR                        pin 8, 9, 13;

" Output pins
QA, QB, QC, QD                      pin 22, 21, 20, 19;

" Set definitions and expressions
INPUT   = [A,   B,   C,   D  ];
LEFTIN  = [B,   C,   D,   LIN];
RIGHTIN = [RIN, A,   B,   C  ];
OUT     = [ QA, QB, QC, QD ];

CTRL  = [S1,S0];
HOLD  = (CTRL == [0,0]);
RIGHT = (CTRL == [0,1]);
LEFT  = (CTRL == [1,0]);
LOAD  = (CTRL == [1,1]);

equations

OUT := HOLD * OUT + RIGHT * RIGHTIN + LEFT * LEFTIN + LOAD * INPUT;
OUT.RE = CLR;

end Four_Bit_Shift_Reg_A
```

the synchronous flip-flop preset signal. (Note that all flip-flops are controlled by the same reset and preset signals.)

SYNC VS. ASYNC

> In most applications, the "clear" inputs of registers and other devices are used only at system reset, to force the device to a known, good state. In most cases, a synchronous clear input works just as well as an asynchronous one, because the clock is already running at reset time.

10.4.3 Counters

The most popular MSI counter is the 74x163 4-bit binary counter, shown in Figure 9–14 on page 596. A glance at this figure shows that the excitation logic for this counter isn't exactly simple, especially considering its use of XNOR gates. Nevertheless, ABEL provides a very simple way of defining counter behavior, which we describe next.

In the absence of the @ALTERNATE directive, ABEL interprets the "+" sym-

Table 10–8 ABEL program for a 4-bit binary counter.

```
module Four_Bit_Counter
title '4-bit Binary Counter'
Z74X163 device 'P16V8R';

" Input pins
CLK, !OE                         pin 1, 11;
A, B, C, D                       pin 2, 3, 4, 5;
!LD, !CLR, ENP, ENT             pin 6, 7, 8, 9;

" Output pins
QA, QB, QC, QD, RCO              pin 19, 18, 17, 16, 15;

" Set definitions
INPUT = [ D,  C,  B,  A ];
COUNT = [QD, QC, QB, QA ];

equations

COUNT := !CLR & (  LD & INPUT
            # !LD &  (ENT & ENP) & (COUNT + 1)
            # !LD & !(ENT & ENP) &  COUNT);

RCO = (COUNT == [1,1,1,1]) & ENT;

end Four_Bit_Counter
```

Table 10–9 MInimized equations for a 4-bit binary counter.

```
QA := (/LD * /CLR *  ENP * ENT * /QA
     + /LD * /CLR * /ENT * QA
     + /LD * /CLR * /ENP * QA
     +  LD * /CLR *  A);

QB := (/LD * /CLR *  ENP * ENT * QA * /QB
     + /LD * /CLR * /ENT * QB
     + /LD * /CLR * /ENP * QB
     + /LD * /CLR * /QA  * QB
     +  LD * /CLR *  B);

QC := (/LD * /CLR *  ENP * ENT * QB * QA * /QC
     + /LD * /CLR * /ENT * QC
     + /LD * /CLR * /ENP * QC
     + /LD * /CLR * /QB  * QC
     + /LD * /CLR * /QA  * QC
     +  LD * /CLR *  C);

QD := (/LD * /CLR *  ENP * ENT * QC * QB * QA * /QD
     + /LD * /CLR * /ENT * QD
     + /LD * /CLR * /ENP * QD
     + /LD * /CLR * /QC  * QD
     + /LD * /CLR * /QB  * QD
     + /LD * /CLR * /QA  * QD
     +  LD * /CLR *  D);

RCO = (ENT * QA * QB * QC * QD);
```

bol as an integer addition operator. When two sets are "added" with this operator, each is interpreted as a binary number; the rightmost set element corresponds to the least significant bit of the number. Thus, the function of a 74x163 can be specified by the ABEL program in Table 10–8, which uses a 16V8.

Table 10–9 shows the minimized logic equations that ABEL generates for the 4-bit counter. Notice that each more significant output bit requires one more product term. As a result, the size of counters that can be realized in a 16V8 or even a 20V8 is generally limited to five or six bits. Other PLDs, the so-called X series, contain an XOR structure that can realize larger counters without increasing product-term requirements. These PLDs and their uses in counter design are described in the optional Section 10.5.

Note that in the absence the @ALTERNATE directive, a program must use ABEL's default symbols for AND, OR, and NOT, namely &, #, and !.	FUNNY SYMBOLS

GALS VS. PALS

> The binary counter in the preceding subsection used one combinational GAL output
> as well as four registered outputs. Alternatively, you could perform the required
> functions in a bipolar PAL device such as the PAL16R4 or PAL16R6, but you'd
> have to be a little more particular about the assignment of signals to pins than we
> were in Table 10–8.

10.4.4   Latches

We showed in Section 7.2 that basic sequential-circuit building blocks such as
latches and flip-flops are normally built using ordinary combinational logic gates
and feedback. For example, the excitation equation for an S-R latch is

$$Q* \ = \ S + R'{\cdot}Q$$

Thus, you could build an S-R latch using one *combinational* output of a PLD.
Furthermore, the S and R signals above could be replaced with more complex
logic functions of the PLD's inputs, limited only by the availability of product
terms (seven per output in a 16V8C or 16L8) to realize the final excitation equation. Note, however, that the Q output must be continuously enabled; otherwise,
the feedback loop would be broken and the latch's state lost. Also, the feedback
loop can be created only when Q is assigned to a bidirectional pin, any I/O pin
in a combinational output logic macrocell of a 16V8R, but limited to one of
IO2–IO7 in a 16V8C or 16L8.

   Probably the handiest sequential circuit to build out of a combinational PLD
is a D latch. The basic excitation equation for a D latch is

$$Q* \ = \ C{\cdot}D + C'{\cdot}Q$$

However, we showed in Section 7.6.1 that this equation contains a static hazard,
and the corresponding circuit does not latch data reliably. To build a reliable D
latch, we must include a consensus term in the excitation equation:

$$Q* \ = \ C{\cdot}D + C'{\cdot}Q + D{\cdot}Q$$

The D input in this equation may be replaced with a more complicated expression,
but the equation's structure remains the same:

$$Q* \ = \ C{\cdot}\textit{expression} + C'{\cdot}Q + \textit{expression}{\cdot}Q$$

It is also possible to use a more complex expression for the C input, as we
showed in Section 7.6.1.

   Probably the most common use of a PLD-based latch is to simultaneously
decode and latch addresses in microprocessor systems. For example, expanding
the memory-decoder example of Table 6–2 on page 407, suppose we have another

input ADDRVALID that indicates that the address is valid and the decoder output should be latched. We can combine the decoding and latching functions in one output with the following equation:

```
ROMCS = ADDRVALID *
          (LARGE * A23 * A22 * A21 * A20 * A19 * A18 * A17 * A16
            + /LARGE * A19 * A18 * A17 * A16 + RTEST)
      + /ADDRVALID * ROMCS
      + ROMCS *
          (LARGE * A23 * A22 * A21 * A20 * A19 * A18 * A17 * A16
            + /LARGE * A19 * A18 * A17 * A16 + RTEST);
```

The equation expands into seven product terms, and thus fits on a 16V8 or 16L8 combinational output.

   After seeing how easy it is to build S-R and D latches using combinational PLDs, you might be tempted to go further and try to build an edge-triggered D flip-flop. Although this is possible, it is expensive because an edge-triggered flip-flop has four internal states and thus two feedback loops, consuming two PLD outputs. Furthermore, the setup and hold times and propagation delays of such a flip-flop would be quite poor compared to those of a discrete flip-flop in the same technology. Finally, as we discussed in Section 7.6.6, the flow tables of all edge-triggered flip-flops contain essential hazards, which may be difficult to mask in a PLD-based design.

* 10.5 COUNTER DESIGN WITH PLDS

Binary counters are good candidates for PLD-based design, for many reasons:

- A large state machine can often be decomposed into two or more smaller state machines where one of the smaller machines is a binary counter that keeps track of how long the other machine should stay in a particular state. This may simplify both the conceptual design and the circuit design of the machine.

- Many applications require almost-binary-modulus counters with special requirements for initialization, state detection, or state skipping. For example, a counter in an elevator controller may skip state 13. Instead of using an off-the-shelf binary counter and extra logic for the special requirements, a designer can put precisely the required functions in a PLD.

- Most standard MSI counters have only 4 bits, while a single 24-pin PLD can be used to create a binary counter with up to 10 bits.

---

*Throughtout this book, optional sections are marked with an asterisk.

In the preceding section we showed how ABEL provides an easy way to specify counter behavior. However, this section describes counter excitation equations in more depth. Understanding this material is necessary if you want to design large counters using X-series PLDs, described later in this section.

* 10.5.1   Counter Behavior

Like most MSI counters, a PLD-based counter is a *synchronous* counter, that is, all of its output bits change at the rising edge of the clock. The basic logic function for one output bit of a synchronous binary up counter is, in words, "Change state if counting is enabled and all the lower-order bits are 1." (To get a binary *down* counter, just change 1 to 0 in this statement.) A corresponding ABEL equation is written below for bit 4 of a synchronous binary up counter:

```
Q4 := /Q4 * (CNTEN * Q3 * Q2 * Q1 * Q0)    " Complement
    +  Q4 * /(CNTEN * Q3 * Q2 * Q1 * Q0); " Preserve
```

The first line complements Q4 if all of the conditions for changing state are true. Otherwise, the second line preserves the state of Q4. This equation requires six product terms to realize in any of the sequential PLDs that we've studied so far, assuming that /Q4 is assigned to a registered output pin. Using ABEL's XOR operator, :+:, we can write the equation in just one line:

```
Q4 := Q4 :+: (CNTEN * Q3 * Q2 * Q1 * Q0); " Toggle if enabled
```

However, after expansion, the equation still requires a total of six product terms.

Other counter features, like synchronous loading and clearing, can be added quite easily to the preceding equation:

```
Q4 := /CLR * /LD * (Q4 :+: (CNTEN * Q3 * Q2 * Q1 * Q0)) "Toggle
    + /CLR * LD * D4; " Load
```

The expanded and reduced equation has eight product terms, which just fits on a registered output of a 16V8R or 16Rx. Higher-order counter bits (Q5, Q6, ...) require even more product terms, and therefore do not fit in the PLDs that we've discussed so far. So, before we look at other modifications to the basic counter equations, we'll introduce a series of PLDs that were specifically designed to reduce the product-term requirements of counter logic.

* 10.5.2   X-Series PLDs

PAL20X8
PAL20X4
PAL20X10
PAL20L10

The "X series" of PLDs uses XOR gates to facilitate the design of binary counters. This series includes the *PAL20X8*, whose logic diagram is shown in Figure 10–13, and three similar devices. The other devices have a different allocation of registered and combinational outputs, as shown in Table 10–10 and Figure 10–14. The *PAL20L10* has *no* registered outputs or XOR gates, but is included in the se-

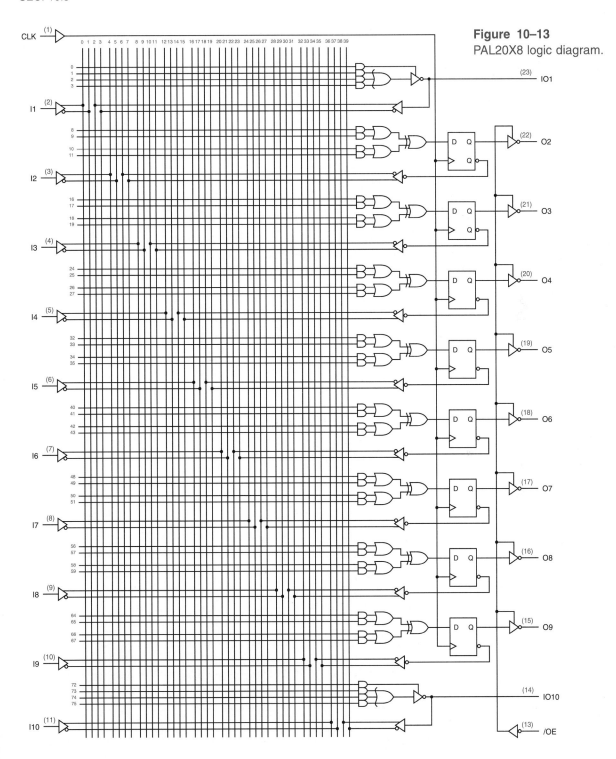

Figure 10–13
PAL20X8 logic diagram.

Table 10–10 Characteristics of "X-series" PLDs.

Part number	Package pins	AND-gate inputs	Inputs to AND array			
			Primary inputs	Bidirectional comb. outputs	Registered outputs	Combinational outputs
PAL20L10	24	20	12	8	0	2
PAL20X4	24	20	10	6	4	0
PAL20X8	24	20	10	2	8	0
PAL20X10	24	20	10	0	10	0

ries because it has the same AND-OR array as the other devices. Lattice Semiconductor also manufactures a GAL device, the GAL20XV10, that can be configured to emulate any device in the table, and more; each output can be individually configured to be combinational or registered, as in the 16V8 and 20V8.

Each X-series output has only four product terms available, which often leaves designers starving for more. The combinational outputs use one of the four terms to control the three-state enable, while the registered outputs perform the XOR of two ORed pairs of product terms. In ABEL, the generic equation for a registered output is

```
OUT := (P1 + P2) :+: (P3 + P4);
```

where P1–P4 are the four product terms.

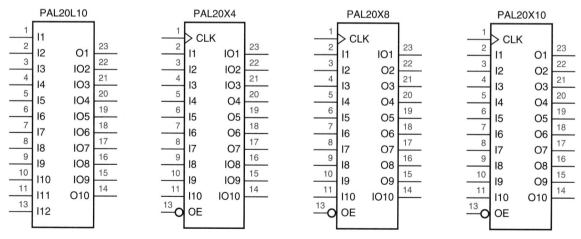

Figure 10–14 Traditional logic symbols for X-series PLDs.

* 10.5.3 Counter Design with X-Series PLDs

With a little factoring, the equation for output Qi of a binary synchronous up counter maps nicely into the structure of X-series PLDs:

```
Qi := (  (/CLR * /LD * Qi)                    " Hold
     +   0 )   :+:                            " Unused
     (  (/CLR * /LD * CNTEN * Qi-1 * ... * Q0)   " Toggle
     + (/CLR *  LD * Di) );                   " Load
```

The same basic equation works for high-order bits as well as low-order ones, with no increase in the number of product terms. In fact, one product term is not even needed (the explicit 0 on the second line).

The Qi signal in the preceding equation is active high at the flip-flop input and output, and therefore active low at the device pin. To obtain an active-high pin output, the flip-flop output must be active-low, /Qi, so the equations must be modified:

```
/Qi := (  (/PR * /LD * /Qi)                   " Hold
     +   0 )   :+:                            " Unused
     (  (/PR * /LD * CNTEN * Qi-1 * ... * Q0)    " Toggle
     + (/PR *  LD * /Di) );                   " Load
```

Note that the functionality of what was the CLR input is now different, since clearing the flip-flop *presets* the active-high output pin to 1. The unused product term can be employed to obtain a true CLR function (see Exercise 10.22). The equations for down counters are identical, except that the toggle term must detect the state /Qi-1*...*/Q0.

If we try to add more features, such as multiple load and clear inputs and nonbinary modulus counting, we quickly run out of product terms in an X-series PLD. Also, PLD compilers aren't smart enough to rearrange an arbitrary expression into the form required by the X-series structure. Let's look at some modifications that *are* possible using at most one extra product term.

In order to obtain a counter that counts from 0 to state N and then returns to state 0 and repeats, we must detect state N and force the counter to return to state 0 on the next count. Suppose that STATEN is a product term that detects state N. Then the expression STATEN·CNTEN is 1 when the counter should be forced into state 0 (unless the /LD input overrides). We can use a combinational output of the PLD to realize this expression:

```
FORCE0 = STATEN * CNTEN;
```

Using FORCE0 as a "helper" term, we can now easily obtain the desired behavior:

```
Qi := (  (/CLR * /LD * Qi * /FORCE0 )            " Hold
     +   0 )   :+:                               " Unused
     (  (/CLR * /LD * CNTEN * Qi-1 *...* Q0 * /FORCE0)" Toggle
     + (/CLR *  LD * Di) );                      " Load
```

If we modified the equation above to use STATEN·CNTEN directly instead of FORCE0, we would run out of product terms—STATEN, a product term, blows up into a sum of many terms when complemented.

Another approach to forcing state 0 does not use a combinational "helper" output, but instead uses at most one extra product term for each Qi output, depending on i and N:

(1) If Qi would be 0 in state $N+1$, do not modify the equation for Qi.

(2) If Qi would normally change from 0 to 1 between states N and $N+1$, add a product term to cancel this change:

```
Qi := (  (/CLR * /LD * Qi)                          " Hold
       + (/CLR * /LD * CNTEN * STATEN) )  :+:       " Cancel
       (  (/CLR * /LD * CNTEN * Qi-1 * ... * Q0)    " Toggle
       + (/CLR *  LD * Di) );                        " Load
```

(3) If Qi would be 1 in both states N and $N+1$, add a product term to toggle it to 0:

```
Qi := (  (/CLR * /LD * Qi)                          " Hold
       + (/CLR *  LD * Di) )  :+:                    " Load
       (  (/CLR * /LD * CNTEN * Qi-1 * ... * Q0)    " Toggle
       + (/CLR * /LD * CNTEN * STATEN) );            " Extra
```

These rules can be extended to force the counter to any desired starting state, M. If Qi has the same value in states M and $N+1$, do nothing. If the values are different, add a "cancel" term if Qi has different values in states N and $N+1$, and an "extra" toggle term otherwise.

In general, if you get stuck with an equation that has more than two product terms on either side of the XOR, you can't perform it in a X-series PLD, unless you use an extra combinational output and two-pass logic. However, sometimes our simple rule of thumb about XOR gates can help:

- Any two signals (inputs or output) of an XOR or XNOR gate may be complemented without changing the resulting logic function.

That is, you can complement both inputs or the output and one input without changing the result. For example, suppose an equation has the form

```
OUT := S1 :+: ((P1) + (P2));
```

where S1 is a sum term and P1 and P2 are product terms. If S1 has more than two variables, the equation can't be realized in one level. However, if you can live with a complemented output signal, you can write

```
/OUT := /S1 :+: ((P1) + (P2));
```

Under DeMorgan's theorem, /S1 is a single product term, so there's no problem.

Another trick is to remember that the XOR gate is equivalent to an OR gate *if* the logic functions on the two inputs are never both 1 simultaneously. So, if you need to perform a logic function that is a sum of four product terms, having nothing to do with XOR, you can still use the X-series output structure, if you can just split the terms into two nonoverlapping pairs.

* 10.5.4 Cascading and Carry Outputs

PLDs can be cascaded to obtain wider counters, by providing each counter stage with a carry output that indicates when it is about to roll over. There are two ways to obtain the carry output:

- *Combinational.* The carry equation indicates that the counter is enabled and is currently in its last state before rollover. For a 5-bit binary up counter, we have

  combinational carry output

  ```
  COUT = CNTEN * Q4 * Q3 * Q2 * Q1 * Q0;
  ```

  Since CNTEN is included, this approach allows carries to be rippled through cascaded counters by connecting each COUT to the next CNTEN.

- *Registered.* The carry equation indicates that the counter is about to enter its last state before rollover. Thus, at the next clock tick, the counter enters this last state and the carry output is asserted. For a 5-bit binary up counter with load and clear inputs, we have

  registered carry output

  ```
  COUT := /CLR * /LD * CNTEN
            * Q4 * Q3 * Q2 * Q1 * /Q0
      + /CLR * /LD * /CNTEN
            * Q4 * Q3 * Q2 * Q1 * Q0
      + /CLR * LD
            * D4 * D3 * D2 * D1 * D0;
  ```

  This approach has the advantage of producing COUT with less delay than the combinational approach. However, external gates are now required between stages, since the CNTEN signal for each stage should be the logical AND of the master count-enable signal and the COUT outputs of all lower-order counters. (These external gates can be avoided if the higher-order counters have multiple enable inputs.)

10.6 SEQUENTIAL-PLD APPLICATIONS

In this section we'll describe a few non-state-machine applications of sequential PLDs. Later sections show how to design state machines with PLDs.

10.6.1 Combining MSI Functions

PLDs can often be used to combine the capabilities of two or more MSI functions into a single IC. One simple example is the combination of a multiplexer and a register. Figure 10–15 shows how a 16V8 could be used as a 4-bit-wide, 2-input multiplexer with a storage register. At each clock tick, the A or B bus is loaded into the register, as selected by SELB. Table 10–11 is an ABEL program that provides the required behavior. Note that the CLK and /OE pin definitions are not used by the compiler to generate the fuse map; they merely serve to make the documentation complete.

In this example, one 16V8 replaces two MSI chips (e.g., a 74x157 and a 74x374), not much of a savings. However, once we have a PLD-based design, we can easily modify it or add to its functionality. For example, the 16V8 in Figure 10–15 has three spare I/O pins. As shown in color, we could use two inputs pins to provide "hold" and "preset" functions, and one output pin to provide a "zero-detect" function. The modified ABEL program requires only the appropriate pin definitions and the following equations:

```
YBUS := /PRESET * /HOLD * (/SELB * ABUS + SELB * BBUS)
      + /PRESET *  HOLD * YBUS
      +  PRESET * [1,1,1,1];

ZERO = (YBUS == 0);
```

Figure 10–15
PLD connections for a 4-bit, 2-input multiplexer with storage; connections for increased functionality are shown in color.

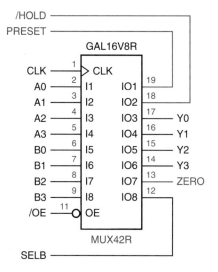

Table 10–11   ABEL program for a 4-bit, 2-input multiplexer with storage.

```
module Mux_with_Storage
title '4-Bit, 2-Input Multiplexer with Storage
J. Wakerly, Marquette University'
MUX42R device 'P16V8R';
@ALTERNATE

" Input pins
CLK, /OE                          pin 1, 11;
A0, A1, A2, A3, B0, B1, B2, B3    pin 2, 3, 4, 5, 6, 7, 8, 9;
SELB                              pin 12;

" Output pins
Y0, Y1, Y2, Y3                    pin 17, 16, 15, 14;

" Set definitions
ABUS = [A0,A1,A2,A3];
BBUS = [B0,B1,B2,B3];
YBUS = [Y0,Y1,Y2,Y3];

equations

YBUS := /SELB * ABUS + SELB * BBUS;

end Mux_with_Storage
```

10.6.2   Shift-Register Counters

In Sections 9.4.5 through 9.4.8, we gave many examples of counter circuits based on shift registers. In a PLD design, shift register operation may be combined with the combinational logic required to achieve the desired counting behavior. For example, Table 10–12 is the program for an 8-bit ring counter. We've used the extra capability of the PLD to add two functions not present in our MSI designs: counting occurs only if CNTEN is asserted, and the next state is forced to S0 if RESTART is asserted.

Ring counters are often used to generate multiphase clocks or enable signals in digital systems, and the requirements in different systems are many and varied. The ability to reprogram the counter's behavior easily is a distinct advantage of a PLD-based design.

Figure 10–16 shows a set of enable signals that might be required in a digital system with six distinct phases of operation. Each phase lasts for two

Table 10–12   Program for 8-bit ring counter.

```
module Eight_Bit_Ring_Counter
title '8-bit Ring Counter
J. Wakerly, Pizza Roma'
RING8 device 'P16V8R';
@ALTERNATE


" Input pins
MCLK, /OE                           pin 1, 11;
CNTEN, RESTART                      pin 2, 3;

" Output pins
S0, S1, S2, S3, S4, S5, S6, S7     pin 19, 18, 17, 16, 15, 14, 13, 12;

equations

S0 := CNTEN * /S0 * /S1 * /S2 * /S3 * /S4 * /S5 * /S6 + /CNTEN * S0 + RESTART;
S1 := CNTEN * /RESTART * S0 + /CNTEN * /RESTART * S1;
S2 := CNTEN * /RESTART * S1 + /CNTEN * /RESTART * S2;
S3 := CNTEN * /RESTART * S2 + /CNTEN * /RESTART * S3;
S4 := CNTEN * /RESTART * S3 + /CNTEN * /RESTART * S4;
S5 := CNTEN * /RESTART * S4 + /CNTEN * /RESTART * S5;
S6 := CNTEN * /RESTART * S5 + /CNTEN * /RESTART * S6;
S7 := CNTEN * /RESTART * S6 + /CNTEN * /RESTART * S7;

end Eight_Bit_Ring_Counter
```

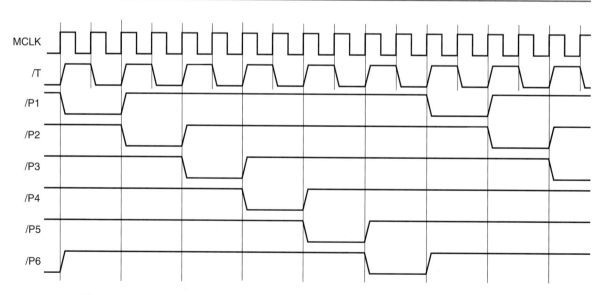

Figure 10–16   Six-phase timing waveforms required in a certain digital system.

Figure 10–17
Using a GAL16V8 as a six-phase
timing generator.

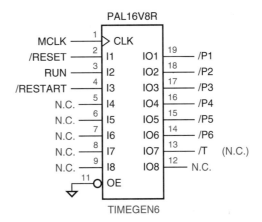

ticks of a master clock signal, MCLK, during which the corresponding active-low
phase-enable signal /Pi is asserted. We can obtain this sort of timing from a ring
counter if we provide an extra flip-flop to count the two ticks of each phase, so
that a shift occurs on the *second* tick of each phase.

The timing generator can be built with a single 16V8 with the connections
shown in Figure 10–17. The three control inputs, whose active levels can be
chosen arbitrarily, are required to have the following behavior:

/RESET   When this input is asserted, no outputs are asserted. The counter
always starts in phase 1 after /RESET is negated.

RUN   This input must be asserted to allow the counter to advance to the
next phase after the second tick of a phase; otherwise, the current
phase is extended.

/RESTART   Asserting this input causes the counter to go back to the first tick of
phase 1, even if RUN is not asserted.

Table 10–13 is a program that creates the required behavior. The equations are
somewhat complicated by the requirements for the /RESET, RUN, and /RESTART
inputs. Some of the requirements are ambiguous; for example, if RUN is negated
during the first tick of a phase, but asserted during the second, is the phase
extended by one tick? You have to decipher the equations to figure out what the
PLD does in this case.

Still, you're better off with PLDs than with "random logic"—you can easily
change the PLD equations if you don't like their behavior (see Exercises 10.18
and 10.19). Going a step further, you can use a state-machine description language (Section 10.8) for an even more understandable design.

Table 10–13 Program for a six-phase timing generator.

```
module Timing_Generator_6
title 'Six-phase Master Timing Generator
J. Wakerly, Unemployed'
TIMEGEN6 device 'P16V8R';
@ALTERNATE

" Input pins
MCLK, /OE                       pin 1, 11;
/RESET, RUN, /RESTART           pin 2, 3, 4;

" Output pins
/P1, /P2, /P3, /P4, /P5, /P6    pin 19, 18, 17, 16, 15, 14;
/T                              pin 13; " not used externally

equations

T  := /RESET * /RESTART * RUN * /T      " Set only for second tick of a phase
   + /RESET * /RESTART * /RUN * T;      " Extend second tick when halted

P1 := /RESET * /P1 * /P2 * /P3 * /P4 * /P5 * /P6     " Start-up after reset
   + /RESET * P1 * /T               " Hold after P1's first tick
   + /RESET * P1 * T * /RUN         " Hold after second tick if not running
   + /RESET * P6 * T * RUN          " Wrap-around after phase 6
   + /RESET * RESTART;             " Assert P1 on restart

P2 := /RESET * /RESTART * P1 * T * RUN  " Assert P2 after P1's second tick
   + /RESET * /RESTART * P2 * /T         " Hold after P2's first tick
   + /RESET * /RESTART * P2 * T * /RUN; " Hold after second tick if not running

P3 := /RESET * /RESTART * P2 * T * RUN  " Assert P3 after P2's second tick
   + /RESET * /RESTART * P3 * /T         " Hold after P3's first tick
   + /RESET * /RESTART * P3 * T * /RUN; " Hold after second tick if not running

P4 := /RESET * /RESTART * P3 * T * RUN  " Assert P4 after P3's second tick
   + /RESET * /RESTART * P4 * /T         " Hold after P4's first tick
   + /RESET * /RESTART * P4 * T * /RUN; " Hold after second tick if not running

P5 := /RESET * /RESTART * P4 * T * RUN  " Assert P5 after P4's second tick
   + /RESET * /RESTART * P5 * /T         " Hold after P5's first tick
   + /RESET * /RESTART * P5 * T * /RUN; " Hold after second tick if not running

P6 := /RESET * /RESTART * P5 * T * RUN  " Assert P6 after P5's second tick
   + /RESET * /RESTART * P6 * /T         " Hold after P6's first tick
   + /RESET * /RESTART * P6 * T * /RUN; " Hold after second tick if not running

end Timing_Generator_6
```

* 10.6.3   X-Series Design Example

Our next PLD design example uses a variation of the waveforms of Figure 10–16 in which each phase signal is asserted for only one clock tick instead of two, as shown in Figure 10–18. Although this is just a small modification to the waveforms, it requires a much different PLD design. In the original design, we effectively used an output-coded state assignment, so that the next state could always be determined from the current outputs and an auxiliary variable T. This is not possible with the new waveforms, since there are six states (corresponding to counts 0, 2, ..., 10 in the figure) that all produce the same output.

A simple approach to generating the new waveforms recognizes that they repeat every 12 clock ticks, and uses a modulo-12 counter and additional logic to assert the proper /Pi output as a function of the counter state indicated in the last line of the figure. Thus, four registered outputs are required for a modulo-12 counter, and six more outputs are needed for /P1–/P6.

The modulo-12 counter can be designed according to the rules of the preceding subsection using an X-series PLD. The PAL20X4 is an obvious choice, since it has four registered outputs for the counter and also has six combinational outputs for /P1–/P6; Figure 10–19 shows the required connections. The ABEL program in Table 10–14 produces the required behavior, and includes /RESET, RUN, and /RESTART inputs similar to the ones in our original six-phase timing generator of Table 10–13.

Since /P1–/P6 are decoded from counter states, they may contain glitches at certain counter transitions, as discussed in Section 9.3.4. Alternatively, we can

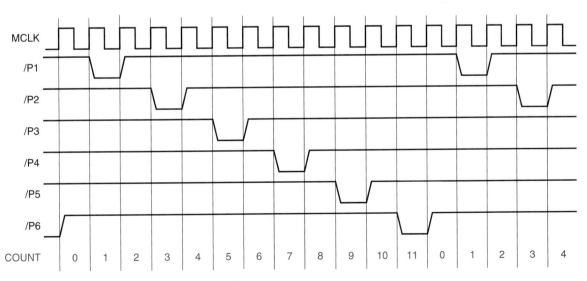

Figure 10–18   Modified timing waveforms for a digital system.

Table 10–14 Program for modified six-phase timing generator.

```
module Pulse_Generator_6
title 'Six-phase Master Timing Generator with 1-Tick Pulses'
PULSGEN6 device 'P20X4';
@ALTERNATE

" Input pins
MCLK, /OE                           pin 1, 13;
/RESET, RUN, /RESTART               pin 2, 3, 4;

" Output pins
/P1, /P2, /P3, /P4, /P5, /P6        pin 23, 22, 21, 16, 15, 14;
/Q3, /Q2, /Q1, /Q0                  pin 20, 19, 18, 17;

" Sets
QSTATE = [Q3,Q2,Q1,Q0];

equations

Q0 := (/RESET * /RESTART * Q0) :+:             " Hold
      (/RESET * /RESTART * RUN);               " Toggle

Q1 := (/RESET * /RESTART * Q1) :+:             " Hold
      (/RESET * /RESTART * RUN * Q0);          " Toggle

Q2 := (/RESET * /RESTART * Q2 +                " Hold
       /RESET * /RESTART * RUN * (QSTATE==11)) :+:   " Cancel
      (/RESET * /RESTART * RUN * Q1 * Q0);     " Toggle

Q3 := (/RESET * /RESTART * Q3) :+:             " Hold
      (/RESET * /RESTART * RUN * Q2 * Q1 * Q0 +      " Toggle
       /RESET * /RESTART * RUN * (QSTATE==11));      " Extra

P1 = /RESET * (QSTATE == 1);

P2 = /RESET * (QSTATE == 3);

P3 = /RESET * (QSTATE == 5);

P4 = /RESET * (QSTATE == 7);

P5 = /RESET * (QSTATE == 9);

P6 = /RESET * (QSTATE == 11);

end Pulse_Generator_6
```

Figure 10–19
PLD connections for modified
six-phase timing generator.

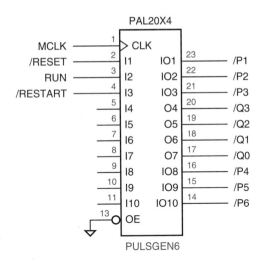

use the *registered* outputs of a PAL20X10 to produce /P1–/P6 without glitches; in this case, we must decode the *predecessor* of the state in which each signal is asserted. In the program of Table 10–14, we need only to change the device declaration and the /P1–/P6 equations:

```
PULSGENR device 'P20X10';
...
P1 := /RESET * (QSTATE == 0);
P2 := /RESET * (QSTATE == 2);
P3 := /RESET * (QSTATE == 4);
P4 := /RESET * (QSTATE == 6);
P5 := /RESET * (QSTATE == 8);
P6 := /RESET * (QSTATE == 10);
```

10.7 STATE-MACHINE DESIGN WITH SEQUENTIAL PLDS

As we showed in the preceding sections, many different sequential circuits can be built with sequential PLDs. However, probably the most important PLD applications are state machines, for two reasons:

(1) An SSI/MSI state-machine design often requires a large number of chips for the next-state and output equations. A PLD saves a lot of PCB area in a board-level design.

(2) An experienced designer can usually get the data paths right the first time in a system design. However, state machines that control the data paths almost always have bugs. Unlike an SSI/MSI design, a PLD-based design can often be salvaged just by changing the PLD equations, without modifying the PCB at all.

In this section we'll show examples of PLD-based state-machine design corresponding to the ad hoc and transition list approaches that we introduced in Sections 8.2 and 8.3. Then, in the next section, we'll describe a preferred method of state-machine design that uses "state-machine description language."

10.7.1 Timing and Packaging of PLD-Based State Machines

Before looking at PLD-based design approaches, we need to examine the timing behavior and packaging considerations for state machines that are built from PLDs. Figure 10–20 shows how a generic PLD with both combinational and registered outputs might be used in a state-machine application. The timing parameters t_{CO} and t_{PD} were explained in Section 10.3; t_{CO} is the flip-flop clock-to-output delay and t_{PD} is the delay through the AND-OR array.

Mealy-type outputs   State variables are assigned to registered outputs, of course, and are stable at time t_{CO} after the rising edge of CLOCK. Mealy-type outputs are assigned to combinational outputs, and are stable at time t_{PD} after any input change that affects them. Mealy-type outputs may also change after a state change, in which case they become stable at time $t_{CO} + t_{PD}$ after the rising edge of CLOCK.

Moore-type outputs   Moore-type outputs may be obtained in either of two ways in a PLD-based state machine. Since a Moore-type output is, by definition, a combinational logic function of the current state, the obvious way is to use a combinational output that is programmed with the output function, for example,

```
Z = /Q1 * Q2 + /Q2 * Q3;
```

Such an output is stable at time $t_{CO} + t_{PD}$ after the rising edge of CLOCK.

A Moore-type output can also be produced on a registered output, since it is supposed to be stable for the entire clock period. However, the registered output must be programmed differently. For example, if the required output function is $Z = Q1' \cdot Q2 + Q2' \cdot Q3$, we can't just substitute the clocked assignment operator in the equation and write

```
Z := /Q1 * Q2 + /Q2 * Q3;
```

lookahead output   Since the output is registered, output values would appear one clock tick later
equation   than in the previous case, where we used a combinational output. Instead, we must write a "lookahead" output equation that predicts what the output should be in the next state, based on the current state and inputs. If the next-state equations in the current example are

```
Q1 := expr1;
Q2 := expr2;
Q3 := expr3;
```

then the required output equation is

```
Z := /(expr1) * (expr2) + /(expr2) * (expr3);
```

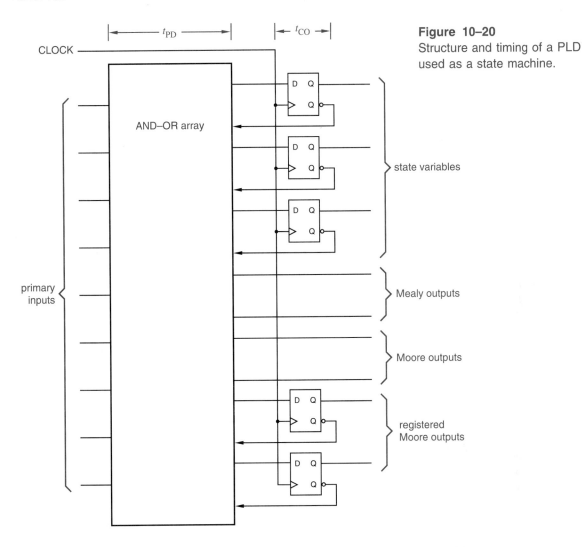

Figure 10–20
Structure and timing of a PLD used as a state machine.

An obvious disadvantage of this approach is that the output equation may "blow up," requiring more product terms than are available. However, it has two advantages: First, the registered outputs are stable sooner, only t_{CO} after the rising edge of the clock; second, they are glitch free. Combinational outputs have an additional delay of t_{PD}, and they may have glitches because of combinational-logic static hazards during single-variable state changes and function hazards during multiple-variable state changes.

In a PLD-based design, it is also possible to use a registered output to generate a signal that is neither a state variable nor a Moore-type output. A good example of such an output is any conventional Mealy or Moore output signal that is fed back into a register to delay it by one or more clock ticks.

PLD-based state-machine designs are often limited by the number of input and output pins available in a single PLD. According to the model of Figure 10–20, one PLD output is required for each state variable and for each Mealy- or Moore-type output. For example, the T-bird tail-lights machine of Table 8–2 on page 556 requires three registered outputs for the state variables and six combinational outputs for the lamp outputs, too many for most of the PLDs that we've described, except for the (expensive) 22V10.

On the other hand, an output-coded state assignment usually requires a smaller total number of PLD outputs. Using an output-coded state assignment, the T-bird tail-lights machine can be built in a single 16V8, as we'll show in Section 10.8.2.

Like any state variable or Moore-type output, an output-coded state variable is stable at time t_{CO} after the clock edge. Thus, an output-coded state assignment improves speed as well as packaging efficiency. In T-bird tail lights, turning on the emergency flashers 10 ns sooner doesn't matter, but in a high-performance digital system, an extra 10 ns of delay could make the difference between a maximum system clock rate of 100 MHz and a maximum of only 50 MHz.

If a design *must* be split into two or more PLDs, the best split in terms of both design simplicity and packaging efficiency is often the one shown in

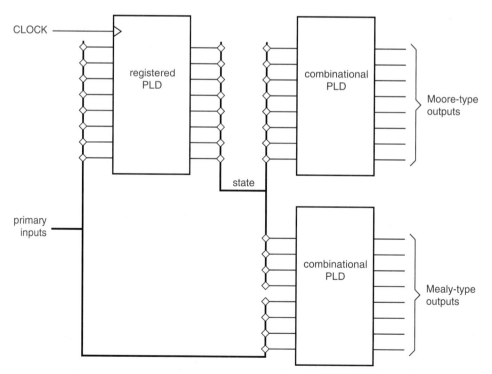

Figure 10–21 Splitting a state-machine design into three PLDs.

Figure 10–21. A single sequential PLD is used to create the required next-state behavior, and combinational PLDs are used to create both Mealy- and Moore-type outputs from the state variables and inputs.

Modern software tools for PLD-based system design can eliminate some of the trial and error that might otherwise be associated with fitting a design into a set of PLDs. To use this capability, the designer enters the equations and state-machine descriptions for the design, without necessarily specifying the actual devices and pinouts that should be used to realize the design. A software tool called a *partitioner* attempts to split the design into the smallest possible number of devices from a given family, while minimizing the number of pins used for inter-device communication. Partitioning can be fully automatic, partially user controlled, or fully user controlled.

Larger-scale PLDs (FPGAs) often have internal, architectural constraints that may create headaches in the absence of expert software assistance. Even if it appears to the designer, based on input, output, and total product-term requirements, that a design will fit in a single FPGA, the design must still be split among multiple macrocells inside that FPGA, where each macrocell has only limited functionality. For example, an output that requires many product terms may have to steal some from physically adjacent macrocells. This may in turn affect whether adjacent pins can be used as inputs or outputs, and how many product terms are available to them. It may also affect the ability to interconnect signals between nearby macrocells and the worst-case delay of these signals. All of these constraints can be tracked by *fitter* software that uses a heuristic approach to find the best split of functions among the macrocells within a single FPGA.

In many FPGA design environments, the partitioner and fitter software work together and interactively with the designer to find an acceptable realization using a set of PLDs and FPGAs.

RELIEF FOR
A SPLITTING
HEADACHE

10.7.2 Ad Hoc Design of PLD-Based State Machines

In Section 8.2, we gave state-machine design examples that followed no particular design methodology, but simply translated the word description of a state machine's desired behavior into a corresponding circuit. The same kind of commonsense approach can be applied to design PLD-based state machines. In fact, PLD-based design is even easier, for two reasons:

- You don't have to be too concerned about the complexity of the resulting excitation equations, as long as they fit in the PLD.

- You may be able to take advantage of PLD language features to make the design easier to express and understand.

Table 10–15   Ad hoc program for example state machine.

```
module Ad_Hoc_Example_1
title 'Ad Hoc State Machine Example'
ADHOC1 device 'P16V8R';
@ALTERNATE

" Input pins
CLOCK, /OE                         pin 1, 11;
RESET, A, B                        pin 2, 3, 4;

" Output pins
LASTA, LLASTA, BOK, Z              pin 12, 13, 14, 15;

equations

LASTA  := A;
LLASTA := LASTA;

BOK := /RESET * B * ((LASTA == LLASTA) + BOK);

Z = /RESET * ((LASTA == LLASTA) + BOK);

END Ad_Hoc_Example_1
```

For example, in Section 8.2 we showed the ad hoc design of a state machine with the following specifications:

> Design a clocked synchronous state machine with two inputs, A and B, and a single output Z that is 1 if:
>
> - A had the same value at each of the two previous clock ticks, *or*
> - B has been 1 since the last time that the first condition was true.
>
> Otherwise, the output should be 0.

This state machine can be realized in a PLD using the ABEL program in Table 10–15. Based on this program, the ABEL compiler generates the excitation and output equations shown in Table 10–16.

Another example from Section 8.2 is a combination-lock state machine (below we omit the HINT output in the original specification):

> Design a clocked synchronous state machine with one input, X, and one output, UNLK. The UNLK output should be 1 if and only if X is 0 and the sequence of inputs received on X at the preceding seven clock ticks was 0110111.

Table 10–16
Reduced equations for example
state machine.

```
LASTA := A;

LLASTA := LASTA;

BOK := /RESET * B * /LASTA * /LLASTA
     + /RESET * B * LASTA * LLASTA
     + /RESET * B * BOK;

Z = /RESET * /LASTA * /LLASTA
  + /RESET * LASTA * LLASTA
  + /RESET * BOK;
```

We used a "finite-memory" approach to design this machine in Figure 8–8 on page 549, and the same thing can be done in a PLD, as shown by the ABEL program in Table 10–17. The program is written using sets to make modifications easy, for example, to change the combination. However, note that the HINT output would be just as difficult to provide in this version of the machine as in the original (see Exercise 10.16).

Table 10–17   Finite-memory program for combination-lock state machine.

```
module Combination_Lock
title 'Combination-Lock State Machine'
COMBLOCK device 'P16V8R';
@ALTERNATE

" Input pins
CLOCK, /OE                      pin 1, 11;
RESET, X                        pin 2, 3;

" Output pins
X1, X2, X3, X4, X5, X6, X7      pin 12, 13, 14, 15, 16, 17, 18;
/UNLK                           pin 19;

" Sets
XHISTORY = [X7,X6,X5,X4,X3,X2,X1];
SHIFTX   = [X6,X5,X4,X3,X2,X1,X ];

equations

XHISTORY := /RESET * SHIFTX;

UNLK = /RESET * (X == 0) * (XHISTORY == [0,1,1,0,1,1,1]);

END Combination_Lock
```

Table 10–18 Ad hoc program for ones-counting state machine.

```
module Ones_Counter
title 'Ones-counting machine'
ONESCNT device 'P16V8R';

" Input pins
CLOCK, !OE                              pin 1, 11;
!RESET, X, Y                            pin 2, 3, 4;

" Output pins
Q1, Q0, Z                               pin 12, 13, 14;

" Definitions
CNT  = [Q1,Q0];                         " Counting variables
CNTX = [ 0, X];
CNTY = [ 0, Y];

equations

CNT :=  RESET & 0
     # !RESET & (CNT + CNTX + CNTY);

Z = (CNT == 0);

END Ones_Counter
```

A final ad hoc example is based on the ones-counting state machine of Section 7.4.6:

> Design a clocked synchronous state machine with two inputs, X and Y, and one output, Z. The output should be 1 if the number of 1 inputs on X and Y since reset is a multiple of 4, and 0 otherwise.

The PLD version of this machine takes advantage of ABEL's addition operator; the program and resulting equations are shown in Tables 10–18 and 10–19. Note that the @ALTERNATE directive is not used. Also note that ABEL "throws away" the carry bit in addition, which is equivalent to performing addition modulo-4 in the example.

10.7.3 Synthesizing PLD-Based State Machines Using Transition Lists

In Section 8.3, we showed how to design state diagrams and convert them into transition lists from which a circuit could be synthesized. Recall that a transition list is a list of current states and transition expressions defining the next state of a machine for every current state. When we drew state diagrams in Section 8.3, we allowed transition expressions to be written in any form. However, when we

Table 10-19
Reduced equations for ones-
counting state machine.

```
Q1 := (/RESET * /Y *  Q1 * /Q0
     + /RESET * /X *  Q1 * /Q0
     + /RESET *  Y * /Q1 *  Q0
     + /RESET *  X * /Q1 *  Q0
     + /RESET *  X *  Y  * /Q1
     + /RESET * /X * /Y  *  Q1);

Q0 := (/RESET *  X * /Y * /Q0
     + /RESET * /X *  Y * /Q0
     + /RESET * /X * /Y *  Q0
     + /RESET *  X *  Y *  Q0);

Z = (/Q1 * /Q0);
```

design the state diagram for a PLD-based state machine, it is convenient to write
transition expressions in sum-of-products form. This makes it possible to write
a transition list whose structure corresponds more closely to that of a PLD's
AND-OR array.

For example, consider the T-bird tail-lights machine, whose transition list
was given in Table 8–2 on page 556. If each transition expression is written as
a sum of products, we can rewrite the transition list as shown in Table 10–20.
Here, the transition-expression column has been replaced with three columns,
corresponding to the machine's three input variables. Each row of the table now

Table 10-20 Sum-of-products transition list for T-bird tail-lights machine.

S	Q2	Q1	Q0	LEFT	RIGHT	HAZ	S*	Q2*	Q1*	Q0*
IDLE	0	0	0	0	0	0	IDLE	0	0	0
IDLE	0	0	0	1	0	0	L1	0	0	1
IDLE	0	0	0	X	X	1	LR3	1	0	0
IDLE	0	0	0	1	1	X	LR3	1	0	0
IDLE	0	0	0	0	1	0	R1	1	0	1
L1	0	0	1	X	X	0	L2	0	1	1
L1	0	0	1	X	X	1	LR3	1	0	0
L2	0	1	1	X	X	0	L3	0	1	0
L2	0	1	1	X	X	1	LR3	1	0	0
L3	0	1	0	X	X	X	IDLE	0	0	0
R1	1	0	1	X	X	0	R2	1	1	1
R1	1	0	1	X	X	1	LR3	1	0	0
R2	1	1	1	X	X	0	R3	1	1	0
R2	1	1	1	X	X	1	LR3	1	0	0
R3	1	1	0	X	X	X	IDLE	0	0	0
LR3	1	0	0	X	X	X	IDLE	0	0	0

Table 10–21   Program for T-bird tail-lights machine.

```
module T_Bird_Tail_Lights_Machine
title 'State Machine for T-Bird Tail Lights'
TBIRDSM device 'P16V8R';
@ALTERNATE

" Input pins
CLOCK, LEFT, RIGHT, HAZ, /OE        pin 1,  2, 3, 4, 11;

" Output pins
/Q0, /Q1, Q2                        pin 17, 16, 15;

equations

Q0   := /Q2 * /Q1 * /Q0 *  LEFT * /RIGHT * /HAZ
      + /Q2 * /Q1 * /Q0 * /LEFT *  RIGHT * /HAZ
      + /Q2 * /Q1 *  Q0                    * /HAZ
      +  Q2 * /Q1 *  Q0                    * /HAZ;

Q1   := /Q2 * /Q1 *  Q0                    * /HAZ
      + /Q2 *  Q1 *  Q0                    * /HAZ
      +  Q2 * /Q1 *  Q0                    * /HAZ
      +  Q2 *  Q1 *  Q0                    * /HAZ;

/Q2  := /Q2 * /Q1 * /Q0 * /LEFT * /RIGHT * /HAZ
      + /Q2 * /Q1 * /Q0 *  LEFT * /RIGHT * /HAZ
      + /Q2 * /Q1 *  Q0                    * /HAZ
      + /Q2 *  Q1 *  Q0                    * /HAZ
      + /Q2 *  Q1 * /Q0
      +  Q2 *  Q1 * /Q0
      +  Q2 * /Q1 * /Q0;

end T_Bird_Tail_Lights_Machine
```

represents a single product term, with 1s, 0s, and x's used to indicate how each variable appears in the product term. Of all the transition expressions in the original table, only one expanded into two product terms, the third and fourth rows of the new table.

We can write ABEL equations for a PLD realization of the state machine directly from the transition list. Recall that the transition/excitation equation for each next-state variable may be written as a sum for transition-list rows for which the variable is 1, or 0 at our option. Each row uses one product term in the PLD.

Table 10–21 shows the ABEL equations corresponding to the transition list of Table 10–20. The Q0* and Q1* equations can be written as a sum of four rows each. Since the Q2* column has more 1s than there are product terms in a standard sequential PLD, we sum 0s instead.

The ABEL compiler can minimize all the equations in Table 10–21; Q0 reduces to three product terms, Q1 to one, and /Q2 to four. For that matter, it wasn't really necessary to write the transition expressions in sum-of-products form in the first place, since the compiler can transform and minimize arbitrary expressions. The sum-of-products method is needed only if we don't have a minimizing PLD compiler, or if for some other reason we want to get a feel for how many product terms we're generating as we go along in a design.

> In Chapter 8 we showed how to generate transition lists from state diagrams and ASM charts. Both methods required quite a bit of "turning the crank"; the fact that a PLD compiler can minimize the resulting equations is little comfort.
>
> State-machine description language, described in the next section, eliminates the tedious part of the process, leaving the designer to do the most interesting parts of the task: designing the equivalent of state diagram or ASM chart, and assigning state-variable combinations to states.

WHERE DO
TRANSITION
LISTS COME
FROM?

10.8   STATE-MACHINE DESCRIPTION LANGUAGE FOR PLDS

The method of the previous subsection is just a mechanical adaptation of the transition-list synthesis procedure that we described in Section 8.5. In it, we simply use the PLD compiler's facilities for "turning the crank" to combine and reduce logic expressions. However, most PLD programming languages have a notation for defining (and documenting!) state machines directly, without ever drawing a state diagram or writing a transition list. Such a notation is called a *state-machine description language*.

state-machine description language

In ABEL, the keyword `state_diagram` indicates the beginning of a state-machine definition. The statements that define a state machine actually resemble the text equivalent of an ASM chart more than they resemble a state diagram. Table 10–22 shows the textual structure of an ABEL "state diagram." Here *state-variables* is an ABEL set that lists the state variables of the machine. If there are n variables in the set, then the machine has 2^n possible states corresponding to the 2^n different assignments of constant values to variables in the set. States are usually given symbolic names in an ABEL program; this makes it easy to try different assignments simply by changing the constant definitions.

`state_diagram`

state-variables

Table 10–22
Structure of a "state diagram" in ABEL.

```
state_diagram state-variables
state state-value 1 : equation ;
state state-value 2 : equation ;
. . .
state state-value n : equation ;
```

An equation for each state variable is developed according to the information in the "state diagram." The keyword `state` indicates that the next states and current outputs for a particular current state are about to be defined; a *state-value* is a constant that defines state-variable values for the current state. An *equation* uses `IF-THEN-ELSE` and `GOTO` statements to define possible next states as a function of the state-machine inputs.

10.8.1 Simple State-Diagram Examples

Our first example using ABEL's "state diagram" capability is based on our first state-machine design example from Section 7.4:

Design a clocked synchronous state machine with two inputs, A and B, and a single output Z that is 1 if:

- A had the same value at each of the two previous clock ticks, *or*
- B has been 1 since the last time that the first condition was true.

Otherwise, the output should be 0.

A state table for this machine was developed in Figure 7–45 on page 485. This state diagram is adapted to ABEL in Table 10–23. Several characteristics of this program should be noted:

Table 10–23
An example of ABEL's state-diagram capability.

```
module State_Machine_Example_1
title 'PLD Version of Example State Machine'
SMEX1 device 'P16V8R';
@ALTERNATE

" Input pins
CLOCK, RESET, A, B, /OE          pin 1, 2, 3, 4, 11;

" Output pins
Q2, Q1, Q0, Z                    pin 12, 13, 14, 15;

" Definitions
QSTATE = [Q2,Q1,Q0];             " State variables
INIT   = [ 0, 0, 0];
A0     = [ 0, 0, 1];
A1     = [ 0, 1, 0];
OK0    = [ 0, 1, 1];
OK1    = [ 1, 0, 0];
XTRA1  = [ 1, 0, 1];
XTRA2  = [ 1, 1, 0];
XTRA3  = [ 1, 1, 1];
```

- The definition of QSTATE uses three variables to encode state.

- The definitions of INIT–XTRA3 determine individual state encodings.

- IF–THEN–ELSE statements are nested. A particular next state may appear in multiple places in one set of nested IF–THEN–ELSE clauses (e.g., see states OK0 and OK1).

- Expressions like "(B==1)*(A==0)" were used instead of equivalents like "B*/A" simply because the program's author felt the former were more readable.

Table 10–23
(continued)

```
state_diagram QSTATE

state INIT:  IF RESET THEN INIT
             ELSE IF (A==0) THEN A0
             ELSE A1;

state A0:    IF RESET THEN INIT
             ELSE IF (A==0) THEN OK0
             ELSE A1;

state A1:    IF RESET THEN INIT
             ELSE IF (A==0) THEN A0
             ELSE OK1;

state OK0:   IF RESET THEN INIT
             ELSE IF (B==1)*(A==0) THEN OK0
             ELSE IF (B==1)*(A==1) THEN OK1
             ELSE IF (A==0) THEN OK0
             ELSE IF (A==1) THEN A1;

state OK1:   IF RESET THEN INIT
             ELSE IF (B==1)*(A==0) THEN OK0
             ELSE IF (B==1)*(A==1) THEN OK1
             ELSE IF (A==0) THEN A0
             ELSE IF (A==1) THEN OK1;

state XTRA1: GOTO INIT;
state XTRA2: GOTO INIT;
state XTRA3: GOTO INIT;

equations

Z = (QSTATE == OK0) + (QSTATE == OK1);

END State_Machine_Example_1
```

NESTING
LIMITATION

> In ABEL, an IF-THEN-ELSE statement should be nested only in the ELSE clause of another IF-THEN-ELSE statement. Unlike Pascal or C, ABEL does not allow BEGIN-END or braces to be used to resolve the ambiguity that results from attempting to nest an IF-THEN-ELSE statement in the IF clause of another.

- The first IF statement in each of states INIT–OK1 ensures that the machine goes to the INIT state if RESET is asserted.

- Just to be safe, next-state equations are given for XTRA1–XTRA3 to ensure that the machine goes to a "safe" state if it somehow gets into an unused state.

- The single equation in the "equations" section of the program determines the behavior of the Moore-type output.

Table 10–24 shows the resulting excitation and output equations produced by ABEL compiler.

Despite the use of "high-level language," the program's author still had to refer to the original, hand-constructed state table in Figure 7–45 to come up with ABEL version in Table 10–23. A different approach is shown in Table 10–25. This program was developed directly from the word description of the state machine, without any intermediate steps.

A key idea the new approach is to remove the last value of A from the state definitions, and to instead have a separate flip-flop that keeps track of it (LASTA). Then only two non-INIT states must be defined: LOOKING (still looking for a match) and OK (got a match or B has been 1 since last match). The Z output is a simple combinational decode of the OK state.

We can go one step further with the example state machine. Notice that Q1 is 1 only in the OK state, so it is equivalent to the Z output. As a result, we can

Table 10–24
Reduced equations for
SMEX1 PLD.

```
Q2 := /RESET * A *   Q2 * /Q1 * /Q0
    + /RESET * A *   B * /Q2 *   Q1
    + /RESET * A * /Q2 *   Q1 * /Q0;

Q1 := /RESET * /A *   B *   Q2 * /Q1 * /Q0
    + /RESET * /B * /Q2 *   Q0
    + /RESET *  A * /Q2 * /Q1
    + /RESET * /A * /Q2 *   Q0;

Q0 := /RESET * /A * /Q1 * /Q0 + /RESET * /A * /Q2;

Z = Q2 * /Q1 * /Q0 + /Q2 * Q1 * Q0;
```

```
module State_Machine_Example_2
title 'Alternate Version of Example State Machine'
SMEX2 device 'P16V8R';
@ALTERNATE

" Input pins
CLOCK, RESET, A, B, /OE          pin 1, 2, 3, 4, 11;

" Output pins
LASTA, Q1, Q0, Z                 pin 12, 13, 14, 15;

" Definitions
QSTATE  = [Q1,Q0];               " State variables
INIT    = [ 0, 0];               " State encodings
LOOKING = [ 0, 1];
OK      = [ 1, 0];
XTRA    = [ 1, 1];

state_diagram QSTATE

state INIT:    IF RESET THEN INIT ELSE LOOKING;

state LOOKING: IF RESET THEN INIT
               ELSE IF (A == LASTA) THEN OK
               ELSE LOOKING;

state OK:      IF RESET THEN INIT
               ELSE IF B THEN OK
               ELSE IF (A == LASTA) THEN OK
               ELSE LOOKING;

state XTRA:    GOTO INIT;

equations

LASTA := A;

Z = (QSTATE == OK);

END State_Machine_Example_2
```

Table 10–25
A more "natural" ABEL program for the example state machine.

eliminate the Q1 state variable and use a registered version of the Z output as a state variable, as shown in Table 10–26. In addition to saving a PLD output, an advantage of this version of the state machine is that Z is valid sooner after the clock edge.

Table 10–26 Example state machine with partially output-coded state assignment.

```
module State_Machine_Example_3
title 'Really Squeezed Version of Example State Machine'
SMEX3 device 'P16V8R';
@ALTERNATE

" Input pins
CLOCK, RESET, A, B, /OE              pin 1, 2, 3, 4, 11;

" Output pins
LASTA, Q0, Z                        pin 12, 13, 14;

" Definitions
QSTATE  = [Z,Q0];                   " State variables
INIT    = [0, 0];                   " State encodings
LOOKING = [0, 1];
OK      = [1, 0];
XTRA    = [1, 1];

state_diagram QSTATE

state INIT:    IF RESET THEN INIT ELSE LOOKING;

state LOOKING: IF RESET THEN INIT
               ELSE IF (A == LASTA) THEN OK
               ELSE LOOKING;

state OK:      IF RESET THEN INIT
               ELSE IF B THEN OK
               ELSE IF (A == LASTA) THEN OK
               ELSE LOOKING;

state XTRA:    GOTO INIT;

equations

LASTA := A;

END State_Machine_Example_3
```

10.8.2 T-Bird Tail Lights

We described and designed a "T-bird tail-lights" state machine in Sections 8.3 and 8.4. Table 10–27 is an ABEL "state diagram" for the T-bird tail-lights machine. There is a close correspondence between this program and the ASM chart in Figure 8–17 on page 560. The program produces exactly the same

Table 10–27 Symbolic program for T-bird tail-lights machine.

```
module T_Bird_Tail_Lights_Machine
title 'State Machine for T-Bird Tail Lights, Symbolic Version'
TBIRDSD device 'P16V8R';
@ALTERNATE

" Input pins
CLOCK, LEFT, RIGHT, HAZ, /OE       pin 1, 2, 3, 4, 11;

" Output pins
/Q0, /Q1, Q2                       pin 17, 16, 15;

" Definitions
QSTATE = [Q2,Q1,Q0];               " State variables
IDLE   = [ 0, 0, 0];               " States
L1     = [ 0, 0, 1];
L2     = [ 0, 1, 1];
L3     = [ 0, 1, 0];
R1     = [ 1, 0, 1];
R2     = [ 1, 1, 1];
R3     = [ 1, 1, 0];
LR3    = [ 1, 0, 0];

state_diagram QSTATE
state IDLE: IF ((HAZ + LEFT*RIGHT) == 1) THEN LR3
            ELSE IF (LEFT == 1) THEN L1
            ELSE IF (RIGHT == 1) THEN R1
            ELSE IDLE;
state L1:   IF (HAZ == 1) THEN LR3 ELSE L2;
state L2:   IF (HAZ == 1) THEN LR3 ELSE L3;
state L3:   GOTO IDLE;
state R1:   IF (HAZ == 1) THEN LR3 ELSE R2;
state R2:   IF (HAZ == 1) THEN LR3 ELSE R3;
state R3:   GOTO IDLE;
state LR3:  GOTO IDLE;

end T_Bird_Tail_Lights_Machine
```

reduced equations as the explicit transition equations resulting from the ASM chart, which we showed in Table 10–21 on page 704.

The program in Table 10–27 handles only the state variables of the tail-lights machine. The output logic requires six combinational outputs, but only four more outputs are available in the 16V8 specified in the program. A larger PLD, such as the (expensive) 22V10, could provide enough outputs for a single-PLD design.

Figure 10–22

A single-PLD design for T-bird tail lights.

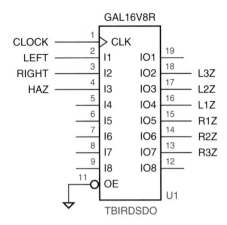

Since the output values of the tail-lights machine are different in each state, we can also use an output-coded state assignment. This requires only six registered outputs and no combinational outputs of a 16V8, as shown in Figure 10–22. Only the device, pin, and state definitions in the previous ABEL

Table 10–28   T-bird tail-lights machine with output-coded state assignment.

```
module T_Bird_Tail_Lights_Machine
title 'Output-Coded State Machine for T-Bird Tail Lights'
TBIRDSDO device 'P16V8R';
@ALTERNATE

" Input pins
CLOCK, LEFT, RIGHT, HAZ, /OE        pin 1, 2, 3, 4, 11;

" Output pins
L3Z, L2Z, L1Z, R1Z, R2Z, R3Z        pin 18, 17, 16, 15, 14, 13;

" Definitions
ZSTATE = [L3Z,L2Z,L1Z,R1Z,R2Z,R3Z];   " State variables
IDLE   = [  0,  0,  0,  0,  0,  0];   " States
L3     = [  1,  1,  1,  0,  0,  0];
L2     = [  0,  1,  1,  0,  0,  0];
L1     = [  0,  0,  1,  0,  0,  0];
R1     = [  0,  0,  0,  1,  0,  0];
R2     = [  0,  0,  0,  1,  1,  0];
R3     = [  0,  0,  0,  1,  1,  1];
LR3    = [  1,  1,  1,  1,  1,  1];

state_diagram ZSTATE
...
end T_Bird_Tail_Lights_Machine
```

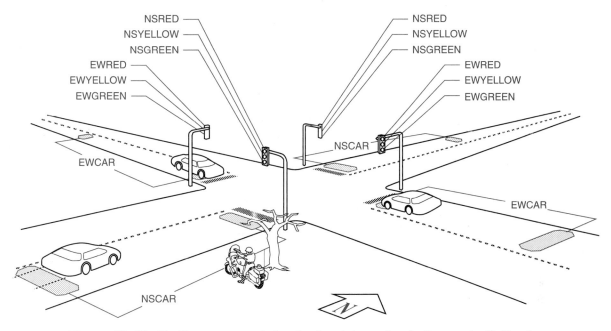

Figure 10–23   Traffic sensors and signals at an intersection in Sunnyvale, California.

program must be changed, as shown in Table 10–28. The six resulting excitation equations use a total of 36 product terms.

10.8.3   Reinventing Traffic-Light Controllers

Our final example is also from the world of cars and traffic. Traffic-light controllers in California, especially in the city of Sunnyvale, are carefully designed to *maximize* the waiting time of cars at intersections. An infrequently used intersection (one that would have no more than a "yield" sign if it were in Chicago) has the sensors and signals shown in Figure 10–23. The state machine that controls the traffic signals uses a 1 Hz clock and a timer and has four inputs:

NSCAR   Asserted when a car on the north-south road is over either sensor on either side of the intersection.

EWCAR   Asserted when a car on the east-west road is over either sensor on either side of the intersection.

TMLONG   Asserted if more than five minutes has elapsed since the timer started; remains asserted until the timer is reset.

TMSHORT   Asserted if more than five seconds has elapsed since the timer started; remains asserted until the timer is reset.

Table 10–29 Sunnyvale traffic-lights program.

```
module Sunnyvale_Traffic_Machine
title 'State Machine for Sunnyvale, CA, Traffic Lights'
SVALETL device 'P16V8R';
@ALTERNATE


" Input pins
CLOCK, /OE                      pin 1, 11;
NSCAR, EWCAR, TMSHORT, TMLONG   pin 2, 3, 8, 9;

" Output pins
/Q0, /Q1, /Q2, /TMRESET         pin 17, 16, 15, 14;

" Definitions
LSTATE  = [Q2,Q1,Q0];           " State variables
NSGO    = [ 0, 0, 0];           " States
NSWAIT  = [ 0, 0, 1];
NSWAIT2 = [ 0, 1, 1];
NSDELAY = [ 0, 1, 0];
EWGO    = [ 1, 1, 0];
EWWAIT  = [ 1, 1, 1];
EWWAIT2 = [ 1, 0, 1];
EWDELAY = [ 1, 0, 0];

state_diagram LSTATE
state NSGO:                         " North-south green
   IF (/TMSHORT) THEN NSGO          " Minimum green is 5 seconds.
   ELSE IF (TMLONG) THEN NSWAIT     " Maximum green is 5 minutes.
   ELSE IF (EWCAR * /NSCAR)         " If E-W car is waiting and no one
         THEN NSGO                  "    is coming N-S, make E-W wait!
   ELSE IF (EWCAR * NSCAR)          " Cars coming in both directions?
         THEN NSWAIT                " Thrash!
   ELSE IF (/NSCAR)                 " Nobody coming N-S and not timed out?
         THEN NSGO                  " Keep N-S green.
   ELSE NSWAIT;                     " Else let E-W have it.

state NSWAIT:  GOTO NSWAIT2;        " Yellow light is on for two ticks for safety.
state NSWAIT2: GOTO NSDELAY;        " (Drivers go 70 mph to catch this turkey green!)
state NSDELAY: GOTO EWGO;           " Red in both directions for added safety.

state EWGO:  " East-west green; states defined analogous to N-S
   IF (/TMSHORT) THEN EWGO
   ELSE IF (TMLONG) THEN EWWAIT
   ELSE IF (NSCAR * /EWCAR) THEN EWGO
   ELSE IF (NSCAR * EWCAR) THEN EWWAIT
   ELSE IF (/EWCAR) THEN EWGO
   ELSE EWWAIT;
```

Table 10–29 (continued)

```
state EWWAIT:  GOTO EWWAIT2;
state EWWAIT2: GOTO EWDELAY;
state EWDELAY: GOTO NSGO;

equations

TMRESET := (LSTATE == NSWAIT2)   " Reset the timer when going into
          + (LSTATE == EWWAIT2);  "   state NSDELAY or state EWDELAY.

end Sunnyvale_Traffic_Machine
```

The state machine has seven outputs:

NSRED, NSYELLOW, NSGREEN Control the north-south lights.

EWRED, EWYELLOW, EWGREEN Control the east-west lights.

TMRESET When asserted, resets the timer and holds TMSHORT and TMLONG negated. The timer starts timing when TMRESET is negated.

A typical, municipally approved algorithm for controlling the traffic lights is embedded in the ABEL program of Table 10–29. This algorithm produces two frequently seen behaviors of "smart" traffic lights. At night, when traffic is light, it holds a car stopped at the light for up to five minutes, unless a car approaches on the cross street, in which case it stops the cross traffic and lets the waiting car go. (The "early warning" sensor is far enough back to change the lights before the approaching car reaches the intersection.) During the day, when traffic is heavy and there are always cars waiting in both directions, it cycles the lights every five seconds, thus minimizing the utilization of the intersection, and maximizing everyone's waiting time and creating public outcry for more taxes to fix the problem.

The equations for the TMRESET output are worth noting. This output is asserted during the "double-red" states, NSDELAY and EWDELAY, to reset the timer in preparation for the next green cycle. The desired output signal could be generated on a combinational output pin by decoding these two states, but we have chosen instead to generate it on a registered output pin by decoding the *predecessors* of these two states.

The ABEL program in Table 10–29 defines only the state variables and one registered Moore output for the traffic controller. Six more Moore outputs are needed for the lights, more than remain on the 16V8. Therefore, a separate combinational PLD is used for these outputs, yielding the complete design shown in Figure 10–24. An ABEL program for the output PLD is given in Table 10–30. We've taken this opportunity to add an OVERRIDE input to the controller. This

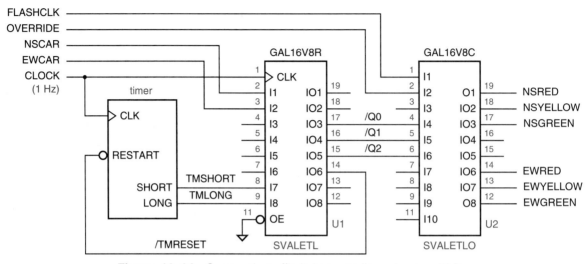

Figure 10–24   Sunnyvale traffic-light controller using two PLDs.

Table 10–30   Output logic for Sunnyvale traffic lights.

```
module Sunnyvale_Traffic_Light_Output
title 'Output logic for Sunnyvale, CA, Traffic Lights'
SVALETLO device 'P16V8C';
@ALTERNATE

" Input pins
FLASHCLK, OVERRIDE, /Q0, /Q1, /Q2   pin 1, 2, 4, 5, 6;

" Output pins
NSRED, NSYELLOW, NSGREEN              pin 19, 18, 17;
EWRED, EWYELLOW, EWGREEN              pin 14, 13, 12;

" Definitions (same as in state machine SVALETL)
...

equations

NSRED = /OVERRIDE * (LSTATE != NSGO) * (LSTATE != NSWAIT) * (LSTATE != NSWAIT2)
      + OVERRIDE * FLASHCLK;
NSYELLOW = /OVERRIDE * ((LSTATE == NSWAIT) + (LSTATE == NSWAIT2))
NSGREEN  = /OVERRIDE * (LSTATE == NSGO);

EWRED = /OVERRIDE * (LSTATE != EWGO) * (LSTATE != EWWAIT) * (LSTATE != EWWAIT2)
      + OVERRIDE * FLASHCLK;
EWYELLOW = /OVERRIDE * ((LSTATE == EWWAIT) + (LSTATE == EWWAIT2))
EWGREEN  = /OVERRIDE * (LSTATE == EWGO);

end Sunnyvale_Traffic_Light_Output
```

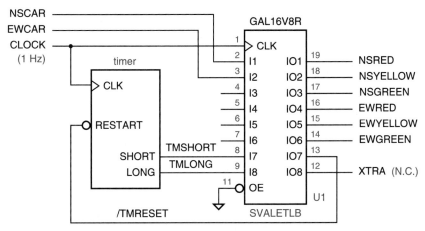

Figure 10–25 Traffic-light state machine using output-coded state assignment in a single PLD.

Table 10–31 Definitions for Sunnyvale traffic-lights machine with output-coded state assignment.

```
module Sunnyvale_Traffic_Machine
title 'State Machine for Sunnyvale, CA, Traffic Lights'
SVALETLB device 'P16V8R';
@ALTERNATE

" Input pins
CLOCK, /OE                     pin 1, 11;
NSCAR, EWCAR, TMSHORT, TMLONG  pin 2, 3, 8, 9;

" Output pins
NSRED, NSYELLOW, NSGREEN       pin 19, 18, 17;
EWRED, EWYELLOW, EWGREEN       pin 16, 15, 14;
/TMRESET, XTRA                 pin 13, 12;

" Definitions
LSTATE  = [NSRED,NSYELLOW,NSGREEN,EWRED,EWYELLOW,EWGREEN,XTRA]; " State vars
NSGO    = [    0,       0,       1,    1,        0,       0,   0]; " States
NSWAIT  = [    0,       1,       0,    1,        0,       0,   0];
NSWAIT2 = [    0,       1,       0,    1,        0,       0,   1];
NSDELAY = [    1,       0,       0,    1,        0,       0,   0];
EWGO    = [    1,       0,       0,    0,        0,       1,   0];
EWWAIT  = [    1,       0,       0,    0,        1,       0,   0];
EWWAIT2 = [    1,       0,       0,    0,        1,       0,   1];
EWDELAY = [    1,       0,       0,    1,        0,       0,   1];
```

Table 10-32 Reduced equations for the output-coded traffic-light machine of Table 10–31.

```
/NSRED := /EWGREEN * EWRED * /EWYELLOW * /NSGREEN * NSRED * /NSYELLOW * XTRA
    + /EWGREEN * EWRED * /EWYELLOW * /NSGREEN * /NSRED * NSYELLOW * /XTRA
    + /EWGREEN * EWRED * /EWYELLOW * NSGREEN * /NSRED * /NSYELLOW * /XTRA;
/NSYELLOW := /EWGREEN * /EWRED * EWYELLOW * /NSGREEN * NSRED * /NSYELLOW
    + EWGREEN * /EWRED * /EWYELLOW * /NSGREEN * NSRED * /NSYELLOW * /XTRA
    + /EWGREEN * EWRED * /EWYELLOW * /NSGREEN * NSRED * /NSYELLOW
    + /EWGREEN * EWRED * /EWYELLOW * /NSGREEN * /NSRED * NSYELLOW * XTRA
    + /EWGREEN * EWRED * /EWYELLOW * /NSCAR * NSGREEN * /NSRED * /NSYELLOW
            * /TMLONG * /XTRA
    + /EWGREEN * EWRED * /EWYELLOW * NSGREEN * /NSRED * /NSYELLOW * /TMSHORT
            * /XTRA;
/NSGREEN := /EWGREEN * /EWRED * EWYELLOW * /NSGREEN * NSRED * /NSYELLOW
    + EWGREEN * /EWRED * /EWYELLOW * /NSGREEN * NSRED * /NSYELLOW * /XTRA
    + /EWGREEN * EWRED * /EWYELLOW * /NSGREEN * NSRED * /NSYELLOW * /XTRA
    + /EWGREEN * EWRED * /EWYELLOW * /NSGREEN * /NSRED * NSYELLOW
    + /EWGREEN * EWRED * /EWYELLOW * NSCAR * NSGREEN * /NSRED * /NSYELLOW
            * TMSHORT * /XTRA
    + /EWGREEN * EWRED * /EWYELLOW * NSGREEN * /NSRED * /NSYELLOW * TMLONG
            * TMSHORT * /XTRA;
/EWRED := /EWGREEN * /EWRED * EWYELLOW * /NSGREEN * NSRED * /NSYELLOW * /XTRA
    + EWGREEN * /EWRED * /EWYELLOW * /NSGREEN * NSRED * /NSYELLOW * /XTRA
    + /EWGREEN * EWRED * /EWYELLOW * /NSGREEN * NSRED * /NSYELLOW * /XTRA;
/EWYELLOW := /EWGREEN * /EWRED * EWYELLOW * /NSGREEN * NSRED * /NSYELLOW * XTRA
    + /EWCAR * EWGREEN * /EWRED * /EWYELLOW * /NSGREEN * NSRED * /NSYELLOW
            * /TMLONG * /XTRA
    + EWGREEN * /EWRED * /EWYELLOW * /NSGREEN * NSRED * /NSYELLOW * /TMSHORT
            * /XTRA
    + /EWGREEN * EWRED * /EWYELLOW * /NSGREEN * NSRED * /NSYELLOW
    + /EWGREEN * EWRED * /EWYELLOW * /NSGREEN * /NSRED * NSYELLOW
    + /EWGREEN * EWRED * /EWYELLOW * NSGREEN * /NSRED * /NSYELLOW * /XTRA;
/EWGREEN := /EWGREEN * EWRED * /EWYELLOW * /NSGREEN * NSRED * /NSYELLOW * XTRA
    + /EWGREEN * /EWRED * EWYELLOW * /NSGREEN * NSRED * /NSYELLOW
    + EWCAR * EWGREEN * /EWRED * /EWYELLOW * /NSGREEN * NSRED * /NSYELLOW
            * TMSHORT * /XTRA
    + EWGREEN * /EWRED * /EWYELLOW * /NSGREEN * NSRED * /NSYELLOW * TMLONG
            * TMSHORT * /XTRA
    + /EWGREEN * EWRED * /EWYELLOW * /NSGREEN * /NSRED * NSYELLOW
    + /EWGREEN * EWRED * /EWYELLOW * NSGREEN * /NSRED * /NSYELLOW * /XTRA;
/XTRA := EWGREEN * /EWRED * /EWYELLOW * /NSGREEN * NSRED * /NSYELLOW * /XTRA
    + /EWGREEN * EWRED * /EWYELLOW * /NSGREEN * NSRED * /NSYELLOW
    + /EWGREEN * EWRED * /EWYELLOW * /NSGREEN * /NSRED * NSYELLOW * XTRA
    + /EWGREEN * EWRED * /EWYELLOW * NSGREEN * /NSRED * /NSYELLOW * /XTRA;
TMRESET := /EWGREEN * /EWRED * EWYELLOW * /NSGREEN * NSRED * /NSYELLOW * XTRA
    + /EWGREEN * EWRED * /EWYELLOW * /NSGREEN * /NSRED * NSYELLOW * XTRA;
```

input may be asserted by the police to disable the controller and put the signals into a flashing-red mode (at a rate determined by FLASHCLK), allowing them to manually clear up the traffic snarls created by this wonderful invention.

A traffic-light state machine including output logic can be built in a single 16V8, shown in Figure 10–25, if we choose an output-coded state assignment. Only the definitions in the original program of Table 10–29 must be changed, as shown in Table 10–31. This PLD does not include the OVERRIDE input and mode, which is left as an exercise (10.30).

Just to give you an appreciation for using a state-machine description language to define and design state machines, Table 10–32 shows the reduced equations that the ABEL compiler generates for the traffic-light machine with the output-coded state assignment of Table 10–31. Now aren't you glad you didn't have to do that by hand?

REFERENCES

Lattice Semiconductor publishes the specifications for their GAL devices in the *GAL Data Book* (Hillsboro, OR 97124, 1992). Advanced Micro Devices specifies their corresponding "PALCE" devices in the *PAL Device Data Book and Design Guide* (Sunnyvale, CA 94088 — ouch, there's that town again!). Both books contain application notes with important information on design topics such as metastability characteristics, ground bounce, latch-up, and transmission-line effects.

Several books are devoted entirely to digital design with PLDs. A particularly good one is *Practical Design Using Programmable Logic* by David Pellerin and Michael Holley, the creators of ABEL (Prentice Hall, 1991). Two more examples are *Digital System Design Using Programmable Logic Devices* by P. K. Lala (Prentice Hall, 1990) and *Digital Systems Design With Programmable Logic* by M. Bolton (Addison-Wesley, 1990).

Although textbooks typically give a good overview of general PLD architectures, they inevitably fall a few to several years behind the state of the art in larger PLD and FPGA architectures. Up-to-date databooks are available from all of the prominent manufacturers of FPGAs, including Actel (Sunnyvale, CA 94086), Advanced Micro Devices (Sunnyvale, CA 94088), Altera (San Jose, CA 95134), Intel (Santa Clara, CA 95051), Lattice Semiconductor (Hillsboro, OR 97124), and Xilinx (San Jose, CA 95124). Besides the manufacturers' literature, a good source of more up-to-date information is the annual PLD Design Conference and Exhibit held each spring in San Jose, CA, chaired by PLD authority Stan Baker and sponsored by *Electronic Engineering Times* (CMP Conference and Exhibit Group, Manhasset, NY 11030).

DRILL PROBLEMS

10.1   Determine the number of fuses in each of the PAL devices in Table 10–1.

10.2   How many fuses are contained in the 16V8 as described in the text? (The commercial device has additional fuses, not described in the text, for user-specific information and security.)

10.3   How many fuses are contained in the 22V10 as described in the text? (The commercial device has additional fuses, not described in the text, for user-specific information and security.)

10.4   Determine f_{max} with external feedback for all of the devices in Table 10–2.

10.5   Determine f_{max} with internal feedback for all of the GAL devices in Table 10–2.

10.6   Describe two different fuse patterns that might result from the 74x374 program in Table 10–3, both of which use just one product term per output.

10.7   Show the changes required in Table 10–4 to realize the 74x377 with a 20R8.

10.8   Show the changes required in Table 10–7 and Figure 10–12(b) to obtain an 8-bit version of the 74x194.

10.9   Show the changes required in Table 10–8 to obtain a 4-bit *down* counter. (*Hint:* In ABEL, "–" works in a way similar to "+".)

10.10 Modify the ABEL program in Table 10–8 to realize the same function as a 74x169 up/down counter.

10.11 Modify the ABEL program in Table 10–8 to realize the same function as a 74x162 decade counter.

10.12 Modify the ABEL program in Table 10–8 to use a PAL16R4. With a one-character change, the same program should work with a PAL16R6.

10.13 How many fusible links are contained in a PAL20X8 PLD?

EXERCISES

10.14 Design an SSI/MSI circuit equivalent to the modified multiplexer of Figure 10–15. Your circuit should have the same input and output signals and active levels as the PLD. Compare the chip count of your circuit with that of the PLD design.

10.15 Design a multiplexer PLD, similar to the ones in Figure 10–15, with four control inputs: LDA, LDB, CLR, and COM. The register should be loaded with the A bus if LDA is asserted and with B if LDB is asserted, and held otherwise. However, if CLR is asserted, then the register is cleared. Finally, if COM is asserted, the next state of the register is the complement of what it would have been otherwise. It is also desirable for the register to be loaded with A⊕B if LDA and LDB happen to be asserted simultaneously. However, you are limited to using a single 16R4 or 16V8 in this design. Draw a logic diagram showing the signal connections to the 16R4 or 16V8, and an ABEL program that produces the required behavior. Also write a truth table that shows the operation performed by your device for all 16 combinations of the control inputs.

10.16 Modify the ABEL program of Table 10–17 to include the HINT output from the original state-machine specification in Section 8.2.

10.17 Determine the number of product terms required for each output of the RING8 PLD in Table 10–12. Does it fit in a 16R8 or 16V8R?

10.18 The timing-generator PLD of Table 10–13 can produce a 1-tick-wide pulse on the /P1–/P6 outputs if /RESET or /RESTART is asserted in the middle of a phase. This may be undesirable, depending on what the outputs are controlling. Modify the equations so that the output pulses are always at least two ticks wide.

10.19 Suppose the timing generator of Table 10–13 is used to control a dynamic memory system, such that all six phases must be completed to read or write the memory. If the timing generator is reset during a write operation without completing all six phases, the memory contents may be corrupted. Modify the equations in Table 10–13 to avoid this problem.

10.20 Modify the timing-generator equations in Table 10–13 so that phase 3 lasts for three ticks of MCLK and phase 4 lasts for four.

10.21 Write an ABEL program to generate the timing waveforms of Exercise 8.16 using a single 16R8 or 16V8R.

10.22 Write a "generic" equation that handles bit i of a synchronous binary up counter with synchronous count-enable, load, clear, and preset inputs in an X-series PLD. Your equation should assume active-high signals at the PLD output pins.

10.23 Write the equations for building, in a 20X8, an 8-bit synchronous binary down counter with count-enable, load, and clear inputs, and an active-low /RCO output that is asserted whenever the counter is in state 00000000. Your equation should assume active-low signals at the PLD output pins.

10.24 Design a binary up/down counter for the elevator controller in a 20-story building, using a single 16R6, 16V8R, or 20X4. The counter should have enable and up/down control inputs. It should stick at state 1 when counting down, stick at state 20 when counting up, and skip state 13 in either mode. Draw a logic diagram and write ABEL equations for your design.

10.25 Modify the ABEL program in Table 10–14 to use a 20X4 with combinational outputs for /P1–/P6. Does this approach have any advantages compared to the original 20X10-based design?

10.26 Design a circuit equivalent to the modulo-193 counter of Figure 9–24 using a single 20X8.

10.27 Redesign the T-bird tail-lights machine to include parking-light and brake-light functions. When the BRAKE input is asserted, all of the lights should go on immediately, and stay on until BRAKE is negated, independent of any other function. When the PARK input is asserted, each lamp is turned on at 50% brightness at all times when it would otherwise be off. This is achieved by driving the lamp with a 100 Hz signal DIMCLK with a 50% duty cycle. Draw a logic diagram for the circuit using one or two PLDs, write an ABEL program for each PLD, and write a short description of how your system works.

10.28 Write an ABEL program for the guessing-game machine described on page 566 using a single 16R6 or 16V8R. Use an output-coded state assignment and ABEL's "state-diagram" notation.

10.29 Design the five-minute timer required traffic-light controller of Figure 10–24 using two 74LS163 counters. (You may reduce the "long" timeout to 255 seconds if you wish.) Use spare PLD outputs and modify the ABEL programs as required to design the entire traffic-light controller without using any additional SSI or MSI parts. Draw a logic diagram for your circuit and write the required ABEL programs.

10.30 Add an OVERRIDE input to the traffic-lights state machine of Figure 10–25, still using just a single 16R8 or 16V8R. When OVERRIDE is asserted, the red lights should flash on and off, changing once per second. Write a complete ABEL program for your machine.

10.31 Write an ABEL program for the state machine described in Exercise 7.26. Use ABEL's "state-diagram" notation.

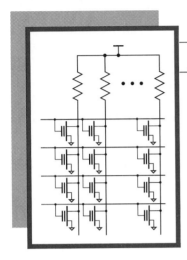

MEMORY

Any sequential circuit has memory of a sort, since each flip-flop or latch stores one bit of information. However, we normally reserve the word "memory" to refer to a bits that are stored in a structured way, usually as a two-dimensional array in which one row of bits is accessed at a time.

This chapter describes several different types of memory organizations and commercially available memory chips. The same kinds of memory may be embedded into larger VLSI chips, where they are combined with other digital or analog circuits to perform a useful function.

The applications of memory are many and varied. In a microprocessor's central processing unit (CPU), a "read-only memory" may be used to define the primitive steps that are performed to execute instruction in the CPU's instruction set. Alongside the CPU, a fast "static memory" may serve as a cache to hold recently used instructions and data. And microprocessor's main-memory subsystem may contain tens to hundreds of millions of bits in "dynamic memory" that store complete operating systems, programs, and data.

Applications of memory are not limited to microprocessors, or even to purely digital systems. For example, equipment in the public telephone system uses read-only memories to perform certain transformations on digitized voice signals, and fast "static memories" as a "switching fabric" to route digitized voice between subscribers. Many portable audio compact-disc players "read ahead" and store a few seconds of audio in a "dynamic memory" so that the unit can keep plating even if it is physically jarred (this requires over 1.4 million bits per second of stored audio). And there are many examples types of modern audio/visual equipment that use memories to temporarily store digitized signals for enhancement through digital signal processing.

This chapter begins with a discussion of read-only memory and its applications, and then describes two different types of read/write memory. Optional subsections discuss the internal structure of the different memory types.

MEMORY—NOT!

> Most types of read-only memory are not really memory in the strictest sense of the word, because they are combinational, not sequential, circuits. They are called "memory" because of the organizational paradigm that describes their function.

11.1 READ-ONLY MEMORY

read-only memory (ROM)
address input
data output

A *read-only memory (ROM)* is a combinational circuit with n inputs and b outputs, as shown in Figure 11–1. The inputs are called *address inputs* and are traditionally named A0, A1, ..., An–1. The outputs are called *data outputs*, and are traditionally named D0, D1, ..., Db-1.

A ROM "stores" the truth table of an n-input, b-output combinational logic function. For example, Table 11–1 is the truth table of a 3-input, 4-output combinational function; it could be stored in a $2^3 \times 4$ (8×4) ROM. Neglecting propagation delays, a ROM's data outputs at all times equal the output bits in the truth-table row selected by the address inputs.

Figure 11–1
Basic structure of a $2^n \times b$ ROM.

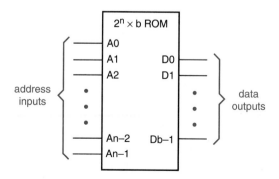

Table 11–1
Truth table for a 3-input, 4-output combinational logic function.

Inputs			Outputs			
A2	A1	A0	D3	D2	D1	D0
0	0	0	1	1	1	0
0	0	1	1	1	0	1
0	1	0	1	0	1	1
0	1	1	0	1	1	1
1	0	0	0	0	0	1
1	0	1	0	0	1	0
1	1	0	0	1	0	0
1	1	1	1	0	0	0

Since a ROM is a *combinational* circuit, you would be correct to say that it's not really a memory at all. In terms of digital circuit operation, you can treat a ROM like any other combinational logic element. However, you can also think of information as being "stored" in the ROM when it is manufactured or programmed (we'll say more about how this is done in Section 11.1.4).

When you think of ROM as being a type of memory, it has a key difference from most other types of integrated-circuit memory. ROM is *nonvolatile memory*;   *nonvolatile memory* that is, its contents are preserved even if no power is applied.

11.1.1 Using ROMs for "Random" Combinational Logic Functions

Table 11–1 is actually the truth table of a 2-to-4 decoder with an output-polarity control, a function that can be built with discrete gates as shown in Figure 11–2. Thus, we have two ways to build the decoder—with discrete gates, or with an 8×4 ROM that contains the truth table, as shown in Figure 11–3.

Figure 11–2
A 2-to-4 decoder with output polarity control.

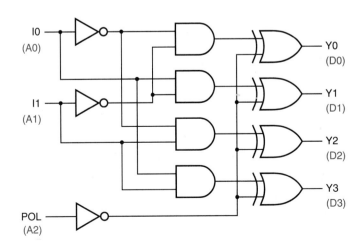

Figure 11–3

Connections to build the 2-to-4 decoder using an 8×4 ROM that stores Table 11–1.

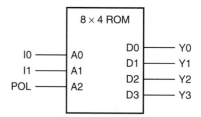

The assignment pattern of decoder inputs and outputs to ROM inputs and outputs in Figure 11–3 is a consequence of the way that the truth table in Table 11–1 is constructed. Therefore, the physical ROM realization of the decoder is not unique. That is, we could write the rows or columns of the truth table in a different order and use a physically different ROM to perform the same logic function, simply by assigning the decoder signals to different ROM inputs and outputs. Another way to look at this is that we can rename the individual address inputs and data outputs of the ROM.

For example, swapping the bits in the D0 and D3 columns of Table 11–1 would give us the truth table for a physically different ROM. However, the new ROM could still be used to build the 2-to-4 decoder simply by swapping the Y0 and Y3 labels in Figure 11–3. Likewise, if we shuffled the data rows of the truth table as shown in Table 11–2, we would get another different ROM, but it can still be used as the 2-to-4 decoder with a rearrangement of the address inputs, A0 = POL, A1 = I0, A2 = I1.

Table 11–2

Truth table with data rows shuffled.

Inputs			Outputs			
A2	A1	A0	D3	D2	D1	D0
0	0	0	1	1	1	0
0	0	1	0	0	0	1
0	1	0	1	1	0	1
0	1	1	0	0	1	0
1	0	0	1	0	1	1
1	0	1	0	1	0	0
1	1	0	0	1	1	1
1	1	1	1	0	0	0

When constructing a ROM to store a given truth table, input and output signals reading from right to left in the truth table are normally assigned to ROM address inputs and data outputs with ascending labels. Each address or data combination may then be read as a corresponding binary integer with the bits numbered in the "natural" way. A data file is typically used to specify the truth table to be stored in the ROM when it is manufactured or programmed. The data file usually gives the address and data values as hexadecimal numbers. For example, a data file may specify Table 11–2 by saying that ROM addresses 0–7 should store the values E, 1, D, 2, B, 4, 7, 8.

Figure 11–4
Connections to perform a 4×4 unsigned binary multiplication using a 256×8 ROM.

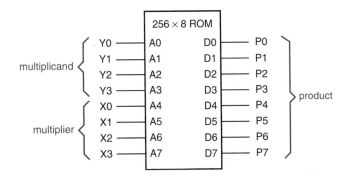

An excellent example of a function that can be built with ROM is 4×4 unsigned binary multiplication, which we performed in Section 5.10 with two chips, the 74284 and the 74285. A $2^8 \times 8$ (256×8) ROM can be used instead, with the connections shown in Figure 11–4.

A ROM's contents are normally specified by a file that contains one entry for every address in the ROM. For example, Table 11–3 is a hexadecimal listing of the 4×4 multiplier ROM contents. Each row gives a starting address in the ROM and specifies the 8-bit data values stored at 16 successive addresses.

Table 11–3
Hexadecimal text file specifying the contents of a 4×4 multiplier ROM.

```
00: 00 00 00 00 00 00 00 00 00 00 00 00 00 00 00 00
10: 00 01 02 03 04 05 06 07 08 09 0A 0B 0C 0D 0E 0F
20: 00 02 04 06 08 0A 0C 0E 10 12 14 16 18 1A 1C 1E
30: 00 03 06 09 0C 0F 12 15 18 1B 1E 21 24 27 2A 2D
40: 00 04 08 0C 10 14 18 1C 20 24 28 2C 30 34 38 3C
50: 00 05 0A 0F 14 19 1E 23 28 2D 32 37 3C 41 46 4B
60: 00 06 0C 12 18 1E 24 2A 30 36 3C 42 48 4E 54 5A
70: 00 07 0E 15 1C 23 2A 31 38 3F 46 4D 54 5B 62 69
80: 00 08 10 18 20 28 30 38 40 48 50 58 60 68 70 78
90: 00 09 12 1B 24 2D 36 3F 48 51 5A 63 6C 75 7E 87
A0: 00 0A 14 1E 28 32 3C 46 50 5A 64 6E 78 82 8C 96
B0: 00 0B 16 21 2C 37 42 4D 58 63 6E 79 84 8F 9A A5
C0: 00 0C 18 24 30 3C 48 54 60 6C 78 84 90 9C A8 B4
D0: 00 0D 1A 27 34 41 4E 5B 68 75 82 8F 9C A9 B6 C3
E0: 00 0E 1C 2A 38 46 54 62 70 7E 8C 9A A8 B6 C4 D2
F0: 00 0F 1E 2D 3C 4B 5A 69 78 87 96 A5 B4 C3 D2 E1
```

The nice thing about ROM-based design is that you can usually write a simple program in a high-level language to calculate what should be stored in the ROM. For example, it took me less than five minutes to write a Pascal program, shown in Table 11–4, that generated the contents of Table 11–3.

Table 11–4
Program to generate the text file specifying the contents of a 4×4 multiplier ROM.

```
PROGRAM MultROM(input, output);
TYPE hex = 0..15;
VAR x, y: integer;

{ Procedure to write d as a hex digit.}
PROCEDURE WriteHexDigit(d: hex);
  BEGIN
    IF d<10 THEN write(char(ord('0')+d))
    ELSE write(char(ord('A')+d-10));
  END;

{ Procedure to write i as two hex digits.}
PROCEDURE WriteHex2(i: integer);
  BEGIN
    WriteHexDigit((i DIV 16) MOD 16);
    WriteHexDigit(i MOD 16);
  END;

BEGIN {Main program.}
  FOR x := 0 UNTIL 15 DO {For all values of x.}
    BEGIN
      WriteHex2(x*16); write(':');
      FOR y := 0 UNTIL 15 DO {For all values of y.}
        BEGIN  write(' '); WriteHex2(x*y)  END;
      writeln;
    END;
END.
```

* 11.1.2   Internal ROM Structure

The mechanism used by ROMs to "store" information varies with different ROM technologies. In most ROMs, the presence or absence of a diode or transistor distinguishes between a 0 and a 1.

Figure 11–5 is the schematic of a primitive 8×4 ROM that you can build yourself using an MSI decoder and a handful of diodes. The address inputs select one of the decoder outputs to be asserted. Each decoder output is called

---

*Throughout this book, optional sections are marked with an asterisk.

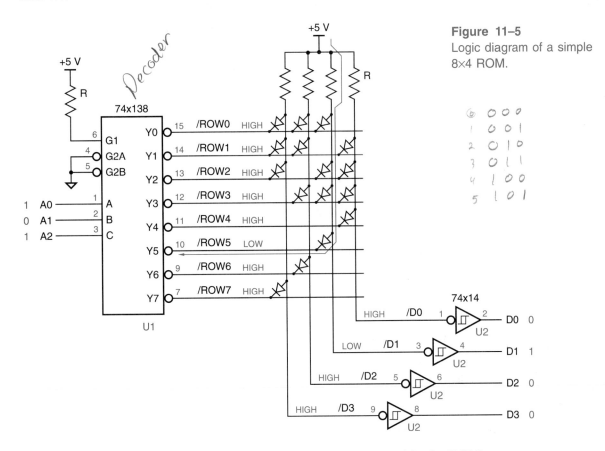

Figure 11–5
Logic diagram of a simple 8×4 ROM.

a *word line* because it selects one row or word of the table stored in the ROM. The figure shows the situation with A2–A0 = 101 and /ROW5 asserted.

    Each vertical line in Figure 11–5 is called a *bit line* because it corresponds to one output bit of the ROM. An asserted word line pulls a bit line LOW if a diode is connected between the word line and the bit line. There is only one diode in row 5, so the corresponding bit line (/D1) is pulled LOW. The bit lines are buffered through inverters to produce the D3–D0 ROM outputs, 0010 for the case shown.

    In the ROM circuit of Figure 11–5, each intersection between a word line and a bit line corresponds to one bit of "memory." If a diode is present at the

word line

bit line

Schmitt-trigger inverters are used in Figure 11–5 to improve the circuit's noise margins, because the 0.7-volt drop across the diodes creates a LOW level on the bit lines that is not really so low. Of course, you'd never really build this circuit except in an academic lab—it's easier just to buy and program a commercial ROM chip.

DETAILS, DETAILS

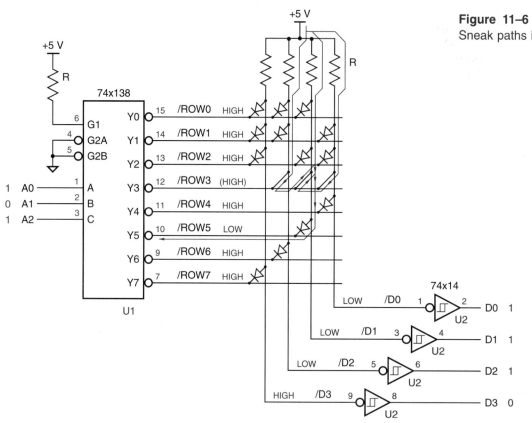

Figure 11–6
Sneak paths in a ROM.

intersection, a 1 is stored; otherwise, a 0 is stored. If you were to build this circuit in the lab, you would "program" the memory by inserting and removing diodes at each intersection. Primitive though it may seem, owners of the DEC PDP-11 minicomputer (circa 1970) made use of similar technology in the M792 32×16 "bootstrap ROM module." The module was shipped with 512 diodes soldered in place, and owners programmed it by clipping out the diode at each location where a 0 was to be stored.

The diode pattern shown in Figure 11–5 corresponds to the 2-to-4-decoder truth table of Table 11–1. This doesn't seem very efficient—we used a 3-to-8 decoder and a bunch of diodes to build the ROM version of a 2-to-4 decoder. We could have just used a subset of the 3-to-8 decoder directly! However, we'll show a more efficient ROM structure and more-useful design examples later.

* 11.1.3 Two-Dimensional Decoding

Suppose you wanted to build a 128×1 ROM using the kind of structure described in the preceding subsection. Have you ever thought about what it would take to

SNEAK PATHS

The ROM circuit of Figure 11–5 must use diodes, not direct connections, at each location where a 1 is to be stored. Figure 11–6 shows what happens if just a few diodes—such as the ones in row 3—are replaced with direct connections. Suppose the address inputs are 101; then /ROW5 is asserted, and only /D1 is supposed to be pulled LOW to create an output of 0010. However, the direct connections allow current to flow along *sneak paths*, so that bit lines /D2 and /D0 are also pulled LOW, and the (incorrect) output is 0111. When diodes are used, sneak paths are blocked by reverse-biased diodes, and correct outputs are obtained.

build a 7-to-128 decoder? Try 128 7-input NAND gates to begin with, and add 14 buffers and inverters with a fanout of 64 each! ROMs with a million bits or more are available commercially; trust me, they do not contain 20-to-1,048,576 decoders. Instead, a different structure, called *two-dimensional decoding*, is used to reduce the decoding complexity to something on the order of the square root of the number of addresses.

two-dimensional decoding

The basic idea in two-dimensional decoding is to arrange the ROM cells in an array that is as close as possible to square. For example, Figure 11–8 shows a possible internal structure for a 128×1 ROM. The three high-order address

Figure 11–7
MOS transistors as storage elements in a ROM.

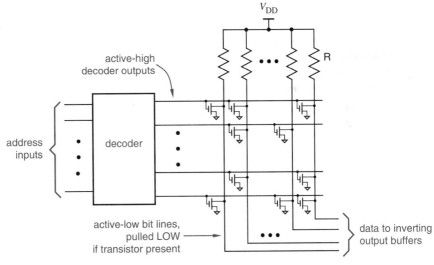

TRANSISTORS AS
ROM ELEMENTS

MOS ROMs actually use a transistor instead of a diode at each location where a bit is to be stored; Figure 11–7 shows the basic idea. The row decoder has active-high outputs. When a row line is asserted, the NMOS transistors in that row are turned on, which pulls the corresponding bit lines low. A similar idea can be used in ROMs built with bipolar transistors.

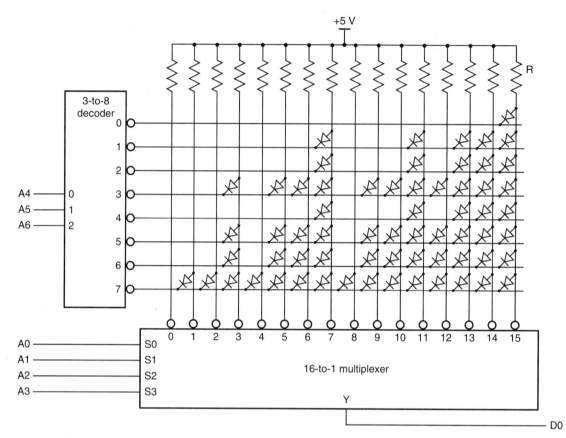

Figure 11–8 Internal structure of a 128×1 ROM using two-dimensional decoding.

bits, A6–A4, are used to select a row. Each row stores 16 bits starting at address A6 A5 A4 0000. When an address is applied to the ROM, all 16 bits in the selected row are "read out" in parallel on the bit lines. A 16-input multiplexer selects the desired data bit based on the low-order address bits.

By the way, the diode pattern in Figure 11–8 was not chosen at random. It performs a very useful 7-input combinational logic function that would require 35 4-input AND gates to build as a minimal two-level AND-OR circuit (see Exercise 11.7). The ROM version of this function certainly saves a lot of engineering effort and space in a board-level design.

Two-dimensional decoding allows the 128×1 ROM to be built with a 3-to-8 decoder and a 16-input multiplexer (whose complexity is comparable to that of a 4-to-16 decoder). A 1M×1 ROM could be built with a 10-to-1024 decoder and a 1024-input multiplexer—not easy, but a lot simpler than the one-dimensional alternative.

Besides reducing decoding complexity, two-dimensional decoding has one other benefit—it leads to a chip with physical dimensions that are close to square, important for chip fabrication and packaging. A chip containing a 1M×1 physical array would be *very* long and skinny and could not be built economically.

In ROMs with multiple data outputs, the storage arrays corresponding to each data output may be made narrower in order to achieve an overall chip layout that is closer to square. For example, Figure 11–9 shows the possible layout of a 32K×8 ROM.

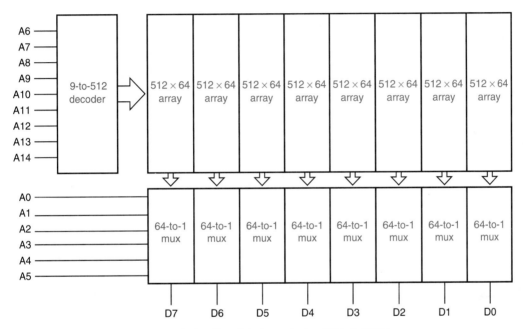

Figure 11–9   Possible layout of a 32K×8 ROM.

11.1.4 Commercial ROM Types

Unless you visit computer museums or surplus stores, it's difficult nowadays to find ROM modules built with discrete diodes. A modern ROM is fabricated as a single IC chip; one that stores two million bits can be purchased for under $10. Various methods are used to "program" the information stored in a ROM, as discussed below and summarized in Table 11–5.

Most of the early integrated-circuit ROMs were *mask-programmable ROMs* (or, simply, *mask ROMs*). A mask ROM is programmed by the pattern of connections and no-connections in one of the *masks* used in the IC manufacturing process. To program or write information into the ROM, the customer gives the manufacturer a listing of the ROM contents, using a floppy disk or other medium.

mask-programmable ROM
mask ROM
mask

Table 11–5 Commercial ROM types.

Type	Technology	Read cycle	Write cycle	Comments
Mask ROM	NMOS, CMOS	20–200 ns	4 weeks	Write once; low power
Mask ROM	Bipolar	<100 ns	4 weeks	Write once; high power; low density
PROM	Bipolar	<100 ns	5 minutes	Write once; high power; no mask charge
EPROM	NMOS, CMOS	25–200 ns	5 minutes	Reusable; low power; no mask charge
EEPROM	NMOS	50–200 ns	10 μs/byte	10,000–100,000 writes/location limit

mask charge

The manufacturer uses this information to create one or more customized masks to manufacture ROMs with the required pattern. ROM manufacturers impose a *mask charge* of several thousand dollars for the "customized" aspects of mask-ROM production. Because of mask charges and the four-week delay typically required to obtain programmed chips, mask ROMs are normally used today only in high-volume applications. For low-volume applications there are more cost-effective choices, discussed next.

programmable read-only memory (PROM)
PROM programmer

fusible link

A *programmable read-only memory (PROM)* is similar to a mask ROM, except that the customer may store data values (i.e., "program the PROM") in just a few minutes using a *PROM programmer*. A PROM chip is manufactured with all of its diodes or transistors "connected." This corresponds to having all bits at a particular value, typically 1. The PROM programmer can be used to set desired bits to the opposite value, typically by vaporizing tiny *fusible links* inside the PROM corresponding to each bit. A link is vaporized by selecting it using the PROM's address and data lines, and then applying a high-voltage pulse (10–30 V) to the device through a special input pin.

Early PROMs had reliability problems. Sometimes the stored bits changed because of incompletely vaporized links that would "grow back," and sometimes intermittent failures occurred because of floating shrapnel inside the IC package. However, these problems have been worked out, and reliable fusible-link technology is used nowadays not only in PROMs, but also in the bipolar PLDs that we introduced in Section 10.1.

erasable programmable read-only memory (EPROM)

floating-gate MOS transistor

An *erasable programmable read-only memory (EPROM)* is programmable like a PROM, but it can also be "erased" to the all-1s state by exposing it to ultraviolet light. No, the light does not cause fuses to grow back! Rather, EPROMs use a different technology, called "floating-gate MOS."

As shown in Figure 11–10, an EPROM has a *floating-gate MOS transistor* at *every* bit location. Each transistor has two gates. The "floating" gate is unconnected and is surrounded by extremely high-impedance insulating material. To program an EPROM, the programmer applies a high voltage to the nonfloating gate at each bit location where a 0 is to be stored. This causes a breakdown in the insulating material and allows a negative charge to accumulate on the floating gate. When the high voltage is removed, the negative charge remains. During subsequent read operations, the negative charge prevents the MOS transistor from turning on when it is selected.

EPROM manufacturers claim that a properly programmed bit will retain 70% of its charge for at least 10 years, even if the part is stored at 125°C, so EPROMs definitely fall into the category of "nonvolatile memory." However, they can also be erased. The insulating material surrounding the floating gate becomes slightly conductive if it is exposed to ultraviolet light with a certain wavelength. Thus, EPROMs can be erased by exposing the chips to ultraviolet   *EPROM erasing* light, typically for 5–20 minutes. An EPROM chip is normally housed in a package with a quartz lid through which the chip may be exposed to the erasing light.

Probably the most common application of EPROMs is to store programs in microprocessor systems. EPROMs are typically used during program development, where the program or other information in the EPROM must be repeatedly changed during debugging. However, ROMs and PROMs usually cost less than EPROMs of similar capacity. Therefore, once a program is finalized, a ROM or PROM may be used in production to save cost. Most of today's PROMs are   *one-time* actually EPROMS housed in inexpensive packages without quartz lids; these are   *programmable* sometimes called *one-time programmable (OTP) ROMs*.   *(OTP) ROM*

An *electrically erasable programmable read-only memory (EEPROM)* is   *electrically erasable* similar to an EPROM, except that individual stored bits may be erased elec-   *programmable* trically. The floating gates in an EEPROM are surrounded by a much thinner   *read-only memory* *(EEPROM)*

Figure 11–10
Storage matrix in an EPROM
using floating-gate MOS
transistors.

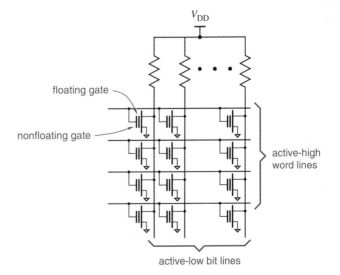

insulating layer, and can be erased by applying a voltage of the opposite polarity as the charging voltage to the nonfloating gate.

As indicated in Table 11–5, programming or "writing" an EEPROM location takes much longer than reading it, so an EEPROM is no substitute for the read/write memories discussed later in this chapter. Also, because the insulating layer is so thin, it can be worn out by repeated programming operations. As a result, EEPROMs can be reprogrammed only a limited number of times, typically 10,000 times per location. Therefore, EEPROMs are typically used for storing data that must be preserved when the equipment is not powered, but that doesn't change very often, such as the default configuration data for a smart terminal.

Popular ROMs for microprocessors and other moderate-speed ROM applications include the *2764*, *27128*, *27256*, and *27512* EPROMs, whose logic symbols are shown in Figure 11–11. The figure also indicates pins that must be connected to a constant signal during normal read-only operation. Inputs labeled V_{CC} must be connected to +5 volts, inputs labeled V_{IH} must be connected to a valid HIGH logic signal, and "N.C." means "no connection." Different input configurations are used to program and test the device; the pin labeled VPP is used to apply the programming voltage.

2764

27128

27256

27512

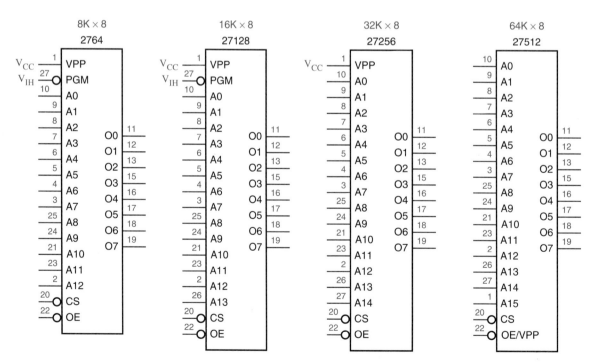

Figure 11–11 Logic symbols for standard EPROMs in 28-pin dual in-line packages.

11.1.5 ROM Control Inputs and Timing

The outputs of a ROM must often be connected to a three-state bus, where different devices may drive the bus at different times. Therefore, most commercial ROM chips have three-state data outputs and an *output-enable (OE) input* that must be asserted to enable the outputs.

output-enable (OE) input

Many ROM applications, especially program storage, have multiple ROMs connected to a bus, where only one ROM drives the bus at a time. Most ROMs have a *chip-select (CS) input* to simplify the design of such systems. In addition to OE, a ROM's CS input must be asserted to enable the three-state outputs.

chip-select (CS) input

Figure 11–12 shows how the OE and CS inputs could be used when connecting four 32K×8 ROMs to a microprocessor system that requires 128 Kbytes of ROM. The microprocessor has an 8-bit data bus and a 20-bit address bus, for a maximum address space of 1 Mbyte (2^{20} bytes). The ROM is supposed to be located in the highest 128K of the address space. Therefore, a NAND gate produces the /HIMEM signal, which is asserted when the address bus contains an

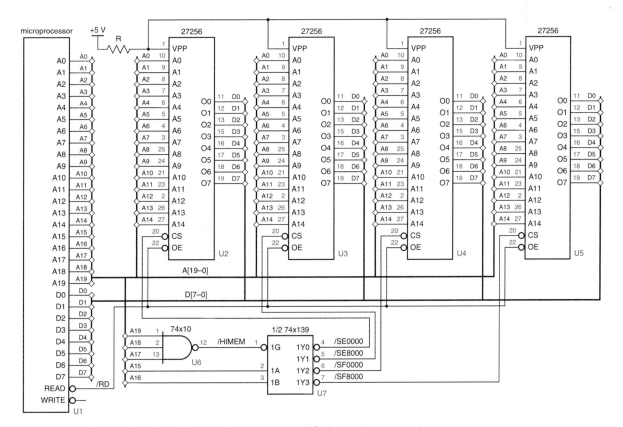

Figure 11–12 Address decoding and ROM enabling in a microprocessor system.

address in the highest 128K (A19–A17 = 111). A 74x139 decoder then selects one of the four 32K×8 ROMs. The selected ROM drives the data bus only when the microprocessor requests a read operation by asserting READ, which is connected to all of the OE inputs.

As we've described it so far, a CS input is no more than a second output-enable input that is ANDed with OE to enable the three-state outputs. However, *power-down input* in many ROMs, CS also serves as a *power-down input*. When CS is negated, power is removed from the ROM's internal decoders, drivers, and multiplexers. *standby mode* In this *standby mode* of operation, a typical ROM consumes less than 10% of *active mode* the power it uses in *active mode* with CS asserted. In Figure 11–12, at most one ROM is selected at any time, so the total power consumption of all the ROM chips is much closer to that of one chip than four.

Figure 11–13 shows how CS and OE inputs are used inside a typical ROM. Figure 11–14 shows typical ROM timing, including the following parameters:

t_{AA}
access time from address

t_{AA} *Access time from address.* The access time from address of a ROM is the propagation delay from stable address inputs to valid data outputs. When designers talk about "a 150-ns ROM," they are usually referring to this parameter.

Figure 11–13
Internal ROM structure, showing use of control inputs.

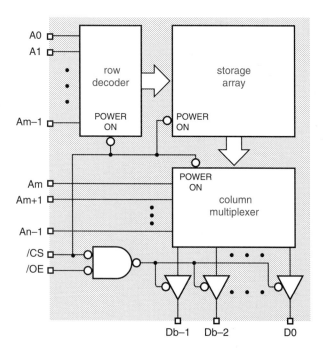

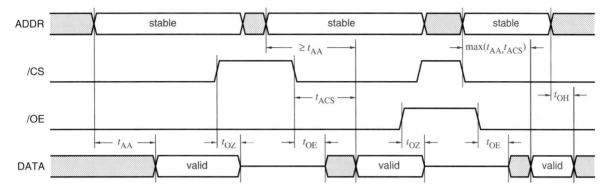

Figure 11–14 ROM timing.

t_{ACS} *Access time from chip-select.* The access time from chip select of a ROM is the propagation delay from the time CS is asserted until the data outputs are valid. In some chips, this is longer than the access time from address, because the chip takes a little while to "power up." In others, this time is shorter because CS controls only output enabling.

t_{ACS}
access time from chip select

t_{OE} *Output-enable time.* This parameter is usually much shorter than access time. The output-enable time of a ROM is the propagation delay from OE and CS both asserted until the three-state output drivers have left the Hi-Z state. Depending on whether the address inputs have been stable long enough, the output data may or may not be valid at that point.

t_{OE}
output-enable time

t_{OZ} *Output-disable time.* The output-disable time of a ROM is the propagation delay from the time OE or CS is negated until the three-state output drivers have entered the Hi-Z state.

t_{OZ}
output-disable time

t_{OH} *Output-hold time.* The output-hold time of a ROM is the length of time that the outputs remain valid after a change in the address inputs, or after /OE or /CS is negated.

t_{OH}
output-hold time

Given the decoding and multiplexing structure in Figure 11–13, you might think that the address access time is shorter from some address inputs than it is from others. In a given application, if some address-input signals were delayed relative to the others, you could recover this delay by connecting the slower signals to the "faster" ROM inputs. After all, any input signal can be connected to any ROM address input if you're willing to rearrange the ROM contents accordingly. However, ROM manufacturers don't specify which, if any, inputs are faster. In fact, the internal electrical characteristics of most ROMs may be such that the difference is not enough to bother with.

NOT ALL INPUTS
ARE CREATED
EQUAL

As with other components, the manufacturer specifies maximum and, sometimes, typical values for all timing parameters. Usually, minimum values are also specified for t_{OE} and t_{OH}. The minimum value of t_{OH} is usually specified to be 0; that is, the minimum combinational-logic delay through the ROM is 0.

UNBELIEVABLY
FAST ROMS

> The actual delay through a ROM is never 0, of course, but it can easily be just slightly less than what you need to meet a nonzero hold-time requirement elsewhere in a design. So it's best to assume that t_{OH} for a ROM is 0, unless you really know what you're doing.

11.1.6 ROM Applications

As we mentioned earlier, the most common application of ROMs is for program storage in microprocessor systems. However, there are many applications where ROM provides a low-cost realization of complex or "random" combinational logic functions. In this section, we'll give a couple of examples of ROM-based circuits that are used in the digital telephone network.

When an analog voice signal enters a typical digital telephone system, it is sampled 8,000 times per second and converted into a sequence of 8-bit bytes representing the analog signal at each sampling point. In Section 9.4.4, we described how these digital voice samples could be converted between parallel and serial formats, but we did not describe the coding of the 8-bit bytes themselves. We'll do that now, and then show how ROM-based circuits can easily deal with this highly encoded information.

The simplest 8-bit encoding of the sign and amplitude of an analog signal would be an 8-bit integer in the two's-complement or signed-magnitude system; this is called a *linear encoding*. However, an 8-bit linear encoding yields a *dynamic range*—the ratio between range of representable numbers and the smallest representable difference—of only 2^8 or 256. For you audiophiles, this corresponds to a dynamic range in signal power of $20 \log 256$ or about 48 dB. By comparison, compact audio discs use a 16-bit linear encoding with a (theoretical) dynamic range of $20 \log 2^{16}$ or about 96 dB.

linear encoding
dynamic range

companded
encoding
μ-law PCM

Instead of a linear encoding, the North American telephone network uses an 8-bit *companded encoding* called *μ-law PCM* ("mu-law" pulse-code modulation). Figure 11–15 shows the format of an 8-bit coded byte, a sort of floating-point representation containing sign (S), exponent (E), and mantissa (M) fields. The analog value V represented by a byte in this format is given by the formula

$$V = (1 - 2S) \cdot [(2^E) \cdot (2M + 33) - 33]$$

Figure 11–15
Format of a μ-law PCM byte.

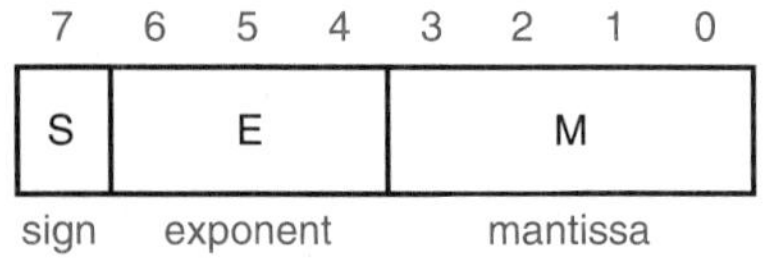

7	6 5 4	3 2 1 0
S	E	M

sign   exponent   mantissa

An analog signal represented in this format can range from $-8159 \cdot k$ to $+8159 \cdot k$, where k is an arbitrary scale factor. The range of signals is $2 \cdot 8159$ and the smallest difference that can be represented is only 2 (when $E = 0$), so the dynamic range is $20 \log 8159$ or about 78 dB, quite an improvement over an 8-bit linear code.

> With μ-law PCM, you don't get something for nothing, of course. In the companded encoding, the difference between successive coded values is greater at high amplitudes than it is at low amplitudes, so the encodings of large analog signals suffer from more quantizing distortion (caused by the difference between the sampled analog value and the nearest available coded representation). Expressed as a percentage of peak signal amplitude, though, quantizing distortion is about constant across the entire range of representable values.

MONEY
FOR NOTHING...

Now let's look at some ROM applications involving μ-law coded voice signals. Believe it or not, in many types of phone connections, your voice is purposely attenuated by a few decibels to make things work better. In an analog phone network, attenuation is performed by a simple, passive analog circuit, but not so in the digital world. Given a μ-law PCM byte, a *digital attenuator* must produce a different PCM byte that represents the original analog signal multiplied by a specified attenuation factor.

digital attenuator

One way to build a digital attenuator is shown in Figure 11–16. The input byte is applied to a μ-law decoder that expands the byte according to the formula given earlier to produce a 14-bit signed-magnitude integer. This 14-bit linear value is then multiplied by a 14-bit binary fraction corresponding to the desired attenuation amount. The fractional bits of the product are discarded, and the result is re-encoded into a new 8-bit μ-law PCM byte. Each block in the figure could be built with perhaps a dozen MSI chips.

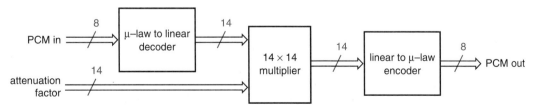

Figure 11–16 Block diagram of a digital attenuator using combinational MSI logic.

Figure 11-17
A digital attenuator.

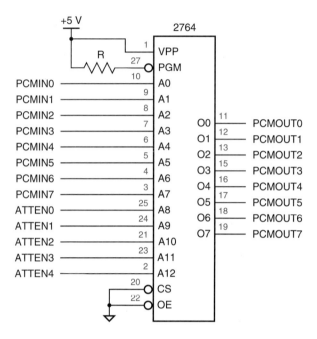

Figure 11–17 is the logic diagram for a digital attenuator that uses a single 8K×8 ROM instead. This ROM can apply any of 32 different attenuation factors to a μ-law input byte—it simply stores 32 different attenuation tables. The high-order address bits select a table, and the low-order address bits select an entry. Each entry is a precomputed μ-law byte corresponding to the given attenuation code and input byte. Table 11–6 is a Pascal program that can be used to generate the ROM contents. The WriteHex procedures are not shown, and details of the UlawToLinear and LinearToUlaw functions are left as an exercise.

Another ROM application in digital telephony is a digital conference circuit. In the analog telephone network, it's fairly easy to make a conference connection between three or more parties. Just connect the analog phone wires together, and you get an analog summing junction, so each person hears everyone else (it's not quite that easy to do it right, but you get the idea). In the digital network,

. . . AND
THE CHECK'S
FOR FREE

The digital attenuator is another good example of the many cost-effective space/time trade-offs that were made possible when the telephone network "went digital." A single analog voice signal produces an 8-bit μ-law PCM sample once every 125 μs, but the digital attenuator ROM can produce a valid output in a few hundred nanoseconds or less. Thus, in a digital telephone system, a single ROM chip can attenuate hundreds of digital voice streams, where hundreds of analog attenuator networks would have been required in the old days.

Table 11–6 Program to generate the contents of an 8K×8, 32-position attenuator ROM for μ-law coded bytes.

```
PROGRAM AttenROM(input, output);
TYPE
  hex = 0..15;
  byte = 0..255;
  linear = -8159..8159;
VAR
  i, j, position : integer;
  pcmIN, pcmOUT: byte;
  atten, attenDB: real;
FUNCTION UlawToLinear(in: byte) : linear;
...
FUNCTION LinearToUlaw(x: linear) : byte;
...
BEGIN {Main program.}
  FOR position := 0 UNTIL 31 DO {Make 32 256-byte tables.}
    BEGIN
      readln{attenDB}; {Get amount in dB from designer,
                    negative for attenuation, positive for gain.}
      atten := exp(ln(10)*attenDB/10); {Convert to fraction.}
      FOR i := 0 to 15 DO {Construct output file in rows of 16.}
        BEGIN
          WriteHex4( (position*256) + i*16);  write(':');
          FOR j := 0 UNTIL 15 DO
            BEGIN
              pcmIN := i*16 + j;
              pcmOUT := LinearToUlaw(atten * UlawToLinear(pcmIN));
              write(' '); WriteHex2(pcmOUT);
            END;
          writeln;
        END;
    END;
END.
```

of course, chaos would result from just shorting together digital output signals. Instead, a digital conference circuit must include a digital *adder* that produces output samples corresponding to sums of the input samples.

You know how to build a binary adder for two 8-bit operands, but binary adders cannot process μ-law PCM bytes directly. Instead, the 8-bit PCM bytes must be converted into 14-bit linear format, then added, and then re-encoded. As in the digital attenuator circuit of Figure 11–16, an MSI design for this function would require many chips. Alternatively, the function could be performed by a single 64K×8 ROM, as shown in Figure 11–18. The ROM has 16 address inputs, accommodating two 8-bit μ-law PCM operands. For each pair of operands,

Figure 11–18
An adder circuit for μ-law coded bytes.

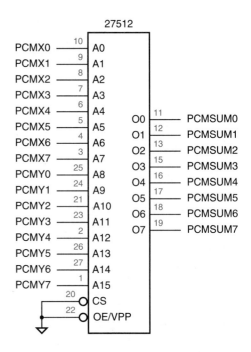

the corresponding ROM address contains the precomputed μ-law PCM sum. Table 11–7 is a Pascal program that can be used to generate the ROM contents.

These examples illustrate the many advantages of building complex combinational functions with ROMs. We usually consider a function to be "complex" if we know that an SSI, MSI, or ad hoc circuit for the function will be a real pain to design. However, most "complex" functions, like the examples in this subsection, have fairly straightforward word descriptions. Such a description can generally be translated into a computer program that "figures out" what the function's outputs should be for every possible input combination, so the function can be built simply by dropping its truth table into a ROM.

Besides ease and speed of design, a ROM-based circuit has other important advantages:

- For a moderately complex function, a ROM-based circuit is usually faster than a discrete SSI/MSI circuit, and often faster than custom LSI in a comparable technology.

- The program that generates the ROM contents can easily be structured to handle unusual or undefined cases that would require additional hardware in any other design. For example, the adder program in Table 11–7 easily handles out-of-range sums. (Also see Exercise 11.22.)

- A ROM's function is easily modified just by changing the stored pattern, usually without changing any external connections to the device. For exam-

Table 11-7 Program to generate the contents of a 64K×8 adder ROM for μ-law coded bytes.

```
PROGRAM AdderROM(input, output);
CONST
  minLinear = -8159;
  maxLinear = 8159;
TYPE
  hex = 0..15;
  byte = 0..255;
  linear = minLinear..maxLinear;
VAR
  i, j, linearSum : integer;
  pcmINX, pcmINY, pcmOUT: byte;
...
BEGIN {Main program.}
  FOR pcmINY := 0 UNTIL 255 DO {For all Y samples...}
    BEGIN
      FOR i := 0 to 15 DO {Construct output file in rows of 16.}
        BEGIN
          WriteHex4( (pcmINY*256) + i*16); write(':');
          FOR j := 0 UNTIL 15 DO {For all X samples...}
            BEGIN
              pcmINX := i*16 + j;
              linearSum := UlawToLinear(pcmINX) + UlawToLinear(pcmINY);
              {The next two lines perform  "clipping" on overflow.}
              IF linearSum < minLinear THEN linearSum := minLinear;
              IF linearSum > maxLinear THEN linearSum := maxLinear;
              write(' '); WriteHex2(LinearToUlaw(linearSum));
            END;
          writeln;
        END;
    END;
END.
```

ple, the PCM attenuator and adder ROMs in this subsection can be changed to use 8-bit "A-law" PCM, the standard digital voice coding in Europe.

- The prices of ROMs and other structured logic devices are always dropping, making them more economical, and the densities are always increasing, expanding the scope of problems that can be solved with a single chip.

There are a few disadvantages of ROM-based circuits, too:

- For simple to moderately complex functions, a ROM-based circuit may cost more or consume more power than a discrete SSI/MSI circuit.

- For functions with more than 16–20 inputs, a ROM-based circuit is impractical because of the limit on ROM sizes that are available. For example, a 16-bit adder would require billions and billions of ROM bits.

11.2 READ/WRITE MEMORY

read/write memory (RWM)

random-access memory (RAM)

The name *read/write memory (RWM)* is given to memory arrays in which we can store and retrieve information at any time. Most of the RWMs used in digital systems nowadays are *random-access memories (RAMs)*, which means that the time it takes to read or write a bit of memory is independent of the bit's location in the RAM. From this point of view, ROMs are also random-access memories, but the name "RAM" is generally used only for read/write random-access memories.

static RAM (SRAM)

dynamic RAM (DRAM)

In a *static RAM (SRAM)*, once a word is written at a location, it remains stored as long as power is applied to the chip, unless the same location is written again. In a *dynamic RAM (DRAM)*, the data stored at each location must be periodically refreshed by reading it and then writing it back again, or else it disappears. We'll discuss both types in this section.

volatile memory
nonvolatile memory

ferroelectric RAM

Most RAMs lose their memory when power is removed; they are a form of *volatile memory*. Some RAMs retain their memory even when power is removed; they are called *nonvolatile memory*. Examples of nonvolatile RAMs are old-style magnetic core memories and modern CMOS static memories in an extra-large package that includes a lithium battery with a 10-year lifetime. Recently, nonvolatile *ferroelectric RAMs* have been introduced; these devices combine magnetic and electronic elements on a single IC chip that retains its state even when power is not applied, just like the old-style core memories.

SERIAL-ACCESS
MEMORY

Random-access memory can be contrasted with *serial-access memories*, in which some location is immediately accessible at any time, but other locations take additional steps to access.

Some early computers used electromechanical serial-access memory devices, such as delay lines and rotating drums. Instructions and data were stored in a rotating medium with only one location under the "read/write head" at any time. To access a random location, the machine would have to wait until the constant rotation brought that location under the head.

In the 1970s, electronic equivalents of serial-access rotating memories were developed, including memories based on charge-coupled devices (CCDs) and others that used magnetic bubbles. Both types of devices were roughly equivalent to very large serial-in, serial-out shift registers with their serial output connected back into the serial input. To read a particular location, you would to clock the shift register until the desired bit appeared at the serial output, and to write the location you would substitute the desired new value at the serial input.

Although they offered higher density (more bits) than DRAMs at the time they were developed, CCD and magnetic-bubble memories never gained much commercial acceptance. One reason for this was the enormous inconvenience of serial access. Another was that they were never more than a few years ahead of DRAMs in the densities that they could achieve.

11.3 STATIC RAM

11.3.1   Static-RAM Inputs and Outputs

Like a ROM, a RAM has address and control inputs and data outputs, but it also has data inputs. The inputs and outputs of a simple $2^n{\times}b$-bit static RAM are shown in Figure 11–19. The control inputs are comparable to those of a ROM, with the addition of a *write-enable (/WE) input*. When /WE is asserted, the data inputs are written into the selected memory location.

write-enable (/WE)
input

Figure 11–19
Basic structure of a $2^n{\times}b$ RAM.

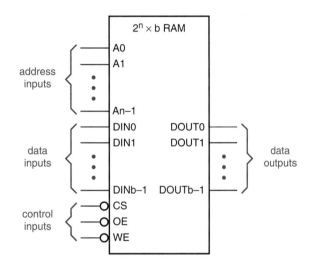

The memory locations in a static RAM behave like D latches, rather than edge-triggered D flip-flops. This means that whenever the /WE input is asserted, the latches for the selected memory location are "open" (or "transparent"), and input data flows into and through the latch. The actual value stored is whatever is present when the latch closes.

Static RAM is typically accessed as follows:

Read An address is placed on the address inputs and /CS and /OE are asserted. The latch outputs for the selected memory location are delivered to DOUT.

Write An address is placed on the address inputs and a data word is placed on DIN; then /CS and /WE are asserted. The latches in the selected memory location open, and the input word is stored.

A certain amount of care is needed when accessing SRAM because it is possible to inadvertently "clobber" one or more locations while writing a selected one, if the SRAM's timing requirements are not met. The next subsection gives details on SRAM internal structure to show why this is so, and the next section explains the actual timing behavior and requirements.

11.3.2 Static-RAM Internal Structure

SRAM cell

Each bit of memory (or *SRAM cell*) in a static RAM has the same functional behavior as the circuit in Figure 11–20. The storage device in each cell is a D latch. When a cell's /SEL input is asserted, the stored data is placed on the cell's output, which is connected to a bit line. When both /SEL and /WR are asserted, the latch is open and a new data bit is stored.

SRAM cells are combined in an array with additional control logic to form a complete static RAM as shown in Figure 11–21 for an 8×4 SRAM. As in a simple ROM, a decoder on the address lines selects a particular row of the SRAM to be accessed at any time.

Although Figure 11–21 is a somewhat simplified model of internal SRAM structure, it accurately portrays several important aspects of SRAM behavior:

- During read operations, the output data is a combinational function of the address inputs, as in a ROM. No harm is done by changing the address lines while the output data bus is enabled. The access time for read operations is specified from the time that the last address input becomes stable.

- During write operations, the input data is stored in *latches*. This means that the data must meet certain setup and hold times with respect to the *trailing* edge of the latch enable signal. That is, input data need not be stable at the moment /WR is asserted internally; it must only be stable a certain time before /WR is negated.

- During write operations, the address inputs *must* be stable for a certain setup time before /WR is asserted internally and for a hold time after /WR is negated. Otherwise, data may be "sprayed" all over the array because of the glitches that may appear on the SEL lines when the address inputs of the decoder are changed.

write cycle

- Internally, /WR is asserted only when both /CS and /WE are asserted. Therefore, a *write cycle* begins when both /CS and /WE are asserted, and ends when either is negated. Setup and hold times for address and data are specified with respect to these events.

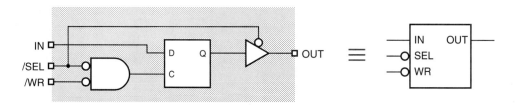

Figure 11–20 Functional behavior of a static-RAM cell.

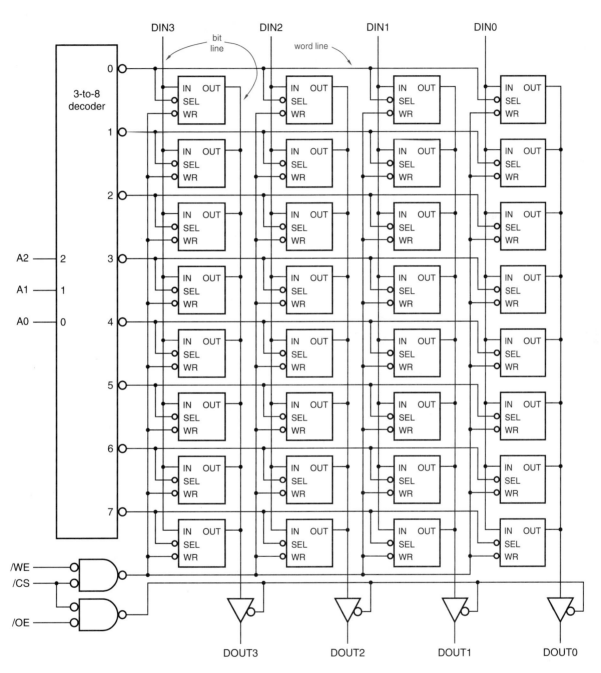

Figure 11–21   Internal structure of an 8×4 static RAM.

11.3.3 Static-RAM Timing

Figure 11–22 shows the timing parameters that are typically specified for read operations in a static RAM; they are described below:

t_{AA}
access time from address

t_{AA} *Access time from address.* Assuming that /OE and /CS are already asserted, or will be soon enough not to make a difference, this is how long it takes to get stable output data after a change in address. When designers talk about a "70-ns SRAM," they're usually referring to this number.

t_{ACS}
access time from chip select

t_{ACS} *Access time from chip select.* Assuming that the address and /OE are already stable, or will be soon enough not to make a difference, this is how long it takes to get stable output data after /CS is asserted. Most often, this parameter is identical to t_{AA}, but it is sometimes longer in SRAMs with a "power-down" mode, and shorter in SRAMs without one.

t_{OE}
output-enable time

t_{OE} *Output-enable time.* This is how long it takes for the three-state output buffers to leave the high-impedance state when /OE and /CS are both asserted. This parameter is normally less than t_{ACS}, so it is possible for the RAM to start accessing data internally before /OE is asserted; this feature is used to achieve fast access times while avoiding "bus fighting" in many applications.

t_{OZ}
output-disable time

t_{OZ} *Output-disable time.* This is how long it takes for the three-state output buffers to enter the high-impedance state after /OE or /CS is negated.

t_{OH}
output-hold time

t_{OH} *Output-hold time.* This parameter specifies how long the output data remains valid after a change in the address inputs.

If you've been paying attention, you may have noticed that the timing diagram and timing parameters for SRAM read operations are identical to what we discussed for ROM read operations in Section 11.1.5. That's the way it is; when

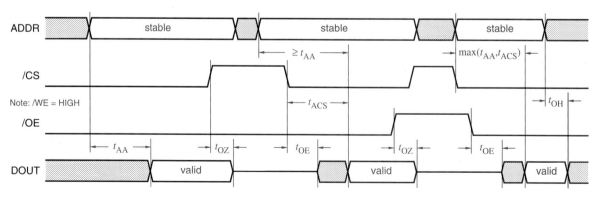

Figure 11–22 Timing parameters for read operations in a static RAM.

they're not being written, SRAMs can be used just like ROMs. The same is not true for DRAMs, as we'll see later.

Timing parameters for write operations are shown in Figure 11–23 and are described below:

t_{AS} *Address setup time before write.* All of the address inputs must be stable at this time before both /CS and /WE are asserted. Otherwise, the data stored at unpredictable locations may be corrupted.

t_{AH} *Address hold time after write.* Analogous to t_{AS}, all address inputs must be held stable until this time after /CS or /WE is negated.

t_{CSW} *Chip-select setup before end of write.* /CS must be asserted at least this long before the end of the write cycle in order to select a cell.

t_{WP} *Write-pulse width.* /WE must be asserted at least this long to reliably latch data into the selected cell.

t_{DS} *Data setup time before end of write.* All of the data inputs must be stable at this time before the write cycle ends. Otherwise, the data may not be latched.

t_{DH} *Data hold time after end of write.* Analogous to t_{DS}, all data inputs must be held stable until this time after the write cycle ends.

Manufacturers of SRAMs specify two types of write cycles, *WE controlled* and *CS controlled*, as shown in the figure. The only difference between these cycles is whether /WE or /CS is the last to be asserted and the first to be negated when enabling the SRAM's internal write operation.

The write-timing requirements of SRAMs could be relaxed somewhat if, instead of using latches, the cells contained edge-triggered D flip-flops with a

t_{AS}
address setup time

t_{AH}
address hold time

t_{CSW}
chip-select setup

t_{WP}
write-pulse width

t_{DS}
data setup time

t_{DH}
data hold time

WE-controlled write
CS-controlled write

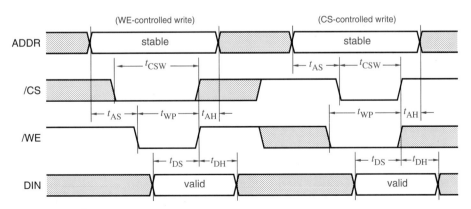

Figure 11–23   Timing parameters for write operations in a static RAM.

common clock input and enable inputs tied controlled by SEL and WR. However, this just isn't done, because it would at least double the chip area of each cell, since a D flip-flop is built from two latches. Thus, the logic designer is left to reconcile the SRAM's latch-type timing with the edge-triggered register and state-machine timing used elsewhere in a system.

A large SRAM does not contain a physical array whose dimensions equal the logical dimensions of the memory. As in a ROM, the SRAM cells are laid out in an almost square array, and an entire row is read internally during read operations. For example, the layout of a 32K×8 SRAM chip might be very similar to that of a 32K×8 ROM shown in Figure 11–9 on page 733. During read operations, column multiplexers pass the required data bits to the output data bus as specified by a subset of the address bits (A5–A0 in the ROM example). For write operations, the write-enable circuitry is designed so that only one column in each subarray is enabled, as determined by the same subset of the address bits.

11.3.4   Standard Static RAMs

Static RAMs are available in many sizes and speeds. In 1993, the largest commonly available SRAMs used CMOS technology, were organized as 512K×8 or 1M×4 bits (4 Mbits), and had access times as fast as 15 ns. The fastest TTL/CMOS compatible SRAMs also used CMOS technology and contained only 1K×4 bits (4 Kbits), but had access times of 2.7 ns!

6164

Shown in Figure 11–24, standard 8K×8 and 32K×8 SRAMs have pinouts that are patterned after the like-size EPROMs in Figure 11–11. The *6164* 8K×8 SRAM is similar to a 2764 EPROM, except that it has a write-enable input and two chip-select inputs instead of one; /CS1 and CS2 both must be asserted to enable the chip. The *5256* 32K×8 SRAM is similar to a 27256 EPROM, with the addition of a write-enable input. Both chips are available from many manufacturers, sometimes with different part numbers, and have access times in the range of 15–150 ns.

5256

The SRAMs in Figure 11–24 have bidirectional data buses, that is, they use the same data pins for both reading and writing. This necessitates a slight change in their internal control logic, shown in Figure 11–25. The output buffer is automatically disabled whenever /WE is asserted, even if /OE is asserted. However, the timing parameters and requirements for read and write operations are practically identical to what we described in the preceding subsection.

A common use of SRAMs is to store data in small microprocessor systems, often in "embedded control" applications (telephones, toasters, electric bumpers, etc.). General-purpose computers more often use DRAMs, discussed in the next subsection, because their density is greater and their cost per bit lower. However,

Figure 11–24
Logic symbols for standard
SRAMs in 28-pin dual in-line
packages.

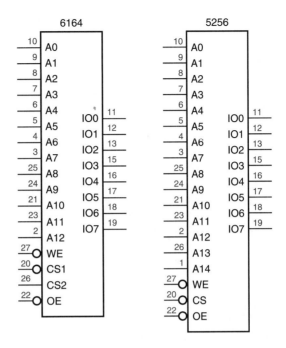

very fast SRAMs are often used in the "cache" memories of high-performance
computers to store frequently used instructions and data.

11.4 DYNAMIC RAM

The basic memory cell in an SRAM, a D latch, requires four gates in a discrete
design, and four to six transistors in a custom-designed SRAM LSI chip. In
order to build RAMs with higher density (more bits per chip), chip designers
invented memory cells that use as little as one transistor per bit.

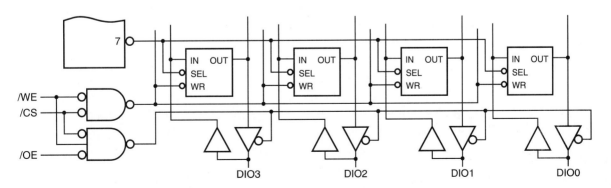

Figure 11–25 Output-buffer control in an SRAM with a bidirectional data bus.

11.4.1 Dynamic-RAM Structure

*dynamic RAM
(DRAM)*

It is not possible to build a bistable element with just one transistor. Instead, the memory cells in a *dynamic RAM (DRAM)* store information on a tiny capacitor accessed through an MOS transistor. Figure 11–26 shows the storage cell for one bit of a DRAM, which is accessed by setting the word line to a HIGH voltage. To store a 1, a HIGH voltage is placed on the bit line, which charges the capacitor through the "on" transistor. To store a 0, LOW voltage on the bit

precharge

line discharges the capacitor. To read a cell, the bit line is first *precharged* to a voltage halfway between HIGH and LOW, and then the word line is set HIGH. Depending on whether the capacitor voltage is HIGH or LOW, the precharged bit

sense amplifier

line is pulled slightly higher or slightly lower. A *sense amplifier* detects this change and recovers a 1 or 0 accordingly. Note that reading a cell destroys the original voltage stored on the capacitor, so that the recovered data must be written back into the cell after reading.

The capacitor in a DRAM cell has a very small capacitance, but the MOS transistor that accesses it has a very high impedance. Therefore, it takes a relatively long time (many milliseconds) for a HIGH voltage to discharge to the point that it looks more like a LOW voltage. In the meantime, the capacitor stores one bit of information.

Naturally, using a computer would be no fun if you had to reboot every few milliseconds because its memory contents disappeared. Therefore, memory

refresh cycle

systems built with DRAMs use *refresh cycles* to update every memory cell every so often, usually once every four milliseconds. This involves sequentially reading the somewhat degraded contents of each cell into a D latch, and writing back a nice solid LOW or HIGH value from the latch. Figure 11–27 illustrates the electrical state of a cell after a write and a sequence of refresh operations.

The first DRAMs, introduced in the early 1970s, contained only 1,024 bits, but modern DRAMs are available containing 16 *megabits* or more. If you had to refresh every cell, one at a time, in four milliseconds, you'd have a real problem—that works out to just 1 ns per cell, and includes no time for useful read and write operations. Fortunately, as we'll show next, DRAMs are organized

Figure 11–26
Storage cell for one bit
in a DRAM.

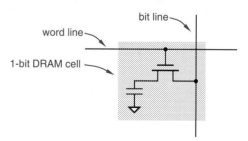

Figure 11–27
Voltage stored in a DRAM
cell after writing and refresh
operations.

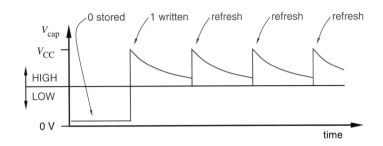

using two-dimensional arrays, and a single operation refreshes an entire row of
the array. Typical DRAM arrays have 256 rows, requiring 256 refresh operations
every four milliseconds, or one about every 15.6 μs. A refresh operation typically
takes less than 200 ns, so the DRAM is available for useful read and write
operations over 98% of the time.

Figure 11–28 is a block diagram of the internal structure of a 64K×1
DRAM. The logical array has 64K×1 bits, but the physical array is square,
containing 256×256 bits. Although the memory has 64K locations, the chip has
only eight *multiplexed address inputs*. A complete 16-bit address is presented
to the chip in two steps controlled by two signals—*/RAS (row address strobe)*
and */CAS (column address strobe)*. Multiplexing the address inputs saves pins,
important for compact design of memory systems, and also fits quite naturally
with the two-step DRAM access methods that we'll describe shortly.

*multiplexed address
  inputs*
*/RAS (row address
  strobe)*
*/CAS (column
  address strobe)*

Figure 11–28
Internal structure of a
64K×1 DRAM.

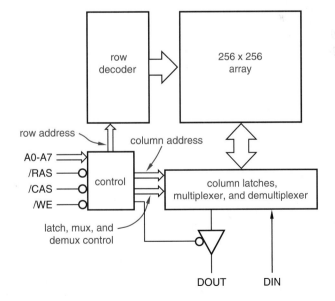

11.4.2   Dynamic-RAM Timing

RAS-only refresh cycle

row-address register

row latch

DRAM timing is quite a bit more complicated than SRAM timing. Timing waveforms for three representative DRAM cycles are shown in Figure 11–29. The first is called a *RAS-only refresh cycle*, and is used to refresh a row of memory without actually reading or writing any data at the external pins of the DRAM chip. The cycle begins when a row address is applied to the multiplexed address inputs (8 bits in the case of a 64K×1 DRAM) and /RAS is asserted. The DRAM stores the row address in an internal *row-address register* on the falling edge of /RAS and reads the selected row of the memory array into an on-chip *row latch*. When /RAS is negated, the contents of the row are written back from

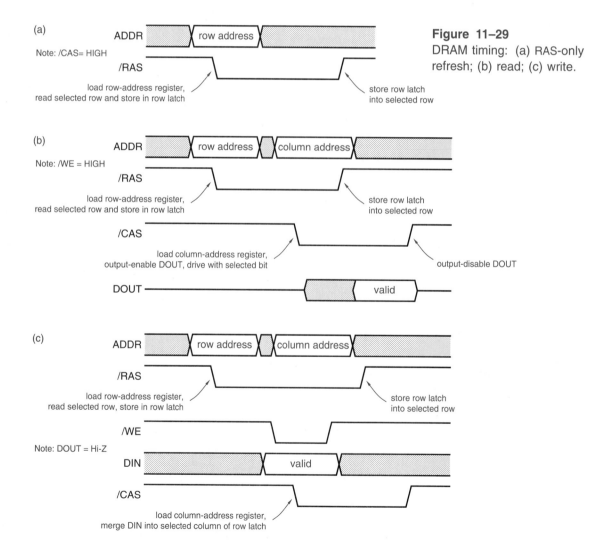

Figure 11–29
DRAM timing: (a) RAS-only refresh; (b) read; (c) write.

the row latch. To refresh an entire 64K×1 DRAM, the logic designer must ensure that 256 such cycles, using all 256 possible row addresses, are executed every four milliseconds. An external 8-bit counter may be used to generate the row addresses, and a timer is used to initiate a refresh cycle once every 15.6 μs.

A *read cycle*, shown in (b), begins like a refresh cycle, where a selected row is read into the row latch. Next, a column address is applied to the multiplexed address inputs, and is stored in an on-chip *column-address register* on the falling edge of /CAS. The column address is used to select one bit of the just-read row, which is made available on the DRAM's DOUT pin. A three-state pin, DOUT is output-enabled as long as /CAS is asserted. In the meantime, the entire row is written back into the array as soon as /RAS is negated.

read cycle

column-address register

A *write cycle*, shown in (c), also begins like a refresh or read cycle. However, */WE (write enable)* must be asserted before /CAS is asserted to select a write cycle. One thing this accomplishes is to disable DOUT for the rest of the cycle, even though /CAS will be asserted subsequently. Once the selected row is read into the row latch, /WE also forces the input bit on DIN to be merged into the row latch, in the bit position selected by the column address. When the row is subsequently written back into the array on the rising edge of /RAS, it contains a new value in the selected column.

write cycle

/WE (write enable)

Several other types of cycles, not shown in the figure, are available in typical DRAMs:

- A *CAS-before-RAS refresh cycle* performs a refresh operation without requiring a row address to be supplied by an external counter. Instead, the DRAM chip maintains an internal row-address counter. If /CAS is asserted before /RAS is asserted, the DRAM refreshes the row selected by the counter and then increments the counter. This simplifies DRAM system design, eliminating the external refresh counter and reducing from three (row, column, refresh) to two the number of multiplexed sources that must drive the DRAM address inputs.

CAS-before-RAS refresh cycle

TRICKY TIMING

You may have noticed that the valid intervals of row and column addresses in Figure 11–29 are skewed to the right with respect to the falling edges of /RAS and /CAS, which load the addresses into on-chip edge-triggered registers. DRAMs typically have a short setup-time requirement for the address inputs (often 0 ns), but a relatively large, positive hold-time requirement (10–20 ns). This can cause some problems in the design of an otherwise fully synchronous system that uses a single clock and flip-flops with zero hold times, and is often solved by distasteful methods involving tapped delay lines or using the opposite edge of the clock. This is one of the reasons that many designers consider DRAM system design a tricky, black art.

read-modify-write cycle
- A *read-modify-write cycle* begins like a read cycle, and allows data to be read on DOUT when /CAS is asserted. However, /WE may be asserted subsequently to write new data back at the same location.

page-mode read cycle
- A *page-mode read cycle* allows up to an entire row ("page") of data to be read without repeating the entire RAS-CAS cycle. Since the entire row is already stored in the row latch, this cycle simply requires /CAS to be pulsed LOW multiple times while /RAS remains asserted continuously. A new column address is provided and a new bit becomes available on DOUT on each falling edge of /CAS. This cycle provides much faster memory access during a sequence of reads to sequential or otherwise nearby addresses; such sequences often occur in computers when they're fetching instructions or filling caches.

page-mode write cycle
- Similarly, a *page-mode write cycle* allows multiple bits of a row to be written with a single /RAS and multiple /CAS cycles.

static-column-mode read cycle
- A *static-column-mode read cycle* is similar to a page-mode read cycle, except that the column address is not stored in an on-chip register; instead it is applied directly to the column multiplexer. Therefore, another bit in the same row may be read at any time simply by changing the column address, *without* pulsing /CAS. This simplifies the external DRAM control logic.

static-column-mode write cycle
- Similarly, a *static-column-mode write cycle* allows the column address to be changed at any time when writing multiple bits in the same row. However, it is necessary to negate either /CAS or /WE when the address is changing, to avoid "spraying" data all over the selected row because of glitches in the column-address decoder.

nibble-mode read cycle
- A *nibble-mode read cycle* is similar to a page-mode read cycle, except that a new column address is not supplied on /CAS pulses after the first. Instead, the DRAM chip generates a sequence of addresses from the address supplied at the initial RAS-CAS cycle. The sequence repeats after four /CAS pulses, hence the name.

nibble-mode write cycle
- Similarly, a *nibble-mode write cycle* allows a sequence of four addresses to be written.

Note that page-mode, static-column-mode, and nibble-mode cycles are mutually exclusive. That is, a given DRAM chip (with a specific part number) supports at most one of these modes.

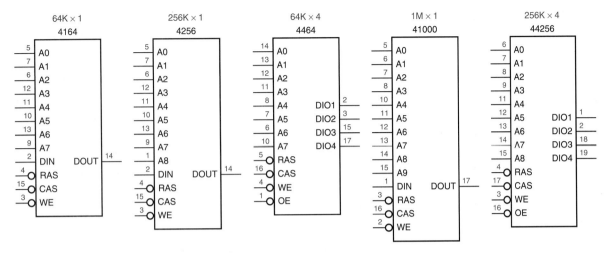

Figure 11–30 Logic symbols for standard DRAMs in dual in-line packages.

Figure 11–30 shows the logic symbols for commonly used DRAMs in dual in-line packages. Note that DRAMs are available in 4-bit-wide as well as in 1-bit-wide configurations. The "size" of a DRAM is the total number of bits it stores; thus, a 44256 256K×4 device is still called a "1-Mbit DRAM." The ×4 configurations are useful in applications that do not require a huge total amount of memory, but can benefit from reduced chip count using high-density DRAMs. In fact, as the 16-Mbit generation of DRAMs emerges, we can expect to see devices introduced with ×8 or even ×16 organizations. Thus, one 1M×16 device can satisfy the entire memory requirement of a low-cost PC clone with 2 Mbytes of memory.

4164
4256
4464
41000
44256

REFERENCES

Manufacturers of ROM and RAM devices publish data books with the specifications for their parts. Intel's data books on their ROM devices are particularly good in that they contain application notes on EPROM and EEPROM programming and use.

Industry publications such as *Computer Design* and *Electronic Design* carry tutorial and update articles from time to time that describe the capabilities and applications of the latest memory chips. For example, Jeffrey Child provided an update on "Ultra-fast SRAMs" in the June 1993 issue *Computer Design* (pp. 101–107). And Dave Bursky reported on "High-Performance DRAMS" in the July 22, 1993, issue of *Electronic Design* (pp. 55–70), describing the capabilities of several new DRAM architectures: enhanced DRAM (EDRAM), cache-DRAM (CDRAM), synchronous DRAM (SDRAM), and RAMBUS DRAM (RDRAM).

DRILL PROBLEMS

11.1 Determine the ROM size needed to realize the combinational logic function in each of the following figures in Chapters 4 and 5: 4–43(b), 5–30, 5–68, and 5–94.

11.2 Determine the ROM size needed to realize the combinational logic function performed by each of the following MSI parts: 74x49, 74x139, 74x153, 74x257, 74x381, 74x682.

11.3 Draw a logic symbol for and determine the size of a ROM that realizes a combinational logic function that can perform the function of either a 74x381 or a 74x382 depending on the value of a single MODE input.

11.4 Draw a logic symbol for and determine the size of a ROM that realizes an 8×8 combinational multiplier.

11.5 Show how to design a 128K×8 SRAM using 5256 SRAMs and a combinational MSI part as building blocks.

EXERCISES

11.6 Our discussion of ROM sneak paths in connection with Figure 11–6 claimed that with A2–A0 = 101, bit lines /D2 and /D0 are pulled LOW through the direct connections. That's not really correct, unless the 74x138 is replaced with a decoder with open-collector outputs. Explain.

11.7 Describe the logic function of seven variables that is performed by the 128×1 ROM in Figure 11–8. Starting with the ROM pattern, one way to describe the logic function is to write the corresponding truth table and canonical sum. However, since the canonical sum has 64 7-variable product terms, you might want to look for a simple but precise word description of the function.

11.8 For the two-output logic function of Figure 4–43(b), compare the number of diodes or transistors needed in the AND-OR, AND, or storage array for PLA, PAL, and ROM realizations.

11.9 Show how to double the number of different attenuation amounts that can be selected in the digital attenuator of Figure 11–17 without increasing the size of the ROM.

11.10 Write the Pascal functions UlawTOlinear and LinearTOulaw that are required in Table 11–6. In LinearTOulaw, you should select the μ-law byte whose theoretical value is closest to the linear input. (*Hint:* The most efficient approach builds, at program initialization, a 256-entry table that is used by both functions.) Both functions should report out-of-range inputs.

11.11   Modify the Pascal program in Table 11–6 to perform clipping in cases where the designer has specified a particular attenuation amount to be a gain, and multiplying an input value by the attenuation factor produces an out-of-range result.

11.12   Write a Pascal program to generate the contents of a 256×8 ROM that converts from 8-bit binary to 8-bit Gray code. (*Hint:* Your program should embody the second Gray-code construction method in Section 2.11.)

11.13   Write a Pascal program to generate the contents of a 256×8 ROM that converts from 8-bit Gray code to 8-bit binary. (*Hints:* You may use the results of Exercise 11.12. It doesn't matter if your program is slow.)

11.14   A certain communication system has been designed to transmit ASCII characters serially on a medium that requires an average signal level of zero, so a 5-out-of-10 code is used to code the data. Each 7-bit ASCII input character is transmitted as a 10-bit word with five 0s and five 1s. Write a Pascal program to generate the contents of a 128×10 ROM that converts ASCII characters into coded words.

11.15   The receiving end of the system in Exercise 11.14 must convert each 10-bit coded word back into a 7-bit ASCII character. Write a Pascal program to generate the contents of a 1K×8 ROM that converts coded words onto ASCII characters. The extra output bit should be an "error" flag that indicates when a noncode word has been received.

11.16   How many ROM bits would be required to build a 16-bit adder/subtracter with mode control, carry input, carry output, and two's-complement overflow output? Be more specific than billions and billions, and explain your answer.

11.17   Repeat Exercise 11.16 assuming that you may use *two* ROMs, so that the delay for a 16-bit addition or subtraction is twice the delay through one ROM. Assume that the two ROMs must be identical in both size and programming. Try to minimize the total number of ROM bits required, and sketch the resulting adder/subtracter circuit. Can the number of ROM bits be further reduced if the two ROMs are allowed to be different?

11.18   Show how, using additional SSI/MSI parts, a 2764 can be used as a 64K×1 ROM. What is the access time of the 64K×1 ROM?

11.19   Show how, using additional SSI/MSI parts, a 2764 can be used as a 2K×32 ROM. You may assume the existence of a free-running clock signal whose period is slightly longer than the access time of the 2764. What is the access time of the 2K×32 ROM?

11.20   Show how to build the μ-law adder circuit of Figure 11–18 with a 32K×8 ROM and two XOR gates. Also, write a Pascal program to generate the ROM contents.

11.21   Determine the ROM size needed to build the fixed-point to floating-point encoder of Figure 5–96. Draw a logic diagram using a commercially available ROM.

11.22 Write a Pascal program to generate the ROM contents for Exercise 11.21. Unlike the original MSI solution, your Pascal program should perform rounding; that is, for each fixed-point number it should generate the nearest possible floating-point number.

11.23 Draw a complete logic diagram for a circuit that performs combinational multiplication of a pair of 8-bit unsigned or two's-complement integers. Signed versus unsigned operation should selected by a single input, SIGNED. You may use any of the commercial ROMs in Figure 11–11, as well as discrete gates.

11.24 Write and test a Pascal or C program that generates the contents of the ROM(s) in Exercise 11.23.

11.25 For each of the timing parameters defined in Section 11.3.3, determine whether its value for the 128K ×8 SRAM you designed in Drill 11.5 is the same as the 5256's. If it is different, indicate the new value. Use the worst-case values in the 74FCT column of Table 5–3 to determine the delays for the MSI part.

synchronous SRAM   11.26 Unlike a conventional SRAM, a *synchronous SRAM* has a clock input, and read and write operations take place on the rising edge of the clock. For read operations, the address and chip-select inputs must be stable a certain time before the rising clock edge, and the output changes after the edge. For write operations, the write-enable and address signals must be stable a certain time before the *falling* clock edge, and the data must be stable a certain time before the rising clock edge; data is stored at the rising edge. Using a 6164 8K×8 SRAM and a handful of MSI parts as building blocks, design an 8K×8 synchronous SRAM.

11.27 Define the relevant timing parameters for the synchronous SRAM as defined in Exercise 11.26.

11.28 Calculate the values of the timing parameters you defined in Exercise 11.27 for your solution to Exercise 11.26.

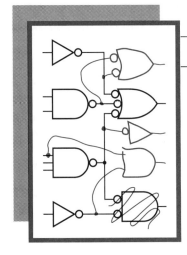

12

ADDITIONAL REAL-WORLD TOPICS

This chapter contains a collection of "real-world" topics that are certain to interest you at some point if you are (or become) a working digital designer. We barely scratch the surface of each of these topic areas, so be sure to consult the References in areas of particular interest.

12.1 COMPUTER-AIDED ENGINEERING TOOLS

"If it wasn't hard, they wouldn't call it hardware."

Many digital designers with twenty years of experience consider this statement to be indisputable. Yet more and more, digital design is being carried out using *software*, and is getting easier as a result.

The terms *computer-aided design (CAD)* and *computer-aided engineering (CAE)* are used to refer to software tools that aid the development of circuits, systems, and many other things. "CAD" is the more general term and applies to tools both inside and outside the electronics area, including architectural and mechanical design tools, for example. Within electronics, "CAD" often refers to *physical-design* tools, such as IC- and PCB-layout programs. "CAE" is used

computer-aided design (CAD)
computer-aided engineering (CAE)

more often to refer to *conceptual-design* tools, such as schematic editors, circuit simulators, and PLD compilers. However, a lot of people in electronics (including the author) tend to use the two terms interchangeably. In this section, we'll introduce some CAD/CAE tools that are often used by digital designers.

<table>
<tr><td>IS HARDWARE
NOW
EASY-WARE?</td><td>Since more and more hardware design and debugging is being carried out using software tools, is it really getting easier? Not necessarily.

    In the author's experience, the increasing use of CAD means that instead of spending time fighting with soldering irons and test clips, many designers can now spend their time fighting with buggy CAD programs running in the wonderful environment of DOS-based PCs. (Just in case you're not a PC user, the above comment is intended as mild sarcasm.)</td></tr>
</table>

12.1.1 Schematic Capture

Except in homework assignments, your first step in designing a circuit usually is to convince someone that your proposed approach is the right one. That means preparing block diagrams and presentation slides, and discussing your ideas with
design review managers and peers in a preliminary *design review*. Once your project has been approved, you can safely start on the "fun stuff," that is, drawing the schematic.

    Schematics used to be drawn by hand, but most are now prepared using
schematic editor *schematic editors*, CAE programs that run on engineering workstations. The
schematic capture process of creating a schematic on a computer is often called *schematic capture*. (Huh? How did it get loose?) This term is used because the schematic editor captures more information than just a drawing. Most information in the schematic has a "type" associated with it, so that selected information can be retrieved automatically as required later in the design process.

    In order to simplify both entry and retrieval of information in a schematic, a
fields schematic editor typically provides *fields* for each information type, either hidden or drawn automatically nearby the associated schematic element. Some typical fields and their uses are described below.

component type
- *Component type.* By specifying a component type (e.g., resistor, capacitor, 74LS00, 74FCT374), the designer can call out a predrawn symbol from a

<table>
<tr><td>WHAT'S A
WORKSTATION?</td><td>An "engineering workstation" nowadays can be anything from a $500 PC clone to a $100,000 parallel processor. The "canonical" engineering workstation in terms of cost (what an average company will spend per engineer) and performance (among the best) is a $10,000 Sun workstation, whatever model you get for that price this year.</td></tr>
</table>

component library. Some components have user-selectable alternate symbols (e.g., DeMorgan equivalents for gates). The component type is used both for manufacturing documentation and for simulation.

component library

- *Component value*. Most analog components have a value that must be specified by this field (e.g., 2.7 kΩ). Additional distinguishing information may appear in a "rating" or "tolerance" field (e.g., 1/4 W, 1%). Digital components may have a speed rating in this field (e.g., -15 for a 15-ns PLD).

component value

- *Approved-part number*. Using all of the distinguishing information from the previous fields, the CAE software may automatically select a part number from the company's "approved parts list" (e.g., 126-10117-0272 for a 2.7-kΩ, 1/4-W, 1% metal-film resistor), or a warning if no such part is available. (No, in large companies you can't use just any part in the catalog; all parts must be approved through a *parts-qualification* process.)

approved-part number

parts qualification

- *Reference designator*. This alphanumeric label distinguishes among multiple instances of similar parts, and is used just about everywhere.

reference designator

- *Component location*. Using a coordinate system, this field may indicate the physical position of the component on an assembled PCB, simplifying the task of finding it during debugging. Of course, the final component location generally is not known during the initial circuit design. However, CAE software with *back-annotation* capability can insert this information into the schematic when the PCB layout is completed.

component location

back annotation

- *Pin numbers*. These specify the assignment of the signal pins on each component, and are usually prespecified in the symbol that is called up from the component library. However, for SSI and other parts that have multiple identical sections in a single package, pin numbers (and even reference designators) might not be specified until the PCB layout is completed; this is another case where back annotation is handy.

pin number

- *Wire type*. Usually the only wire types are "ordinary" and "bus," where a bus is just a set of ordinary wires that are grouped for drawing purposes. However, other wire types, such as extra-wide PCB traces for analog signals that carry high currents, may be indicated in some systems; such information is passed to the PCB-layout program.

wire type

- *Signal name*. The user may (and should) associate a name with each signal. The name is especially useful in simulation and debugging, but it does not affect the physical PCB layout.

signal name

connection flag
- *Connection flag.* This drawing element indicates the connection of one signal line to another one on the same or a different page. Software can ensure that all outgoing connections are matched by incoming ones. In a hierarchical schematic, software can use the connection flags to generate a "logic symbol" that represents the entire page.

Once you've "captured" a circuit's schematic in a CAE system, there are lots of things you can do with it besides printing it. At least two documents can be generated for manufacturing purposes:

parts list
- *Parts list.* This is a list of components and their reference designators.

net
net list
- *Net list.* A *net* is a set of pins that are all connected to the same electrical node or signal. A *net list* specifies all such connections that are required by the schematic, typically using an alphabetically sorted list of signal names. For each signal name, it lists the device pins (identified by reference designator and pin number) where that signal appears.

pin list
Alternatively, the net list may be sorted by reference designator and pin number, and show the signal name connected to each pin (sometimes this is called a *pin list*).

The parts list and net list are the main inputs to the PCB-layout process. The designer may also use the parts list to estimate the cost and reliability of the circuit, and the manufacturing department obviously can use it to order the parts.

Other interesting things you can do once you have a schematic are to analyze and simulate the circuit, as discussed in the next subsection.

12.1.2 Circuit Analysis and Simulation

component model
The component library of a sophisticated CAE system contains more than just a symbol for each component. For ICs, the library may contain a *component model* describing the device's logical and electrical operation. Component models, along with the interconnection information in the schematic, can be used to find design errors.

design-rule checker
As a minimum, the model for an IC indicates whether each pin is an input or output. With just this information, a *design-rule checker* can detect some of the more common "dumb errors" in a design, such as shorted outputs and floating inputs. With the addition of input loading and output driving characteristics for each pin, the program can also determine whether any output's fanout capability is exceeded.

timing verification
The next step is *timing verification.* Even without a detailed model of an IC's logical behavior, the component library can provide a worst-case delay value for each input-to-output path, and setup and hold times for clocked devices.

Using this information, a *timing verifier* can find the worst-case delay paths in   *timing verifier*
the overall circuit, so the designer can determine whether the timing margins are
adequate.

Finally, the library may contain a detailed model of each component's logi-
cal behavior, in which case *simulation* may be used to predict the overall circuit's   *simulation*
behavior for any given sequence of inputs. The designer provides an input sequence, and a *simulator* program determines how the circuit would respond to   *simulator*
that sequence. The simulator output is usually displayed graphically, in the form
of timing waveforms that the designer could see on an oscilloscope or logic analyzer if the same inputs were applied to a real circuit. In such an environment, it
is possible to debug an entire circuit without "breadboarding" it, and to assemble
a PCB that works on the very first try.

Simulators use different component models depending on the type of component. Analog simulators like SPICE may use mathematical models with just
one parameter for a resistor, and with anywhere from a few to dozens of parameters for a bipolar transistor. Based on these models and the circuit's structure,
the simulator develops and solves equations to determine the circuit's behavior.

Digital simulators normally use one of the following models for each logic
element:

- *Behavioral model.* In such a model, the software "understands" the basic   *behavioral model*
  function of the logic element. For example, an AND gate is understood to
  produce a 1 output with delay t_{pLH} after all inputs are 1, and a 0 output
  with delay t_{pHL} after any input is 0.

- *Structural model.* Larger elements may be modeled as a collection of   *structural model*
  smaller elements that have behavioral models. For example, a 74x138
  decoder may be modeled as a set of gates connected according to the internal
  logic diagram of Figure 5–28 on page 304. Alternatively, the '138 could
  have a behavioral model consisting of a truth table and timing information.

Both of the foregoing models are called *software models* because the chip's op-   *software model*
eration is simulated entirely by software. Some chips, including microprocessors
and other LSI parts, require huge software models. In most cases, the chip
manufacturers don't provide such models (they're considered to be proprietary
information), and the models would take months or years for a user to develop.
To get around this problem, innovative CAE companies have developed a third
type of model:

- *Physical model*, sometimes called a *hardware model*. A real, working copy   *physical model*
  of the chip is connected to the computer that's running the simulation. The   *hardware model*
  simulator applies inputs to the real chip and observes its outputs, which are
  then used as inputs to the other models in the simulation.

event queue

Regardless of the model type(s), the simulator's operation is driven by an *event queue*, a time-ordered list of signal transitions. For example, when the simulator determines that a 0-to-1 transition occurs at time t at the input of an inverter with a 5 ns delay, it inserts a 1-to-0 transition at the inverter's output at time $t+5$. Eventually, after processing all of the other events that occur between times t and $t+5$, the simulator propagates the change on the inverter output to the inputs that it affects.

A digital designer's input to a logic simulator is a schematic and a sequence of input vectors to be applied to the circuit in simulated time. The first output that the designer usually sees is "XXXXX," meaning that the simulator can't figure out what the output will be. This usually occurs because the circuit contains sequential circuits (flip-flops and latches) that are not initialized. To obtain meaningful simulator output, the designer must ensure that all circuits are explicitly reset to a known value, even if reset is not required for correct operation in the real system. For example, the self-correcting ring counters of Figure 9–46 on page 622 cannot be simulated, because the simulator has no way of knowing their starting state.

uninitialized sequential circuits

Once the simulated circuit has been initialized, the designer's ability to probe its operation is limited only by the simulator's speed. Simulation of complex circuits can be very slow, compared to the speed of the real thing. Depending on the level of detail being simulated, the simulation time can be 10^3 to 10^8 times longer than the interval being simulated. A strictly functional simulation, in which all components are assumed to have zero delay, is fastest. A much more realistic simulation, one that calculates and considers a worst-case range of delays for all signal paths, is the slowest.

simulation inefficiency

12.1.3 Hardware Description Languages

hardware description language

Most logic design is performed graphically, using block diagrams and schematics, but many different textual languages, called *hardware description languages*, have been developed for describing logic circuits and systems. PLD programming languages like ABEL are examples of hardware description languages for a predefined, limited collection of circuit structures. General state-machine description languages with features similar to ABEL's "state diagrams" can be used to describe arbitrary clocked, synchronous state machines. Other languages can describe an even wider range of circuit structures.

register-transfer language

Register-transfer languages have been used for decades to describe the operation of synchronous systems, including the simple structure of Figure 8–1 on page 541. Such a language combines the control-flow notation of a state-machine description language with a means for defining and operating on multibit registers. Register-transfer languages have been especially useful in computer design, where individual machine-language instructions are defined as a sequence of more primitive steps involving loading, storing, combining, and testing registers.

In the mid-1980s, the U.S. Department of Defense and the IEEE sponsored the development of a highly capable hardware-description language called *VHDL* that includes the following features: *VHDL*

- Designs may be decomposed hierarchically.

- Each design element has both a well-defined interface (for connecting it to other elements) and a precise behavioral specification (for simulating it).

- Behavioral specifications can use either an algorithm or an actual hardware structure to define an element's operation. An element may initially be described by an algorithm, to enable design verification of the higher-level elements that use it; the algorithm can be replaced with a hardware structure later.

- Concurrency, timing, and clocking can all be modeled. VHDL handles asynchronous as well as synchronous circuit structures.

- Circuit operation of a design can be simulated.

- Advanced VHDL-based compilers can synthesize hardware structures directly from algorithms, without designer intervention. For example, it's possible to synthesize an entire VLSI chip from a VHDL description.

Since VHDL is a government standard, you'd naturally expect to see its use grow. More important, VHDL circuit descriptions are attractive because they can provide the input to CAD tools that automatically synthesize the corresponding chip or board-level circuit and fulfill documentation, manufacturing, and testing requirements.

"VHDL" stands for "VHSIC Hardware Description Language." VHSIC, in turn, stands for "Very High Speed Integrated Circuit," a U.S. Department of Defense program to encourage the development of high-performance IC technology (using Very Healthy Sums of Instant Cash?).

THE MEANING
OF VHDL

12.1.4 Printed-Circuit-Board Layout

Printed-circuit-board layout can be carried out by a program or by a person (a PCB designer), or by a combination of the two. A PCB designer is usually someone who likes to solve puzzles. The PCB designer's job is to fit all of the components in the schematic onto a board of a given size, and then to hook up all the connections as required by the schematic, and usually to do so with the smallest possible PCB size and number of interconnection layers.

PCB design can be done automatically by software, but the results aren't always so good. Both the PCB designer and the circuit designer normally must guide the initial component-placement steps, taking into account many practical concerns, including the following:

mechanical constraints

- *Mechanical constraints.* Certain connectors, indicators, and other components may have to be placed at a particular location on the PCB because of a predetermined mechanical requirement.

critical signal paths

- *Critical signal paths.* Although an automatic placement program can attempt to place components in a way that globally minimizes the average length of interconnections, the designers usually have a much better idea of what areas of the circuit and which specific signals have the most critical requirements. In fact, the circuit designer normally provides the PCB

floorplan

  designer with a *floorplan*, a suggested placement for groups of components that reflects the designer's ideas about which things need to be close to each other.

thermal concerns

- *Thermal concerns.* Some areas of the PCB may have better cooling than others when the board is installed in a system, and certainly some components run hotter than others. The designers must ensure that the operating temperature range of the components is not exceeded.

  The temperature anywhere on a well-designed PCB in a properly cooled system typically will not exceed 10°C above ambient, but a power-hungry chip (such as a microprocessor) placed in an area with no air flow could create a temperature rise of 30°C or more. If the ambient temperature is 40°C, the actual temperature at the microprocessor would be 70°C, which is at the limit of the commercial operating range for most chips.

radio-frequency emissions

- *Radio-frequency emissions.* Electronic equipment radiates radio-frequency energy as a side effect. Government agencies in the U.S. and other countries specify the maximum permissible level of emissions; in the U.S. the limits

FCC Part 15

  are spelled out in *FCC Part 15* regulations.

  A circuit's emissions levels usual depend not only on component performance and circuit configuration, but also on the physical layout of the PCB. During layout, special attention must therefore be paid to component placement, grounding, isolation, and decoupling, to minimize unwanted (and illegal!) radio-frequency emissions.

- *Stupid mistakes.* Probably the most important step in PCB design, especially   *stupid mistakes*
  when using new or unfamiliar components, is checking for stupid mistakes.
  Almost every digital designer, even successful ones, can tell you stories of
  laying out a PCB using the mirror image of a new part's pinout, putting
  mounting holes or connectors in the wrong place, using the wrong version
  of an IC package, using the wrong sex of a connector, leaving out the inner
  connection layers when the board is fabricated, and so on. Remember,
  "Computers don't make mistakes, humans do!"

Once component placement is done, all of the signals must be hooked up;
this is called *routing*. This phase of PCB layout is most easily automated; in fact,   *routing*
a decent "auto-router" can route a moderate-sized PCB overnight. Still, design
engineers may have to give special attention to the routing of critical-path traces
and to other issues such as placement of signal terminations and test points,
which are discussed later in this chapter.

12.2 DESIGN FOR TESTABILITY

When you buy a new piece of equipment, you expect it to work properly "right
out of the box." However, even with modern automated equipment, it's impossible for an equipment manufacturer to guarantee that every unit produced will be
perfect. Some units don't work because they contain individual components that
are faulty, or because they were assembled incorrectly or sloppily, or because
they were damaged in handling. Most manufacturers prefer to discover these
problems in the factory, rather than get a bad reputation with their customers.
This is the purpose of *testing*.   *testing*

The most basic test, a *go/no-go test* yields just one bit of information—   *go/no-go test*
is the system 100% functional or not? If the answer is yes, the system can
be shipped. If not, the next action depends on the size of the system. The
appropriate response for a digital watch would be to throw it out and try again.
If many watches are turning up faulty, it's more efficient to repair what must be
a problem in the assembly process than to try to salvage individual units.

On the other hand, you wouldn't want to toss out a $10,000 computer that
fails to boot. Instead, a more detailed *diagnostic test* may be run to locate the   *diagnostic test*
particular subsystem that is failing. Depending on cost, the failed subsystem
may be either repaired or replaced. A typical digital subsystem is a PCB whose
assembled cost with components ranges from $50 to $1,000. The repair/replace
decision is an economic trade-off between the cost of the assembled PCB and
the estimated time to locate and repair the failure.

design for testability (DFT)

This is where *design for testability (DFT)* comes in. "DFT" is a general term applied to design methods that lead to more thorough and less costly testing. Many benefits ensue from designing a system or subsystem so that failures are easy to detect and locate:

- The outcome of a go/no-go test is more believable. If fewer systems with hidden faults are shipped, fewer customers get upset, which yields obvious economic as well as psychological benefits.

- Diagnostic tests run faster and produce more accurate results. This reduces the cost of salvaging a subsystem that fails the go/no-go test, making it possible to manufacture more systems at lower cost.

- Both go/no-go and diagnostic tests require less test-engineering time to develop.

- Although the savings in test-engineering time may be offset by added design-engineering effort to include DFT, any increase in overall product development cost can be expected to be outweighed by decreased manufacturing cost.

12.2.1 Testing

test vector

Digital circuits are tested by applying *test vectors* which consist of inputs combinations and expected output combinations. A circuit "passes" if its outputs match what's expected. In the worst case, an n-input combinational circuit requires 2^n test vectors. However, if we know something about the circuit's physical realization and make some assumptions about the type of failures that can occur, the number of vectors required to test the circuit fully can be greatly reduced.

single stuck-at fault

The most common assumption is that failures are *single stuck-at faults*, that is, that they can be modeled as a single input or output signal stuck at logic 0 or logic 1. Under this assumption, an 8-input NAND gate, which might otherwise require 256 test vectors, can be fully tested with just nine—11111111, 01111111, 10111111, ..., 11111110.

Under the single-fault assumption, it's easy to come up with test vectors for individual logic elements. However, the problem in practice is *applying* the test vectors to logic elements that are buried deep in a circuit, and *seeing* the results. For example, suppose that a circuit has a dozen combinational and sequential logic elements between its primary inputs and the inputs of an 8-input NAND gate that we want to test. It's not at all obvious what primary-input vector, or sequence of primary-input vectors, must be applied to generate the test vector 11111111 at the NAND-gate inputs. Furthermore, it's not obvious what else might be required to propagate the NAND gate's output to a primary output of the circuit.

Sophisticated *test generation programs* deal with this complexity and try to create a *complete test set* for a circuit, that is, a sequence of test patterns that fully tests each logic element in the circuit. However, the computation required can be huge, and it's quite often just not possible to generate a complete test set.

test generation program

complete test set

DFT methods attempt to simplify test-pattern generation by enhancing the "controllability" and "observability" of logic elements in a circuit. In a circuit with good *controllability*, it's easy to produce any desired values on the internal signals of the circuit by applying an appropriate test-vector input combination to the primary inputs. Similarly, good *observability* means that any internal signal can be easily propagated to a primary output for comparison with an expected value by the application of an appropriate primary-input combination. The most common method of improving controllability and observability is to add *test points*, additional primary inputs and outputs that are used during testing.

controllability

observability

test points

12.2.2 Bed-of-Nails and In-Circuit Testing

In a digital circuit that is fabricated on a single PCB, the "ultimate" in observability is obtained by using every pin of every IC as a test point. This is achieved by building a special *test fixture* that matches the underside (solder side) of the PCB and contains a spring-loaded pin (*nail*) at each IC-pin position. The PCB is placed on this *bed of nails*, and the nails are connected to an *automatic tester* that can monitor each pin as required by a test program.

test fixture
nail
bed of nails
automatic tester
in-circuit testing

Going one step further, *in-circuit testing* also achieves the "ultimate" in controllability. This method not only monitors the signals on the bed of nails for observability, but also connects each nail to a very low impedance driving circuit in the tester. In this way, the tester can override (or *overdrive*) whatever circuit on the PCB normally drives each signal, and directly generate any desired test vector on the internal signals of the PCB. Overdriving an opposing gate output causes excessive current flow in both the tester and the opposing gate, but the tester gets away with it for short periods (milliseconds).

overdrive

To test an 8-input NAND gate (e.g., 74LS30), an in-circuit tester needs only to provide the nine test vectors mentioned previously, and can ignore whatever values the rest of the circuit is trying to drive onto the eight input pins. And the NAND-gate output can be observed directly on the output pin, of course. With in-circuit testing, each logic element can be tested in isolation from the others.

Although in-circuit testing greatly enhances the controllability and observability of a PCB-based circuit, logic designers must still follow a few DFT guidelines for the approach to be effective. Some of these are listed below.

- *Initialization.* It must be possible to initialize all sequential circuit elements to a known state.

The advent of high-density surface-mount packaging has made bed-of-nails testing considerably more difficult. Since components may be mounted on both sides of the PCB, a special test fixture called a clam shell may be needed which connects nails to both sides of the PCB.

Furthermore, the pins of many surface-mount devices are so small and their spacing is so tight (25 mils or less), that it may be impossible to reliably land a test pin on them. In such cases, the PCB designer may have to explicitly provide test pads, extra copper patches that are large enough for a test pin to contact (e.g., 50 mils in diameter). A separate test pad must be provided for each signal that does connect to a larger (e.g., 62-mil through-hole) component pad somewhere on the PCB.

Since the preset and clear input pins of registers and flip-flops are available to an in-circuit tester, you would think that this is no problem. However, Figure 12–1(a) shows a classic case of a circuit (a Gray-code counter) that cannot be initialized, since the flip-flops go to an unpredictable state when /PR and /CLR are asserted simultaneously. The correct way to handle the preset and clear inputs is shown in (b).

Figure 12–1
Flip-flops with pull-up resistors
for unused inputs: (a) untestable;
(b) testable.

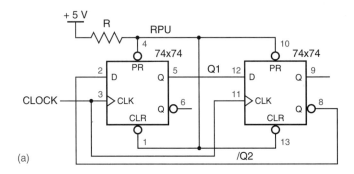

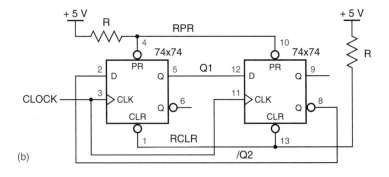

- *Clock generation.* The tester must be able to provide its own clock signal *without* overdriving the on-board clock signals.

Testers usually must override an on-board clock, for several reasons: The speed at which they can apply test vectors is limited; they must allow extra time for overdriven signals to settle; and sometimes they must stop the clock. However, overdriving the clock is a no-no. An overdriven signal may "ring" and make several transitions between LOW and HIGH before finally settling to the level that the tester wants. On a clock signal, such transitions can create unwanted state changes.

Figure 12–2 shows a recommended clock driver circuit. To inject its own clock, the tester pulls CLKEN LOW and inserts its clock on /TESTCLK. Since the tester is not overdriving any gate outputs, the resulting CLOCK signal is clean. In general, any normally glitch-free signal that is used as a clock input or other asynchronous input must not be overdriven by the tester, and would have to be treated in a way similar to Figure 12–1. This is another reason why synchronous design with a single clock is so desirable.

Figure 12–2
Clock driver circuit that allows
a tester to cleanly override the
system clock.

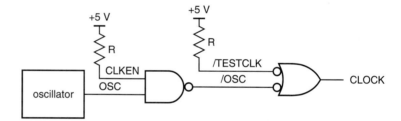

- *Grounded inputs.* In general, ground should not be used as a constant-0 logic source.

The in-circuit tester can overdrive *most* signals, but it can't overdrive ground. Therefore, signals that require a constant-0 input during normal operation should be tied to ground through a resistor, which allows the tester to set them to 1 as required during testing. A good example of this situation occurs when a PLD is used to create a set of clock signals, as in Figure 10–17 on page 691. Pin 11 of the GAL16V8 should be connected to ground through a resistor, so that the tester can float the PLD outputs and provide its own clock signals on /P1–/P6. Although the tester could theoretically overdrive /P1–/P6 directly, it shouldn't if they are used as clocks.

- *Bus drivers.* In general, it should be possible to disable the drivers for wide buses so that the tester can drive the bus without having to overdrive all the signals on the bus.

That is, it should be possible to output-disable all of the three-state drivers on a bus, so that the tester drives a "floating" bus. This reduces electrical stress both on the tester and on multibit drivers (e.g., 74x244) that might otherwise be overheated and damaged by having all of their outputs overdriven simultaneously.

12.2.3   Scan Methods

In-circuit testing works fine, up to a point. It doesn't do much good for custom VLSI chips and ASICs, because the internal signals simply aren't accessible. Even in board-level circuits, high-density packaging technologies such as surface mounting greatly increase the difficulty of providing a test point for every signal on a PCB. As a result, an increasing number of designs are using "scan methods" to provide controllability and observability.

scan method
scan-path method
        A *scan method* attempts to control and observe the internal signals of a circuit using only a small number of test points. A *scan-path method* considers any digital circuit to be a collection of flip-flops or other storage elements interconnected by combinational logic, and is concerned with controlling and observing the state of the storage elements. It does this by providing two operating
scan mode
modes: a normal mode, and a *scan mode* in which all of the storage elements are reorganized into a giant shift register. In scan mode, the state of the circuit's *n* storage elements can be read out by *n* shifts (*observability*), and a new state can be loaded at the same time (*controllability*).

        Figure 12–3 shows a circuit designed using a scan-path method. Each
two-port D flip-flop
storage element in this circuit is a *two-port D flip-flop* that can be loaded from
scan enable (SE)
one of two sources. The *scan enable (SE)* selects the source—normal data (D) or
scan data (SD)
scan data (SD). The SD inputs are daisy-chained to create the scan path shown in color. By asserting ENSCAN for 11 clock ticks, a tester can read out the current state of the flip-flops and load a new state. The test engineer is left with the job of deriving test sets for the individual combinational logic blocks, which can be fully controlled and observed using the scan path and the primary inputs and outputs.

        Scan-path design is used most often in custom VLSI and ASIC design, because of the impossibility of providing a large number of conventional test points. The two-port flip-flops used in scan-path design add chip area. For example, in LSI Logic Corp.'s LCA10000 series of CMOS gate arrays, an FD1 D flip-flop macrocell uses 7 "gate cells," while an FD1S two-port D flip-flop macrocell uses 9 gate cells, almost a 30% increase in silicon area. However, the overall increase in chip area is much less, since flip-flops are only a fraction of the chip, and large "regular" memory structures (e.g., RAM) may be tested by other means. In any case, the improvement in testability may actually *reduce* the cost of the packaged chip when the cost of testing is considered.

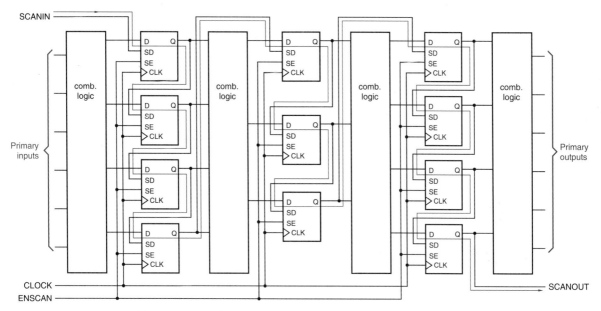

Figure 12–3   Circuit containing a scan path, shown in color.

12.3   ESTIMATING DIGITAL SYSTEM RELIABILITY

Qualitatively, the *reliability* of a digital system is the likelihood that it works   *reliability*
correctly when you need it. Marketing and sales people like to say that the
systems they sell have "high reliability," meaning that the systems are "pretty
likely to keep working." However, savvy customers ask questions that require
more concrete answers, such as, "If I buy 100 of these systems, how many will
fail in a year?" To provide these answers, digital design engineers are often
responsible for calculating the reliability of the systems they design, and in any
case they should be aware of the factors that affect reliability.

Quantitatively, reliability is expressed as a mathematical function of time:

$R(t)$ = Probability that the system still works correctly at time t

Reliability is a real number between 0 and 1; that is, at any time $0 \leq R(t) \leq 1$.
We assume that $R(t)$ is a monotonically decreasing function; that is, failures are
permanent and we do not consider the effects of repair. Figure 12–4 is a typical
reliability function.

Figure 12–4

Typical reliability function for a system.

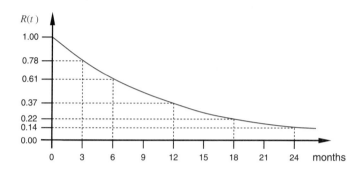

The foregoing definition of reliability assumes that you know the mathematical definition of probability. If you don't, reliability and the equivalent probability are easiest to define in terms of an experiment. Suppose that we were to build and operate N identical copies of the system in question. Let $W_N(t)$ denote the number of them that would still be working at time t. Then,

$$R(t) = \lim_{n \to \infty} W_N(t)/N$$

That is, if we build *lots* of systems, $R(t)$ is the fraction of them that are still working at time t. When we talk about the reliability of a single system, we are simply using our experience with a large population to estimate our chances with a single unit.

It would be very expensive to compute $R(t)$ if the only way to do it was to perform an experiment—to build and monitor N copies of the system. Worse, for any t, we wouldn't know the value of $R(t)$ until time t had elapsed in real time. Thus, to answer the customer's question posed earlier, we'd have to build a bunch of systems and wait a year; by then, our potential customer would have purchased something else.

Instead, we can estimate the reliability of a system by combining reliability information for its individual components, using a simple mathematical model. The reliability of mature components (e.g., 74FCT CMOS chips) may be known and published based on actual experimental evidence, while the reliability of new components (e.g., a Sexium microprocessor) may be estimated or extrapolated from experience with something similar. In any case, a component's reliability is typically described by a single number, the "failure rate" described next.

12.3.1 Failure Rates

failure rate

λ

The *failure rate* is the number of failures that occur per unit time in a component or system. In mathematical formulas, failure rate is usually denoted by the Greek letter λ. Since failures occur infrequently in electronic equipment, the failure rate

Figure 12–5
The "bathtub curve" for
electronic-component
failure rates.

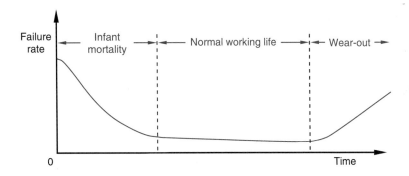

is measured or estimated using many identical copies of a component or system. For example, if we operate 10,000 microprocessor chips for 1,000 hours, and eight of them fail, we would say that the failure rate is

$$\lambda = \frac{8 \text{ failures}}{10^4 \text{ chips } \cdot \text{ } 10^3 \text{ hours}} = (8 \cdot 10^{-7} \text{ failures}/\text{hour})/\text{chip}$$

That is, the failure rate of a single chip is $8 \cdot 10^{-7}$ failures/hour.

The actual process of estimating the reliability of a batch of chips is not nearly as simple as we've just portrayed it; for more information, see the References. However, we can use individual component failure rates, derived by whatever means, in a straightforward mathematical model to predict overall system reliability, as we'll show later in this section.

Since the failure rates for typical electronic components are so small, there are several scaled units that are commonly used for expressing them: percent failures per 10^3 hours, failures per 10^6 hours, and failures per 10^9 hours. The last unit is called a *FIT*:

FIT

$$1 \text{ FIT} = 1 \text{ failure}/10^9 \text{ hours}$$

In our earlier example, we would say that $\lambda_{\text{microprocessor}} = 800$ FITs.

The failure rate of a typical electronic component is a function of time. As shown in Figure 12–5, a typical component has a high failure rate during its early life, during which most manufacturing defects make themselves visible; failures during this period are called *infant mortality*. Infant mortality is why manufacturers of high-quality equipment perform *burn-in*—operating the equipment for 24 to 168 hours before shipping it to customers. With burn-in, most infant mortality occurs at the factory rather than at the customer's premises. Even without a thorough burn-in period, the seemingly stingy 90-day warranty offered by most electronic equipment manufacturers does in fact cover most of the failures that occur in the first few *years* of the equipment's operation (but if it fails on the 91st day, that's tough!). This is quite different from the situation

infant mortality
burn-in

with an automobile or other piece of mechanical equipment, where "wear and tear" increases the failure rate as a function of time.

wear-out

Once an electronic component has successfully passed the burn-in phase, its failure rate can be expected to be pretty much constant. Depending on the component, there may be *wear-out* mechanisms that occur late in the component's life, increasing its failure rate. In the old days, vacuum tubes often wore out after a few thousand hours because their filaments deteriorated from thermal stress. Nowadays, most electronic equipment reaches obsolescence before its solid-state components start to experience wear-out failures. For example, even though it's been over 15 years since the widespread use of EPROMs began, many of which were guaranteed to store data for only 10 years, we haven't seen a rash of equipment failures caused by their bits leaking away. (Do you know anyone with a 10-year-old PC or VCR?)

Thus, in practice, the infant-mortality and wear-out phases of electronic-component lifetime are ignored, and reliability is calculated on the assumption that failure rate is constant during the normal working life of electronic equipment. This assumption, which says that a failure is equally likely at any time in a component's working life, allows us to use a simplified mathematical model to predict system reliability, as we'll show later in this section.

There *are* some other factors that can affect component failure rates, including temperature, humidity, shock, vibration, and power cycling. The most significant of these for ICs is temperature. Many IC failure mechanisms involve chemical reactions between the chip and a contaminant, and these are accelerated by higher temperatures. Likewise, electrically overstressing a transistor, which heats it up too much and eventually destroys it, is worse if the device temperature is high to begin with. Both theoretical and empirical evidence support the following widely used rule of thumb:

- An IC's failure rate roughly doubles for every 10°C rise in temperature.

This rule is true to a greater or lesser degree for most other electronic parts.

Note that the temperature of interest in the foregoing rule is the *internal* temperature of the IC, not the ambient temperature of the surrounding air. A power-hogging component in a system without forced-air cooling may have an internal temperature as much as 40–50°C higher than ambient. A well-placed fan may reduce the temperature rise to 10–20°C, reducing the component's failure rate by perhaps a factor of 10.

12.3.2 Reliability and MTBF

For components with a constant failure rate λ, it can be shown that reliability is a simple exponential function of time:

$$R(t) \;=\; e^{-\lambda t}$$

The reliability curve in Figure 12–4 is such a function; it happens to use the value $\lambda = 1$ failure / year.

Another measure of the reliability of a component or system is the *mean time between failures (MTBF)*, the average time that it takes for a component to fail. For components with a constant failure rate λ, it can be shown that MTBF is simply the reciprocal of λ:

mean time between failures (MTBF)

$$\text{MTBF} \;=\; 1/\lambda$$

Actually, the abbreviation *MTTF* (mean time *to* failure) should be used to talk about IC failures. Strictly speaking, "MTBF" should only be used when speaking about repairable equipment; ICs are never repaired and given the opportunity to fail again. However, most people ignore this minor semantic nit.

MTBF VERSUS MTTF

12.3.3 System Reliability

Suppose that we build a system with m components, each with a different failure rate, $\lambda_1, \lambda_2, \ldots, \lambda_m$. Let us assume that for the system to operate properly, *all* of its components must operate properly. Basic probability theory says that the system reliability is then given by the formula

$$\begin{aligned}
R_{\text{sys}}(t) &\;=\; R_1(t) \cdot R_2(t) \cdot \;\cdots\; \cdot R_m(t) \\
&\;=\; e^{-\lambda_1 t} \cdot e^{-\lambda_2 t} \cdot \;\cdots\; \cdot e^{-\lambda_m t} \\
&\;=\; e^{-\lambda_1 t - \lambda_2 t - \;\cdots\; -\lambda_m t} \\
&\;=\; e^{-\lambda_{\text{sys}} t}
\end{aligned}$$

where

$$\lambda_{\text{sys}} \;=\; \lambda_1 + \lambda_2 + \cdots + \lambda_m$$

Thus, system reliability is also an exponential function, using a composite failure rate λ_{sys} that is the sum of the individual component failure rates.

The constant-failure-rate assumption makes it very easy to determine the reliability of a system—simply add the failure rates of the individual components to get the system failure rate. Individual component failure rates can be obtained from manufacturers, reliability handbooks, or company standards.

For example, a portion of one company's standard is listed in Table 12–1. Since failure rates are just estimates, this comapny simplifies the designer's job

Table 12–1
Typical component failure rates at 55°C.

Component	Failure Rate (FITs)
SSI IC	90
MSI IC	160
LSI IC	250
VLSI microprocessor	500
Resistor	10
Decoupling capacitor	15
Connector (per pin)	10
Printed-circuit board	1000

by considering only broad categories of components, rather than giving "exact" failure rates for each component. Other companies use more detailed lists, and some CAE systems maintain a component failure-rate database, so that a circuit's composite failure rate can be calculated automatically from its parts list.

Suppose that we were to build a single-board system with a VLSI microprocessor, 16 memory and other LSI ICs, 2 SSI ICs, 4 MSI ICs, 10 resistors, 24 decoupling capacitors, and a 50-pin connector. Using the numbers in Table 12–1, we can calculate the composite failure rate,

$$\lambda_{sys} = 1000 + 16 \cdot 250 + 2 \cdot 90 + 4 \cdot 160 + 10 \cdot 10 + 24 \cdot 15 + 50 \cdot 10 + 500 \text{ FITs}$$

$$= 7280 \text{ failures} / 10^9 \text{ hours}$$

The MTBF of our single-board system is $1 / \lambda$, or about 15.6 years.

12.4 TRANSMISSION LINES, REFLECTIONS, AND TERMINATION

Nothing happens instantly, especially where digital circuits are concerned. In particular, consider the fact that the "speed-of-light" propagation delay of electrical signals in *wire* is on the order of 1.5–2 ns per foot (the exact delay depends on characteristics of the wire). When wire delays are on the same order as the transition times of the signals that they carry, we must treat wires not as zero-delay, perfect conductors, but as the "transmission lines" that they really are.

Transmission-line behavior includes signal changes that would not be predicted by a "DC" analysis of circuit operation. The most significant changes occur for an interval of approximately $2T$ after an output changes state, where T is the delay from the output to the far end of the wire that it drives.

Transmission-line behavior usually doesn't affect the logical operation of devices whose signal transition times and propagation delays are much longer than $2T$. Thus, transmission-line behavior typically is considered only for 74LS connections over a couple feet in length, for 74AS, 74AC and 74ACT connections over a foot, and for 74FCT and ECL connections over 3–6 inches. Naturally, you can sometimes get away with longer connections depending on the application, but there are also situations where transmission-line theory must be applied to even shorter connections. Even some VLSI chips achieve high enough speeds that transmission-line behavior must be considered in the internal chip design.

12.4.1 Basic Transmission-Line Theory

Two conductors in parallel constitute the simplest *transmission line*. Consider a pair of conductors with infinite length, as shown in Figure 12–6(a). If we instantaneously place a voltage source across the pair, a certain current flows to create a voltage wave that travels along the pair. The ratio of voltage to current, V_{out}/I_{out} depends on the physical characteristics of the conductors, and is called the *characteristic impedance* Z_0 of the conductor pair.

transmission line

characteristic impedance Z_0

The magnitude of V_{out} is determined by viewing the series combination of R_{src} and Z_0 as a voltage divider, so that

$$V_{out} = V_{src} \cdot \frac{Z_0}{R_{src} + Z_0}$$

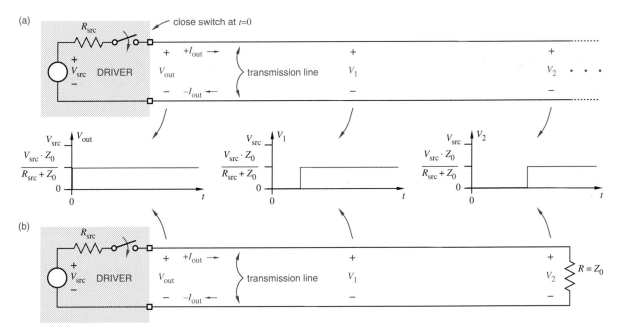

Figure 12–6   Transmission lines: (a) with infinite length; (b) with finite length, terminated with characteristic impedance.

Of course, we don't have any infinitely long conductors. However, suppose that we have a 5-foot-long pair of conductors, and place a resistor equal to Z_0 across the far end. If we instantaneously place a voltage source across the pair as shown in (b), the same current flows forever as in the infinite-length case. Thus, when a line is terminated by its characteristic impedance, we needn't consider transmission-line effects any further.

The situation is different for a transmission line of finite length that is not terminated in its characteristic impedance. An extreme case, in which the far end is short-circuited, is shown in Figure 12–7(a). For simplicity, we assume that $R_{src} = Z_0$ in this example. Initially, all the driver sees is the characteristic impedance of the line, and a voltage wave with amplitude $V_{src}/2$ happily propagates down the line. However, when the wave hits the far end at time T, it sees the short-circuit. In order to satisfy Kirchhoff's laws, a voltage wave of the *opposite* polarity propagates back down the line, canceling the original wave.

reflection   The far end has *reflected* the original wave, and the driver sees the short-circuit at time $2T$.

Another extreme case, in which the far end is open-circuited, is shown in (b). Everything starts out as before. However, when the initial voltage wave hits the far end, the current has nowhere to go, and so a voltage wave of the *same* polarity propagates back up the line, adding to the original voltage. When the reflected wave reaches the driver at time $2T$, the voltage everywhere is V_{src}, and nothing more happens.

reflection   In the general case, the amplitude of the wave reflected at the end of a
coefficient, ρ   transmission line is determined by the *reflection coefficient, ρ* (rho). The value
termination   of ρ depends on Z_0 and Z_{term}, the *termination impedance* that appears at the end
impedance   of the line:

$$\rho = \frac{Z_{term} - Z_0}{Z_{term} + Z_0}$$

When a voltage wave with amplitude V_{wave} hits the end of a transmission line, a wave with amplitude $\rho \cdot V_{wave}$ is reflected. Note that three simple cases match our previous discussion:

$Z_{term} = Z_0$ When a transmission line is terminated in its characteristic impedance, the reflection coefficient is 0.

$Z_{term} = 0$ The reflection coefficient of a short-circuited line is −1, producing a reflection of equal magnitude and opposite polarity.

$Z_{term} = \infty$ The reflection coefficient of an open-circuited line is +1, producing a reflection of equal magnitude and the same polarity.

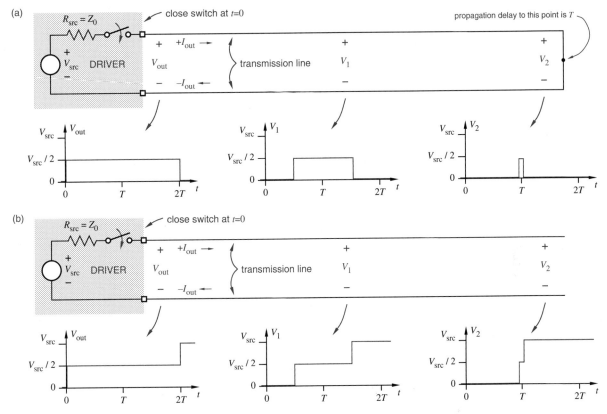

Figure 12–7　Transmission lines: (a) short-circuited; (b) open-circuited.

If the source impedance (R_{src}) does not equal Z_0, then reflections occur at the near end of the line as well as at the far end. Each end of the line has its own value of ρ. The *principle of superposition* applies, so that voltage at any point on the line and instant in time is the sum of that point's initial condition and all waves that have passed it so far. The end of a transmission line is said to be *matched* if it is terminated with its characteristic impedance.

principle of superposition

transmission-line matching

Figure 12–8 shows the behavior of a transmission line that is not matched at either end. (Note that the driver's ideal voltage source V_{src} is considered to have a resistance of 0 Ω for this analysis.) Reflections occur at both ends of the line, with smaller and smaller waves reflecting back and forth. The voltage everywhere on the line asymptotically approaches $0.9V_{src}$, the value that would be predicted by a "DC" analysis of the circuit using Ohm's law.

12.4.2   Logic-Signal Interconnections as Transmission Lines

So, how does transmission-line theory affect logic signals? Let us consider the case of a TTL driver producing a LOW-to-HIGH transition on a transmission line consisting of a signal line and ground, as illustrated in Figure 12–9. An LS-TTL output in the HIGH state is roughly equivalent to a 4 V source in series with a 120-Ω resistor. The characteristic impedance of a typical PCB trace with respect to ground is on the order of 100–150 Ω; let's assume 120 Ω for convenience, so the reflection coefficient on the driving end of the line is 0. An LS-TTL input has an impedance of about 20 kΩ, so the reflection coefficient on the receiving end is very nearly +1.

When the TTL driver switches from LOW to HIGH, the 4 V source in the driver sees the 120-Ω resistance of the driver in series with the 120-Ω Z_0 of the line, so a 2-volt wave propagates down the line. After time T, this wave reaches the receiving gate U2 on the far end and is reflected. After time $2T$, the reflected wave reaches the sending end and is absorbed without a reflection because $\rho = 0$ at that end.

Everything works fine as far as receiving gate U2 is concerned—it sees an instantaneous transition from 0 V to 4 V at time T after the driver switches. However, consider the waveform seen by another receiving gate U1 positioned halfway between the driver and U2. As shown in the figure, U1 sees an input of only 2 V for an interval T. A receiving gate positioned closer to the driver would see this input voltage even longer. This is a problem, because 2 V is right on the borderline for a valid HIGH signal. If the driver resistance were a little higher, or if the line impedance were a little lower, or if a little noise occurred,

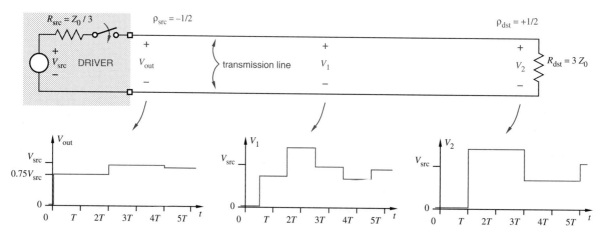

Figure 12–8   A transmission line that is not matched at either end.

the voltage could be even lower, which could cause the receiver to oscillate or produce a nonlogic-voltage output.

Any logic-signal connection, no matter how long or how short, is a transmission line. However, the foregoing analysis is quite idealized. In practice, transmission-line behavior causes no problem if T is somewhat shorter than the transition times of logic signals or the propagation delays of logic gates. In such a case, the reflections typically settle out before the receivers on the line have a chance to notice them.

A somewhat nastier case occurs when a TTL output switches from HIGH to LOW, as illustrated in Figure 12–10. In the LOW state, an LS-TTL output looks roughly like a 10-Ω resistor to ground, so the reflection coefficient on the sending end is about -0.85. When the output is first switched LOW it sees a voltage divider consisting of the 10-Ω resistor in series with the line impedance of 120 Ω. Since the line was originally at 4 V, the new voltage at the output of the divider must be

$$V_{out} = \frac{10\ \Omega}{120\ \Omega + 10\ \Omega} \cdot 4\ V = 0.3\ V$$

Therefore, a voltage wave $(0.3 - 4.0) = -3.7$ V is propagated down the line. When this wave hits the receiving end at time T, it produces a reflection of equal sign and magnitude (since $\rho = 1$). Thus, the voltage at the receiving end is now $V_2 = 4.0 - 3.7 - 3.7 = -3.4$ V, a negative voltage! This is called *undershoot*.   *undershoot*

When the -3.7 V reflection gets back to the sending end at time $2T$, yet another reflection occurs. This time, since ρ is negative, the reflected wave has a positive amplitude, $-0.85 \cdot -3.7 = +3.15$ V. The output voltage is now

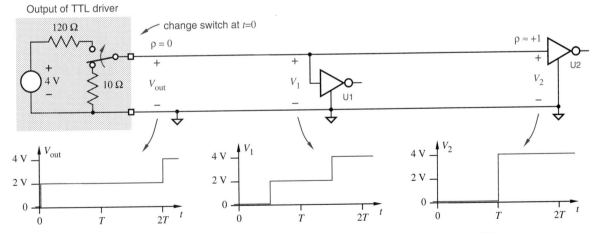

Figure 12–9 Reflections on a TTL signal line changing from LOW to HIGH.

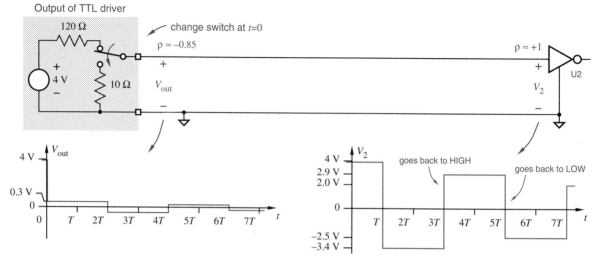

Figure 12–10   Reflections on a TTL signal line changing from HIGH to LOW.

modified by the sum of the incoming reflection and the new outgoing reflection: $V_{out} = 0.3 - 3.7 + 3.15 = -0.25$ V. This is no problem. However, when the +3.1 V reflection reaches the receiving end at time $3T$, the ensuing positive reflection yields $V_2 = -3.4 + 3.15 + 3.15 = 2.9$ V, back to positive again!

As shown in the figure, the reflections continue, with the voltage at both the sending and receiving ends asymptotically approaching 0 V, the value one would predict with a "DC" analysis of the circuit. This oscillating pattern is called *ringing*.

ringing

The large magnitude of the ringing on the receiving end may be a problem, since V_2 does not settle to a range that is less than the LOW-state input voltage (0.8 V) until time $17T$. Thus, the effective propagation delay of the circuit has been increased by many times the wire delay. Worse, if the signal is a clock, extra edges will be detected on the ringing transitions, falsely clocking the flip-flops at the receiving end.

Once again, the transmission-line behavior that we've described causes no problems if T is much shorter than the transition times of logic signals or the propagation delays of logic gates. In addition, TTL input structures include *clamping diodes*, the normally reverse-biased diodes from each input to ground that we showed in Figure 3–76 on page 154. Since these diodes change the receiver's input impedance from very high to very low for negative voltages, they limit the negative excursion at time T to about 1 V. This reduces the reflection back to the sending end, which in turn reduces the excursion at time $3T$ to less than 1 V.

clamping diodes

REFERENCES

CAE tools are just tools, and you won't find any scholarly references on how to use them. However, *IEEE Design & Test of Computers* (*D&T*) and *High-Performance Systems* magazines are excellent ongoing sources of information about them. The best source of information for a particular CAE tool is, naturally, the supplier's documentation. When a particular tool is used *very* widely, an independent tutorial guide may appear; a good example is *SPICE: A Guide to Circuit Simulation and Analysis Using Pspice®* by Paul W. Tuinenga (Prentice Hall, 1988).

Register-transfer languages are discussed in most computer design books, including *Computer Architecture and Organization*, 2nd ed., by John P. Hayes (McGraw-Hill, 1988). *IEEE Computer* magazine had a special issue on hardware description languages in February 1985, and *D&T* had a special issue on VHDL in April 1986.

The emergence of VHDL has led to the publication of several descriptive books, such as *Chip-Level Modeling with VHDL* by James R. Armstrong (Prentice Hall, 1989) and *The VHDL Handbook* by David R. Coelho (Kluwer Academic Publishers, 1989). You know that VHDL "has arrived" when you see that some books are already in their second editions, including *VHDL*, 2nd ed., by Douglas L. Perry (McGraw-Hill, 1994).

The authoritative treatment of design for testability in the context of logic design is in Edward J. McCluskey's *Logic Design Principles* (Prentice Hall, 1986). A thorough treatment of the subject from the point of view of manufacturing and test engineers is *Digital Test Engineering* by J. Max Cortner (Wiley, 1987). Once again, *D&T* is an excellent ongoing information source. In the August 1989 special issue on built-in self-test methods, an article by Kenneth P. Parker, "The Impact of Boundary Scan on Board Test," gives a very practical perspective on the growing importance of scan methods in board-level design.

The required mathematical background and general methods for analyzing system reliability can be found in a handy little book by J. C. Cluley, *Electronic Equipment Reliability* (Halsted Press/Wiley, 1974). To see how ICs actually fail, and how their failure rates can be predicted in less than a lifetime (yours or an IC's), read *Failure Mechanisms in Semiconductor Devices* by E. A. Amerasekera and D. S. Campbell (Wiley, 1987).

An excellent introduction to transmission-line theory and practice from the point of view of a digital designer can be found in Thomas R. Blakeslee's *Digital Design with Standard MSI and LSI*, 2nd ed. (Wiley, 1979).

Highly recommended reading and an authoritative treatment of the analog aspects of digital design that I love to harp on is *High-Speed Digital Design* by

Howard W. Johnson and Martin Graham (Prentice Hall, 1993). Although the book's subtitle is *A Handbook of Black Magic*, the authors go a long way toward demystifying the analog behavior that all high-speed digital system designers must cope with.

APPENDIX A

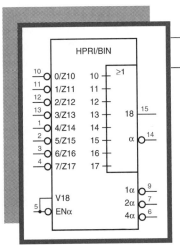

IEEE STANDARD SYMBOLS

Together with the American National Standards Institute (ANSI), the Institute of Electrical and Electronic Engineers (IEEE) has developed a standard set of logic symbols. The most recent revision of the standard is ANSI/IEEE Std 91-1984, *IEEE Standard Graphic Symbols for Logic Functions*. It is compatible with standard 617 of the International Electrotechnical Commission (IEC), and must be used in all logic diagrams drawn for the U.S. Department of Defense.

ANSI
IEEE

A.1 GENERAL DEFINITIONS

The IEEE standard supports the notion of bubble-to-bubble logic design with the following definitions:

- An *internal logic state* is a logic state assumed to exist inside a symbol outline at an input or an output.

  internal logic state

- An *external logic state* is a logic state assumed to exist outside a symbol outline either (1) on an input line prior to any external qualifying symbol at that input, or (2) on an output line beyond any external qualifying symbol at that output.

  external logic state

791

qualifying symbol   A *qualifying symbol* is graphics or text added to the basic outline of a device's logic symbol to describe the physical or logical characteristics of the device. The "external qualifying symbol" mentioned above is typically an inversion bubble, which denotes a "negated" input or output, for which the external 0-state corresponds to the internal 1-state. This concept is illustrated in Figure A–1. When

internal 1-state   the standard says that a signal is in its *internal 1-state*, we would say that the
internal 0-state   signal is asserted. Likewise, when the standard says that a signal is in its *internal 0-state*, we would say that the signal is negated.

Figure A–1
Internal and external logic states.

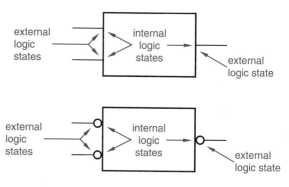

The IEEE standard provides two different types of symbols for logic gates.
distinctive-shape   One type, called *distinctive-shape symbols*, is what we've been using all along.
symbols   The other type, called *rectangular-shape symbols*, uses the same shape for all
rectangular-shape   the gates, along with an internal label to identify the type of gate. Figure A–2
symbols   compares the two types. According to the IEEE standard, "the distinctive-shape symbol is not preferred." Some people think this statement means that rectangular-shape symbols *are* preferred. However, all the standard really says is that it gives no preference to distinctive-shape symbols compared to rectangular-shape symbols. On the other hand, since most digital designers, authors, and computer-aided design systems prefer the distinctive-shape symbols, that's what we use in this book.

Before the promulgation of the IEEE standard, logic symbols for larger-scale logic elements were drawn in an ad hoc manner; the only standard rule was to use rectangles with inputs on the left and outputs on the right. Although the logic symbol might contain a short description of the element (e.g., "3–8

ANOTHER KIND
OF BUBBLE

In addition to the familiar bubble, the IEEE standard also allows an external, triangular "polarity symbol" to be used to specify active-low inputs and outputs, for which the external LOW level corresponds to the internal 1-state. However, under a positive-logic convention, the bubble and the triangular polarity symbol are equivalent, so we use the more traditional bubble in this appendix.

Figure A–2
Distinctive- and rectangular-
shape logic symbols.

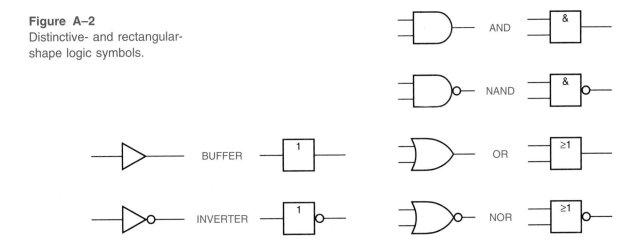

decoder," "2–1 multiplexer"), it was usually necessary to refer to a separate
table to determine the element's logic function. However, the IEEE standard
contains a rich set of concepts, such as bit grouping, common control blocks,
and dependency notation, that allow some or all of a larger-scale logic element's
function to be displayed in the symbol itself. We'll introduce these concepts
as appropriate as we cover the symbols for various categories of devices in the
sections that follow.

A.2 DECODERS

Chapter 5 used "traditional" logic symbols for decoders and other MSI logic
elements. Although traditional symbols show the active levels of the inputs and
outputs, they do not indicate the logical function of the device—you already have
to know what a 74x138 or 74x139 does when you read its symbol.

 The IEEE standard for logic symbols, on the other hand, allows a decoder's
logic function to be displayed as part of the symbol, shown in Figure A–3. These
symbols use several concepts of the standard:

- *Internal qualifying symbols.* Individual input and output signals may be   *internal qualifying*
  labeled with qualifying symbols inside the logic-symbol outline to describe   *symbol*
  the signals' characteristics. In this book, we call such symbols *qualifying* *qualifying label*
  labels for short.

- *General qualifying symbols.* The top of a logic symbol may contain an   *general qualifying*
  alphanumeric label to denote the general function performed by the device.   *symbol*
  Decoders and encoders (called *coders*) use the general qualifying symbol   *coder*
  X/Y to indicate the type of coding performed, where X is the input code

Figure A–3
IEEE standard symbols for decoders: (a) 74x138; (b) 74x139.

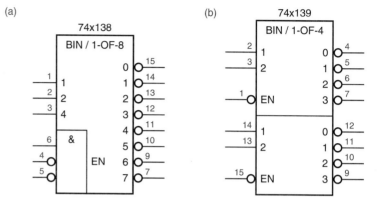

and Y is the output code. For example, a 3-to-8 decoder may be labeled
BIN/1-OF-8.

internal value

- *Internal values.* Each input combination of a coder produces an "internal
  value" that is displayed by the coder's outputs. The internal values for a
  3-to-8 decoder are 0–7.

input weight

- *Input weights.* The inputs of a coder may have qualifying labels indicating
  the numerical weights associated with those inputs. In this case, the internal
  value at any time is the sum of the weights of the asserted inputs. The input
  weights for a 3-to-8 decoder are 1, 2, and 4.

output value

- *Output values.* Each output may have a qualifying label listing the internal
  values that cause that output to be asserted. In a binary decoder, each output
  is asserted for just one internal value.

enable input

- *Enable input.* An enable input has the qualifying label EN and permits
  action when asserted. When negated, an enable input imposes the external high-impedance state on three-state outputs, and the negated state on
  other outputs. The 74x138 and 74x139 have active-low outputs, which are
  therefore 1 when the enable input is negated.

embedded element
abutted element

- *Embedded and abutted elements.* The outlines of individual logic elements
  may be embedded or abutted to form a larger composite symbol. There
  is at least one common logic connection when the dividing line between
  two outlines is perpendicular to the direction of signal flow, as shown in
  Figure A–4. For example, the 74x138 symbol has an embedded 3-input
  AND gate that drives the internal EN input. There is no connection between
  the elements when the dividing line is in the direction of signal flow, as
  shown in Figure A–5. For example, the 74x139 contains two separate 2-
  to-4 decoders.

Figure A–4
A composite symbol with one or
more logic connections between
its elements.

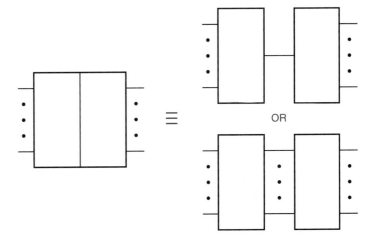

The ability to embed individual logic elements in a larger symbol is probably the most useful feature of the IEEE standard. For example, Figure A–6 shows the symbol for a 3-to-8 decoder with a different set of enable inputs than the 74x138. The fictitious *74x328* decoder is enabled when pin 6 is 0 or both pins 4 and 5 are 1.

74x328

Figure A–5
A composite symbol with no
connection between its elements.

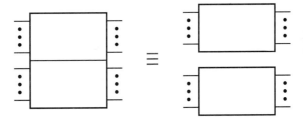

Figure A–6
IEEE standard symbol for a
fictitious decoder, the 74x328.

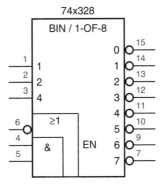

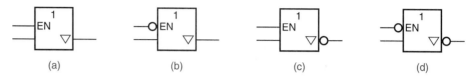

Figure A–7   IEEE standard rectangular-shape symbols for three-state buffers: (a) noninverting, active-high enable; (b) noninverting, active-low enable; (c) inverting, active-high enable; (d) inverting, active-low enable.

A.3   THREE-STATE BUFFERS

Three-state-buffer symbols use one more feature of the IEEE standard:

downward-pointing triangle

- *Downward-pointing triangle.* This denotes a three-state output.

Also recall that an enable input, labeled EN, is specifically understood to force all affected outputs to a disabled state. For three-state outputs, the disabled state is Hi-Z. Thus, the three-state buffers of Figure 5–35 are drawn as shown in Figure A–7. Another important feature of the IEEE standard is introduced in the symbols for MSI three-state buffers:

common control block

- *Common control block.* This concept, illustrated in Figure A–8, may be used with an array of related elements. Inputs to the common control block are understood to affect all the elements of the array.

Figure A–8
Common control block in an IEEE standard symbol.

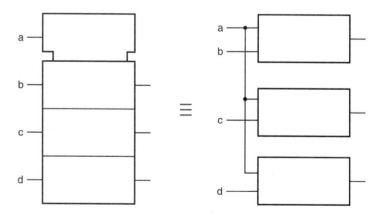

Thus, symbols may be drawn for the 74x541 and 74x245 as shown in Figure A–9. The enable and direction inputs (pins 1 and 19) apply to all elements of the device. The '541 and '245 symbols introduce several other features of the standard:

Figure A–9
IEEE standard logic symbols for
the 74x541 and 74x245.

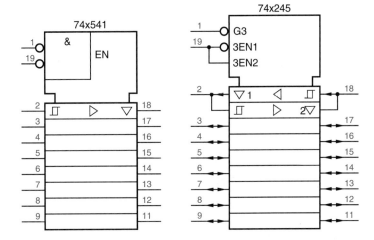

- *Hysteresis symbol.* Inputs bearing this symbol have hysteresis.   *hysteresis symbol*

- *Right-pointing or left-pointing triangle.* These symbols are used to denote   *right-pointing*
  "amplification"; in the case of three-state buffers, they indicate an output   *triangle*
  circuit that has more fanout and can drive a heavier load than an ordinary   *left-pointing*
  output circuit.   *triangle*

- *Arrows.* These denote the direction of signal flow when it is not strictly   *arrows*
  left to right, as in the '245 symbol.

- *Identical elements.* Only the first of two or more identical elements in an   *identical elements*
  array must be drawn in detail. Thus, the bottom seven elements of the '245
  symbol are understood to be identical to the top element (which in this case
  happens to be divided into two subelements, the three-state buffers for the
  two directions).

The '245 symbol also introduces *dependency notation*, a means of display-   *dependency notation*
ing some of the more common logical relationships among input and output
signals. A few more concepts are needed to make this notation fly:

- *Affecting signals.* An input signal (or, occasionally, an output signal) may   *affecting signal*
  affect other inputs or outputs in a way that can be displayed on the symbol.
  Such a signal has a qualifying label *Li*, where *L* is a letter that indicates the
  type of relationship or effect, and i is an integer that identifies the affected
  signals. The '245 internal signal labeled G3 is such a signal.

- *Affected signals.* Inputs or outputs that are affected by a signal *Li* bear the   *affected signal*

WHOOPS!

> The discussion of "affected signals" said "G3" when it should have said "the signal labeled G3." In the IEEE standard, G3 is a qualifying label, and other signals may have the same label. The standard does not specify unique names for internal and external signals. However, in the rest of this appendix, we'll take the liberty of using an internal signal's qualifying label as its name if no ambiguity results.

qualifying label i. If an affected signal requires a qualifying label of its own, then i is used as a prefix to that label. Thus, the signal labeled 3EN1 in the '245 symbol affects signals labeled 1, and is itself affected by G3.

AND dependency
Gi

- *AND dependency.* This relationship is denoted by Gi, and is a sort of enable function. Affected signals perform their "normal" functions only if Gi is asserted; otherwise, they are negated. In the '245 symbol, inputs 3EN1 and 3EN2 can "do their thing" only if G3 is asserted. Figure A–10 shows an equivalent notation for the dependent signals. Notice that dependency is defined in terms of internal, not external, signal values.

Figure A–10
Illustration of AND dependency.

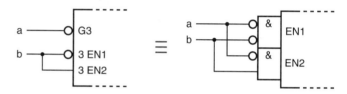

enable dependency
ENi

- *Enable dependency.* This dependency is denoted by ENi, and has the same effect as EN inputs defined earlier. Thus, in the '245 symbol, the internal signal 3EN1, which is asserted only if pins 1 and 19 are LOW, enables the three-state drivers for pins 2 through 9.

Whewww! You might be thinking that this is a lot of trouble for a few lousy three-state buffers, but just wait until you see some of the IEEE-standard counter and shift-register symbols later in this appendix!

A.4   PRIORITY ENCODERS

Figure A–11 shows the IEEE standard symbol for a 74x148. Still more features of the standard are used in this symbol:

solidus

- *Solidus.* The slash in a label such as 0/Z10, called a *solidus* in the standard, separates multiple functions of a single internal signal. An equivalent notation is shown in Figure A–12.

Figure A–11
IEEE standard logic symbol
for the 74x148 8-input priority
encoder.

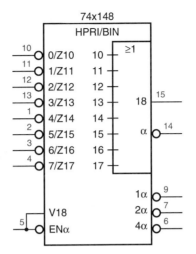

- *OR dependency.* This relationship is denoted by Vi; the Vi signal is OR'ed with affected signals. Thus, affected signals perform their "normal" functions only if the Vi signal is negated; otherwise, they are asserted.

  OR dependency
  Vi

- *Virtual inputs and outputs.* These are internal signals, denoted by a horizontal line going nowhere, such as the ones labeled 10–17 in the '148 symbol. These signals have no external connection but affect or are affected by other signals via dependency notation.

  virtual input
  virtual output

- *Interconnection dependency.* This relationship is denoted by Zi, and indicates an internal connection. Affected signals equal the Zi signal, unless modified by additional dependency notation. Think of the Z as a zig-zag internal wire.

  interconnection
  * dependency*
  Zi

- *Greek letters.* A Greek letter may be used instead of an integer in qualifying labels to avoid ambiguity when the affected signals have a numeric function label, as in the inputs or outputs of a coder.

  Greek letters

If you understand these and the previously introduced features of the standard, you can "read" the functional behavior of a '148 right from its symbol. However, most people do the opposite—already knowing how a '148 works, they try to deduce how the standard works from the '148 symbol!

Figure A–12
Equivalent notation for solidi.

a ──○| 0 / Z10   ≡   a ──●──○| 0
                                    ○| Z10

A.5 MULTIPLEXERS AND DEMULTIPLEXERS

The IEEE standard provides a special notation for multiplexers and demultiplexers. For example, Figure A–13 shows the IEEE symbols for multiplexer ICs that we discussed in Chapter 5. The general qualifying symbol MUX identifies a *bit-grouping symbol* multiplexer. The bracket is called a *bit-grouping symbol* and indicates that the *internal value* grouped inputs produce an *internal value* that is a weighted sum. The weights are given by the qualifying labels on the individual inputs; if the weights are all powers of 2, they may be replaced by the corresponding exponents, and all but the first and last exponents may be omitted "if no confusion is likely." Thus, the weights of pins 9, 10, and 11 of the 74x151 are 2^2, 2^1, and 2^0, respectively; if the input signal on these pins is 110, the internal value is 6.

The internal value produced by bit grouping affects other internal values or outputs according to the qualifying label written to the right of the grouping *range notation* symbol. In the 74x151 multiplexer symbol, the notation $G\frac{0}{7}$ indicates AND dependency with signals whose labels are in the range 0–7. In other words, input *i* is selected (and transferred to the output) if and only if the internal value is *i*. There are two outputs, normally equal to the selected input and its complement. However, these outputs are asserted only if the enable input EN is asserted.

The 74x153 symbol also uses bit grouping for the select inputs, and it uses a common control block to indicate that the select inputs affect both sections of

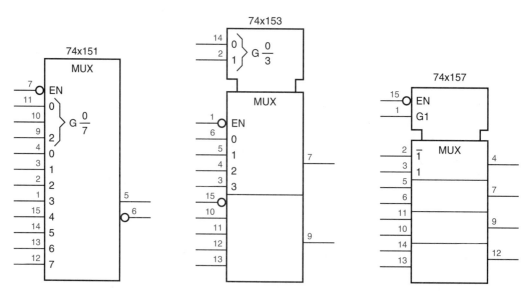

Figure A–13   IEEE standard symbols for multiplexers.

the multiplexer. Since the bottom half of the multiplexer is identical in function to the top, its qualifying labels are not repeated. Notice that each half has an independent EN input.

The symbol for the 74x157 does not use bit grouping, because it has only one select input. Instead, the select input is labeled G1, indicating that it has an AND dependency with signals bearing the identifier "1." Thus, pin 3 is selected only if G1 is asserted. Pin 2, on the other hand, bears the identifier "$\overline{1}$"; the overbar indicates that pin 2 is selected only if G1 is negated. All four sections are controlled by G1 in this way.

Figure A–14
IEEE standard symbols
for demultiplexers.

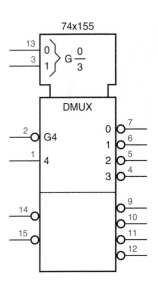
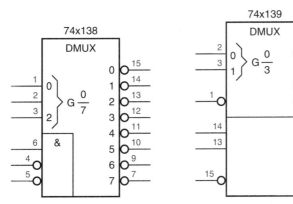

Figure A–14 shows the IEEE standard demultiplexer symbols for the MSI decoder/demultiplexer ICs that we discussed in Chapter 5. The general qualifying symbol DMUX identifies a demultiplexer. The notation $G\frac{0}{7}$ indicates AND dependency with outputs whose labels are in the range 0–7. In other words, output i may be asserted only if the internal value is i. In addition, all of the other inputs (pins 4–6) must be asserted.

A similar notation is used in the 74x139, which contains two independent demultiplexers. Since the second demultiplexer is identical in function to the first, the qualifying labels are not repeated.

The symbol for the 74x155 uses a common control block to show that the same select inputs (and internal value produced by bit grouping) are used for both sections of the demultiplexer. Also, the input labeled "4" has an AND dependency on the input labeled "G4," so the selected output is asserted only if both of these inputs are asserted. Now, is that all clear?

A.6 EXCLUSIVE-OR AND PARITY FUNCTIONS

The IEEE standard symbols for EXCLUSIVE OR and EXCLUSIVE NOR gates are shown in Figure A–15. Either of two different notations may be used, depending on how you're thinking about the gate's function. The top symbols, with "=1" inside the symbol outline, assert their outputs when exactly one of the inputs is asserted. The bottom symbols, with "=" inside the symbol outline, assert their outputs when their inputs are equal.

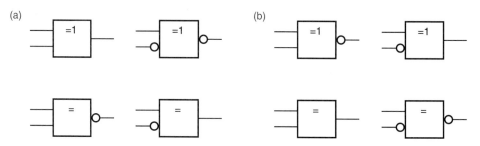

Figure A–15   IEEE standard symbols: (a) EXCLUSIVE OR gates; (b) EXCLUSIVE NOR gates.

The IEEE standard symbol for 74x280 9-bit parity generator is shown in Figure A–16(a). The general qualifying symbol "2k" at the top of the symbol indicates that the outputs are asserted if $2k$ of the inputs are asserted for some integer k. Thus, pin 5, an active-high output, and pin 6, an active-low output, both indicate even parity. The nature of the function, of course, is such that pin 6 could instead be viewed as an active-high output that denotes odd parity. The standard symbol for this interpretation of the device function is shown in (b).

Figure A–16
IEEE standard symbols for the 74x280 9-bit parity generator: (a) normal symbol; (b) both outputs active high.

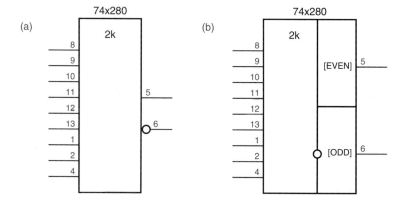

A.7 COMPARATORS

IEEE standard symbols for MSI comparators are shown in Figure A–17. Like the select inputs of multiplexers, the data inputs have qualifying labels indicating their arithmetic weights in powers of 2 (0–3 corresponding to 2^0–2^3 in the '85). The cascading inputs and outputs are labeled with the appropriate arithmetic function. In the '682 symbol, the right-pointing triangle indicates that the outputs have high fanout capability, and hysteresis is indicated on the data inputs.

Figure A–17
IEEE standard symbols
for MSI comparators.

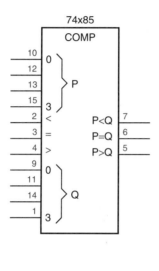

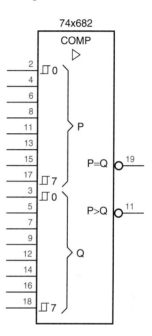

A.8 ADDERS

The IEEE standard uses the general qualifying symbol "Σ" to identify an adder or addition function. Figure A–18(a) shows the symbol for a 74x283 4-bit adder. The numbers on the addend inputs and sum output indicate the weight of each pin, as a power of 2.

The IEEE symbol for a 74x181 4-bit ALU is shown in Figure A–18(b). The first five inputs of the common control block form a "mode control" word, which the standard designates by the letter M. The weights of the mode control bits are shown as powers of 2, and they designate a mode number in the range 0–31. According to the standard, a separate table accompanies the logic symbol to define the functions performed in each mode. The CP, CG, and CO outputs are enabled in modes 0–15. The four individual ALU blocks are labeled with the weights of the bits they process. The IEEE symbol for the 74x182 carry lookahead circuit is much simpler, and is shown in (c).

Figure A–18
IEEE standard symbols:
(a) 74x283; (b) 74x181;
(c) 74x182.

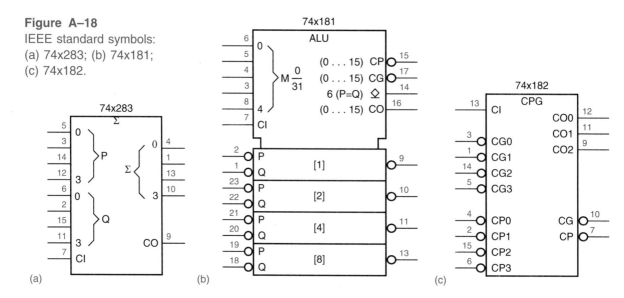

(a)   (b)   (c)

A.9   LATCHES, FLIP-FLOPS, AND REGISTERS

There are a few differences between traditional and IEEE symbols for latches and flip-flops. Figure A–19 shows IEEE standard symbols for SSI latches and flip-flops. A major difference is that the asynchronous preset and clear inputs are drawn on the left, not on the top and bottom. The names for these inputs are different, too: S *(set)* and R *(reset)*. Also, a clock input is simply named Ci, where i is an integer and all other inputs labeled with i (e.g., iD) are controlled by Ci; this is just another instance of dependency notation:

S (set)
R (reset)

control dependency
Ci

- *Control dependency.* This relationship is denoted by Ci and, like enable dependency, is a sort of enable function. It is intended to be used only for clock or timing inputs. Affected signals are enabled when the Ci input is in the internal 1-state or, if the Ci input has a dynamic indicator, on the 0-to-1 change in internal state.

The symbols for the '74, '109, and '112 follow the usual IEEE convention that only the first of two or more identical elements must be drawn in detail when they are abutted in an array. Alternatively, the two independent sections of each SSI package in Figure A–19 may be drawn separately, as are other traditional and IEEE symbols for devices with independent sections, such as the 74x139.

Figure A–19
IEEE standard symbols for SSI
latches and flip-flops.

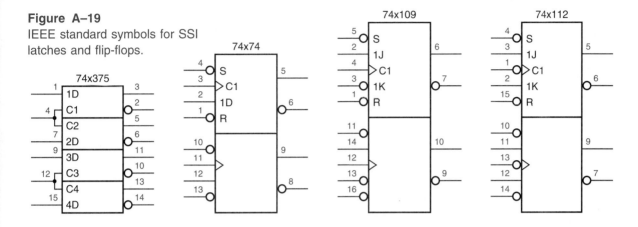

This is especially useful if the logic functions performed by the two sections in a particular application are unrelated.

Figure A–20 shows the IEEE standard symbols for MSI latches and registers. All of these symbols make good use of the IEEE notation for common control blocks. In the '373 and '374 symbols, the label C1 on the clock input indicates that it controls all inputs that bear the label 1, that is, all of the 1D inputs. The downward-pointing triangles indicate three-state outputs; by convention, an input labeled EN enables such outputs.

In the '377 symbol, the input labeled G1 is an enable for inputs bearing the label 1, that is, for the clock input 1C2. The clock input in turn controls all of the inputs bearing the label 2, that is, data inputs 2D. As usual, only the first of the eight identical flip-flop elements is drawn in detail.

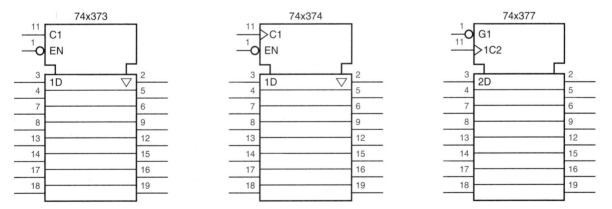

Figure A–20   IEEE standard symbols for MSI latches and registers.

A.10 COUNTERS

IEEE standard symbols for popular counters are shown in Figure A–21. Like the IEEE symbols for other MSI devices, the counter symbols make good use of the common-control-block and array features of the standard. The general qualifying symbol CTRDIV16 indicates that the device is a divide-by-16 counter, and labels [1]–[8] indicate the arithmetic weight of each counter bit. However, the additional notation used within the common control blocks to describe the devices' functions, though precise, is hardly intuitive. We'll have to describe a few more features of the standard to understand these symbols:

content input
CT=m input

- *Content input.* When an input signal bearing the label CT=m is asserted, the value m is loaded into the device. In the counter symbols, you might read "CT" as "count," but in general it means "content."

The only difference between the '161 and '163 symbols in Figure A–21 is that the 163's 5CT=0 has a control dependency on the clock input (C5), and is therefore a synchronous clear; the '161 has an asynchronous clear.

content output
CT=m output

- *Content output.* An output bearing the label CT=m is asserted when the content of the device is m.

In the '161 and '163 symbols, the output 3CT=15 is asserted when the counter is in state 15 and G3 is asserted.

mode dependency
Mi

- *Mode dependency.* This dependency is denoted by Mi and, like enable dependency, is a sort of enable function. Affected signals perform their

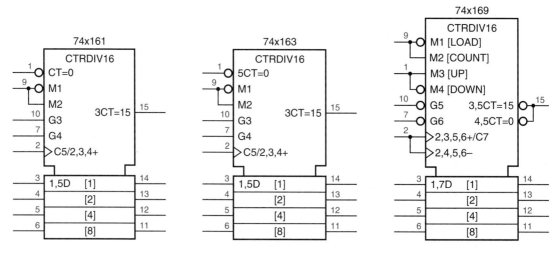

Figure A–21   IEEE standard symbols for counters.

"normal" functions only if Mi is asserted; otherwise, the affected signals have no effect on the device's function and are ignored.

- *Multiple dependencies.* A signal may be affected by several other signals. The identifiers of the affecting signals are listed, separated by commas, in the qualifying label of the affected signal. The dependencies are applied in the order that they are written, from left to right. If all of the dependencies are of the same type (e.g., AND dependency), then the order is irrelevant.

  multiple dependencies

- *Multifunction outputs.* A single external output signal, such as pin 15 in the '169 symbol, may have several sets of qualifying labels corresponding to multiple modes of operation. Such a signal may be represented by multiple outputs that are connected together externally. Such outputs normally have a functional OR relationship—the external signal is asserted if any internal signal is asserted.

  multifunction output

In the '161 and '163 symbols, the label 1,5D indicates that the data will be stored when M1 and C5 are asserted (but C5 is "asserted" only on an edge, because of its dynamic indicator). In the '169 symbol, the output 3,5CT=15 is asserted if M3 and G5 are asserted and the counter is in state 15; and 4,5CT=0 is asserted if M4 and G5 are asserted and the counter is in state 0; the external signal (pin 15) is asserted (LOW) if either of these internal signals is asserted.

- *Counting inputs.* When asserted, an input labeled with a + causes the device to count up once. An input labeled with a − causes the device to count down once.

  counting input
  +
  −

According to the '161 and '163 symbols, the device counts up on the rising edge of the signal on pin 2 if M2, G3, and G4 are asserted. The '169 counts up if M2, M3, G5, and G6 are asserted, and down if M2, M4, G5, and G6 are asserted. In each device, the qualifying labels for two separate functions of pin 2—load and count up—are drawn on the same input line, separated by a solidus; they could also have been drawn on two separate input lines, as we showed in Figure A–12 on page 799.

- *Nonstandardized information.* Descriptive function names and other non-standardized (i.e., helpful) information may be written in brackets next to the qualifying labels in a symbol.

  nonstandardized information

The '169 symbol has four "nonstandard" labels to describe the traditional /LD and /ENP inputs. Theoretically, you don't need such labels if you understand the standard. Conversely, if signals are given meaningful names (as in traditional logic symbols), then you don't need the standard!

A.11 SHIFT REGISTERS

Figure A–22 shows the IEEE standard symbols for popular shift registers. The general qualifying symbol SRGn denotes an *n*-bit shift register. We've used most of the other notation in these symbols previously; there's just one new thing:

shifting input

$\rightarrow$

$\leftarrow$

- *Shifting inputs.* When asserted, an input labeled with a $\rightarrow$ causes the device to shift its information one position in the direction from left to right or from top to bottom. An input labeled with a $\leftarrow$ causes a shift in the opposite direction.

Figure A–22
IEEE standard symbols for shift registers.

A.12 PROS AND CONS OF IEEE STANDARD SYMBOLS

The absence of a standardized method of giving unique *names* to pins and displaying the names on the logic symbol is probably the biggest defect of the IEEE standard. For example, how do we name the internal signals on pins 4, 5, and 6 of the 74x328 in Figure A–6? Worse, how can we distinguish between the internal signals on pins 3 and 11, which are both labeled "4" on the logic symbol? Apparently, the standard makers expected us to refer to these signals by pin number only. But this is awkward and inconvenient, not only in textbooks, but also in design and debugging. It is far more natural to refer to a signal by a functional name than by a pin number. (Indeed, in ASIC design there are no pin numbers for internal signals!)

Still, the standard has several strengths:

- It's a standard.

- It's consistent.

- It supports, indeed, defines, the symbology used in bubble-to-bubble design.

- It is very complete. In addition to the basic devices covered in this book, the full IEEE standard covers many other less commonly used devices and structures.

- In most cases, the standard allows the function of a logic device to be defined unambiguously by the device symbol. For example, the '299 symbol in Figure A–22 conveys as much information as the function table in Table 9–3, and more.

That's the good news; now here's more bad news, if you hadn't already noticed:

- Although the standard allows you to figure out, with moderate effort, the precise function of an unfamiliar symbol, it does not provide any standard way to remind you quickly of the function of a familiar symbol. That is, it does not provide any standard, descriptive names for internal signals. Such items are relegated to the status of "nonstandard information," and their use is not encouraged.

- Many signals in IEEE symbols have no qualifying labels, while others have duplicates. This makes it awkward to talk about signals during design and debugging (e.g., "Connect X to the third bit up from the bottom...").

- The standard encourages the omission of qualifying labels "if no confusion will result." That's like saying "Don't use your turn signal if nobody's nearby to see it." It's precisely when someone is driving in your blind spot that the habit of *always* using your turn signal can avert an accident. IEEE symbols have lots of "blind" inputs and outputs that may be hooked up incorrectly while drawing a schematic, or misapplied during debugging.

- Many possible symbols exist for moderately complex devices. For example, in the '194 symbol in Figure A–22, pin 11 could be handled as shown, or could be drawn as a single input line with the label C4/1→/2← or as three lines with the labels C4, 1→, and 2←. If used, these variations make it even more difficult to "eyeball" the symbol for a familiar device and recognize its functions.

What logic designers really need is a standard set of descriptive, functional, naming conventions for the inputs and outputs of MSI and LSI devices, one that is consistently followed by all data books and CAD systems. For example, it's ridiculous that in the present environment, a single manufacturer can produce 2-, 4-, and 8-input multiplexers with descriptive input names ranging from A–B to C0–C3 to D0–D7! Unfortunately, the industry has too much invested in documentation and CAD software for both inconsistent traditional symbols and unhelpful IEEE standard symbols, for a consistent, helpful naming standard to be deployed anytime soon. At least, today's ASIC designers can work to ensure a modicum of naming consistency and helpfulness in the logic "macrocells" that they define and use in their own designs.

E X E R C I S E S

A.1 Draw and explain the IEEE standard symbol for a 74x49 seven-segment decoder.

A.2 Draw the IEEE standard symbol for the device in Exercise 5.43.

A.3 Explain all of the notation used in the IEEE standard symbol for a 74x148. (Some research is required to answer this one.)

A.4 Draw the IEEE standard symbol for the 74x155 used as a dual 2-to-4 decoder.

A.5 Draw the IEEE standard symbol for a 74x251 multiplexer.

A.6 Draw the IEEE standard symbol for the circuit in Exercise 5.56.

A.7 How does the meaning of the label 1D differ between the traditional and the IEEE symbols for registers like the 74x374? (*Hint:* How might pin 4 of a '374 be labeled in the IEEE standard?)

A.8 Draw an IEEE-standard symbol for the modified multiplexer of Figure 10–15.

INDEX

Boldface type is used to indicate page numbers on which the corresponding index terms are defined or used in an important way. The index terms usually appear in the outside margin on such pages.

Sub-entries for terms that have an acronym appear under the acronym, not under the corresponding spelled-out entry.

826

INDEX

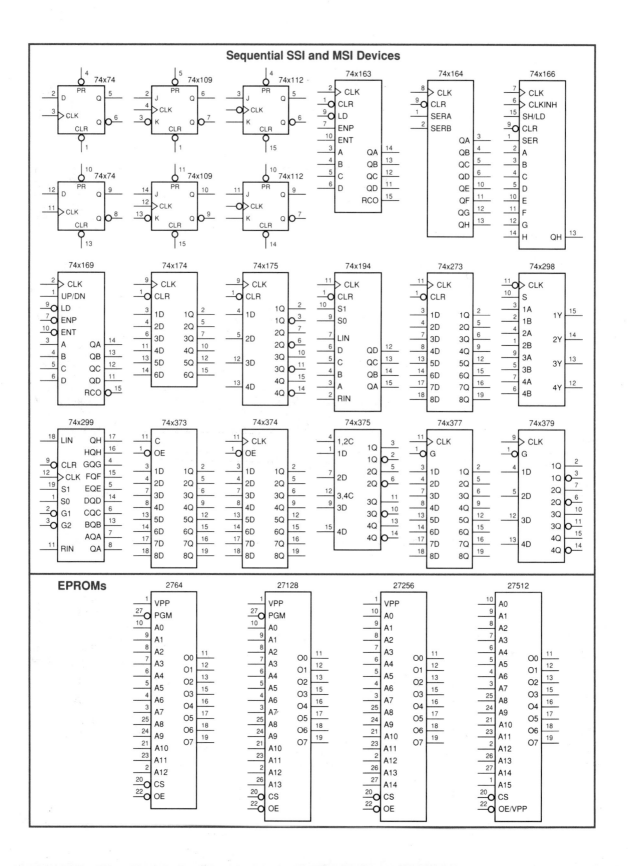